Population Ecology in Practice

Population Ecology in Practice

Edited by

Dennis L. Murray
Trent University
Peterborough, Ontario, Canada

Brett K. Sandercock
Norwegian Institute for Nature Research
Trondheim, Trøndelag, Norway

This edition first published 2020

Registered Offices
John Wiley & Sons, Inc., 111 River Street, Hoboken, NJ 07030, USA
John Wiley & Sons Ltd, The Atrium, Southern Gate, Chichester, West Sussex, PO19 8SQ, UK

Editorial Office
9600 Garsington Road, Oxford, OX4 2DQ, UK

For details of our global editorial offices, customer services, and more information about Wiley products visit us at www.wiley.com.

Wiley also publishes its books in a variety of electronic formats and by print-on-demand. Some content that appears in standard print versions of this book may not be available in other formats.

Library of Congress Cataloging-in-Publication Data

Names: Murray, Dennis, 1966– editor. | Sandercock, Brett K. (Brett Kevin), 1966– editor.
Title: Population ecology in practice / edited by Dennis Murray and Brett K. Sandercock.
Description: First Edition. | Hoboken : Wiley-Blackwell, 2019. | Includes bibliographical references and index.
Identifiers: LCCN 2019013320 (print) | LCCN 2019980192 (ebook) | ISBN 9780470674147 (Paperback) | ISBN 9781119574620 (eBook) | ISBN 9781119574644 (PDF)
Subjects: LCSH: Population–Environmental aspects. | Population ecology. | Nature–Effect of human beings on. | Sustainable development.
Classification: LCC HB849.415 P6585 2019 (print) | LCC HB849.415 (ebook) | DDC 304.2–dc23
LC record available at https://lccn.loc.gov/2019013320
LC ebook record available at https://lccn.loc.gov/2019980192

Cover Design: Wiley
Cover Image: © Copyright Michael Cummings/Getty Images

Set in 10/12pt Warnock by SPi Global, Pondicherry, India
Printed and bound in Singapore by Markono Print Media Pte Ltd

10 9 8 7 6 5 4 3 2 1

To Dylan, and to Nate, Eilish, and Tiernan

Contents

Contributors

Christopher R. Ayers
Wildlife, Fisheries, and Aquaculture
Mississippi State University
Starkville, MS, USA

Guillaume Bastille-Rousseau
School of Biological Sciences
Southern Illinois University
Carbondale, IL, USA

Lynne E. Beaty
Department of Biology
Penn State Erie
Erie, PA, USA

Jerrold L. Belant
Camp Fire Program in Wildlife Conservation
College of Environmental Science and Forestry
State University of New York
Syracuse, NY, USA

Catherine M. Bodinof Jachowski
Department of Forestry and
Environmental Conservation
Clemson University
Clemson, SC, USA

Thomas W. Bonnot
Department of Fisheries and
Wildlife Sciences
University of Missouri
Columbia, MO, USA

Luca Börger
Department of Biosciences
College of Science
Swansea University
Swansea, United Kingdom

Justin M. Calabrese
Department of Biology
University of Maryland College Park
College Park, MD, USA

Clément Calenge
Direction de la Recherche et de l'Expertise
Office National de la Chasse et de la Faune Sauvage
Saint Benoist, Auffargis, France

Sarah J. Converse
U.S. Geological Survey
Washington Cooperative Fish and
Wildlife Research Unit
School of Environmental and Forest Sciences (SEFS) and
School of Aquatic and Fishery Sciences (SAFS)
University of Washington
Seattle, WA, USA

Steven Delean
School of Biological Sciences and
the Environment Institute
The University of Adelaide
Adelaide, South Australia, Australia

David A. Eads
U.S. Geological Survey
Fort Collins Science Center
Fort Collins, CO, USA

John Fieberg
Department of Fisheries
Wildlife and Conservation Biology
University of Minnesota
St. Paul, MN, USA

Chris H. Fleming
Smithsonian Conservation Biology Institute
Front Royal, VA, USA

Marie-Josée Fortin
Department of Ecology and Evolutionary Biology
University of Toronto
Toronto, Ontario, Canada

Robert A. Gitzen
Forestry and Wildlife Sciences
Auburn University
Auburn, AL, USA

Jon S. Horne
Idaho Department of Fish and Game
Boise, ID, USA

Megan L. Hornseth
Borealis Ecology
Thunder Bay, Ontario, Canada

Dylan C. Kesler
The Institute for Bird Populations
Point Reyes Station, CA, USA

Charles J. Krebs
Department of Zoology
University of British Columbia
Vancouver, British Columbia, Canada

Stéphane Legendre
Institut de Biologie de
l'Ecole Normale Supérieure (IBENS)
Le Centre National de la Recherche
Scientifique (CNRS)
Paris, France

Stephen C. Lougheed
Department of Biology
Queen's University
Kingston, Ontario, Canada

Brett T. McClintock
NOAA National Marine Mammal Laboratory
Alaska Fisheries Science Center
National Marine Fisheries Service
Seattle, WA, USA

Joshua J. Millspaugh
Wildlife Biology Program
Department of Ecosystem and
Conservation Sciences
University of Montana
Missoula, MT, USA

Robert A. Montgomery
Department of Fisheries and Wildlife
Michigan State University
East Lansing, MI, USA

Dennis L. Murray
Department of Biology
Trent University
Peterborough, Ontario, Canada

Michael J.L. Peers
Department of Biological Sciences
University of Alberta
Edmonton, Alberta, Canada

Thomas A.A. Prowse
School of Mathematical Sciences
The University of Adelaide
Adelaide, South Australia, Australia

Janet L. Rachlow
Department of Fish and Wildlife Sciences
University of Idaho
Moscow, ID, USA

Eloy Revilla
Department of Conservation Biology
Estación Biológica de Doñana CSIC
Sevilla, Spain

Joshua V. Ross
School of Mathematical Sciences
The University of Adelaide
Adelaide, South Australia, Australia

Christopher T. Rota
Wildlife and Fisheries Resources Program
School of Natural Resources
West Virginia University
Morgantown, WV, USA

Jeffrey R. Row
School of Environment, Resources
and Sustainability, University of Waterloo
Waterloo, Ontario, Canada

J. Andrew Royle
U.S. Geological Survey
Patuxent Wildlife Research Center
Laurel, MD, USA

Brett K. Sandercock
Department of Terrestrial Ecology
Norwegian Institute for Nature Research
Trondheim, Norway

Michael Schaub
Swiss Ornithological Institute
Sempach, Switzerland

Len Thomas
Centre for Research into Ecological and Environmental Modelling
School of Mathematics and Statistics
University of St. Andrews
St. Andrews, United Kingdom

Daniel H. Thornton
School of Environment
Washington State University
Pullman, WA, USA

Jonathan Tuke
School of Mathematical Sciences
The University of Adelaide
Adelaide, South Australia, Australia

Preface

Our motivation as editors for assembling a book on current methods in population ecology arose from our ongoing interactions with graduate students and professionals in the fields of ecology, conservation biology, and wildlife management. Over the past several decades, research in population ecology has developed at a rapid pace, from a largely descriptive field dominated by observation and description, to a mature discipline that emphasizes innovative and robust analyses of ecological patterns and processes. Many recent advances have been driven by persistent knowledge gaps, not the least of which are urgent questions about the key drivers of population dynamics and their ecological relevance in the face of ongoing global environmental change. Increasingly, population ecologists have recognized that key questions in ecology and evolutionary biology must be investigated using the data and analytical methods that allow researchers to make robust inferences about causality. At the same time, advances in satellite or GPS-based telemetry, noninvasive genetic sampling, automated field photography, and other new technologies have revolutionized our ability to collect new data on the occurrence, abundance, and distributions of rare or elusive organisms under natural conditions. Emerging technologies have opened up new possibilities for data collection, but many have also required development of innovative approaches for data analysis. In some cases, new quantitative approaches have been adopted directly from other fields of research but some types of data have required the development of entirely novel analytical tools. Improvements in the capacity of personal and cloud-based computing, availability of Program R and other freeware statistical packages, and online resources for learning and troubleshooting new statistical procedures have led to tremendous improvements in the potential capacity for data analysis in population ecology. Developments in population ecology have paralleled improvements in data quality and analysis in genomics, data sciences, and other scientific disciplines. Still, population ecology has been transformed in recent decades so that our current ability to answer longstanding and elusive questions greatly surpasses what could have been imaginable even a short time ago.

Development of new tools for ecological analysis has been exciting to witness but presents a challenge for both new and seasoned ecologists who would like to stay current with available technologies and analytical approaches. During our own formative years as graduate students a few decades ago, the prevailing quantitative methods for data analysis consisted mainly of statistical tests in a frequentist framework that were originally designed for analysis of data from controlled experiments and balanced study designs. Basic tests like analysis of variance and regression were familiar because of their extensive coverage in undergraduate courses, or else they were readily adopted following focused reading or trial and error. Even specialized techniques like population estimation or habitat selection analysis were mostly accessible using off-the-shelf analytical approaches. Accordingly, at the time most ecologists were not unduly challenged to conduct data analysis that met contemporary standards. However, ecological systems are rarely governed by factors that conform to controlled conditions, and therefore ecological research rarely yields field data that truly fits standard assumptions of independence, normality, and lack of bias. Moreover, the sheer volume, structure, and complexity of ecological data collected in many field studies preclude standard statistical approaches. New quantitative methods in ecology often deviate substantially from the standard approaches that form the basis of undergraduate training in statistics, and ecologists may be left scrambling to correctly identify and implement an appropriate analytical technique. The correct application of contemporary methods for data analysis is increasingly a prerequisite for publication and for implementation of effective management policy in ecology.

Our edited volume is primarily aimed at graduate students and early career professionals who may be embarking on their first attempt to analyze ecological data using contemporary methods. We aimed to assemble a series of chapters that review the state of knowledge in the core areas of population ecology, and our selection of topics and authors was purposeful to cover the main areas by experts in the field. Our final submissions included

16 manuscripts from 39 contributors working in 8 different countries. Our aim was for every chapter to serve as a stand-alone assessment for different topics in population ecology, including pros and cons of related quantitative methods, basic assumptions and limitations when deriving inference from a given approach, and some of the potential pitfalls in the application of available techniques. Perhaps unavoidably, the chapters include some bias toward methods that are especially relevant for use with wildlife species and for using data that have been collected through newer monitoring technologies. Nevertheless, many of the general concepts and approaches covered in our contributed chapters have broad relevance to a diversity of research questions and study systems.

The 16 chapters of our book are organized into five sections. The first section begins with two chapters that provide a framework for asking relevant questions in ecology, including how research studies can be best designed to derive robust inference. The second section assembles five chapters covering a variety of analytical approaches in population demography and population time series analysis; these topics normally form the requisite basis of most investigations into population status and trend. The analytical approaches differ in whether they are based on closed or open population models, use encounter histories from marked or unmarked individuals, or control for situations where detection may be perfect or imperfect. The third section highlights population-level analysis, including newer approaches that use integrative and individual-based models to understand population drivers and forecast their potential change. The fourth section includes five chapters that address genetic and spatial approaches in population analysis, covering topics like home range and resource selection analysis and species distribution modeling.

This volume is intended to provide an overview for researchers using a variety of analytical tools and platforms. Importantly, the R statistical software platform has been transformative to data analysis in ecology, and to that end the final chapter provides an ecologically focused overview of basic nomenclature and data management using R software. Chapters are supported by a compendium of online exercises in R that provide worked through examples that reinforce topics covered in individual chapters. The intent is for exercises to provide readers with both the necessary background to implement more common analytical approaches, as well as sample code in R that can be adapted to start their own data analysis. All online exercises can be accessed from the publisher's website (www.wiley.com/go/MurrayPopulationEcology).

Our edited book would not have seen the light of day without the significant efforts of a number of people to whom we are indebted. We thank Guillaume Chapron, who began this journey with us and helped start the editorial process of selecting topics for the different chapters and inviting contributors. We thank all of the contributors who contributed their work to this volume for sharing with us a vision for the book, mostly adhering to our editorial requests, peer-reviewing each other's chapters, and for working hard to improve the quality of their chapters. Working on an edited volume can provide new appreciation of the old adage that a caravan is only as fast as the slowest camel. We thank the contributors for their sustained efforts and commitment, but especially for their patience in graciously accepting delays that arose while two slow camels worked to keep the editorial process on track. Special thanks to all of the external reviewers who provided anonymous reviews of chapters, including the many graduate students who served as test groups for the chapters and the online exercises. The students provided many useful comments that helped calibrate the volume for its intended audience. We also highlight the valuable contribution by Pat Heney, who standardized and tested all the online exercises prior to their release. Likewise, a debt of gratitude is owed to Sam Sonnega for help with indexing the complete volume. The staff at Wiley-Blackwell, especially Anupama Sreekanth, Kavitha Chandrasekar and Emma Cole, provided valuable assistance in support of our vision for the book. Last, we thank H. Dean Cluff for being an initial source of inspiration and for an exploding can of sardines.

Our hope is that our edited book will contribute to a growing body of literature that guides researchers in the rigorous analysis of ecological data. The current state of our planet, and of the species and ecosystems that have captivated the fascination of population ecologists for decades, are under grave peril. The quantitative methods described in this volume provide a valuable set of tools for addressing some of the current and emerging environmental problems that will command humanity's attention for the foreseeable future. Our book will be a success if it provides a new generation of early career researchers with the necessary tools to tackle some of these problems.

In recognition of the daunting environmental challenges facing this and future generations, the editors are pleased to donate royalties from the book to conservation activities of Wildlife Conservation Society Canada. For more information about this organization, please visit www.wcscanada.org.

Dennis L. Murray
Trent University
Peterborough, Ontario, Canada

Brett K. Sandercock
Norwegian Institute for Nature Research
Trondheim, Trøndelag, Norway

About the Companion Website

This book is accompanied by a companion website:

www.wiley.com/go/MurrayPopulationEcology

The website includes:

Exercises related to each chapter. The exercises are meant to reinforce the application of the main concepts covered in this volume, and therefore include datasets, coding, and explanatory details to help guide users in their own analyses.

Part I

Tools for Population Biology

1

How to Ask Meaningful Ecological Questions

Charles J. Krebs

Department of Zoology, University of British Columbia, Vancouver, British Columbia, Canada

Summary

I present and discuss four rules for asking good ecological questions:

Rule No. 1. Understand the successes and failures from ecological history but do not let this knowledge become a straitjacket.
Rule No. 2. Develop and define a series of multiple alternative hypotheses and explicitly state what each hypothesis predicts and what it forbids.
Rule No. 3. Seek generality from your hypotheses and experiments but distrust it.
Rule No. 4. If your research has policy implications, read the social science literature about how scientific information and policy decisions interface.

Meaningful questions in population ecology address theoretical issues or management questions that demand a solution. The solution should be looked for among a set of multiple working hypotheses. If you have only one hypothesis with no alternatives, there is nothing to do. The classical null hypothesis in a statistical sense is not an alternative hypothesis in which population ecology is interested. Given a question, the possible outcomes of the study should be noted before any field work is carried out, and an interpretation given of what each possible result means in terms of basic theory or applied management. The most useful questions often have multiple dimensions and apply to more than one taxonomic group. Once you have an important question formulated with alternative hypotheses, you must discuss the critical aspects of the experimental design – replication, randomization, treatments, and controls. How many replicates are needed over what landscape units? How long a study is required? How often do you need to sample? Will the confidence limits of any estimates be narrow or wide? If the proposed steps are not followed, it is possible to get lost in the mechanical details of a study without knowing clearly how the outcome will reflect back on the original questions. Serendipity may rescue poorly conceived studies, but the probability of this event may be less than $P < 0.01$. Management and conservation problems demand both good data and effective policy development. Ecologists need to become more proactive in providing solutions to politicians and business leaders who develop policy options with ecological consequences.

1.1 What Problems Do Population Ecologists Try to Solve?

Every ecological question comes down to a question of *population ecology*, and hence it is useful to start by asking how one goes about asking meaningful ecological questions in population ecology. Implicitly the starting point must involve answering the flip question of: How does one avoid questions that yield information that do not help in solving an ecological problem? The first and simplest guide is to look at the historical literature in population ecology, which is littered with questions that have led nowhere in terms of increased understanding of ecological dynamics or improving sustainable land management (Hartway and Mills 2012; Walsh et al. 2012). The second guide must be that a historical search is not sufficient, because it will not tell you about future research questions. Thus, it is possible to make a mistake and to spend time exploring alleys that are dead ends. But it is useful to realize that setbacks are not a scientific defeat because these explorations will show the next generation of ecologists what to avoid. So this advice might be coded as the first rule of asking meaningful questions:

Population Ecology in Practice, First Edition. Edited by Dennis L. Murray and Brett K. Sandercock.

Companion website: www.wiley.com/go/MurrayPopulationEcology

Rule No. 1. Understand the successes and failures from ecological history but do not let this knowledge become a straitjacket.

A simple example will illustrate this point. The management of Northern Bobwhites (*Colinus virginianus*) in the USA involved a controversial issue of whether quail populations could be limited by the lack of water and thus would benefit from managers providing free water, such as a pond, in their habitat. Guthery (1999) examined the competing hypotheses about water limitation and showed that even in southern Texas quail did not need free water to survive and thus that water sources were not required as a management tool. Whether this conclusion will hold under climate change is an important issue for managers in the future.

At a general level, philosophers of science provide a set of guidelines on how to develop general theory. Ask general questions rather than particular ones. General questions will apply to a variety of species and habitats, particular questions will involve only one or a few species in a restricted environmental space. Formulate your research questions as *testable hypotheses*, and if possible develop multiple working hypotheses with *alternative predictions* that are mutually exclusive (Platt 1964; Chapter 2; Chamberlin 1897).

The two major questions that population ecologists address involve the *distribution* and *abundance* of organisms. This focus for population ecology was clearly stated by Charles Elton (1927) and rigorously re-stated by Andrewartha and Birch (1954). Knowing the factors that limit the distribution of an organism can assist in analyzing problems with introduced pests (Urban et al. 2007), as well as giving some indication of how organisms might change their distributions in light of climate change or other anthropogenic stressors like habitat loss (Thomas et al. 2006; Flockhart et al. 2015, Chapter 15). Knowing the factors that affect changes in the abundance of an organism can be even more critical if the species is a keystone in the community or if it is endangered and declining in numbers. It is with these kinds of problems that this book grapples, and as methods of approach are continually improved, we ecologists hope to answer pressing questions more rapidly and more accurately.

We must recognize at the start that population ecologists should not pretend to solve every ecological problem or solve every management question. In particular, ecologists try to answer *scientific questions* and not policy issues. If a songbird is declining in abundance, the job of the population ecologist is to find out why it is declining and to recommend what might be done to reverse the observed decline. Our political systems and society at large make the *policy decisions*, for example the decision either to set aside arable grasslands to protect this bird population or to use the grassland area to produce more crops for human consumption. Ecologists will have strong views about the value of *biodiversity conservation*, and will press for policy decisions that favor biodiversity, but their role as scientists is to make estimates of the probable course of events under policy A vs. policy B. So let us begin with a clear understanding that we ecologists do not run the world and do not make policy, but rather we provide evidence-based recommendations from the science we are able to do. The separation of policy options and research questions is central to this approach to global issues to which ecological data on populations are relevant (Sutherland et al. 2010).

Many ecological questions are posed with no clear connection to population ecology. For example, increasing levels of carbon dioxide (CO_2) in the atmosphere are affecting the acidity of sea water and potentially affecting the geochemical carbon cycle (Dybas 2006; Ruttimann 2006; Boyd et al. 2010). On the surface the problem appears to be one for chemical ecologists, but quickly the question become exactly which species of phytoplankton are being affected by changes in seawater acidity, and how this disruption of population growth affects predators or competitors in the community that either feed on the particular phytoplankton species or compete with it for nutrients. Problems of this type, once broken down in a reductionist manner, quickly fall into the basket of population dynamics.

There is a temptation to ask questions about community or ecosystem ecology with the implicit belief that we can reach an understanding of the problem, and in particular to be able to recommend policy alternatives to alleviate the problem, without getting buried in population dynamics. Neither *community ecology* nor *ecosystem ecology* have solved ecological problems without delving into the details of population dynamics to sort out mechanisms. *Macroecology* is also useful for recognizing ecological patterns that require explanations at the level of both community and population ecology (Trebilco et al. 2013; Borrelli et al. 2015).

Given the broad questions about distribution and abundance, there are many more steps that have to be decided before one has posed a good ecological question. The first step is to choose the species of interest. Research priorities may be dictated to you by your employer if you work for a wildlife agency, or may be decided by funding options if you are a graduate student. Financial support would seem to be a major constraint for a new scientist, but in fact there are important and interesting questions to be asked for every species. Important questions are either general questions that apply to many species, or conservation questions that have a direct bearing on management decisions. Important questions always have at least two and possibly three or more potential answers which are not presently known. To confirm potential knowledge gaps, you will have to know the literature on your species and closely related species very well.

Studies of single-species populations could be considered ecological stamp-collecting, but this would be an error. Some species can be considered model organisms whose results can be generalized to many species in their group. Studies of single species are necessary to answer broad questions regarding, for example, the types of *numerical responses* predicted by predator–prey theory (Sundell et al. 2013; Bowler et al. 2014). Designing single-species studies to test broad ecological models is an essential way to refine *ecological theory* (Chapters 8–10).

Theoretical ecologists put forward many different models of the ecological universe; some of these are useful and important, while others are irrelevant and unanswerable. The theoretical literature in ecology abounds with concepts like density dependence, competitive exclusion, chaos, resilience, and stability that are potentially useful if they can be defined rigorously and are available for empirical measurement. *Resilience*, for example, is a useful word, but its ecological measurement is fraught with problems (Carpenter et al. 2001; Myers-Smith et al. 2012). Even if a theoretical concept can be measured, it may not have much utility, so it pays a young investigator to ask where the concept leads. A key example is the idea of *direct density dependence* in reproductive and survival rates. The concept is clearly presented in every textbook as the foundation for understanding population changes; the means of measuring it are fairly straightforward for many species, but having done so leads one to a dead end. The concept has the illusion of precision but suffers from two problems. It provides no predictability, so the observed density dependence in one population will not allow one to predict quantitatively the relationship in other populations of the same species (Krebs 2002). The second problem is that it does not define mechanisms that can be manipulated for wildlife management or conservation questions. Without mechanisms like predation, food shortage or disease, managers have no potential levers to use to solve the problems they face. Consequently, demonstrating density dependence in population dynamics is useful only as a first step toward the much more difficult goal of finding mechanisms involving births, deaths, and movements that drive density changes (Strong 1986). Birth rates may not automatically increase (or death rates decrease) as a population declines in abundance, and *Allee effects* may doom some populations to local extinction (Courchamp et al. 2008). Too many ecological concepts lead one to unanswerable questions or questions that once answered have no utility for management or conservation (Peters 1991).

Two major empirical processes stare ecologists in the face at this time in history and should demand our attention – *climate change* and *habitat loss*. Both factors are having and will have major impacts on distribution and abundance, and when they are both occurring together, they may be difficult to disentangle. These two processes raise general questions that are applicable to many species: how is global warming changing the distribution of species? Are most geographical distributions limited by climatic factors? Will alpine and subalpine species be driven to extinction? How quickly can a species adapt to temperature shifts? Will top-down systems be affected by climate change? Habitat loss is universal in the era of rising human populations, and the effects of habitat loss and habitat fragmentation are key issues that may have general effects or individual species-specific effects (Stephens et al. 2003; Hanski 2011).

Two aspects of these global problems complicate ecological investigations in this century. First, both problems are strongly affected by human actions. While in past decades ecologists could argue that human influences were relatively minor and large ecosystems were relatively intact over much of the globe, now the rapidly growing human population and the need for resources for human consumption have plundered the natural world and set up new situations for organisms. Thus, the justification that one was studying a system in a *steady state* that was at least, in ecological time, unchanging is now gone. Given these new realities, past observed population dynamics are only an approximate guide to future population dynamics for any particular species. Second, the adaptations that organisms have made to their environment may now be antiquated (Conroy et al. 2011). A confronting example is the lack of anti-predator adaptations in native Australian mammals and birds to introduced predators such as the red fox (*Vulpes vulpes*, Short et al. 2002) or toxic cane toads (*Rhinella marina*). Cane toads were introduced into northern Australia, and because they are toxic to predators, initial concerns were that generalist predators would be devastated by this potential prey species. Fortunately, many species have learned to avoid eating cane toads, and to date, the predicted devastation of the predator guild has been minimal for most species (Shine 2010). Both these aspects mean that ecologists must rely more on empirical studies of our current ecosystems than on predictions based on past observations. One alternative is to rely on predictions from studies on *model species*, with the assumption that the dynamics of the chosen model species is general and applies to all similar species. This alternative assumes a generality of mechanistic understanding that is only slowly accumulating in population ecology. An excellent example is the general observation of strong population declines in migratory species of insectivorous birds (Benton et al. 2002). The associated habitat change has been an intensification of agricultural production, with the presumption that the mechanism involved was food shortage on the nesting ground. The hypothesis of food limitation was tested and rejected for Tree Swallows (*Tachycineta bicolor*, Rioux Paquette et al. 2014), with the suggestion that the responsible mechanism might be operating during migration or on over-wintering sites that are yet unstudied. We do not yet know if these general declines in bird populations have many different causes or whether there is a common mechanism which could then provide a common solution.

There has always been a tension in ecology between those who argue that basic studies of population distribution and abundance are the key to progress, versus others who feel that all of ecology is now a crisis discipline and ecologists should study nothing but solutions to practical management and conservation issues (Fleishman et al. 1999). I doubt that this debate is fruitful, since there is no way of knowing the future needs of biodiversity conservation. There is a pressing need to study immediate conservation and management problems, but there is also a pressing need to develop more general understanding of population dynamics, research that can often be done with abundant species of no great economic value or conservation concern.

1.2 What Approaches Do Population Ecologists Use?

The factors limiting geographic distributions of organisms have been well dissected for more than 100 years (Chapter 15), but studies of distributional limitations have become important only since the era of climate change was recognized during the 1980s. *Species distributions* can be difficult to map because they are scale-dependent (Forman 1964; Gaston 1991). The scale dependency of studies of distributions has produced a strange literature of limitations that are contradictory. As a simple example, dispersal abilities may limit a species range at a continental scale, but have little relevance to understanding why species X appears in patch Y but not in patch Z only a few meters away (Kroiss and HilleRisLambers 2014). Given that you have mapped a geographic range at an appropriate scale, the two possible approaches you can take are *observational*, "watch and wait, hope something happens," or *experimental* manipulations. Studies of distributional limitation are hampered by the long time frame needed to see changes and the confounding of human actions, climate change, and organismal adaptations to these changes.

There are more studies of changes in the abundance of particular organisms than studies of distributional limitations. Changes in abundance have been particularly attractive to theoretical ecologists, and we can surmise that the number of models now exceeds the number of empirical studies explicitly designed to test the models (Chapter 5). In general, the theoretical literature has directed population ecologists to two paradigms or two approaches to answering the question of what determines the rate of population growth of species X (Sibly and Hone 2002). I have called these the *density paradigm* and the *mechanistic paradigm* (after Kuhn 1970). If you do not like the word "paradigm" replace it with "approach."

The *density paradigm* instructs us to plot population growth rate against population density. At this point, we should become suspicious because the variables on the X- and the Y-axis are not independent. But we are assured by some biometricians that this is not a problem (Griffiths 1998), so we might forget about this potential problem. If the density data are a time series of one or more plots, much now depends on the trend shown by the data (Chapter 4). If density is monotonically falling (or rising), it will not be possible to estimate an "equilibrium" density, except by assuming independence between points or the use of formal time series methods. If a population does not change much in density, the relationship may well look like a shotgun pattern (Strong 1986). Experimental manipulations of density are needed in many cases.

A *decision tree* illustrates how to proceed (Figure 1.1). If there is a negative relationship between population growth rate and density, the next question is which of the demographic components drive this relationship. Given that data are available to answer this question, the next step is to find out which factors or combinations of factors cause changes in births and deaths, as well as movements if the population is open. All this is what I will call the standard analysis procedure of the *density paradigm*. However, what happens if there is no pattern in the plot of growth rate against density? Does this mean that the population is not subject to density-dependent constraints on growth, or merely that there is too much noise in the data or that there are other factors at play, such as tradeoffs among demographic rates, that obfuscate the relationship?

We are assured by both theoreticians and empiricists that there *must* be a negative relationship between population growth rate and density (Nicholson 1933; Sinclair 1989; Turchin 1999). If this is true, it raises an interesting question of the relationship of theory in ecology to empirical data. If there must be a relationship, the problem of the field ecologist is to describe this relationship in terms of its slope and intercept, and to determine if Allee effects occur at low density. The problem is not to ask if indeed such a relationship exists (Murray 1999, 2000). There is no alternative hypothesis to test.

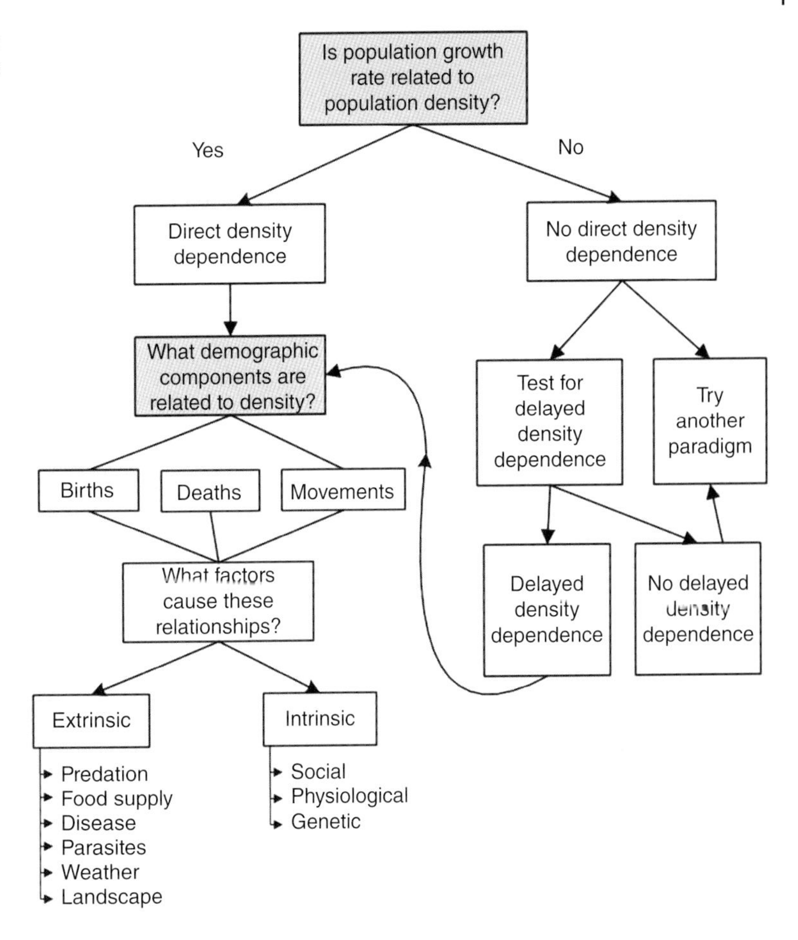

Figure 1.1 Decision tree for the density paradigm for explaining changes in population density (after Krebs 2002). The gray boxes indicate the key questions in which the density paradigm differs from the mechanistic paradigm shown in Figure 1.2.

The first strategy that is adopted after finding that there is no relationship between population growth rate and population density is to invoke *delayed density dependence* (Turchin 1990). This is a reasonable strategy because virtually every interaction in population ecology involves some time delays. But this strategy opens Pandora's Box because data analysis begins to take on the form of *data dredging* if we have no a priori way of knowing the duration of the critical time delays. Fortunately, we have independent natural history data for many systems that can set limits for what are biologically reasonable time delays, which permits us to define the limits for time series analyses. There are elegant methods of time series analysis that can be applied to population data to estimate the integrated time lags in a series of density estimates (Stenseth et al. 1998), but it is less clear how to translate these estimated time lags into ecological understanding. Do predators respond to changes in prey abundance quickly via *dispersal movements* (Korpimäki 1994), or more slowly via *recruitment processes* (O'Donoghue et al. 1997; Eberhardt and Peterson 1999)? Data constraints, such as only annual census data, affect our ability to draw biology out of statistics at this point.

If delayed density dependence can be identified in a time series of population densities, we can proceed in the same manner as the standard analysis procedure of the density paradigm, and try to determine what causes these time lags. The remaining problem is what to do with cases in which no direct or delayed density dependence can be identified in a time series. In theory, this situation cannot occur, but it seems to arise frequently enough to cause endless arguments in the literature about the means of testing for direct and delayed density dependence (den Boer and Reddingius 1989; Dennis and Taper 1994). Many ecologists in this situation would not give up studying population regulation, but would switch to the *mechanistic paradigm* discussed by Sibly and Hone (2002).

The *mechanistic paradigm* can be viewed in two different ways. Sibly and Hone (2002) consider it an elaboration of the density paradigm (Figure 1.1), and indicate that one can proceed to this level of analysis for populations that are well studied in a reductionist manner. Note that the key variable in the density paradigm illustrated in Figure 1.1 is always *population density*. Krebs (1995), by contrast, postulated that the key variable should be *population growth rate*, and suggested that the mechanistic paradigm is an alternative to the conventional approach that proceeds via the density paradigm. The mechanistic paradigm

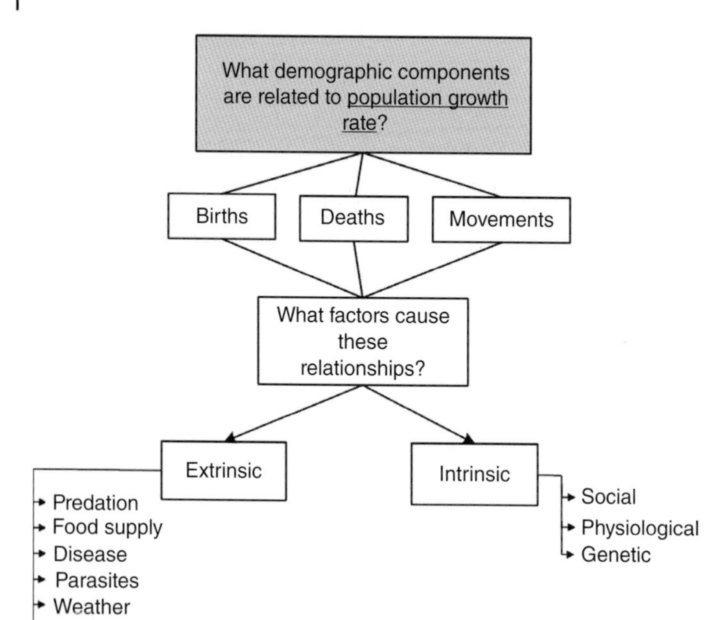

Figure 1.2 Decision tree for the mechanistic paradigm for explaining population growth rate changes (after Krebs 2002). The gray box indicates the key question in which this paradigm differs from that of the density paradigm shown in Figure 1.1.

short-circuits the search for density dependence on the assumption that no predictive science of population dynamics could be founded on describing relationships between vital rates and population density without specifying the ecological mechanisms driving these rates. Density is not a mechanism but a surrogate for a variety of mechanisms that require study.

The key question seems to be whether or not any density-dependent relationships are repeatable in time or space. I have been able to find few ecologists who have asked this question. For those I have found, none of the studies have shown repeatable patterns in time or space (Krebs 2002) and the explanation for this failure lies in the fact that density is not a mechanism. When density dependence can be found, it is not quantitatively repeatable in space or in time. Both (1998) showed this in his elegant work on populations of Great Tits (*Parus major*) in the Netherlands. Annual mean clutch size in this species declined strongly with density but the regression lines between population density and clutch size differed greatly in his six study areas, probably because of food supplies. The pattern was clear, but the process not. The conclusion I reached was that density-dependent relationships occur often (as most ecologists believe) but are not repeatable (as is rarely tested) and are an unreliable basis for a predictive ecology. Thus, spending valuable research time on showing density dependence is rarely needed and should not be the central focus for solving problem in wildlife management and conservation.

The flow diagram for the mechanistic paradigm (Figure 1.2) is similar to that of the density paradigm (Figure 1.1), but has one significant difference (outlined in gray): instead of asking what demographic components are related to *population density*, it asks which components are related to the *population growth rate.* In cases where density is closely related to population growth rate, there will be no difference between these two approaches. But in every nonequilibrial system, the differences can be large. The critical assumption again depends on whether there is an equilibrium point for the system under study. A mechanistic approach is best adapted to short-term considerations in which questions about ultimate equilibrium states are not particularly relevant or interesting because the world is changing too rapidly. It is closely related to the approach to population dynamics typified by the Leslie matrix (Caswell 2001; Chapter 8), and is particularly well suited to our current ecological situation in which climate change is rewriting many ecological interactions. The mechanistic approach does not concern itself with asymptotic properties, but takes into account the fact that asymptotic properties may not capture transient, short-term dynamics. An example of the use of the mechanistic paradigm is given from our long-term studies of the population cycles of snowshoe hares (*Lepus americanus*) in the Yukon (Box 1.1).

The mechanistic paradigm asks how individual animals are influenced by the factors affecting density, and recognizes that individuals vary in their responses to predators, food supplies, parasites, and weather, as well as in their social standing within the population. Behavioral ecology has made a particularly strong contribution to our understanding of individual differences, and is pushing strongly to utilize this understanding to enrich population dynamics.

Is the mechanistic approach better than the density approach? Both approaches rely on precise estimates of

Box 1.1 An Example of Hypothesis Development and Testing – The Kluane Project

The Kluane Project began in 1976 as a research program aimed at providing an explanation for the 9–10 year cycles in population numbers of snowshoe hares (*Lepus americanus*) in the boreal forests of Canada. Jamie Smith, Tony Sinclair, and I started the first 10 years of study with the aim of testing the three most likely explanations for the hare cycle – winter food shortage, excess predation pressure, or social interactions. We wanted to test social interactions since this was such an important part of the explanation of the three to four year vole and lemming cycle, and if we could produce a general explanation for both these mammalian cycles, it would be a valuable achievement. Alas it was not to be because we learned very early that snowshoe hares have no social behavior that is of any consequence for population limitation (since they are not territorial and do not commit infanticide), and our hypothesis list was reduced to two likely mechanisms – food shortage and predation mortality. It is critical to note that there was a long history of detailed work on snowshoe hares before we began, particularly by Lloyd Keith and his associates but also by many other ecologists, so we knew much about the natural history of their population fluctuations when we started in 1976. We knew that cycles were regionally (500 km) synchronous, so movements of animals could not be the explanation for increases and declines. Lloyd Keith had shown that hare reproductive rates varied dramatically with the cycle in a delayed density-dependent manner so that the number of litters was reduced from four to two as hares reached peak density and entered the decline and low phase. Everything in the boreal forest eats snowshoe hares – Canada lynx (*Lynx canadensis*), coyotes (*Canis latrans*), Great-horned Owls (*Bubo virginianus*), and Northern Goshawks (*Accipiter gentilis*) – and if you include the predators of juvenile hares you can add another long list from Gray Jays (*Perisoreus canadensis*), to red squirrels (*Tamiasciurus hudsonicus*) and arctic ground squirrels (*Spermophilus parryii*), and several of the smaller raptors. It was known from the early work of Lloyd Keith, Jerry Wolff, and John Bryant that in some cycles severe over-browsing of winter shrubs occurred, which could suggest food shortage.

Given all this information, how can one proceed? We first tested the simplest hypothesis of winter food shortage from 1976 to 1985, predicting that if we provided winter food experimentally to specific populations, we could stop the cycle or at least slow it down. We provided commercial rabbit chow to three areas over each winter, and because of a complaint about unnatural high-quality chow for food, we also supplied natural food over three winters by cutting down large trees. We knew from natural history observations that hares devoured white spruce and aspen trees that blew over in wind storms, and spruce needles from tall trees were in fact their favorite winter food in cafeteria tests. The results were unequivocal – adding winter food increased the local hare density by immigration but all the fed grids collapsed at the same time and at the same rate as the control grids with no added food. It took 10 years to establish this, and the suggestion from these results was that we should study predation as a more likely cause of the cycle.

Consequently in 1986 we began the Kluane Boreal Forest Ecosystem Project. The entire project was very much a team effort. In addition to Jamie Smith and Tony Sinclair, we were fortunate to have Rudy Boonstra, Stan Boutin, Susan Hannon, Kathy Martin, Roy Turkington, and Mark Dale on the team, along with many first-class graduate students. We adopted a conventional statistical design of control areas (60 ha) and manipulated areas (100 ha) with the treatments being fertilizer addition (nutrient bottom up), food addition (bottom up), predator reduction (top down) and a combined treatment of simultaneous food addition and predator reduction. Predator reduction involved electric fences around 1 km^2 to keep out mammal predators. We attempted to eliminate avian predation with fishing net strung between trees but it was unsuccessful due to snow accumulation. For each treatment we wrote down before the studies were done what the predictions were and what the alternative hypotheses would predict for all the major species in the ecosystem.

The strongest impact on hare densities was achieved by the combined food addition – predator reduction treatment, and we were left after 10 years to try to explain this interaction of food and predation. We had determined by radio telemetry that the immediate cause of death of >90% of all hares was predation, so clearly predation was a critical driver of the mortality component of hare dynamics. But we were puzzled by the interaction with food supplies since food limitation seemed the most likely cause of reproductive curtailment yet at the same time by no other measures could we find evidence of food shortage in hares on control areas.

Rudy Boonstra saved the day by suggesting that reproductive curtailment might arise from stress, with the stressing agent being unsuccessful predator chases. If this was correct, predation could be the cause of both the mortality changes as well as the reproductive changes that occurred to drive the cyclic dynamics. By the 1990s new methods had been developed by physiologists to measure stress levels in individuals by the metabolic breakdown products released in fecal pellets. After much work the stress hypothesis was validated by Michael

Sheriff in 2009 with the important addition that there was a maternal effect of stress – stressed mothers reduced their reproductive rate and also produced stressed offspring, raising the issue of how long stress effects might be passed on from generation to generation by nongenetic means. The end point now is that maternal effects are seen as a potentially critical variable in the population ecology of many vertebrate species.

So all this research over 40 years has validated the hypothesis that predation is the dominant mechanism driving the changes in reproduction and mortality in the snowshoe hare cycle. This abbreviated synopsis is examined in more detail in the book by Krebs et al. (2001) and the publications cited in it. The whole process is neatly summarized in Popper (1963) – conjectures (= hypotheses) and refutations (= testing).

birth, death, and movement rates (Chapters 6, 7, and 13). At present, the bulk of the population literature now favors the density approach, probably because it is simpler to estimate population density with narrow confidence limits than it is to estimate population growth rate. But the question we need to answer is whether or not the density approach has led to a rapid rate of progress in understanding the problems ecologists face, whether of overabundant wildlife or endangered species. Two good examples to investigate the utility of these two paradigms would be the issues of how to conserve polar bears (*Ursus maritimus*; Sahanatien and Derocher 2012), and how to manage overabundant ungulates (Bradford and Hobbs 2008). The mechanistic approach involves much more effort to analyze mechanisms, whether they be predation, disease, food supplies, or weather. I think that we would solve more problems and have less controversy if we adopted the mechanistic approach with clear alternative hypotheses, but only the future will tell if that is correct.

1.2.1 Generating and Testing Hypotheses in Population Ecology

Many philosophers of science define science by the *hypothetico-deductive method* (Popper 1963; Mentis 1988; O'Connor 2000). An alternative approach is a completely empirical method often called the *inductive approach*. Induction operates by gathering data and then trying to decipher what they mean. Both the hypothetico-deductive and inductive methods are further discussed in Chapter 2. Hypotheses can be generated after all the data are collected, and further data collection can test these hypotheses. In a sense, all ecologists have been using inductive methods by observing patterns in nature, and certainly one cannot begin any study without some natural history knowledge and some idea of patterns that you wish to investigate. But for many ecological systems the patterns are clear but the explanations are not known. It is at this point that the hypothetico-deductive methods of Popper (1963), Platt (1964), and Chamberlin (1897) become most useful.

Induction is often defended in ecology by the necessity of having information to generate hypotheses. This viewpoint is certainly correct. Without some data or understanding it is impossible to generate hypotheses to explain any ecological problem. Consequently, there is much discussion at cross purposes in ecology about these issues, and the simple advice is: (i) develop a hypothesis, (ii) make some predictions, and (iii) test the predictions. The method of *multiple alternative hypotheses* comes into play here (Chamberlin 1897; Chapter 2). A single hypothesis is quite useless in science and pairing a single hypothesis with the statistical null hypothesis is probably the most common error in ecological science (Anderson et al. 2000). A simple example will illustrate this problem, which appears regularly in graduate student theses. A single hypothesis that a herbivore species is selective in its diet could be tested against the null hypothesis that its diet is completely nonselective. The test would be a complete waste of time since no ecologist on earth would question the idea that all herbivores are selective foragers. But the basic idea could be turned into a set of interesting questions, for example with multiple alternative hypotheses that this species is a generalist or a specialist, that it selects food plants high in protein, or alternatively high in carbohydrates, or alternatively low in feeding deterrents, and whether the diet varies seasonally. The second approach would be informative and predictive of how this herbivore species operates in its foraging universe. And this simple example illustrates how most forage ecologists now operate (Bryant 1981).

Rule No. 2. Develop and define a series of multiple alternative hypotheses and explicitly state what each hypothesis predicts and what it forbids.

There are a whole set of assumptions buried in the hypothetico-deductive framework that few ecologists tend to discuss. A list of multiple alternative hypotheses may not include the correct hypothesis or the best fitting statistical model. There is no way to correct for this problem except by having excellent natural history information and knowing the background of studies in your particular field. The optimistic view that ecological science progresses in a linear way toward correct ideas is far from reality. Progress in understanding is slow and

Box 1.2 Rules of Thumb for Judging Ecological Models

(1) Compare the number of parameters with the number of data points. When a model uses 10 parameters to fit to a time series of 25 data points, chances are that it can fit almost any 25 data points.

(2) Compare the complexity of the proposed model with the complexity of the phenomenon that it seeks to explain. Often, proposed models turn out to be dramatically more complex than the ecological problems that they seek to solve. If one can state the ecological phenomenon in fewer words than it takes to formulate the model, the theory is probably not useful.

(3) Beware of meaningless caveats confessing oversimplification. Eager for their work to be embraced by ecologists, theoreticians like to conclude that their models are oversimplified. An already complex model that "admits" that there are more mechanisms to be taken into account (read: more parameters) betrays a tendency toward further unjustified complexity

(4) Beware of being given what you expect. As ecologists, we have come to expect that our data will be "messy," and many theoreticians will go out of their way to meet this expectation. One way to make the curves look "less perfect" is to simply add environmental noise and observational error (each variance adding one more parameter). Suspect that rhetoric is at work when models that are fully capable of producing a perfect fit are tweaked to show a more palatable near-perfect fit.

Source: Adapted from Ginzburg and Jensen (2004).

shows reverses not infrequently. The most graphic new illustration is the technique of multimodal inference based on *information theory* (AIC [Akaike's Information Criterion] analysis) of alternative statistical models (Anderson et al. 2000; Chapter 2). Anderson and Burnham (2002) discussed four issues that limit the utility of this approach. The four issues include: (i) poor scientific questions, (ii) too many models, (iii) the true model is not in the set evaluated, and (iv) using AIC methods as equivalent to standard statistical tests that use p-values. If the number of models exceeds the number of data points, more data are required. We should not think, Anderson and Burnham (2002) argue, that we will find the "true" model or hypothesis but rather one or more models in a set of alternative models that best fit the data. The most important point these authors recognize is to think carefully about the models or hypotheses you wish to test, and define alternative hypotheses as precisely as possible before you gather or analyze data (Chapter 2).

Much controversy in the past has emerged between the view that population ecologists should do *field experiments* versus the view that *observational studies* are equally valid (Diamond 1986). This discussion is of limited use, as the most important item on the agenda is to gather the necessary data that will test a series of hypotheses. In some cases, this can be done only by observational methods; in other cases some experimental manipulation can be carried out. There are many reasons to pick one or the other approach – financial, political, time available – but we must not lose the critical aspect of defining the hypotheses precisely and specifying what data will reject each particular hypothesis.

There is also much discussion in ecology about the value of *simulation models* for testing alternative hypotheses (Aber 1997). Ginzburg and Jensen (2004) pointed out that many mathematical models used in population ecology are overfitted with too many parameters. The authors presented a series of *rules of thumb* for judging ecological models (Box 1.2). Ginzburg et al. (2007) suggest that if a hypothesis (or a model representing that hypothesis) is so general that data cannot be used to test it, the hypothesis or model must be made more specific. Simulation models should contain parameters that can be measured in the real world, and if that restriction is widely accepted, the value of simulation models could be greatly improved (Chapter 10).

1.3 Generality in Population Ecology

We search for generality in ecology but are constrained by a series of problems that are not easily resolved. Typically, we assume in our studies some type of spatial and temporal invariance. For example, if nematode intestinal parasites limit density of Red Grouse (*Lagopus lagopus*) in Yorkshire, they should also limit density in Scotland (cf. Redpath et al. 2006; Moss et al. 2010). Moreover, if this idea was correct in 1990, it will also be correct in 2020, or 2120. Ecologists now recognize the role of climate change in affecting populations but tend to avoid discussing all the issues associated with the assumption of spatial and temporal invariance. Because of habitat changes due to agricultural intensification, and habitat loss due to land degradation, the assumption of spatial and temporal invariance is particularly worrisome. This task is further complicated by concerns about how natural selection will affect populations in relation to climate change as well as human land-use change (Henden et al. 2009; Conroy et al. 2011).

Rule No. 3. Seek generality from your hypotheses and experiments but distrust it.
For the present, we do not seem to have any recourse for this problem but it has consequences for what population research is to be done. Typically, ecologists do not wish to repeat studies done in the past, which is exactly what we ought to be doing in the face of climate and land-use changes to test the generality of our findings. Funding agencies stress the importance of novel research without acknowledging the value of repetition in space, time, or across study taxa. But in our current situation, ecologists have more questions than we have money or person-power to answer. The answer we have to this problem is to design optimal monitoring programs at a large scale, and there is much discussion of the design of these programs at a regional and national level (Stadt et al. 2006; Lindenmayer et al. 2012). At the level of species distributions, monitoring is clearly effective in detecting spatial and temporal changes (Fujisaki et al. 2008; Snäll et al. 2011; Sullivan et al. 2014, Chapters 12 and 15). Monitoring of cryptic mammals is more difficult and can only rarely be done by citizen groups (Harris and Yalden 2004; Sutherland 2006), although remote networks of camera traps are providing a new approach for achieving monitoring goals (Meek et al. 2014).

Replication is an important issue in experimental design, but the conflict here for population studies is that the advice of "use a larger study area" conflicts with the recommendations of statisticians to "take many replicates." If one can take replicates of some experimental study, it is possible to do power analyses to detect what effect sizes can be measured at a given level of replication (Schindler 1998; Johnson 2002; Field et al. 2007), although one must be wary of post-hoc power tests and such assessment is best conducted prior to data collection. In many cases, the ecological needs of a study must trump the statistical needs, and hence the rule "$n = 1$ is better than $n = 0$". In our Kluane Lake experiments (Box 1.1) we could not replicate an experimental treatment that combined an electric fence + food additions because of the financial costs involved. And the standard recommendations to randomize treatments could not always be followed in our project for logistical reasons (Krebs 2010). Replication can be done in space as well as in time.

Meta-analyses are one possible solution to the problems of replication and limited spatial coverage (Stewart 2010). The basic assumption of meta-analyses is that a given hypothesis has been investigated many times by different research groups, and by pooling statistical results we obtain a more precise estimate of the effect size for a given treatment. Problems arise when individual studies utilize different methods of measurement so that the average effect size may be a poor estimate of the correct but unknown effect size. Ecological studies always differ and judgment is required about how similar they must be for pooled effects to be meaningful. Meta-analysis may overestimate effect size because of publication bias due to failure to publish studies with nonsignificant or negative results (Jennions and Møller 2002), or may be biased in favor of the prevalent paradigm (Koricheva 2003). Careful interpretation is required to avoid spuriously precise estimates of effect size (Slavin 1995; Whittaker 2010). Stewart (2010) rejected these criticisms as unwarranted, but Vetter et al. (2013) then evaluated 133 papers that used meta-analyses in biodiversity conservation, and found that few of them were reliable. The authors provided recommendations for improving the use of meta-analyses in ecology. Their warning is still relevant: meta-analyses can be useful if done properly, with the caveat that both the analyses and studies replicated in time are subject to the types of environmental changes that are now under way.

1.4 Final Thoughts

Ecological knowledge may be restricted to the scientific literature and read by the academic community, but the implications of findings are often lost when dealing with management issues or the development of policy alternatives (Lawton 2007; Sutherland et al. 2010). As Lawton (2007) has succinctly put it:

> Many scientists hold to the "deficit model" of turning science into policy, the view that if only politicians are told what the science reveals, "correct" policies will automatically follow. Nothing could be further from the truth. Politicians have all kinds of reasons, some valid, some less valid, not to adopt what often seem to us to be common sense policies to protect the environment.

The litany of reasons why ecological information may not be used in policy and management issues is unfortunately long. Possible reasons include poor communication by scientists, an overabundance of information, a lack of public support for the changes required, conflicts with other financial or political interests, and ambiguous science. Another important issue is time scale – ecologists wish to have years to study a problem, but politicians wish to have an answer today so they can decide what to do.

Ecologists should not only identify and research problems in natural resource management, but also suggest solutions.

At different times, national governments of Canada, Australia, and the USA have forbidden government scientists from any discussions of policy. Scientific research in ecology, geological, and medicine often has implicit policy implications, and restrictions would seem to ignore science in favor of political opinions. Fortunately, university scientists are not bound by government rules and are usually able to speak out in matters of science that affect public policy. The implications of science for policy decisions can lead ecologists into the field of social science.

A good example of the clash between ecological understanding and public policy occurs in the management of wolves (*Canis lupus*) in North America and Europe. Wolves are apex predators in many ecosystems (Ripple et al. 2014), and wolf populations have increased in both North America and Europe in recent years after more than a hundred years of decline (Jacobs et al. 2014). In the USA and parts of Canada, governments have conducted lethal control of wolves to benefit livestock producers and to enhance populations of moose (*Alces* spp.), caribou (*Rangifer tarandus*), and other native ungulates. The ecological consequences of large predator reductions are now well known but the problem has rested in public opinion about the utility of wolf control (Bergstrom et al. 2014). Much of wolf management is strongly affected by public opinion, and rather less by ecological information.

Rule No. 4. If your research has policy implications, read the social science literature about how scientific information and policy decisions interface.
Linking science to policy decisions is not easy, as illustrated by the global problems of dealing with climate change, or addressing competing interests of industrial development and landscape protection. One important approach is to identify policies for conservation and land management and to map the research needs relevant to each policy option. Sutherland et al. (2010) have completed these tasks for the United Kingdom and have listed 25 policy areas for conservation that affect the UK as well as many other countries in Europe. The key to success is to increase the interactions among ecologists, government, and business policy makers so that mutual understanding and tolerance are well recognized. By focusing on the twin issues of *biodiversity conservation* and *ecosystem services*, ecologists may be able to build public support for measures that require some realignment of our current economic system. Opposition to change will be strong, as it has been in the past (Oreskes and Conway 2010), and consequently ecologists should not be discouraged by a lack of immediate success. Public education is the key.

Particular issues in land management require a balanced approach that extends beyond ecological science to social and political science. Braysher et al. (2012) identified five key principles for successful efforts in pest management, and I have restated these broad principles as a five-point template for conservation researchers:

1) All key stakeholders need to be actively engaged and consulted for effective conservation plans.
2) Land-use management for biodiversity needs to focus on the outcome, not just on effort and dollars expended.
3) A whole ecosystem approach is required for managing conservation programs.
4) Most conservation management occurs in ecosystems in which our knowledge is imperfect.
5) An effective monitoring and evaluation strategy is essential for all management interventions designed to protect biodiversity.

Not all population ecologists will be directly concerned with policy and management for biodiversity, since we need to base all these policies on the best science available. But more and more the public are asking ecologists for information and recommendations on issues of concern, whether the issues involve the protection of threatened whales or the ecological consequences of climatic warming. We need to be ready.

References

Aber, J.D. (1997). Why don't we believe the models? *Bulletin of Ecological Society of America* 78: 232–233.

Anderson, D.R. and Burnham, K.P. (2002). Avoiding pitfalls when using information-theoretic methods. *Journal of Wildlife Management* 66: 912–918.

Anderson, D.R., Burnham, K.P., and Thompson, W.L. (2000). Null hypothesis testing: problems, prevalence, and an alternative. *Journal of Wildlife Management* 64: 912–923.

Andrewartha, H.G. and Birch, L.C. (1954). *The Distribution and Abundance of Animals.* Chicago: University of Chicago Press.

Benton, T.G., Bryant, D.M., Cole, L., and Crick, H.Q.P. (2002). Linking agricultural practice to insect and bird populations: a historical study over three decades. *Journal of Applied Ecology* 39: 673–687.

Bergstrom, B.J., Arias, L.C., Davidson, A.D. et al. (2014). License to kill: reforming federal wildlife control to restore biodiversity and ecosystem function. *Conservation Letters* 7: 131–142.

Borrelli, J.J., Allesina, S., Amarasekare, P. et al. (2015). Selection on stability across ecological scales. *Trends in Ecology and Evolution* 30: 417–425.

Both, C. (1998). Experimental evidence for density dependence of reproduction in Great Tits. *Journal of Animal Ecology* 67: 667–674.

Bowler, B., Krebs, C.J., O'Donoghue, M., and Hone, J. (2014). Climatic amplification of the numerical response of a predator population to its prey: the case of coyotes and snowshoe hares. *Ecology* 95: 1153–1161.

Boyd, P.W., Strzepek, R., Fu, F., and Hutchins, D.A. (2010). Environmental control of open-ocean phytoplankton groups: now and in the future. *Limnology and Oceanography* 55: 1353–1376.

Bradford, J.B. and Hobbs, N.T. (2008). Regulating overabundant ungulate populations: an example for elk in Rocky Mountain National Park, Colorado. *Journal of Environmental Management* 86: 520–528.

Braysher, M., Saunders, G., Buckmaster, T., and Krebs, C.J. (2012) Principles underpinning best practice management of the damage due to pests in Australia. *Proceedings of the 25th Vertebrate Pest Conference,* Monterey, California. pp. 300–312.

Bryant, J.P. (1981). Phytochemical deterrence of snowshoe hare browsing by adventitious shoots of four Alaskan trees. *Science* 213: 889–890.

Carpenter, S.R., Walker, B.H., Anderies, J.M., and Abel, N. (2001). From metaphor to measurement: resilience of what to what? *Ecosystems* 4: 765–781.

Caswell, H. (2001). *Matrix Population Models: Construction, Analysis, and Interpretation,* 2e. Sunderland, Mass: Sinauer Associates.

Chamberlin, T.C. (1897). The method of multiple working hypotheses. *Journal of Geology* 5: 837–848. (reprinted in *Science* 148, 754–759 in 1965).

Conroy, M.J., Runge, M.C., Nichols, J.D. et al. (2011). Conservation in the face of climate change: the roles of alternative models, monitoring, and adaptation in confronting and reducing uncertainty. *Biological Conservation* 144: 1204–1213.

Courchamp, F., Berec, L., and Gascoigne, J. (2008). *Allee Effects in Ecology and Conservation.* Oxford, UK: Oxford University Press.

den Boer, P.J. and Reddingius, J. (1989). On the stabilization of animal numbers. Problems of testing. 2. Confrontation with data from the field. *Oecologia* 79: 143–149.

Dennis, B. and Taper, M.L. (1994). Density dependence in time series observations of natural populations: estimation and testing. *Ecological Monographs* 64: 205–224.

Diamond, J. (1986). Overview: laboratory experiments, field experiments, and natural experiments. In: *Community Ecology* (eds. J. Diamond and T.J. Case), 3–22. New York: Harper and Row.

Dybas, C.L. (2006). On a collision course: ocean plankton and climate change. *BioScience* 56: 642–646.

Eberhardt, L.L. and Peterson, R.O. (1999). Predicting the wolf-prey equilibrium point. *Canadian Journal of Zoology* 77: 494–498.

Elton, C. (1927). *Animal Ecology.* London, UK: Sidgwick and Jackson.

Field, S.A., O'Connor, P.J., Tyre, A.J., and Possingham, H.P. (2007). Making monitoring meaningful. *Austral Ecology* 32: 485–491.

Fleishman, E., Wolff, G.H., Boggs, C.L. et al. (1999). Conservation in practice: overcoming obstacles to implementation. *Conservation Biology* 13: 450–452.

Flockhart, D.T.T., Pichancourt, J.-B., Norris, D.R., and Martin, T.G. (2015). Unravelling the annual cycle in a migratory animal: breeding-season habitat loss drives population declines of monarch butterflies. *Journal of Animal Ecology* 84: 155–165.

Forman, R.T.T. (1964). Growth under controlled conditions to explain the hierarchical distributions of a moss, *Tetraphis pellucida. Ecological Monographs* 34: 1–25.

Fujisaki, I., Pearlstine, E., and Miller, M. (2008). Detecting population decline of birds using long-term monitoring data. *Population Ecology* 50: 275–284.

Gaston, K.J. (1991). How large is a species' geographic range? *Oikos* 61: 434–438.

Ginzburg, L.R. and Jensen, C.X.J. (2004). Rules of thumb for judging ecological theories. *Trends in Ecology and Evolution* 19: 121–126.

Ginzburg, L.R., Jensen, C.X.J., and Yule, J.V. (2007). Aiming the "unreasonable effectiveness of mathematics" at ecological theory. *Ecological Modelling* 207: 356–362.

Griffiths, D. (1998). Sampling effort, regression method, and the shape and slope of size-abundance relations. *Journal of Animal Ecology* 67: 795–804.

Guthery, F.S. (1999). The role of free water in bobwhite management. *Wildlife Society Bulletin* 27: 538–542.

Hanski, I. (2011). Habitat loss, the dynamics of biodiversity, and a perspective on conservation. *Ambio* 40: 248–255.

Harris, S. and Yalden, D.W. (2004). An integrated monitoring programme for terrestrial mammals in Britain. *Mammal Review* 34: 157–167.

Hartway, C. and Mills, L.S. (2012). A meta-analysis of the effects of common management actions on the nest success of North American birds. *Conservation Biology* 26: 657–666.

Henden, J.-A., Ims, R.A., and Yoccoz, N.G. (2009). Nonstationary spatio-temporal small rodent dynamics: evidence from long-term Norwegian fox bounty data. *Journal of Animal Ecology* 78: 636–645.

Jacobs, M., Vaske, J., Dubois, S., and Fehres, P. (2014). More than fear: role of emotions in acceptability of lethal control of wolves. *European Journal of Wildlife Research* 60: 589–598.

Jennions, M.D. and Møller, A.P. (2002). Relationships fade with time: a meta-analysis of temporal trends in

publication in ecology and evolution. *Proceedings of the Royal Society of London B* 269: 43–48.
Johnson, D.H. (2002). The importance of replication in wildlife research. *Journal of Wildlife Management* 66: 919–932.
Koricheva, J. (2003). Non-significant results in ecology: a burden or a blessing in disguise? *Oikos* 102: 397–401.
Korpimäki, E. (1994). Rapid or delayed tracking of multi-annual vole cycles by avian predators? *Journal of Animal Ecology* 63: 619–628.
Krebs, C.J. (1995). Two paradigms of population regulation. *Wildlife Research* 22: 1–10.
Krebs, C.J., Boutin, S., and Boonstra, R. (2001). *Ecosystem Dynamics of the Boreal Forest: The Kluane Project.* New York: Oxford University Press.
Krebs, C.J. (2002). Two complementary paradigms for analyzing population dynamics. *Philosophical Transactions of the Royal Society of London B* 357: 1211–1219.
Krebs, C.J. (2010). Case studies and ecological understanding. In: *The Ecology of Place: Contributions of Place-Based Research to Ecological Understanding* (eds. I. Billick and M.V. Price), 283–302. Chicago: University of Chicago Press.
Kroiss, S.J. and HilleRisLambers, J. (2014). Recruitment limitation of long-lived conifers: implications for climate change responses. *Ecology* 96: 1286–1297.
Kuhn, T. (1970). *The Structure of Scientific Revolutions.* Chicago, Illinois: University of Chicago Press.
Lawton, J.H. (2007). Ecology, politics and policy. *Journal of Applied Ecology* 44: 465–474.
Lindenmayer, D.B., Gibbons, P., Bourke, M. et al. (2012). Improving biodiversity monitoring in Australia. *Austral Ecology* 37: 285–294.
Meek, P., Fleming, P., Ballard, G. et al. (2014). *Camera Trapping: Wildlife Management and Research.* Melbourne, Victoria: CSIRO Publishing.
Mentis, M.T. (1988). Hypothetico-deductive and inductive approaches in ecology. *Functional Ecology* 2: 5–14.
Moss, R., Storch, I., and Müller, M. (2010). Trends in grouse research. *Wildlife Biology* 16: 1–11.
Murray, B.G.J. (1999). Can the population regulation controversy be buried and forgotten? *Oikos* 84: 148–152.
Murray, B.G.J. (2000). Dynamics of an age-structured population drawn from a random numbers tables. *Austral Ecology* 25: 297–304.
Myers-Smith, I.H., Trefry, S.A., and Swarbrick, V.J. (2012). Resilience: easy to use but hard to define. *Ideas in Ecology and Evolution* 5: 44–53.
Nicholson, A.J. (1933). The balance of animal populations. *Journal of Animal Ecology* 2: 132–178.
O'Connor, R.J. (2000). Why ecology lags behind biology. *The Scientist* 14: 35.
O'Donoghue, M., Boutin, S., Krebs, C.J., and Hofer, E.J. (1997). Numerical responses of coyotes and lynx to the snowshoe hare cycle. *Oikos* 80: 150–162.
Oreskes, N. and Conway, E.M. (2010). *Merchants of Doubt: How a Handful of Scientists Obscured the Truth on Issues from Tobacco Smoke to Global Warming.* New York: Bloomsbury Press.
Peters, R.H. (1991). *A Critique for Ecology.* Cambridge, England: Cambridge University Press.
Platt, J.R. (1964). Strong inference. *Science* 146: 347–353.
Popper, K.R. (1963). *Conjectures and Refutations: The Growth of Scientific Knowledge.* London: Routledge and Kegan Paul.
Redpath, S.M., Mougeot, F., Leckie, F.M. et al. (2006). Testing the role of parasites in driving the cyclic population dynamics of a gamebird. *Ecology Letters* 9: 410–418.
Rioux Paquette, S., Pelletier, F., Garant, D., and Bélisle, M. (2014). Severe recent decrease of adult body mass in a declining insectivorous bird population. *Proceedings of the Royal Society of London B* 281: 20140649.
Ripple, W.J., Estes, J.A., Beschta, R.L. et al. (2014). Status and ecological effects of the world's largest carnivores. *Science* 343: 1241484.
Ruttimann, J. (2006). Sick seas. *Nature* 442: 978–980.
Sahanatien, V. and Derocher, A.E. (2012). Monitoring sea ice habitat fragmentation for polar bear conservation. *Animal Conservation* 15: 397–406.
Schindler, D.W. (1998). Replication versus realism: the need for ecosystem-scale experiments. *Ecosystems* 1: 323–334.
Shine, R. (2010). The ecological impact of invasive cane toads (*Bufo marinus*) in Australia. *Quarterly Review of Biology* 85: 253–291.
Short, J., Kinnear, J.E., and Robley, A. (2002). Surplus killing by introduced predators in Australia – evidence for ineffective anti-predator adaptations in native prey species? *Biological Conservation* 103: 283–301.
Sibly, R.M. and Hone, J. (2002). Population growth rate and its determinants: an overview. *Philosophical Transactions of the Royal Society of London B* 357: 1153–1170.
Sinclair, A.R.E. (1989). Population regulation in animals. In: *Ecological Concepts* (ed. J.M. Cherrett), 197–241. Oxford: Blackwell Scientific.
Slavin, R.E. (1995). Best evidence synthesis: an intelligent alternative to meta-analysis. *Journal of Clinical Epidemiology* 48: 9–18.
Snäll, T., Kindvall, O., Nilsson, J., and Pärt, T. (2011). Evaluating citizen-based presence data for bird monitoring. *Biological Conservation* 144: 804–810.
Stadt, J.J., Schieck, J., and Stelfox, H.A. (2006). Alberta biodiversity monitoring program – monitoring effectiveness of sustainable forest management planning. *Environmental Monitoring and Assessment* 121: 33–46.

Stenseth, N.C., Falck, W., Chan, K.S. et al. (1998). From patterns to processes: phase and density dependencies in the Canadian lynx cycle. *Proceedings of the National Academy of Sciences USA* 95: 15430–15435.

Stephens, S.E., Koons, D.N., Rotella, J.J., and Willey, D.W. (2003). Effects of habitat fragmentation on avian nesting success: a review of the evidence at multiple spatial scales. *Biological Conservation* 115: 101–110.

Stewart, G. (2010). Meta-analysis in applied ecology. *Biology Letters* 6: 78–81.

Strong, D.R. (1986). Density-vague population change. *Trends in Ecology and Evolution* 1: 39–42.

Sullivan, B.L., Aycrigg, J.L., Barry, J.H. et al. (2014). The eBird enterprise: an integrated approach to development and application of citizen science. *Biological Conservation* 169: 31–40.

Sundell, J., O'Hara, R.B., Helle, P. et al. (2013). Numerical response of small mustelids to vole abundance: delayed or not? *Oikos* 122: 1112–1120.

Sutherland, W.J. (2006). *Ecological Census Techniques: A Handbook*. Cambridge, UK: Cambridge University Press.

Sutherland, W.J., Albon, S.D., Allison, H. et al. (2010). The identification of priority policy options for UK nature conservation. *Journal of Applied Ecology* 47: 955–965.

Thomas, C.D., Franco, A.M.A., and Hill, J.K. (2006). Range retractions and extinction in the face of climate warming. *Trends in Ecology and Evolution* 21: 415–416.

Trebilco, R., Baum, J.K., Salomon, A.K., and Dulvy, N.K. (2013). Ecosystem ecology: size-based constraints on the pyramids of life. *Trends in Ecology and Evolution* 28: 423–431.

Turchin, P. (1990). Rarity of density dependence or population regulation with lags? *Nature* 344: 660–663.

Turchin, P. (1999). Population regulation: a synthetic view. *Oikos* 84: 153–159.

Urban, M.C., Phillips, B.L., Skelly, D.K., and Shine, R. (2007). The cane toad's (*Chaunus [Bufo] marinus*) increasing ability to invade Australia is revealed by a dynamically updated range model. *Proceedings of the Royal Society B* 274: 1413–1419.

Vetter, D., Rücker, G., and Storch, I. (2013). Meta-analysis: a need for well-defined usage in ecology and conservation biology. *Ecosphere* 4: art74.

Walsh, J.C., Wilson, K.A., Benshemesh, J., and Possingham, H.P. (2012). Unexpected outcomes of invasive predator control: the importance of evaluating conservation management actions. *Animal Conservation* 15: 319–328.

Whittaker, R.J. (2010). Meta-analyses and mega-mistakes: calling time on meta-analysis of the species richness – productivity relationship. *Ecology* 91: 2522–2533.

2

From Research Hypothesis to Model Selection

A Strategy for Robust Inference in Population Ecology

Dennis L. Murray[1], *Guillaume Bastille-Rousseau*[1], *Lynne E. Beaty*[1], *Megan L. Hornseth*[1], *Jeffrey R. Row*[2] and *Daniel H. Thornton*[3]

[1] Department of Biology, Trent University, Peterborough, Ontario, Canada
[2] School of Environment, Resources, and Sustainability, University of Waterloo, Waterloo, Ontario, Canada
[3] School of Environment, Washington State University, Pullman, WA, USA

Summary

Study design and data analysis in population ecology are becoming more sophisticated, rigorous, and insightful. This transition involves development of research hypotheses that are mindful of the complexity of field data, combined with judicious evaluation of statistical models serving as direct measures of these hypotheses. Yet, the analytical philosophy and methods designed to deal with ecological complexity and data integrity are poorly covered in statistical texts and courses, often giving rise to weak inference despite good intentions. Research hypotheses should reflect natural relationships between response and predictor variables, which are complex and inadequately described by a simple statistical null hypothesis. Translating research hypotheses to candidate statistical models is a nontrivial exercise requiring careful consideration and creative structuring. Once data are collected, diagnostic tests should verify whether transformation, imputation, or variable reduction are required to best test explicit predictions. Model selection via information-theoretic methods ranks models that are weighted relative to their fit and degree of parsimony, and multimodel inference incorporates model weights in parameter and variance estimation. A similar approach can be adopted using Bayesian statistics. An important concluding step involves model validation to confirm the biological relevance of the findings. When properly applied, these steps promote full integration of complex field data into a robust analytical context, providing stronger inference and evaluation of research questions than is possible using traditional statistical methods. This is an important point as research in population ecology assumes a new level of complexity, relevance, and immediacy in response to ongoing environmental change.

2.1 Introduction

The last decades have witnessed a remarkable shift in the analytical philosophy and degree of sophistication used for deriving inference from ecological research. Largely gone are the days of blindly running a series of ANOVAs (analysis of variances) or linear regressions in the quest for a statistically significant result and using a rigid and arbitrary probability value to determine whether a null hypothesis should be rejected. Through time and plenty of questionable findings, researchers and journal editors now recognize that robust inference should come mainly through data collection and analysis that considers the variety of factors affecting natural populations; capturing the essence of this complexity is normally beyond the scope of traditional approaches based on frequentist tests of a simple null hypothesis and dichotomous *P*-value (Johnson and Omland 2004). Indeed, *frequentist statistical methods* (hereafter also referred to as "traditional") were developed in the first half of the twentieth century specifically to deal with data from studies with repeated observations where experimental design and sample randomization were under the researcher's strict control (Stephens et al. 2007). In this context, *hypothesis testing* evaluates whether a given treatment elicits a statistically significant response, based on a specified probability (usually $\alpha = 0.05$). At the same time that traditional methods were being developed for application in experimental research, quantitative analysis in ecology and other field-based disciplines was in its formative stages, and distinctions between analytical contexts for different research designs and questions were not immediately obvious. This led to widespread adoption of the null hypothesis as the foundation for statistical analysis in field research,

Population Ecology in Practice, First Edition. Edited by Dennis L. Murray and Brett K. Sandercock.

Companion website: www.wiley.com/go/MurrayPopulationEcology

even despite earlier calls for a more pluralistic approach (Chamberlin 1890, see also Elliott and Brook 2007). Indeed, the observational nature of most field studies sets the stage for examining how a set of factors may influence a given response, rather than whether manipulation of a specific factor drives a predicted response. The former line of investigation should force consideration of a broader array of predictors than is typically entertained in experimental research (Garamszegi 2011).

2.1.1 Inductive Methods

Until recently, traditional methods served as the mainstay of quantitative analysis in ecology, making many important contributions to our current understanding of patterns and processes governing natural systems. Given this, new researchers logically may ask why a different analytical philosophy is necessary, but consider the following example: An ecologist is interested in how pollinating insects affect plant productivity and therefore designs a study to document the relationship between number and diversity of pollinators and seed production (Box 2.1). The researcher predicts that more pollinator visits will result in greater seed yields and thus she might design an experiment directly manipulating the number of pollinators that reach plants during the floral season. Traditional statistical methods such as ANOVA would evaluate whether differences in seed production occur between flowers that are exposed to experimental treatment versus unmanipulated controls (i.e. natural visitation vs. no visitation), but the test would focus exclusively on whether the null model was rejected irrespective of the particular research hypothesis conceived by the researcher. Accordingly, the statistical test would never directly inform the broader ecological question being posed, that is the role of pollinator visitation rate

Box 2.1 Comparing Analytical Frameworks

Using the pollinator system described above, let us break down how – and what type of – questions may be asked and answered following inductive, hypothetico-deductive, information-theoretic, and Bayesian analytical frameworks. For each framework, we outline how the hypothesis(es) may be framed, how an experiment can be designed to test the hypothesis(es), what statistical test may be performed, and the conclusions that can be drawn from the results of the experiment.

Inductive Methods:

- Hypothesis: The number of visiting pollinators affects seed production of a flowering plant.
- Experimental Design: Manipulate number of pollinators visiting plants and measure seed yield.
- Statistical Analysis: Likely an ANOVA
- Conclusions: The null hypothesis (that seed production does not differ between treatments with different numbers of pollinators) is rejected or not.

Hypothetico-deductive Methods:

- Hypotheses: The number of visiting pollinators, or the size of the floral patch, or the distance between patches, influences seed production by a flowering plant.
- Experimental Design: Field observations of pollinator number, patch size, distance between patches, and the number of seeds produced by each plant.
- Statistical Analysis: Multiple or stepwise regression
- Conclusions: Pollinator number and/or patch size and/or distance have a statistically significant relationship with seed yield.

Information-theoretic Methods:

- Hypotheses: The number of visiting pollinators and/or the size of the floral patch and/or the distance between patches influences seed production by a flowering plant.
- Experimental Design: Field observations of pollinator number, patch size, distance between patches, and the number of seeds produced by each plant.
- Statistical Analysis: Model selection and, if necessary, model averaging, of multiple regression models
- Conclusions: The most-supported model contains the number of visiting pollinators and/or the size of the floral patch and/or the distance between patches.

Bayesian Methods:

- Hypotheses: The number of visiting pollinators and/or the size of the floral patch and/or the distance between patches influences seed production by a flowering plant.
- Experimental Design: Field observations of pollinator number, patch size, distance between patches, and the number of seeds produced by each plant.
- Statistical Analysis: Decide on prior distributions for model parameters and conduct Bayesian analysis.
- Conclusions: Given what is known about this system and the observed data, the model containing the number of visiting pollinators and/or the size of the floral patch and/or the distance between patches has the highest posterior probability.

on seed production (Gerrodette 2011). The statistical significance of the pollinator variable, based on the *P*-value of the test, would be directly related to sample size and thus potentially irreproducible by future researchers attempting to corroborate the findings (Halsey et al. 2015). From a practical perspective, the coefficient describing the magnitude of seed production difference according to treatment would have little biological relevance or predictive value because it would refer to the unnatural (dichotomous) condition of pollination vs. no pollination, rather than a more realistic gradient in visitation and pollination rates. This study would be inductive in the sense that over the longer term the researcher would sequentially modify or refine the initial research hypothesis to better fit new information, and then retest the hypothesis using variations on the initial study design. Eventually, the inductive approach should lead to a statistically significant test of a hypothesis that is derived from the original (Figure 2.1a).

The above approach is pervasive in ecology but it is easy to see how it may not be best-suited for field research (Halsey et al. 2015; Betini et al. 2017). Depending on the study design and treatment, it could give rise to testing "pet" hypotheses in an inefficient, repetitive, and, to some extent, contrived manner. Returning to the question of seed production and pollinator visits, the researcher would realistically look at a variety of different experiments and probably conduct the study over several years before a satisfactory answer to the question could be obtained. Indeed, apart from manipulating the number of pollinators, one might imagine companion experiments on pollinator travel distance, floral density, nectar yields, and a range of other factors, before the researcher would have a comprehensive understanding of the mechanisms underlying pollination. In addition, variation in the importance of the above measures doubtless would arise because of annual variation in temperature, precipitation, and other factors that should not be ignored.

2.1.2 Hypothetico-deductive Methods

Alternatively, the pollination researcher may adopt a more observational approach by relating natural insect visitation rate to the number of seeds produced. The study could use the model coefficient for pollinator visitation rate to describe how natural seed production varies across a realistic range of visits. This research context differs substantially from the earlier experimental study, and the single null hypothesis and significance test

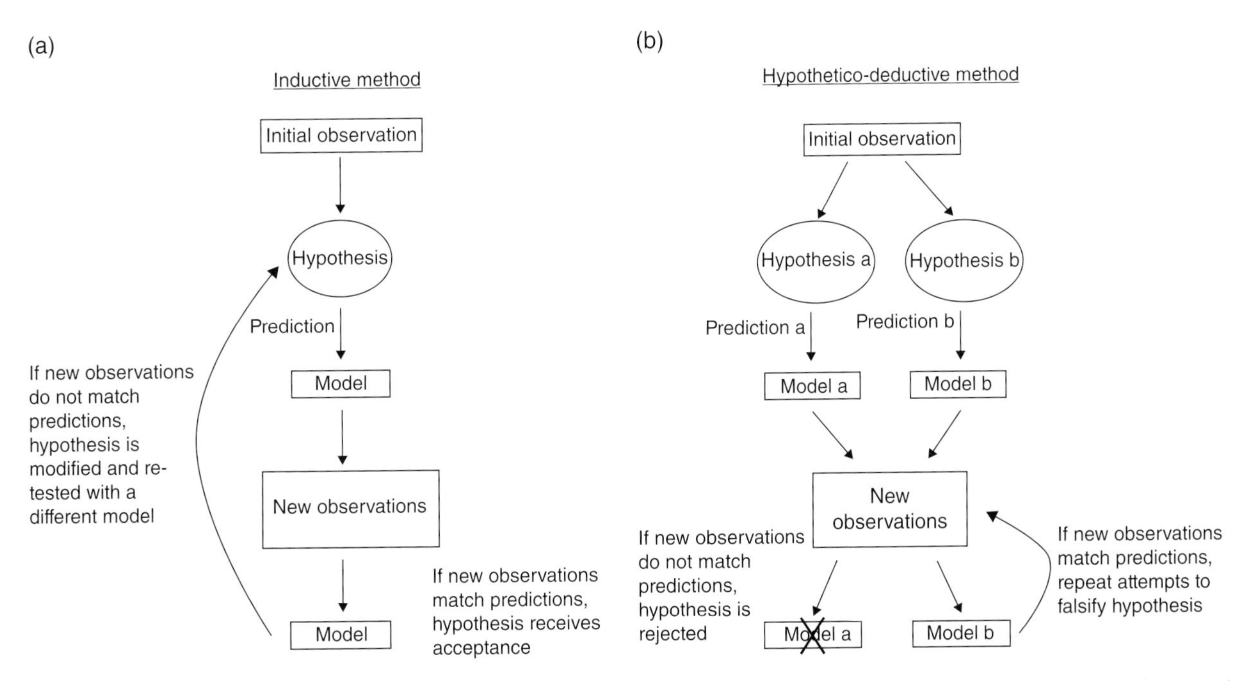

Figure 2.1 Flowcharts for the inductive (a) and hypothetico-deductive (b) methods. Initial observations are obtained and are used to develop research hypotheses and their predictions, which are then tested through statistical models. Models are mathematical versions of research hypotheses. Through induction, the cycle of hypothesis, prediction, model, and new observation is conducted repeatedly, with ultimate acceptance of a modified hypothesis being the endpoint of the process. The hypothetico-deductive method involves simultaneous evaluation of multiple research hypotheses, with emphasis on falsification rather than verification of individual hypotheses. Using the hypothetico-deductive approach, the correct hypothesis is the one that withstands multiple attempts at falsification. Source: Adapted from Gotelli and Ellison (2013).

from before would be of little relevance to the more comprehensive investigation. Instead, the researcher would adopt an analysis that is fully sensitive to biological complexity while surely recognizing that including other ecological factors like pollinator visits, size of the floral patch, distance of the patch to other patches, identity and abundance of pollinators, or nearby occurrence of other flowering plant species, also may influence pollinator behavior. The researcher may have suspected some of these relationships at the outset of the study and could have collected relevant data to test these ideas. However, while inclusion of these extra variables might improve model fit and provide a more informed understanding of the factors influencing pollination success, a single model including all variables could generate biased parameter estimates depending on the strength of individual predictors and their interactions. Even if the number of variables was reduced sequentially, as is typically the case when using a *stepwise* procedure (Hegyi and Garamszegi 2011), the researcher still would base variable retention on an arbitrary *P*-value and without consideration for the diversity of hypotheses being investigated (Box 2.1). Because the multiple hypotheses that the researcher has in mind probably involve different combinations of the predictor variables, the stepwise model would not be useful for testing predictor relationships against each other (Whittingham et al. 2006). Accordingly, a single model would poorly describe the complex relationships affecting pollination rates, and a more careful consideration of a variety of hypotheses should provide additional insight into the biological relationship between pollination and seed production. In fact, consideration of multiple hypotheses simultaneously conforms to a *hypothetico-deductive approach* (Figure 2.1b), where evaluation relies upon multiple hypotheses from the same set of observations. It follows that lack-of-fit with the data leads to rejection of one or more hypotheses. This process is repeated to identify the hypothesis that best withstands repeated attempts at falsification.

2.1.3 Multimodel Inference

The recent philosophical metamorphosis in quantitative analysis of ecological data is supported primarily by *information-theoretic* (IT) *methods*, which are based on Kullback–Leibler (KL) information that identifies the model within a set of comparable models that best approximates reality (Burnham et al. 2011; Figure 2.2). Instead of focusing on effect size and statistical precision associated with the study design, the emphasis is on *relative support* for individual candidate models across a wider set of plausible models. The IT method encourages researchers to adopt a more nuanced approach by allowing for inclusion of information from *multiple hypotheses* when evaluating the role of a given variable or the strength of an individual model. Practically speaking, the shift in perspective makes sense because while field experiments are often highly regarded for robust inference in ecology (MacNab

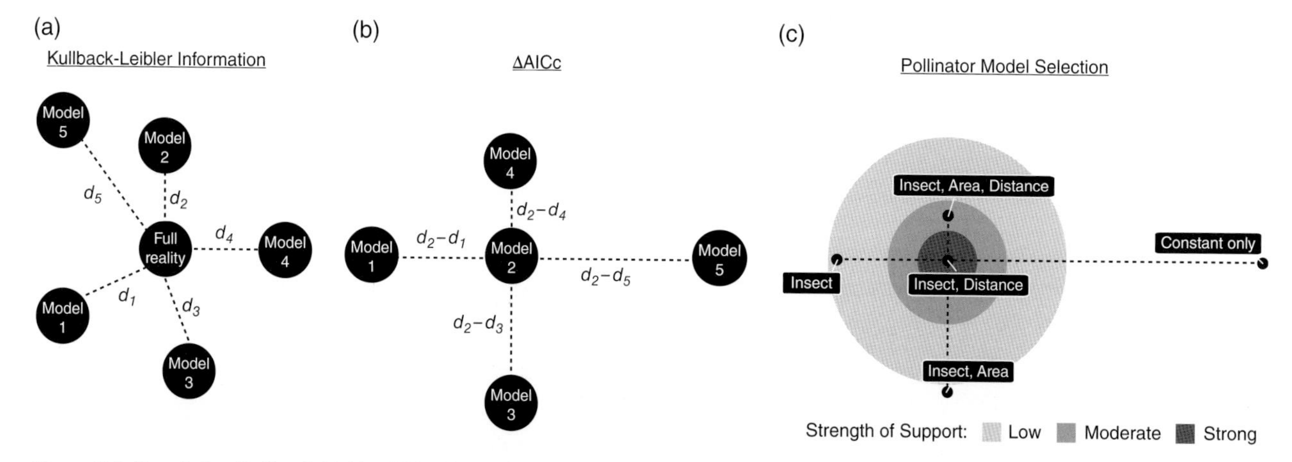

Figure 2.2 Translating Kullback-Liebler (KL) information to model selection via Akaike's Information Criterion (AIC). (a) All plausible models from a candidate set are different distances (*d*) from true reality. (b) Given that we do not know what full reality is, the model that is the shortest distance from reality given its fit to observed data and degree of parsimony becomes the most-supported model of the candidate set and is the model against which all other models are compared (Model 2, in this case). The distance between the most-supported model and the other models is known as ΔAIC (ΔAICc = AIC corrected for small sample sizes). (c) For the pollinator example, the model with the most support contains the number of visiting pollinators and distance between floral patches ("Insect, Distance"). The degree of support for the other candidate models depends on their relative distance from the Insect, Distance model. Those within 2 ΔAICc units have strong support and are a plausible alternative hypothesis while those within 4 and 8 ΔAICc units receive moderate and low support, respectively. Models that are >8 ΔAICc units are generally not supported. Note that other information criteria (e.g. Bayes Information Criterion (BIC); Schwartz Information Criterion) are functionally similar to AIC.

1983; Carpenter et al. 1995), most natural systems are simply too complex or difficult to study to provide reliable inference from traditional, dichotomous approaches. In fact, the loss of information associated with ignoring hypotheses that may have partial support should concern anyone familiar with ecological complexity and the subtleties involved in seeking reliable knowledge from field observations. In our example of seed production versus pollinator visits, the array of factors potentially affecting the response would be integrated into the analysis through models having different combinations of the variables, ultimately with each candidate providing a weighted measure of the predicted relationship (Figure 2.3). Accordingly, valuable information obtained from fitting a broader set of models would be considered explicitly, and hypotheses having different levels of support would be weighted according to their relative support (Box 2.1).

In as little as 30 years, IT methods have become common in field-based ecological research and are making significant inroads into other disciplines where multivariate complexity is the norm (Johnson and Omland 2004; Lindberg et al. 2015). However, the transition has not been without some debate. Some researchers have voiced calls to either stave off the current trend in IT-based analysis or to adopt a more measured transition to these new

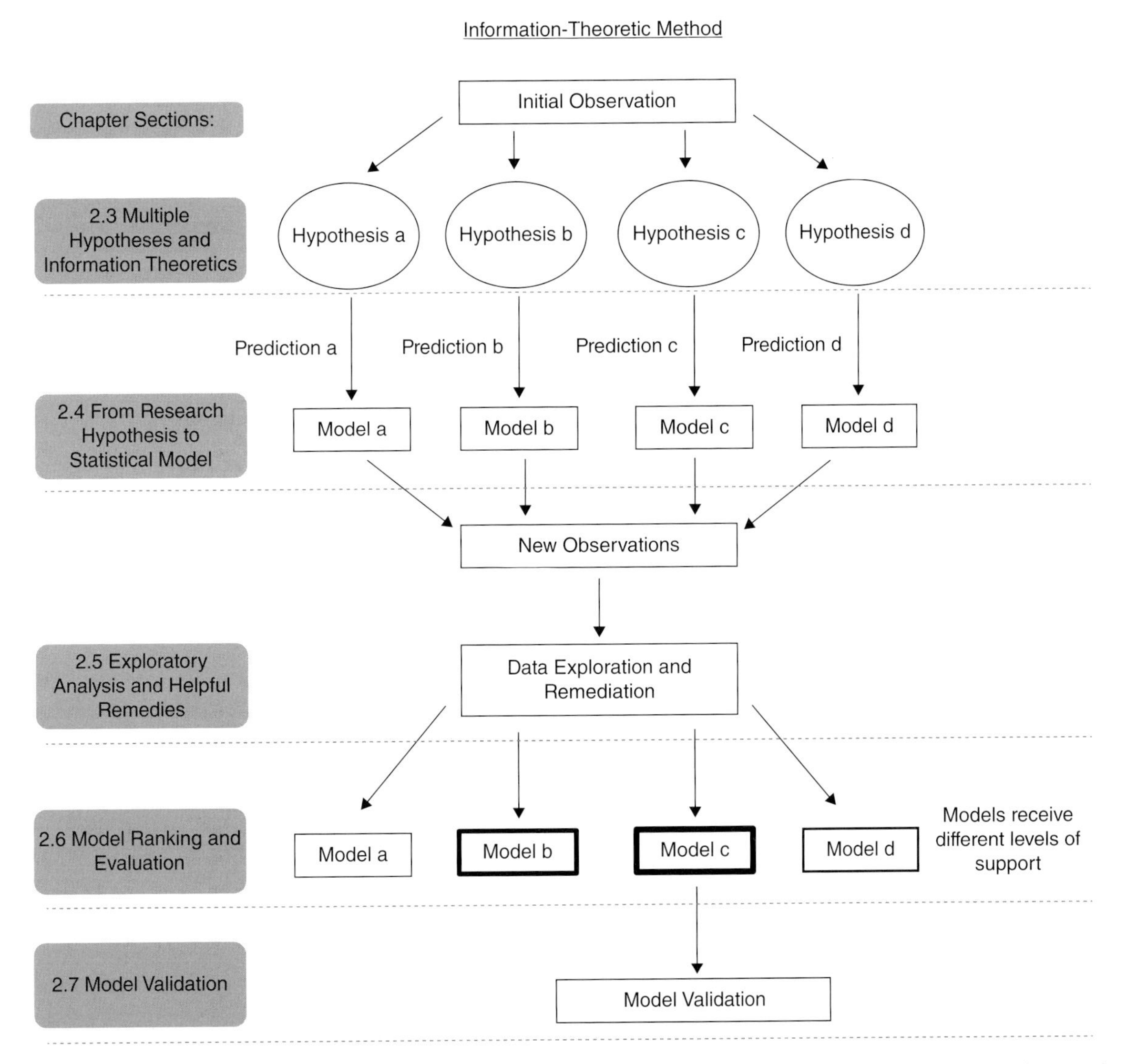

Figure 2.3 The method of evaluating multiple research hypotheses and the relevant sections of this chapter for guidance in the completion of this process. Initial observations obtained from background literature, modeling, or theory are used to develop hypotheses that serve as basis for one or more unique predictions that are then evaluated when new observations are fit to models corresponding to each hypothesis. Here, rectangle thickness represents the level of support for each model and corresponding hypothesis, from high-to-low (Hypothesis c > Hypothesis b > Hypothesis d > Hypothesis a).

approaches (Guthery et al. 2005; Stephens et al. 2005; Spanos 2014; Madden et al. 2015), which may not be completely new methods (Murtaugh 2014; de Valpine 2014). To a large extent, these arguments are based on the perception that null hypothesis testing remains more rigorous especially in cases of inferring causality in simple systems or when field experiments have been conducted (Fieberg and Johnson 2015; Betini et al. 2017). In contrast, calls to jettison null hypothesis testing and stepwise approaches have been similarly voiced by advocates of the IT philosophy (Anderson et al. 2000; Lukacs et al. 2007; Burnham and Anderson 2014), leading to a high degree of polarization among ecologists. Not surprisingly, traditional approaches based on the null hypothesis and *P*-value remain at the foundation of laboratory research and statistical teaching in the biological sciences (Whitlock and Schluter 2009; Zar 2010).

2.1.4 Bayesian Methods

Bayesian methods complement the suite of approaches that are available for statistical analysis, and they have recently received increasing attention from population ecologists (Gotelli and Ellison 2013; Hooten and Hobbs 2015; McCarthy 2015). Bayesian methods are based on determining the probability that particular conditions exist, given a set of observations, and the approach contrasts sharply with traditional frequentist methods that aim to determine the probability of observing the data (or obtaining a significant *P*-value), given the conditions. Bayesian methods involve development of a prior probability distribution that is based on initial knowledge; then, given the likelihood of the observations, a posterior probability distribution is generated and serves as the basis for inference (Box 2.1). Bayesian methods constitute a fundamental shift in statistical philosophy over traditional approaches, and their benefits for ecologists include making better use of ancillary data when designing a field study (Martínez-Abraín et al. 2014), and the potential to consider model parameters as random instead of fixed factors. Consideration of multiple research hypotheses and multimodel inference are also possible in a Bayesian context, allowing these methods to be adapted to the complexity that is inherent in natural systems. The downside of Bayesian methods is that they are conceptually foreign to many ecologists and normally require substantial computing power; to date they have been under used across all sciences. Yet, Bayesian methods are becoming increasingly familiar and accessible, and are now covered in some undergraduate and graduate statistics courses. Recognizing the importance of Bayesian methods to population ecology, we highlight several excellent guides to help users apply these techniques in their research (Ellison 2004; King et al. 2009; Beaumont 2010; Hooten and Hobbs 2015). While the remainder of this chapter focuses on hypothesis development and testing, data diagnostic approaches, and multimodel inference using IT methods, we draw links to Bayesian methods where appropriate.

2.2 What Constitutes a Good Research Hypothesis?

In Chapter 1, Charles Krebs provided an outline of the development of hypotheses in ecology (see also Ford 2009); here we elaborate on the linkage between *research hypotheses* and *statistical models*. Research hypotheses help distill our understanding of natural phenomena, and the development of clear and testable hypotheses is part of the foundation of any sound scientific investigation. Whereas the *null hypothesis* is simply a statement implying no relationship between predictor and explanatory variables, a research hypothesis seeks to explain *why* patterns occur; developing an effective research hypothesis can be challenging when there is limited baseline knowledge of the study system and the various complexities that drive its dynamics (Box 2.2). The early science philosopher, Karl Popper, was explicit that a hypothesis cannot be accepted or proved, but as an individual research hypothesis withstands repeated attempts at falsification, it gains firmer support. Normally, it is best to begin with a simple *research hypothesis* that can be readily conjured from available background information. For instance, returning to our example of seed production and insect pollination, a starting point based on prior knowledge of this biological system (Jennersten and Nilsson 1993), combined with a dose of common sense, could give rise to the hypothesis that: *Plants visited at a higher frequency by pollinating insects have higher seed production*. The simple statement captures two ingredients that are key to any effective research hypothesis: (i) an established *causality* between the observation and response; and (ii) *testability* with an indication of the necessary data that should be collected. It should be evident that even this simple hypothesis is clearly more informative than the corresponding null alternative, and another advantage is that its structure can be extended to accommodate additional factors that may elaborate on the hypothesized role of pollinators.

The research hypothesis assumes a greater importance in the context of IT compared to traditional statistical methods, and accordingly it requires careful thought and insight into the underlying mechanisms driving the relationship of interest. Likewise, development of the prior probability distribution in Bayesian statistics also

Box 2.2 Asking the Right Questions to Answer the Right Question

Developing alternative hypotheses and translating them into statistical models can be difficult as it requires a thorough understanding of your study system and your question(s) of interest. Before beginning to brainstorm hypotheses, ask yourself the following questions about your study system:

- What are the key players (abiotic and biotic) in your system?
- What are the secondary players in your system (i.e. those that may influence one or more of your key players)?
- What ecological theories are relevant to your study system? Do they provide any additional insight into the dynamics of your system?
- Are there gaps in our understanding of how your system works? If so, what information or data are missing? How might this gap influence your experiment?
- What conclusions do you want to be able to draw by the end of your study? Can you make those conclusions (i.e. are they feasible and are you collecting the right data) given your current experimental design?

After you have asked yourself these questions about your study system, try to form alternative hypotheses. Consider the following questions during hypothesis development:

- Which of your key players are likely to independently and singly influence your response variable of interest? What is the directionality of this influence (i.e. is it positive or negative)?
- Which primary and secondary players cannot logically be paired together? That is, the pairing of which factors does not make biological sense?
- Which factors may work in combination to influence your response variable? Are they likely to interact or will their effect be additive?
- Which factors are likely to interact with another factor to influence your response variable?
- Go through each of the hypotheses you have developed – if this hypothesis/model is the most supported at the end of your experiment and analysis, how do you interpret the results?

demands careful understanding of the study system (Ellison 2004). Predictions accompany hypotheses and serve as the direct link to the data being collected, so they must be testable and ultimately challenge the integrity of the hypothesis. For example, the prediction that there is *positive correlation between pollinator visits and seed production* is completely testable (and falsifiable) by an observant researcher collecting appropriate data. It follows that although this particular prediction may seem self-evident and somewhat trivial, in the context of the pollination study it helps identify the expected outcome of the investigation. Naturally, predicting the outcome of a hypothesis test becomes increasingly challenging in complex study systems where outcomes are dependent on a range of factors, as may be the case especially for research addressing ecosystem function, community dynamics, or cumulative effects. Regardless, a sound, explicit, and parsimonious prediction should help direct relevant data collection in even the most complex research situations. To the fullest extent possible, predictions should be *unequivocally falsifiable*, leaving no uncertainty from the collected data as to whether they are supported or not.

Dochtermann and Jenkins (2011) provide three general approaches for developing research hypotheses, namely: (i) collation of previous research results through literature review; (ii) prediction derived from theory and modeling; and (iii) exploratory analysis of existing data. Using previous research findings or theory to generate hypotheses and predictions is standard protocol across the majority of ecological research. For example, Levi and Wilmers (2012) reviewed the extensive literature on competition among mammalian carnivores, and surmised that natural restoration of wolf populations (*Canis lupus*) should lead to suppression of coyotes (*Canis latrans*), which in turn, would release red foxes (*Vulpes vulpes*) from competition with coyotes. Using the second approach, Fussmann et al. (2000) developed a mathematical model predicting dynamical relationships in simple predator–prey systems and tested the model by showing that population dynamics of rotifers (*Brachyonus* spp.) and algae (*Chlorella* spp.) correspond to specific predictions. The third approach cited by Dochtermann and Jenkins (2011), exploratory analysis, requires more critical evaluation as it rarely should form the sole or primary basis for developing research hypotheses. Here, we distinguish between *exploratory analysis* (sensu Dochtermann and Jenkins 2011), as a separate concept from *data exploration* where the purpose is remediation in advance of formal statistical analysis. First, exploratory analysis supports data collection that lacks the focus usually imposed by a hypothesis and associated prediction, and thereby can lead to omissions or excesses in acquired information. Rather, the hypothesis and prediction normally should drive data collection and thus be stated a priori. Second, data exploration with the purpose of generating a research hypothesis promotes *data dredging* (also known as *data fishing* or *data snooping*), which increases the likelihood of identifying associations between variables through chance alone (Anderson et al. 2001). For example, if the aforementioned plant pollination study was conducted in the absence of an underlying hypothesis, it is conceivable that data snooping could reveal correlation with unrelated factors like day of the week that seeds were counted or the color of the researcher's t-shirt when pollinators were observed; obviously such relationships have no logical basis for driving the observed patterns. The point is that spurious associations tend to be minimized when data collection and analysis are guided by a research hypothesis and related predictions that are formulated a priori, which consequently promotes broader generality and legitimacy of the analytical results. Stepwise regression is a form of data dredging because of the circularity in having the single best-fit model serve as a test of one or more unspecified hypotheses (Hegyi and Garamszegi 2011).

Notwithstanding these points, there are instances when it is inevitable that data are collected before a research hypothesis is developed. This situation may lead to the temptation that data should be explored statistically before the hypothesis and prediction(s) are established, but usually there is sufficient information available from background literature, theory, or model systems to help maintain an independent approach to hypothesis development. However, if in exceptional circumstances it is essential to explore data first, the researcher should recognize this compromise and conduct the actual analysis using a separate and independent dataset to reduce the likelihood that spurious associations are transferred. Note that a researcher alternatively might divide a dataset into two subsets, one for exploration and the other for analysis, but this strategy is not ideal because it does not fully establish independence between datasets and may aggravate sample size limitations that are common in ecological studies (Quinn and Keough 2002). Regardless, in this age of growth and maturation in our understanding of ecological systems, there should be few instances when hypotheses and predictions cannot be generated prior to data collection and analysis.

2.3 Multiple Hypotheses and Information Theoretics

Multiple hypotheses offer the advantage of more efficiently addressing complex research questions by contrasting several plausible alternatives simultaneously. To illustrate this point, consider the relationship between the occurrence of snowshoe hares (*Lepus americanus*) relative to habitat features of forested landscapes of North America. Chapter 1 provided the framework for testing hypotheses related to the role of snowshoe hares in the boreal forest ecosystem, and here we elaborate on the link between local habitat features and hare occupancy. A thorough literature review on this topic reveals several competing ideas regarding what drives site occupancy by hares: (i) food limitation during winter, due to the hare's reliance on woody browse as food (Pease et al. 1979); (ii) structural cover in the form of dense vegetation, allowing hares to avoid predator detection and capture (Wirsing et al. 2002); (iii) large patches of suitable habitat, providing hares with their minimum spatial needs (Ausband and Baty 2005); and (iv) linkages between habitat patches, allowing hares to travel between suitable areas (Griffin and Mills 2009). Other factors also influence patterns of hare occurrence on the landscape (Murray 2003), but several of these can be excluded because they operate at a different spatial or temporal scale than is relevant to the study. Other hypotheses may need to be excluded simply because they are untestable given the study design or other constraints.

Hares have a dynamic relationship with their natural environment and it is unlikely that any single factor can fully explain their habitat occupancy. To reflect real-life complexity, additional hypotheses beyond the four univariate relationships described above are added to the candidate set (Table 2.1). Some hypotheses involve pairs or groups of factors that, when considered together, may provide additional insight into hare occupancy dynamics. For example, areas that have an abundance of both food and cover might be especially attractive to hares (Hypothesis 5), or sites with cover may only be occupied when connected to other patches having food (Hypothesis 10). If we expect that the relationship between food and cover, or between food, cover, and patch connectivity, is consistent across the range of conditions (in other words that the effect of one factor remains proportional to the other), then developing the corresponding multi-factor hypothesis may be straightforward because the factors are *additive*. However, if we hypothesize that factors have inconsistent relationships with the predicted outcome (that one factor varies disproportionately relative to the other) then we can expect multiplicative effects, as indicated by additional terms for *interactions* among factors. Possible interactions was the rationale underlying Hypothesis 12, as it may be that the abundant vegetation that provides adequate cover also contains high-quality food (Table 2.1). Alternatively, we might consider that cover density and patch size are highly correlated and thus avoid including both together in the same model. Elsewhere, Burnham et al. (2011) provide a helpful example concerning the development of multiple hypotheses to explain extra-pair copulations in birds.

Table 2.1 Candidate hypotheses, predictions, and models associated with patterns of occurrence of snowshoe hares in forested landscapes.

Number	Hypothesis	Predicted association (direction)	Model
1	Food[a]	Food density (+)	$\beta_0 + \beta_1 X_1$
2	Cover[b]	Cover density (+)	$\beta_0 + \beta_2 X_2$
3	Patch size[c]	Patch size (+)	$\beta_0 + \beta_3 X_3$
4	Patch connectivity[d]	Nearest patch distance (−)	$\beta_0 + \beta_4 X_4$
5	Food, Cover	Food density (+); Cover density (+)	$\beta_0 + \beta_1 X_1 + \beta_2 X_2$
6	Food, Patch size	Food density (+); Patch size (+)	$\beta_0 + \beta_1 X_1 + \beta_3 X_3$
7	Food, Patch connectivity	Food density (+); Nearest patch distance (−)	$\beta_0 + \beta_1 X_1 + \beta_4 X_4$
8	Cover, Patch size	Cover density (+); Patch size (+)	$\beta_0 + \beta_2 X_2 + \beta_3 X_3$
9	Food, Cover, Patch size	Food density (+); Cover (+); Patch size (+)	$\beta_0 + \beta_1 X_1 + \beta_2 X_2 + \beta_3 X_3$
10	Food, Cover, Patch connectivity	Food density (+); Cover (+); Nearest patch distance (+)	$\beta_0 + \beta_1 X_1 + \beta_2 X_2 + \beta_4 X_4$
11	Food, Patch size, Patch connectivity	Food density (+); Patch size (+); Nearest patch distance (−)	$\beta_0 + \beta_1 X_1 + \beta_3 X_3 + \beta_4 X_4$
12	Food, Cover interaction	Food density (+); Cover density (+); Interaction (+)	$\beta_0 + \beta_1 X_1 + \beta_2 X_2 + \beta_{1,2}(X_1 * X_2)$
13	Food, Patch size interaction	Food density (+); Patch size (+); Interaction (+)	$\beta_0 + \beta_1 X_1 + \beta_3 X_3 + \beta_{1,3}(X_1 * X_3)$
14	Food, Patch connectivity interaction	Food density (+); Nearest patch distance (−); Interaction (+)	$\beta_0 + \beta_1 X_1 + \beta_4 X_4 + \beta_{1,4}(X_1 * X_4)$
15	Cover, Patch connectivity interaction	Cover density (+); Nearest patch distance (−); Interaction (+)	$\beta_0 + \beta_2 X_2 + \beta_4 X_4 + \beta_{2,4}(X_2 * X_4)$
16	Constant only	none	β_0

[a] Pease et al. (1979).
[b] Wirsing et al. (2002).
[c] Ausband and Baty (2005).
[d] Griffin and Mills (2009).
Researchers count the number of fecal pellets on transects, convert fecal pellet counts to binary hare occupancy (present[1] – absent[0]), and relate occupancy to food density (vegetation consumption on transects), cover density (visual obstruction of a spherical densiometer), forest patch size (estimated size of the forest stand), and forest patch connectivity (abundance of forest surrounding the transect; see Thornton et al. 2012). Hypotheses 1–4 were developed through review of the literature and 5–15 reflect potential multivariate processes governing hare spatial dynamics. Predictions were developed a priori, assuming standardized variables (see text). We include the Constant-only model as a gauge of the relevancy of the candidate set.

2.3.1 How Many Are Too Many Hypotheses?

How many research hypotheses should be developed for a given study? There is no simple answer to this question but researchers often consider too many hypotheses at once, even if some have little prior biological support (Anderson and Burnham 2002). The number of potential hypotheses increases geometrically with the number of variables under consideration, even when interactions between variables are not included (Dochtermann and Jenkins 2011), so that ultimately the number of candidate models should be considered against the anticipated sample size of observations. Burnham et al. (2011) state that the upper limit on the number of hypotheses should be less than the sample size of the dataset, which in our view seems obvious. In fact, some authors have suggested that a statistical penalty should be imposed for each additional hypothesis under consideration (Taper 2004; Forstmeier and Schielzeth 2011), although this idea has yet to receive broad support.

In our snowshoe hare example, we developed 16 hypotheses (including the null) based on the four starting hypotheses and several more complex alternatives that are consistent with our broader understanding of hare ecology. However, we did not use all possible combinations in developing our set of candidates. For example, we noted earlier that cover density and patch size could be correlated and thus avoided this redundancy. In fact, the inclusion of all variable combinations when developing the set of candidates presents a problem that is analogous to the aforementioned issue of data dredging when developing hypotheses: Fitting all combinations of variables or *model dredging* favors models that lack generality and discourages the careful thinking that is expected when developing research hypotheses. Unfortunately, many software packages offer all-subsets regression as a default, which implements model dredging and inevitably propels the researcher down a path of uninformed, post hoc hypothesis generation which can lead to weak inference. Instead, hypothesis development should rely on an understanding of how multiple variables may relate to each other, which is achieved through careful thought and understanding of likely relationships.

Last, we need to mention that in the multiple hypothesis framework there is some debate as to whether predictions must be *mutually exclusive* (Dochtermann and Jenkins 2011). It is important to recognize that greater differentiation between predicted outcomes helps ensure that weak hypotheses are unseated through the hypothetico-deductive or model selection methods. In the case of our snowshoe hare example, it should be relatively straightforward to weigh the support for each of the different hypotheses relative to the differentiation spelled out in the predictions (Table 2.1). Of course, in some study situations it may be necessary to accept more overlap between hypotheses and predictions, which will make IT statistics particularly relevant for weighing the relative merit of each candidate model.

2.4 From Research Hypothesis to Statistical Model

Statistical models are mathematical depictions of research hypotheses, and developing an appropriate formulation of the hypothesis requires insight into the nature of the relationship between *predictor* and *response* variables. This important point highlights an additional reason why hypotheses and predictions should be stated simply and clearly, so they can be translated to models with relative ease and accuracy. Some research hypotheses are straightforward and easily converted to mathematical notation to provide a direct reflection of the stated question. Here, the outcome of model fitting logically constitutes an evaluation of the strength of support for the prediction (Table 2.1). In other cases, translation from hypothesis to model is more challenging, and the researcher may grapple with proper formulation of the hypothesis. The following sections highlight what we consider to be important challenges faced when developing statistical models that seek to effectively and efficiently reflect stated hypotheses.

2.4.1 Functional Relationships Between Variables

Most hypotheses in ecology are adequately represented through *linear* or *logit* (i.e. logistic) models, which adequately reflect the error distribution of much ecological data. However, sometimes the relationship between predictor and response variables are *nonlinear*, and substantial additional information would be missed if linearity was assumed (Anderson and Burnham 2002). We predicted a linear relationship between snowshoe hare occupancy and habitat patch size (Table 2.1), but in fact the relationship could be nonlinear given that the area occupied by the hare represents two-dimensional space. Instead, we could have modeled the hare occupancy vs. patch size relationship (i.e. Model 3, Table 2.1) as: $\beta_0 + \beta_3 X_3 + \beta_3 {X_3}^2$, where X_3 is the variable for area. Typically, variables that are modeled in a nonlinear context are multidimensional units such as mass, density, or age that should not vary monotonically with the response. Newer methods for addressing complex nonlinearity include multivariable fractional polynomial functions and a range of spline techniques (Sauerbrei et al. 2007), but there probably are few instances in population ecology where these methods are necessary for revealing more common nonlinear associations between variables. Again, we emphasize that careful thinking about biological relationships will help generate the best hypotheses, implying that nonlinear relationships may normally arise only once coarser linear relationships have been modeled successfully. For instance, snowshoe hare populations can be estimated using fecal pellet counts on fixed transects and there is often a strong linear relationship between the two variables (Krebs et al. 1987, 2001). However, if hare or fecal pellet counts are somehow biased according to pellet density, or if hares have differential patterns of feeding or activity at high or low population density, a more precise description of the pellet number–hare density relationship may involve a nonlinear function (Murray et al. 2002). Appreciation for this alternative may only arise upon further reflection after the first dataset is collected and analyzed. Alternatively, there are cases when researchers are most interested in developing a generalizable model that can be applied to a range of sites or circumstances, meaning that a complex nonlinear relationship between predictor and response variables may be adequately represented by a more parsimonious linear fit.

2.4.2 Interactions Between Predictor Variables

In field-based research, the role of individual predictor variables often depends on a range of other environmental factors, and codependence between variables is an important consideration when developing models designed to reflect real world complexity. Yet, while interaction terms representing variable codependence are key to traditional statistics like ANOVA, in observational studies they are often perceived more as a nuisance and frequently not given adequate attention (Aiken and West 1991). In the context of multiple linear regression, an interaction between two variables reflects that the partial slope of the regression is not independent of the value of other variables, implying that ignoring such dependency as an explicit component of statistical models can lead to imprecise parameter estimates and poor fit. In contrast, including interaction terms complicates the interpretation of main effects due to the conditionality of predictor variables upon multiple terms in the model (Engqvist 2005). It follows that weak interaction terms lead to parameter bias and imprecision. An important challenge associated with interactions between variables is that to the fullest extent possible potential codependence should be recognized a priori and thus directly incorporated as part of the hypothesis statement. In practice, proper hypothesis development will usually consider potential dependency between predictors.

Surprisingly, many ecological studies either do not consider interactions at all, or develop an initial set of models that is restricted to main effects only with interactions only included in later modeling. Ignoring potential interactions is clearly contrary to the IT philosophy and reflects a quest for improved model fit without an underlying hypothesis. Along the same line, researchers sometimes include models having interaction terms while excluding main predictors associated with the interaction, which is counter to the *principle of marginality* (Nelder 1977) and suggests both weak hypothesis development and stepwise variable reduction. Thus, by default researchers become "model dredgers" if they allow themselves to deviate from conducting analysis that is restricted to the initial set of models.

Our hypotheses of snowshoe hare occupancy include four cases where we considered at the outset that variables may be codependent; the corresponding statistical models include appropriate interaction terms to reflect these possible dependencies (Table 2.1). Our interaction terms are restricted to those with two factors, but sometimes researchers develop models consisting of three or more interacting variables even though such complexity can rarely be conceived a priori. When we developed our set of hypotheses explaining snowshoe hare occupancy, we could readily imagine how food and cover density might be inter-related (Figure 2.4), or that food or cover each could be dependent upon patch connectivity. On the other hand, it was challenging to envision the outcome of a three-way interaction between food density, cover density, and patch connectivity, despite our relative familiarity with the snowshoe hare study system. Thus, barring rare insight into the inner workings of an organism's interaction with its environment, it is more realistic for ecologists to restrict interaction term complexity even if this limits the strength of model fit. However, note that while later we discuss IT methods as a useful means of inference, problems arise when calculating model-averaged parameter estimates because those that show up in both main effects and interactions cannot be easily weighted (Dochtermann and Jenkins 2011).

2.4.3 Number and Structure of Predictor Variables

Ecologists often fit statistical models having too many predictors, which inevitably leads to overly complex models that lack generality even if IT-based methods are used to promote model parsimony. To avoid this problem, researchers should restrict research hypotheses to a manageable number of components that need to be modeled as variables. The key point in assessing whether a particular model can be fit with reliable certainty is the *number of observations* and how this affects accuracy of the parameter estimates. As a rule of thumb, Vittinghoff and McCullough (2006) advised that reasonable inference can be drawn from logistic regression with a single predictor using as few as five observations, but most researchers use a higher threshold. In a simulation exercise, Peduzzi et al. (1996) showed that logistic regression parameter estimates are biased when there are <10 observations of the least frequent outcome per variable, which is a general rule of thumb for either continuous or discrete covariates having a balanced distribution. Given typical sample sizes in ecological studies, most models should responsibly include no more than four to five covariates, including interaction terms. It is reasonable to suggest that this number of covariates corresponds with the upper limit of factors that a researcher can realistically include in a research hypothesis.

Our final point regarding the translation from research hypothesis to statistical model involves the actual structure of the predictor variables. Data can be collected as *continuous* or *categorical* variables and it is generally good practice to retain the original format for analysis. One common practice is to dichotomize continuous variables by grouping observations into two or more categories, to allow for improved interpretability of the fitted model (Owen and Froman 2005). However, this approach has been heavily criticized due to information lost when reclassifying original data and its effect on parameter estimates and residuals (Royston et al. 2006). An additional concern relates to the need to select arbitrary cutpoints when designing categories, which usually is accomplished after some degree of data snooping to detect appropriate thresholds. Thus, in most cases there should be neither need nor merit in reclassifying data. However, sometimes simplifying continuous or categorical variables into dummy (binary) variables can be beneficial if this relates specifically to the stated research hypothesis (Farrington and Loeber 2000). For example, snowshoe hares are known to require a minimum amount of vegetative cover before they will colonize an area (Murray 2003, p. 155). In our field study of snowshoe hares, we could have estimated density of plant stems in areas that were occupied versus unoccupied by hares, and then dichotomized the cover measurement into a binary variable according to the known plant stem density threshold. Alternatively, if we hypothesized that food and cover density would affect hare occupancy only when the baseline hare population was high and under strong competition for food and space, we might conduct a multiyear study and code the time variable according to a hare density threshold. In both examples, restructuring the continuous variable could refine our evaluation of the research hypothesis and improve model fit, although in either case the justification to reclassify should be based on a goal of improved inference rather than by the lure of a simplified predictor. The point is that whether the potentially harmful effects of dichotomizing variables are fully realized will depend to some extent on data structure and conversion needs, as well as the appropriateness of the modification with respect to the research hypothesis.

2.5 Exploratory Analysis and Helpful Remedies

After hypotheses are developed and formulated into statistical models, and data are collected to parameterize models and gauge support for predictions, researchers can proceed to data manipulation and analysis. Before a dataset is ready for formal statistical analysis, a preliminary assessment of its properties should identify any underlying problems that may violate basic assumptions or otherwise compromise the integrity of the analysis. *Data exploration* is distinguished from the aforementioned "data snooping" because it serves specifically to remedy problems rather than detect relationships that guide post-hoc hypothesis development. Data exploration is the only part of the analysis that does not need to be hypothesis-driven, and typically should demand a substantial portion of time, perhaps up to 50% of the total time devoted to analysis (Zuur et al. 2010). Although this comprises more time investment than probably is afforded by most ecologists, proper data exploration can have a major impact on the integrity of the statistical analysis and thus should be conducted thoroughly.

2.5.1 Exploratory Analysis and Diagnostic Tests

During recent decades, statistical analysis in ecology has undergone a major transformation following the advent of powerful analytical tools allowing researchers to address questions in ways that were previously unimaginable. For example, new techniques based on traditional capture-mark-recapture methodology provide new tools for estimation of population abundance (Chapter 5) and powerful new methods in genetic analysis have expanded our insights into population processes (Chapter 11). Many newer analytical methods are accessible to ecologists through the availability of free software combined with improved computing speed and power (Zuur et al. 2010). Yet, despite their novelty and sophistication, newer methods retain the need for judicious pre-analysis data treatment to meet statistical requirements. Indeed, given the broad range of structural problems commonly plaguing ecological datasets (Schielzeth 2010; Nakagawa and Freckleton 2011), exploratory data analysis, and if necessary, data remediation, should form an integral component of any robust statistical analysis. Yet, the lure of quick analytical results using state-of-the-art statistical tools sometimes preempts careful data exploration as the first step in the analysis. Further, some improvements in data exploration and remediation are recent additions to the statistical toolbox and therefore remain obscure. Accordingly, to help support robust inference in population ecology, exploratory data analysis and remediation should constitute a necessary component of statistical analysis.

There are two broad categories of data problems that emerge through exploratory analysis; (i) irregularities in the distribution or completeness of observations; and (ii) inter-relationships between individual observations. A variety of graphical and quantitative methods have been designed to check and remedy the more straightforward problems with data (summarized in Table 2.2), and more extensive coverage of each of these procedures is available from other sources (Quinn and Keough 2002; Zuur et al. 2010). One point, however, is that the listed quantitative tests are largely frequentist (i.e. based on P-values) and thus do not conform to the philosophical shift toward statistical inference that we are advocating in this chapter. Several procedures require examination of model residuals, and it is inevitable that they sometimes are conducted post hoc. Nevertheless, we highlight three important areas of remediation that have been prioritized by the statistical community.

2.5.2 Missing Data

Ecologists often work with incomplete datasets, where one or more predictors are not complete for all observations. Missing observations arise because data are not properly collected during the field study or if exploratory analysis reveals outliers or errors that are removed prior to statistical analysis. *Missing data* present a general problem with multivariate datasets and are an increasing concern in ecology (Nakagawa and Freckleton 2008).

Table 2.2 Summary of problems detected through exploratory analysis and possible remedies.

Problem	Variable(s) affected	Diagnostics	Remedy
Outliers	Response & Predictor	Graphical: Boxplot; Cleveland dot plot Quantitative: Grubb's test; Peirce's criterion; Dixon's Q test; Mahalanobis distance	Data transformation or analysis based on a different distribution. Outlier deletion is discouraged.
Homogeneity of variance	Response	Graphical: Conditional boxplot Quantitative: Levene's test	Data transformation or analysis based on a different distribution. Nonparametric approaches as a last resort.
Normality	Response	Graphical: Probability plot Quantitative: Shapiro-Wilks test	Data transformation or analysis based on a different distribution. Nonparametric approaches as a last resort.
Preponderance of zeros	Response	Graphical: Frequency plot Quantitative: Skewness test	Zero-inflated models are available for continuous or discrete variables. Two-class models partition observations into prevalence and frequency.
Independence	Response	Graphical: Autocorrelation function; Partial-autocorrelation function Quantitative: Durbin-Watson test	Time-series methods.
Missing data	Response & Predictor	Graphical: Frequency plot Quantitative: Descriptive statistics with sample sizes	Multiple imputation is generally recommended. Case deletion is discouraged.
Collinearity	Predictor	Graphical: Frequency plot Quantitative: Correlation; Variance inflation factors	Removal of redundant variables or develop synthetic variables through principal components analysis.
Different data scales	Predictor	Graphical: Frequency plot Quantitative: Descriptive statistics with mean and variance	Re-scale variables by centering on zero and dividing by standard deviation.

Source: Adapted from Quinn and Keough (2002) and Zuur et al. (2010).

Consider a dataset comprising of morphometric measurements collected from live-trapped small mammals, including body mass, body length, skull size, foot length, and other variables. If several samples provide only partial observations for some of the metrics of interest, it will be difficult to establish reliable ordinal relationships between variables. In the context of model selection and multimodel inference, such omissions are particularly worrisome because values of all models in the set of candidates are compared together but information criteria and measures of goodness-of-fit require that each model includes the same data. We suspect that many first-time users of IT methods may not be aware of this important limitation, and that improper model selection exercises due to missing data are more prevalent than expected (Nakagawa and Freckleton 2011).

The most common approaches for dealing with missing data are to either *omit incomplete variables* or to *remove cases* with incomplete observations; these tactics are widespread, admittedly even in papers published by authors of this chapter. However, removing incomplete predictors leads to information loss whereas deletion of incomplete records lowers sample size and reduces precision. For example, if our dataset of small mammal morphometrics has 10 explanatory variables each with 5% of the cases missing, deletion of all entries with missing data will reduce the sample size by up to 40%. Many studies in population ecology already suffer from small sample sizes and can ill afford unnecessary information loss. More importantly, deleting records can result in biased parameter estimates if observations are not missing at random (Little and Rubin 2002). In the case of our small mammal example, missing observations can be especially recurrent among animals that are more resistant to handling, or individuals that are so small that precise measurements are not possible. In the first case, the sample may under-represent squirmy animals whereas the second will favor larger individuals; in both cases data are *missing not at random* (MNAR) and parameter estimates will be biased. As a rule of thumb in other ecological datasets, shy individuals are poorly represented in behavioral syndrome datasets (Biro and Dingemanse 2009), highly mobile individuals are undersampled in habitat selection studies (Smith et al. 2010), and individuals with predisposing traits are poorly surveyed in survival estimation work (Zens and Peart 2003). Often, it is not immediately obvious how removing cases with missing observations can invoke sampling bias, but as a general rule any consideration of selective case deletion should be approached carefully. In fact, we recommend that unless there is clear evidence that missing observations clearly are random with no underlying pattern determining which data are missing, researchers should assume that selective case deletion has occurred and will promote bias.

Understanding the source and patterns of missing data is important when faced with situations involving incomplete information, and proper reporting of such problems should be a priority. There is increasing recognition that the appropriate remedy for treating missing data is to substitute omissions through *imputation* or *augmentation*. Several techniques are available to do this, but the simpler approaches like substituting blanks with mean values or fitting a regression model to predict missing observations ignore uncertainty in the missing data and lead to artificially low variability. Alternatively, *multiple imputation* is a simulation-based approach that allows for sampling variability by generating multiple complete datasets; once multiple datasets are simulated, analysis proceeds using standard statistical techniques and then results are pooled across datasets (Little and Rubin 2002). Proper imputation should involve a model that is as general as possible and provides unbiased parameter estimates and confidence intervals that achieve nominal coverage when randomized over the imputed models; diagnostic tests can be applied to confirm that the imputation model is correctly specified (Abayomi et al. 2008). Diagnostics can be best achieved by including as many variables as possible (including the response) in the imputation model. Ultimately, the number of imputations should depend on the amount of missing information as well as the analysis model itself, and may be relatively small in most applications (i.e. 3–20; Nakagawa and Freckleton 2011). White et al. (2011) provide guidelines for determining the appropriate number of imputations, and in general it is good practice to vary the number in exploratory analyses to confirm that the imputation model is not unduly sensitive to combinations of variables. An important point to consider is that variables with extensive omissions will be more difficult to impute reliably and in some cases will warrant being dropped from the analysis. Nakagawa and Freckleton (2011) provide a useful overview of how to deal with model imputation in an IT context, which poses a unique challenge given that multiple datasets are generated and a weighted statistic from this composite group is being sought.

2.5.3 Inter-relationships Between Predictors

Correlation between two or more predictor variables, termed *collinearity* or *multicollinearity*, is a common feature of many observational datasets and can be a problem in regression-type analyses because it causes unstable parameter estimates and inflated variance (Quinn and Keogh 2002). Collinearity also hinders the ability to identify separate effects among correlated variables, and hence precludes detection of causal predictors. When extrapolating the results of a model outside the study area, collinearity will cause a decrease in predictive performance of models because inter-relationships often change through time and space. Considering multiple predictors that are different manifestations of the same underlying process is a common source of collinearity with ecological datasets using predictors that are indirectly linked to the response. Although there is no specific way to "solve" problems of collinearity, there are ways to minimize correlations between variables before analysis, or to perform subsidiary analysis to gain a better understanding of independent and joint effects of variables. Due caution must be exercised when interpreting the output of models with correlated predictors, particularly when attempting to identify causal links (Dormann et al. 2012).

Exploratory analysis may reveal collinearity between predictors through Pearson correlation coefficients or nonparametric alternatives such as Spearman's tau. Alternatively, the *variance inflation factor* (VIF) is a regression-based approach specifically quantifying change in a variable's coefficient owing to collinearity. As a rule of thumb, $VIF > 5$ or $VIF > 10$ indicate that the level of correlation warrants concern, although increasingly there is recognition that these thresholds are arbitrary and perhaps too conservative (O'Brien 2007). In our snowshoe hare dataset, correlation coefficients and VIFs indicated overall weak correlation between variables (Table 2.3), allowing us to proceed with minimal concern regarding collinearity. Indeed, the degree of change in standard error (SE) for the coefficient from a correlated variable is calculated as: $\sqrt{VIF}$. For the snowshoe hare study, cover density had a $VIF = 1.03$,

Table 2.3 Correlation coefficients between predictor variables used to model snowshoe hare occupancy.

Predictor variable	Food density	Cover density	Patch area	Patch connectivity
Food density				
Cover density	0.083			
Patch area	0.096	0.150		
Patch connectivity	−0.052	0.009	0.060	

Note the low level of correlation between variables, which is ideal for fitting and interpreting parameter estimates. Variance inflation factors (VIF) were low for all variables: Food density (1.01); Cover density (1.03); Patch area (1.04); Patch connectivity (1.01). Cover density and patch area were assumed a priori to be correlated and therefore not subject to correlation analysis.

meaning that the SE of the coefficient is $\sqrt{1.03}$ = 1.02 times higher than in the absence of collinearity. Another common measure of a variable's collinearity is its *tolerance*, which is simply: tolerance = 1 / VIF.

Once collinearity is identified, it can be addressed either by reducing the set of variables until only an uncorrelated set remains, or the number of predictors can be reduced by using new synthetic variables. If doing the former, a general rule of thumb is that pairs of predictor variables with a correlation greater than 0.7 should not be used together. Although this cut-off is arbitrary, model performance decreases around that level of correlation (Dormann et al. 2012). Generally, the predictor that is more biologically meaningful, or that is expected to have a stronger causal link to the response, should be the one that is retained, but it remains important to ensure that eliminating variables does not diminish assessment of particular research hypotheses as they have been stated. If no decision on variable elimination can be made a priori, use of univariate tests to identify the most important predictor to retain is recommended (Murray and Conner 2009). Alternatively, sets of correlated predictors can be combined together using Principle Components Analysis or other clustering techniques, thereby generating uncorrelated synthetic variables that can be used. However, potential drawbacks of this approach are that synthetic variables may be difficult to develop a priori in the context of a research hypothesis, and their interpretation can be problematic because they represent combinations of several variables that are non-repeatable and specific to a particular dataset. Alternatively, it may be better to select whichever original predictors best fit the synthetic variables and exclude the remainder from the analysis. Other possible approaches for dealing with collinearity may be applicable to particular circumstances. For example, if there is good a priori reason to expect that one of a pair of correlated predictor variables is causally linked to the response, but it is desirable to retain both variables in the analysis, sequential regression techniques can be applied (Dormann et al. 2012).

In some cases, even after the data preparation steps outlined above, correlations will remain between predictors. Therefore, caution is necessary in making interpretations about their relative importance based on standardized parameter estimates or information criteria, which can perform poorly in the presence of collinearity (Murray and Connor 2009). Recently, several techniques have been developed to quantify the relative importance of variables even in the presence of collinearity, such that we can now identify the independent versus joint effects of groups of predictors to help determine important drivers of a particular response (Mac Nally 2000; Murray and Connor 2009).

2.5.4 Interpretability of Model Output

Before going through the model selection and fitting process, the predictor variables may need to be *standardized* to facilitate comparison of the relative influence of individual predictors, provide meaningful estimates of main effects in models with interaction terms, or avoid problems of collinearity in models containing polynomial terms (Schielzeth 2010). Standardization involves two steps: *centering* and *scaling*. Centering a variable subtracts the sample mean from each individual value of the variable, and scaling divides each value by the sample standard deviation. A continuous predictor variable that has been standardized therefore has a mean of zero and a standard deviation of one.

Although not always necessary, standardization is particularly helpful when comparing the relative influence of predictors, especially when they are measured on markedly different scales. As an example, suppose we model the influence of patch area and surrounding food availability on snowshoe hare population density: Patch area might vary between 50 and 300 ha, and food availability could range from 0.40 to 0.60 of the landscape. We obtain the following model: hare density = $0.2 + 0.004X_1 + 1.2X_2$, where X_1 and X_2 represent patch area and food availability, respectively. Parameter estimates from a fitted model with unstandardized variables indicate the change in hare density given a one-unit change in the predictor. But how does one compare the influence of a one-unit change in patch area (0.004) to that of a one-unit change in tree cover (1.2) when they are measured on such different scales? Standardization allows a direct comparison of the relative influence of predictors. The same model of hare density fitted with standardized predictors gives the following: hare density = $1.49 + 0.31X_1 + 0.074X_2$. The parameter estimates from this model (0.31 and 0.074) indicate the increase in hare density given a *one standard deviation change* in patch area or food availability, respectively. We conclude from this hypothetical example that patch area has a greater influence on hare density than surrounding food availability, but this relationship is most easily discerned when the analysis involves standardized variables. However, for graphical purposes standardized variables should be back-transformed to their original value to improve interpretability.

Although researchers typically standardize continuous variables, similar procedures are available for categorical variables, although their influence on continuous variables can be problematic. Scaling continuous predictors by dividing by two standard deviations may more accurately compare the influence of standardized categorical and continuous predictors (Schielzeth 2010). Notably, standardized parameter estimates may be biased when there is a high degree of correlation among predictor variables. In such cases, other techniques for estimating relative influence may be preferred (Murray and Conner 2009).

Standardizing or centering predictor variables before model fitting also allows easier interpretation of main effects of parameter estimates from models with interaction terms. In a model with two-way interactions, main effects indicate the influence of each variable on the response, when the value of the other predictors involved in the interaction is zero. This is problematic when using unstandardized predictors because many ecological variables cannot have a zero value, or if zero lies outside of the range in sample data. For example, if we fit a model of snowshoe hare density with an interaction term between patch area and food availability, the main effect estimate for food availability indicates its influence on hare density when patch area is zero. However, this is a meaningless estimate because patch area can never be zero, indicating that parameter estimates of main effects and associated tests of statistical significance from models with unstandardized predictors are often uninformative. Use of standardized predictors solves this problem, and in a model with standardized values of patch area and food availability, the parameter estimate for the main effect of food indicates the influence of a one standard deviation change in food availability when patch area is at its *average* value (which is equal to 0, for a standardized variable). Thus, main effects become meaningful, and given the parameter estimate of the interaction term, we can easily determine the influence of food availability on hare density when patch area is at its average, high, and low value.

Standardization also can be helpful in mitigating problems of collinearity between predictors and polynomial terms. The squared and nonsquared version of unstandardized predictor variables, particularly all-negative or all-positive predictors, will be highly correlated with each other (Schielzeth 2010), and such correlations result in large SE estimates and problems in interpretation of independent effects (Section 2.5.3). Standardization (or more accurately, centering) of a variable before the creation of polynomial terms eliminates this collinearity and therefore allows accurate estimation and interpretation of the independent effects of the squared and nonsquared versions of the variable of interest.

In summary, there are a number of exploratory tests and remedies that should be applied to observational datasets in order to mitigate sampling bias, ensure independence between variables, and improve model interpretation. Ultimately, improving the structure and functionality of statistical models will allow researchers to conduct more rigorous tests of their research hypotheses.

2.6 Model Ranking and Evaluation

2.6.1 Model Selection

Once exploratory analysis has been used to check for the various data integrity and structure issues, and after appropriate data remediation measures have been conducted, the researcher can proceed to evaluate how the set of *candidate models* fits the data. The maximized log-likelihood [$\log(L)$], a fundamental quantity in mathematical statistics, uses KL information to express the difference between each candidate model versus full reality (Burnham and Anderson 2002). Notably, while KL information identifies the model that is closest to full reality, a common misconception regarding the application of IT methods is that the correct model should be part of the candidate set. The relationship between KL information and full reality across a set of models is such that the best model has least distance but does not necessarily (or likely) achieve full reality. In the absence of full reality, the best model receives the strongest support and serves as basis for comparison with other candidates (Figure 2.2). The maximized log-likelihood is the starting point for determining the distance between the best versus other candidate models, and is converted to one of several information criteria for calculating relative distance. The asymptotic bias in maximum log-likelihood is corrected for large samples by the total number of estimable parameters in the model (K); $\log(L)-K$, which leads to the well-known *Akaike Information Criterion* (AIC) as a measure of distance between two models:

$$\text{AIC} = -2\log(L) + 2K. \tag{2.1}$$

AIC allows us to compute the distance of a model relative to other candidate models and thereby provides a measure of intermodel deviance (Box 2.3). Of particular note is the inclusion of $2K$ in the AIC calculation, which increases the AIC score according to the number of parameters in the model and essentially serves as a penalty against models that are overfit with uninformative parameters. *Overfit models* are problematic for a variety of reasons, not the least of which is that they lack precision when sample size is low. For example, for every additional parameter in the model, the information remaining to fit the next parameter is proportionally less. Overfit models tend to match the noise in a given dataset and thus are prone to instability and spurious association, and therefore have poor performance when applied to a different study. Thus, AIC and other information criteria balance model *bias* from including too many parameters against the *variance* of the fitted model parameters and the undesirable situation of having too much unexplained variance. However, we note that AIC is not an actual

Box 2.3 AIC Calculation and Model Selection

A researcher generates a set of linear models evaluating seed production for a hypothetical plant species. Thirty sites are monitored (n = 30) and the number of seeds produced (Success) is compared relative to insect visitation rate (Insect) as well as other variables like size of the floral patch (Area) and distance of the floral patch to other floral patches (Distance). The dataset is provided below.

Record	Success	Insect	Area	Distance
1	19	8	10	11
2	32	15	22	74
3	35	19	32	42
4	17	6	17	39
5	21	11	11	21
6	7	3	11	19
7	57	11	54	91
8	27	11	44	62
9	87	32	50	91
10	45	49	45	57
11	31	43	40	43
12	28	10	51	34
13	35	15	31	65
14	74	33	18	22
15	22	15	3	34
16	11	5	32	17
17	7	2	3	11
18	7	17	16	50
19	86	19	22	91
20	34	14	41	54
21	45	22	40	31
22	75	29	33	65
23	64	33	16	32
24	54	17	27	93
25	29	12	36	32
26	60	24	34	17
27	43	34	51	78
28	56	22	44	97
29	88	41	52	143
30	27	15	11	46

There are five hypotheses under consideration (related principally to the role of insect visitation), including a constant-only model. Models include main effects only. The RSS for the model with Insect is: 9783.518, and the number of estimated parameters (K) in the model is 3 (Insect, Constant, model error). The Akaike's Information Criterion for the model is:

$$\mathrm{AIC} = n \cdot \mathrm{Log_e}(\mathrm{RSS}/\mathrm{n}) + (2 \cdot K)$$
$$= 30 \cdot \mathrm{Log_e}(9783.519/30) + (2 \cdot 3) = 179.618$$

Because AIC is biased when the sample size is <40, we apply a correction factor to calculate unbiased model distance:

$$\mathrm{AIC_c} = \mathrm{AIC} + (2K(K+1))/(n-K-1)$$
$$= 179.618 + (2 \cdot 3(3+1))/(30-3-1) = 180.541$$

Using the RSS and K specific to each model, we calculate AIC and $\mathrm{AIC_c}$ for each candidate in the set:

Model	K	RSS	AIC	$\mathrm{AIC_c}$
Insect	3	9783.519	179.618	180.541
Insect, Distance	4	6976.855	171.475	173.075
Insect, Area	4	9261.158	179.972	181.572
Insect, Area, Distance	5	6971.622	173.452	175.952
Constant only	2	16 786.702	193.814	194.259

We see that although the Insect, Area, Distance model has lowest log-likelihood, it has the second-lowest AIC score, owing to the penalty due to its larger number of parameters.

Next, we determine the difference in $\mathrm{AIC_c}$ scores between the best versus alternate models. We reorder models from low-to-high $\mathrm{AIC_c}$ and calculate Δ_i such that the best model (lowest $\mathrm{AIC_c}$) is the base model. Because $\mathrm{AIC_c}$ is lowest for the Insect, Distance model, it serves as reference. Thus, for the Insect, Distance model, we obtain: Δ_i = 173.075 − 173.075 = 0. The next best model, which includes Insect, Area, and Distance, has: Δ_i = 175.952–173.075 = 2.877. The same distance measure is calculated for all models in the set:

Model	$\mathrm{AIC_c}$	Δ_i
Insect, Distance	173.075	0
Insect, Area, Distance	175.952	2.877
Insect	180.541	7.466
Insect, Area	181.572	8.497
Constant	194.259	21.184

Δ_i is useful for evaluating the level of support for a given model relative to the best model in the set of i models. Δ_i also is necessary for determining the relative weight of each model. To obtain model weights, first we calculate the relative likelihood of each model, which is simply:

exp (−1 / 2 Δ_i). For the Insect, Distance model, we get exp (−1 / 2 · 0) = 1; for the Insect, Area, Distance model, we get exp (−1 / 2 · 2.877) = 0.237. We complete the calculations for the entire set of models.

Model	Δ_i	Relative Likelihood	w_i
Insect, Distance	0	1	0.784
Insect, Area, Distance	2.877	0.237	0.186
Insect	7.466	0.024	0.019
Insect, Area	8.497	0.014	0.011
Constant	21.184	0.000	0.000
		$\sum = 1.275$	

The proportion of the relative likelihood for individual models against the sum of relative likelihood for the entire set (1.275) is the model weight. For the Insect, Distance model, w_i = 1 / 1.275 = 0.784. The appropriate interpretation of w_i is that on a scale of 0–1, it provides the proportion of evidence for a given model. The weight of evidence is 0.784 (from 0 to 1) that the Insect, Distance model is the best approximating model, given the data and set of candidates. There is considerably less weight of evidence that either the Insect, Area, Distance (w_i = 0.186), the Insect (w_i = 0.019), or the Insect, Area (w_i = 0.011) models are the best in the set (Figure B2.3.1).

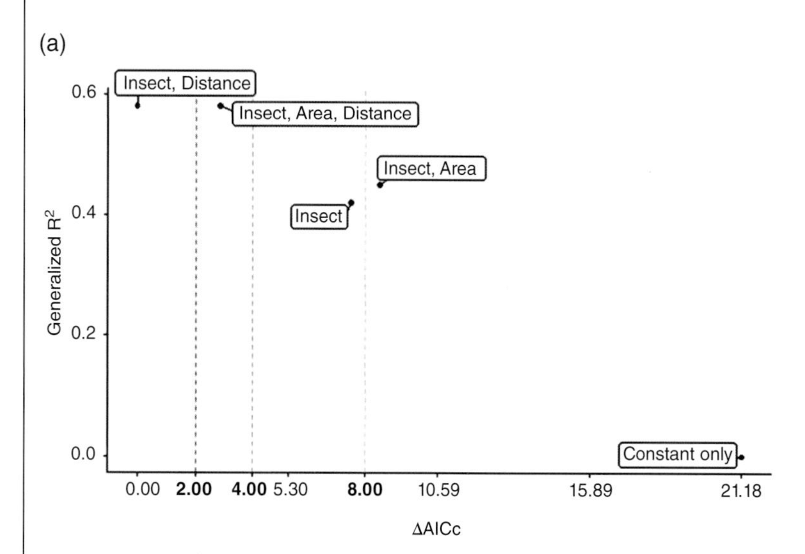

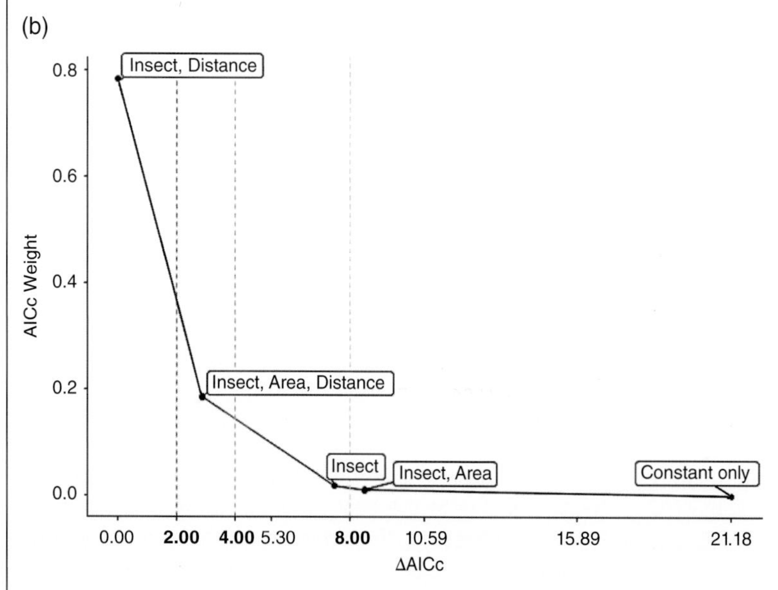

Figure B2.3.1 Plot of the relationship between ΔAICc vs. generalized R^2 (a) and AICc weight (b) for the sample pollinator dataset. Note that some models that have little support through model selection (>8 ΔAICc units; e.g. the model containing visiting number of pollinating insects and patch area – "Insect, Area") may still fit the data fairly well.

measure of model fit and the "best" model from a set of candidates may have low AIC but still suffer from large distance from full reality (Figure 2.2).

The log-likelihood and its derivatives are a product of logistic regression and hazard models, and we need a different measure of model distance to calculate AIC for least squares models:

$$\text{AIC} = n\log_e(\text{RSS}/n) + 2\text{K} \quad (2.2)$$

where RSS is the residual sum of squares (RSS) from the fitted model (Anderson et al. 2000). Therefore, using either of the two AIC expressions (Eqs. 2.1 and 2.2) depending on the nature of the analysis, we can determine the distance among candidate models across a broad range of statistical analyses. The special case of overdispersion in the response variable requires calculation of a modified unit, QAIC (Burnham and Anderson 2002).

Ecological studies often are beset with small sample sizes, which can pose problems when using AIC and other information criteria. A correction factor is applied (AIC_c) to adjust for sample size (n):

$$\text{AIC}_c = \text{AIC} + (2K(K+1))/(n-K-1). \quad (2.3)$$

Note that AIC_c and AIC converge when $n/K \approx 40$, but for simplicity many researchers determine model distance using the more conservative AIC_c irrespective of sample size. Because the AIC units themselves are meaningless and relative differences between models are preserved whether AIC or AIC_c is used, there is no harm in consistently using AIC_c as the unit of model distance.

To interpret values from AIC units they must be compared to units from competing models. The *AIC difference* (Δ_i) is the difference between the AIC value for a given model against the value for the best model (i.e. lowest AIC_c) in the set of i models; Δ_i values are on a continuous scale and provide an actual measure of the distance of each candidate from the best model in the set. For example, we show that for snowshoe hares, Model 12 (Food, Cover interaction) has the lowest overall distance (i.e. lowest AIC_c), and Δ_i ranges from 2.16 for Model 5 (Food, Cover availability) to 82.01 for Model 4 (Patch connectivity) (Table 2.4).

What Δ_i constitutes strong support for a model? The specific thresholds for qualifying model fit may vary depending on a number of factors, but generally models with $\Delta_i < 2$ are considered to have strong support, those with $2 < \Delta_i < 4$ have moderate support, and those with

Table 2.4 Model selection results from an analysis of food density, cover density, patch size, and patch connectivity influences on snowshoe hare occupancy.

Number	Covariates	K	Log-Likelihood	AICc	Δ_i	w_i	Generalized R^2	AUC
12	Food, Cover interaction	4	−194.95	398.01	0.00	0.59	0.29	0.78
5	Food, Cover	3	−197.05	400.17	2.16	0.20	0.28	0.78
9	Food, Cover, Patch size	4	−196.53	401.16	3.15	0.12	0.28	0.78
10	Food, Cover, Patch connectivity	4	−196.77	401.64	3.63	0.10	0.28	0.78
6	Food, Patch size	3	−207.75	421.58	23.57	0.00	0.21	0.75
1	Food	2	−209.32	422.67	24.66	0.00	0.20	0.71
11	Food, Patch size, Patch connectivity	4	−207.55	423.21	25.20	0.00	0.21	0.75
14	Food, Patch connectivity interaction	4	−207.74	423.60	25.59	0.00	0.21	0.71
13	Food, Patch size interaction	4	−207.75	423.61	25.60	0.00	0.21	0.75
7	Food, Patch connectivity	3	−209.18	424.43	26.42	0.00	0.20	0.71
8	Cover, Patch size	3	−222.70	451.47	53.46	0.00	0.11	0.67
2	Cover	2	−223.92	451.87	53.86	0.00	0.10	0.65
15	Cover, Patch connectivity interaction	4	−223.69	455.48	57.48	0.00	0.10	0.65
3	Patch size	2	−235.46	474.96	76.95	0.00	0.02	0.60
16	Constant only	1	−238.12	478.25	80.24	0.00	0	0.50
4	Patch connectivity	2	−237.99	480.02	82.01	0.00	0	0.51

Number corresponds to model number from Table 2.1, K represents the number of parameters in the model, log-likelihood is a measure of the probability of the observed outcome, AICc is the Akaike Information Criterion corrected for small sample size, Δ_i is the model difference from the best model, and w_i is the probability of the model, given the data. Generalized R^2 and area under the curve (AUC) are measures of goodness-of-fit (Table 2.6).

$4 < \Delta_i < 8$ have low support. Models with $\Delta_i > 9$–11 are widely recognized as having little support and often are not considered further (Burnham and Anderson 2002). These thresholds have been justifiably criticized as being arbitrary (Murtaugh 2014). For snowshoe hares, the respective Δ_i values indicate that Model 12 (Food, Cover interaction) has strongest support, Models 5 (Food, Cover availability), 9 (Food, Cover, Patch Size), and 10 (Food, Cover, Patch connectivity) have moderate support, and the remaining models have virtually no support (Table 2.4).

Notably, our decision as to whether individual models had support was based not only on Δ_i but also from model weights and evidence ratios. The *model weight*, w_i, representing the probability of each model i, given the data, is:

$$w_i = \frac{\exp\left(-\frac{1}{2}\Delta_i\right)}{\sum_{r=1}^{R}\exp\left(-\frac{1}{2}\Delta_r\right)} \qquad (2.4)$$

where R is the full set of models, $r = 1, 2,\ldots R$. Note that model weights reveal the proportional support for individual models and are normalized to one (Table 2.4). Model weights are necessary for calculating *evidence ratios*, which correspond to the magnitude of difference in support between two models. For example, from Table 2.4 we see that the weight of evidence supporting the best-fitting model (Model 12) is 0.59 compared to the model with food and density alone (Model 5 : 0.20). In other words, the weight of evidence in support for Model 12 is almost three times stronger than for the next best model in the set (Burnham et al. 2011).

We can also calculate the *sum of weights* for models that include a common variable. The collective weight of evidence for either the Food or Cover variable is: 0.59 + 0.20 + 0.12 + 0.10 = 1.0, whereas the collective weight of evidence for the Patch size and Patch connectivity variables is 0.12 and 0.10, respectively. Note, however, that there is emerging evidence that the sum of weights may not be particularly reliable as an indicator of predictor variable importance and thus should be used cautiously (Galipaud et al. 2014, but see Giam and Olden 2016). The confidence set of candidate models, which is analogous to a confidence interval for a mean estimate, includes models with AIC weights that are >10% of the highest w_i, which in the case of Table 2.4 includes Models 12 and 5.

Collectively, model likelihoods, evidence ratios, and model weights provide a sound basis for comparing statistical models in a manner that is superior to traditional statistics involving probabilities and single candidate models. Notably, Bayesian methods also can be used for similar purposes (Ellison 2004; Hooten and Hobbs 2015), and on a side note, in some instances *Bayes Information Criterion* (BIC) performs better than AIC (Aho et al. 2014, 2017). However, as we pointed out earlier, it is important to recognize that the units derived from KL information never actually reveal how well a model actually fits the data and the "best" in a set of candidates can still have large distance from full reality. Therefore, it is important to provide an additional metric of model fit, which can be R^2 in the case of linear regression and analogous units for logit functions (Dochtermann and Jenkins 2011; Table 2.4; Figure B2.3.1). The generalized R^2 is a measure of model fit which is corrected for number of parameters, and therefore is philosophically consistent with AIC.

2.6.2 Multimodel Inference

The above methods offer a superior approach for obtaining more judicious inference from observational studies compared to traditional statistical methods. In population ecology, the goal of a study is often to obtain parameter estimates that will serve in risk analysis or numerical projection, so it is important that such estimates be unbiased. *Multimodel inference* uses information from a set of weighted models to provide estimates with high accuracy and precision. Sometimes, there is a single viable contender in the set (i.e. $w_i = 1.0$; all others $w_i = 0$) and parameter estimates can be calculated directly from the best model. More likely, several models provide varying levels of support, and estimates for the same predictors will differ between models. To make full use of the information available from the full set of models we should derive model-weighted estimates (Buckland et al. 1997) (Box 2.4). *Model-averaged estimates* $(\bar{Y})$ can be computed as a sum of the estimates obtained for each model (Y_i), weighted by the mean weight for that model (w_i),

$$\hat{\bar{Y}} = \sum_{i=1}^{R} w_i \hat{Y}_i. \qquad (2.5)$$

Note that model-averaged estimates also can be calculated across only the set of models containing the parameter of interest, but to do this one must recalculate w_i using only the candidates that include that parameter. Of course, this basic approach assumes that model-averaged estimates also should include an estimate of uncertainty that corrects for dependency of the estimated model variance on the model itself, and that weighs the estimated error according to the weight of each model. Model-averaged variance is usually reported as the *unconditional standard error* (Box 2.4).

Research studies should report the appropriate output from model selection exercises, including log-likelihood value [log(L)] for logit models or RSS for least squares, number of estimable parameters (K), value for the

Box 2.4 Model-averaged Parameter Estimation and Uncertainty

To make full use of the benefits of model-based approaches, we can compute parameter estimates and corresponding uncertainty that is weighted by the set of candidate models rather than simply from the best-fit model. Using the example of seed production developed in Box 2.3, the fitted models for the set of five candidates provides the following parameter estimates (± SE):

Model	Constant	Insect	Area	Distance
Insect, Distance	5.505 (6.482)	0.918 (0.267)	–	0.338 (0.103)
Insect, Area, Distance	5.904 (7.193)	0.930 (0.283)	–0.034 (0.244)	0.346 (0.118)
Insect	16.214 (6.523)	1.272 (0.284)	–	–
Insect, Area	10.530 (7.936)	1.102 (0.313)	0.301 (0.244)	–
Constant	41.102 (4.393)	–	–	–

To obtain weighted parameter estimates, we return to Δ_i values for each model and calculate adjusted relative likelihoods and weights; in this particular case we chose to sum using only the models containing the parameter of interest. For the Area parameter estimate we calculate the relative likelihood for the Insect, Area, Distance model ($\exp(-1/2 \cdot 2.877) = 0.237$) and the Insect, Area model ($\exp(-1/2 \cdot 8.497) = 0.014$), as well as the total for the two models ($0.237 + 0.014 = 0.251$). The relative likelihood for the individual model divided by the new sum is the adjusted model weight, such that for Insect, Area, Distance the weight is: $0.237 / 0.251 = 0.944$ and for Insect, Area the weight is: $0.014 / 0.251 = 0.056$. For the entire model set, the adjusted relative likelihood and model weights are:

The next step involves multiplying parameter estimates by the new model weight, and summing the values to provide a model-averaged estimate. For the Area variable, we get: $(-0.034 \cdot 0.944) + (0.301 \cdot 0.056) = -0.015$ as a model-averaged estimate. The model-averaged estimates for all variables are:

Variable	Estimate
Insect	0.929
Area	–0.015
Distance	0.340
Constant	5.837

Thus, the composite model for seed production is:

$$Y = 5.837 + (0.929 \cdot X_1) - (0.015 \cdot X_2) + (0.340 \cdot X_3) + error$$

where estimates for the predictors are: X_1 = insect, X_2 = area, and X_3 = distance.

The model-averaged parameter estimates should be presented with estimates of uncertainty, but the SE obtained from the fitted models are conditional on the candidate model so they must also include an additional source of variance. Model selection variance (MSV) is obtained from the difference between the model-averaged parameter estimate and the raw parameter estimate from the fitted model:

$$\text{MSV} = (\text{model-averaged estimate} - \text{raw parameter estimate})^2$$

		Adjusted relative likelihood and model weight							
		Constant		Insect		Area		Distance	
Model	Δ_i	rel *L*	w_i	rel *L*	w_i	rel *L*	w_i	rel *L*	w_i
Insect, Distance	0	1.000	0.784	1.000	0.784	–	–	1.000	0.808
Insect, Area, Distance	2.877	0.237	0.186	0.237	0.186	0.237	0.944	0.237	0.192
Insect	7.466	0.024	0.019	0.024	0.019	–	–	–	–
Insect, Area	8.497	0.014	0.011	0.014	0.011	0.014	0.056	–	–
Constant	21.184	0.000	0.000	–	–	–	–	–	–
	$\sum$ =	1.275		1.275		0.251		1.237	

For the Insect variable in the Insect, Distance model, we calculate: MSV = $(0.929–0.918)^2$ = 0.00012. We repeat the calculation for each combination of variable and model:

Model	Model selection variance (MSV)			
	Constant	Insect	Area	Distance
Insect, Distance	0.110	0.000	–	0.000
Insect, Area, Distance	0.004	0.000	0.000	0.000
Insect	107.682	0.118	–	–
Insect, Area	22.024	0.030	0.100	–
Constant	1243.620	–	–	–

To calculate the unconditional SEs, we combine MSV and the conditional variance (SE^2). Then, we take the square root of this sum and weight it according to the adjusted w_i. The sum of these values is the unconditional SE for the parameter. The variance for the Distance variable in the Insect, Distance model is $0.103^2 = 0.011$, so we get $\sqrt{(0.000 + 0.011)} = 0.103$ as the new SE and $(0.103 \cdot 0.808) = 0.083$ as the weighted SE. For the Insect, Area, Distance model the new SE is $\sqrt{(0.000 + 0.118^2)} = 0.118$ and the weighted SE is $(0.118 \cdot 0.092) = 0.011$. Therefore, we can sum these two values to get the unconditional SE for parameter Distance: $(0.083 + 0.011) = 0.094$. By repeating the calculation for each parameter and summing, we obtain error estimates for each parameter:

Model	Unconditional SE			
	Constant	Insect	Area	Distance
Insect, Distance	5.089	0.210	–	0.083
Insect, Area, Distance	1.338	0.053	0.230	0.023
Insect	0.233	0.009	–	–
Insect, Area	0.101	0.004	0.023	
Constant	0.000	–	–	–
	$\sum = 6.761$	0.275	0.253	0.106

Therefore, we conclude that the model-averaged composite, with unconditional SEs, is:

$$Y = 5.837\,(6.761) + [0.929\,(0.275) \cdot X_1] - [(0.015\,(0.253)X_2] + 0.340\,(0.106)X_3] + error$$

where estimates for the predictors are: X_1 = Insect, X_2 = Area, and X_3 = Distance.

information criterion used, differences in information criterion (Δ_i), model weights (w_i), and a measure of goodness-of-fit (Table 2.4). Proper reporting also should include parameter estimates resulting from the model-fitting exercise; note that in the snowshoe hare example there is variability in the estimates provided for Food and Cover between the best versus remaining models (Table 2.5). On the other hand, the Food, Cover interaction in Model 12 (Figure 2.4) clearly shows that the two factors have multiplicative effects on hare occupancy and thus the interaction term should be considered. Table 2.5 does not include parameter estimates that have been weighted through model averaging because interaction terms preclude our ability to derive parameter estimates that are not conditioned by such terms. This issue is problematic because, as we stated earlier, complexity is par for the course in field research and ecologists should embrace this fact by explicitly modeling interacting factors where appropriate; our inability to reconcile model complexity with unbiased parameter estimates currently represents an important challenge in the application of IT methods in population ecology.

To conclude this section, we would be remiss if we failed to mention some ongoing concerns related to multimodel inference. Consideration of multiple models should not be necessary when dealing with fixed states such as sampling design optimization or when the true model is known (Ver Hoef and Boveng 2015). Likewise, in the quest to infer causality between predictor and response variables, multimodel inferencing should not trump robust estimation of effect sizes (Fieberg and Johnson 2015). In terms of the multimodel inferencing process itself, parameter estimates may not hold equivalent interpretations across all models that include a given predictor, highlighting the need to conduct model development with an eye to interactions between variables (Banner and Higgs 2017). Indeed, Cade (2015) warns that multimodel inferencing does not yield valid estimates when there is multicollinearity between predictors because correlated variables do not have common scales across different models. It follows that averaging models with correlated variables can lead to flawed statistical interpretation, and variable scaling is not sufficient in addressing this concern. Although it is currently unclear exactly how much correlation between predictors is tolerable in multimodel inferencing, we remind of the crucial need to assess multicollinearity and other aspects of exploratory analysis (Section 2.5.3) prior to conducting model selection and multimodel inferencing.

Table 2.5 Logit parameter estimates (± SE) from models relating snowshoe hare occupancy to food density, cover density, patch size, and patch connectivity.

Number	Covariates	Inter	Food	Cover	PS	PC	Cover * Food	Food * PS	Food * PC	Cover * PC
12	Food, Cover interaction	1.05	1.67 (0.33)	0.77 (0.15)			0.51 (0.23)			
5	Food, Cover	0.94	1.45 (0.28)	0.58 (0.12)						
9	Food, Cover, Patch size	0.95	1.44 (0.28)	0.56 (0.12)	0.13 (0.13)					
10	Food, Cover, Patch connectivity	0.95	1.45 (0.28)	0.59 (0.12)		−0.09 (0.12)				
6	Food, Patch size	0.93	1.46 (0.28)		0.22 (0.13)					
1	Food	0.92	1.46 (0.27)							
11	Food, Patch size, Patch connectivity	0.94	1.46 (0.28)		0.23 (0.13)	−0.08 (0.12)				
14	Food, Patch connectivity interaction	1.02	1.65 (0.32)			−0.30 (0.20)			−0.57 (0.35)	
13	Food, Patch size interaction	0.93	1.46 (0.27)		0.20 (0.19)			−0.04 (0.31)		
7	Food, Patch connectivity	0.92	1.47 (0.27)			−0.06 (0.11)				
8	Cover, Patch size	0.63		0.57 (0.12)	0.20 (0.13)					
2	Cover	0.63		0.59 (0.12)						
15	Cover, Patch connectivity interaction	0.63		0.59 (0.12)		−0.06 (0.12)				0.05 (0.13)
3	Patch size	0.61			0.28 (0.03)					
16	Constant only	0.60								
4	Patch connectivity	0.60				−0.06 (0.11)				

PS: Patch size; PC: Patch connectivity; Inter: Intercept.
Models are ordered according to their weight (higher weighted models at top) from model selection (see Table 2.4). Model averaging could not be employed due to the presence of interaction terms. Predictors were standardized prior to analysis to improve interpretability of the coefficients.

2.7 Model Validation

Model selection techniques identify the best model among a set of candidates, but if the best model fails to provide a good fit, biological processes of interest may be inadequately explained. A best-fitting model that does a poor job explaining variation in the response may be caused by predictors that are not strongly correlated with the response, or generally "noisy" datasets. It is good practice to evaluate goodness-of-fit or predictive performance of models, and a variety of methods are available for comparing fitted versus predicted values in a model (Table 2.6). *Model validation* seems to be an often-overlooked concluding step in model selection, and when conducted model validation tends to focus on the best model rather than on the suite of weighted candidates obtained through model selection.

Three general approaches exist for evaluating the *predictive performance* of a best-fitting model. For small datasets, it may be most appropriate to *calibrate the model* and estimate parameters, and then test predictive accuracy using the same set of observations (Guisan and Zimmermann 2000). Calibration is perhaps the most common approach in ecology, likely due to the preponderance of small datasets. For larger sample sizes, it is possible to *partition observations* into training and test data, where training data serve to calibrate the models and test data evaluate predictive performance (Hastie et al. 2009; Heikkinen et al. 2012). An approach based on partitioning of datasets provides a more conservative and accurate estimate of the model's predictive abilities. Training and test data can be split randomly into two samples, typically using a 50/50 split, although other splits may be more appropriate based on sample size

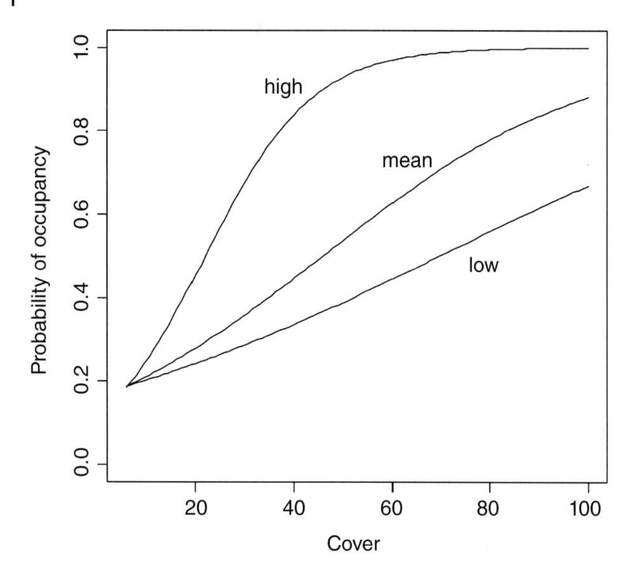

Figure 2.4 Plot of the interaction between food and cover on snowshoe hare occupancy. Food level is binned into low, medium, and high depending on the intensity of browsing on woody twigs, and cover ranges from low to high based on the visual obstruction of a spherical densitometer. The fitted lines represent the predicted relationship between food and cover based on Model #12. Coefficients and confidence intervals are omitted for clarity.

and the number of parameters in the model (see Fielding and Bell 1997). A trade-off exists between retaining enough training data to calibrate the model properly, and enough test data to get an accurate assessment of model performance (Guisan and Zimmermann 2000). However, we should note that splitting data for training and testing purposes is subject to the same concerns as those expressed when developing and evaluating robust hypotheses using the same dataset (Section 2.2). Indeed, the choice of data used for training versus testing is mostly arbitrary and ad hoc, and to date there lacks a standardized approach.

Recently, repeated subsampling procedures have become more common for dividing a dataset into training and test samples, including *cross-validation* and *bootstrapping*. In cross-validation procedures, the dataset is divided into k subsets (typically 5 or 10 subsets, called "folds") with model fitting and validation repeated k times with all of the folds minus one being used to calibrate the model and the held-out fold serving as a test for model validation (Hastie et al. 2009). Each distinct fold is used exactly once as the test data. In this manner, the entire dataset is used for testing the model, and because of

Table 2.6 Some commonly used metrics for testing the predictive performance of models.

Metric	Description
Continuous response variables	
Correlation coefficients	Measures correlation between predicted and observed values; Pearson correlations can be used for normally distributed variables, nonparametric correlations such as Kendall's tau are used for non-normally distributed variables
Root mean squared error	Measures the mean squared difference between predicted and observed values for normally distributed variables
Continuous, count, or binary variables	
Generalized R^2 (coefficient of determination)	For continuous response variables, the generalized R^2 is equivalent to traditional R^2 and indicates the amount of explained variation. Although the generalized R^2 also can be applied to count and binary data, care in analysis and interpretation is necessary
Pearson goodness-of-fit statistic	Indicates whether the model provides a good fit to the data. Large test statistics and small P-values indicate poor fit
Binary response variables	
Biserial correlation coefficient	Measures the correlation between predicted probabilities and observed binary response
Sensitivity	Measures the probability that the model will correctly classify a positive response (e.g. presence of a species at a location). Requires that predicted probabilities from the model are converted to a 1/0 binary response using a user-defined threshold.
Specificity	Measures the probability that the model will correctly classify a negative response (e.g. absence of a species at a location). Requires that predicted probabilities from the model are converted to a 1/0 binary response using a user-defined threshold.
Cohen's Kappa	Measures the overall accuracy of model predictions, corrected for the accuracy expected to occur by chance. Requires that predicted probabilities from the model are converted to a 1/0 binary response using a user-defined threshold.
AUC (area-under-the-curve)	Measures the discriminative ability of a model and is especially useful for models fit to binary data. It is a non-threshold-dependent measure of a model's predictive ability with values ranging from 0.5 to 1.0.

See Fielding and Bell (1997) and Guisan and Zimmerman (2000) for additional details.

the repeated assessment of performance, average values and measures of variability (e.g. R^2) can be calculated. N-fold cross-validation is a modification where n is the sample size and a single datapoint is used as the test data (equivalent to a jackknife procedure; Steyerberg et al. 2001). Bootstrapping procedures are similar to cross-validation but the test data are randomly selected with replacement from the training data, and the sample size of the test and training data are equivalent (Hastie et al. 2009).

If the goal of model validation is to identify a general model that can predict responses in different places or times, the data splitting approach for generating training and testing data is flawed and a better estimate of predictive performance is derived from a completely independent set of observations taken at a different place and time (Wenger and Olden 2012). For example, we could develop a best-fit model for snowshoe hare occupancy in a fragmented landscape of Idaho, and test the predictive performance of that model on a dataset of hare occupancy from a fragmented landscape in Minnesota. A model may perform well when predicting "in-sample" test data, where training and test data come from the same overall dataset, but perform poorly on "out-of-sample" test data from a completely independent dataset. The decline in predictive ability may correspond to overfitting in the original dataset, as the best performing models for predicting in-sample test data tend to be more complex and thus less transferable to other landscapes or times. Heterogeneity in relationships between predictors and response variables across space or time is another possible reason why model performance on completely independent datasets often declines, as we might expect to be the case in terms of snowshoe hare occupancy patterns across the vastly different landscape features in Idaho versus Minnesota.

The exact way in which a model should be validated will depend on sample size and the question that is being addressed, and requires decisions regarding using the same dataset for calibration and evaluation, splitting a single dataset randomly into test and training samples versus testing the model on a completely independent dataset. Regardless, some form of model validation should be presented as a final complement to the model selection process. This validation gives us confidence that our models are indeed useful for explaining patterns in the distribution, abundance, or population dynamics of species.

2.8 Software Tools

Most statistical programs have functions in their default installation for exploratory analyses, including the production of figures for diagnostics such as scatterplots and pairwise correlations. Default installations also generally offer functions for general statistical tests. The output of these functions often include diagnostics tests regarding whether the assumptions of the model are respected or not. More advanced functions are sometimes provided in additional packages, for example, a function to calculate the VIF is available in the package `car` in R. An additional package, `bbmle`, has an `ICtab` function that computes information criteria for a series of models, optionally giving information about Δ_i, w_i and other relevant metrics. The package speeds up the model selection process but the calculations are black-boxed, which is not helpful for those first learning model selection approaches.

Log-likelihood, AIC, and other information criteria are frequently provided in the output of regression models, but can also be extracted using specific functions such as the `logLik` and `AIC` functions in R. Model selection can be performed manually once the log-likelihood or information criteria are extracted, but thankfully, this process is made easier in R with packages such as `AICcmodavg` and `MuMIn`. Both packages offer similar outputs in term of model selection, but `MuMIn` is more flexible in how models included in the selection process are specified. Both packages can also calculate model-averaged coefficients with unconditional confidence intervals across a candidate set of multiple models.

2.9 Online Exercises

The online exercises are designed to reinforce concepts presented in this chapter and to provide the user with an understanding of the steps toward robust analysis in population ecology. Exercise 1 uses a study on snowshoe hares to develop a model assessing factors affecting body condition of individual hares. Exercise 2 uses a related dataset to assess the fit of multiple models predicting hare cause of death relative to body condition and environmental factors. Last, Exercise 3 extends this approach by using the snowshoe hare dataset described in this chapter to conduct a model selection and multimodel inferencing exercise.

2.10 Future Directions

A scientific approach that has become strongly established in population ecology consists of three elements: (i) developing multiple research hypotheses and predictions at the outset of a study; (ii) testing hypotheses through corresponding statistical models; and (iii) applying IT (or Bayesian) methods and multimodel weighting to evaluate the strength of inference. Further, mindful data exploration and remediation, as well as model validation, are increasingly recognized as important complements to model-based inference. We think that these procedures will continue to lead to improved inference about population ecology, although frequentist statistical methods should retain a place in the ecologist's statistical toolbox especially when conducting experiments or developing sampling protocols or adaptive management strategies. The advantages of the IT approach in an observational research context are resoundingly clear, and similar arguments can be made for Bayesian methods especially when robust prior information is available.

There are key areas that need work to facilitate the further integration of multimodel approaches into the ecological toolbox. We mentioned previously the challenges of multiple imputation in the context of model selection, as well as the current uncertainty concerning multimodel inferencing with models having interaction terms or multicollinearity. We also illustrated difficulties with model validation when predictive data are not truly independent or when the focus of validation is a single best-fit model rather than the weighted suite of candidates. Likewise, model selection for integrated population models (IPM) remains a challenge, even though IPM have become increasingly popular in population ecology (Chapter 9).

From a more philosophical perspective, there remains the thorny issue of mixing modeling paradigms and whether it is even appropriate to combine IT with frequentist approaches in the same research framework. To this point we offer that researchers should keep an open mind and entertain the full set of tools that are available for robust inference, as it may be that the research questions are served using a variety of techniques. Indeed, as with all new or transformational approaches in science, we should not be surprised by resistance or caution in the widespread adoption of model-based methods, and more thorough coverage in undergraduate statistics courses and ecological texts will help raise their profile.

More broadly, we underscore that while students can be trained in the proper use of new statistical tools and software to conduct robust inference, ultimately a greater appreciation for the development of research hypotheses and predictions that reflect real-world relationships between variables, as well as a more explicit understanding of the direct link between hypotheses, predictions, and models in a philosophical as well as statistical context, is crucial. Such advances will help ensure that population ecologists are in position to rigorously and effectively address the analytical challenges that doubtless will arise in response to forthcoming environmental changes.

References

Abayomi, K., Gelman, A., and Levy, M. (2008). Diagnostics for multivariate imputations. *Journal of the Royal Statistical Society C* 57: 273–291.

Aho, K., Derryberry, D., and Peterson, T. (2014). Model selection for ecologists: the worldviews of AIC and BIC. *Ecology* 95: 631–636.

Aho, K., Derryberry, D., and Peterson, T. (2017). A graphical framework for model selection criteria and significance tests: refutation, confirmation and ecology. *Methods in Ecology and Evolution* 8: 47–56.

Aiken, L.S. and West, S.G. (1991). *Multiple Regression: Testing and Interpreting Interactions*. Newbury Park, NJ: Sage Publications.

Anderson, D.R. and Burnham, K.P. (2002). Avoiding pitfalls when using information-theoretic methods. *Journal of Wildlife Management* 66: 912–918.

Anderson, D.R., Burnham, K.P., Gould, W.R. et al. (2001). Concerns about finding effects that are actually spurious. *Wildlife Society Bulletin* 29: 311–316.

Anderson, D.R., Burnham, K.P., and Thompson, W.L. (2000). Null hypothesis testing: problems, prevalence, and an alternative. *Journal of Wildlife Management* 64: 912–923.

Ausband, D.E. and Baty, G.R. (2005). Effects of precommercial thinning on snowshoe hare habitat use during winter in low-elevation montane forests. *Canadian Journal of Forest Research* 35: 206–210.

Banner, K.M. and Higgs, M.D. (2017). Considerations for assessing model averaging of regression coefficients. *Ecological Applications* 27: 78–93.

Beaumont, M.A. (2010). Approximate Bayesian computation in evolution and ecology. *Annual Review of Ecology and Systematics* 41: 379–406.

Betini, G.S., Avgar, T., and Fryxell, J.M. (2017). Why are we not evaluating multiple competing hypotheses in ecology and evolution? *Royal Society Open Science* 4: 160756.

Biro, P.A. and Dingemanse, N.J. (2009). Sampling bias resulting from animal personality. *Trends in Ecology and Evolution* 24: 66–67.

Buckland, S.T., Burnham, K.P., and Augustin, N.H. (1997). Model selection: an integral part of inference. *Biometrics* 53: 603–618.

Burnham, K.P. and Anderson, D.R. (2002). *Model Selection and Multimodel Inference*. New York, NY: Springer-Verlag.

Burnham, K.P. and Anderson, D.R. (2014). *P* values are only an index to evidence: 20th vs. 21st-century statistical science. *Ecology* 95: 627–630.

Burnham, K.P., Anderson, D.R., and Huyvaert, K.P. (2011). AIC model selection and multimodel inference in behavioural ecology: some background, observations, and comparisons. *Behavioural Ecology and Sociobiology* 65: 23–35.

Cade, B.S. (2015). Model averaging and muddled multimodel inferences. *Ecology* 96: 2370–2382.

Carpenter, S.R., Chisholm, S.W., Krebs, C.J. et al. (1995). Ecosystem experiments. *Science* 269: 324–327.

Chamberlin, T.C. (1897). The method of multiple working hypotheses. *Journal of Geology* 5: 837–848. (reprinted in Science 148, 754–759 in 1965).

De Valpine, P. (2014). The common sense of *P* values. *Ecology* 95: 617–621.

Dochtermann, N.A. and Jenkins, S.H. (2011). Developing multiple hypotheses in behavioural ecology. *Behavioural Ecology and Sociobiology* 65: 37–45.

Dormann, C.F., Elith, J., Bacher, S. et al. (2012). Collinearity: a review of methods to deal with it and a simulation study evaluating their performance. *Ecography* 35: 1–20.

Elliott, L.R. and Brook, B.W. (2007). Revisiting Chamberlin: multiple working hypotheses of the 21st century. *BioScience* 57: 608–614.

Ellison, A.M. (2004). Bayesian inference in ecology. *Ecology Letters* 7: 509–520.

Engqvist, L. (2005). The mistreatment of covariate interaction terms in linear model analyses of behavioural and evolutionary ecology studies. *Animal Behaviour* 70: 967–971.

Farrington, D.P. and Loeber, R. (2000). Some benefits of dichotomization in psychiatric and criminological research. *Criminal Behaviour and Mental Health* 10: 100–122.

Fieberg, J. and Johnson, D.H. (2015). MMI: multimodel inference or models with management implications? *Journal of Wildlife Management* 79: 708–718.

Fielding, A.H. and Bell, J.F. (1997). A review of methods for the assessment of prediction errors in conservation presence/absence models. *Environmental Conservation* 24: 38–49.

Ford, E.D. (2009). The importance of a research data statement and how to develop one. *Annales Zoologici Fennici* 46: 82–92.

Forstmeier, W. and Schielzeth, H. (2011). Cryptic multiple hypotheses testing in linear models: overestimated effect sizes and the winner's curse. *Behavioural Ecology and Sociobiology* 65: 47–55.

Fussmann, G.F., Ellner, S.P., Shertzer, K.W. et al. (2000). Crossing the Hopf bifurcation in a live predator-prey system. *Science* 290: 1358–1360.

Galipaud, M., Gillingham, M.A.F., David, M., and Dechaume-Moncharmong, F.-X. (2014). Ecologists overestimate the importance of predictor variables in model averaging: a plea for cautious interpretations. *Methods in Ecology and Evolution* 5: 983–991.

Garamszegi, L.Z. (2011). Information-theoretic approaches to statistical analysis in behavioural ecology: an introduction. *Behavioural Ecology and Sociobiology* 65: 1–11.

Gerrodette, T. (2011). Inference without significance: measuring support for hypotheses rather than rejecting them. *Marine Ecology* 32: 404–418.

Giam, X. and Olden, J.D. (2016). Quantifying variable importance in a multimodel inference framework. *Methods in Ecology and Evolution* 7: 388–397.

Gotelli, N.J. and Ellison, A.M. (2013). *A Primer of Ecological Statistics*. Sunderland, MA: Sinauer Associates, Inc.

Griffin, P.C. and Mills, L.S. (2009). Sinks without borders: snowshoe hare dynamics in a complex landscape. *Oikos* 118: 1487–1498.

Guisan, A. and Zimmermann, N.E. (2000). Predictive habitat distribution models in ecology. *Ecological Modelling* 135: 147–186.

Guthery, F.S., Brennan, L.A., Peterson, M.J. et al. (2005). Information theory in wildlife science: critique and viewpoint. *Journal of Wildlife Management* 69: 457–465.

Halsey, L.G., Curran-Everett, D., Vowler, S.L., and Drummond, G.B. (2015). The fickle *P* value generates irreproducible results. *Nature Methods* 12: 179–185.

Hastie, T., Tibshirani, R., and Friedman, J. (2009). *The Elements of Statistical Learning: Data Mining, Inference, and Prediction*. New York, NY: Springer Science +Business Media, LLC.

Hegyi, G. and Garamszegi, L.Z. (2011). Using information theory as a substitute for stepwise regression in ecology and behaviour. *Behavioural Ecology and Sociobiology* 65: 69–76.

Heikkinen, R.K., Marmion, M., and Luoto, M. (2012). Does the interpolation accuracy of species distribution models come at the expense of transferability? *Ecography* 35: 276–288.

Hooten, M.B. and Hobbs, N.T. (2015). A guide to Bayesian model selection for ecologists. *Ecological Monographs* 85: 3–28.

Jennersten, O. and Nilsson, S.G. (1993). Insect flower visitation frequency and seed production in relation to patch size of *Vicaria vulgaris* (Caryophyllaceae). *Oikos* 68: 283–292.

Johnson, J.B. and Omland, K.S. (2004). Model selection in ecology and evolution. *Trends in Ecology and Evolution* 19: 101–108.

King, R., Morgan, B.J.T., Gimenez, O. et al. (2009). *Bayesian Analysis for Population Ecology*. Boca Raton, FL: Chapman & Hall/CRC.

Krebs, C.J., Boonstra, R., Nams, V. et al. (2001). Estimating snowshoe hare population density from pellet counts: a further evaluation. *Canadian Journal of Zoology* 79: 1–4.

Krebs, C.J., Gilbert, B.S., Boutin, S. et al. (1987). Estimation of snowshoe hare population density from turd transects. *Canadian Journal of Zoology* 65: 565–567.

Levi, T. and Wilmers, C.C. (2012). Wolves-coyotes-foxes: a cascade among carnivores. *Ecology* 93: 921–929.

Lindberg, M.S., Schmidt, J.H., and Walker, J. (2015). History of multimodel inference via model selection in wildlife science. *Journal of Wildlife Management* 79: 704–707.

Little, R.J.A. and Rubin, D.B. (2002). *Statistical Analysis with Missing Data*. Hoboken, NJ: Wiley.

Lukacs, P.M., Thompson, W.L., Kendall, W.L. et al. (2007). Concerns regarding a call for pluralism of information theory and hypothesis testing. *Journal of Applied Ecology* 44: 456–460.

Mac Nally, R.M.A.C. (2000). Regression and model-building in conservation biology, biogeography and ecology: the distinction between – and reconciliation of – "predictive" and "explanatory" models. *Biodiversity and Conservation* 9: 655–671.

MacNab, J. (1983). Wildlife management as scientific experimentation. *Wildlife Society Bulletin* 11: 397–401.

Madden, L.V., Shah, D.A., and Esker, P.D. (2015). Does the *P* value have a future in plant pathology? *Phytopathology* 105: 1400–1407.

Martínez–Abraín, A., Conesa, D., and Forte, A. (2014). Subjectivism as an unavoidable feature of ecological statistics. *Animal Biodiversity and Conservation* 37 (2): 141–143.

McCarthy, M.A. (2015). Approaches to statistical inference. In: *Ecological Statistics: Contemporary Theory and Application* (eds. G.A. Fox, S. Negrete-Yankelevich and V. J. Sosa), 15–43. Oxford, UK: Oxford University Press.

Murray, D.L. (2003). Snowshoe hare and other hares. In: *Wild Mammals of North America*, vol. II (eds. G.A. Feldhamer, B. Thompson and J. Chapman), 147–175. Baltimore, MD: John Hopkins University Press.

Murray, D.L., Roth, J.D., Ellsworth, E. et al. (2002). Estimating low density snowshoe hare populations using fecal pellet counts. *Canadian Journal of Zoology* 80: 771–781.

Murray, K. and Conner, M.M. (2009). Methods to quantify variable importance: implications for the analysis of noisy ecological data. *Ecology* 90: 348–355.

Murtaugh, P.A. (2014). In defense of *P* values. *Ecology* 95: 611–617.

Nakagawa, S. and Freckleton, R.P. (2008). Missing inaction: the danger of ignoring missing data. *Trends in Ecology and Evolution* 11: 592–596.

Nakagawa, S. and Freckleton, R.P. (2011). Model averaging, missing data and multiple imputation: a case study for behavioural ecology. *Behavioural Ecology and Sociobiology* 65: 103–116.

Nelder, J.A. (1977). A reformulation of linear models. *Journal of the Royal Statistical Society* 140: 48–77.

O'Brien, R.M. (2007). A caution regarding rules of thumb for variance inflation factors. *Quality and Quantity* 7: 673–690.

Owen, S.V. and Froman, R.D. (2005). Why carve up your continuous data? *Research in Nursing and Health* 28: 496–503.

Pease, J.L., Vowles, R.H., and Keith, L.B. (1979). Interaction of snowshoe hares and woody vegetation. *Journal of Wildlife Management* 43: 43–60.

Peduzzi, P., Concato, J., Kemper, E. et al. (1996). A simulation study of the number of events per variable in logistic regression analysis. *Journal of Clinical Epidemiology* 49: 1373–1379.

Quinn, G.P. and Keough, M.J. (2002). *Experimental Design and Analysis for Biologists*. Cambridge, UK: Cambridge University Press.

Royston, P., Altman, D.G., and Sauerbrei, W. (2006). Dichotomizing continuous predictors in multiple regression: a bad idea. *Statistics in Medicine* 25: 127–141.

Sauerbrei, W., Royston, P., and Binder, H. (2007). Selection of important variables and determination of functional form for continuous predictors in multivariable model building. *Statistics in Medicine* 26: 5512–5528.

Schielzeth, H. (2010). Simple means to improve the interpretability of regression coefficients. *Methods in Ecology and Evolution* 1: 103–113.

Smith, D.W., Bangs, E.E., Oakleaf, J.O. et al. (2010). Survival of colonizing wolves in the northern Rocky mountains of the United States, 1982–2004. *Journal of Wildlife Management* 74: 620–634.

Spanos, A. (2014). Recurring controversies about *P* values and confidence intervals revisited. *Ecology* 95: 645–651.

Stephens, P.A., Buskirk, S.W., Hayward, G.D. et al. (2005). Information theory and hypothesis testing: a call for pluralism. *Journal of Applied Ecology* 42: 4–12.

Stephens, P.A., Buskirk, S.W., and Martínez del Rio, C. (2007). Inference in ecology and evolution. *Trends in Ecology and Evolution* 22: 192–197.

Steyerberg, E.W., Harrell, F.E., Borsboom, G.J. et al. (2001). Internal validation of predictive models: efficiency of some procedures for logistic regression analysis. *Journal of Clinical Epidemiology* 54: 774–781.

Taper, M.L. (2004). Model identification from many candidates. In: *The Nature of Scientific Evidence: Statistical, Philosophical, and Empirical Considerations* (eds. M.L. Taper and S.R. Lele), 488–508. Chicago, IL: University of Chicago Press.

Thornton, D.H., Wirsing, A.J., Roth, J.R. et al. (2012). The asymmetric role of snowshoe hare population density on extinction and colonization processes in variegated landscapes. *Ecography* 36: 610–621.

Ver Hoef, J.M. and Boveng, P.M. (2015). Iterating on a single model is a viable alternative to multimodel inference. *Journal of Wildlife Management* 79: 719–729.

Vittinghoff, E. and McCulloch, C.E. (2006). Relaxing the rule of ten events per variable in logistic and Cox regression. *American Journal of Epidemiology* 165: 710–718.

Wenger, S.J. and Olden, J.D. (2012). Assessing transferability of ecological models: an underappreciated aspect of statistical validation. *Methods in Ecology and Evolution* 3: 260–267.

White, I.R., Royston, P., and Wood, A.M. (2011). Multiple imputation using chained equations: issues and guidance for practice. *Statistics in Medicine* 30: 377–399.

Whitlock, M.C. and Schluter, D. (2009). *The Analysis of Biological Data*. Greenwood, CO: Roberts and Company.

Whittingham, M.J., Stephens, P.A., Bradbury, R.B. et al. (2006). Why do we still use stepwise modeling in ecology and behaviour? *Journal of Animal Ecology* 75: 1182–1189.

Wirsing, A.J., Steury, T.D., and Murray, D.L. (2002). A demographic analysis of a southern snowshoe hare population in a fragmented habitat: evaluating the refugium model. *Canadian Journal of Zoology* 80: 169–177.

Zar, G.H. (2010). *Biostatistical Analysis*, 5e. Upper Saddle River, NJ: Prentice Hall.

Zens, M.S. and Peart, D.R. (2003). Dealing with death data: individual hazards, mortality and bias. *Trends in Ecology and Evolution* 18: 366–373.

Zuur, A.F., Ieno, E.N., and Elphick, C.S. (2010). A protocol for data exploration to avoid common statistical problems. *Methods in Ecology and Evolution* 1: 3–14.

Part II

Population Demography

3

Estimating Abundance or Occupancy from Unmarked Populations

Brett T. McClintock[1] *and Len Thomas*[2]

[1] *NOAA National Marine Mammal Laboratory, Alaska Fisheries Science Center, National Marine Fisheries Service, Seattle, WA, USA*
[2] *Centre for Research into Ecological and Environmental Modelling, School of Mathematics and Statistics, University of St. Andrews, St. Andrews, United Kingdom*

Summary

When it comes to collecting field data, "you can't always get what you want" in population ecology. We are often interested in understanding the vital rates that drive population dynamics, such as fecundity or survival, but the realities of fieldwork under logistical and financial constraints often preclude us from collecting the data we would most like to have. Despite these challenges, there are plenty of ways to "get what you need" and make meaningful inferences about population status, trends, and habitat associations using relatively simple and inexpensive field sampling methods. Here, we focus on the estimation of abundance and species occurrence from unmarked population data. We provide an overview of common sampling and analysis methods for squeezing the most information out of unmarked population data while accounting for imperfect detection and other obstacles. These unmarked population methods include plot sampling, distance sampling, spatially replicated counts, removal sampling, and presence/absence sampling.

3.1 Introduction

From the perspective of an applied population ecologist, an ideal study population would consist of individually identifiable or "marked" organisms that could be observed at any time and for as long as we wish without disturbing them. Whether for a long-term investigation about the effects of global climate change or the shorter-term studies typical of graduate student projects, reliable inference about population patterns and dynamics from observations of a population of *marked individuals* would be considerably easier than observations of a study population consisting of indistinguishable and difficult-to-observe (or to manually mark) individuals. Some wildlife species possess easily distinguishable marks, including natural pelage coloration of some felids or acquired scars in some cetaceans, but otherwise this ideal study population rarely exists outside the artificial constructs of a well-designed field experiment. Instead, we must often contend with the reality that our study population may contain relatively few to no marked individuals, and that these individuals may be difficult to observe when conducting data collection surveys. This chapter focuses on using data from populations of *unmarked individuals* to estimate two quantities fundamental to understanding population dynamics and species distributions: *population abundance* and *species occurrence.*

3.1.1 Why Collect Data from Unmarked Populations?

Unmarked populations are not as informative for estimation of demographic parameters, such as survival, compared to marked populations. However, status, trends, and habitat associations can be inferred from unmarked populations, and there are entire books devoted to use of data from unmarked individuals for estimation of population abundance (Buckland et al. 2001, 2004, 2015) and species occurrence (MacKenzie et al. 2006). In deciding whether to collect and analyze data from an unmarked population, there are many issues to consider. When scientific hypotheses concern the role of demographic rates (i.e. recruitment, survival) and movements (i.e. immigration, emigration) as drivers of changes in abundance, then investment in capture-recapture methods for marked populations may be necessary (Chapters 7 and 9). However, a key advantage of unmarked population studies is that time and money need not be invested

Population Ecology in Practice, First Edition. Edited by Dennis L. Murray and Brett K. Sandercock.

Companion website: www.wiley.com/go/MurrayPopulationEcology

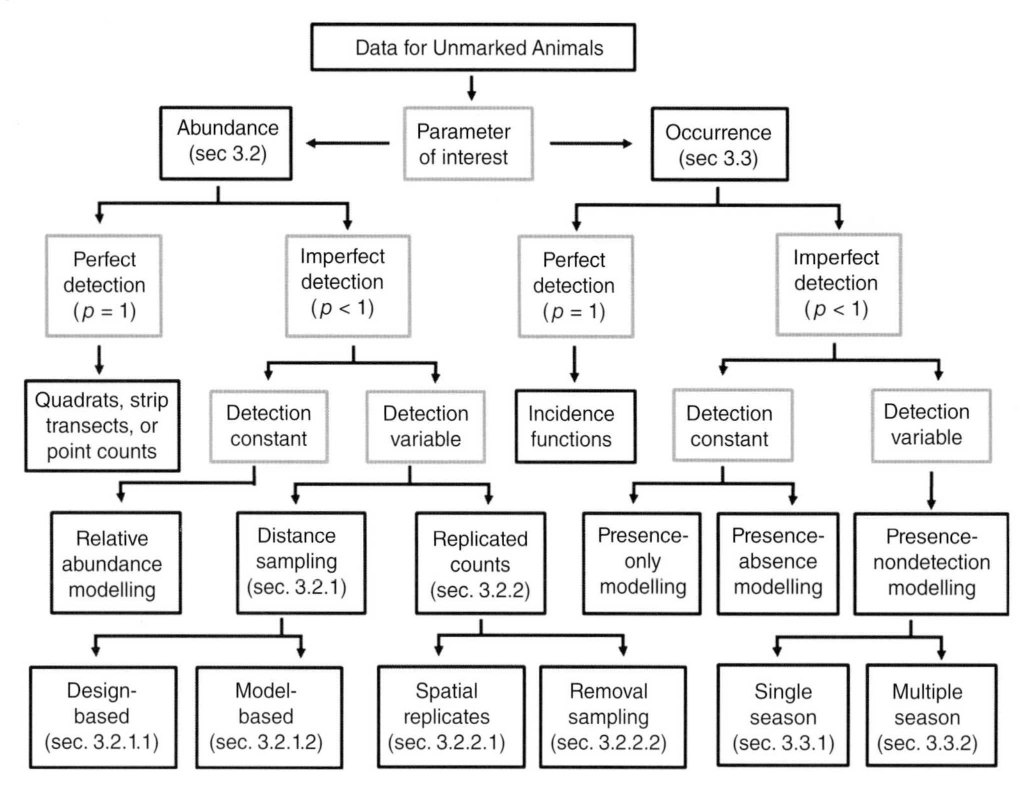

Figure 3.1 Decision tree for Chapter 3. Some of the most important considerations for the design and analysis of unmarked population studies include the item of ecological interest and detection probability (p).

in capturing and marking animals or, in the case of non-invasive genetic sampling, the collection and analysis of DNA samples (Lukacs and Burnham 2005, Bravington et al. 2016; Fewster et al. 2016; Chapter 5). Thus, limited resources can be invested in other aspects of the study design, thereby enabling a broader geographic or temporal scale of monitoring. Not only can capturing and marking animals be expensive in time and money, but it is also typically stressful and potentially harmful to wild animals. For species of conservation concern, unmarked population studies may be necessary because the capturing and marking of individuals is not politically or socially acceptable. Despite the limited information contained in unmarked data, in this chapter we will demonstrate how meaningful inferences about population abundance and species occurrence probability can be obtained using relatively simple and inexpensive field sampling methods for unmarked populations. A guide to this material is provided in Figure 3.1.

3.1.2 Relative Indices and Detection Probability

3.1.2.1 Population Abundance

Suppose one is interested in the size of an endangered population after protective management policies have been initiated. If a population of animals is endangered, physical capture of individuals for marking or genetic sampling procedures may be prohibited. Inconveniently, many populations of individuals are not distinguishable from natural markings. As part of a monitoring plan, surveys of unmarked individuals are conducted throughout the study area each year. Whenever an individual is encountered during a survey, this datum is recorded.

If these surveys had perfect detection and all individuals in the population were encountered each year, raw counts of individuals would constitute annual estimates of population size or *abundance* (N). This example of a sampling scenario is often referred to as a *population census*. Population censuses are rare in wild animal populations because study areas are often too large (or inaccessible) to sample completely, and individuals present in the portion of the study area subject to sampling often go undetected. However, some populations are regularly censused with high success, such as nesting penguins. Typically, species that can be readily surveyed with complete detection are those where individuals are highly visible, occupy a discrete location, and can be distinguished by the observer. However, for the majority of populations where a population census is not possible, it may seem natural to assume that the number of individuals seen each year still reflects the underlying population size. In other words, if more individuals are seen each year, one may be inclined to conclude that the population increased during the study period. Under this assumption, the raw counts (C) constitute an index of *relative abundance* for each year. However, this index is in fact a product of the *true abundance* (N) and the *probability of detection* (p):

$$E(C) = Np, \tag{3.1}$$

where $E(C)$ is the expected value for C. For example, suppose the rate of change in abundance (h) from year i to year $i + 1$ is $h = N_{i+1}/N_i$. From Eq. 3.1, we have:

$$E(h) = \frac{E(C_{i+1})}{E(C_i)} = \frac{N_{i+1}p_{i+1}}{N_i p_i} = h\frac{p_{i+1}}{p_i}. \tag{3.2}$$

Reliable inferences about changes in abundance from relative indices therefore depend on the critical assumption that detection probability is constant across surveys (i.e. $p_i = p_{i+1}$). It is now widely acknowledged that a constant probability of detection is generally an unrealistic assumption (Anderson 2001; Mazerolle et al. 2007; Johnson 2008; Archaux et al. 2012). The probability of detection during any given survey can depend on observer ability, environmental variables such as time of day, season, precipitation, habitat type, wind speed, and human disturbance, as well as species characteristics such as behavior, group size, or calling intensity. For example, Archaux et al. (2012) demonstrated by simulation that a detection probability difference between count surveys as small as 4–8% can lead to a 50–90% risk of a type I error with false rejection of the null hypothesis under sampling conditions that are commonplace in ecological studies. In a seven-year study of four species of forest birds in Utah, USA, Norvell et al. (2003) showed that the assumption of constant proportionality was violated and lead to differences in estimated trends for relative abundance compared to population size.

When individuals in a population are marked, it is possible to estimate p and make inferences about N and other demographic parameters using capture-recapture methods (Chapters 5 and 7). When the object of interest is unmarked and immobile, such as individual plants, nests, or sessile animals, spatial location can take the place of the mark in providing an identification, and detection probability can be estimated with capture-recapture protocols based on *repeated sampling* or *multiple observers* (Nichols et al. 2000). Things are not so easy for unmarked, mobile populations. However, with appropriate data collection and analysis, one can also estimate detection probability and reliably estimate N (or h) using a variety of different methods (Seber 1982; Borchers et al. 2002; Williams et al. 2002; Nichols et al. 2009).

3.1.2.2 Species Occurrence

Instead of abundance, suppose that one is interested in the spatial patterns and dynamics of *species occurrence*. In this case, so-called *presence/absence* surveys can be used to make inferences about the spatial distributions of species, and these methods do not require marked individuals. The sample units in presence/absence surveys are individual sites within a larger study region of interest. Surveys consist of visits to each site, where the species of interest is either encountered or not encountered, and the number of individuals encountered at each site may or may not be recorded. Assuming species occurrence is detected perfectly at each site, the *probability of species occurrence* for the study area may be estimated as x/n, where x is the number of sites where the species was detected and n is the total number of sites. Assuming perfect detection, one may infer patterns and dynamics in the species' distribution using *presence-only* or presence/absence modeling, such as incidence functions and related approaches (Hanski 1992, 1999; He and Gaston 2003; Phillips et al. 2006; Yackulic et al. 2012a). However, it is now widely recognized that presence/absence data are also subject to imperfect detection, and presence/absence data are perhaps more appropriately described as *detection/nondetection* data (MacKenzie et al. 2002; He and Gaston 2003; Tyre et al. 2003; MacKenzie 2005).

Similar to using raw counts to infer abundance, problems arise when attempting to make inferences about species occurrence under imperfect detection. The issue is that the number of sites with detections x is a function of both the *probability of site occupancy* (ψ) and the *probability of detection* for the species at a site (p):

$$E(x) = pn\psi. \tag{3.3}$$

Failing to account for $p < 1$ will therefore result in underestimation of species occurrence. Now suppose the finite rate of change in occupancy from year i to year $i + 1$ is $h = \psi_{i+1}/\psi_i$. From Eq. 3.3, we have:

$$E(h) = \frac{E(x_{i+1})}{E(x_i)} = \frac{\psi_{i+1}p_{i+1}}{\psi_i p_i} = h\frac{p_{i+1}}{p_i}. \tag{3.4}$$

Attempting to infer changes in occupancy from an index of *relative occupancy* therefore depends on the assumption that p is constant across surveys. If one is willing to assume a constant detection probability, then the relative patterns and dynamics of species distribution can be investigated using presence-only and presence/absence modeling without explicitly accounting for detection probability (Phillips et al. 2006; Royle et al. 2012; Yackulic et al. 2012a). However, for many animals, and even plants (Kéry and Gregg 2004; Chen et al. 2012), it will more often be the case that detection is imperfect (i.e. $p < 1$) and is not constant among observers, sampling occasions, or sites. By using repeated sampling, one can use detection/nondetection data to estimate both site-level probabilities of detection and occurrence, as well as local extinction and colonization rates.

3.1.3 Hierarchy of Sampling Methods for Unmarked Individuals

A complete population census represents the most informative sampling method for an unmarked population. More commonly, study areas are too large to sample completely but individuals that are exposed to sampling are perfectly detected, allowing the use of uncorrected counts from sampled portions of the study area to estimate regional abundance (or density). Such *plot sampling* methods include quadrats (square plots), strip transects (rectangular plots), and point counts (circular plots, Seber 1982; Buckland et al. 2001; Borchers et al. 2002; Williams et al. 2002).

Given R plots each of size a, randomly selected from within a study area of size A, and containing c individuals, then abundance $(\hat{N})$ may be estimated as

$$\hat{N} = \frac{c}{Ra}A \tag{3.5}$$

and density $(\hat{D})$ as

$$\hat{D} = \frac{\hat{N}}{A} = \frac{c}{Ra}. \tag{3.6}$$

A simulated strip transect example is shown in Figure 3.2. Strip transects are often used for aerial surveys of conspicuous, common animals. For strip transects, we can modify the above formulae to accommodate the area sampled by each strip. In the example, R = 10 strips of half-width w = 150 m and length l_i = 20 km (i = 1, …, R) were placed according to a systematic random design in a study area of size A = 560 km^2. Hence, we now have

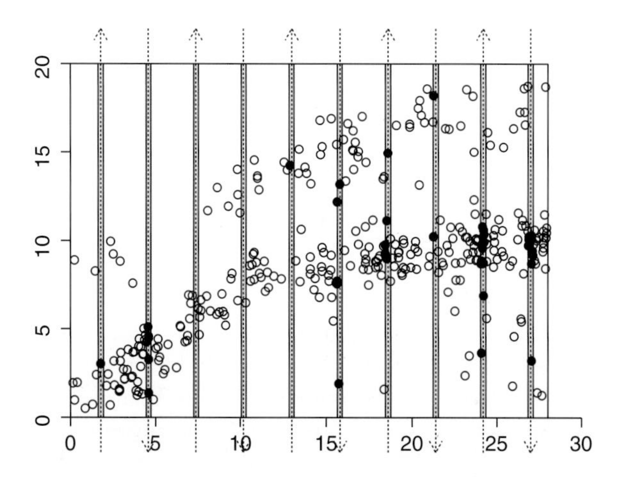

Figure 3.2 A simulated example strip transect survey that was created using the `wisp` package in `R`; the population comes from the `seal.pop` dataset. The surveyed strips are shown in gray with dashed lines indicating the tracklines taken during the survey. Animal locations occur within the simulated landscape (white circles), and all animals within the surveyed strips are assumed to be detected (black circles).

$$\hat{N} = \frac{c}{2wL}A, \tag{3.7}$$

and

$$\hat{D} = \frac{\hat{N}}{A} = \frac{c}{2wL}, \tag{3.8}$$

where $L = \sum_{i=1}^{R} l_i$. Assuming all c = 40 individuals in the strips were counted, we obtain $\hat{N} = 373$ and $\hat{D} = 0.67$ individuals km^{-2}, and because the analysis is based on a simulation, we know the true values were 345 and 0.62, respectively. Variance is calculated treating transect as the sampling unit; methods for *systematic random designs* such as this are discussed in Fewster et al. (2009). Applying the "O2" estimator of Fewster et al., the coefficient of variation (CV – standard error divided by estimate) on the above estimates is 0.23. Systematic random survey designs are generally more efficient than *completely random designs*, in the sense that they produce lower variance for the same survey effort.

When detection in the areas exposed to sampling is not perfect, one must account for $p < 1$ to reliably estimate abundance or species occurrence. Detection probability is often considered a "nuisance" parameter that is of little interest from an ecological perspective. However, because the detection process must be properly accounted for to make reliable inferences about underlying ecological processes, detection probability shall be the central theme of our chapter. Fortunately, numerous study designs facilitate simultaneous estimation of detection probability and ecological parameters of interest from unmarked populations. If distances to detected individuals can be measured, then *distance sampling methods* can be used to simultaneously estimate detection probability and abundance. However, if it is not possible to estimate distances to detected individuals, but surveys are spatially or temporally replicated, then *repeated counts* can be used to estimate detection probability and abundance.

Some animal populations are not immediately amenable to the above approaches, but nevertheless may reliably be surveyed using modifications of the methods. For example, cottontop tamarin monkeys (*Saguinus oedipus*) are endemic to Colombian forests, but are not suitable for standard count-based methods, being highly cryptic and showing strong avoidance of observers. However, monkeys can be induced to approach observers with playbacks of conspecific vocalizations; exploiting this behaviour led to the development of a *lure strip transect* by Savage et al. (2010) to derive the first range-wide population estimate. Similarly, Buckland et al. (2006) used acoustic lures in a point-based sampling scheme to derive an estimate of detection probability of Scottish Crossbills (*Loxia scotica*).

One option that is often useful when the animals themselves are hard to survey is to survey animal signs such as whale surface blows, carnivore tracks, songbird calls, primate nests, or hare pellets. The density of these alternative, easier to measure, objects (often called "cues") can then be linked to animal density by using one or more *multipliers* or conversion factors, which usually require separate estimation using secondary surveys. For example, Laing et al. (2003) discuss methods for estimating deer dung deposition and decomposition rates, for estimation of deer density using deer dung collected along transect surveys. Some species that are hard to detect visually make frequent, loud vocalizations that can be transmitted for long distances through the environment (e.g. many cetacean species, forest elephants). Passive acoustic density estimation is a rapidly expanding field for monitoring of secretive species (Marques et al. 2013).

Multipliers are also useful when some proportion of the population can be surveyed using a standard method, and this proportion can be estimated. For example, *aerial strip transect surveys* have become more widely used now that it is possible to replace a human observer on the vehicle with a high-definition camera or video system (Buckland et al. 2012; Conn et al. 2014). For marine mammals, some proportion of animals will be underwater when the survey vehicle passes over, but if this proportion can be estimated from auxiliary information, such as a sample of tagged animals, then correction factors can be included as multipliers in the denominator of the standard strip transect formula (Eq. 3.7).

In the absence of counts of detected individuals, presence/absence surveys represent the least-informative sampling method for unmarked populations. These data represent a coarse summary of population structure and dynamics that is also subject to potential biases induced by imperfect detection at the site level. However, using repeated sampling, detection/nondetection data can be used to investigate complex hypotheses about the patterns and dynamics of species occurrence, while still accounting for imperfect detection. As we shall see, there is a fundamental relationship between abundance and species occurrence, and under certain conditions, one may even be able to estimate abundance from detection/nondetection data.

In the rest of this chapter, we focus on developing a conceptual understanding of the various methods used in abundance and occupancy estimation for unmarked animals (Figure 3.1); we limit our treatment of issues associated with survey design and field methods. However, good design and execution are key to the success of all of these methods. Anyone planning on putting these methods into practice would be well advised to consider these issues carefully, and undertake pilot surveys and pilot analyses before starting in earnest. Poor analyses can typically be redone at low cost; the same is not true if a repeat collection of field data is required!

3.2 Estimating Abundance (or Density) from Unmarked Individuals

3.2.1 Distance Sampling

Distance sampling methods were originally developed to estimate abundance (or density) for unmarked populations where $p < 1$. A comprehensive review of the history and methods of distance sampling, as well as practical considerations for the design and implementation of these studies, is provided by Buckland et al. (2001, 2004, 2015). A large and growing literature describes the extension or modification of these methods for nearly all imaginable sampling situations, but only a fraction of these can be covered here. We focus on the fundamentals of estimating abundance with *distance sampling data* collected from line and point transects. We do not cover the practicalities of survey design and field methods, although getting these right is absolutely key to obtaining reliable results – the reader is referred to chapters 2 and 4 in Buckland et al. (2015), chapter 7 in Buckland et al. (2001), or chapter 7 in Buckland et al. (2004).

We first describe *design-based* (or *classical*) distance sampling analysis methods (Buckland et al. 2001; Buckland et al. 2015, Chapter 5). We then consider recent *model-based* distance sampling analysis methods (Hedley and Buckland 2004; Royle et al. 2004; Buckland et al. 2015, chapters 7 and 8 of Kéry and Royle 2015; see review by Miller et al. 2013a). The two terms are convenient for distinguishing different approaches to distance sampling, but design-based distance sampling is a bit of a misnomer. Both approaches use model-based methods for describing variation in detection probability within the sampled areas. The key difference is that design-based distance sampling methods use assumptions about random transect placement to extrapolate from estimated abundance in the sampled area to abundance over the entire survey region, whereas model-based distance sampling methods use a spatial model of animal distribution for extrapolation. The key advantage of design-based methods is that we can ensure the assumptions are met through good survey design, whereas for model-based methods we cannot be sure that the required assumptions about animal distribution are correct. On the other hand, model-based methods have relaxed requirements for the sampling design, and may therefore be better able to accommodate nonrandom sampling designs such as opportunistic sampling. However, flexibility does not mean that model-based approaches can salvage a poorly designed study. Relative to the desired level of inference, reliable use of model-based approaches will necessarily depend on both the appropriateness of the model and on the quality of available data, which in turn depends on the sampling design. For a detailed discussion of model-based versus design-based inference in the context of distance sampling and animal abundance estimation in general, see Borchers et al. (2002).

3.2.1.1 Classical Distance Sampling

3.2.1.1.1 Line and Point Transects

Consider the *strip transect scenario* described in Section 3.1.3 and shown in Figure 3.2. To meet the assumption that all individuals in the strip are detected (i.e. $p = 1$), the strips have to be very narrow, so that they can be searched thoroughly as the observer moves down the trackline. Detecting all animals out to the boundary of the strip implies that there will be many animals just outside the strip that are also detected, but cannot be counted. Hence strip transect (and other plot sampling) methods are inefficient in the sense that they do not use all of the data potentially available. The survey design is fine for animals that occur at high density, where a reasonable sample can be obtained even using small plots, but it is not optimal for animals that occur at lower density, or are harder to detect, so that the plots must be prohibitively small to ensure $p = 1$. Distance sampling methods relax the assumption that all individuals must be detected, thereby allowing a larger area to be included in the sample, and hence more data to be collected (compare Figures 3.2 and 3.3a). The penalty for this is that we must now account for the animals missed. From Eqs. 3.1, 3.5, and 3.6, we can estimate abundance (N) and density (D) corrected for imperfect detection ($p < 1$):

$$\hat{N} = \frac{c}{2wL\hat{p}}A, \tag{3.9}$$

and

$$\hat{D} = \frac{\hat{N}}{A} = \frac{c}{2wL\hat{p}}. \tag{3.10}$$

Hence, to reliably estimate abundance (or density) when $p < 1$, one must be able to reliably estimate p.

Line transect surveys are the distance sampling equivalent of a strip transect (Figure 3.3a). Observers travel along randomly placed lines within the study area, and whenever an individual is detected, the perpendicular distance from the line to the individual is measured and recorded. Alternatively, the radial distance and sighting angle can be recorded in the field and then used to calculate the perpendicular distance. The other common type of distance sampling survey is a *point transect* (also called a *variable circle plot*), which is an extension of the point count plot sampling method. In point transect surveys, observers visit randomly placed points within the study area and record the radial distance from the point to any detected individuals (Figure 3.3b).

Distance sampling theory allows for individuals in the sampled areas to go undetected by exploiting the tendency for detectability to decrease with increasing distance from the line or point. Buckland et al. (2001, chapters 4 and 6) discuss some of the relative advantages of line and point transects, but where feasible, line transect distance sampling methods are generally considered more efficient, accurate, and robust (Bollinger et al. 1988). However, sampling techniques are often linked to habitat structure. For example, line transects are typically easier to implement in open habitats like grasslands or ocean,

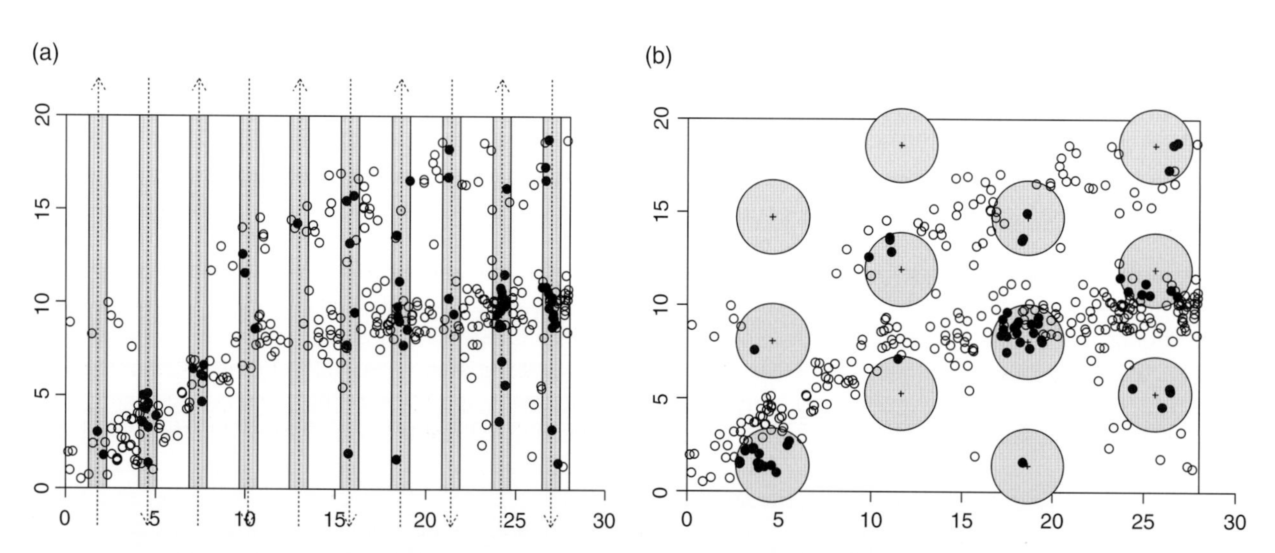

Figure 3.3 Examples of (a) line transect and (b) point transect sampling. Detected and undetected animals are shown as black and white circles, respectively. Note that compared with the strip transect survey of Figure 3.2, the line transect covers a larger area, but some individuals within the covered area are missed. Individuals closer to the line or point are more likely to be detected.

whereas point transects are more often used in closed or rugged habitats such as forests or mountains.

In cases where animals occur in distinct *clusters* (groups, flocks, herds, etc.), it is often easier to treat the cluster as the object that is sampled, recording the distance to the center of the cluster and, separately, the cluster size. Eqs. 3.9 and 3.10 then give the abundance and density of clusters; to convert to abundance and density of individuals one needs to include an additional term for population mean cluster size. In some cases, this can be reliably estimated by the mean of the observed cluster sizes. More often, however, cluster size is only recorded accurately at short distances; various methods are available to correct for any *size bias* for observations of clusters at longer distances (see Buckland et al. 2001, Section 3.5).

The density estimators using line or point transects are necessarily different because the geometry of the sampling frame differs, but both approaches have three key assumptions in common: (i) all individuals located on the line or point are detected; (ii) animal movements are negligible; and (iii) all distances (and angles, if necessary) are measured accurately. A fourth assumption is that detections are independent, but the classical methods are robust to violations of this assumption.

Based on the observed distances $\boldsymbol{x} = (x_1, x_2, \ldots, x_c)$ to c detected individuals from line or point transect surveys, the basic strategy underlying classic distance sampling methods is a combination of design- and model-based approaches. Model-based methods are used to describe the detection process as a decreasing function of distance, and, given a model for the detection process, design-based methods are used to make inferences about density (or abundance). The *detection function*, $g(x \mid \theta)$, is a model for the detection process that generated the observed distance data ($\boldsymbol{x}$) as a function of (unknown) detection process parameter(s), θ. Specifically, $g(x \mid \theta)$ is the probability of detecting an object at distance x from the line or point, conditional on the model parameters θ. Hence, $g(0 \mid \theta) = 1$ based on assumption 1. Careful specification of $g(x \mid \theta)$ is paramount to reliable estimation of θ and D (or N) from the observed distances $\boldsymbol{x}$. Regardless of the exact form of the model, we can estimate the average detection probability, p, from the fitted detection function for line transects:

$$\hat{p} = \frac{\int_0^w \hat{g}(x \mid \theta) dx}{w}, \tag{3.11}$$

or the detection function for point transects:

$$\hat{p} = \frac{2\int_0^w x\hat{g}(x \mid \theta) dx}{w^2} \tag{3.12}$$

(for full derivations of these estimators, see chapter 2 of Buckland et al. 2001). Here, w is the *truncation distance*, or the distance beyond which detections are ignored in the estimation of density; proper truncation of outliers can greatly aid in the modeling of the detection function (see Buckland et al. 2001, pp. 15–17). The estimates of p can then be used to obtain estimates of density (or abundance) from Eqs. 3.9 and 3.10 (or the equivalent for point transects). Fortunately, given a specified form for the detection function $g(x \mid \theta)$, free software tools are available for estimating θ and therefore N or D (Section 3.4), thereby precluding the need for us to dust off our undergraduate calculus texts.

3.2.1.1.2 Specification of the Detection Function $g(x \mid \theta)$

Histograms of distance sampling data can help provide some initial insights about the purpose of the detection function. Figure 3.4a shows the distance to detected individuals from the example in Figure 3.3a, in 50 m intervals. Note the tendency for fewer individuals to be detected at greater distances from the line. We have described the detection function as a mathematical model for detection probability as a function of the measured distances to individuals detected from line or point transect surveys. There are many different models that could be used for this purpose, but decades of research have converged on a relatively small, but flexible, set of models to choose from.

Buckland (1992) synthesized much previous work and developed a unified formulation for specifying a suite of detection functions that are commonly used by practitioners. The strategy is to select a few models for $g(x \mid \theta)$ that tend to exhibit desirable qualities, such as generality, flexibility, and efficiency. In terms of shape, it is desirable for detection functions to be monotonically decreasing with a "shoulder" near the line or point. The width of the shoulder indicates the distance from the line or point to which detection remains nearly certain – this is something that is typically, at least partly, under the control of the observers; a wider shoulder leads to more robust inference. Truncation is often used to remove outlier observations recorded at unusually large distances, further improving inference (Buckland et al. 2001, pp. 15–17).

Many models for the detection function $g(x \mid \theta)$ could exhibit these qualities, but Buckland (1992) proposed detection functions using the following conceptual form:

$$g(x \mid \theta) \propto \text{key}(x \mid \theta_k)[1 + \text{series}(x \mid \theta_s)],$$

where $\text{key}(x \mid \theta_k)$ is a *key function* that serves as baseline for the model, and $\text{series}(x \mid \theta_s)$ is a *series expansion* used to adjust the key function. The key function alone can often suffice, but a series expansion can help improve the fit of

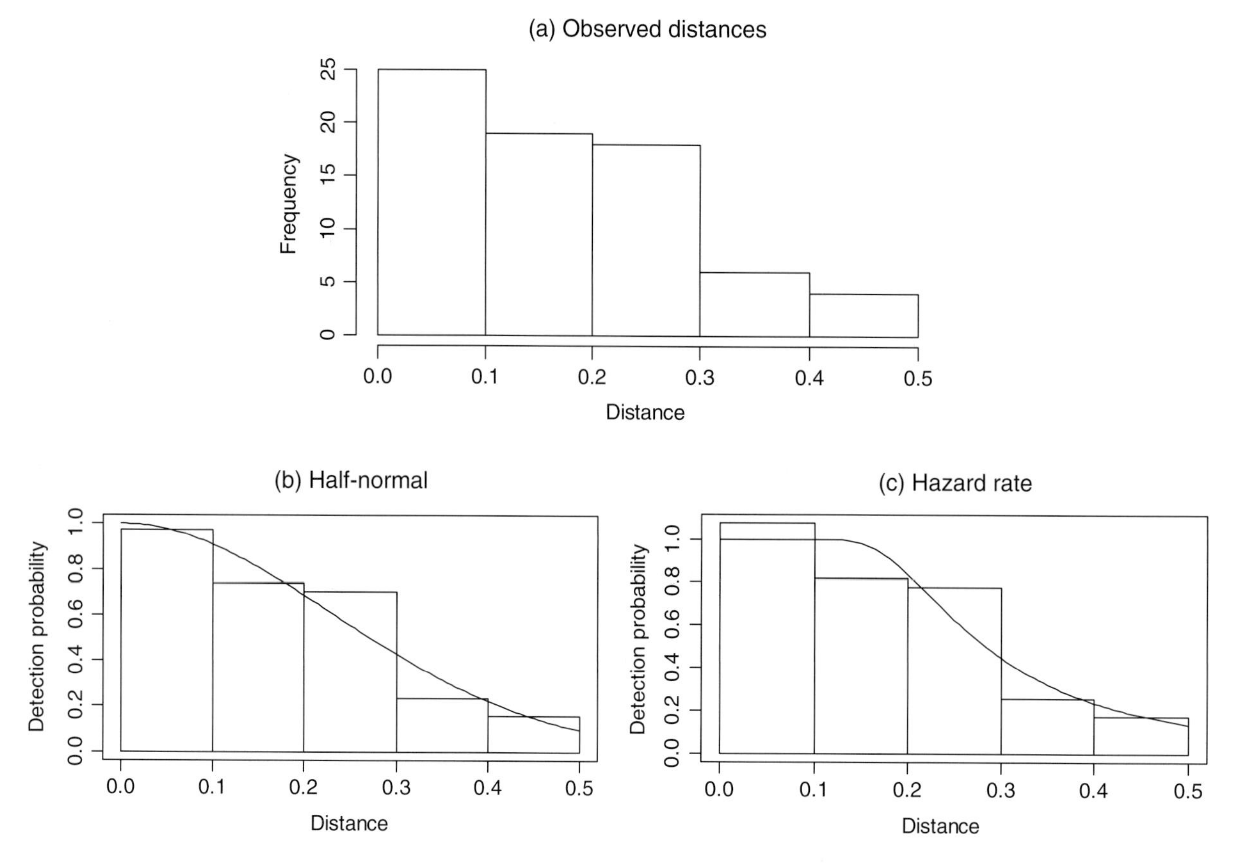

Figure 3.4 (a) Histogram of observed distances to detected individuals in the line transect example from Figure 3.3a. (b) and (c) Detection functions fitted to these data using half-normal (b) and hazard-rate (c) models without series expansions. Histograms of the distances are also shown in (b) and (c), scaled so that the area of the histogram is the same as the area under the fitted functions. The half-normal model has a lower AIC value and so is the preferred model, although both AIC values and estimated average detection probability, $\hat{p}$, are similar (see text).

Table 3.1 Recommended models for detection functions of the form $g(x \mid \theta) = \text{key}(x \mid \theta_k)[1 + \text{series}(x \mid \theta_s)]$, where x is distance from the line, θ_k are the key function parameters, θ_s are the series expansion parameters, and w is the truncation point (i.e. distances exceeding w are either not recorded or removed before analysis). The series expansion terms depend on a scaled value of x, $x_s = x/w$ or $x_s = x/\sigma$.

		Series expansion			
Key function	θ_k	**Cosine**	**Simple polynomial**	**Hermite polynomial**	θ_s
Uniform, $1/w$		$\sum_{j=1}^{m} a_j \cos(j\pi x_s)$	$\sum_{j=1}^{m} a_j (x_s)^{2j}$		a_j
Half-normal, $\exp(-x^2/2\sigma^2)$	σ	$\sum_{j=2}^{m} a_j \cos(j\pi x_s)$		$\sum_{j=2}^{m} a_j H_{2j}(x_s)$	a_j
Hazard-rate, $1 - \exp(-(x/\sigma)^{-b})$	σ, b	$\sum_{j=2}^{m} a_j \cos(j\pi x_s)$	$\sum_{j=2}^{m} a_j (x_s)^{2j}$		a_j

Source: Adapted from Buckland et al. (2001, p. 47).

the key function to the distance data. Buckland et al. (2001) recommend three key functions and three mth-order series expansions (Table 3.1). Figure 3.4b and c show detection functions for the *half-normal* and *hazard-rate models* without series expansions, fit to example data from Figure 3.4a. Standard likelihood-based analysis methods can be used to fit models and estimate parameters of the detection function, and selection among candidate models can be accomplished using likelihood-based model selection procedures (Chapter 2). Additional model structures explaining variation in detection probability, such as observer, environmental, or species covariates, can also be incorporated into this framework (Marques and Buckland 2004; Marques et al. 2007).

Estimating variance of $\hat{N}$ and $\hat{D}$ involves combining estimated variances from the random components that make up the abundance and density estimates: number of observations (c), detection probability (p), and where applicable, population mean cluster size. Details of these procedures and calculation of related quantities such as coefficient of variation and confidence intervals, are given in Buckland et al. (2001, Section 3.6), although the standard method of calculating variance in c has changed based on recommendations of Fewster et al. (2009).

For the example line transect data, $\hat{p}$ is 0.56 (SE 0.05) for the half-normal and 0.62 (SE 0.07) for the hazard rate models; AIC (Akaike's Information Criterion) is −121.53 and −119.13 respectively, and so the half-normal model would be selected based on lowest AIC. Given this model, and using Eqs. 3.9 and 3.10 with $c = 72$ individuals and $w = 0.5$ km, we estimate that $\hat{N} = 360$ and $\hat{D} = 0.64$ individuals km^{-2}, which are close to the true values of 345 and 0.62 respectively. It is instructive to compare the estimates of uncertainty from the strip and line transect surveys: the former had a CV of 0.23 while the latter was somewhat more precise, with a CV of 0.20. Hence, even though line transect analysis requires estimating the detection function as an additional quantity, the additional uncertainty that this generates in the final estimate was more than compensated for by the reduction in uncertainty from being able to survey a larger area by having a longer truncation distance (w).

We were also rather generous in assuming all individuals in the strips were detected for the strip transect survey, given that the estimated detection probability at the strip transect boundary of 150 m was estimated, from the line transect data, to be 0.8. Hence, even when undertaking a strip transect survey, it is worthwhile to record the perpendicular distance to detected individuals, as a way to check the assumption that all animals are detected (see Buckland et al. 2001, pp. 335 for another example).

The above example is somewhat artificial, being based on simulated data. Numerous real-world worked examples exist, including seven case studies in chapter 8 of Buckland et al. (2001), many of which have accompanying data available as sample datasets in Program `Distance`. Three other examples in order of increasing complexity include: Williams and Thomas (2007), Marques et al. (2007) and Durant et al. (2011). A simplified version of an analysis from the latter paper is given in Box 3.1. Last, Buckland et al. (2015) provide numerous examples, with data and `R` code on an associated website.

Much of the modern research in classic distance sampling seeks to eliminate, relax, or better cope with various assumptions about the detection process. Extended models include *multiple-observer* or *mark-recapture distance sampling* to allow for $g(0 \mid \theta) < 1$ (Laake and Borchers 2004; Borchers et al. 2006; Buckland et al. 2010), animals moving in response to observers (Fewster et al. 2008), transect placement that is not fully random (Marques et al. 2010), and measurement error in detection distance (Marques 2004; Borchers et al. 2010). Avoidance of the observer is more common in point transects than line transects, and the issue often requires left-truncation of the distance data, thereby creating a point "donut." General recommendations for point transect surveys were reviewed by Sólymos et al. (2013) and Matsuoka et al. (2014).

3.2.1.2 Model-Based Distance Sampling

Classic distance sampling methods have strict requirements about the design of sampling surveys, thereby avoiding assumptions about spatial distribution of individuals within the study area. It is not required that individuals be *randomly distributed* within the study area (e.g. Poisson), instead that transects are randomly placed within the study area. This requirement ensures the individual-to-transect distances are random, from a known distribution (e.g. uniform in the case of a line transect), and hence allows the distance to *observed individuals* to be used to draw inferences about detectability. Randomly placed lines or points also ensure that the estimated density from the sampled area also applies to the entire study area.

In contrast, model-based distance sampling methods do not require these key assumptions about the sampling design. Model-based methods instead rely on assumptions about the spatial distribution of individuals within the sampled region. By including a model for local abundance, covariates for habitat type, region, or other factors may be used to explain variability in both abundance and detection in sampled areas, as well as make predictions about abundance in unsampled areas. Model-based methods are therefore better suited for *opportunistic sampling*. Another potential advantage is that the model may explain spatial variation in density through use of appropriate covariates, in contrast to design-based methods where spatial variability within strata contributes to estimator variance. Hence model-based methods may produce more precise estimates. One strong disadvantage, however, is that model-based methods rely on a model for animal distribution within the study area – if this model is incorrect (and remember that all models are wrong) then estimates may be biased, and variances may be wrong. We do not usually control the animal distribution, so we cannot know completely whether our models are right, or nearly so. By contrast, in classic, design-based methods, the assumptions are about the design, and we are in control of that.

Box 3.1 Estimating Density from a Line Transect Survey Using Classical Methods

Durant et al. (2011) estimated the seasonal and habitat-specific density, as well as temporal trends, of seven carnivore species using line transects surveys performed within a 3000 km^2 section of Serengeti National Park, Tanzania. Here, we focus on one species, the spotted hyaena (*Crocuta crocuta*). Our analysis is a simplified version of the one by Durant et al. using the four surveys performed in the wet and dry season in 2002–3 (hereafter 2002) and the wet and dry season of 2005 (see paper for complications that are inherent in use of earlier survey data). Approximately 30 transects, placed according to a systematic random design with 2 km transect spacing and totaling approximately 1200 km, were surveyed on each occasion, where the exact number varied by occasion depending on random start point and orientation. Sightings were of clusters (or groups) of individuals, and for each sighting the species and cluster size was recorded, as well as the estimated distance from the center of the cluster to transect, in one of nine intervals: 0–10, 10–50, 50–100, 100–150, 150–200, 200–300, 300–400, 400–500, and >500 m.

For spotted hyaenas, 494 groups were sighted. We follow Durant et al., in truncating the data at 200 m, leaving 389 groups sighted within this distance. We fit the detection function models given in Table B3.1.1, allowing up to two series expansion terms. The model would be sufficient if our interest were only in an overall density estimate, however, a goal of the analysis was to estimate density for two habitat strata within the surveys (long grass plains, LGP, and short grass plains, SGP), and it is possible that hyaena detectability varied by habitat or survey. Hence, we also fit multiple covariate models with habitat, year, and season as factor covariates (see Marques et al. 2007 for an accessible introduction to multiple covariate distance sampling). Last, we tried models that included cluster size as a continuous covariate, in case larger clusters were easier to spot. The AIC-best models (ΔAIC < 2) were as follows (HR means hazard rate and HN half-normal; cos(2) means cosine series expansion of order 2):

We observe that the AIC-best model is the HR with no series expansion terms, but that all of the other models that are close in terms of AIC, and have similar estimated

Table B3.1.1 Model selection for detection functions for counts of spotted hyaenas.

	No. of parameters	AIC	ΔAIC	$\hat{p}$	$CV(\hat{p})$
HR (no series expansion)	2	1142.83	0.0	0.48	0.10
HN cos(2)	2	1142.93	0.10	0.46	0.07
HN cos(2) + cluster size	3	1143.98	1.15	0.46	0.04
HN cos(2) + season	3	1144.40	1.57	0.46	0.04
HN cos(2) + habitat	3	1144.47	1.64	0.46	0.04
HN cos(2) + year	3	1144.53	1.70	0.46	0.04
HR + cluster size	3	1144.72	1.89	0.50	0.04

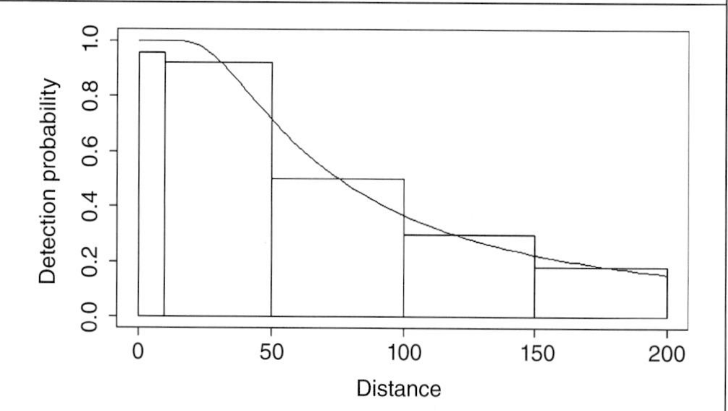

Figure B3.1.1 Fit of the hazard-rate model to detections at 50 m intervals to groups of spotted hyenas.

average detection probabilities ($\hat{p}$). These results are typical for "good" distance sampling data with a wide shoulder of high detectability at short distances. Goodness-of-fit for the selected model can be assessed by comparing the fitted detection function to a scaled histogram of the distances (see plot) as well as via a χ^2-test (other tests are available for data not collected in intervals). The fit of the hazard-rate model is good ($\chi^2 = 0.34$ with 2 d.f.; $p = 0.84$, Figure B3.1.1). Durant et al. looked for evidence of variation in cluster size between habitats or surveys and found no differences, so we follow the authors in estimating population mean cluster size as the mean of the observed cluster sizes (1.91; CV 0.05).

Density estimates were calculated by habitat and survey using Eq. 3.10, with an additional multiplier for mean cluster size. Variances and confidence intervals were derived using the standard methods described by Buckland et al. (2001, Section 3.6), except that when calculating variance in counts, *c*, a variance estimator was used that accounts for the systematic random design (estimator O2 of Fewster et al. 2009). The results are shown below (Figure B3.1.2). The results clearly demonstrate that density in the wet season is higher than the dry in the SGP habitat but not in the LGP. The findings are consistent with a "commuter" system, where Serengeti hyaenas have been observed to commute long distances from their clan territory to areas, like SGP in the wet season, where the density of migratory prey is high.

Analyses reported here were performed in the `Distance` software but almost identical results were obtained using the `Distance` package in `R`, which was also used to produce the plots.

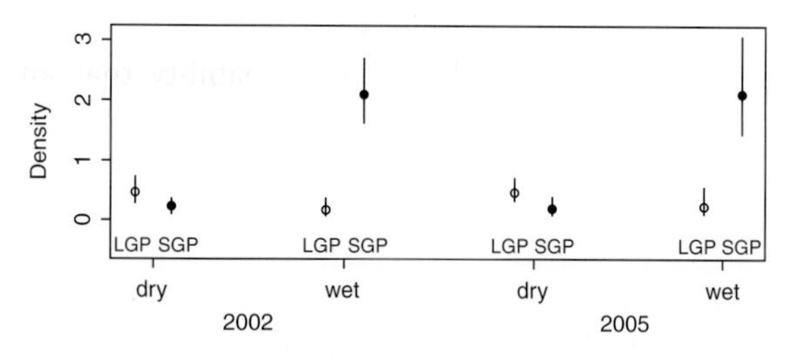

Figure B3.1.2 Estimates of density for spotted hyenas during the dry and wet seasons in the long grass plains (LGP) and short grass plains (SGP). Dots are the estimates and vertical lines are 95% confidence intervals.

Relative to classic methods, development of model-based distance sampling methods is still at an early stage. As such, there remains room for much theoretical and practical development. Here, we will briefly describe some of the notable developments in model-based distance sampling.

Hedley and Buckland (2004) and Royle et al. (2004) were among the earlier developments in model-based distance sampling methods, and both seek to account for imperfect detection while incorporating environmental covariates that explain local variation in abundance within the study region of interest. Hedley and Buckland (2004) approached the problem from a classic distance sampling perspective and proposed a model conditioned on detection: the conditional distribution of x given that the observation appeared in the sample c. However, they found it simpler in practice to attack the problem in a *two-stage approach*: first to fit a detection function model to the distance data, and then conditional on this model, to fit a spatial density surface model to the corrected count data. Two-stage approaches are analytically more tractable than direct modeling of both density and observation processes. Because the transects are usually narrow compared to the size of the study area, there is also little to be learned about the large-scale spatial distribution of animals from the distances of observed animals from the transects – this provides further justification for a two-stage approach. One disadvantage of such an approach, however, is that it is not straightforward to propagate uncertainty from the first stage based on detection function modeling through to the second stage of density surface modeling. A review of model-based methods, with an emphasis on practical two-stage approaches, is given by Miller et al. (2013a), who also provide an example application with accompanying computer code. For the remainder of this section, we focus mainly on *one-stage approaches*.

Royle et al. (2004) approached the problem from a hierarchical modeling perspective and used an *unconditional* model that is the joint distribution of x and c. The approach groups the distance data for each of R transects into J distance classes, such that the distance data for sample unit (or site) i are $\boldsymbol{x}_i = (x_{i1}, x_{i2}, \ldots, x_{iJ})$ for $i = 1, \ldots, R$, with distance group end points (e_0, e_1), (e_1, e_2), ..., (e_{J-1}, e_J). Typically, $e_0 = 0$ and $e_J = w$. For these grouped distance data, the probability that an individual is present at site i and detected in distance class j is calculated by the following function for line transects:

$$\pi_{ij} = \frac{\int_{e_{j-1}}^{e_j} g(x \mid \theta)dx}{e_j - e_{j-1}}, \tag{3.13}$$

and by a similar function for point transects:

$$\pi_{ij} = \frac{2\pi \int_{e_{j-1}}^{e_j} xg(x \mid \theta)dx}{\pi\, e_J^2}. \tag{3.14}$$

Similar to classic distance sampling, the detection function $g(x \mid \theta)$ can incorporate site-level covariates to explain variability in π_{ij}.

It is then assumed that the $\boldsymbol{x}_i$ follow a multinomial distribution with an unknown index N_i, the local population size at site i. A simple model for N_i is:

$$N_i \sim \text{Poisson}(\lambda a_i), \tag{3.15}$$

where we assume the $\boldsymbol{x}_i$ are derived from an underlying (homogeneous) Poisson point process for the distribution of individuals in the study area, and the intensity parameter λ is the expected local density of individuals in each site (of area a_i). For line transects of length l_i, $a_i = 2e_J l_i$, and for point transects, $a_i = \pi e_J^2$. After integrating the multinomial likelihood for $\boldsymbol{x}_i$ over the random effects distribution for N_i, the resulting likelihood for $\boldsymbol{x}_i$ is

$$L(\lambda, \theta \mid \boldsymbol{x}_i) = \prod_{j=1}^{J} \text{Poisson}\left(x_{ij}; \lambda a_{ij} \pi_{ij}\right), \tag{3.16}$$

where $a_{ij} = 2l_i(e_j - e_{j-1})$ for line transects, and $a_{ij} = \pi e_J^2$ for point transects. Density (and its variance) can then be estimated for each of the R sites using standard likelihood-based analysis methods.

From an ecological perspective, this formulation is quite interesting because it allows local abundances to be related to site-specific covariates such as habitat. The underlying distribution of animals in the study area need not be the same across sites, and Eqs. 3.15 and 3.16 can be relatively easily extended to an *inhomogeneous point process* that is a function of the covariates believed to influence local abundance (Box 3.2). Here, "inhomogeneous" refers to distributions that are not spatially uniform. For example, the degree of habitat fragmentation within a site may be related to lower local densities. The relationship can be investigated using the log link function:

$$\log(\lambda_i) = \alpha + f_i \beta, \tag{3.17}$$

where f_i is some measure of fragmentation for site i, β is a slope parameter describing the relationship between f_i and local abundance, and α is an intercept parameter. In this manner, if measurable covariates can adequately explain variability in local densities, and these covariates can be measured in areas not exposed to distance sampling (e.g. from GIS records), it is possible to make predictions about local densities in unsampled areas. Conn et al. (2014) provide an example where aerial transect

Box 3.2 Covariates and Link Functions

Ecological hypotheses about factors that influence detection probability, abundance, or species distribution can be investigated through the use of link functions (Table B3.2.1) that relate the parameter of interest (θ) to measurable covariates (X). When θ is a probability (i. e. $0 \le \theta \le 1$), a common link function is the logit link:

$$\theta = \frac{\exp(\boldsymbol{X\beta})}{1+\exp(\boldsymbol{X\beta})} \tag{3.18}$$

or, equivalently,

$$\text{logit}(\theta) = \log\left(\frac{\theta}{1-\theta}\right) = \boldsymbol{X\beta}, \tag{3.19}$$

where

$$\boldsymbol{X\beta} = \sum_{j=1}^{r} x_j\beta_j. \tag{3.20}$$

The parameter vector $\boldsymbol{\beta}$ therefore describes the relationship(s) between θ and r measurable covariates x_j ($j = 1, \ldots, r$). Note that it is common for β_1 to be designated as an intercept term by setting $x_1 = 1$. When θ is non-negative (i. e. $\theta \ge 0$), the log link function is often used:

$$\theta = \exp(\boldsymbol{X\beta})$$

or, equivalently,

$$\log(\theta) = \boldsymbol{X\beta}.$$

Table B3.2.1 Common link functions and covariates relevant to models of abundance or species occurrence under imperfect detection.

Parameter	Link function	Example covariates (x)
Detection probability (p)	logit	observer, weather conditions, habitat, effort
Mean density (λ)	log	habitat, elevation, region
Occupancy probability (ψ)	logit	habitat, elevation, region

surveys were used in conjunction with spatial covariates for sea surface ice, temperature, and other factors to estimate local densities for three species of seals that are associated with sea ice in the Bering Sea – a study area the size of Texas!

The N_i were integrated out of the likelihood (Eq. 3.16), and estimates for N_i cannot be obtained directly from the likelihood. However, N_i can be estimated as a derived parameter conditional on $\hat{\boldsymbol{\pi}}_i$ and $\hat{\lambda}_i$ using an empirical Bayes procedure[1]:

$$\Pr\left(N_i = k \mid \boldsymbol{x}_i, \hat{\lambda}_i, \hat{\boldsymbol{\pi}}_i\right) = \frac{\Pr(\boldsymbol{x}_i \mid N_i = k, \hat{\boldsymbol{\pi}}_i)\Pr\left(N_i = k \mid \hat{\lambda}_i\right)}{\sum_{j=0}^{\infty}\Pr(\boldsymbol{x}_i \mid N_i = j, \hat{\boldsymbol{\pi}}_i)\Pr\left(N_i = j \mid \hat{\lambda}_i\right)}, \tag{3.21}$$

which, for the Poisson local population model, provides the basis for the so-called *best unbiased predictor* of N_i (Royle et al. 2004):

$$E(N_i \mid \boldsymbol{x}_i) = \sum_{j=1}^{J} x_{ij} + \hat{\lambda}_i a_i\left(1 - \sum_{j=1}^{J}\hat{\pi}_{ij}\right). \tag{3.22}$$

1 This is based on the famous conditional probability theorem attributed to the Rev. Thomas Bayes (1701–1761), $\Pr(A \mid B) = \frac{\Pr(B \mid A)\Pr(A)}{\Pr(B)}$.

For simplicity, we have focused on an inhomogeneous Poisson point process model for explaining variability in local abundance, but there are many alternative models available for N_i that allow even greater flexibility in specifying various forms of spatial variation (Royle and Dorazio 2006; Royle et al. 2007; chapter 24 of Kéry and Royle 2015). However, it should be noted that although model-based approaches do not require strict random placement of line or point transects with respect to the distribution of individuals within the study area, reliable inferences to unsampled areas *still* require that the distribution of individuals in the sampled areas is representative of that in the unsampled areas. Careful study design is therefore still a requisite for reliable predictions about local abundance in unsampled areas when using model-based distance sampling methods.

Example datasets and R code utilizing these methods can be found in Royle et al. (2004) and extensions (Chandler et al. 2011; Chelgren et al. 2011; Sillett et al. 2012). Sillett et al. (2012) estimated local and total population densities for Island Scrub-Jays (*Aphelocoma insularis*) as a function of habitat-level covariates, finding higher jay densities in low-elevation chaparral habitat. Johnson et al. (2010) generalized the approach of Hedley and Buckland (2004) by implementing a full likelihood-based approach for the simultaneous estimation of detection probability and an (in)homogeneous spatial point process, and Conn et al. (2012, 2013) developed other promising model-based extensions for multiple-observer distance sampling surveys that can accommodate species misidentification.

3.2.2 Replicated Counts of Unmarked Individuals

In distance sampling, the detection distances supply the information needed to estimate detection probability from (typically unreplicated) counts of unmarked individuals. The collection of distance sampling data is not always feasible or appropriate, and logistics often depend on the specific species of interest and sampling conditions in the field. For example, even with modern laser rangefinders, estimating distances to all individuals seen or heard during surveys can be difficult, and both accuracy and precision may depend on whether the cue for initial detection is visual or auditory. Auditory cues can be especially challenging for distance sampling because both distance *and* location must usually be estimated. However, when count surveys are temporally replicated, repeated surveys allow detection probability and therefore abundance to be estimated from unmarked counts without any need for distance data. Examples of study designs include *spatially replicated counts* (Royle 2004a) and *removal models* (Farnsworth et al. 2002; Royle 2004b). Many of these methods share similarities with model-based distance sampling approaches.

3.2.2.1 Spatially Replicated Counts

In the case of spatially replicated counts, we consider counts of unmarked individuals detected during T visits to R sample units (or sites). Let y_{it} denote the number of distinct individuals counted at site i ($i = 1, \dots, R$) on sampling occasion t ($t = 1, \dots, T_i$). Note that the number of sampling occasions, T_i, is allowed to vary among sites. The model has three assumptions: (i) the local population in each site is closed to birth, death, immigration, and emigration during the sampling period; (ii) counts at each site are independent; and (iii) individuals are not double-counted within a single sampling occasion.

Utilizing information about detection probability and local abundance afforded by repeated sampling, it is natural to assume that the counts for each site, $\boldsymbol{y}_i = (y_{i1}, y_{i2}, \dots, y_{iT_i})$ for $y_{it} \in \{0, 1, 2, \dots, N_i\}$, are binomial random variables with an unknown index N_i, the local population size at each site, and detection probability p:

$$L(N_i, p \mid \boldsymbol{y}_i) = \prod_{t=1}^{T_i} \binom{N_i}{y_{it}} p^{y_{it}} (1-p)^{N_i - y_{it}}. \quad (3.23)$$

Even with a constant detection probability, this likelihood is notoriously unstable and sensitive to small perturbations in the data. Instability is exacerbated by a tendency for repeated count data to be sparse, with some sites having few or no detections (Box 3.3). In fact, until recently, this repeated count sampling protocol was not widely used. Royle (2004a) proposed a hierarchical modeling solution to the problem by specifying a model for the local abundance at each site (N_i). There are many options for these so-called *N-mixture* models, but a natural choice in this case is the Poisson distribution:

$$N_i \sim \text{Poisson}(\lambda), \quad (3.24)$$

with local abundance rate parameter λ, defined as the density per site or mean local abundance. Maximum likelihood analysis may then proceed using the integrated likelihood:

$$L(p, \lambda \mid \boldsymbol{y}_1, \boldsymbol{y}_2, \dots, \boldsymbol{y}_R) = \prod_{i=1}^{R} \left[\sum_{N_i = \max(\boldsymbol{y}_i)}^{\infty} \left(\prod_{t=1}^{T_i} \text{Binomial}(y_{it}; N_i, p) \right) \text{Poisson}(N_i; \lambda) \right]. \quad (3.25)$$

Note that a similar integrated likelihood approach was used to derive Eq. 3.16, but in this case the integrated likelihood is not of a standard form (e.g. Poisson). Given a maximum likelihood estimate (MLE) for λ, an estimate of total abundance across the R sites in the sampled area is

$$\hat{N} = R\hat{\lambda}, \quad (3.26)$$

with variance approximated by the delta method:

$$\text{var}(\hat{N}) = R^2 \text{var}(\hat{\lambda}). \quad (3.27)$$

Similar to distance sampling (Eq. 3.21), the N_i are integrated out of the likelihood (Eq. 3.25), and estimates for N_i are not obtained from the likelihood. However, these again can be estimated (conditional on $\hat{p}$ and $\hat{\lambda}$) using an empirical Bayes procedure:

$$\Pr(N_i = k \mid \boldsymbol{y}_i, \hat{\lambda}, \hat{p}) = \frac{\Pr(\boldsymbol{y}_i \mid N_i = k, \hat{p}) \Pr(N_i = k \mid \hat{\lambda})}{\sum_{j=0}^{\infty} \Pr(\boldsymbol{y}_i \mid N_i = j, \hat{p}) \Pr(N_i = j \mid \hat{\lambda})}. \quad (3.28)$$

Conveniently, this expression also allows estimation of the probability of species occurrence for sites with no detections:

$$\hat{\psi}_i = 1 - \Pr(N_i = 0 \mid \boldsymbol{y}_i, \hat{\lambda}, \hat{p}) \quad (3.29)$$

(Royle et al. 2005). We will return to this relationship between local abundance and site occupancy in Section 3.3.

One of the advantages of the modeling approach above is an ability to incorporate *covariate information* about

Box 3.3 Estimating Abundance from Spatially Replicated Counts

Kéry et al. (2005) estimated abundances for eight species of bird in Switzerland using the spatially replicated count model of Royle (2004a). Here, we demonstrate this modeling approach using data from one of these species, Mallards (*Anas platyrhynchos*). Survey data were collected in 2002 during $T = 2$ or 3 visits to 235 sites throughout Switzerland. Mallards were detected at 40 sites, but counts tended to be quite low, with 87% of visits detecting no Mallards and 8% of visits detecting a single individual. In addition to counts, several temporal and site-level covariates that were believed to influence detection probability or local abundance were recorded. The covariates included date, elevation, and percent forest cover. Kéry et al. (2005) suspected detection probability might vary with date and elevation because activity associated with breeding was expected to decline during the study, but less so at higher elevations due to later breeding at higher altitudes. Based on the natural history of Mallards, they also suspected Mallards would have higher densities at lower elevations and in areas with less percent forest cover. Based on AIC, they found the best-supported models included

$$\text{logit}(p_{it}) = \alpha_0 + (\text{date})_{it}\alpha_1 + (\text{date})^2_{it}\alpha_2 + (\text{date})_{it}(\text{elev})_i\alpha_3 + (\text{date})^2_{it}(\text{elev})_i\alpha_4$$

for detection probability at site i during visit t, and

$$\log(\lambda_i) = \beta_0 + (\text{elev})_i\beta_1 + (\text{forest})_i\beta_2$$

for mean local abundance at site i.

Estimates of the coefficients were based on standardized values for the set of covariates (Table B3.3.1). The estimates indicate that surveys occurring later in the season and at higher elevations tended to have lower detection probabilities. Perhaps more interesting from an ecological perspective, the estimates for β_1 and β_2 supported the hypothesis that sites at higher elevations

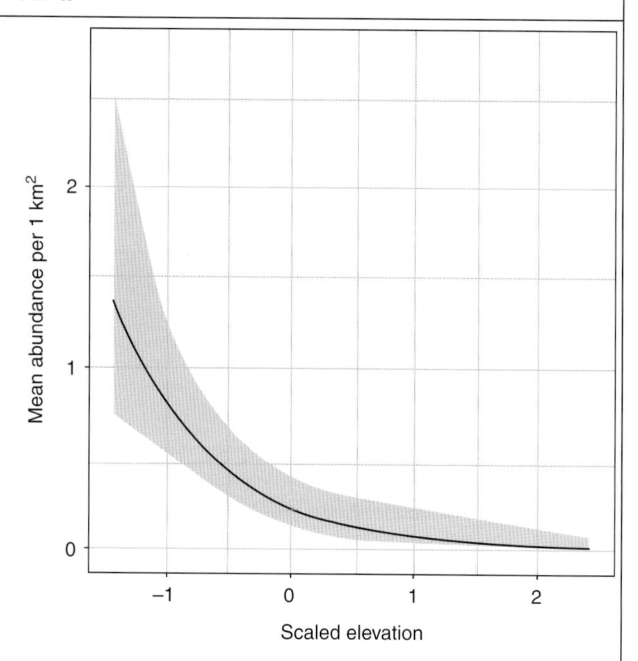

Figure B3.3.1 Estimates of density for Mallards (per km^2) as a function of scaled elevation. 95% confidence intervals are shaded.

(Figure B3.3.1) or with more forest cover (Figure B3.3.2) tended to have lower densities of Mallards. Across all sites, Kéry et al. (2005) estimated the total population size as $\hat{N} = 104$ (95% CI: 67–152), with a mean estimated density of 0.43 (95% CI: 0.28–0.64) Mallards per 1 km^2.

Table B3.3.1 Estimates of slope coefficients for the effects of standardized covariates on abundance of Mallards.

Parameter	Estimate	SE
α_0	0.01	0.27
α_1	−0.37	0.25
α_2	−0.33	0.17
α_3	−0.14	0.25
α_4	−0.38	0.17
β_0	−1.47	0.30
β_1	−0.92	0.30
β_2	−0.81	0.22

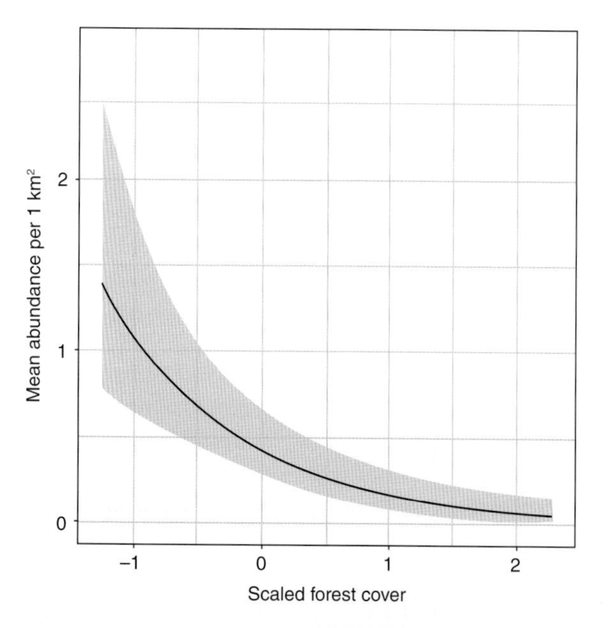

Figure B3.3.2 Estimates of density for Mallards (per km^2) as a function of scaled forest cover. 95% confidence intervals are shaded.

detection and local abundance. Either objective can be easily achieved using standard link functions:

$$\log(\lambda_i) = \alpha + \sum_{j=1}^{r} x_{ij}\beta_j, \tag{3.30}$$

or

$$\text{logit}(p_{it}) = \alpha + \sum_{j=1}^{r} x_{ij}\beta_j + \sum_{k=1}^{q} z_{tk}\delta_k, \tag{3.31}$$

where x_{ij} ($j = 1, \ldots, r$) are r measurable covariates for site i, and z_{tk} ($k = 1, \ldots, q$) are q measurable covariates for sampling occasion t (Box 3.3). Note that spatial variation in local abundance that is not explained by the measurable covariates can be incorporated using additional distributional assumptions. For example, Royle and Dorazio (2006) accounted for substantial spatial heterogeneity when estimating local abundance for a species of stream fish, the Okaloosa darter (*Etheostoma okaloosae*), by assuming $\log(\lambda_i)$ is a normally distributed random variable with unknown mean (μ_i) and variance (σ^2). When there is spatial variation in local abundance, an estimate of total abundance for the sampled area is $\hat{N} = \sum_{i=1}^{R} \hat{\lambda}_i$. Assuming the r covariates can be measured over a larger region, including unsampled sites with no count data, a total population estimate can be obtained by summing all $\hat{\lambda}_i = \hat{\alpha} + \sum_{j=1}^{r} x_{ij}\hat{\beta}_j$ over the entire region of interest.

Spatially replicated count methods have been applied to many vertebrate species, including amphibians (Dodd and Dorazio 2004; Mazerolle et al. 2007), birds (Royle 2004a; Kéry and Royle 2010; Riddle et al. 2010), and mammals (Zellweger-Fischer et al. 2011). These count methods have also been extended to open populations, thereby relaxing the closure assumption (Kéry et al. 2009; Chandler et al. 2011; Dail and Madsen 2011; Zipkin et al. 2014). When double counting occurs within sampling occasions, for example because of animal movements, Chandler and Royle (2013) use a design-induced spatial dependence among counts to estimate density, although, in practice, this approach is best suited to studies with at least some marked individuals in the population.

Although spatially replicated count methods have received a great deal of attention in recent years, a major problem with these methods is that unless detection probability is constant across sites, then a correct model for detectability must be specified. When the factors driving detectability can be identified and reliably measured, then some limited modeling of detectability is possible (Eq. 3.31). However, if there is any correlation between detection-related covariates and density-related covariates, then reliable inference is not possible because the two sets of parameters are confounded. In our opinion, these methods should therefore be relied upon more as a last resort because other survey methods based on distance sampling or capture-recapture allow detectability to be estimated independently of density. There is no free lunch in population ecology, and repeated count methods require strong and largely untestable assumptions about detectability (Barker et al. 2018, Link et al. 2018).

3.2.2.2 Removal Sampling

A popular technique for estimating the size of exploited populations such as fisheries, *removal sampling methods* also involve counts of unmarked individuals that are detected during T visits to R sample units (or sites). As the name implies, removal methods were originally developed to estimate abundance when individuals are trapped and removed from the population (Hilborn and Walters 1992). However, if captured individuals can be temporally removed and then released after sampling is completed (Jung et al. 2005), applications of removal sampling methods need not be limited to harvested populations. In fact, if one can keep track of individuals after they are initially detected, then no physical removal is required (Farnsworth et al. 2002).

There are numerous models for estimating abundance using counts arising from removal sampling protocols (Zippin 1958; Otis et al. 1978; Farnsworth et al. 2002; Williams et al. 2002, pp. 320–325; Royle 2004b), but each uses the decline in numbers of individuals detected for the first time across the T sampling occasion to inform the estimation of detection probability. All of these removal approaches make two assumptions: (i) the population in each site is closed to birth, mortality, and movement during the sampling period of interest; and (ii) there is no double-counting of individuals across the T sampling occasions.

The general removal design under consideration involves counts for the number of individuals first detected during each sampling occasion, $\boldsymbol{y}_i = (y_{i1}, y_{i2}, \ldots, y_{iT})$, for site $i = 1, \ldots, R$. For illustration, if we assume $T = 2$, $y_{i1} > y_{i2}$ and a constant detection probability, a simple removal estimator is:

$$\hat{N}_i = \frac{y_{i1}^2}{y_{i1} - y_{i2}} \tag{3.32}$$

(Zippin 1958). This model has been generalized for $T > 2$ with time variation in detection probability (Otis et al. 1978; White et al. 1982), as well as for $T > 3$ with individual heterogeneity in detection probability (Pledger 2000). The familiar closed population capture-recapture models (e.g. models "M_b," "M_{tb}," "M_{bh}," and "M_{tbh}") are discussed elsewhere in Chapter 5. As with capture-recapture models for marked animals, removal models for unmarked animals that do not account for individual heterogeneity or other sources of variability in detection probability can yield biased estimates of abundance.

Motivated by avian point counts, Farnsworth et al. (2002) developed a practical removal sampling protocol for unmarked populations that allows modeling of individual heterogeneity in detection probability when $T = 3$. The authors proposed partitioning a single visit (of duration K minutes) to R points within the study area into $T = 3$ intervals of length k_t, such that $\sum_{t=1}^{3} k_t = K$. For each point, the $\boldsymbol{y}_i = (y_{i1}, y_{i2}, y_{i3})$ then consists of the number of individuals first detected in each of the intervals. *Individual heterogeneity* in detection probability is characterized by partitioning N into two groups; group 1 consists of individuals that are easily detected (with probability of detection during the first interval equal to 1), and group 2 includes those that are more difficult to detect (with probability of detection within one minute p). All individuals in group 1 (and some individuals from group 2) are therefore detected in the first interval (of length k_1). Defining c as the expected proportion of the population in group 2, $q = 1 - p$, $y_{.t} = \sum_{i=1}^{R} y_{it}$, and $y_{..} = \sum_{t=1}^{3} y_{.t}$, the model likelihood is multinomial:

$$L(c,p \mid \boldsymbol{y}_1, \boldsymbol{y}_2, \ldots \boldsymbol{y}_R)$$
$$= \frac{y_{..}!}{y_{.1}! y_{.2}! y_{.3}!} \left[\frac{1 - cq^{k_1}}{1 - cq^K}\right]^{y_{.1}} \left[\frac{cq^{k_1}(1 - q^{k_2})}{1 - cq^K}\right]^{y_{.2}} \left[\frac{cq^{k_1 + k_2}(1 - q^{k_3})}{1 - cq^K}\right]^{y_{.3}}. \quad (3.33)$$

Abundance for the sampled area can then be estimated as:

$$\hat{N} = \frac{y_{..}}{\hat{p}}. \quad (3.34)$$

When the size of the sampled area is known (e.g. from fixed-radius point counts), then density can be estimated as:

$$\hat{D} = \frac{\hat{N}}{A} \quad (3.35)$$

with variance calculated as:

$$\operatorname{var}(\hat{D}) = \frac{y_{..}^2 \operatorname{var}(\hat{p})}{A^2 \hat{p}^4} + \frac{y_{..}(1 - \hat{p})}{A^2 \hat{p}^2}, \quad (3.36)$$

where A is the total area sampled (e.g. the sum of the areas within each fixed-radius point).

Farnsworth et al. (2002) used this approach to demonstrate strong differences in detectability for 15 bird species in Great Smoky Mountains National Park, USA, that could be attributed to variation in call intensity, time of day, and observer ability. From a practical perspective, the design can be useful because physical removal is not required. However, one must be able to distinguish newly detected individuals from those previously detected, which can be difficult for mobile or nonterritorial species.

We also note that for $T > 3$, the removal models allowing individual heterogeneity proposed by Norris and Pollock (1996) and Pledger (2000) may be preferable because these models do not assume that any group has detection probability equal to 1.

Using a similar integrated likelihood approach to that already described for model-based distance sampling (Eq. 3.16) and spatially replicated counts (Eq. 3.25), Royle (2004b) proposed a model of local abundance for removal data that assumes the $\boldsymbol{y}_i$ $(i = 1, \ldots, R)$ from each site are multinomial random variables with local population index N_i and cell probabilities $\boldsymbol{\pi}_i = (\pi_{i1}, \pi_{i2}, \ldots, \pi_{iT_i})$, where

$$\pi_{it} = p_{it}(1 - p_{it})^{t-1}, \quad (3.37)$$

and p_{it} is the probability of detection for site i on sampling occasion t. Therefore, π_{it} is the probability that an individual in site i is detected for the first time and removed during the tth sampling occasion $(t = 1, \ldots, T_i)$. The number of sampling occasions, T_i, is allowed to vary among sites. As before, a natural model for local abundance is

$$N_i \sim \text{Poisson}(\lambda_i), \quad (3.38)$$

which yields a now familiar integrated likelihood:

$$L(\lambda_i, \boldsymbol{p}_i \mid \boldsymbol{y}_i) = \prod_{t=1}^{T_i} \text{Poisson}(y_{it}; \lambda_i \pi_{it}). \quad (3.39)$$

Both λ_i and p_{it} may be modeled using site-level (i) and temporal (t) covariates as in Eqs. 3.30 and 3.31. As before, the local abundances (conditional on $\hat{\boldsymbol{\pi}}_i$ and $\hat{\lambda}_i$) can be estimated using an empirical Bayes procedure (Eq. 3.21) which, for the Poisson local population model, yields the best unbiased predictor of N_i:

$$E(N_i \mid \boldsymbol{y}_i) = \sum_{t=1}^{T_i} y_{it} + \hat{\lambda}_i \left(1 - \sum_{t=1}^{T_i} \hat{\pi}_{it}\right). \quad (3.40)$$

The removal sampling methodology of Royle (2004b) has been applied to a diverse range of wildlife species, including fish (Royle and Dorazio 2006), amphibians (Royle and Dorazio 2008, pp. 291–294), and birds (Royle 2004b).

3.3 Estimating Species Occurrence under Imperfect Detection

Despite its relative infancy, the past two decades have seen an explosion in the development and application of methods for estimating *species occurrence* under imperfect detection. Prior to these developments, presence/absence data were typically used to infer patterns and dynamics in occupancy through incidence functions (Hanski 1992) or other methods that do not account for detection probability (He and Gaston 2003). In some situations, it may be perfectly reasonable to assume species are detected without error, which is analogous to a complete population census as described previously. For example, if relatively small sample units or sites within a study area are visited to determine the presence or absence of a particular species of vascular plant, then it may be unlikely that the species would go undetected in any site that is occupied (but then again, see Kéry and Gregg 2004; Chen et al. 2012). Unfortunately, this scenario does not apply to most animals, and species that are relatively small, or have cryptic or elusive behaviour, are especially problematic. Even potential proxies for species presence, such as tracks, scats, or nests, can be difficult to detect. Much recent research has therefore focused on the estimation of patterns and dynamics in site occupancy when species detection is not perfect (i.e. $p < 1$).

Recent reviews of the history and methods of occupancy models, as well as practical considerations for the design of these studies, are provided by MacKenzie et al. (2006), Royle and Dorazio (2008), Bailey et al. (2014), and Guillera-Arroita (2016). Here, we focus on the basics of occupancy estimation for a single species when $p < 1$, as well as some exciting recent developments. All of these methods use repeated sampling protocols to inform the detection process, thereby allowing inferences about species occurrence from detection/nondetection data.

3.3.1 Single-Season Occupancy Models

We begin with the *single-season occupancy model* originally presented by MacKenzie et al. (2002), which provides a foundation for extensions that follow. The sampling scenario under consideration involves surveys for a species at R distinct sample units (or sites), and each site is surveyed on T sampling occasions. We assume the occupancy status of each site is closed, such that there is no site colonization or extinction during the sampling period. The species is either detected or not detected during each of the T surveys at each of the R sites. We assume the species is never falsely detected when the species is absent from a site, i.e., there are no *false positive species detections*. When the species is present at a site, the species may be detected (with probability p) or may not be detected (with probability $1 - p$). We also assume that the detection of the species at one site is independent of detecting the species at any other site.

Similar to capture-recapture methods for marked animals (Chapters 5 and 7), we can summarize the detection data for each site using an *encounter history*. For illustration, consider an occupancy study of a calling anuran species with $T = 3$ sampling occasions at each of $R = 30$ ponds. If the species is heard during the tth visit at the ith site, we shall denote this with a "1". If the species is not detected, we denote this with a "0". The encounter history for each of the ponds is then a vector of 1s and 0s (Table 3.2). For example, the encounter history "101" indicates the species was detected at a site on the first and third sampling occasions, but was not detected on the second. In a closed population, a site with an encounter history containing at least one detection is occupied by the species. If ψ denotes the probability of species occurrence across the R sites, and p_t is the probability of detecting the species at an occupied site on occasion t, we observe the encounter history $\boldsymbol{h}_i = (h_{i1}, h_{i2}, h_{i3}) = 101$ at site i with probability

$$\Pr(\boldsymbol{h}_i = 101) = \psi p_1(1 - p_2)p_3. \tag{3.41}$$

Things are a bit more tricky when a site has the detection history $\boldsymbol{h}_i = 000$. In this case, we do not know if the species was *truly absent* from the site (with probability $1 - \psi$), or if the species was indeed present (with probability ψ) but observers *failed to detect* the species on all three sampling occasions [with probability $(1 - p_t)^3$]:

$$\Pr(\boldsymbol{h}_i = 000) = \left[\psi \prod_{t=1}^{T} (1 - p_t)\right] + (1 - \psi). \tag{3.42}$$

The formulation allows the specification of a general model likelihood for ψ and $\boldsymbol{p} = (p_1, p_2, \ldots, p_T)$ given the encounter histories summarizing the $T_i\,(i = 1, \ldots, R)$ sampling occasions for all R sites:

$$L(\psi, \boldsymbol{p} \mid \boldsymbol{h}_1, \boldsymbol{h}_2, \ldots, \boldsymbol{h}_R) = \prod_{i=1}^{R} \left\{ \left[\psi \prod_{t=1}^{T_i} p_t^{h_{it}} (1 - p_t)^{1 - h_{it}}\right] + (1 - \psi) I\left(\sum_{t=1}^{T_i} h_{it} = 0\right) \right\}, \tag{3.43}$$

where $I\left(\sum_{t=1}^{T_i} h_{it} = 0\right)$ is an indicator function taking the value 1 when $\sum_{t=1}^{T_i} h_{it} = 0$ (i.e. the species was never

Table 3.2 The $2^T = 8$ possible encounter histories ($\boldsymbol{h}_i$) and their respective probabilities, $\Pr(\boldsymbol{h}_i \mid \psi, \boldsymbol{p}_i)$, for a single-season occupancy study with $T = 3$ sampling occasions. Here we allow the conditional (on presence) probability of detection (p_{it}) to vary by both site (i) and occasion (t).

$\boldsymbol{h}_i$	$\Pr(\boldsymbol{h}_i \mid \psi, \boldsymbol{p}_i)$
000	$\psi \prod_{t=1}^{T} (1 - p_{it}) + 1 - \psi$
100	$\psi p_{i1}(1 - p_{i2})(1 - p_{i3})$
010	$\psi(1 - p_{i1})p_{i2}(1 - p_{i3})$
110	$\psi p_{i1}p_{i2}(1 - p_{i3})$
001	$\psi(1 - p_{i1})(1 - p_{i2})p_{i3}$
101	$\psi p_{i1}(1 - p_{i2})p_{i3}$
011	$\psi(1 - p_{i1})p_{i2}p_{i3}$
111	$\psi p_{i1}p_{i2}p_{i3}$

detected at site i), and 0 otherwise. Likelihood-based methods may therefore be used to simultaneously estimate occupancy and detection probability.

One of the advantages of this occupancy modeling approach is its ability to incorporate covariates that may help explain site occupancy or detection probability (Box 3.4). Using the logit link function, covariate modeling of ψ enables a broad range of ecological hypotheses to be investigated:

$$\text{logit}(\psi_i) = \alpha + \sum_{j=1}^{r} x_{ij}\beta_j, \tag{3.44}$$

where $\boldsymbol{x}_i = (x_{i1}, x_{i2}, \ldots, x_{ir})$ is a collection of r measurable covariates that are believed to influence the probability of occupancy at site i (ψ_i), α is an intercept parameter, and the β_j are slope parameters describing the relationship between x_{ij} and site occupancy. As in Eq. 3.31, covariates may also be used for modeling detection probability:

$$\text{logit}(p_{it}) = \alpha + \sum_{j=1}^{r} x_{ij}\beta_j + \sum_{k=1}^{q} z_{ik}\delta_k. \tag{3.45}$$

Similar to repeated count methods (Section 3.2.2.1), if there is correlation between species occupancy and detectability, then the two become confounded. There are therefore limitations as to what relationships can be properly accounted for, and careful consideration of these limitations is important in the design and analysis of occupancy studies (see MacKenzie et al., 2006).

3.3.2 Multiple-Season Occupancy Models

Single-season occupancy models are useful for examining occupancy during the snapshot of time spanning the T sampling occasions. However, we are often interested in patterns and dynamics of species occurrence over time. Suppose that instead of a single sampling period of T occasions where we assume that sites are closed to changes in occupancy status, we conduct M sampling periods of $T_m (m = 1, \ldots, M)$ occasions. We will still assume that the sites are closed to changes in occupancy *within* each of the M sampling periods, but not *between* these periods. A *multiseason occupancy model* allows us to investigate changes in site occupancy through time, as previously unoccupied sites become occupied and previously occupied sites become unoccupied.

To investigate species occurrence dynamics under imperfect detection, some additional parameters are required. In addition to ψ_m and p_{mt} for $m = 1, \ldots, M$ and $t = 1, \ldots, T_m$, multiseason occupancy models also estimate the probabilities of local colonization and local extinction. The *probability of local colonization* (γ_m) is the probability that a site unoccupied during sampling period m is occupied during period $m + 1$. The *probability of local extinction* (ϵ_m) is the probability that a site occupied during sampling period m is unoccupied during period $m + 1$. With these additional occupancy dynamics parameters, probabilistic arguments may still be used to model encounter histories and estimate ψ, p, γ, and ϵ for multiple seasons (MacKenzie et al. 2003).

Consider the case for $M = 2$, $T_1 = T_2 = 3$, and the complete encounter history $\boldsymbol{h}_i = (\boldsymbol{h}_{i1}, \boldsymbol{h}_{i2}) = 010\ 000$. The probability of observing the detection history $\boldsymbol{h}_{i1} = (h_{i11}, h_{i12}, h_{i13}) = 010$ during the first sampling period is:

$$\Pr(h_{i1} = 010) = \psi_1(1 - p_{11})p_{12}(1 - p_{13}). \tag{3.46}$$

In this case, site i was clearly occupied during the first period of sampling. For the second period of sampling, the site could have remained occupied (with probability $1 - \epsilon_1$) or become unoccupied (with probability ϵ_1). The probability of observing $\boldsymbol{h}_{i2} = (h_{i21}, h_{i22}, h_{i23}) = 000$ for the second sampling period is therefore:

$$\Pr(\boldsymbol{h}_{i2} = 000) = (1 - \epsilon_1)\prod_{t=1}^{3}(1 - p_{2t}) + \epsilon_1. \tag{3.47}$$

Hence, the probability of observing the complete detection history, $\Pr(\boldsymbol{h}_i = 010\ 000)$, is simply the product of Eqs. 3.46 and 3.47.

A more complicated case arises when $\boldsymbol{h}_i = 000\ 000$. Here, the occupancy status of site i is never known with certainty, and all combinations of occupancy, local extinction, and local colonization are possible: (i) the site was occupied during both sampling periods, but not detected, with probability $\psi_1\prod_{t=1}^{3}(1 - p_{1t})(1 - \epsilon_1)\prod_{t=1}^{3}(1 - p_{2t})$; (ii) the site was occupied, but not detected, during the first sampling period and unoccupied during the second period, with probability $\psi_1\prod_{t=1}^{3}(1 - p_{1t})\epsilon_1$; (iii) the site was unoccupied during the first sampling period and occupied, but not detected, during the second sampling period, with probability $(1 - \psi_1)\gamma_1\prod_{t=1}^{3}(1 - p_{2t})$; or (iv) the site was unoccupied during both sampling periods, with probability $(1 - \psi_1)(1 - \gamma_1)$. Hence, the expression must combine all four possible scenarios:

$$\begin{aligned}\Pr(\boldsymbol{h}_i = 000\ 000 \mid \psi_1, \epsilon_1, \gamma_1, \boldsymbol{p}) \\ = \psi_1\prod_{t=1}^{3}(1 - p_{1t})\left[(1 - \epsilon_1)\prod_{t=1}^{3}(1 - p_{2t}) + \epsilon_1\right] \\ + (1 - \psi_1)\left[\gamma_1\prod_{t=1}^{3}(1 - p_{2t}) + (1 - \gamma_1)\right].\end{aligned} \tag{3.48}$$

Occupancy for the second sampling period may then be derived as $\psi_2 = \psi_1(1 - \epsilon_1) + (1 - \psi_1)\gamma_1$. Clearly, the probability statements for large M can become complicated

Box 3.4 A Single-Season Occupancy Analysis

Returning to the Mallard data of Kéry et al. (2005) that was described in Box 3.3, we will now use these data to estimate the probability of species occurrence using the standard single-season occupancy model of MacKenzie et al. (2002). During T = 2 or 3 visits to R = 235 sites throughout Switzerland, Mallards were detected at 40 sites. If only a single visit had been made to each site, then the x = 31 sites where Mallards were detected on the first visit would yield the naive estimate of occurrence probability $\hat{\psi} = \frac{x}{R} = \frac{31}{235} = 0.13$. Using the single-season occupancy model that accounts for imperfect detection, we can model both the probability of occurrence and detection probability as a function of covariates using the logit link function. Here are AIC rankings for a candidate model set including elevation (elev), percent forest cover (forest), or intercept-only (.) effects on ψ, as well as date, elevation, and intercept-only effects on detection probability (p) (Table B3.4.1).

We see there is some evidence of a seasonal effect on detection probability, but based on the AIC weights, this evidence is not overwhelming. Consistent with the original analysis of Kéry et al. (2005), both elevation and forest cover were found to affect the probability of species occurrence. Unconditional on elevation or forest cover, the mean estimate for ψ from model $\psi(.)p(date)$ was 0.20 (95% CI: 0.14–0.26), which is significantly higher than the naive estimate of 0.13. The mean estimate for p from model $\psi(elev + forest)p(.)$ was 0.67 (95% CI: 0.57–0.76), indicating that on average there was a 67% chance that a Mallard would be detected during a survey of an occupied site. When plotting site occupancy as a function of elevation or forest cover, the estimated probability of a site being occupied declines with increasing elevation (Figure B3.4.1) or increasing percent forest cover (Figure B3.4.2).

Table B3.4.1 Model selection for the effects of standardized covariations on the probability of occupancy for Mallards.

Model	No. of parameters	AIC	ΔAIC	AIC weight
$\psi(elev + forest)p(date)$	5	314.4	0.0	0.59
$\psi(elev + forest)p(.)$	4	315.1	0.8	0.40
$\psi(elev)p(date)$	4	323.8	9.5	0.01
$\psi(elev)p(.)$	3	324.4	10.1	0.00
$\psi(forest)p(date)$	4	341.6	27.3	0.00
$\psi(forest)p(.)$	3	350.7	36.4	0.00
$\psi(.)p(date)$	3	355.5	41.2	0.00
$\psi(.)p(.)$	2	360.3	45.9	0.00

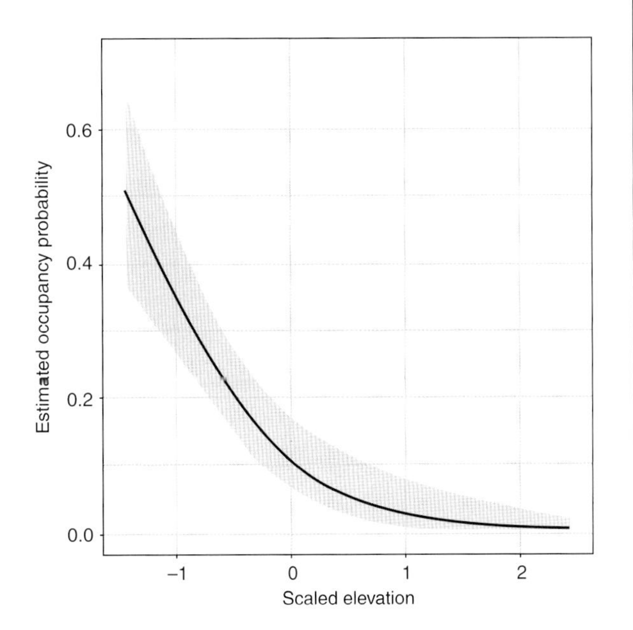

Figure B3.4.1 Estimates of the probability of occupancy for Mallards as a function of scaled elevation. 95% confidence intervals are shaded.

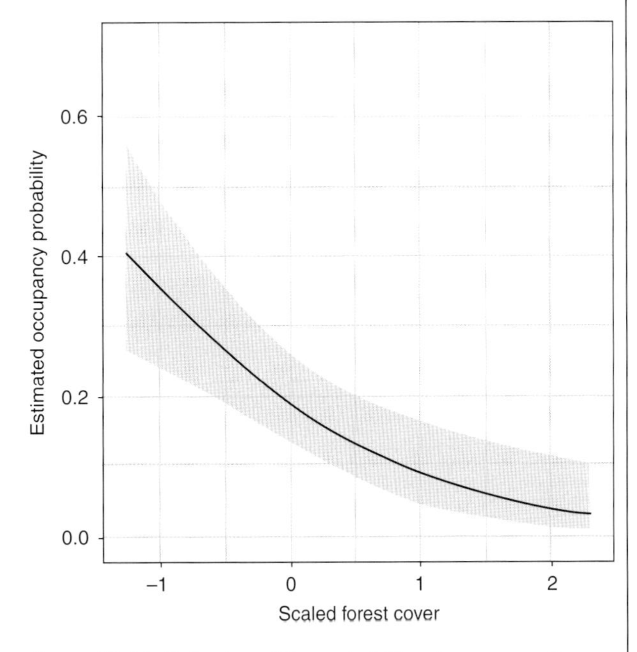

Figure B3.4.2 Estimates of the probability of occupancy for Mallards as a function of scaled forest cover. 95% confidence intervals are shaded.

and tedious, but matrix notation makes the problem tractable (MacKenzie et al. 2003, 2006).

For general M, the multiple-season occupancy model likelihood is:

$$L(\psi_1, \gamma, \epsilon, \boldsymbol{p} \mid \boldsymbol{h}_1, \boldsymbol{h}_2, \ldots, \boldsymbol{h}_R) = \prod_{i=1}^{R} \Pr(\boldsymbol{h}_i). \quad (3.49)$$

Likelihood-based analysis methods may be used to estimate parameters and derive the probability of occupancy during sampling period m using the recursive relationship

$$\psi_m = \psi_{m-1}(1 - \epsilon_{m-1}) + (1 - \psi_{m-1})\gamma_{m-1}. \quad (3.50)$$

Similar to the single-season occupancy model, site-level or temporal covariates for probabilities of site occupancy (ψ_{im}), local colonization (γ_{im}), local extinction (ϵ_{im}), and detection (p_{imt}) can be incorporated into the model parameters using logit link functions (Box 3.2).

3.3.3 Other Developments in Occupancy Estimation

The basic single- and multiseason occupancy models described above have been extended to accommodate more complicated hypotheses about species occurrence and the species detection process. The methodological explosion cannot be covered in its entirety here, but many of these developments are covered in MacKenzie et al. (2006), Royle and Dorazio (2008), and Bailey et al. (2014). Here, we briefly review a few of these extensions.

3.3.3.1 Site Heterogeneity in Detection Probability

Similar to other methods described in this chapter, heterogeneity in detection probability among sites can bias estimators of species occurrence (McClintock et al. 2010a; Miller et al. 2015). One of the earliest extensions of occupancy estimation methods sought to accommodate site heterogeneity in detection probability beyond that explained by measurable covariates believed to influence detection. These "generic" individual heterogeneity models are often needed for reliable inference about species occurrence when it may not be possible to identify, measure, or control for all important sources of variation in detection probability through study design. Occupancy models allowing *heterogeneity in detection* assume that the probability of detection probability at each site (p_i) is a random variable from some distribution, usually referred to as a *mixture distribution* (Royle 2005).

3.3.3.2 Occupancy and Abundance Relationships

In Section 3.2.2.1, we touched on a fundamental relationship between local abundance and the probability of site occupancy:

$$\psi_i = 1 - \Pr(N_i = 0). \quad (3.51)$$

Occupancy probability is therefore the probability that there is *at least* one individual present at a site. It is also seems reasonable to suspect there may be a positive relationship between the probability of detecting a species and local abundance. In other words, sites with higher local densities may be more likely to be detected as occupied simply because there are more individuals available for detection. One may describe heterogeneity in detection probability that is induced by variability in abundance among sites by placing a mixture distribution on the unknown abundance at each site. Under certain conditions, use of mixtures may allow inference about abundance (or density) from detection/nondetection data. Royle and Nichols (2003) proposed that the detection probability for site i at sampling occasion t (p_{it}) is determined by its local population size (N_i) and the detection probability of an individual in the population (r_{it}):

$$p_{it} = 1 - (1 - r_{it})^{N_i}. \quad (3.52)$$

In other words, the probability of detection for site i is the probability that *at least* one of the N_i individuals is detected.

Similar to models of local abundance described in Section 3.2, a model for N_i must also be specified, such as Poisson or negative binomial. The model for N_i provides an estimate of mean density (λ_i) of animals for equal-sized plots under the following four assumptions: (i) every individual in the population has the same probability of detection; (ii) individual detections must be independent; (iii) each local population is closed and N_i must be constant across all surveys of a plot; and (iv) any other sources of variability in detection can be adequately explained by measurable covariates.

In practice, these strong and largely untestable assumptions are unlikely to be valid, and violation of any of these assumptions may lead to questionable inferences about abundance (or density). For discussions on the interpretability of local abundance parameters using this approach, see Royle and Nichols (2003), MacKenzie et al. (2006, pp. 140–141), and Royle and Dorazio (2008, pp.139–140). Regardless, this formulation can still be a useful model for generic site-level heterogeneity in detection probability (Royle 2006).

3.3.3.3 Multistate and Multiscale Occupancy Models

In previous sections, we focused on two states of species occurrence, whether the site is occupied or not occupied, and at a single spatial scale of the site. However, extensions of the single- and multiple-season occupancy models can accommodate additional states and scales of occupancy. *Multistate occupancy models* allow occupied sites to be characterized by additional attributes (or categories of occupancy), such as occupied sites where reproduction occurred (Nichols et al. 2007; MacKenzie et al. 2009), or occupied sites containing diseased or parasitized individuals (Kendall 2009; McClintock et al. 2010c). For example, suppose that anurans were repeatedly sampled at ponds and swabbed for a pathogenic fungus suspected in recent global amphibian declines. Each visit to a pond could fall under one of three possible states: (i) unoccupied by the host anuran species; (ii) occupied by the host with no infected individuals; or (iii) occupied by the host with infected individuals. If we assume no false positive host or pathogen detections, then there is no uncertainty about the state of a site when an infected individual is detected. However, due to nondetection, there remains uncertainty about the disease state of occupied ponds if the pathogen was not detected, and also about the occupancy state of ponds if the host was not detected.

Royle and Link (2005) and Nichols et al. (2007) proposed a multiple-state extension of the single-season occupancy model for this type of situation. These models may be considered a special case of *species co-occurrence models* (not covered here; see MacKenzie et al. 2006, pp. 225–247; Richmond et al. 2010). As with standard occupancy estimators, these approaches allow the estimation of occupancy and detection probability, but they also provide a means to estimate the probability that an occupied site has particular attributes such as infected or not infected, while still accounting for imperfect detection of such attributes such as infected but not detected. MacKenzie et al. (2009) further extended these methods to multiple-season models, thereby allowing investigation of species occurrence dynamics with multiple states. Many interesting ecological hypotheses about the patterns and dynamics of species occurrence may therefore be addressed using multistate occupancy modeling. For example, MacKenzie et al. (2009) showed that California Spotted Owls (*Strix occidentalis occidentalis*) did not tend to colonize a new territory and successfully reproduce within the same season.

Multistate models can also be used to investigate relationships among local abundance, occupancy, and detection probability. For example, it can be difficult to count individuals during calling anuran surveys, but calling intensity is often used as an index of breeding population size. In their analysis of calling survey data for green frogs (*Rana clamitans*), Royle and Link (2005) used a multistate occupancy model with four types of observations that were imperfectly detected: nondetection ($h_{it} = 0$), nonoverlapping calls ($h_{it} = 1$), discrete overlapping calls ($h_{it} = 2$), and a full chorus of continuous overlapping calls ($h_{it} = 3$). Each site was assumed to have a maximum potential calling index that could be observed given the underlying (latent) breeding population size. They authors found that ca. 47% of sites were unoccupied, 33% were capable of generating nonoverlapping calls (corresponding to the lowest abundance level), 15% were capable of generating discrete overlapping calls (intermediate abundance level), and only 4% of sites had abundance levels capable of generating a full chorus of continuous overlapping calls. In an investigation of an assemblage of threatened stream fishes, Falke et al. (2010) used a similar approach that incorporated two relative abundance states (high or low), and modeled detection probability as a function of time, relative abundance, and depth of spawning habitat. They found spawning habitat area, depth, or type were important predictors of occupancy, with spawning habitat size being an important predictor of both larval occurrence and relative abundance.

By incorporating additional levels of repeated sampling, occupancy models can also be extended to multiple spatial scales (Nichols et al. 2008; Kendall 2009). For example, suppose interest is in occupancy across M geographic regions of interest, and each of these regions contains R_i sites for $i = 1, \ldots, M$. Nichols et al. (2008) developed a model for this scenario that allows occupancy estimation at both the region and site level while accounting for imperfect detection. Motivated by large-scale wildlife disease monitoring, McClintock et al. (2010c) extended the multi-state, multiple-season model of MacKenzie et al. (2009) to accommodate multiple spatial scales, and the model awaits further testing with empirical data.

3.3.3.4 Metapopulation Occupancy Models

The occupancy models described thus far have not explicitly accounted for the relative spatial locations of sites. However, the relative locations of sites could be an important factor driving patterns and dynamics of site occupancy. For example, one might expect similarity in the occupancy status of neighboring sites as a function of distance or connectivity; dispersal from an occupied site could lead to the colonization of neighboring sites, or the spread of a disease could lead to local extinction in neighboring sites. Less mechanistic approaches could utilize spatially correlated random effects (Magoun et al. 2007), but spatial models for binary response data or *autologistic models* (Besag 1972) can be used to model correlations in the occupancy status of neighboring sites.

Sargeant et al. (2005) used an autologistic model to estimate the spatial distribution of swift foxes (*Vulpes velox*) under imperfect detection. By dividing the study area into a spatial lattice of Q sites (R of which were surveyed for foxes), each site was assigned a set of s_i neighboring sites (G_i) that shared a boundary with site i. The occupancy status of each site, z_i, was then assumed to have a conditional Bernoulli distribution

$$z_i \mid \boldsymbol{z}_{-i} \sim \text{Bernoulli}(\psi_i), \qquad (3.53)$$

where $\mathbf{z}_{-i}$ is a vector indicating the occupancy status (1 = occupied, 0 = unoccupied) of all sites except site i. Inclusion of spatial structure allowed site occupancy to be modeled as a function of the number of occupied neighbors for site i:

$$\text{logit}(\psi_i) = \alpha + x_i\beta, \tag{3.54}$$

where the auto-covariate

$$x_i = \frac{1}{s_i}\sum\nolimits_{j \in G_i} z_j. \tag{3.55}$$

Hoeting et al. (2000) and Royle and Dorazio (2008, pp. 314–321) describe some alternative autologistic formulations under imperfect detection. McClintock et al. (2010c) describe the use of autologistic methods in multi-state occupancy models to explain correlations in the disease infection state of neighboring sites. Bled et al. (2011) and Yackulic et al. (2012b) extended the multiple-season occupancy model to accommodate spatiotemporal autologistic models. Using a different approach, Sutherland et al. (2014) extended classical stochastic patch occupancy models (Hanski 1999) to investigate metapopulation dynamics and persistence of water voles (*Arvicola amphibius*), while accounting for imperfect detection and missing data.

3.3.3.5 False Positive Occupancy Models

It is now widely acknowledged that false negatives or nondetections are an important source of observation error that must be accounted for when making inferences about occupancy from detection/nondetection data. The false negative detection process can be adequately explained using the occupancy models described thus far, and the assumption of no false positive detections, that unoccupied sites are never falsely detected as occupied, is likely reasonable in many cases. For example, it seems reasonable to assume that studies relying on the physical capture of individuals for species identification would have low false positive error rates. However, studies relying on visual or auditory detections for species identification may be more susceptible to *false positive errors* where misidentification can lead to apparent detection of a species that is actually absent (Simons et al. 2007; McClintock et al. 2010b; Miller et al. 2012b; McClintock et al. 2015). Until recently, little attention has been focused on accounting for false positive errors in the estimation of species occurrence (but see Royle and Link 2006). If not accounted for, false positive detections can lead to overestimation of occupancy probability (Royle and Link 2006) and subsequently bias estimators for both local extinction and local colonization (McClintock et al. 2010a). Even when false positive detections are thought to be relatively rare, they are an important issue to consider. For example, McClintock et al. (2010a) demonstrated that false positive errors constituting ≤1% of all detections in calling anuran surveys can cause severe overestimation of site occupancy, colonization, and extinction probabilities.

Miller et al. (2011) extended occupancy models to accommodate both false negative and false positive detections by utilizing additional information about the false positive detection process. The approach relies on study designs that enable occupancy status of some sites to be determined with certainty. Some species detections are known to be true positive detections, but all other (less certain) detections are susceptible to false positive errors. One way this can be accomplished is by adopting sampling protocols where observers categorize species detections as certain or uncertain. For example, calling intensities of anurans are often recorded during auditory surveys, and it may be reasonable to assume that detections of many calling individuals in a "chorus" at a site may be more reliable than a single individual detection. Another study design that can inform the false positive detection process utilizes two detection methods at each site. One sampling method may be intensive and not susceptible to false positive errors such as physical capture of individuals, whereas the second sampling method may be less intensive, but susceptible to false positive errors, such as auditory or visual surveys. Using either study design, the true underlying state for each site is either occupied or not occupied, but the observations can fall under three categories: (i) species not detected; (ii) species detected, but not with certainty; and (iii) species detected with certainty. The approach of Miller et al. (2011) uses these three types of observations to simultaneously estimate occupancy, false negative detection, and false positive detection probabilities.

Sutherland et al. (2013) extended the approach of Royle and Link (2006) to accommodate multiple seasons and transients moving through the study area. In this case, *false positives* were attributed to transient movements through otherwise unoccupied sites. Based on detections of highly distinctive latrines used to mark territories of established colonies, Sutherland et al. (2013) used their approach to separate *true occupancy* of residents from *apparent occupancy* of transients in a highly dispersive population of water voles (*Arvicola amphibious*).

3.4 Software Tools

A growing number of software tools are available for implementing the methods described in this chapter. One of the most useful and versatile is the R package `unmarked` (Fiske and Chandler, 2011). For abundance and related demographic parameters, unmarked includes the distance sampling models of Royle et al. (2004) and Chandler et al. (2011), replicated count models of local abundance and open population extensions (Royle 2004a; Kéry et al. 2005; Chandler et al. 2011; Dail and Madsen 2011), and removal (or multiple-observer) sampling models (Royle 2004b). For species occurrence, the unmarked package includes the single- and multiple-season models of MacKenzie et al. (2002, 2003), the Royle-Nichols model relating detection probability and abundance (Royle and Nichols 2003), and single-season models accounting for both false negative and false positive detections (Royle and Link 2006; Miller et al. 2011). Kéry and Royle (2015) include examples of the use of unmarked as well as equivalent Bayesian models implemented in the BUGS language.

Program Distance (Thomas et al. 2009) is the most popular and comprehensive stand-alone software for the design and analysis of distance sampling data, including a built-in GIS for survey design, the classical analysis methods described above, and the model-based two-stage approach of Hedley and Buckland (2004). There are also R packages dedicated to distance sampling analyses, including Distance (Miller 2013), mrds (Laake et al. 2012), and dsm (Miller et al. 2013a). The dsm package implements the two-stage model-based approach of Hedley and Buckland (2004), with some extensions. Johnson et al. (2010) implemented a one-stage approach in the R package DSpat (Johnson et al. 2014). Conn et al. (2012, 2013) developed other promising model-based extensions for multiple-observer distance sampling surveys that can accommodate species misidentification and are available in the R package hierarchicalDS (Conn 2014). An R-based simulation package, wisp, is also available (www.ruwpa.st-and.ac.uk/estimating.abundance/WiSP), and was designed primarily as a teaching aid to accompany the text of Borchers et al. (2002), but covers many methods other than distance sampling. However, wisp is no longer under active development; instead we recommend the R-package DSsim for simulation studies to examine optimal survey design, test the effect of assumption violations, and so forth.

Programs PRESENCE (MacKenzie et al. 2006) and MARK (White and Burnham 1999) are popular stand-alone software for detection/nondetection and replicated count data. Both PRESENCE and MARK include local abundance models for spatially replicated counts (Royle 2004a), single- and multiple-season occupancy models (MacKenzie et al. 2002, 2003) and multiple-state extensions (Royle and Link 2005; Nichols et al. 2007; MacKenzie et al. 2009), finite mixture occupancy models for site heterogeneity in detection probability (Royle 2005), the Royle-Nichols occupancy model relating detection probability and abundance (Royle and Nichols 2003), the multiscale occupancy model of Nichols et al. (2008), single-and multiple-season occupancy models accounting for both false negatives and false positives (Miller et al. 2011, 2013b), and single-season species co-occurrence models (MacKenzie et al., 2006). PRESENCE also includes the autologistic occupancy model of Yackulic et al. (2012b), and several other species occurrence models not described here. Program MARK also includes removal models (White et al. 1982; Pledger 2000), multiple-season species co-occurrence models (Richmond et al. 2010; Miller et al. 2012a), and many different capture-recapture models (Chapters 5 and 7). We note that R users can implement most of the models featured in Program MARK using the package RMark (Laake 2013).

Most of the approaches described in this chapter are implemented in some form in stand-alone software or R packages. However, the objectives underlying the design and analysis of unmarked population studies often necessitate custom computer code or model-fitting algorithms. Loaded with fully worked examples and custom code, Royle and Dorazio (2008) and Kéry and Royle (2015) provide an excellent foundation for motivated ecologists whose analysis needs are not covered by existing software.

3.5 Online Exercises

The online R exercises for chapter 3 include four exercises intended to acquaint the reader with several different types of analyses using unmarked population data. Exercises 1 and 2 utilize the dolphin data from Miller et al. (2013a) to provide comparative examples of classical and model-based distance sampling analyses, respectively. The goal of Exercise 3 is to recreate the analysis of spatially replicated counts of Mallards from Kéry et al. (2005) as described in Box 3.3. Exercise 4 uses the same dataset from Mallards to fit the single-season occupancy models as described in Box 3.4.

3.6 Future Directions

Over the past two decades, we have witnessed an explosion of new methods for the analysis of unmarked population data to make inferences about population abundance and species occurrence under imperfect detection. Classical distance sampling methods continue to be extended, and the more recent model-based distance sampling methods continue to be refined. Spatially replicated count and removal methods have been developed further and are regularly applied by population ecologists. Fortunately, user-friendly software is helping to facilitate increased use of these abundance estimation methods by practitioners (Section 3.5). A recent review by Dénes et al. (2015) covers some of the recent "bleeding edge" developments in the estimation of abundance from unmarked individuals that we were unable to cover in detail here, including open population and spatially explicit models.

Classical distance sampling is partly model-based with detection function modeling, and partly design-based by using design properties to extrapolate beyond the surveyed area. Recent developments in detection function modeling have been aimed at increasing robustness through using different classes of flexible models (Miller and Thomas 2015) or relaxing assumptions such as perfect detectability (Borchers and Cox 2016) or no animal movements (Glennie et al. 2015). One particularly interesting development has been a move to unify distance sampling and capture-recapture in a common conceptual framework (Borchers et al. 2015). For model-based distance sampling, one ongoing research thrust is to embed detection function estimation within the well-established framework of point process modeling (Yuan et al. 2017), thereby leveraging that extensive literature. Another development has been to include distance sampling within the general framework of hierarchical modeling (chapters 7–8 of Kèry and Royle 2015; chapter 24 of Kèry and Royle 2016).

Developments in occupancy estimation are appearing so regularly that it can sometimes be difficult to keep up. Bailey et al. (2014) provide a review of recent advances in occupancy estimation, including many approaches not covered here. As with abundance estimation, the continued development of freely available occupancy estimation software is helping practitioners track and apply these developments as they become available. Metapopulation occupancy modeling approaches have not yet received a great deal of attention, but we anticipate the use and development of autologistic and related spatial models to rapidly increase in the future (Bled et al. 2011; Yackulic et al. 2012b; Sutherland et al. 2014). Most occupancy models are for a single species, but species co-occurrence models (MacKenzie et al. 2006; Richmond et al. 2010) have the potential to accommodate a variety of different trophic interactions and may have broad utility for questions in ecology. Species richness and community composition can also be investigated using data collected from unmarked populations in ways that are similar to the occupancy models described above. MacKenzie et al. (2006, pp. 249–264) and Royle and Dorazio (2008, pp. 379–400) provide accessible introductions to these methods, and community-level occupancy modeling is becoming more commonplace (Zipkin et al. 2009; Dorazio et al. 2010). Further development and application of methods for inferring species richness, interactions, and community dynamics from unmarked population data remains a very promising avenue for future research.

Despite mostly separate treatment of the topics in this chapter, there are clear relationships between abundance and species occurrence. We touched on some of these relationships in Sections 3.2.2.1 and 3.3.3.2. In an interesting "marriage" of the two concepts, Wenger and Freeman (2008) combined the spatially replicated count model of Royle (2004a) with the occupancy model of MacKenzie et al. (2002) for the simultaneous estimation of occupancy probability and abundance while accounting for imperfect detection. We anticipate the joint modeling of abundance and species occurrence will continue to be a focus of much future research (Royle et al. 2005).

Much of the material we have covered attempts to account for imperfect detection. One source of variability in detection probability is attributable to differences among individual animals due to behavior or appearance, and among sites due to size or habitat characteristics. Distance sampling methods are typically robust to such heterogeneity, a property of the estimators that is generally referred to as *pooling robustness* (Section 11.12 of Buckland et al. 2004). However, accommodating additional covariates that affect detectability can sometimes be useful (Section 3.2.1). By contrast, many of the other methods covered in this chapter are highly nonrobust to unmodelled heterogeneity. If not properly accounted for, individual heterogeneity can result in biased estimators of abundance because only the "most detectable" individuals are encountered. Unmarked population data contain little, if any information about individual heterogeneity in detection probability. We recommend using capture-recapture methods or other approaches for estimating abundance (Chapter 5), when individual heterogeneity is non-negligible and distance sampling methods cannot be used.

Similarly, unmarked population data contain little information about the vital rates driving population dynamics. If interest lies in estimation of demographic parameters such as fecundity, survival, or movement, then one should consider using capture-recapture or other methods that can directly examine these processes (Chapters 5, 7, and 9). If more informative methods based on marked individuals are not an option, Dail and Madsen (2011) and Zipkin et al. (2014) describe open population models for estimating demographic parameters from data on unmarked individuals. Although this may seem like squeezing juice from a turnip, these new and relatively untested approaches could be reasonable under certain conditions (but see Knape and Korner-Nievergelt 2016), particularly when used in an integrated population model. Integrated population models perhaps hold the most promise for extracting more reliable inferences from unmarked population data (Chapter 9), and we anticipate much development in this area in the coming years (Chandler and Clark 2014).

Besides individual heterogeneity in detection probability, other issues can make reliable inference about abundance difficult to obtain from unmarked populations. Distance sampling methods are not always feasible when a species of interest is highly mobile or when it is difficult to obtain accurate distance measurements to detected individuals. Spatially replicated counts or removal methods that do not involve physical capture are not always feasible with highly mobile species because it can be difficult to avoid double counting of previously detected individuals. Similarly, removal methods involving physical removal are not always feasible for large animals or species of concern. In these circumstances, capture-recapture, or less-invasive methods, may be necessary to reliably estimate abundance (Chapter 5).

When the abundance estimation methods described in this chapter are not feasible or satisfactory, mark-recapture distance sampling (Borchers et al. 1998) and mark-resight models (White and Shenk 2001) are two examples of methodologies that can often be less expensive and less invasive alternatives to conventional capture-recapture (Chapters 5 and 7). Mark-recapture distance sampling can be particularly useful when detection at distance zero is not perfect, and mark-resight can be useful when individual heterogeneity or other sources of variability in detection probability make it difficult to reliably estimate abundance based solely on counts of unmarked individuals. By utilizing encounter data from both marked and unmarked individuals, mark-recapture distance sampling and mark-resight constitute hybridizations of capture-recapture and the unmarked population abundance estimation methods described in this chapter. Borchers et al. (1998) and Laake and Borchers (2004) describe mark-recapture distance sampling, and these methods have seen increased use since their implementation in Program `Distance` (Thomas et al. 2009) and the R package `mrds` (Laake et al. 2012). White and Shenk (2001) review mark-resight methods for estimating abundance, and McClintock and White (2012) describe some more recent approaches to estimating abundance and other demographic parameters from mark-resight data that are implemented in Program `MARK` (White and Burnham 1999). Chandler and Royle (2013) and Sollmann et al. (2013) have developed spatial models of abundance that accommodate this hybrid study design. In our chapter, we have focused on estimation of occupancy and abundance from unmarked individuals, and other book chapters address estimation of abundance and related demographic parameters from data consisting entirely of marked individual encounter (Chapters 5–7). In the future, we predict more development of hybrid models that can exploit the potentially advantageous trade-off between "inexpensive" unmarked versus "expensive" marked population data.

References

Anderson, D.R. (2001). The need to get the basics right in wildlife field studies. *Wildlife Society Bulletin* 29: 1294–1297.

Archaux, F., Henry, P.-Y., and Gimenez, O. (2012). When can we ignore the problem of imperfect detection in comparative studies? *Methods in Ecology and Evolution* 3: 188–194.

Bailey, L.L., MacKenzie, D.I., and Nichols, J.D. (2014). Advances and applications of occupancy models. *Methods in Ecology and Evolution* 5: 1269–1279.

Barker, R.J., Schofield, M.R., Link, W.A., and Sauer, J.R. (2018). On the reliability of N-mixture models for count data. *Biometrics* 74: 369–377.

Besag, J. (1972). Nearest-neighbour systems and the auto-logistic model for binary data. *Journal of the Royal Statistical Society B* 34: 75–83.

Bled, F., Royle, J.A., and Cam, E. (2011). Hierarchical modeling of an invasive spread: the Eurasian Collared-Dove *Streptopelia decaocto* in the United States. *Ecological Applications* 21: 290–302.

Bollinger, E.K., Gavin, T.A., and McIntyre, D.C. (1988). Comparison of transects and circular-plots for estimating Bobolink densities. *Journal of Wildlife Management* 52: 777–786.

Borchers, D.L., Buckland, S.T., and Zucchini, W. (2002). *Estimating Animal Abundance*. New York: Springer.

Borchers, D.L. and Cox, M.J. (2016). Distance sampling: 2D or not 2D? *Biometrics* 73: 593–602.

Borchers, D.L., Laake, J.L., Southwell, C., and Paxton, C.G.M. (2006). Accommodating unmodeled heterogeneity in double-observer distance sampling surveys. *Biometrics* 62: 372–378.

Borchers, D.L., Marques, T.A., Gunnlaugsson, T., and Jupp, P.E. (2010). Estimating distance sampling detection functions when distances are measured with errors. *Journal of Agricultural, Biological, and Environmental Statistics* 15: 346–361.

Borchers, D.L., Stevenson, B.C., Kidney, D. et al. (2015). A unifying model for capture-recapture and distance sampling surveys. *Journal of the American Statistical Association* 110: 195–204.

Borchers, D.L., Zucchini, W., and Fewster, R.M. (1998). Mark-recapture models for line transect surveys. *Biometrics* 54: 1207–1220.

Bravington, M.V., Skaug, H.J., and Anderson, E.C. (2016). Close-kin mark-recapture. *Statistical Science* 31: 259–274.

Buckland, S.T. (1992). Fitting density functions using polynomials. *Applied Statistics* 41: 63–76.

Buckland, S.T., Anderson, D.R., Burnham, K.P. et al. (2001). *Introduction to Distance Sampling*. Oxford, UK: Oxford University Press.

Buckland, S.T., Anderson, D.R., Burnham, K.P. et al. (2004). *Advanced Distance Sampling*. Oxford, UK: Oxford University Press.

Buckland, S.T., Burt, M.L., Rexstad, E. et al. (2012). Aerial surveys of seabirds: the advent of digital methods. *Journal of Applied Ecology* 49: 960–967.

Buckland, S.T., Laake, J.L., and Borchers, D.L. (2010). Double-observer line transect methods: levels of independence. *Biometrics* 66: 169–177.

Buckland, S.T., Rexstad, E.A., Marques, T.A., and Oedekoven, C.S. (2015). *Distance Sampling: Methods and Applications*. Switzerland: Springer International Publishing.

Buckland, S.T., Summers, R.W., Borchers, D.L., and Thomas, L. (2006). Point transect sampling with traps or lures. *Journal of Applied Ecology* 43: 377–384.

Chandler, R.B. and Clark, J.D. (2014). Spatially explicit integrated population models. *Methods in Ecology and Evolution* 5: 1351–1360.

Chandler, R.B. and Royle, J.A. (2013). Spatially explicit models for inference about density in unmarked or partially marked populations. *Annals of Applied Statistics* 7: 936–954.

Chandler, R.B., Royle, J.A., and King, D.I. (2011). Inference about density and temporary emigration in unmarked populations. *Ecology* 92: 1429–1435.

Chelgren, N.D., Samora, B., Adams, M.J., and McCreary, B. (2011). Using spatiotemporal models and distance sampling to map the space use and abundance of newly metamorphosed western toads (*Anaxyrus boreas*). *Herpetological Conservation and Biology* 6: 175–190.

Chen, G., Kéry, M., Plattner, M. et al. (2012). Imperfect detection is the rule rather than the exception in plant distribution studies. *Journal of Ecology* 101: 183–191.

Conn, P.B. 2014. `hierarchicalDS`: Functions For Performing Hierarchical Analysis of Distance Sampling Data. R package version 2.9. https://CRAN.R-project.org/package=hierarchicalDS

Conn, P.B., Laake, J.L., and Johnson, D.S. (2012). A hierarchical modeling framework for multiple observer transect surveys. *PLoS One* 7: e42294.

Conn, P.B., McClintock, B.T., Cameron, M.F. et al. (2013). Accommodating species identification errors in transect surveys. *Ecology* 94: 2607–2618.

Conn, P.B., Ver Hoef, J.M., McClintock, B.T. et al. (2014). Estimating multispecies abundance using automated detection systems: ice-associated seals in the Bering Sea. *Methods in Ecology and Evolution* 5: 1280–1293.

Dail, D. and Madsen, L. (2011). Models for estimating abundance from repeated counts of an open metapopulation. *Biometrics* 67: 577–587.

Dénes, F.V., Silveira, L.F., and Beissinger, S.R. (2015). Estimating abundance of unmarked animal populations: accounting for imperfect detection and other sources of zero inflation. *Methods in Ecology and Evolution* 6: 543–556.

Dodd, J.R. and Dorazio, R.M. (2004). Using counts to simultaneously estimate abundance and detection probabilities in a salamander community. *Herpetologica* 60: 468–478.

Dorazio, R.M., Kéry, M., Royle, J.A., and Plattner, M. (2010). Models for inference in dynamic metacommunity systems. *Ecology* 91: 2466–2475.

Durant, S., Craft, M., Hilborn, R. et al. (2011). Long term trends in carnivore abundance using distance sampling in Serengeti national park, Tanzania. *Journal of Applied Ecology* 48: 1490–1500.

Falke, J.A., Fausch, K.D., Bestgen, K.R., and Bailey, L.L. (2010). Spawning phenology and habitat use in a Great Plains, USA, stream fish assemblage: an occupancy estimation approach. *Canadian Journal of Fisheries and Aquatic Sciences* 67: 1942–1956.

Farnsworth, G.L., Pollock, K.H., Nichols, J.D. et al. (2002). A removal model for estimating detection probabilities from point-count surveys. *Auk* 119: 414–425.

Fewster, R.M., Buckland, S.T., Burnham, K.P. et al. (2009). Estimating the encounter rate variance in distance sampling. *Biometrics* 65: 225–236.

Fewster, R.M., Southwell, C., Borchers, D.L. et al. (2008). The influence of animal mobility on the assumption of uniform distances in aerial line transect surveys. *Wildlife Research* 35: 275–288.

Fewster, R.M., Stevenson, B.C., and Borchers, D.L. (2016). Trace-contrast models for capture–recapture without capture histories. *Statistical Science* 31: 245–258.

Fiske, I.J. and Chandler, R.B. (2011). `unmarked`: an `R` package for fitting hierarchical models of wildlife occurrence and abundance. *Journal of Statistical Software* 43: 1–23.

Glennie, R., Buckland, S.T., and Thomas, L. (2015). The effect of animal movement on line transect estimates of abundance. *PLoS One* 10: e0121333.

Guillera-Arroita, G. (2016). Modelling of species distributions, range dynamics and communities under imperfect detection: advances, challenges and opportunities. *Ecography* 40: 281–295.

Hanski, I. (1992). Inferences from ecological incidence functions. *American Naturalist* 139: 657–662.

Hanski, I. (1999). *Metapopulation Ecology*. Oxford, UK: Oxford University Press.

He, F. and Gaston, K.J. (2003). Occupancy, spatial variance, and the abundance of species. *American Naturalist* 162: 366–375.

Hedley, S.L. and Buckland, S.T. (2004). Spatial models for line transect sampling. *Journal of Agricultural, Biological, and Environmental Statistics* 9: 181–199.

Hilborn, R. and Walters, C.J. (1992). *Quantitative Fisheries Stock Assessment: Choice, Dynamics, and Uncertainty.* New York, NY: Chapman and Hall.

Hoeting, J.A., Leecaster, M., and Bowden, D. (2000). An improved model for spatially correlated binary variables. *Journal of Agricultural, Biological, and Environmental Statistics* 5: 102–114.

Johnson, D.H. (2008). In defense of indices: the case of bird surveys. *Journal of Wildlife Management* 72: 857–868.

Johnson, D.S., Laake, J.L., and Ver Hoef, J.M. (2010). A model-based approach for making ecological inference from distance sampling data. *Biometrics* 66: 310–318.

Johnson, D.S., Laake J.L., and Ver Hoef J.M. (2014). `DSpat`: Spatial Modelling for Distance Sampling Data. R package version 0.1.6. https://CRAN.R-project.org/package=DSpat

Jung, R.E., Royle, J.A., Sauer, J.R. et al. (2005). Estimation of stream salamander (Plethodontidae, Desmognathinae, and Plethodontinae) populations in Shenandoah National Park, Virginia, USA. *Alytes* 22: 72–84.

Kendall, W.L. (2009). One size does not fit all: adapting mark-recapture and occupancy models for state uncertainty. In: *Modeling Demographic Processes in Marked Populations*, Environmental and Ecological Statistics 3 (eds. D.L. Thomson et al.), 765–780. New York, NY: Springer.

Kéry, M., Dorazio, R.M., Soldaat, L. et al. (2009). Trend estimation in populations with imperfect detection. *Journal of Applied Ecology* 46: 1163–1172.

Kéry, M. and Gregg, K.B. (2004). Demographic estimation methods for plants with dormancy. *Animal Biodiversity and Conservation* 27: 129–131.

Kéry, M. and Royle, J.A. (2010). Hierarchical modelling and estimation of abundance and population trends in metapopulation designs. *Journal of Animal Ecology* 79: 453–461.

Kéry, M. and Royle, J.A. (2015). *Applied Hierarchical Modeling in Ecology: Analysis of Distribution, Abundance and Species Richness in R and BUGS. Volume 1: prelude and static models.* London: Academic Press.

Kéry, M. and Royle, J.A. (2016). *Applied Hierarchical Modeling in Ecology: Analysis of Distribution, Abundance and Species Richness in R and BUGS. Volume 2: Dynamic and Advanced Models.* London: Academic Press.

Kéry, M., Royle, J.A., and Schmid, H. (2005). Modeling avian abundance from replicated counts using binomial mixture models. *Ecological Applications* 15: 1450–1461.

Knape, J. and Korner-Nievergelt, F. (2016). On assumptions behind estimates of abundance from counts at multiple sites. *Methods in Ecology and Evolution* 7: 206–209.

Laake, J.L. 2013. `RMark`: An R Interface for Analysis of Capture-Recapture Data with MARK. AFSC Processed Rep 2013-01, 25p. Alaska Fish. Sci. Cent., NOAA, Natl. Mar. Fish. Serv., 7600 Sand Point Way NE, Seattle WA 98115 USA.

Laake, J.L. and Borchers, D.L. (2004). Methods for incomplete detection at distance zero. In: *Advanced Distance Sampling* (eds. S.T. Buckland et al.), 108–188. Oxford: Oxford University Press.

Laake, J., Borchers D., Thomas L., et al. 2012. `mrds`: mark-recapture distance sampling (`mrds`). R package version 2.0.5. http://CRAN.R-project.org/package=mrds.

Laing, S.E., Buckland, S.T., Burn, R.W. et al. (2003). Dung and nest surveys: estimating decay rates. *Journal of Applied Ecology* 40: 1102–1111.

Link, W.A., Schofield, M.R., Barker, R.J. and Sauer, J.R. (2018). On the robustness of N-mixture models. *Ecology* 99: 1547–1551.

Lukacs, P.M. and Burnham, K.P. (2005). Review of capture–recapture methods applicable to noninvasive genetic sampling. *Molecular Ecology* 14: 3909–3919.

MacKenzie, D.I. (2005). What are the issues with presence-absence data for wildlife managers. *Journal of Wildlife Management* 69: 849–860.

MacKenzie, D.I., Nichols, J.D., Hines, J.E. et al. (2003). Estimating site occupancy, colonization, and local extinction when a species is detected imperfectly. *Ecology* 84: 2200–2207.

MacKenzie, D.I., Nichols, J.D., Lachman, G.B. et al. (2002). Estimating site occupancy rates when detection probabilities are less than one. *Ecology* 83: 2248–2255.

MacKenzie, D.I., Nichols, J.D., Royle, J.A. et al. (2006). *Occupancy Estimation and Modeling.* San Diego, CA: Associated Press.

MacKenzie, D.I., Nichols, J.D., Seamans, M.E., and Gutierrez, R.J. (2009). Modeling species occurrence dynamics with multiple states and imperfect detection. *Ecology* 90: 823–835.

Magoun, A.J., Ray, J.C., Johnson, D.S. et al. (2007). Modeling wolverine occurrence using aerial surveys of tracks in snow. *Journal of Wildlife Management* 71: 2221–2229.

Marques, T.A. (2004). Predicting and correcting bias cause by measurement error in line transect sampling using multiplicative error models. *Biometrics* 60: 757–763.

Marques, F.F.C. and Buckland, S.T. (2004). Covariate models for the detection function. In: *Advanced Distance Sampling* (eds. S.T. Buckland et al.), 31–47. Oxford, UK: Oxford University Press.

Marques, T.A., Buckland, S.T., Borchers, D.L. et al. (2010). Point transect sampling along linear features. *Biometrics* 66: 1247–1255.

Marques, T.A., Thomas, L., Fancy, S.G., and Buckland, S.T. (2007). Improving estimates of bird density using multiple-covariate distance sampling. *Auk* 124: 1229–1243.

Marques, T.A., Thomas, L., Martin, S.W. et al. (2013). Estimating animal population density using passive acoustics. *Biological Reviews* 88: 287–309.

Matsuoka, S.M., Mahon, C.L., Handel, C.M. et al. (2014). Reviving common standards in point-count surveys for broad inference across studies. *Condor* 116: 599–608.

Mazerolle, M.J., Bailey, L.L., Kendall, W.L. et al. (2007). Making great leaps forward: accounting for detectability in herpetological field studies. *Journal of Herpetology* 41: 672–689.

McClintock, B.T., Bailey, L.L., Pollock, K.H., and Simons, T. R. (2010a). Unmodeled observation error induces bias when inferring patterns and dynamics of species occurrence via aural detections. *Ecology* 91: 2446–2454.

McClintock, B.T., Bailey, L.L., Pollock, K.H., and Simons, T. R. (2010b). Experimental investigation of observation error in anuran call surveys. *Journal of Wildlife Management* 74: 1882–1893.

McClintock, B.T., Moreland, E.E., London, J.M. et al. (2015). Quantitative assessment of species identification in aerial transect surveys for ice-associated seals. *Marine Mammal Science* 31: 1057–1076.

McClintock, B.T., Nichols, J.D., Bailey, L.L. et al. (2010c). Seeking a second opinion: uncertainty in disease ecology. *Ecology Letters* 13: 659–674.

McClintock, B.T. and White, G.C. (2012). From NOREMARK to MARK: software for estimating demographic parameters using mark-resight methodology. *Journal of Ornithology* 152 (Suppl 2): S641–S650.

Miller, D.L. 2013. `Distance`. R package version 0.9.4. https://cran.r-project.org/web/packages/Distance.

Miller, D.A., Bailey, L.L., Campbell Grant, E.H. et al. (2015). Performance of species occurrence estimators when basic assumptions are not met: a test using field data where true occupancy status is known. *Methods in Ecology and Evolution* 6: 557–565.

Miller, D.A., Brehme, C.S., Hines, J.E. et al. (2012a). Joint estimation of habitat dynamics and species interactions: disturbance reduces co-occurrence of non-native predators with an endangered toad. *Journal of Animal Ecology* 81: 1288–1297.

Miller, D.L., Burt, M.L., Rexstad, E.A., and Thomas, L. (2013a). Spatial models for distance sampling data: recent developments and future directions. *Methods in Ecology and Evolution* 4: 1001–1010.

Miller, D.A., Nichols, J.D., Gude, J.A. et al. (2013b). Determining occurrence dynamics when false positives occur: estimating the range dynamics of wolves from public survey data. *PLoS One* 8: e65808.

Miller, D.A., Nichols, J.D., McClintock, B.T. et al. (2011). Improving occupancy estimation when two types of observational error occur: non-detection and species misidentification. *Ecology* 92: 1422–1428.

Miller, D.L. and Thomas, L. (2015). Mixture models for distance sampling detection functions. *PLoS One* 10: e0118726.

Miller, D.A., Weir, L.A., McClintock, B.T. et al. (2012b). Experimental investigation of false positive errors in auditory species occurrence surveys. *Ecological Applications* 22: 1665–1674.

Nichols, J.D., Bailey, L.L., O'Connell, A.F. et al. (2008). Multi-scale occupancy estimation and modelling using multiple detection methods. *Journal of Applied Ecology* 45: 1321–1329.

Nichols, J.D., Hines, J.E., MacKenzie, D.I. et al. (2007). Occupancy estimation and modeling with multiple states and state uncertainty. *Ecology* 88: 1395–1400.

Nichols, J.D., Hines, J.E., Sauer, J.R. et al. (2000). A double-observer approach for estimating detection probability and abundance from point counts. *Auk* 117: 393–408.

Nichols, J.D., Thomas, L., and Conn, P.B. (2009). Inferences about landbird abundance from count data: recent advances and future directions. In: *Modeling demographic processes in marked populations*, Environmental and Ecological Statistics 3 (eds. D.L. Thomson et al.), 201–235. New York, NY: Springer.

Norris, J.L. and Pollock, K.H. (1996). Nonparametric MLE under two closed capture-recapture models with heterogeneity. *Biometrics* 52: 639–649.

Norvell, R.E., Howe, F.P., and Parrish, J.R. (2003). A seven-year comparison of relative-abundance and distance-sampling methods. *Auk* 120: 1013–1028.

Otis, D.L., Burnham, K.P., White, G.C., and Anderson, D.R. (1978). Statistical inference from capture data on closed animal populations. *Wildlife Monographs* 62: 1–133.

Phillips, S.J., Anderson, R.P., and Schapire, R.E. (2006). Maximum entropy modeling of species geographic distributions. *Ecological Modelling* 190: 231–259.

Pledger, S. (2000). Unified maximum likelihood estimates for closed capture-recapture models using mixtures. *Biometrics* 56: 434–442.

Richmond, O.M., Hines, J.E., and Beissinger, S.R. (2010). Two-species occupancy models: a new parameterization applied to co-occurrence of secretive rails. *Ecological Applications* 20: 2036–2046.

Riddle, J.D., Pollock, K.H., and Simons, T.R. (2010). An unreconciled double-observer method for estimating detection probability and abundance. *Auk* 127: 841–849.

Royle, J.A. (2004a). *N*-mixture models for estimating population size from spatially replicated counts. *Biometrics* 60: 108–115.

Royle, J.A. (2004b). Generalized estimators of avian abundance from count survey data. *Animal Biodiversity and Conservation* 27: 375–386.

Royle, J.A. (2005). Site occupancy models with heterogeneous detection probabilities. *Biometrics* 62: 97–102.

Royle, J.A. (2006). Site occupancy models with heterogeneous detection probabilities. *Biometrics* 62: 97–102.

Royle, J.A., Chandler, R.B., Yackulic, C., and Nichols, J.D. (2012). Likelihood analysis of species occurrence probability from presence-only data for modeling species distributions. *Methods in Ecology and Evolution* 3: 545–554.

Royle, J.A., Dawson, D.K., and Bates, S. (2004). Modeling abundance effects in distance sampling. *Ecology* 85: 1591–1597.

Royle, J.A. and Dorazio, R.M. (2006). Hierarchical models of animal abundance and occurrence. *Journal of Agricultural, Environmental, and Ecological Statistics* 11: 249–263.

Royle, J.A. and Dorazio, R.M. (2008). *Hierarchical Modeling and Inference in Ecology: The Analysis of Data from Populations, Metapopulations and Communities*. New York, NY: Academic Press.

Royle, J.A., Kéry, M., Gautier, R., and Schmid, H. (2007). Hierarchical spatial models of abundance and occurrence from imperfect survey data. *Ecological Monographs* 77: 465–481.

Royle, J.A. and Link, W.A. (2005). A general class of multinomial mixture models for anuran calling survey data. *Ecology* 86: 2505–2512.

Royle, J.A. and Link, W.A. (2006). Generalized site occupancy models allowing for false positive and false negative errors. *Ecology* 87: 835–841.

Royle, J.A. and Nichols, J.D. (2003). Estimating abundance from repeated presence-absence data or point counts. *Ecology* 84: 777–790.

Royle, J.A., Nichols, J.D., and Kéry, M. (2005). Modelling occurrence and abundance of species when detection is imperfect. *Oikos* 110: 353–359.

Sargeant, G.A., Sovada, M.A., Slivinski, C.C., and Johnson, D. H. (2005). Markov chain Monte Carlo estimation of species distributions: a case study of the swift fox in western Kansas. *Journal of Wildlife Management* 69: 483–497.

Savage, A., Thomas, L.K.A., Leighty, L.H. et al. (2010). Novel survey method finds dramatic decline of wild cotton-top tamarin population. *Nature Communications* 1: 30.

Seber, G.A.F. (1982). *The Estimation of Animal Abundance and Related Parameters*, 2e. New York, USA: MacMillian Publication Co.

Sillett, T.S., Chandler, R.B., Royle, J.A. et al. (2012). Hierarchical distance sampling models to estimate population size and habitat-specific abundance of an island endemic. *Ecological Applications* 22: 1997–2006.

Simons, T.R., Alldredge, M.W., Pollock, K.H., and Wettroth, J.M. (2007). Experimental analysis of the auditory detection process on avian point counts. *Auk* 124: 986–999.

Sollmann, R., Gardner, B., Parsons, A.W. et al. (2013). A spatial mark-resight model augmented with telemetry data. *Ecology* 94: 553–559.

Sólymos, P., Matsuoka, S.M., Bayne, E.M. et al. (2013). Calibrating indices of avian density from nonstandardized survey data: making the most of a messy situation. *Methods in Ecology and Evolution* 4: 1047–1058.

Sutherland, C.S., Elston, D.A., and Lambin, X. (2013). Accounting for false positive detection error induced by transient individuals. *Wildlife Research* 40: 490–498.

Sutherland, C.S., Elston, D.A., and Lambin, X. (2014). A demographic, spatially explicit patch occupancy model of metapopulation dynamics and persistence. *Ecology* 95: 3149–3160.

Thomas, L., Buckland, S.T., Rexstad, E.A. et al. (2009). Distance software: design and analysis of distance sampling surveys for estimating population size. *Journal of Applied Ecology* 47: 5–14.

Tyre, A.J., Tenhumberg, B., Field, S.A. et al. (2003). Improving precision and reducing bias in biological surveys by estimating false negative error rates in presence-absence data. *Ecological Applications* 13: 1790–1801.

Wenger, S.J. and Freeman, M.C. (2008). Estimating species occurrence, abundance, and detection probability using zero-inflated distributions. *Ecology* 89: 2953–2959.

White, G.C., Anderson D.R., Burnham K.P., and Otis D.L. (1982). Capture-recapture and removal methods for sampling closed populations. Los Alamos National Laboratory Rep. LA-8787-NERP, Los Alamos, NM.

White, G.C. and Burnham, K.P. (1999). Program MARK: survival estimation from populations of marked animals. *Bird Study* 46: S120–S139.

White, G.C. and Shenk, T.M. (2001). Population estimation with radiomarked animals. In: *Radio Tracking and Animal Populations* (eds. J. Millspaugh and J.M. Marzluff), 329–350. San Diego, CA: Academic Press.

Williams, B.K., Nichols, J.D., and Conroy, M.J. (2002). *Analysis and Management of Animal Populations*. San Diego, CA: Associated Press.

Williams, R. and Thomas, L. (2007). Distribution and abundance of marine mammals in the coastal waters of British Columbia, Canada. *Journal of Cetacean Research and Management* 9: 15–28.

Yackulic, C.B., Chandler, R.B., Zipkin, E.F. et al. (2012a). Presence-only modelling using MAXENT: when can we trust the inferences? *Methods in Ecology and Evolution* 4: 236–243.

Yackulic, C.B., Reid, J., Davis, R. et al. (2012b). Neighborhood and habitat effects on vital rates: expansion of the Barred Owl in the Oregon Coast Ranges. *Ecology* 93: 1953–1966.

Yuan, Y., Bachl, F.B., Lindgren, F. et al. (2017). Point process models for spatio-temporal distance sampling data from a large-scale survey of blue whales. *Annals of Applied Statistics* 11: 2270–2297.

Zellweger-Fischer, J., Kéry, M., and Pasinelli, G. (2011). Population trends of brown hares in Switzerland: the role of land-use and ecological compensation areas. *Biological Conservation* 144: 1364–1373.

Zipkin, E.F., DeWan, A., and Royle, J.A. (2009). Impacts of forest fragmentation on species richness: a hierarchical approach to community modelling. *Journal of Applied Ecology* 46: 815–822.

Zipkin, E.F., Thorson, J.T., See, K. et al. (2014). Modeling structured population dynamics using data from unmarked individuals. *Ecology* 95: 22–29.

Zippin, C. (1958). The removal method of population estimation. *Journal of Wildlife Management* 22: 82–90.

4

Analyzing Time Series Data

Single-Species Abundance Modeling

Steven Delean[1], Thomas A.A. Prowse[2], Joshua V. Ross[2] and Jonathan Tuke[2]

[1] School of Biological Sciences and the Environment Institute, The University of Adelaide, Adelaide, South Australia, Australia
[2] School of Mathematical Sciences, The University of Adelaide, Adelaide, South Australia, Australia

Summary

The quest to understand if, how, and why the abundance of a wildlife population varies over time, and how it is likely to change in the future, is central to population ecology. To address these questions requires information on population abundance over time, yielding sequential time series data that pose particular statistical challenges because the population size at any given time is not independent of that at previous times. Whereas traditional statistical approaches to time series analysis were developed for data-rich research fields such as econometrics and focused on forecasting, ecological time series are typically short for statistical purposes and are more often used to explore population-level processes than for prediction. Time series for wildlife populations are also "noisy" in that they can be considered to result from the combined effects of a deterministic population dynamics process along with "process error" (including components from demographic and environmental stochasticity) and "measurement error" (including imperfect detection probabilities). The range of time series analysis techniques available to ecologists make different assumptions about what is process and what is noise, and the ability to tease apart these components depends critically on the length and temporal resolution of the time series and the availability of covariate information. In this chapter, we explore a range of techniques for analyzing population time series that might exhibit trends or cycling and be affected by intrinsic and extrinsic factors. First, we consider classical statistical approaches based on autoregressive linear models that assume population size is a linear function of one or more lagged values of itself, and then generalize these techniques to nonlinear responses and non-Gaussian error structures. Second, we explore phenomenological models that directly model variation in a population growth rate that results from hidden birth and death processes, and illustrate methods for incorporating density dependence and environmental drivers. Whereas these model classes assume that all noise in the data is due either to observation or process error, we go on to illustrate the application of state-space models to population time series, which formally partition process and observation error components. Throughout, we provide advice on how to negotiate the myriad methods available and to tailor time series analysis to particular research needs and the characteristics of the ecological data available.

4.1 Introduction

One of the most absorbing challenges in ecology is to understand how and why the abundance of wildlife populations varies over time. In some cases, the primary motivation might simply be to detect trends in population size or density (Gerrodette 1987). On the other hand, more complex analyses might ask what is causing those changes (Bjornstad and Grenfell 2001; Knape and de Valpine 2011; Boggs and Inouye 2012), whether the population exhibits cycling or density dependence (Turchin 1990; Stenseth 1999; Louca and Doebeli 2015), or seek to determine what harvesting effort could be sustained by a managed stock (Jacobson and Maccall 1995). The raw materials we need to address such questions are time series data that consist of sequential abundance observations for a population. Within a time series, observations that are closer together in time are more likely to be related and such *autocorrelation* violates one of the core assumptions of classical statistical tests such as linear regression – that the data are independent. Special techniques are therefore required to account for the fact that a population size at time t is some function of the population size at previous times. The quest to understand how present abundance relates to past abundance and other biotic and abiotic variables is central to time series analysis in ecology, but this task is far from trivial (Kendall et al. 1999; Bjornstad and Grenfell 2001; Turchin 2003; Knape and de Valpine 2012a).

Population Ecology in Practice, First Edition. Edited by Dennis L. Murray and Brett K. Sandercock.

Companion website: www.wiley.com/go/MurrayPopulationEcology

When confronted with *ecological time series* for the first time, one cannot escape the overriding impression that they bounce around a lot. There are many reasons why time series might be variable, but these can be split usefully into process and noise components (de Valpine and Hastings 2002; Clark and Bjornstad 2004). The *process* reflects the way in which the average fertility and survival rates, or *vital rates*, of the population change over time as a function of intrinsic or extrinsic factors, and thereby determine the population growth rate. Process-driven variation in population size could result from a range of mechanisms such as density dependence or a relationship between a time-varying environmental variable and a vital rate. In general, the goal of time series analysis is to characterize this process, but in the ecological context this can be difficult because our data are typically "noisy" (Ellner and Turchin 1995; Reuman et al. 2008; Lindén et al. 2013).

Traditional time series analysis assumes that temporal variation in a response variable is governed by a deterministic process, and that all noise is due to *observation error* (Lundberg et al. 2000; de Valpine and Hastings 2002). In truth, however, even if all observation error were eliminated, then time series for wildlife populations would still be "noisy" due to *process error* that includes contributions from both demographic and environmental stochasticity (Bartlett 1960; May 1973; Shaffer 1981; Bjornstad and Grenfell 2001). *Demographic stochasticity* refers to variation in realized survival and fertility rates because a population is composed of a finite number of individuals, so that even if the per-individual survival rate is constant over time, the actual proportion of individuals surviving will vary from year to year. *Environmental stochasticity* means variation in the vital rates due to extrinsic factors that are not included in the analysis such as the density of predators or competitors and climatic variables. Modern time series methods can estimate the contributions of both observation and process error provided there is sufficient data to do so (de Valpine and Hastings 2002; Clark and Bjornstad 2004; Knape 2008).

Although noise can obscure the processes behind population dynamics (Dennis et al. 2006), disturbance from perturbations are crucial to developing a full understanding of population-level responses (Lundberg et al. 2000). For example, time series data for a population that remains close to carrying capacity yield little information about density-dependent processes. In contrast, time series data for another population that has been affected by harvesting, extreme climatic variation, or other disturbances, and has been reduced to low density will often offer much more information about intrinsic and extrinsic drivers of population growth (Polansky et al. 2009; Clark et al. 2010). In many cases, it is only through disturbing a population and observing its growth response at different abundances that the underlying dynamics are revealed (Lundberg et al. 2000).

As we shall see, there are many decisions to make when choosing an appropriate model for time series data. In this respect, *time series modeling* in ecology is something of an "art" in that there seldom exists a single best approach to modeling any given dataset (Ziebarth et al. 2010). Rather, the possible approaches will be to some extent determined by the length and temporal resolution of the time series, the goal of the analysis, and the attitudes of the individual researcher as to what should be treated as process and what as noise. In our chapter, we focus on these different approaches and decisions, by considering time series methods for analyzing abundance data for animal populations.

4.1.1 Principal Approaches to Time Series Analysis in Ecology

The two primary aims of any time series analysis are: (i) to identify the process underlying the sequential observations; and (ii) to predict future values of the variable of interest, which is also termed *forecasting* (Hyndman and Athanasopoulos 2013). In our chapter, we focus on the first goal because, with the possible exception of threatened species research, time series for wildlife populations are usually used to explore the population dynamics process rather than to develop future predictions (Sibly et al. 2005; Ziebarth et al. 2010). Ecological theory suggests that many intrinsic and extrinsic processes could cause populations to fluctuate in size over time, but the challenge lies in confronting these data with models (Lundberg et al. 2000).

To illustrate this point, Figure 4.1 presents real time series data for four example populations. In the first two cases, a single population count or index has been recorded each year (Figure 4.1a and b) yet the time series produced are clearly quite different. The first example illustrates the population size of beavers (*Castor canadensis*) in the USA and Canada between 1931 and 1981 (NERC Centre for Population Biology, Imperial College 2010) (Figure 4.1a). The population size and variance grow through time, and although we might estimate the average rate of increase, little information about carrying capacity is included in the data for beavers in this environment. The second example illustrates the growth of a population of Tasmanian sheep (*Ovis aries*) between 1818 and 1836 (Davidson 1938) (Figure 4.1b). The sheep population grew initially before appearing to stabilize from the mid-1850s, largely because pasture availability became a limiting factor, although economic factors also played a role (Renshaw 1991). In contrast to the beaver example, one could assume a density-dependent model for these data and estimate the maximum annual rate

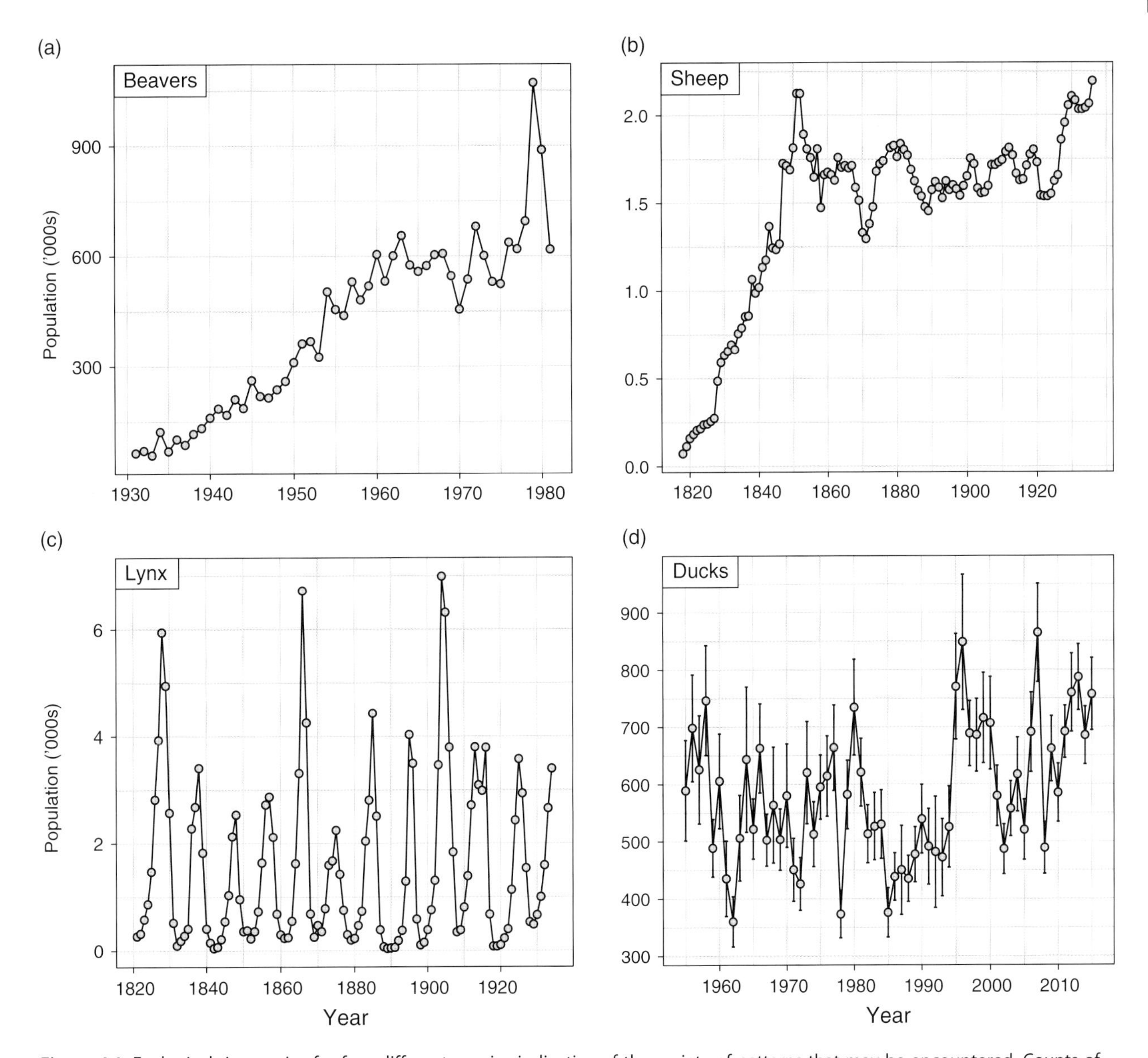

Figure 4.1 Ecological time series for four different species indicative of the variety of patterns that may be encountered. Counts of population size are recorded annually for all species; error bars in (d) are 95% confidence intervals (CI).

of increase for sheep as the annual rate of population growth when sheep density is low, and also the carrying capacity of the environment.

The third example shows the annual numbers of Canada lynx (*Lynx canadensis*) trapped in Canada between 1821 and 1934 (Campbell and Walker 1977; Brockwell and Davis 1991) (Figure 4.1c). Assuming we are prepared to use trap counts as an index of abundance, it is clear that the lynx population cycles through time, and so we might want to estimate the amplitude and periodicity of the cycles and explore the mechanisms driving this pattern. The final example presents more detailed data on the population size of Canvasbacks (*Aythya valisineria*) between 1955 and 2015 (U.S. Fish and Wildlife Service 2015; Figure 4.1d). Each year the duck population was surveyed at a number of sites, so that estimates of the total population at each sampling time are possible. There is no clear trend in the duck population size over the period, yet considerable variability exists over time and there is strong evidence of autocorrelation in the series. As we shall see, the availability of replicate data at each sampling time affords the opportunity to disentangle process and observation error and refine estimates of the parameters governing population processes.

These four examples are far from exhaustive, but they illustrate the breadth of possibilities one faces when considering how to proceed with a time series analysis. In some cases, the goal of a research project might be clear from the outset such as: Is the population size of a threatened species declining? In many other cases, an iterative approach might be required in that characteristics of the time series data obtained will inform the choice of modeling conducted. Many different approaches can be taken to analyzing ecological time series but these can be classified into three main methods: statistical, phenomenological, and mechanistic.

Statistical methods for time series analysis were originally developed within the field of econometrics, with the primary goal to generate useful predictions of economic variables, such as share prices or wages, based on historical data. The standard statistical approaches adopted for this purpose usually rely on *autoregressive integrated moving average* (ARIMA) and related models. ARIMA models are a class of regression model that assume the response variable is a linear function of one or more lagged values of itself as the *autoregressive component*, and a weighted sum of one or more recent errors as *the moving average component*. The models are commonly used to reveal population cycles and characterize the strength of density-dependent processes (Lundberg et al. 2000; Ziebarth et al. 2010), but assume all error is observation error. ARIMA models were developed as linear models for data with normally distributed errors, but as we will see they can be extended to account for nonlinear responses and nonstandard error distributions through a flexible suite of correlated-error models. In contrast, *phenomenological models* directly model the intrinsic rate of increase (r) of a population. Phenomenological models usually assume some density-dependent modulation of the population growth rate but they can also incorporate the effects of other environmental covariates (Post et al. 2009). Although these models have typically been fit under the assumption that all error is observation error, they are readily extended to a state-space model framework that allows a process error component.

One advantage shared by statistical and phenomenological models is that they only require data on the population size at each time step, which is often a year. If, however, data on survival and fertility rates are available instead, then a "mechanistic" model could be used to estimate the specific effects of population density or external factors on these vital rates and the emergent population growth rate (Pelletier et al. 2012; Letcher et al. 2015). In this chapter, we focus on statistical and phenomenological models for ecological time series, while readers interested in mechanistic models are directed elsewhere (Newman et al. 2014, Chapters 8 and 9).

4.1.2 Challenges to Time Series Analysis in Ecology

Ecological time series pose a number of challenges that are rarely encountered by other disciplines. One of the most obvious technical difficulties is that a time series that might be considered "long" by ecologists (e.g. more than 10 years) are still relatively "short" for statistical purposes (Ives et al. 2010). When a time series is short, parameter estimates from a fitted model will have high variance and may not be informative, which is the usual small-sample problem (Ives et al. 2010; Dennis and Ponciano 2014). Similarly, spurious trends or relationships between the population size and environmental covariates are to be expected for short time series (Knape and de Valpine 2011). Ecological time series data also come with a host of familiar challenges – the data can be *continuous* (e.g. density) or *discrete* (e.g. survey counts or harvest numbers), and might contain *many zeros*. Therefore, different data transforms or distributional assumptions may be required to model the data adequately. The logistical challenges of monitoring wild populations in the field also dictate that *missing values* are commonplace. The importance of missing values is linked to the temporal autocorrelation in the time series – when autocorrelation is high, a few missing values might be unimportant due to redundancy or because they can be interpolated; otherwise such data gaps can present a serious problem. With the possible exception of time series for harvested species, large intervals between sampling times at a seasonal or annual time step are typical. Monitoring wildlife populations is expensive and time-consuming and surveys are normally conducted periodically. As a result, discrete-time models are far more common in the literature than continuous-time approaches, particularly where data are collected at regular intervals and species undergo at least some demographic change like reproduction and dispersal on a seasonal basis. However, as much as possible, ecological time series models should be formulated to reflect the dynamics being examined and can then be calibrated against any available data.

In this chapter, we will introduce three modeling approaches used for analyzing single-species time series data. The first type are classical *autoregressive moving average* (ARMA) time series models, where past trends, rather than particular underlying mechanistic processes, are used to predict future behaviour; this can really be viewed as a statistical model of the data. We then introduce biological models of population dynamics in the sense that they parameterize population growth over time in terms of the intrinsic growth rate and the feedback effects of past population density. We consider *ordinary differential equation* models (ODE), which are the classical models used in dynamic modeling of population processes where the emphasis is on explaining the average expected behaviour of a species' dynamics. We also discuss the approximation of these models in discrete time using difference equations, with integration over specified time periods. Last, we consider *state-space models* that describe population time series as partially-observed

Markov processes where the true, but unknown, population state is estimated, assuming that the observed population counts are measured with error.

4.2 Time Series (ARMA) Modeling

4.2.1 Time Series Models

As a fundamental starting point, we will consider an empirical statistical approach to estimating the dynamics of ecological time series using ARIMA models. ARIMA models provide a flexible structure for estimating the characteristics of ecological time series to assess changes in population sizes of species for management and conservation (Box 4.1). We refer interested readers to the coverage of ARMA models for ecological time series analysis in Ives et al. (2010) and to Cowpertwait and Metcalfe (2009) for general development of the ideas.

Consider the set of n observations of the population size (or density) N made at discrete time points t. We will denote this time series data by $\{N_t : t = 1, \dots, n\}$: which may be abbreviated as $\{N_t\}$. Our goal for this introduction is to understand the underlying dynamics of the population and forecast future values of N_t given the observed data. We will denote the forecast for a future value at time $t + k$ made at time t by $\hat{N}_{t+k|t}$. Population sizes (or densities) are generally log-transformed for analysis as $x_t = \log(N_t)$ to represent log-linear dynamics.

Fundamentally, time series models can be split into three components: a *trend*, a *seasonal effect*, and a *stochastic error*. A simple model is:

$$x_t = m_t + s_t + \varepsilon_t, \tag{4.1}$$

where m_t is the trend component, s_t is the seasonal component, and ε_t is an appropriate stochastic error term at time t. This model can be refered to as the *additive decomposition model*, however, our focus in this chapter is on models for the trend and error components and we do not address the seasonal component models further (see Cowpertwait and Metcalfe [2009] for details; also see Kendall et al. [1998] and Louca and Doebeli [2015] for reviews of estimating cyclicity in ecological time series; Box 4.2). To aid in the understanding of how to model the error term, we will first consider some measures of the properties of time series that will be useful. Usually in statistical modeling, we assume that the error component is independent and identically distributed from a prespecified distribution. In the case of time series, this may not be the case: often consecutive error terms will be correlated. For example, we would expect that the population size for any one year will be highly correlated with the population size from the previous year. A key aim in time series analysis is to model and then estimate this correlation structure.

A fundamental assumption to modeling autocorrelation in time series data is that of *stationarity*. We assume that the time series is stationary in the mean if the sample mean abundance $\bar{x}$ does not change with time, and stationary in the variance if the sample variance σ^2 is constant for all times. The correlation between consecutive observations can be measured by the *sample autocovariance function (acvf)* $\hat{\gamma}_k = n^{-1}\sum_{t=1}^{n-k}(x_t - \bar{x})(x_{t+k} - \bar{x})$, which is the covariance between time points separated by a time lag of k. The sample autocorrelation function, which we will use later for evaluating correlation lags, is then defined as $\hat{\rho}_k = \hat{\gamma}_k / \hat{\gamma}_0$, where $\hat{\gamma}_0 = \hat{\sigma}^2$.

So far we have used the term stationary to describe both time series that are stationary in the mean (constant mean over time) and variance (constant variance over time). Consider a time series as a whole; a model $\{x_t\}$ is strictly stationary if the joint statistical distribution of $x_{t_1}, \dots, x_{t_n}$ is the same as the joint distribution of $x_{t_1+m}, \dots, x_{t_n+m}$ for all $t_1, \dots, t_n$ and m. This is a strict requirement, and so instead we often consider time series models that are second-order stationary. *Second-order stationary* time series models have mean and variance that are constant in time, and an autocovariance that only depends on the time lag k.

The most basic time series model that represents an observation in terms of the previous observation is the random walk; it is defined for a time series $\{x_t\}$ as:

$$x_t = x_{t-1} + \varepsilon_t, \tag{4.2}$$

where ε_t is white noise (i.e. the values of ε_t are independent and identically distributed with a mean of zero). If, additionally, the distribution is normal, then the series is called Gaussian white noise.

4.2.2 Autoregressive Moving Average Models

Extending the idea of expressing the observation at time t in terms of the previous observation, we can define a time series $\{x_t\}$ as an *ARMA* process of order (p, q), denoted ARMA(p, q), in general terms as follows:

$$(x_t - \mu) = \sum_{i=1}^{p} \beta_i (x_{t-i} - \mu) + \sum_{j=0}^{q} \phi_j \varepsilon_{t-j} \tag{4.3}$$

where μ is the mean of the stationary process, β_i are the autoregressive parameters of order p, ε_t is white noise, and ϕ_j are the parameters of a moving average (MA) process (Box et al. 1994; Ives et al. 2010). The autoregressive process defines x_t in terms of past observations and thus detects delayed effects of past population size on current dynamics. The MA process defines x_t in terms of previous

Box 4.1 Modeling the Population Dynamics of Beavers Using ARIMA

The data are a time series of annual population counts of beavers in the USA and Canada from 1931 to 1981 (NERC Centre for Population Biology, Imperial College 2010; Figure B4.1.1a). Examination of the plot shows an increasing beaver population size over time with increased variability in population size for later years, which indicates that the series is clearly not stationary in the mean, and possibly not in the variance. We could formally test for stationarity using, for example, the augmented Dickey-Fuller test for unit roots (Dickey and Fuller 1979). Given the increasing variance in population size counts that is evident from the plot, we will log-transform the series prior to further analysis. The resulting series is clearly nonlinear (Figure B4.1.1b).

The next step is to determine the nature of the trend over time, and particularly to determine whether the series is *trend-stationary* where the residuals from a linear trend fitted to the logged counts are stationary, or alternatively *difference-stationary* where the first-differences of logged counts are stationary. In this example, the augmented Dickey-Fuller test applied to each series indicates that the logged beaver population size is difference-stationary, but not trend-stationary. The ecological interpretation would be that there is nonlinearity in the temporal trend, associated with the population size apparently reaching a plateau since the 1960s, which may reflect a carrying capacity, but that the magnitude of annual population changes are relatively consistent over time. We could proceed in one of two ways: (i) examine the dynamics around the differenced series of population size transitions; or (ii) identify a satisfactory model for the nonlinear trend (for example, using piece-wise-polynomial splines) and then examine the dynamics around the residual deviations from that trend.

The next step is to use sample autocorrelations to infer the order of autoregressive and MA processes for our time series. The *autocorrelation function* measures the cross-correlation of a time series with lagged series up to an arbitrary maximum lag. In contrast, a *partial autocorrelation* is the amount of correlation between a series and a given lag of the series that is not explained by correlations at all lower-order lags of shorter duration. Generally, 95% confidence limits for an independent time series with $\rho_0 = 1$ are shown on autocorrelation function plots (Venables and Ripley 2002), which can be used to identify sample autocorrelations that are significantly different from zero.

The first-differenced beaver population series exhibits a negative lag-1 autocorrelation (Figure B4.1.2), which would suggest that a first-order MA term should be included in the model. The lag-1 partial autocorrelation can indicate the order of the AR process, but where the correlation is negative, as it is here, it provides additional evidence for an MA process of order 1. Therefore, we would fit an ARMA ($p = 0$, $q = 1$) model to first-differenced time series.

Whilst the autocorrelation functions provide a guide to determine the order of the correlations, we can also use an information-theoretic approach to determine the appropriate ARMA model. To do so, we would fit ARMA models with all pairwise combinations of AR and MA orders and select the model that minimizes the information criterion. Using the AIC criterion corrected for small-sample bias (AIC_c; Hurvich and Tsai 1989; Chapter 2) as a parameter-penalized log likelihood for the beaver example, the ARMA(0,1) model is ranked highest. This indicates that the errors ε_t can be modeled by a MA process with $\varepsilon_t = -0.35 \times \varepsilon_{t-1}$, where ε_{t-1} is white noise with $\hat{\sigma}^2 = 0.042$.

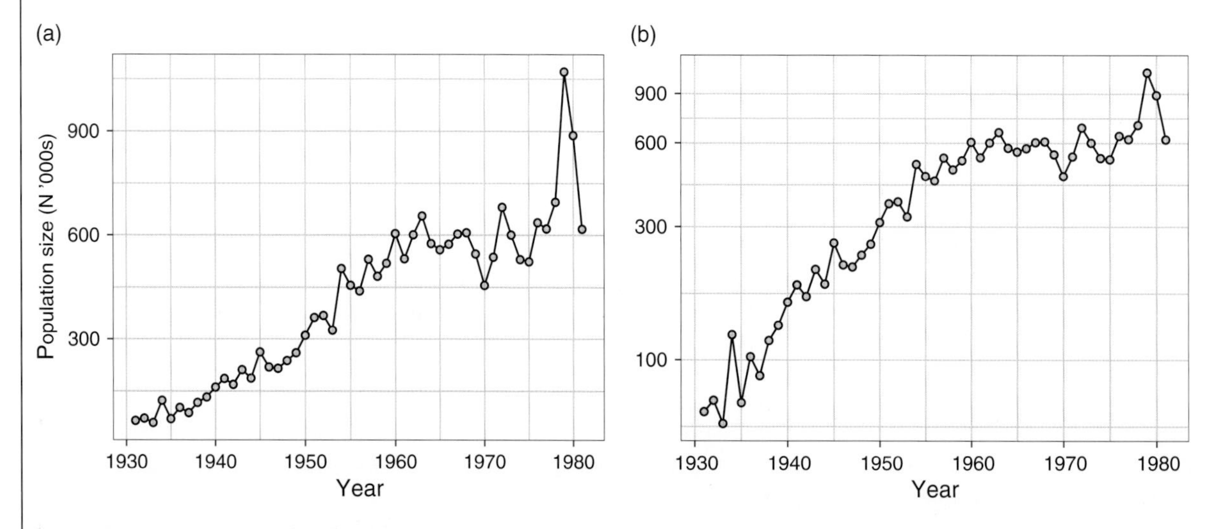

Figure B4.1.1 Time series plot of yearly beaver population size (in thousands) from 1931 to 1981 on the (a) raw and (b) log scales.

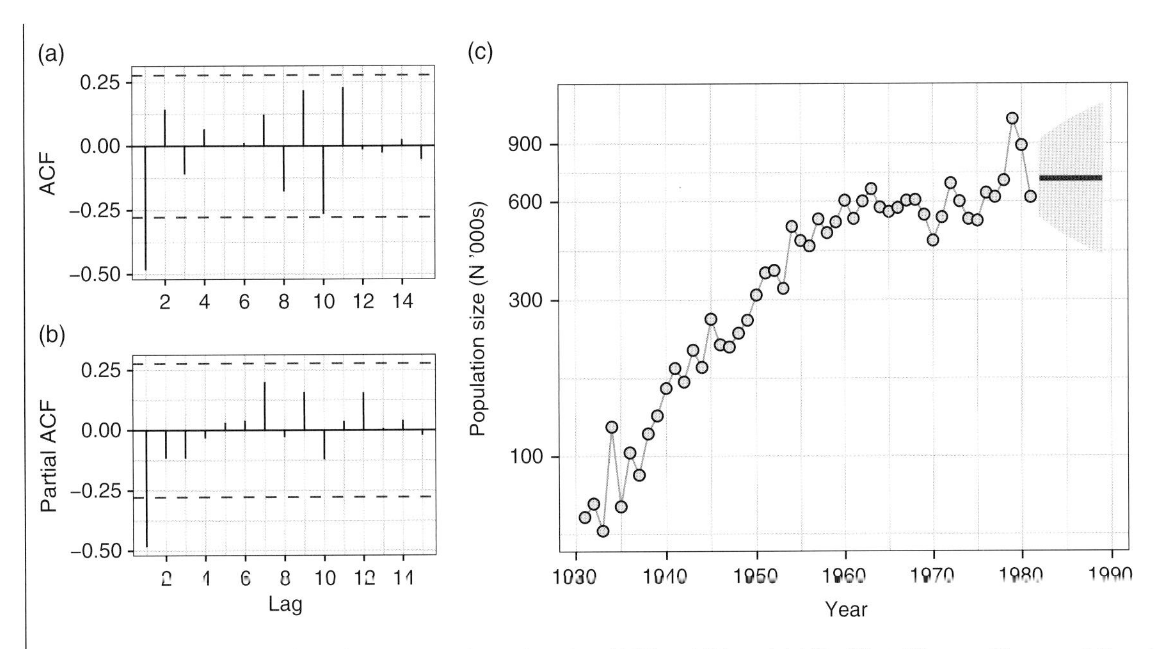

Figure B4.1.2 Time series plots of (a) autocorrelation functions (ACF) and (b) partial ACF of first-differenced log population size, and (c) logged population size time series and forecast population size over 10 years (solid line) based on an ARMA(0,1) model with 95% confidence intervals (CI) on the predictions (shaded region). Dashed lines in (a) and (b) indicate 95% CI cut-offs for significant autocorrelations.

Box 4.2 Population Cycles of Canadian Lynx

Ecological time series often exhibit cyclicity, so a common goal is to understand these underlying patterns in conjunction with trends and environmental relationships (Kendall et al. 1998; Louca and Doebeli 2015). To illustrate these types of analyses, we use a classical dataset of the annual numbers of lynx trapped from 1821 to 1934 in Canada (Figure B4.2.1a; initially described in Elton and Nicholson 1942). The data may be used as an index of lynx population size to examine the underlying dynamics that explain substantial temporal fluctuations in population numbers. In the absence of direct information about the driver(s) of the apparently cycling population numbers, we may proceed to model this stochastic process.

The distribution of trapping numbers is right-skewed, suggesting a log-transformation of the counts would be appropriate prior to further analysis (Figure B4.2.1b). In visualizing the relationship, there seems to be no obvious linear trend, so we do not detrend the data. There is a strong positive autocorrelation in the series at a lag of 10 and a strong negative correlation at a lag of 5, which is indicative of a periodic effect of 10 years (Figure B4.2.2). The partial autocorrelation function indicates strong first- and second-order lags, and smaller but potentially important correlations at greater lags (e.g. lags 4, 7, and 11). Based on the autocorrelation functions, a parsimonious approach could be to fit an ARMA(2,0) model, and in fact this is the lag that has highest support (from the range 1–11 lags) based on model rankings with the Bayesian Information Criterion (BIC; Schwarz 1978). A cumulative periodogram (Diggle 1990) suggests that the residuals from that model are white noise, however the autocorrelation functions reveal some small higher-order correlations are present. In contrast, an ARMA(11,0) model is ranked highest based on AIC_c, and this model captures the periodic dynamics (i.e. there are no remaining peaks in the autocorrelation functions), and results in independent residuals, but at the expense of parsimony compared to the model ranked highest using BIC (Note: BIC penalizes the model likelihood by $\log(n)K$, where K is the number of parameters in the model, so for $K = 1$ and sample size $n > 12$, BIC places a greater penalty on model complexity than AIC_c).

An alternative approach to ARMA that could be explored in search of a more parsimonious model is to fit a linear (in the parameters) statistical model to capture the cyclic population dynamics. One example is to specify a harmonic model for the 10-year cycle (based on the ACF plot) of the form:

$$x_t = m_t + s_i \sin(2\pi t/10) + c_i \cos(2\pi t/10) + z_t \quad (4.4)$$

where m_t is the trend, z_t is the error term, and $[s_i, c_i]$ are unknown constants. However, it transpires that there is still some partial autocorrelation at lags 1 (positive) and 2 (negative), as well as some weaker higher-order correlation in the residuals from this model. Refitting the model using generalized least squares allows accounting for any remaining ARMA structure in the residuals; an ARMA(2,0) process for the residual correlation has highest-rank using BIC, although this is less parsimonious than our previous AR(2) model. Various authors have examined the lynx trapping data (for example, Moran 1953; Stenseth et al. 1997; Stenseth et al. 1998), and most conclude that it is related to a predator–prey (Lynx-Hare) model. Actually, nonlinear autoregressive models (e.g. SETAR; Tong 1990) are generally recognized to afford improved predictions over an AR(2) model for these data.

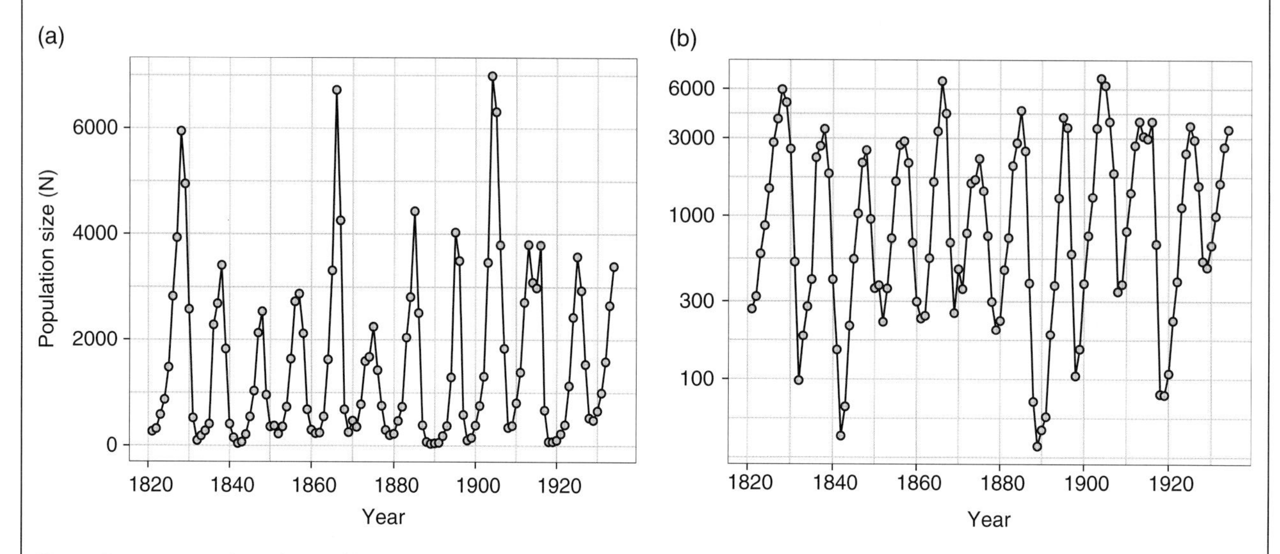

Figure B4.2.1 Annual numbers of lynx trappings for 1821–1934 in Canada on the (a) raw and (b) log scales.

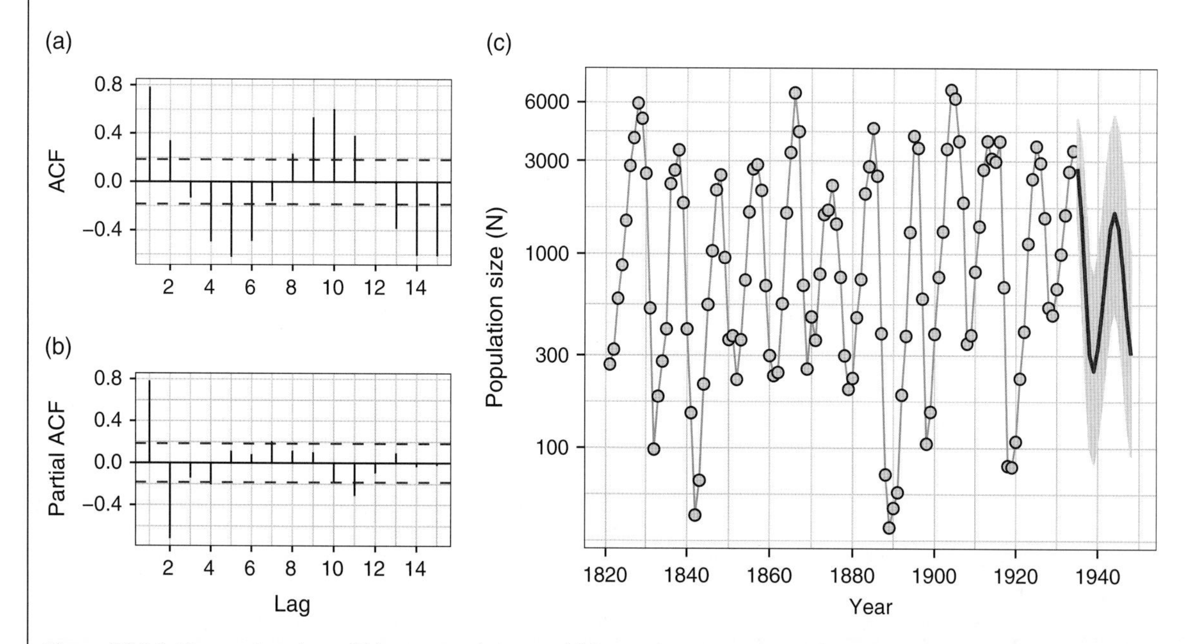

Figure B4.2.2 Time series plots of (a) autocorrelation and (b) partial autocorrelation functions of log population size, and (c) logged population size time series and forecast population size over 12 years (solid line) based on an ARMA(2,0) model with 95% confidence intervals (CI) on the predictions (shaded region). Dashed lines in (a) and (b) indicate 95% CI cut-offs for significant autocorrelations.

white noise terms. Further, we can partition this model to obtain an AR (p) as the special case ARMA(p, 0), and a MA(q) as the special case ARMA(0, q). Autoregressive lags can be generated in age- or stage-structured populations where reproductive maturity is delayed relative to age (or stage) intervals. The delayed effects can also occur due to species interactions, such as predator–prey relationships, and are amplified where both these processes occur together.

Rather than detrending the data series prior to analysis, we can difference the series directly within an ARIMA framework; here the order of the process extends to (p, d, q), denoted ARIMA(p, d, q), where d indicates the degree of differencing. The random walk model defined above is the special case of an ARIMA of order (p = 0, d = 1, q = 0) where $\mu = 0$; and in the case where $\mu \neq 0$ the model is a random walk with drift parameter μ. The ARIMA framework opens up a rich array of models for the dynamics of ecological time series, including a random walk with autocorrelated errors (p = 1, d = 1, q = 0), or with smoothed errors (p = 0, d = 1, q = 1), as well as models with density dependence (p = 1, d = 0, q = 0) plus a constant term (Dennis et al. 2006), which we will expand on elsewhere in the chapter (Box 4.1). Low-dimensional autoregressive models have been shown to give accurate forecasts of future population states across a wide range of taxa (Ward et al. 2014).

Stability in the dynamics of a population can be measured by the rate at which the population returns to its stationary distribution. Known as the *characteristic return time* (Box et al. 1994), this measure depends on the AR coefficients of an ARMA model and is calculated as the magnitude of the inverse of the minimum root of the characteristic equation of Eq. 4.3 (Box et al. 1994). Ives et al. (2010) argue that this summary measure of population dynamics may be particularly useful as it is robust to the imprecision in ARMA model parameters that results from the characteristically short lengths of ecological time series. Return time has been used as a key metric of population regulation, particularly in showing that return times are generally long in ecological time series, thus providing partial evidence for the conclusion that density dependence is generally weak (Ziebarth et al. 2010).

4.3 Regression Models with Correlated Errors

To this point, we have considered traditional statistical approaches to time series analysis that include some combination of an autoregressive model and an error term modeled as a linear combination of current and previous errors. When the goal of a time series analysis is to explore trends or the effects of environmental covariates on population size, an alternative strategy is to treat all temporal autocorrelation as noise and fit a regression model without any lag terms while assuming some correlation structure for the errors. For example, assuming we have a time series of population counts N_t of length n and we are interested in whether there is evidence of a trend in abundance over time t or whether the mean population size is stationary, we could fit the following regression model:

$$N_t = \beta_0 + \beta_1 t + \varepsilon_t \qquad (4.5)$$

where β_0 and β_1 are the ordinary least squares estimates of the intercept and slope parameters, respectively, and $\varepsilon_t \sim \mathcal{N}(0,\sigma^2)$, meaning that the residual error at time t is drawn from a normal distribution with a mean of zero and variance σ^2. A key assumption of this model is that the errors are independent, and for time series data we want ε to be white noise. We can express the distribution of the vector of errors from the model in matrix form as $\varepsilon \sim \mathcal{N}(0,\mathcal{I}\sigma^2)$ where $\mathcal{I}$ is a $n \times n$ identity matrix. The ordinary least squares approach to fitting linear models can be extended by incorporating a more general covariance matrix than $\mathcal{I}\sigma^2$ using generalized least squares (Box et al. 1994). For time series data where we can not necessarily assume that the errors are independent of one another, we then have the flexibility to construct this covariance matrix (Σ) to account for the fact that errors clustered close together in time are likely to be correlated.

To see how this works, let us assume that the errors of this model are stationary. In other words, they are derived from a process with a mean of zero and variance σ^2 (i.e. neither mean nor variance change over time), and that their covariance depends only on the lag k between them. The latter requirement ensures that the covariance of two errors is simply equal to $\sigma^2\rho_k$, which is the product of the error variance and the correlation between errors separated by lag k. Therefore, the error covariance matrix Σ would be:

$$\Sigma = \sigma^2 \begin{bmatrix} 1 & p_1 & p_2 & p_3 & \cdot & p_{n-1} \\ p_1 & 1 & p_1 & p_2 & \cdot & p_{n-2} \\ \cdot & \cdot & \cdot & \cdot & \cdot & \cdot \\ p_{n-1} & p_{n-2} & p_{n-3} & p_{n-4} & \cdot & 1 \end{bmatrix} \qquad (4.6)$$

where the diagonal entries correspond to the constant error variances and the off-diagonal entries reflect the autocorrelated errors. We then need to use the data to estimate both the error variance σ^2 as well as all the ρ_k values. To achieve this, we assume some relationship between the ρ_k values that reduces the number of parameters requiring estimation. If the time series data are collected at regular intervals, we might assume an ARMA structure that is now familiar to us from Section 4.2.2.

Consider, for example, a first-order autoregressive model for the errors, equivalent to an ARMA(1,0) process, such that the error at time t is a function of the error from the previous time step:

$$\varepsilon_t = \phi\varepsilon_{t-1} + \nu_t \tag{4.7}$$

where the ν_t are Gaussian white noise with mean zero and variance σ_ν^2. It follows that $\rho_k = \phi^k$ and some mathematical reasoning can prove that $\sigma^2 = \frac{\sigma_\nu^2}{1-\phi^2}$. This has simplified our problem dramatically because only two parameters need to be estimated to characterize the autocorrelated error component of our regression model. These techniques are extendable to more complex ARMA formulations and other autocorrelation structures. We believe that many ecologists will be attracted to this approach to time series analysis because it is extendable to generalized linear or nonlinear models (Box 4.3). Whereas traditional ARIMA methods assume that the response variable is normally distributed, models with correlated error structures can be fitted using link functions to non-standard data types, including count or zero-inflated data that commonly arise when populations are monitored.

4.4 Phenomenological Models of Population Dynamics

Choosing the type of model to adopt in analyzing ecological time series depends upon the questions you wish to answer. The statistical models described in Section 4.3 use past trends, rather than a particular underlying mechanistic process, to predict future behavior. Statistical time series models are often better at handling seasonality. However, if the question is to understand what factors are most important to a species' abundance, and explore population responses to perturbations, then a model representing the biological phenomenon is required. We now introduce a particular type of deterministic model which is common to biological modeling, called ODE. The parameters of these models represent biological processes and are used when the emphasis is on explaining the average expected behaviour of a species' dynamics. Here, population dynamics are modeled as evolving continuously in time, and also as a continuous state variable, the latter meaning that fractions of animals are assumed to exist. We then provide examples of fitting discrete-time models using difference equations, which

Box 4.3 Models with Correlated Error Structures Using a Simulated Example

A common application of regression models with correlated errors to modeling ecological time series is to relate temporal patterns in the population dynamics of a species with environmental variables that may explain the patterns. Suppose that we are concerned with the viability of a mammal population that has declined over the last 50 years. Over the same period, air temperatures have risen due to climate change and it is suspected that warmer temperatures lead to a reduction in juvenile survival rates. We want to estimate the relationship between average annual air temperature and the population abundance of the species. Since the maximum lifespan of the mammal is around 10 years, the population size is likely to be correlated from one year to the next. To illustrate such a scenario, we simulated a population time series arising from a first-order autoregressive process as follows:

$$x_t = \log(N_t) = \beta_0 + \beta_1 T_t + \phi\varepsilon_{t-1} + \nu_t \tag{4.8}$$

where T_t is the logarithm of the annual temperature, ϕ is the correlation between successive errors, ε_{t-1} is the error from the previous year, and ν_t is Gaussian white noise. Since ε_{t-1} is simply the difference between the true value of x_{t-1} and its expected value $\beta_0 + \beta_1 T_{t-1}$, we can rewrite this equation as:

$$x_t = \beta_0 + \beta_1 T_t + \phi[x_{t-1} - (\beta_0 + \beta_1 T_{t-1})] + \nu_t \tag{4.9}$$

which clearly shows that this is an AR(1) model formulation because the population size at time t is a function of that at time $t-1$. The simulated covariate series is shown in Figure B4.3.1b, together with one realization of the population time series derived according to the autoregressive process described above (Figure B4.3.1a).

To illustrate an analysis that assumes a correlated error structure, we analyzed one realization of the population time series using generalized least squares. We assumed a first-order autoregressive error structure and used either ML or REML for the model fitting. Parameter estimates and 95% CIs from both approaches are presented in Table B4.3.1, as are the true values used to produce the simulated data.

Both fitting methods produce accurate estimates of the intercept β_0 and slope β_1 parameters, correctly identifying the negative relationship between temperature and population abundance. However, the ML estimate of ϕ is low compared to the true value (0.71 cf. 0.90) which suggests that this estimate is biased, while the REML estimate is slightly better (0.75) but still biased low. In fact, the bias of both ML and REML estimates of ϕ is a known issue (Ives et al. 2010).

Although, in this simulated example, we focused on identifying the relationship between population size and an environmental covariate, it is worth stressing that analyses of this kind are correlative and cannot demonstrate causality. Spurious correlations between population size and extrinsic variables can result because we only have access to data for one realization of a noisy population process, or simply because both the population and covariate time series are trending. One approach that can be used to evaluate whether relationships that are identified are an artifact of trends in both series is to detrend (or take first differences) prior to analysis, and this method should be employed routinely. Another approach is to fit phenomenological models of population growth that incorporate density dependence and environmental covariates to time series of ecological populations (Knape and de Valpine 2011). However, care must be taken to ensure that the correct form of density dependence is identified, and that errors in observed population counts are properly accounted for, as these factors can lead to biased estimates of relationships with environmental effects (Lindén and Knape 2009; Lindén et al. 2013). Identifying the best ways of inferring causality between environmental and population time series is an active field of research.

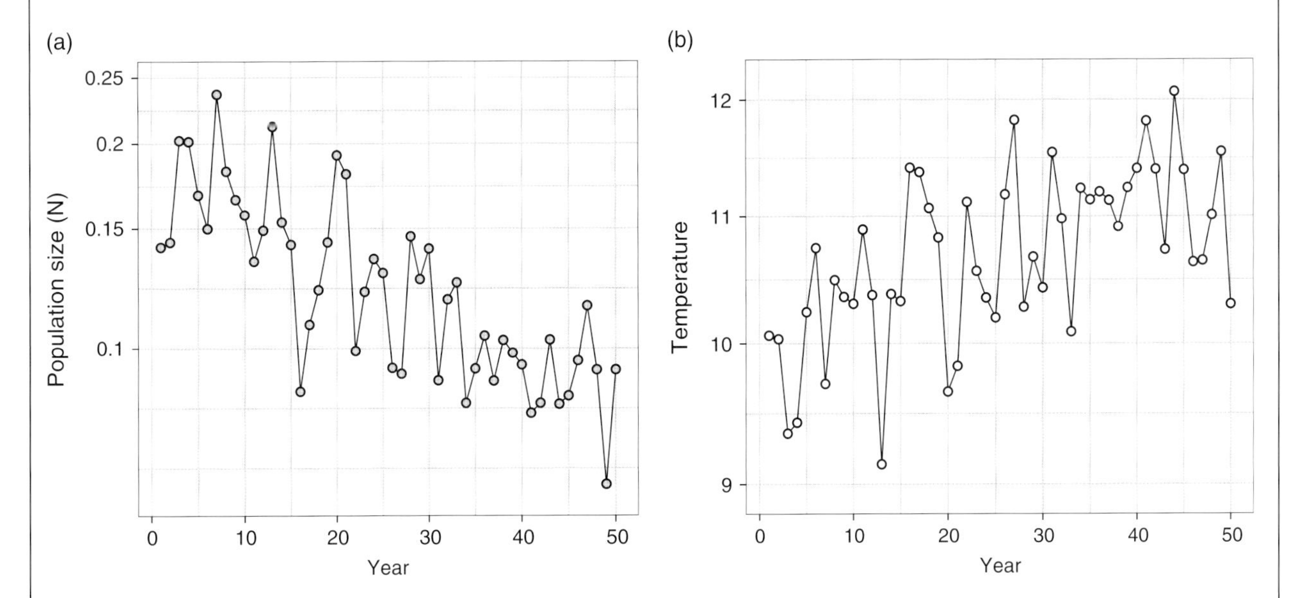

Figure B4.3.1 Plot showing simulated time series of (a) population size of a mammal species, and (b) temperature over 50 annual time steps.

Table B4.3.1 Generating true values used to simulate the time series, and estimated parameter values of the fitted linear model with autocorrelated errors.

Parameter	True value	ML Estimate	95% CI	REML Estimate	95% CI
β_0	5.0	5.11	(3.88,6.32)	5.09	(3.87,6.31)
β_1	−3.0	−3.02	(−3.53, −2.50)	−3.01	(−3.52, −2.50)
ϕ	0.9	0.71	(0.46,0.86)	0.75	(0.46,0.89)
σ	0.1	0.11	(0.08,0.16)	0.12	(0.08,0.19)

ML, maximum likelihood; REML, restricted maximum likelihood; CI, confidence interval.

can be viewed as approximations of the continuous-time population dynamics, including stochastic features representing extrinsic environmental forces that are unrelated to population processes.

4.4.1 Deterministic Models

4.4.1.1 Exponential Growth

The most famous deterministic model is perhaps the model giving rise to exponential growth. Let N be the population size at time t, and assume that each individual is growing at per-unit-time constant rate r. Hence, we have the approximate dynamics:

$$\frac{dN}{dt} = rN. \tag{4.10}$$

This differential equation states that the rate of change in population size with respect to time (left-hand side) is equal to r times the current population size (right-hand size). Importantly, r measures the *instantaneous rate of increase*, which is the per capita rate of population increase over a short time interval, which reflects the difference between the instantaneous birth (b) and death (d) rates. Given the importance of terminology here, it is worth mentioning that r is sometimes called the *realized intrinsic rate of increase*, or alternatively, the Malthusian parameter. Consequently, changes in population size depend on r, such that the population remains constant where $r = 0$, increases exponentially where $r > 0$ (i.e. instantaneous birth rate exceeds death rate), and decreases exponentially for $r < 0$. The model assumes constant birth and death rates and that growth is continuous in time, as well as population closure with no immigration or emigration, and that there is no age or stage structure in demographic performance.

The differential equation above specifies the population growth rate, so to calculate population size at any given time, we need to integrate; the solution of this equation can be given explicitly as:

$$N_t = N_0 e^{rt} \tag{4.11}$$

where N_0 is the initial population size (at time $t = 0$) and e is Euler's constant ($e \approx 2.718...$).

4.4.1.2 Classic ODE Single-Species Population Models that Incorporate Density Dependence

When resources are limited, so that birth and death rates depend on the population size in some way, the population growth rate will be constrained by population size. Perhaps the most ubiquitous single species ODE model in population modeling is the *logistic model*:

$$\frac{dN}{dt} = r_m N \left[1 - \frac{N}{K}\right], \tag{4.12}$$

where N is the population size at time t, r_m is the *maximum intrinsic rate of increase* (in contrast to the realized rate of increase $r = r_m\left[1 - \frac{N}{K}\right]$) and K is *the carrying capacity* (Box 4.4). In the limit as t tends to infinity, we have that the population size approaches 0 if $r < 0$ and approaches K if $r > 0$. Note that if $r = 0$ the population will not change from its initial population size N_0. Once again, to calculate population size at any given time, we need to integrate to provide the solution:

$$N_t = \frac{K}{1 + \left(\frac{(K - N_0)}{N_0}\right) e^{-r_m t}}. \tag{4.13}$$

An equivalent way to formulate this model is, for example, to take the birth rate as constant and introduce a density-dependent mortality rate. Analogous to the exponential growth case where population rate of change is equal to $rN = (b - d)N$, we can then express the rate of change of the population size as a birth rate minus a death rate, assuming population closure:

$$\frac{dN}{dt} = bN - dN^2, \tag{4.14}$$

where b is the per-capita *birth rate parameter* and d is a *friction coefficient* (death rate parameter; Verhulst 1838; Gabriel et al. 2005; Ross 2010). These steps result in the carrying capacity as $K = b/d$ and the growth rate $r_m = b$ in terms of the previous formulation (Eq. 4.12). A motivation for considering this alternate formulation (Eq. 4.14) is to enhance the exposition of a popular generalization of the logistic model, namely the theta-logistic model (Gilpin and Ayala 1973; Stacey and Taper 1992; Sæther et al. 2000, 2002; 2008; Gerber et al. 2004; Chamaillé-Jammes et al. 2008; Ross 2010).

The *theta-logistic model* may be parameterized as:

$$\frac{dN}{dt} = bN - dN^{\gamma}, \tag{4.15}$$

being a clear generalization of the logistic model (Eq. 4.14), where b is the per-capita birth rate and d is a friction coefficient provided $\gamma > 1$; if $0 < \gamma < 1$ then the model can still model density dependence provided b and d are chosen to be negative; in such a case, $|b|$ is the per-capita death rate and $|d|$ is a birth rate parameter (where $|a| = a$ if $a \geq 0$ and $|a| = -a$ if $a < 0$; Ross 2010). The theta-logistic model is more commonly presented in the form:

$$\frac{dN}{dt} = r_m N \left[1 - \left(\frac{N}{K}\right)^{\theta}\right], \tag{4.16}$$

where the parameter θ gives rise to the name of the model and describes the curvature of the relationship between population growth rate and population abundance. The correspondence between parameters in the two different representations (Eqs. 4.15 vs. 4.16) is such that $r_m = b$, $\theta = \gamma - 1$ and $K = (b/d)^{1/\theta}$. When using the second parameterisation (Eq. 4.16), it is recommended to restrict $\theta > -1$ (corresponding to $\gamma > 0$ in Eq. 4.15, see Ross 2010). We note that when $\theta > 0$ (or equivalently $\gamma > 1$) the parameter r_m is the per-capita birth rate, whilst if $\theta < 0$ (or equivalently $\gamma < 1$) the parameter r_m is the per-capita growth rate of the population as the population size tends to infinity (Ross 2010). The θ parameter describes the curvature of

Box 4.4 Density-Dependent Population Dynamics of Sheep

To illustrate the examination of density-dependent population growth, we will use a dataset on the population size of sheep that were introduced on the Australian island state of Tasmania in 1818 (Davidson 1938). The population increased steadily following the introduction and reached a plateau from around 1860 onwards, though there were substantial fluctuations in the population between 1850 and 1936 (Figure B4.4.1).

To calibrate ODE models of population growth, one approach is to minimize the sum of squared errors between the model predicted population sizes $N(t_i)$ and the data N_{t_i}, where the model predicted population sizes are evaluated by integrating forward the ODE(s). Suppose that there is a parameter (or vector of parameters) v (for example, we might have $v = (r_m, K)$ for the logistic model (1.9)). Then, we evaluate the (point) estimates $\hat{v}$ as:

$$\hat{v} = \underset{v}{\operatorname{argmin}} \sum_{i=1}^{T} (N(t_i) - N_{t_i})^2$$

using an optimization algorithm (given some set of initial parameter estimates). Last, we solve the ODE(s) given a population starting value N_0 to get the simulated population trajectory. The estimated trajectory for the sheep population based on the intrinsic growth rate (r) and the carrying capacity (K) for the logistic ($\hat{r}_m = 0.18$; $\hat{K} = 1712$) and Gompertz ($\hat{r} = 0.69$; $\hat{K} = 1732$) models are shown in Figure B4.4.2.

We can also estimate the parameters for the theta-logistic model fitted to the sheep data ($\hat{r}_m = 0.17$; $\hat{K} = 1714; \hat{\theta} = 1.16$). The population trajectory based on these parameters was similar (not shown) to that for the logistic model (as would be expected given that the θ parameter estimate was close to 1). The method we used for estimating the parameters of the theta-logistic model is described in Byrd et al. (1995), which allows for constraints on the parameters. Such constraints are necessary because the function fails to converge if all parameters are allowed to vary freely due to the correlation between the r_m and θ parameters, as described in Polansky et al. (2009).

Another commonly used approach to modeling population growth is to use discrete-time models. Using the Ricker model as a discrete-time approximation of the logistic model, it is straight-forward to regress $\log(N_{t+1}/N_t)$ on $r_m[1 - (N_t/K)]$, using nonlinear regression in some form (Figure B4.4.2). The model implicitly assumes a constant growth rate (the per-capita growth rate) over the interval $[t_i, t_{i+1}]$. If the time interval is very small then this approximation can be justified; in general, the accuracy of the approximation depends on the actual size of the per-capita growth rate and the width of the time interval between observations. However, for most populations and datasets (where population estimates are, say, yearly) this approximation may not be suitable and therefore should be checked on a case-by-case basis.

Estimated model parameters for the Ricker ($\hat{r}_m = 0.18$, 95% CI = [0.14, 0.23]; $\hat{K} = 1722$, [1570, 1888]) and Gompertz ($\hat{r}_{N=1} = 0.69$, [0.54, 0.88]; $\hat{K} = 1754$, [1434, 2073]) models are comparable to those estimated using continuous-time models (above) and translate to the estimated relationships shown in Figure B4.4.2. The strength of density dependence can be calculated as a derived parameter $1 - c$ from the Gompertz model (Section 4.4.2); here the strength parameter was $1 - \hat{c} = 0.09$, [0.07, 0.11], providing support for density dependent dynamics (as the 95% CI excludes 0; but note that this interval ignores uncertainty in the observed population sizes).

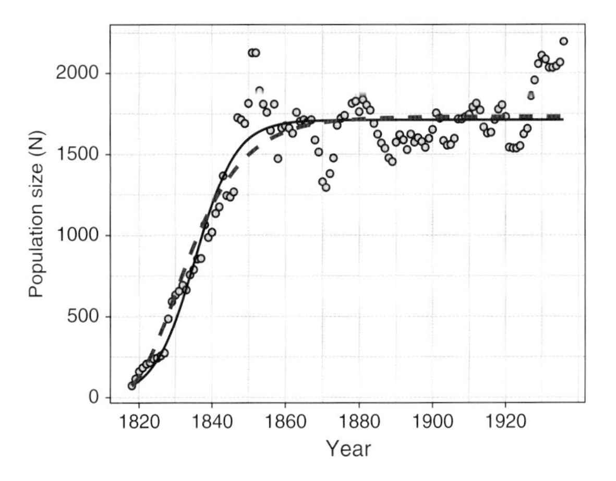

Figure B4.4.1 Time series plot of sheep population size (points) from 1818 to 1936. Population size estimated from the solution of ODEs of continuous-time logistic (solid line) and Gompertz models (dashed line).

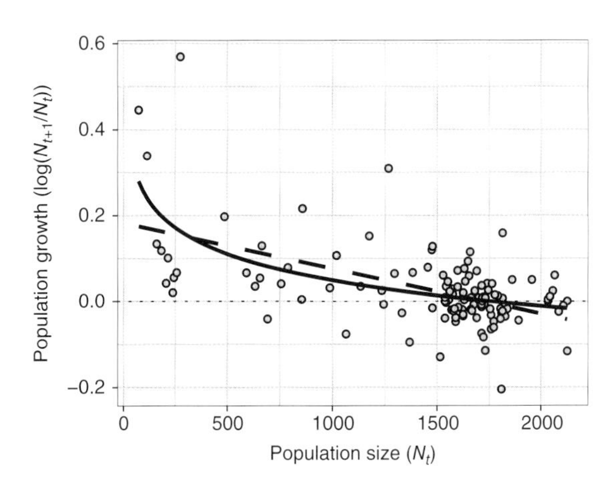

Figure B4.4.2 Scatterplot of sheep population growth rate (differences in log(N) between years) and population size. The fitted relationships are shown for the discrete-time Ricker (dashed line) and Gompertz models (solid line). The intersection of the fitted relationships and the (dotted) zero line indicates the estimated carrying capacity $\hat{K}$, whilst the y-intercept gives the maximum rate of increase $\hat{r}_m$ for the Ricker model (for the Gompertz $\hat{r}$ is the rate of increase when $N = 1$).

the relationship; when $\theta < 1$ there is a concave relationship between population size and population growth rate, and when $\theta > 1$ a convex relationship exists.

Another model that has appeared in the literature but has not been widely adopted is the *basic Savageau model*:

$$\frac{dN}{dt} = bN^{\theta_1} - dN^{\gamma}, \tag{4.17}$$

with per-capita birth rate b and per-capita death rate d (both positive) and $\theta_1 < \gamma$. Furthermore, we advise the requirement that $\theta_1 > 0$ (Ross 2010). This model was called "basic" by Savageau (1979). Once again, this model may be reformulated into the form more akin to the earlier presentations:

$$\frac{dN}{dt} = r_m N^{\theta_1}\left[1 - \left(\frac{N}{K}\right)^{\theta_2}\right], \tag{4.18}$$

where $r_m = b$, $\theta_2 = \gamma - \theta_1$ and $K = (b/d)^{1/\theta_2}$. We believe this double-theta-logistic model has been largely overlooked in the ecological community. Whereas the theta-logistic model imposes linear changes in either the population birth or population death rate, this model allows for nonlinear changes in both of these rates. Hence, it allows for much greater flexibility in the modeling of single species dynamics.

Another commonly used model for the analysis of population dynamics is the *Gompertz model* (Gompertz 1825), which has the deterministic form:

$$\frac{dN}{dt} = \alpha N[\log(K) - \log(N)], \tag{4.19}$$

where α is a measure of the rate of return to equilibrium and K is the carrying capacity. The solution to this equation is given by:

$$N_t = Ke^{e^{-\alpha t}\log\left(\frac{N_0}{K}\right)}, \tag{4.20}$$

where N_0 is the initial population size.

4.4.2 Discrete-Time Population Growth Models with Stochasticity

The choice between a discrete-time versus a continuous-time model should be made based upon the underlying demography of the population being modeled, as opposed to the data that are available. If the species evolves continuously in time, such that breeding and deaths take place at essentially any time of year, then a continuous-time model should be used. The fact that data are only collected at discrete time points is irrelevant. On the other hand, for species with life histories that result in discrete generations, such that births are pulses that only take place at a relatively small window of the year, or those where there are discrete seasonal pulses in population growth, demographic change may not necessarily be considered as continuous in time. In these situations, discrete difference equations might be used to approximate population dynamics (Box 4.4). Indeed, the dynamics of any population for which population counts are observed at equal time intervals could be approximated by an appropriate discrete difference equation for population growth. However, the accuracy of the approximation will decrease with the length of time step.

The *Gompertz population model* (Royama 1992) is one of several models that are commonly used to explain the dynamics of ecological populations (Reddingius 1971; Dennis and Taper 1994; de Valpine and Hastings 2002; Dennis et al. 2006; Knape 2008). The discrete form of the model that is generally used is derived by specifying a relationship between the rate of return to equilibrium and the generation time (May et al. 1974). The continuous-time model is re-expressed by setting $a = \alpha\log(K)$ and the generation time to one such that the dynamics are dependent only on a single estimated parameter (b, see below). Using the notation employed by Dennis et al. (2006) and Knape and de Valpine (2012a), a discrete time model for population size at time t, N_t, is defined as:

$$N_{t+1} = N_t e^{a + b\log(N_t) + \varepsilon_t}, \tag{4.21}$$

where N_t is the population size (or density) at time t, a measures the maximum rate of increase when population size equals one (because the growth rate when population size equals zero is negative infinity), b is the parameter measuring density dependence, and ε_t is Gaussian-distributed error, usually termed "process error", with mean zero and variance σ^2. We introduce this stochasticity to represent unexplained fluctuations in the per capita population growth rate between sampling periods caused by environmental heterogeneity that is not explained by the density feedbacks (Dennis and Taper 1994). Generally, such data are analyzed on a log scale (appropriately adjusted if there are zeros), so by transforming such that $x_t = \log(N_t)$ the equation can be simplified as follows:

$$\begin{aligned} x_{t+1} &= x_t + a + bx_t + \varepsilon_t, \\ \Rightarrow x_{t+1} &= a + (b+1)x_t + \varepsilon_t, \\ \Rightarrow x_{t+1} &= a + cx_t + \varepsilon_t, \end{aligned} \tag{4.22}$$

where $c = b + 1$ is the strength of autocorrelation at lag 1. This interpretation of an AR(1) process assumes stationarity, which only occurs where $|c| < 1$ (Dennis et al. 2006; Knape and de Valpine 2012a). The return tendency of the population to equilibrium is measured by the strength of density dependence $1 - c$ (Royama 1992).

The (log) abundance time series is therefore density independent when $c = 1$ (i.e. $b = 0$), and this case links back to a density-independent random walk with drift (as estimated by parameter a). The Gompertz model estimates the strength of compensatory dynamics; so $c = 0$ (i.e. $b = 1$) indicates perfectly compensatory responses of population size, whereas $c > 0$ ($|b| < 1$) indicates (negative) density dependence. The Gompertz model exhibits the greatest density dependent effect at small population sizes, but as the population size increases, the effect of density becomes less pronounced. This form of density feedback can approximate a wide range of population dynamics (Sibly et al. 2005).

A common alternative to the Gompertz is the *Ricker population model* (Ricker 1954; Dennis and Taper 1994) for estimating the dynamics of fish stocks and recruitment. The Ricker model differs in using N_t rather than $\log(N_t)$ in the exponential function of the Gompertz model (Eq. 4.21). Consequently, the Ricker model of population abundances on the log scale is not linear in its parameters. This model does imply a linear density feedback relationship between log scale population growth rate ($x_{t+1} - x_t$) and N_t:

$$N_{t+1} = N_t e^{r_m\left(1 - \frac{N_t}{K}\right) + \varepsilon_t}, \tag{4.23}$$

$$\Rightarrow N_{t+1} = N_t e^{r_m(1 - dN_t) + \varepsilon_t}, \text{ where } d = 1/K.$$

Transforming to the log scale gives:

$$x_{t+1} = x_t + r_m(1 - de^{x_t}) + \varepsilon_t, \tag{4.24}$$

where r_m is the maximal growth rate per time step when population size is small and d is the equilibrium point. Both stochastic versions of the Ricker and Gompertz models presented here assume stochasticity arises solely from environmental (process) noise. Knape and de Valpine (2012a) use a first-order Taylor expansion of the above form of the Ricker model to show that $1 - \hat{r}_m$ from the Ricker model and parameter $\hat{c}$ from the autoregressive form of the Gompertz model both measure autocorrelation of first-order linearizations, and therefore are comparable measures of the strength of density dependence from these two models.

The predicted dynamics of the Ricker model are similar to the continuous-time logistic model when the intrinsic population growth rate is relatively small ($r_m < 0.5$). However, the approximation deteriorates as population growth rate increases, and where growth rate approaches or exceeds 2 the model exhibits complex dynamics, including damping oscillations ($r_m = 1 - 2$), cycles ($r_m = 2 - 2.692$), or chaos ($r_m > 2.692$; May 1976). Following from the derivation of the Ricker model as a discrete-time approximation of the logistic model, the discrete form of the theta-logistic density feedback model (sometimes called the theta-Ricker model) takes the form:

$$N_{t+1} = N_t e^{r_m\left[1 - \left(\frac{N}{K}\right)^\theta\right]}. \tag{4.25}$$

Predicting r_m and K from the theta-Ricker model, however, gives biased and imprecise estimates because of the inherent trade-off between r_m and θ and the generally flat likelihood for this model (Polansky et al. 2009; Clark et al. 2010), particularly where population abundance fluctuates around K and thus contains little information about the true value of r_m. For these reasons, a strong emphasis should be placed on convergence diagnostics and parameter precision when deciding to use this model for predicting dynamics.

Moreover, it is now well known that there are significant biases associated with tests for density dependence based on population growth models that do not account for uncertainty in the observed population sizes in a time series (Lebreton 2009), and that the bias increases with the magnitude of the uncertainty (Lebreton and Gimenez 2013), providing misleading results. Lebreton and Gimenez (2013) advocate that approaches ignoring uncertainty in population size should be abandoned and that state-space methods should be preferred instead (Section 4.5).

4.5 State-space Modeling

Population abundance data are likely to contain errors for a variety of reasons, including imperfect detection associated with habitat heterogeneity, sampling method, and observer experience. Often these problems lead to the use of indirect counting methods as the basis for estimating population sizes. The modeling approaches we have presented so far contain only one level of error; however there are two main components of variance in population abundance time series data: observation or measurement errors; and process errors (de Valpine and Hastings 2002; Calder et al. 2003; Clark and Bjornstad 2004; Buckland et al. 2004). *Observation errors* therefore represent variation in the observed population counts associated with the survey methodology that lead to imperfect detection as described above, as well as variation in environmental conditions under which data are collected that affect accuracy of measurement, logistical issues, and funding constraints that change over time, as well as human error (Freckleton et al. 2006). In contrast, *process errors* encompass the natural variability in the true population size associated with the biotic and abiotic drivers of population fluctuations. Failure to account for observation error in population models risks incorrect inferences about the processes influencing dynamics, for example overestimating the strength of density dependence (Shenk et al. 1998; Freckleton et al. 2006; Knape 2008; Lebreton 2009) and inflating estimates of process variance.

State-space models (SSM; including *Hidden Markov models* or HMM) provide an avenue to analyzing population fluctuations and their potential drivers, while simultaneously accounting for both observation and process errors (de Valpine and Hastings 2002; de Valpine 2003; Clark and Bjornstad 2004; Knape and de Valpine 2012b). SSM are hierarchical models that explain the true size or "state" of the population in terms of parameters explaining population growth as an autocorrelated latent process with stochasticity on one level, and observation errors (usually) as log-Normal deviations around the true, but unobserved, population size. Accounting for errors in the observation process can provide more accurate estimates of the underlying population states, as well as the variables that may be driving fluctuations in those states, such as climate variability. State-space approaches are now commonly employed to model the dynamics of animal populations (Zeng et al. 1998; Wang et al. 2006; Wilson et al. 2011; Ahrestani et al. 2013).

State-space models can model linear and nonlinear relationships and can potentially incorporate a variety of statistical distributions. However, most classical applications of state-space methods apply to models that are linear in the parameters of the population growth model and assume that the error components are Gaussian-distributed. These constraints allow the model parameters to be estimated with the *Kalman Filter* (Kalman 1960) based on various algorithms (Koopman et al. 1999; Shumway and Stoffer 2006).

The key ingredients in defining a state-space model are that: (i) the observed population size at each time point N_t is said to be independent of the past observations of population size, conditional on the estimated true population state x_t (termed "conditional independence"); and (ii) the current estimated state x_t is dependent only on the state at the previous time x_{t-1}, termed the *Markov property*.

4.5.1 Gompertz State-space Population Model

The Gompertz population model provides a simple form to illustrate a state-space model that includes density dependence. As described above, the Gompertz model on the log scale can be re-expressed as a lag-1 autoregressive process when stationarity is assumed. If we allow for sampling errors that lead to uncertainty in the observed counts of population abundances, N_t, and assume that these errors follow a log-Normal distribution, then the observation errors can be incorporated into a state-space formulation of the Gompertz population model on the log scale $y_t = \log(N_t)$ as:

$$\begin{aligned} y_t &= x_t + \eta_t \quad \text{(Observation model)}, \\ x_{t+1} &= a + cx_t + \varepsilon_t \quad \text{(Process model)}, \end{aligned} \tag{4.26}$$

where $y_t = \log(N_t)$ and $\eta_t \sim \mathcal{N}(0, \sigma_O^2)$ (Dennis et al. 2006; Knape and de Valpine 2012a). We use the Gompertz state-space model to estimate the underlying dynamics in population size of Canvasbacks (Box 4.5).

Small sample bias leads to overestimates of the strength of density dependence for short time series ($n < 40$) if the underlying dynamics are *undercompensatory* (Knape and de Valpine 2012a). One possible test for density dependence under this framework that avoids the problems of bias outlined for process-error-only models is a likelihood

Box 4.5 State-space Models of Population Dynamics of Canvasbacks

To illustrate the application of SSM to examine density-dependent population growth while accounting for both process and observation errors, we use annual estimates of the population size of Canvasbacks in North America, recorded on their breeding grounds from 1955 to 2015 (U.S. Fish and Wildlife Service 2015). Population size estimates varied substantially over the years, and the annual surveys provided an estimate of sampling variability for each population estimate (Figure B4.5.1a).

The relationship between population growth rate and population size appears to be roughly similar for both the Gompertz and Ricker models over the observed range of Canvasback population sizes (Figure B4.5.1b; although this initial visualization ignores the presence of observation errors in the annual estimates). We then fit the Gompertz state-space model to the time series of Canvasback population sizes (Figure B4.5.2), and aimed to estimate all the model parameters, including both the process (σ_P^2) and observation (σ_O^2) error variances. The estimated strength of density dependence is relatively imprecise ($1-\hat{c}=0.30, [-0.08, 0.69]$), however the likelihood ratio test for density dependence ($H_0: b = 0$) is significant ($\chi^2_{df=1} = 5.1, P = 0.024$). Also, estimates of the model variance components σ_P and σ_O have low precision (Coefficient of Variation $CV_P = 0.40$ and $CV_O = 0.27$, respectively) when the model is fitted to the observed time series without any other information about the magnitude of the observation error variance (Lebreton 2009).

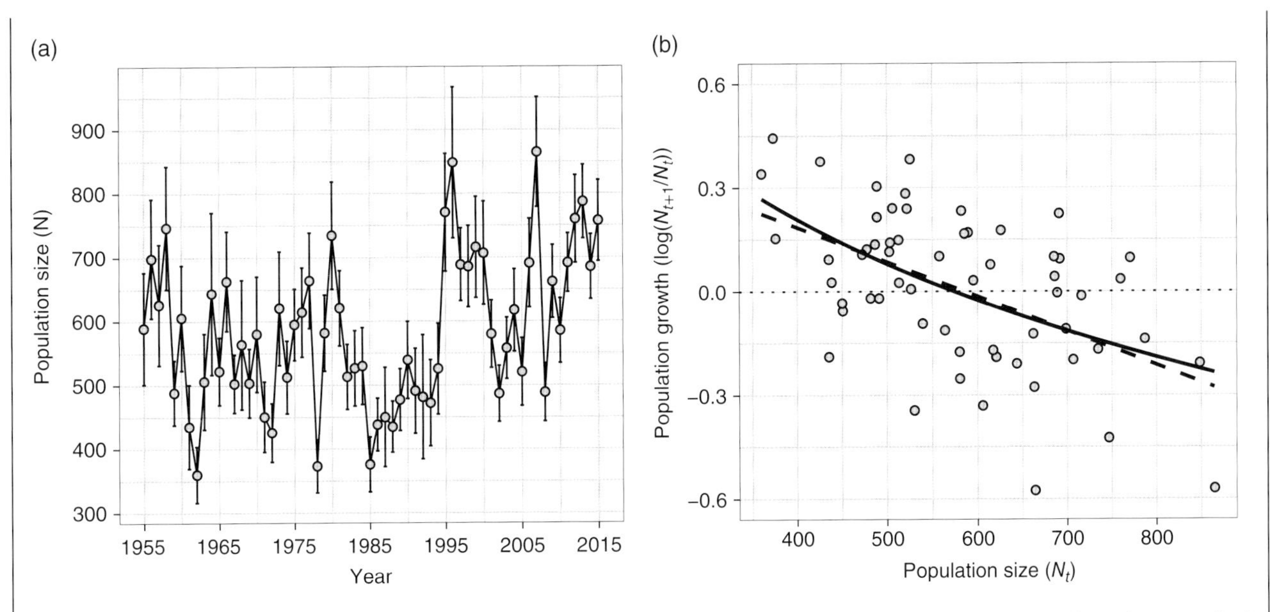

Figure B4.5.1 (a) Time series plot of estimated population size of Canvasbacks (points) from 1955 to 2015; error bars show standard errors of estimates associated with sampling variability. (b) Scatterplot of population growth rate and population size; the fitted relationships are shown for the discrete-time Ricker (dashed line) and Gompertz models (solid line).

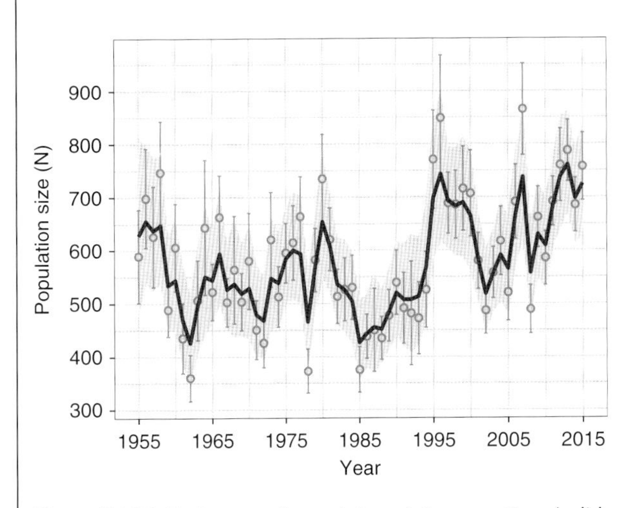

Figure B4.5.2 Trajectory of population states over time (solid line) estimated from a Gompertz state space model. Shaded region shows 95% confidence intervals (CI) for the latent states. The observed population sizes (and standard errors) are shown in background.

However, in this case we have additional data that could improve our estimates. Thus, we proceed by specifying the standard deviation of the Gaussian distribution from which the observation errors are drawn (σ_O) to equal the standard error of population size in each sampling year. Using known estimates of sampling error to specify the amount of observation error in the model does not substantially change the point estimate for σ_P. However, there are two important consequences that result from incorporating the additional information: (i) a more precise estimate of the strength of density dependence ($1-\hat{c}=0.35, [0.06, 0.63]$); and (ii) markedly greater precision in the estimation of process error σ_P (CV_P = 0.19). The likelihood ratio test for density dependence (H_0: b = 0) provides evidence for a feedback response ($\chi^2_{df=1}=8.4, P=0.0038$). However, this test is anti-conservative (Knape and de Valpine 2012a; Lebreton and Gimenez 2013), and comparison with an empirical distribution of the likelihood ratio based on a parametric bootstrap provides lower confidence in the conclusion of density dependence (P = 0.062).

In contrast to the state-space model, the *process-error-only* Gompertz model that assumes the population counts are observed without error gives (upwardly) biased estimates of the strength of density dependence ($1-\hat{c}= 0.58, [0.34, 0.81]$) and the magnitude of process error is greater by ~40%. This example therefore demonstrates the importance of accounting for observation error when modeling population growth and evaluating evidence for density dependence, and the importance of having additional information that informs on the relative magnitudes of process and observation error variances.

To facilitate the comparisons of estimates from both the Gompertz (linear) and Ricker (nonlinear) SSM, parameters were estimated using a method that treats the latent population states as random effects (Fournier et al. 2012; Kristensen et al. 2016; Pedersen et al. 2011). It is important to note that we could have used the Kalman Filter for fitting the linear and Gaussian Gompertz state-space model (Lindley 2003; Knape and de Valpine 2012a), or indeed we could have taken a fully Bayesian approach that would allow comparisons involving nonlinear and/or non-Gaussian models (Buckland et al. 2004; Clark and Bjornstad 2004; Newman et al. 2006).

ratio test comparing, for example, the Gompertz state-space model with a reduced model that assumes density independence (i.e. setting $b = 0$) using a parametric bootstrap approach (Knape and de Valpine 2012a).

A problem that is encountered when fitting SSM, such as those described above, is that the likelihood surfaces are ridge-like, indicating correlation in the parameter space, or have multiple local maxima. For example, high correlations can occur between the population model parameters $\hat{r}$ and $\hat{b}$ in the Gompertz state-space model. Alternatively, the process $\hat{\sigma}_P$ and observation $\hat{\sigma}_O$ error components trade off against one another such that one component shrinks to close to zero and all variance is captured in the other term (Dennis et al. 2006; Knape 2008). Essentially, in the absence of other information these model parameters may be unidentifiable. This phenomenon occurs frequently for short time series (Dennis and Ponciano 2014), and in cases where the population does not depart substantially from equilibrium (Dennis and Taper 1994). Model diagnostics are critical in these cases to assess the validity of parameter estimates and consequent model inferences.

Employing a Bayesian approach to model fitting and incorporating informative priors on the population growth rate parameter from demographic data has the potential to alleviate problems of identifiability (Delean et al. 2013; Lebreton and Gimenez 2013), and where possible should be encouraged. Another avenue is to obtain estimates of sampling error through replicated population sampling, such as repeat counts, to inform the magnitude of observation error (Dennis et al. 2010). The resulting improvement to the estimate of process error can increase the precision of the estimated strength of density dependence (Box 4.5).

4.5.2 Nonlinear and Non-Gaussian State-space Population Models

Linear Gaussian state-space models tend not to be suitable for modeling small populations where demographic stochasticity plays a more substantial role in the population fluctuations, and where random variability in survival and reproduction between individuals inflates process error. Models for these types of data that do not account for errors in the observation process can predict negative estimates of population size. Greater flexibility is obtained when fitting state-space models using Markov chain Monte Carlo methods. Calder et al. (2003) used Gibbs sampling to fit a Ricker state-space model, where the system of equations is similar to that described above for the Gompertz state-space model but also requires specification of prior distributions for all model parameters, including the initial population state X_0. The Bayesian framework also allows nonlinear functional forms of density feedback (Wang 2007) and non-Gaussian probability density functions (Knape et al. 2011).

Other approaches that are available to fit nonlinear models for density-dependent feedbacks and multiple sources of error that follow non-Gaussian distributions include: Bayesian state-space models (de Valpine and Hastings 2002; Clark and Bjornstad 2004), data cloning (Lele et al. 2007), and a class of models for partially observed Markov processes encompassing iterated filtering (Ionides et al. 2015), approximate Bayesian Computation (Toni et al. 2009), particle Markov chain Monte Carlo (Andrieu et al. 2010; Peters et al. 2010), and synthetic likelihood (Wood 2010). This highly flexible class of models are computationally intensive approaches to estimate true population abundances as unobserved latent variables using hierarchical models (King et al. 2016).

There has been a recent focus toward estimation of SSM for population abundance time series using sequential Monte Carlo or particle filter methods (Peters et al. 2010; Knape and de Valpine 2012b; Hosack et al. 2012a). These methods involve a combination of particle filters to explore the latent (or hidden) states such as the true, but unknown, population size, and Metropolis-Hastings Markov chain Monte Carlo to estimate the parameters of the population growth models. Efficient sampling of the parameter space means that these approaches are superior to standard Metropolis-Hastings or Gibbs sampling for high-dimensional problems, which are the case for long time series. In addition, they can be simple to use and the application of an adaptive proposal for the static population model parameters means that only minimal tuning is required (Peters et al. 2010). Hosack et al. (2012b) introduce the use of the normal inverse Gaussian distribution for the observation error variance in particle filter state-space models. This particular distribution is heavy-tailed to accommodate outlying values and captures a variety of distributions, including the Gaussian, log-Normal, Student's t and Gamma, making it a flexible choice when no direct information about the distribution of observation errors is available.

4.6 Software Tools

Software for time series modeling are widely available, including several packages available in `R` (R Core Team 2018) and other programming languages (e.g. `Python`). With specific relevance to the approaches and examples presented herein, Ives et al. (2010) provide descriptions and both `MATLAB` (MathWorks 2018) and `R` code (R Core Team 2018) for estimation of the parameters of ARMA models using both maximum likelihood (ML) and restricted maximum likelihood (REML). The authors show that REML estimates are generally less biased than ML estimates. The likelihood functions presented in Ives et al. (2010) can be used with alternative maximization routines that are suited to optimization problems where there are multiple local maxima, and the authors suggest simulated annealing as one possible example (Kirkpatrick et al. 1983).

As mentioned, most classical applications of state-space methods apply to models that are linear in the parameters of the population growth model and assume that the error components are Gaussian-distributed. The constraints allow the model parameters to be estimated with the Kalman Filter (Kalman 1960) based on various algorithms (Koopman et al. 1999; Shumway and Stoffer 2006), which can be fitted easily using freely available software packages such as the `dlm` and `KFAS` packages in Program `R`. Dennis and Ponciano (2014) provide `R` code to calculate ML and REML estimates of parameters of the Ornstein Uhlenbeck (OU) state space model. Other alternatives for fitting time series state-space models include the packages `pomp` (King et al. 2016) and `TMB` (Kristensen et al. 2016), or using a Bayesian approach through `rjags` (Plummer 2016).

4.7 Online Exercises

The online R exercises for Chapter 4 provide examples that apply the concepts presented in this chapter to real-world time series data sets. In Exercise 1 we step through the code to fit ARMA models to the time series of beaver population abundances described in Box 4.1. In Exercise 2, we outline approaches to fit continuous ordinary differential equations to the time series of sheep abundances presented in Box 4.4, and then go on to show how to maximize the likelihood of discrete-time density dependence models that include a stochastic term to the same time series, which in this example provides a good approximation to the continuous-time model. In Exercise 3, we use a time series of Cackling Goose abundance to introduce the fitting of state-space models for density dependence (Box 4.5) to time series data for a single species. We provide the code to estimate model parameters using a Laplace approximation to the likelihood, and provide a parametric bootstrap approach to evaluate evidence for density dependence. We additionally show how to apply a full Bayesian analysis to the estimation of the model parameters for this data set.

4.8 Future Directions

Our major focus here has been on the dynamics of single populations of a given species, however biological and environmental effects might be expected to vary in space and time. For example, the strength of density dependence can differ among spatially separated populations due to environmental differences (Wang et al. 2006). Importantly, there have been recent developments that allow estimation of spatial variation in population growth rates (Thorson et al. 2015) and the explicit estimation of spatial variation in the parameters for population growth rate and strength of density dependence using a spatially varying coefficients approach (Roy et al. 2016) using state-space models.

Another approach to population growth state-space models incorporates imperfect detection into the likelihood explicitly, and can model variation in detection as a function of spatial and temporal covariates (Hostetler and Chandler 2015). These models, based on hierarchical N-mixture models for repeated counts from metapopulations (Dail and Madsen 2011), can also account for other common processes in ecological count data such as excess zeros and spatial variation in dynamics. Here, population growth is modeled as a Poisson process and the mean can be expressed in terms of a Gompertz or Ricker model for density dependence, and the parameters of these models can be allowed to vary as functions of environmental covariates (Hostetler and Chandler 2015). Alternatively, environmental stochasticity can be modeled as log-Normal variation in population growth, and this process can be constrained to differ spatially (Hostetler and Chandler 2015). Commensurate with the flexibility of these models, the precision with which parameters are estimated depends on the information available on each process within the data. As such, the precision of state parameters depends on accurate estimation of the detection probability, and therefore replicated count data that allow the assumption of population closure are recommended (Hostetler and Chandler 2015).

A common limitation when fitting models incorporating density-dependent dynamics to population time series using discrete-time formulations of the models occurs when the time intervals between sampled observations are not equal. Unequal intervals may be a consequence of environmental conditions that affect the timing of sampling, or simply the logistical constraints of field sampling programs. To address this scenario, Dennis and Ponciano (2014) extend the density-dependent Gompertz state-space model to allow for unequal time intervals. The model is based on a stochastic version of the continuous-time, deterministic Gompertz model, which, when transformed to the log-scale, describes as a continuous-time diffusion process where density-dependent growth is perturbed by environmental noise (Dennis and Ponciano 2014). On the log-scale, the model is an OU process, and this process can be extended to a state-space model by adding a normally distributed observation error component, as described above for the Gompertz state-space model.

Of course, given the complexity and dimensionality of state-space models, great care should be taken when fitting these models to population time series. Consideration should be placed on whether, despite the best intentions of capturing all of the components of variation in the observed data, there is indeed enough "information" in a particular time series to support inferences from such complex models. Problems in the estimation of both model parameters and state variables have been shown to give misleading results, even for simple linear Gaussian state-space models (Auger-Méthéé et al. 2016). There is a high cost associated with the estimation of complex lag structures and observation errors when using short and variable time series, yet the return, in terms of predictive capacity, may be low (Ward et al. 2014). These issues highlight the fact that valid inferences can only follow from a careful evaluation of model results to determine whether accurate model parameters are actually achieveable for a given state-space model.

References

Ahrestani, F.S., Hebblewhite, M., and Post, E. (2013). The importance of observation versus process error in analyses of global ungulate populations. *Scientific Reports* 3: 3125.

Andrieu, C., Doucet, A., and Holenstein, R. (2010). Particle Markov chain Monte Carlo methods. *Journal of the Royal Statistical Society B* 72: 269–342.

Auger-Méthéé, M., Field, C., Albertsen, C.M. et al. (2016). State-space models' dirty little secrets: even simple linear Gaussian models can have estimation problems. *Scientific Reports* 6: 26677.

Bartlett, M.S. (1960). *Stochastic Population Models in Ecology and Epidemiology*. London: Methuen.

Bjornstad, O.N. and Grenfell, B.T. (2001). Noisy clockwork: time series analysis of population fluctuations in animals. *Science* 293: 638–643.

Boggs, C.L. and Inouye, D.W. (2012). A single climate driver has direct and indirect effects on insect population dynamics. *Ecology Letters* 15: 502–508.

Box, G.E.P., Jenkins, G.M., and Reinsel, G.C. (1994). *Time Series Analysis: Forecasting and Control*, 3e. Englewood Cliffs, NJ: Prentice Hall.

Brockwell, P.J. and Davis, R.A.G. (1991). *Time Series and Forecasting Methods*, 2e. G. Springer.

Buckland, S.T., Newman, K.B., Thomas, L., and Koesters, N.B. (2004). State-space models for the dynamics of wild animal populations. *Ecological Modeling* 171: 157–175.

Byrd, R.H., Lu, P., Nocedal, J., and Zhu, C. (1995). A limited memory algorithm for bound constrained optimization. *SIAM Journal on Scientific Computing* 160: 1190–1208.

Calder, C.C., Lavine, M.L., Muller, P., and Clark, J.S. (2003). Incorporating multiple sources of stochasticity into dynamic population models. *Ecology* 84: 1395–1402.

Campbell, M.J. and Walker, A.M. (1977). A survey of statistical work on the Mackenzie river series of annual Canadian lynx trappings for the years 1821–1934 and a new analysis. *Journal of the Royal Statistical Society A* 140: 411–431.

Chamaillé-Jammes, S., Fritz, H., Valeix, M. et al. (2008). Resource variability, aggregation and direct density dependence in an open context: the local regulation of an African elephant population. *Journal of Animal Ecology* 77: 135–144.

Clark, F., Brook, B.W., Delean, S. et al. (2010). The theta-logistic is unreliable for modelling most census data. *Methods in Ecology and Evolution* 1: 253–262.

Clark, J.S. and Bjornstad, O.N. (2004). Population time series: process variability, observation errors, missing values, lags, and hidden states. *Ecology* 85: 3140–3150.

Cowpertwait, P.S.P. and Metcalfe, A.V. (2009). *Introductory Time Series with R*. Use R! Springer. ISBN: 9780387886985.

Dail, D. and Madsen, L. (2011). Models for estimating abundance from repeated counts of an open metapopulation. *Biometrics* 67: 577–587.

Davidson, J. (1938). On the growth of the sheep population in Tasmania. *Transactions of the Royal Society of South Australia* 32: 342–346.

de Valpine, P. (2003). Better inferences from population-dynamics experiments using Monte Carlo state-space likelihood methods. *Ecology* 84: 3064–3077.

de Valpine, P. and Hastings, A. (2002). Fitting population models incorporating process noise and observation error. *Ecological Monographs* 72: 57–76.

Delean, S., Brook, B.W., and Bradshaw, C.J.A. (2013). Ecologically realistic estimates of maximum population growth using informed Bayesian priors. *Methods in Ecology and Evolution* 4: 34–44.

Dennis, B. and Ponciano, J.M. (2014). Density-dependent state-space model for population abundance data with unequal time intervals. *Ecology* 95: 2069–2076.

Dennis, B., Ponciano, J.M., Lele, S.R. et al. (2006). Estimating density dependence, process noise, and observation error. *Ecological Monographs* 76: 323–341.

Dennis, B., Ponciano, J.M., and Taper, M.L. (2010). Replicated sampling increases efficiency in monitoring biological populations. *Ecology* 91: 610–620.

Dennis, B. and Taper, B. (1994). Density dependence in time series observations of natural populations: estimation and testing. *Ecological Monographs* 64: 205–224.

Dickey, D.A. and Fuller, W.A. (1979). Distribution of the estimators for autoregressive time series with a unit root. *Journal of the American Statistical Association* 74: 427–431.

Diggle, P. (1990). *Time Series: A Biostatistical Introduction*. Oxford science publications. Clarendon Press. ISBN: 9780198522263.

Ellner, S. and Turchin, P. (1995). Chaos in a noisy world: new methods and evidence from time series analysis. *American Naturalist* 145: 343–375.

Elton, C. and Nicholson, M. (1942). The ten-year cycle in numbers of the lynx in Canada. *Journal of Animal Ecology* 11: 215–244.

Fournier, D.A., Skaug, H.J., Ancheta, J. et al. (2012). AD Model Builder: using automatic differentiation for statistical inference of highly parameterized complex nonlinear models. *Optimization Methods and Software* 27: 233–249.

Freckleton, R.P., Watkinson, A.R., Green, R.E., and Sutherland, W.J. (2006). Census error and the detection of density dependence. *Journal of Animal Ecology* 75: 837–851.

Gabriel, J.-P., Saucy, F., and Bersier, L.-F. (2005). Paradoxes in the logistic equation? *Ecological Modelling* 185: 147–151.

Gerber, L.R., Buenau, K.E., and Vanblaricom, G. (2004). Density dependence and risk of extinction in a small population of sea otters. *Biodiversity and Conservation* 13: 2741–2757.

Gerrodette, T. (1987). A power analysis for detecting trends. *Ecology* 68: 1364–1372.

Gilpin, M.E. and Ayala, F.J. (1973). Global models of growth and competition. *Proceedings of the National Academy of Sciences United States of America* 70: 3590–3593.

Gompertz, B. (1825). On the nature of the function expressive of the law of human mortality and on a new model of determining life contingencies. *Philosophical Transactions of the Royal Society of London* 115: 513–585.

Hosack, G.R., Peters, G.W., and Hayes, K.R. (2012a). Estimating density dependence and latent population trajectories with unknown observation error. *Methods in Ecology and Evolution* 3: 1028–1038.

Hosack, G.R., Peters, G.W., and Hayes, K.R. (2012b). Estimating density dependence and latent population trajectories with unknown observation error. *Methods in Ecology and Evolution* 3: 1028–1038.

Hostetler, J.A. and Chandler, R.B. (2015). Improved state-space models for inference about spatial and temporal variation in abundance from count data. *Ecology* 96: 1713–1723.

Hurvich, C.M. and Tsai, C.-L. (1989). Regression and time series model selection in small samples. *Biometrika* 76: 297–307.

Hyndman, R.J. and Athanasopoulos, G. (2014). *Forecasting: Principles and Practice*. OTexts.com http://otexts.com/fpp2.

Ionides, E.L., Nguyen, D., Atchadé, Y. et al. (2015). Inference for dynamic and latent variable models via iterated, perturbed Bayes maps. *Proceedings of the National Academy of Sciences United States of America* 112: 719–724.

Ives, A.R., Abbott, K.C., and Ziebarth, N.L. (2010). Analysis of ecological time series with ARMA(p,q) models. *Ecology* 91: 858–871.

Jacobson, L.D. and Maccall, A.D. (1995). Stock-recruitment models for Pacific sardine (*Sardinops sagax*). *Canadian Journal of Fisheries and Aquatic Sciences* 52: 566–577.

Kalman, R. (1960). A New Approach to Linear Filtering and Prediction Problems. *ASME Journal of Basic Engineering* 82: 35–45.

Kendall, B.E., Briggs, C.J., Murdoch, W.W. et al. (1999). Why do populations cycle? A synthesis of statistical and mechanistic modeling approaches. *Ecology* 80: 1789–1805.

Kendall, B.E., Prendergast, J., and Bjørnstad, O.N. (1998). The macroecology of population dynamics: taxonomic and biogeographic patterns in population cycles. *Ecology Letters* 1: 160–164.

King, A.A., Nguyen, D., and Ionides, E.L. (2016). Statistical inference for partially observed Markov processes via the R Package pomp. *Journal of Statistical Software* 69: 1–43.

Kirkpatrick, S., Gelatt, C.D., and Vecchi, M.P. (1983). Optimization by simulated annealing. *Science* 220: 671–680.

Knape, J. (2008). Estimability of density dependence in models of time series data. *Ecology* 89: 2994–3000.

Knape, J. and de Valpine, P. (2011). Effects of weather and climate on the dynamics of animal population time series. *Proceedings of the Royal Society of London B* 278: 985–992.

Knape, J. and de Valpine, P. (2012a). Are patterns of density dependence in the Global Population Dynamics Database driven by uncertainty about population abundance? *Ecology Letters* 15: 17–23.

Knape, J. and de Valpine, P. (2012b). Fitting complex population models by combining particle filters with Markov chain Monte Carlo. *Ecology* 93: 256–263.

Knape, J., Jonzén, N., and Sköld, M. (2011). On observation distributions for state space models of population survey data. *Journal of Animal Ecology* 80: 1269–1277.

Koopman, S.J., Shephard, N., and Doornik, J.A. (1999). Statistical algorithms for models in state space using SsfPack 2.2. *The Econometrics Journal* 2: 107–160.

Kristensen, K., Nielsen, A., Berg, C.W. et al. (2016). TMB: automatic differentiation and Laplace approximation. *Journal of Statistical Software* 70: 1–21.

Lebreton, J.-D. (2009). Assessing density dependence: where are we left? Modeling demographic processes in marked populations. In: *Modeling demographic processes in marked populations* (eds. D.L. Thomson, E.G. Cooch and M.J. Conroy), 19–42. New York, USA: Springer.

Lebreton, J.-D. and Gimenez, O. (2013). Detecting and estimating density dependence in wildlife populations. *Journal of Wildlife Management* 77: 12–23.
Lele, S.R., Dennis, B., and Lutscher, F. (2007). Data cloning: Easy maximum likelihood estimation for complex ecological models using Bayesian Markov chain Monte Carlo methods. *Ecology Letters* 10: 551–563.
Letcher, B.H., Schueller, P., Bassar, R.D. et al. (2015). Robust estimates of environmental effects on population vital rates: an integrated capture-recapture model of seasonal brook trout growth, survival and movement in a stream network. *Journal of Animal Ecology* 84: 337–352.
Lindén, A., Fowler, M.S., and Jonzén, N. (2013). Mischaracterising density dependence biases estimated effects of coloured covariates on population dynamics. *Population Ecology* 55: 183–192.
Lindén, A. and Knape, J. (2009). Estimating environmental effects on population dynamics: consequences of observation error. *Oikos* 118: 675–680.
Lindley, S.T. (2003). Estimation of population growth and extinction parameters from noisy data. *Ecological Applications* 13: 806–813.
Louca, S. and Doebeli, M. (2015). Detecting cyclicity in ecological time series. *Ecology* 96: 1724–1732.
Lundberg, P., Ranta, E., Ripa, J., and Kaitala, V. (2000). Population variability in space and time. *Trends in Ecology and Evolution* 15: 460–464.
MATLAB (2018) Version 9.0.5.0. The MathWorks Inc. Natick, Massachusetts.
May, R.M. (1973). *Stability and Complexity in Model Ecosystems*. Princeton, NJ: Princeton University Press.
May, R.M. (1976). Simple mathematical models with very complicated dynamics. *Nature* 261: 459–466.
May, R.M., Conway, G.R., Hassell, M.P., and Southwood, T.R.E. (1974). Time delays, density dependence and single-species oscillations. *Journal of Animal Ecology* 43: 747–770.
Moran, P.A.P. (1953). The statistical analysis of the Canadian lynx cycle. *Australian Journal of Zoology* 10: 291–298.
NERC Centre for Population Biology, Imperial College (2010) *The Global Population Dynamics Database*, Version 2. http://www.sw.ic.ac.uk/cpb/cpb/gpdd.html.
Newman, K., Buckland, S.T., Lindley, S. et al. (2006). Hidden Process Models For Animal Population Dynamics. *Ecological Applications* 16: 74–86.
Newman, K., Buckland, S.T., Morgan, B. et al. (2014). *Modelling Population Dynamics: Model Formulation, Fitting and Assessment Using State-Space Methods*. New York: Springer.
Pedersen, M.W., Berg, C.W., Thygesen, U.H. et al. (2011). Estimation methods for nonlinear state-space models in ecology. *Ecological Modeling* 222: 1394–1400.
Pelletier, F., Moyes, K., Clutton-Brock, T.H., and Coulson, T. (2012). Decomposing variation in population growth into contributions from environment and phenotypes in an age-structured population. *Proceedings of the Royal Society of London B* 279: 394–401.
Peters, GW, Hosack, GR, Hayes, KR (2010) Ecological non-linear state space model selection via adaptive particle Markov chain Monte Carlo (AdPMCMC). Preprint *arXiv:1005.2238v1*. Cornell University Library, Ithaca, New York, USA.
Plummer, M (2016) `rjags`: Bayesian Graphical Models using MCMC. R package version 4–6. https://CRAN.R-project.org/package=rjags
Polansky, L., de Valpine, P., Lloyd-Smith, J.O., and Getz, W. M. (2009). Likelihood ridges and multimodality in population growth rate models. *Ecology* 90: 2313–2320.
Post, E., Brodie, J., Hebblewhite, M. et al. (2009). Global population dynamics and hot spots of response to climate change. *Bioscience* 59: 489–497.
R Core Team (2016). *R: A Language and Environment for Statistical Computing*. Vienna, Austria: R Foundation for Statistical Computing URL https://www.R-project.org.
Reddingius, J. (1971). Gambling for existence: a discussion of some theoretical problems in animal population ecology. *Acta Biotheoretica* 20 (Supplement): 1–208.
Renshaw, E. (1991). *Modelling Biological Populations in Space and Time*. Cambridge, UK: Cambridge University Press.
Reuman, D.C., Costantino, R.F., Desharnais, R.A., and Cohen, J.E. (2008). Colour of environmental noise affects the nonlinear dynamics of cycling, stage-structured populations. *Ecology Letters* 11: 820–830.
Ricker, W.E. (1954). Stock and recruitment. *Journal of the Fisheries Research Board of Canada* 11: 559–623.
Ross, J.V. (2010). A note on density dependence in population models. *Ecological Modelling* 220: 3472–3474.
Roy, C., McIntire, E.J.B., and Cumming, S.G. (2016). Assessing the spatial variability of density dependence in waterfowl populations. *Ecography* 39: 942–953.
Royama, T. (1992). *Analytical population dynamics*. London: Chapman & Hall.
Sæther, B.-E., Engen, S., Grøtan, V. et al. (2008). Forms for density regulation and (quasi-) stationary distributions of population sizes in birds. *Oikos* 117: 1197–1208.
Sæther, B.-E., Engen, S., Lande, R. et al. (2000). Estimating the time to extinction in an island population of song sparrows. *Proceedings of the Royal Society of London B* 267: 621–626.
Sæther, B.-E., Engen, S., and Matthysen, E. (2002). Demographic characteristics and population dynamical patterns of solitary birds. *Science* 295: 2070–2073.
Savageau, M.A. (1979). Growth of complex systems can be related to the properties of their underlying determinants. *Proceedings of the National Academy of Sciences United States of America* 76: 5413–5417.
Schwarz, G.E. (1978). Estimating the dimension of a model. *Annals of Statistics* 6: 461–464.

Shaffer, M.L. (1981). Minimum population sizes for species conservation. *Bioscience* 31: 131–134.

Shenk, T.M., White, G.C., and Burnham, K.P. (1998). Sampling-variance effects on detecting density dependence from temporal trends in natural populations. *Ecological Monographs* 68: 445–463.

Shumway, R.H. and Stoffer, D.S. (2006). *Time Series Analysis and Its Applications with R Examples*. New York: Springer.

Sibly, R.M., Barker, D., Denham, M.C. et al. (2005). On the regulation of populations of mammals, birds, fish, and insects. *Science* 309: 607–610.

Stacey, P.B. and Taper, M. (1992). Environmental variation and the persistence of small populations. *Ecological Applications* 2: 18–29.

Stenseth, N.C. (1999). Population cycles in voles and lemmings: density dependence and phase dependence in a stochastic world. *Oikos* 87: 427–461.

Stenseth, N.C., Falck, W., Bjørnstad, O.N., and Krebs, C. J. (1997). Population regulation in snowshoe hare and Canadian lynx: asymmetric food web configurations between hare and lynx. *Proceedings of the National Academy of Sciences USA* 940: 5147–5152.

Stenseth, N.C., Falck, W., Chan, K.-S. et al. (1998). From patterns to processes: phase and density dependencies in the Canadian lynx cycle. *Proceedings of the National Academy of Sciences USA* 950: 15430–15435.

Thorson, J.T., Skaug, H.J., Kristensen, K. et al. (2015). The importance of spatial models for estimating the strength of density dependence. *Ecology* 96: 1202–1212.

Tong, H. (1990). *Non-Linear Time Series: A Dynamical System Approach*. Oxford University Press.

Toni, T., Welch, D., Strelkowa, N. et al. (2009). Approximate Bayesian computation scheme for parameter inference and model selection in dynamical systems. *Journal of the Royal Society Interface* 6: 187–202.

Turchin, P. (1990). Rarity of density dependence or population regulation with lags. *Nature* 344: 660–663.

Turchin, P. (2003). *Complex Population Dynamics*. Princeton, NJ: Princeton University Press.

US Fish and Wildlife Service (2015). *Waterfowl Population Status, 2015*. Washington, DC USA: US Department of the Interior.

Venables, W.N. and Ripley, B.D. (2002). *Modern Applied Statistics with S*. Statistics and Computing. Springer. ISBN: 9780387954578.

Verhulst, P.F. (1838). Notice sur la loi que la population suit dans son accroisement. *Correspondence Mathematique et Physique* 10: 113–121.

Wang, G. (2007). On the latent state estimation of nonlinear population dynamics using Bayesian and non-Bayesian state-space models. *Ecological Modelling* 200: 521–528.

Wang, G.M., Hobbs, N.T., Boone, R.B. et al. (2006). Spatial and temporal variability modify density dependence in populations of large herbivores. *Ecology* 87: 95–102.

Ward, E.J., Holmes, E.E., Thorson, J.T., and Collen, B. (2014). Complexity is costly: a meta-analysis of parametric and non-parametric methods for short-term population forecasting. *Oikos* 123: 652–661.

Wood, S.N. (2010). Statistical inference for noisy nonlinear ecological dynamic systems. *Nature* 466: 1102–1104.

Zeng, Z., Nowierski, R.M., Taper, M.L. et al. (1998). Complex population dynamics in the real world: modeling the influence of time-varying parameters and time lags. *Ecology* 79: 2193–2209.

Ziebarth, N.L., Abbott, K.C., and Ives, A.R. (2010). Weak population regulation in ecological time series. *Ecology Letters* 13: 21–31.

5

Estimating Abundance from Capture-Recapture Data

J. Andrew Royle[1] and Sarah J. Converse[2,3]

[1] U.S. Geological Survey, Patuxent Wildlife Research Center, Laurel, MD, USA
[2] U.S. Geological Survey, Washington Cooperative Fish and Wildlife Research Unit, University of Washington, Seattle, WA, USA
[3] School of Environmental and Forest Sciences (SEFS) and School of Aquatic and Fishery Sciences (SAFS), University of Washington, Seattle, WA, USA

Summary

Capture-recapture models are a foundational tool in population ecology, because we recognize that the individuals we observe at any point in time represent only a sample from some larger population of interest. Conceptually, estimating population abundance requires us to model and estimate detection probability, the proportion of individuals in the population that occur in a sample. In this chapter, we consider the use of capture-recapture models for estimating abundance, and explore how insights into sampling and ecological processes allow us to model detection across time, space, and individuals, thus improving abundance estimates. We review standard capture-recapture models for estimation of abundance, including their assumptions and applications. We include in our review models where populations are assumed demographically and geographically closed during sampling, and models that relax that assumption. We also review recent advances arising from the development of hierarchical capture-recapture models and spatial capture-recapture (SCR) models.

5.1 Introduction

Capture-recapture methods represent perhaps the most widely used statistical methodology in population ecology. The basic principles have existed for decades, and a large number of synthetic monographs or texts have been produced on these techniques (Otis et al. 1978; Seber 1982; White 1982; Williams et al. 2002; Borchers et al. 2002; Royle et al. 2014). The methods are based on encounter data from marked animals and the models enable inference to be made about population parameters, including *demographic rates* such as survival, movement, and population change, *state parameters* such as population size or density, and in the presence of *imperfect detection* of individuals. Conceptually, we understand the effect of imperfect detection in the context of binomial sampling. If we sample a population and observe n individuals, we suppose that n represents a binomial sample of some population of size N:

$$n \sim Binomial(N, \bar{p}) \tag{5.1}$$

where $\bar{p}$ is the probability that an individual appears in the sample. Therefore, heuristically, we can think of constructing an estimator of N by equating the observed sample size of individuals, n, to its expected value $\bar{p}N$, and then solving for N:

$$\hat{N} = \frac{n}{\bar{p}}. \tag{5.2}$$

When we substitute an estimate, $\hat{\bar{p}}$, into this equation, it is a standard population size estimator referred to as the *conditional estimator* (Sanathanan 1972). The estimator illustrates one of the most important concepts in ecological sampling, and thus many authors refer to it as the *canonical estimator* of abundance because it forms the conceptual basis of many models and procedures (Williams et al. 2002, Kéry and Schaub 2012).

What makes capture-recapture models interesting and useful is not a simple heuristic for estimating N but rather that a wide variety of models are available for explaining how individually identifiable animals are observed in the field and how features of both sampling design and ecological processes affect these observations. In particular, the parameter $\bar{p}$ is really a function of a number of more fundamental parameters that determine the probability of detection of individuals during specific *sampling occasions*. Capture-recapture models therefore allow researchers to model specific features of the sampling scheme or biology that vary by individual or over time.

Population Ecology in Practice, First Edition. Edited by Dennis L. Murray and Brett K. Sandercock.

Companion website: www.wiley.com/go/MurrayPopulationEcology

The data obtained from CR studies are often much richer than just the observed sample of size n. In particular, the data can be referenced in time and space, and individual characteristics can be incorporated in order to more realistically model the process of observing animals. In this chapter, we review some of the standard capture-recapture models that are widely used for estimating population size and density, and we discuss recent advances in this field based on the use of hierarchical capture-recapture models.

Capture-recapture models for inference about *population size* have been the cornerstone of statistical population ecology for decades. However, estimating and modeling population size allows for development of an understanding of how populations vary in space and time, and this is of fundamental interest in both population ecology and wildlife management. The theoretical and empirical bases of these models are more relevant now than ever before, due to the advent of new technologies both for obtaining individual encounter history information (e.g. DNA, camera trapping), but also due to new statistical methods that make more efficient use of encounter history data and allow researchers to build more realistic models of population structure and dynamics.

5.2 Genesis of Capture-Recapture Data

Essentially all capture-recapture models make use of *encounter history* data obtained from repeated sampling of individuals in a population. Encounter histories in basic form are a sequence of 0s and 1s indicating whether an individual was captured or not during each of several sampling occasions. We denote by y the binary variable of capture state so that $y = 1$ represents the event *captured* and $y = 0$ is the event *not captured.*

Individual encounter history data are often obtained by classical field methods such as live trapping and tagging small mammals in trapping grids or, in the case of birds, using mist-nets and numbered leg bands. However, more recently developed methods such as identification of individuals with natural marks using camera trapping (O'Connell et al. 2010), and genetic identification from various sources, including scat surveys carried out by detector dogs, are becoming more common (Long et al. 2008). The explosion of new technology for detecting and obtaining individual identity has made capture-recapture methods more useful now than ever before.

Use of such field methods gives rise to two basic classes of capture-recapture models: models for which the population is assumed to be *closed* to additions or subtractions (deaths, births, or migration), and conversely, models for populations that are *open* to demographic change, such as the Jolly–Seber class of models (Jolly 1965; Seber 1965). From a data collection standpoint, the major difference between the two classes is the time elapsed between consecutive sampling occasions. Closed population models apply to situations where the sampling occasions occur in rapid succession, so that absence of mortality, recruitment, or movement may be reasonable to expect. Conversely, open population models apply when sampling occasions are far enough apart that such processes may be operative. *Robust design models* are unique in that they combine features of closed and open models (Pollock 1982; Kendall et al. 1999), however, for abundance estimation, robust design models function essentially as closed models. We focus primarily in this chapter on closed models, as they are the most widely used and diverse set of models for abundance estimation. However, we will visit Jolly–Seber models toward the end of the chapter.

5.3 The Basic Closed Population Models: M_0, M_t, M_b

At the core of the *closed capture-recapture models* is a simple binomial or logistic regression model. The response variable is whether an individual was captured during a particular sample occasion (and perhaps at a particular location; more to come on that). Denote this binary outcome by $y_{ik} = 1$ if individual i was captured during occasion (sample) k and $y_{ik} = 0$ if not, for all $i = 1, 2,\ldots, N$ individuals in the population. The essence of closed population models is that the y_{ik} are Bernoulli trials which we express symbolically as:

$$y_{ik} \sim \text{Bernoulli}(p_{ik}), \qquad (5.3)$$

where $p_{ik} = \Pr(y_{ik} = 1)$ is the capture or encounter probability of individual i during sample occasion k. For now, we assume that the y_{ik} observations are *independent* within individuals and among individuals. That is, we assume that capture of individuals does not affect the capture of others, and their capture at some occasion does not affect their *subsequent capture.* These two assumptions can be relaxed in certain situations, some of which we discuss later. Three additional basic assumptions of closed capture-recapture models that we will assume throughout are that:

1) The population is *closed* geographically to movement and demographically to births and deaths.
2) Marks are not lost, overlooked, or *misread.*
3) Each sample of individuals is a *random sample* of the population of interest.

Williams et al. (2002) provide a thorough discussion of these model assumptions, testing them and the effects of their violation. In general, when important assumption violations are suspected, the strategy should be to extend the model to accommodate those violations. For example, if the population is suspected to be open due to recruitment and survival, then open population models should be considered. Alternatively, if the population is not geographically closed, then spatial capture-recapture (SCR) models should be considered.

Closed population models are concerned with modeling the *probability of encounter* (p_{ik}) and, principally, with estimating abundance (N). We discuss a few common model structures for p_{ik} and the strategies for estimating N. When detection probability is constant for all individuals and sample occasions, i.e. $p_{ik} = p_0$, this is called "model M_0" in the capture-recapture literature (Otis et al. 1978). There are several basic extensions of this model, some of which we discuss shortly, but see also Otis et al. (1978), Williams et al. (2002), and Cooch and White (2008) for more detail. The recent book by Kéry and Schaub (2012) discusses the standard suite of models from a Bayesian perspective analyzed in the `WinBUGS` software.

For cases where detection or encounter probability depends on any number of *fixed covariates* that are completely known for all individuals and all sample occasions, the probabilities p_{ik} depend on covariates according to some model, such as a logit model:

$$logit(p_{ik}) = \alpha_0 + \alpha_1 x_{ik}, \tag{5.4}$$

where x_{ik} is the value of some covariate measured for individual i at sampling occasion k. Two standard examples of models with fixed covariates are time-specific models and behavioral response models. Time-specific encounter probability models, so-called "model M_t" account for variation in detection over time. Time-specific variation can arise due to variation in weather or other environmental conditions that cause variation in detection probability. A general version of model M_t allows for occasion-specific detection probabilities, one for each occasion. Models that allow for behavioral response to capture, "model M_b," are valuable in situations where animals respond negatively with *trap shyness* or positively with *trap happiness* to the experience of being captured, thus inducing a nonindependence of encounters of the same individual that must be modeled. One obvious example is when food is used to bait animals into a trap – this can result in a trap-happy effect. Behavior models can be developed by introducing x_{ik} as an indicator variable of previous capture (i.e. $x_{ik} = 1$ if individual i was captured prior to occasion k. Such a behavioral response is sometimes called a *persistent* behavioral response, because the effect of initial capture lasts for the duration of the study. Alternatively, we might consider a *Markovian response,* in which the effect of previous capture is ephemeral and lasts only until the next (or a few more) capture occasions (Yang and Chao 2005). A common protocol for estimating abundance in fisheries applications is *removal sampling* where animals are temporarily or permanently removed from the population. Removal sampling is a special and extreme case of a behavior model in that once an animal is captured, it will never be captured again. The decline in encounter frequency over consecutive rounds of removals is informative about encounter probability.

We consider some further extensions of closed population models shortly but we use the basic models with fixed effects here to introduce a number of widely used strategies for inference.

5.4 Inference Strategies

As models for a binary response, closed capture-recapture models are closely related to standard *logistic regression models*. Indeed, if N were known, or if it is conditioned out of the likelihood (Section 5.4.1), capture-recapture is precisely a logistic regression model. As a result, some strategies for analyzing closed population models closely parallel the way that standard logistic regression models are analyzed, and one may analyze capture-recapture models using strategies based on either *maximum likelihood* or *Bayesian inference.*

The main challenge in analyzing closed capture-recapture models is that N is unknown and, as a result, our observed data are only the positive encounter histories for individuals that were captured at least once, where the outcome *not captured at all* is not observable. Thus, capture-recapture produces a sort of biased sampling, and this must be accounted for formally in developing estimators of N (and other quantities) from encounter history data.

5.4.1 Likelihood Inference

Two classical approaches have been widely used for inference about population size (N): one based on the *conditional likelihood*, and a second based on the *unconditional* or *full likelihood* (Borchers et al. 2002; Cooch and White 2008). Following standard ideas of unequal probability sampling (Horvitz and Thompson 1952), the conditional likelihood is formed by expressing the probability distribution for the data conditional on the event that the individual appears in the sample. In the context of capture-recapture, "appears in the sample" is equivalent to "is captured at least one time." Conversely, the full likelihood includes a contribution to account for the fact that "is captured at least one time" is a random event which also depends on the model parameters.

We clarify the nature of the conditional and full likelihood approaches using the simplest model M_0 in which encounter probability, p, is constant. In that case, we can aggregate the data to form a binomial likelihood for the individual totals (encounter frequencies) $y_i = \sum_{k=1}^{K} y_{ik}$. Then, if N were known, the data would be a sample of binomial observations where each y_i is a binomial response with number of trials equal to K, and where the goal is to estimate the binomial probability parameter p. That is, the probability distribution for the observations from which we construct the likelihood is the standard binomial probability mass function (pmf). In the conditional likelihood approach, to account for the fact that we can only observe positive values, we need to condition on the event that $y > 0$ – that is, we need to find out what the pmf is for the observed counts, which are strictly positive, i.e. each observed frequency is at least 1. The resulting pmf is used to construct the *conditional likelihood*, which we can compute using the law of total probability. In particular, the probability of the event $y = k$ for any value k is the sum of two parts: $\Pr(y = k) = \Pr(y = k|y > 0)\Pr(y > 0) + \Pr(y|y = 0)\Pr(y = 0)$. For $k = 1, 2, \ldots, K$, clearly the second part of this expression evaluates to 0, i.e. for the observable values of y. Therefore, the conditional probability that $y = k$, given that $y > 0$, is:

$$\Pr(y = k \mid y > 0) = \Pr(y = k)/\Pr(y > 0). \tag{5.5}$$

For the binomial encounter frequency model of model M_0 and related models, the conditional probability distribution has the following form:

$$\Pr(y \mid y > 0) = \frac{\binom{K}{y} p^y (1-p)^{K-y}}{1-(1-p)^K}. \tag{5.6}$$

This probability mass function is also called a *zero-truncated* binomial. Note that the combinatorial term, which is not a function of any model parameter, can be ignored in maximizing the likelihood to obtain the maximum likelihood estimates (MLEs). Therefore, the likelihood for the n observed encounter frequencies $y_1, y_2, \ldots, y_n$ is therefore the product of n components based on the zero-truncated binomial pmf:

$$L(p \mid y_1, y_2, \ldots, y_n) = \prod_{i=1}^{n} \frac{p^{y_i}(1-p)^{K-y_i}}{1-(1-p)^K}. \tag{5.7}$$

The conditional maximum likelihood estimator of p is obtained by maximizing Eq. 5.7.

You may have noticed that there is no N in the conditional likelihood for the zero-truncated binomial (Eq. 5.7) – only p. So, what do we do to estimate N? The so-called conditional estimator of N derives from the canonical estimator, by noting that the expected value of the observed sample size n is:

$$E(n) = \bar{p}N. \tag{5.8}$$

where $\bar{p} = 1 - (1-p)^K$. Then, the conditional estimator of N, $\hat{N}_c$, can be defined according to:

$$\hat{N}_c = n/\hat{\bar{p}}, \tag{5.9}$$

where $\hat{\bar{p}}$ is obtained by plugging-in the MLE of p into the expression for $\bar{p}$. This idea of conditioning on capture can be generalized to other models, though the formulation of the marginal pmf can become more complex in such cases.

An alternative framework for inference in capture-recapture models is the joint or *full likelihood*, which is developed by computing the joint likelihood of the encounter histories and n, under the binomial assumption for n. Sanathanan (1972) proves the asymptotic equivalence of the conditional and full likelihood estimators. The unconditional (full or joint) likelihood is the product of the conditional likelihood and the binomial contribution from n:

$$L(p, N \mid y_1, y_2, \ldots, y_n, n) = \left\{ \prod_{i=1}^{n} \frac{p^{y_i}(1-p)^{K-y_i}}{1-(1-p)^K} \right\} \left\{ \frac{N!}{n!(N-n)!} \right\} \bar{p}^n (1-\bar{p})^{N-n}, \tag{5.10}$$

which if we combine terms we wind up with a more familiar binomial-looking expression for the model M_0 likelihood (Eq. 14.6 of Williams et al. 2002). The full likelihood for many capture-recapture models, including model M_0, can be derived as a multinomial distribution, for distributing individuals into $K + 1$ classes (Royle and Dorazio 2008), including the class representing $y = 0$ (not captured).

Likelihood inference extends directly for cases where p depends on any number of fixed covariates. When there are covariates thought to influence detection probability, this requires that the likelihood for each of the n observed individuals be expressed in terms of probabilities p_{ik} which may depend on covariates according to some model, such as a logit model as discussed in Section 5.3, and also the parameters α_0 and α_1 (see Eq. 5.4) and possibly additional parameters depending on how many covariate effects are being modeled. To indicate that p now depends on the parameters α_0 and α_1, we will write $p(\boldsymbol{\alpha})$. When covariates are modeled, it is convenient to analyze a general form of the likelihood formulated in terms of individual level encounter histories:

$$L(\alpha_0,\alpha_1,N \mid \{y_{ik}\}) = \left\{\frac{N!}{n!(N-n)!}\right\} \left\{\prod_{i=1}^{n}\prod_{k=1}^{K} p_{ik}(\boldsymbol{\alpha})^{y_{ik}}\left(1-p_{ik}(\boldsymbol{\alpha})^{1-y_{ik}}\right)\right\} \left\{\prod_{k=1}^{K}(1-p_k(\boldsymbol{\alpha}))\right\}^{N-n}, \quad (5.11)$$

where here $p_k(\boldsymbol{\alpha})$ is the probability of not encountering an individual in sample occasion k which will, in general, depend on the fixed covariate values for sample occasion k.

5.4.2 Bayesian Analysis

Parameter estimation and inference based on likelihood methods remains the standard approach in applications involving inference about N. *Bayesian analysis* has become more popular due to the advent of accessible and efficient computing platforms, the appealing nature of Bayesian inference, which allows for the direct characterization of uncertainty about parameters using probability distributions, and the flexibility in model development that Bayesian analysis permits. In principle, Bayesian analysis of either the conditional or full likelihood of the model is straightforward, and several developments and applications have appeared in the literature (Castledine 1981; Smith 1991; George 1992; Basu and Ebrahimi 2001; King and Brooks 2001). Any of the closed population capture-recapture models which contain a number of encounter probability parameters and, for analysis based on the full likelihood, the parameter N, can easily be analyzed by Bayesian methods. To conduct a Bayesian analysis requires specification of prior distributions for each unknown parameter. Given the likelihood and set of prior distributions, inference about N or other model parameters is then based on the posterior distribution, which is the probability distribution of N *given* the data.

For some models or in some applications, Bayesian analysis is facilitated by the use of *data augmentation* (Royle et al. 2007a; Kéry and Schaub 2012; Royle and Dorazio 2012), which is based on a reformulation of the model in terms of individual encounter histories under a certain prior specification for N. As a practical matter, analysis based on data augmentation proceeds by augmenting the observed encounter history matrix with a large number of *all-zero* encounter histories which correspond to individuals that were not captured (Figure 5.1). The analysis must be based on adding a sufficient number

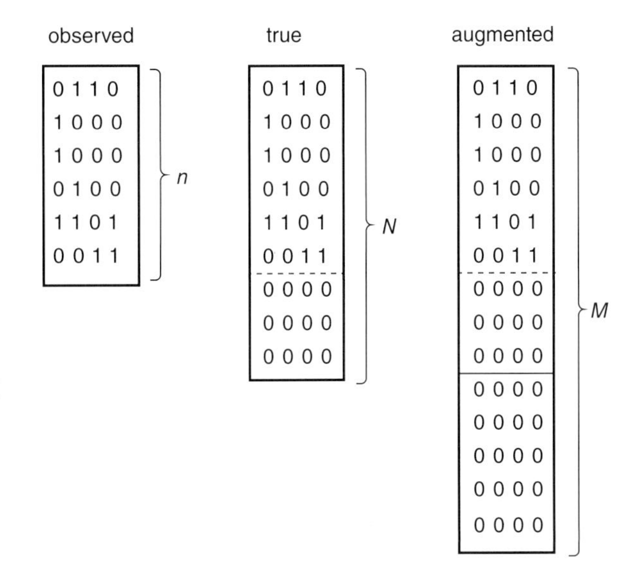

Figure 5.1 Data augmentation schematic showing the observed data set (left), the unobserved "true" data set (middle), and the augmented data set (right). Source: Borrowed with permission from Kéry and Schaub (2012, figure 10.2).

of these all-zero encounter histories such that the total size of the augmented data set is certain to be larger than the true value of N. The key idea is that the model for the augmented data set is a zero-inflated version of the known-N model. For example, considering model M_0, if N were known, then the model of the encounter frequencies is a simple binomial model with parameter p and sample size K. However, under data augmentation, where the data set is augmented up to a size of M, then the model for the augmented encounter frequencies is a zero-inflated binomial with additional parameter ψ which is related to the quantity of excess zeros beyond that which would be accounted for given N. In fact, the relationship between unknown N and the parameter ψ derives from the binomial distribution, $E(N) = \psi M$. Because M is fixed, data augmentation essentially shifts the inference problem from one of estimating N to one of estimating ψ.

We illustrate the concept of data augmentation applied to the simplest model – model M_0. For this model, the likelihood under data augmentation – that is, for the data set of size M – is a simple zero-inflated binomial likelihood for the observed encounter frequencies (out of K samples) $y_1, y_2, \ldots, y_n$ augmented with all-zero encounter frequencies $y_{n+1}, \ldots, y_M$. The zero-inflated binomial model can be described hierarchically, by introducing a set of binary latent variables – the data-augmentation variables, $z_1, z_2, \ldots, z_M$ – to indicate whether each individual i is ($z_i = 1$) or is not ($z_i = 0$) a member of the population of N individuals exposed to sampling. We assume that z_i~Bernoulli(ψ) where ψ is the probability that an individual in the data set of size M is a member of the sampled population. The zero-inflated binomial model which arises under data augmentation can be formally expressed by the following set of assumptions where we include typical prior distributions for the parameters p and ψ:

$$yi|zi = 1 \sim \text{Binomial}(K, p) \tag{5.12}$$

$$yi|zi = 0 \sim I(y = 0) \tag{5.13}$$

$$zi \sim \text{Bernoulli}(\psi) \tag{5.14}$$

$$\Psi \sim \text{Uniform}(0, 1) \tag{5.15}$$

$$p \sim \text{Uniform}(0, 1) \tag{5.16}$$

for $i = 1,\ldots, M$, where $I(y = 0)$ is a point mass at $y = 0$. This is model M_0 when formulated by the use of data augmentation, which partitions the zeros of the augmented data set into real individuals (sampling zeros) versus excess augmentation (structural zeros). It is convenient in some of the standard Bayesian software packages (BUGS/JAGS) to express the conditional-on-z observation model concisely in just one step: $y_i \mid z_i \sim \text{Binomial}(K, z_i p)$, which we understand to mean, if $z_i = 0$ then y_i is necessarily 0 because its success probability is $z_i p = 0$.

5.4.3 Other Inference Strategies

There are some other general approaches to inference in closed population capture-recapture models. One approach is to use *Poisson-integrated likelihood* (Sandland and Cormack 1984), in which the full likelihood having parameter N is integrated over a Poisson prior distribution for N. The Poisson-integrated likelihood also extends readily to structured populations (Converse and Royle 2012) and so it is handy for modeling variation in abundance among strata or groups. A related idea is to use a binomial-integrated likelihood (chapter 6 of Royle et al. 2014) in which the full likelihood is integrated over a binomial prior: $N \sim \text{Binomial}(M, \psi)$ where M is fixed at some large value and ψ is estimated instead of N. The model is consistent with the data augmentation approach but it can be readily applied in a likelihood estimation framework. Royle and Dorazio (2008) provide an application of the binomial-integrated likelihood to distance sampling.

5.5 Models with Individual Heterogeneity in Detection

The basic models described above make some restrictive assumptions about the manner in which detection probability varies among individuals in the population. However, *individual heterogeneity* in detection is ubiquitous for reasons related both to biology and sampling. For example, individuals use space differently and therefore might be exposed to different numbers of traps, or our sampling method might favor certain demographic classes (e.g. age-, sex-, or size-biased sampling). The existence of heterogeneity in detection probability that is not accounted for by the modeling can result in underestimation of population size. One can see this intuitively by considering the extreme case where some individuals have no chance of being detected, i.e. $p = 0$, such as young animals that are not yet mobile. In this case, the population that would be estimated would not include that group of animals with $p = 0$. In general, we see that the sample of individuals favors those with high detection probability, and thus our sample will tend to overestimate the population level p and underestimate N, because $\hat{N} = n/\hat{\bar{p}}$.

The problem of heterogeneity is well-known, and several types of capture-recapture models that control for individual level effects in detection probability have been developed and are in widespread use. Examples include models with fixed effects measured at the individual level, which are called *individual covariate models*, and models that contain unstructured or latent variation in the form of an individual random effect, including *finite mixture models*.

5.5.1 Model M_h

Models with individual heterogeneity are models which posit that the detection probability parameter p varies by individual, say p_i for individual i. The class of models and estimators that accommodate such variation are referred to collectively as "model M_h." In reality, this is a broad class of models, each being distinguished by the specific distribution assumed for p_i. There are many different varieties of model M_h including parametric and various nonparametric approaches. Burnham and Overton (1978) proposed a nonparametric jackknife estimator,which formed the basis for estimating population size in the presence of heterogeneity for many years. Chao (1984, 1987) also developed a nonparametric estimator in the presence of heterogeneity that has been widely adopted. A formal model for heterogeneity in p was proposed by Norris and Pollock (1996), and generalized by Pledger (2000) and subsequent papers. The mixture model assumes that p_i, the probability of detection for individual i, is a latent variable belonging to a finite number of latent classes say pc for $c = 1, 2,\ldots, C$ with probabilities fc where $\sum_c f_c = 1$. In practice, two or three classes is usually sufficient to explain heterogeneity in realistic data sets. In addition, from a practical standpoint, it is difficult to estimate more complex latent class models because the number of parameters increases rapidly. The number of parameters in a finite mixture model is $2C$ where C is the number of latent classes. As a result, the likelihood can be difficult to explore due to its complex form and potentially multiple modes.

Fully parametric models of heterogeneity have also been proposed. For example, a standard model is the *logit-normal mixture model* (Coull and Agresti 1999) which supposes that the encounter probability p_i for individual i varies according to: $logit(p_i) = \mu + \eta_i$ where $\eta i \sim \text{Normal}(0, \sigma^2)$ or, equivalently, written as:

$$\text{logit}(p_i) \sim \text{Normal}(\mu, \sigma^2). \tag{5.17}$$

This model is a natural extension of the basic model with constant p, as a mixed generalized linear model GLMM, and similar models occur throughout statistics. It is also natural to consider a beta prior distribution for p_i (Burnham 1972; Dorazio and Royle 2003).

Inference about model parameters, including N, in model M_h is slightly more involved than in other standard models, because the model involves *individual random effects*. The standard method in all of statistics for inference in random effects is based on the so-called *marginal likelihood* or *integrated likelihood* (Coull and Agresti 1999; Dorazio and Royle 2003). The basic idea is that the marginal likelihood for the observations is computed by removing the random effect from the conditional-on-η likelihood by integration, yielding a marginal likelihood that is only a function of the structural parameters and not the random effect, and which can be maximized to obtain the MLE of the model parameters. The technical details of analyzing the marginal likelihood are widely discussed, for examples Coull and Agresti (1999), Dorazio and Royle (2003), and chapter 6 of Royle and Dorazio (2008). Bayesian analysis of models with random effects is straightforward using standard methods of Markov chain Monte Carlo (MCMC). Examples of this abound in the literature (Fienberg et al. 1999; Royle and Dorazio 2008; Link and Barker 2009; King et al. 2010; Royle et al. 2014).

One important practical matter is that estimates of N can be extremely sensitive to the choice of heterogeneity model (Fienberg et al. 1999; Link 2003). Indeed, Link (2003) showed that in some cases it is possible to fit different models that yield precisely the same log-likelihood, yet produce wildly different estimates of N. In that sense, N for most practical purposes is not identifiable across classes of mixture models, which should be considered before adopting model M_h as a definitive solution for individual heterogeneity. An alternative solution to the problem of individual heterogeneity is to seek to model explicit factors that contribute to heterogeneity, such as using individual covariates or spatially explicit models.

5.5.2 Individual Covariate Models

The models described in Section 5.5.1 accommodate heterogeneity in detection probability that is not explainable by explicit covariates. However, closed population models can be extended to accommodate *explicit individual-level covariates*. These models are similar in structure to model M_h, except that some individual-level effect is observed for the n individuals that appear in the sample. In particular, the model is of the form:

$$y_i \sim \text{Binomial}(K, p_i), \tag{5.18}$$

where p_i is functionally related to a covariate x_i. A standard model is the logit-linear model:

$$\text{logit}(p_i) = \alpha_0 + \alpha_1 x_i. \tag{5.19}$$

In many studies, the covariate x might be related to body size or age, for example. Following the model notation of chapter 6 of Kéry and Schaub (2012), we will call this model M_x. There are two different approaches to inference about N under model M_x (section 11.3 of Borchers et al. 2002). The first approach is to put a model on the covariate x, which allows us to deal with the fact that x is missing for the $N - n$ individuals never captured, and then we can either do likelihood analysis with the "full likelihood" approach (Borchers et al. 2002), or we can conduct a Bayesian analysis using MCMC (Royle 2009). An alternative approach in common use is based on a Horvitz-Thompson unequal probability sampling estimator, and is called the Huggins-Alho estimator (Huggins 1989; Alho 1990; Williams et al. 2002).

5.5.2.1 The Full Likelihood

The individual covariate is unobserved for the $N - n$ uncaptured individuals and therefore we require a model to describe variation among individuals so that the sample can be extrapolated to the population. For example, we might suppose that the individual covariate has a normal distribution:

$$x_i \sim \text{Normal}(\mu, \sigma^2). \tag{5.20}$$

This assumption, together with Eqs. 5.18 and 5.19, provides a fully specified model which can be analyzed using standard methods. A standard example of an individual covariate model is that in which animal size or body mass is thought to influence detectability. Another standard example is "group size" detected in aerial surveys, such as flocks of waterfowl (Royle 2008).

Analysis of the full likelihood for individual covariate models can be challenging because the covariate values for the $N - n$ individuals that were never captured are missing data, and therefore the marginal probability of $y = 0$ must be computed by numerical integration. Alternatively, Bayesian analysis of individual covariate models is relatively straight forward in the BUGS language and similar MCMC algorithms using data augmentation (Royle 2009).

5.5.2.2 Horvitz-Thompson Estimation

Traditionally, estimation of N in individual covariate models is achieved using methods based on ideas of unequal probability sampling. This approach leads to an estimator of N of the following form:

$$\hat{N} = \sum_{i}^{n} \frac{1}{\bar{p}_i(\boldsymbol{\alpha})}, \tag{5.21}$$

where $\bar{p}_i(\boldsymbol{\alpha})$ is the probability that individual i appeared in the sample, which depends on the parameters $\boldsymbol{\alpha} = (\alpha_0, \alpha_1)$ through Eq. 5.19. The quantity $\bar{p}_i(\boldsymbol{\alpha})$ is the probability that an individual is captured, i.e. $\Pr(y_i > 0; \boldsymbol{\alpha})$ which in closed population capture-recapture models is given by:

$$\bar{p}_i(\boldsymbol{\alpha}) = \Pr(y_i > 0; \boldsymbol{\alpha}) = 1 - (1 - p_i(\boldsymbol{\alpha}))^K. \tag{5.22}$$

In practice, parameters $\boldsymbol{\alpha}$ are estimated from the conditional likelihood of the observed encounter histories which is,

$$L_c(\boldsymbol{\alpha} \mid y_1, \dots, y_n) = \prod_{i=1}^{N} \frac{\text{Binomial}(y_i \mid \boldsymbol{\alpha})}{\bar{p}_i(\boldsymbol{\alpha})}. \tag{5.23}$$

The conditional likelihood is maximized to obtain the MLE of $\boldsymbol{\alpha}$, which is then used in Eq. 5.21 to obtain an estimator of N.

5.5.3 Distance Sampling

Distance sampling is not traditionally considered a capture-recapture method, but it is perhaps illuminating to note that conceptually, distance sampling is a type of individual covariate model with only a single sample, $K = 1$. Therefore, animals are not marked in the traditional sense but, rather, they are individually identified in the sampling process, and individual-specific information is recorded, including location, perpendicular distance to the transect line, and perhaps group size. Inference is traditionally based on a conditional estimator, although estimation based on the full likelihood is also straightforward (section 7.2 of Borchers et al. 2002).

The model underlying distance sampling is precisely the same as that which applies to the individual-covariate models, except that observations are made at only $K = 1$ sampling occasions. However, in distance sampling we pay for having only a single sample (i.e. $K = 1$) by requiring constraints on the model of detection probability. In particular, it cannot have an intercept parameter in the basic model, which leads to the often-stated assumption that distance sampling assumes perfect detection on the transect line. A standard model for encounter probability is:

$$p_i = exp\left(-\frac{x_i^2}{2\sigma^2}\right), \tag{5.24}$$

where x_i denotes the distance at which the ith individual is detected relative to some reference location where perfect detectability ($p = 1$) is assumed. The parameter σ determines the rate of change in detection probability as a function of distance. The function is usually called the *half-normal* detection probability model. An additional assumption is that of a uniform distribution for the individual covariate x_i. The customary choice is $x_i \sim \text{Uniform}(0, B)$ where $B > 0$ is specified, and corresponds to the limit of observation: either a count radius, or a transect width.

5.5.4 Spatial Capture-Recapture Models

Capture-recapture models are enormously popular in population ecology. However, the central focus on population size, N, in closed models has long been recognized as a deficiency in practical applications. The reason is that, in practice, no population is truly closed because geographic closure can never be satisfied in practice and, as a result, a strict sample area cannot be defined. As a result, it is not possible to convert N to density (a summary of abundance that is invariant to sample area) in the absence of additional information. Or, to put the problem another way, the area to which the N estimate applies is unknown. To understand this intuitively, imagine a grid of live traps (Figure 5.2). Some of the animals caught in the live traps may have home ranges that only partly overlap the grid – illustrated by the home range on the left of the trapping grid. Thus, the size of the home

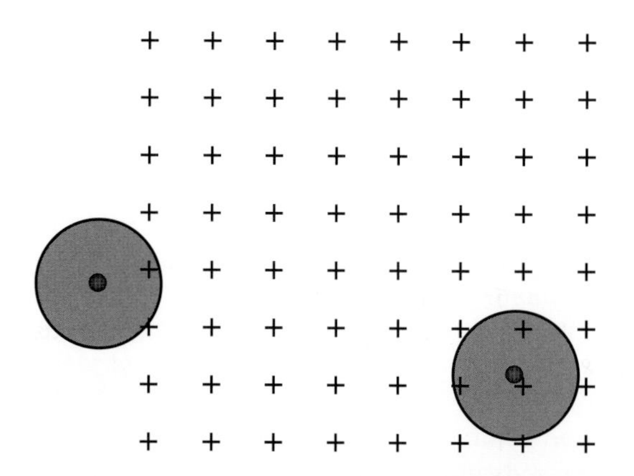

Figure 5.2 A trapping grid with two individual home ranges superimposed. Home-range centers are indicated by black dots. The effective sample area of a trapping grid depends on the typical home ranges of individuals. Individuals with home ranges located predominantly off-grid may still be captured.

ranges inflates the effective area sampled. If the home ranges are large, we are effectively sampling a larger area because animals that use areas relatively far from the grid may be caught on the grid. If the home ranges are small, we are effectively sampling a smaller area. But how do we know? It is the spatial information in SCR that helps us, conceptually, to resolve this.

Historically, the fact that area is unknown in capture-recapture studies, thereby complicating the conversion of N to density, has been addressed by the prescription of buffer strips or nested trapping grids based on an ad hoc treatment of the movement information in trapping data (reviewed by section 14.3 of Williams et al. 2002), or by using telemetry data to compute an adjustment to estimated detection probability based on a model of temporary emigration (White and Shenk 2001; Ivan et al. 2013).

Another problem that arises in the application of classical capture-recapture models is that individual heterogeneity is induced by the spatial juxtaposition of individuals in a population with the array of traps or sampling devices (Figure 5.2). For example, individuals whose home range is on the edge of a study area or trap array are exposed to fewer traps and therefore likely have lower probabilities of detection compared to individuals whose home range is more centrally located within the trap array. Thus, the spatial nature of capture-recapture studies induces individual heterogeneity that should be accounted for. Historically, this problem of heterogeneity has been accounted for by fitting standard types of heterogeneity models M_h (Karanth and Nichols 1998; Section 5.5.1), or alternatively, model M_x with observed location of capture as an individual covariate (Boulanger and McLellan 2001; Section 5.5.2).

New classes of *spatially explicit capture-recapture models* (SECR) or just SCR provide a formal resolution to the problems of undefined sample area, and of heterogeneity in detection probability (Efford 2004; Royle et al. 2014; Borchers 2012), by taking advantage of auxiliary information about location of capture that exists in virtually all capture-recapture data sets. SCR models are essentially a version of ordinary individual covariate models described above, but with imperfect information about the individual covariate. Spatial models are being rapidly adopted in practice, with many recent applications to DNA sampling, camera trapping, and other situations (Dawson & Efford 2009; Gardner et al. 2010a; Gopalaswamy et al. 2012).

The basic idea with SCR models is to extend the ordinary capture-recapture model by introducing a latent variable, **s**, for each individual in the population. The variable **s** represents the individual's home range or *activity center*. The collection of latent variables for the population of N individuals can then be treated as a spatial point process, which can be modeled explicitly. Then, encounter probability is expressed as a function of that latent variable. In a typical problem, suppose an array of J camera traps is used to observe a population of individuals that are individually identifiable based on natural marks over some period of time (K nights). The array of traps produces individual- and trap-specific encounter frequencies y_{ij} for individual i and trap j. For camera traps, a sensible encounter probability model is the binomial model:

$$y_{ij} \sim \text{Binomial}(K, p_{ij}), \quad (5.25)$$

where the trap- and individual-specific encounter probability is assumed to be a function of distance between an individual's home-range center and the traps. For example, a plausible model is:

$$logit(p_{ij}) = \alpha_0 + \alpha_1 \times dist(\mathbf{s}_i, \mathbf{x}_j)^2, \quad (5.26)$$

where $dist(\mathbf{s}_i, \mathbf{x}_j)$ is the (unobserved) distance between the location of trap j, $\mathbf{x}_j$, and a summary of individual location $\mathbf{s}_i$. Conceptually, we view $\mathbf{s}_i$ as the center of activity or home-range center for individual i. That $dist(\mathbf{s}_i, \mathbf{x}_j)$ is not observed may appear problematic, but this is analogous to any random effect in classical random effects models. Therefore, formal inference for SCR models is achieved by regarding the activity centers $\mathbf{s}i$ as latent variables or random effects. The model is extended then by specifying a probability model for these latent variables. In particular, the customary assumption (used in almost all recent applications of SCR models) is, for $i = 1, 2, \ldots, N$,

$$\mathbf{s}_i \sim \text{Uniform}(S), \quad (5.27)$$

where S is a region of the plane in the vicinity of the trap locations. You can think of this as the area which contains the population exposed to sampling by the traps. The common view of SCR models is that the activity centers $\mathbf{s}_1, \ldots, \mathbf{s}_N$ represent a realization of a statistical point process, and the region of the plane S is usually called the state-space of the point process. Given the state-space for any particular model, the parameter N of the SCR model is the population size of individuals as the number of activity centers which exist in that state-space. Therefore, SCR models provide a direct linkage between the data and density, the number of individuals per unit area, $D = N/\text{area}(S)$.

In applications of SCR models, the logit model of Eq. 5.26 is rarely used, even though it is a perfectly reasonable and valid model. Instead, standard models are those which are common in distance sampling. The similarity is sensible because distance from a trap to a home range center is analogous to the distance between an observer and an animal if the data were collected under distance sampling protocols. Therefore, a distance sampling model such as the half-normal model can be used:

$$p_{ij} = p_0 \exp\left(-\text{dist}(\mathbf{s}_i, \mathbf{x}_j)^2 / (2\sigma^2)\right), \quad (5.28)$$

where the parameters p_0 and σ are estimated (note: σ here is unrelated to σ of the logit-normal model M_h described previously). Many other models are possible, and the `secr` package of Efford (2011) contains about a dozen candidate models (see Table 7.1 in Royle et al. 2014).

One of the key concepts underlying SCR models is that the encounter probability model corresponds to an explicit model of home range or space use. In the half-normal model, the parameter σ is the parameter of a bivariate normal home range model. Models can be developed which include explicit elements of resource selection (Royle et al. 2013), latent movement models (Royle et al. 2016), or that accommodate non-Euclidean space use (Sutherland et al. 2015; Fuller et al. 2016).

5.5.4.1 The State-space

In Bayesian analysis of SCR models, it is necessary to prescribe S because the analysis proceeds by simulating realizations of $\mathbf{s}_1, \ldots, \mathbf{s}_N$ from the required posterior distribution. Individuals have to reside somewhere. In developing the problem this way, N is sensitive to the choice of S – as its area increases, so too does N and vice versa. However, as S becomes large, then p_{ij} diminishes to zero rapidly, at least under well-behaved models, and additional increases in S are inconsequential. In particular, N will continue to increase but *density* will become invariant to the size of S, which is a consequence of the *model*, and which implies a constant density of individuals. The same phenomenon is relevant to the likelihood-based analyses of Borchers and Efford (2008). The region over which the likelihood is integrated must be defined explicitly, but construction of the state-space often happens "under the hood" when likelihood analysis is conducted, and so it may not seem like an explicit element of the model in such cases.

5.5.4.2 Inference in SCR Models

Both likelihood and Bayesian analyses of SCR models are straightforward using standard methods. Bayesian analysis of the model specified in terms of the latent activity center variables is easily achieved using standard methods of MCMC (Royle et al. 2014). Likelihood analysis of SCR models is based on the marginal likelihood in which the random effects $\mathbf{s}i$ are removed from the conditional-on-$\mathbf{s}$ likelihood by integration over the state-space S. The basic approach applies directly to analysis of three alternative models: (i) the full likelihood which includes N; (ii) the conditional likelihood which is conditional on n; and (iii) the Poisson-integrated likelihood where the full likelihood is integrated over a Poisson prior distribution for N (Borchers and Efford 2008; chapter 6 of Royle et al. 2014). The basic approach of computing the marginal likelihood is standard in other classes of capture-recapture model (model M_h, individual covariate models, distance sampling) with no additional technical or conceptual considerations.

5.6 Stratified Populations or Multisession Models

A situation that arises often in practice is that in which a stratified or group-structured population is sampled, so that the resulting encounter history data fall naturally into $g = 1, 2, \ldots, G$ groups or strata (organized in space and/or time). A common situation occurs in studies of small mammal populations, which might involve establishing a set of G trap arrays. In these types of problems, it is usually of interest to model variation in the unknown population sizes, N_g, among the trapping arrays, perhaps in response to landscape/habitat (Royle et al. 2007b) or treatment effects (Converse and Royle 2012). Another common situation occurs in bird surveys which are often conducted using capture-recapture counting protocols, such as double-observer sampling or time-removal sampling. In these situations, it is natural to formulate the model in terms of local population sizes N_g for each point count location g. Then, the observed data for point count location g is naturally described as conditional on N_g. In both situations for small mammal and bird counts, the groups were spatial groups and the population size parameters N_g related to the spatial populations being sampled. However, the groups do not have to be spatial groups: It would be natural also to define them to be temporal groups, where N_g is the population size in year $g = 1, 2, \ldots, G$, or different combinations of spatial units and temporal periods. Because it is common to define temporal strata in capture-recapture models, the term *multisession model* is often used to describe these scenarios and the resulting models.

5.6.1 Nonparametric Estimation

If there is not explicit interest in modeling patterns in abundance across the g groups or strata, there are substantial benefits to combining data across groups into a single analysis for abundance estimation; the benefits accrue in terms of efficiency in estimating p and therefore N for each group. For example, a behavioral effect may be reasonably modeled in common across individuals of the same species, existing in different places. If 10 different spatial strata had been sampled, a shared effect of behavior would save nine parameters compared to an independent analysis for each strata. Converse and Royle (2012) examined this issue explicitly, and some efficient approaches include sharing information across groups to estimate fixed effects, and also exploiting the group structure to estimate random effects for the detection parameters. Converse et al. (2006) demonstrated use of such estimates in a weighted regression analysis to model the abundance estimates, while accounting for the non-zero sampling covariances. However, advances in hierarchical estimation models (Converse and Royle 2012; Royle et al. 2012) make this ad hoc approach less attractive. One can imagine taking a non-hierarchical approach to produce estimates of N across strata within the context of a monitoring program, where the primary interest is the abundance estimates themselves. With sufficient data on each population of interest, this approach should be reasonably efficient from a statistical standpoint, and avoids having to make explicit assumptions about N.

5.6.2 Hierarchical Capture-Recapture Models

A flexible approach to building capture-recapture models for stratified populations can be achieved using *hierarchical capture-recapture models* based on developing explicit statistical models for population size among the groups (Royle et al. 2012). In this way, we can move beyond simply estimating abundance or density to modeling variation in abundance or density over space and time, which is almost always more interesting from both a management and research perspective. From a practical standpoint this is advantageous because it makes efficient use of data from all groups for estimating shared parameters, which can be useful when some of the groups have small sample sizes.

To motivate the hierarchical capture-recapture model, suppose capture-recapture data are collected on $g = 1, 2, \ldots, G$ populations, each having population size N_g. To devise a model for this situation we might consider any of our standard capture-recapture observation models (e.g. models M_0, M_b, etc.), and thus the model for the observations for each group is one of the standard models, having one or more detection parameters and a population size parameter N_g. In addition, we assume a model for the population size parameters N_g. For example, we might assume that $N_g \sim \text{Poisson}(\lambda_g)$ where:

$$log(\lambda_g) = \beta_0 + \beta_1 x_g, \tag{5.29}$$

and where x_g is some covariate measured at the population level. Other abundance models could be considered (e.g. negative binomial), but the point here is that, using a hierarchical capture-recapture model, we analyze the joint model which includes component models that describe the observation of the capture-recapture data – the observation model – and that which describes variation in abundance among the strata.

Hierarchical capture-recapture models for stratified populations have not appeared widely in the literature although the data structure is somewhat common in practice. Royle et al. (2007b) analyzed a version of these models based on the capture-recapture summary statistics, which have a multinomial distribution with index N_g. The main objective of their analysis was to model spatial variation in N_g and they considered a Poisson type of model with over-dispersion. Likelihood analysis of hierarchical capture-recapture models is possible. Several detailed case studies of such models applied to bird population studies can be found in chapter 7 of Kéry and Royle (2016). A general Bayesian analysis of stratified capture-recapture models was given by Converse and Royle (2012) and Royle and Converse (2014) in the context of SCR models.

5.7 Model Selection and Model Fit

When considering alternative models for the probability of detection, such as whether or not a mixture appropriately captures heterogeneity, or if an individual covariate should be used, there will be interest in identifying the model(s) that are the best fit for the data. *Model selection* provides the tools for ordering or ranking the models, and maybe choosing a "best" model, for obtaining *model-averaged estimates* across several candidate models with support, and for evaluating whether a given model is appropriate for inference to answer the question, does it "fit" the data? Here, we review some basic strategies for carrying out model selection and evaluating goodness-of-fit in models for estimating population size. Readers are referred to Chapter 2 for a more detailed treatment of model selection and multimodel inference.

5.7.1 Model Selection

Using classical analysis based on likelihood, model selection is easily accomplished using the *Akaike Information Criterion* (AIC) (Burnham and Anderson 2002; also Chapter 2). The AIC of a model is simply twice the negative log-likelihood evaluated at the MLE, penalized by the number of parameters (K) in the model:

$$\text{AIC} = -2\log\text{L}\left(\hat{\boldsymbol{\theta}} \mid \mathbf{y}\right) + 2K \tag{5.30}$$

Models with small values of AIC are preferred. It is common to use a "corrected" version of AIC, referred to as AICc, that has a correction term to account for small sample sizes:

$$\text{AIC}_c = -2\text{logL}\left(\hat{\boldsymbol{\theta}} \mid \mathbf{y}\right) + \frac{2K(K+1)}{n-K-1} \tag{5.31}$$

where n is the sample size. While the use of AIC and similar criteria have been widely adopted in ecology, and especially in the analysis of capture-recapture data, researchers are increasingly recognizing practical difficulties in their application. For example, they do not apply directly to hierarchical models that contain random effects (Grueber et al. 2011), unless they are computed directly from the marginal likelihood. Moreover, it is not clear what should be the effective sample size n in calculation of AICc, as there can be covariates that affect individuals that vary over time, or space. Nevertheless, the use of AIC is straightforward in many problems and, as a result, it has gained widespread use in capture-recapture models since the publication of Burnham and Anderson (2002).

Bayesian model selection is somewhat less straightforward than that based on AIC or AICc. O'Hara and Sillanpää (2009) provide a general review of *variable selection* ideas from a Bayesian standpoint, and other candidate approaches were evaluated by Royle et al. (2014) in the context of SCR models (see also Kéry and Royle 2016). The area of Bayesian model selection has undergone rapid development and synthesis, with many recent papers which should be consulted for additional guidance including Millar (2009), Link and Barker (2009), Tenan et al. (2014), and Hooten and Hobbs (2015).

A good heuristic approach that is easy to apply when the model consists of a small number of fixed effects is a conventional approach based on *hypothesis testing*. That is, if the posterior distribution for a parameter overlaps zero substantially, then it is probably reasonable to discard that effect from the model. A second approach, which can be applied in many situations, is direct calculation of posterior model probabilities for a given set of models. One idea for achieving this is the *indicator variable selection method* of Kuo and Mallick (1998). To illustrate how Bayesian variable selection can be applied, consider a simple model which has a behavioral response covariate $C_{ik} = 1$ if individual i is captured prior to occasion k, and $C_{ik} = 0$ otherwise. The behavioral response model posits that:

$$\text{logit}(p_{ik}) = \alpha_0 + \alpha_1 C_{ik}. \tag{5.32}$$

Kuo and Mallick (1998) suggest expanding the model to include a binary latent variable $w \sim \text{Bernoulli}(0.5)$ and expressing the model for p according to:

$$\text{logit}(p_{ik}) = \alpha_0 + w\alpha_1 C_{ik}. \tag{5.33}$$

Then, the variable w is estimated along with the other parameters of the model. The importance of the covariate C is then measured by the posterior probability that $w = 1$. These ideas are demonstrated by Royle et al. (2014), and covered by Royle and Dorazio (2008). Link and Barker (2009) have good coverage of computing posterior model probabilities and general issues related to Bayesian model selection.

An AIC-like approach that has gained considerable traction in Bayesian analysis is the *Deviance Information Criterion* or DIC (Spiegelhalter et al. 2002). Although DIC is widely used, there seem to be several variations, and a consistent version is not always reported across computing platforms. Even statisticians do not have general agreement on practical issues related to the use of DIC (Millar 2009). Royle et al. (2014) evaluated the use of DIC for SCR models and suggest that it can produce sensible results if the hierarchical structure of the models is the same but only the fixed effects are being varied. The authors also suggested that attempts to evaluate or calibrate the use of DIC for any specific problem should be made. The main advantage of DIC as a model selection criterion is ease of calculation. *Model deviance* is defined as $-2log(\text{L})$; i.e. for a given model with parameters θ: Dev $(\theta) = -2 * \text{logL}(\theta|\mathbf{y})$. The DIC is defined as the posterior mean of the deviance, Dev(θ), plus a measure of model complexity, p_D: DIC = Dev(θ) + p_D. The standard definition of p_D is $p_D = \text{Dev}(\theta) - \text{Dev}(\bar{\theta})$ where the second term is the deviance evaluated at the posterior mean of the model parameter(s), $\bar{\theta}$. The p_D term here is interpreted as the effective number of parameters in the model. Gelman et al. (2004) suggest a different version of p_D based on one-half the posterior variance of the deviance: $p_V = \text{Var}(\text{Dev}(\theta) \mid \mathbf{y})/2$. While DIC is easy to compute, care should be taken in its application, especially in the context of models with latent variables (Millar 2009).

5.7.2 Goodness-of-Fit

In practical settings, we estimate parameters of a desirable model, or maybe several models, with the idea of using one or all of them to produce an estimate of N along with some statement of uncertainty. For that to be a valid statement of uncertainty we would like to be confident that our model is a reasonable approximation to truth. Or, in other words, does the model appear to be an adequate description of our data? Formal assessment of model adequacy or *goodness-of-fit* is a challenging problem and there are no all-purpose algorithms for doing this in either frequentist or Bayesian paradigms (Chapter 2). However, in many cases, whether Bayesian or frequentist, the main idea for assessing model fit is the same: we compare data sets from the model we are interested in with the data set we have in hand. If they appear to be consistent with one another, then our faith in the model increases, at least to some extent, and we say that the model fits.

In simple cases, using classical inference methods, it is sometimes possible to identify a test statistic of theoretical merit, perhaps with a known asymptotic distribution. For classes of models for which the sufficient statistic is a count statistic, we can apply conventional fit testing ideas based on chi-square statistics using the Pearson residual, based on the difference between the observed and expected frequencies. Otis et al. (1978, appendix K) provide some examples of this for the basic models such as M_b and M_t. These classical ideas provide a satisfactory approach for simple models where the sufficient statistic is a vector of count frequencies but, in general, this will not be the case, including any model containing individual level effects, such as in SCR models, or individual covariate models. In these cases, it is not possible to reduce the data to sensible count statistics, or the choice of a fit statistic may not be obvious. There is little guidance in the literature on goodness-of-fit in the context of capture-recapture models based on individual encounter histories. Therefore, only tentative guidance can be given. For this, we draw on analogies with site-occupancy models which have a similar data structure, in that they are composed of encounter histories for individual sites. MacKenzie and Bailey (2004) suggest a goodness-of-fit testing framework based on aggregating data into unique encounter histories, and Kéry and Royle (2016, section 10.8) adapted this strategy based on aggregating data into row or column totals which, in the context of capture-recapture, would be individual encounter frequencies or sample occasion encounter frequencies. It seems reasonable to use a parametric bootstrap to characterize the null distribution of such a statistic. To our knowledge, no one has investigated this approach for capture-recapture models.

For Bayesian analyses, the most effective practical approach for evaluating model fit seems to be based on the Bayesian p-value (Gelman et al. 1996) which has not been widely applied in capture-recapture models (but see chapter 8 of Royle et al. 2014). Using this approach, data sets are simulated from the posterior distribution, and some fit statistic is computed for each simulated data set, usually based on the discrepancy of the observed data from its expected values. The same fit statistic for the actual data set at hand is then compared to the posterior simulated values of the fit statistic. The Bayesian p-value is the proportion of times the fit statistic for the simulated data is greater than that of the observed data. A model that is adequate for a given data set should have a Bayesian p-value that is not too close to 0 or 1. For practical purposes, we might judge this to be greater than 0.10 and less than 0.90.

5.7.3 What to Do When Your Model Does Not Fit

Invariably with large and complex data sets, situations may arise in which the *global model* as the most complex model of possible interest, or a reduced model of interest, simply does not fit the data set at hand. In this case, a standard suggestion is to calculate a *variance inflation factor* or *c-hat*, an adjustment parameter that is also referred to as a *lack-of-fit ratio* or a *quasi-likelihood adjustment for over-dispersion*. For some background on this in the context of capture-recapture models see Williams et al. (2002; p. 323) and Cooch and White (2008, chapter 5). The *c*-hat statistic is the ratio of the fit statistic computed for the actual data to its expected value. In classical capture-recapture applications of goodness-of-fit assessment, inference for nonfitting models is dealt with by inflating the resulting standard errors of the nonfitting model by the square-root of *c*-hat (as done in Program `Mark`, White and Burnham 1999). Kéry and Schaub (2012, p. 401) apply an idea similar to the *c*-hat adjustment but in the context of Bayesian analysis and Bayesian p-values. Kéry and Schaub compute a *c*-hat-like statistic as the ratio of the fit statistic computed for the actual data to that of the replicate data sets, and interpret the parameter as analogous to the classical *c*-hat for likelihood-based inference. Additional guidance and examples of assessment of model fit are given throughout Kéry and Royle (2016).

5.8 Open Population Models

In extending capture-recapture models to demographically open populations, the major change is that individuals may enter (by birth or immigration), or exit (death and emigration) between sampling occasions. A relatively large class of open models allows individuals only to exit the population, such as *time-since-marking models* which condition on first capture in the traditional Cormack–Jolly–Seber model. Alternatively, the model of Jolly (1965) and Seber (1965) also allows for entry into the population as well via recruitment or immigration processes. *Open population models* allow for the estimation of abundance at each capture occasion, along with population dynamics parameters such as survival and recruitment (Chapter 7). In many applications of open population models, it is conventional or even necessary to interpret survival and recruitment as *effective survival* and recruitment, which includes both true birth/death and also immigration/emigration. A critical difference between the closed models and open models that allow abundance estimation is that behavioral effects, for example trap happiness or trap avoidance, *cannot* be accommodated in traditional parameterizations of open population models, and will result in bias in N, if the effects are present. It is therefore important that newly encountered individuals have the same detection probability as previously encountered individuals. If this assumption holds, Jolly–Seber models can be useful for estimating abundance.

Historically, a number of specific formulations have been developed of open population models, which we collectively describe as Jolly–Seber type models. Cooch and White (2008; chapter 12) provide a good technical development and literature review of these different formulations of the model. One of the key formulations is the *super-population model* in the Schwarz and Arnason (1996) formulation. In this parameterization, the model is parameterized in terms of distinct survival probabilities, and the total population size over the duration of the study, called the "super-population," say N^{super}. Then, the *recruitment model* allocates the N^{super} individuals among the T years using a multinomial model based on when they recruited to the population.

A limitation of classical formulations of Jolly–Seber type models is that they are *not* formulated in terms of individual encounter histories, but instead are based on sufficient statistic summaries of the data such as encounter history frequencies. As a result, modeling individual level effects, for example individual covariates, heterogeneity, or latent spatial location as in SCR models, is not possible. However, Bayesian analysis of Jolly–Seber type models based on data augmentation (Royle and Dorazio 2008; Kéry and Schaub 2012; Royle and Dorazio 2012) permits general formulations of open population models that can easily be extended to models with individual effects. Bayesian analysis of the Jolly–Seber model using data augmentation is based on a convenient state-space formulation of the model involving an individual "alive state", $z_{i,t}$, a binary variable indicating whether individual i is alive ($z_{i,t} = 1$) or dead ($z_{i,t} = 0$) during time period t. The state model allowing for recruitment and survival is described by two component models, one for the initial state

$$z_{i,t} \sim \text{Bernoulli}(\gamma_1), \tag{5.34}$$

and a model describing subsequent state transitions:

$$z_{i,t+1} \mid z_{i,1}, \ldots, z_{i,t} \sim \text{Bernoulli}\left(z_{i,t}\phi_{i,t} + \gamma_{t+1} \prod_{k=1}^{t} (1 - z_{i,k}) \right), \tag{5.35}$$

where $\phi_{i,t}$ is the apparent survival probability of individual i during the interval $(t, t+1)$ and γ_t are effective "recruitment" parameters. In the context of Bayesian analysis using the data augmentation formulation, the data set is augmented up to a large size M, and then γ_1 is the probability that $z_1 = 1$ in period 1 relative to a super population of size M and $\gamma_2, \ldots, \gamma_{T-1}$ are then probabilities relative to individuals that have not yet been recruited. The γ_t parameters are related to recruitment in the sense that the number of births during period t, B_t, is a function of the γ_t parameters. For $t = 1$, $E(B_1) = M\gamma_1$ and for $t > 1$,

$$E(B_t) = M \prod_{k=1}^{t-1} (1 - \gamma_k)\gamma_t. \tag{5.36}$$

The observation model for this formulation of the Jolly–Seber model is specified conditional on the latent state variable $z_{i,t}$. For example, when there is only a single sampling of the population during each period t, the observation model is

$$y_{i,t} \mid (z_{i,t} = 1) \sim \text{Bernoulli}(p_{i,t}) \tag{5.37}$$

and, if $z_{i,t} = 0$, $y_{i,t} = 0$ with probability 1.

An advantage of this state-space formulation of the Jolly–Seber model is that it is based on individual observations and individual states, and therefore one can develop general models in which parameters depend on individual. As an example, a model with individual heterogeneity in $\phi_{i,t}$ is

$$\text{logit}(\phi_{i,t}) = \mu_t + \eta_i, \tag{5.38}$$

where $\eta_i \sim \text{Normal}(0, \sigma^2)$. It also easily accommodates a SCR observation model (Gardner et al. 2010b) in which the observations also depend on trap locations $\mathbf{x}_j$ (the location of trap j), and individual activity centers $\mathbf{s}_i$ as in the basic SCR models described previously.

We noted at the beginning of this section that there are several classical non-Bayesian formulations of the Jolly–Seber model which are not based on individual level encounter data. Conversely, there are also a number of distinct Bayesian formulations of individual-level Jolly–Seber models. Each set of models are effectively different parameterizations for the recruitment process. See Kéry and Schaub (2012) for details and examples of the available models.

5.9 Software Tools

Capture-recapture models described in this chapter are easily analyzed using both Bayesian and classical likelihood methods. Most applied Bayesian analyses are conducted with general-purpose software for carrying out MCMC, such as `WinBUGS` (Lunn et al. 2000) or `JAGS`

(Plummer 2003). Given what amounts to a pseudo-code description of the model, each of these pieces of software comes up with an algorithm for conducting the MCMC sampling of the posterior distribution. These BUGS engines can be called from R using convenient packages R2WinBUGS (Sturtz et al. 2005), Rjags (Plummer 2003), or jagsUI (Kellner 2014). Most of the existing BUGS implementations of capture-recapture models for estimating N are based on the use of data augmentation (Royle et al. 2007a; section 4.1). Extensive coverage of implementations based on data augmentation using the BUGS language to fit closed and open population models can be found in Royle and Dorazio (2008), Kéry (2010), and Kéry and Schaub (2012). The general considerations for choosing priors, and conducting MCMC, are not unique to capture-recapture models and so we avoid a detailed discussion here, but see Kéry (2010), Link and Barker (2009), or King et al. (2010), and especially Kéry and Schaub (2012) which has quite a bit of material on analysis of both closed and open population capture-recapture models.

The most frequently used software today for likelihood-based abundance estimation and inference about population size is Program MARK (White and Burnham 1999; Cooch and White 2008), which has largely replaced earlier packages for fitting closed population models (primarily CAPTURE; White 1982). Within Program MARK, one can fit a wide variety of model types, including most broadly, either the conditional likelihood (Huggins 1989; Alho 1990) or full likelihood models (section 14.3 of Cooch and White 2008. The design matrix within MARK provides flexibility to the user in terms of effects to include (time and behavior). Heterogeneity can be accounted for in several ways: within the conditional likelihood models, individual covariates can be integrated to model heterogeneity in detection. The finite mixture model of Pledger (2004) can also be fit. Program MARK also allows users to fit the Jolly–Seber models, other classes of open population models, and multistate models. An R package, Rmark, is also available to facilitate use of Program MARK for R users.

A number of specialized R packages exist for inference about N in capture-recapture like data. The package secr can fit a wide range of SCR models including those relevant to camera trapping, live trapping, sampling by acoustic detectors, and many other classes of models (Efford 2011). The R package unmarked was designed specifically to deal with data from unmarked individuals (Fiske and Chandler 2011), and implements likelihood analysis of certain classes of multinomial-mixture models described by Chandler et al. (2011). See also chapter 7 of Kéry and Royle (2016). The model structure which can be handled by unmarked allows for multinomial observations (which includes basic capture-recapture models) based on population sizes N_g and prior distributions for N_g such as Poisson or negative binomial, where the mean may depend on explicit covariates.

5.10 Online Exercises

We illustrate the application of some of the standard capture-recapture models as well as a SECR using captures of deer mice (*Peromyscus maniculatus*, Box 5.1). An R script to process the data and reproduce the analyses is available in the R package oSCR (Sutherland et al. 2019 see the help file? peromyscus; the oSCR package can be installed from github [Box 5.1]).

Box 5.1 Example of *Peromyscus* Trapping Grid Data

We illustrate the application of some of the standard capture-recapture models as well as a SECR using captures of deer mice (*Peromyscus maniculatus*). Mice were captured over a five-day period on a trapping grid comprised of 121 live traps arranged in a square with 25-m spacing between traps (Figure B5.1.1). For this specific trapping grid, $n = 66$ individuals were captured a total of 185 times over the $K = 10$ sampling occasions (five morning and five afternoon occasions). The data represent a subset of a larger dataset from an experiment using spatially and temporally replicated trapping grids to evaluate impacts of forest fuels management practices (Converse et al. 2006; Royle and Converse 2014). An R script to process the data and reproduce the analyses here is available in the R package oSCR (see the help file? peromyscus; the oSCR package can be installed from here: https://github.com/jaroyle/oSCR).

For this data set, we considered versions of models M_t, M_b, and M_h and various combinations of these models. Time of day is thought to affect encounter probability due to the nocturnal habits of deer mice. Thus, we model this time effect as a binary "dummy variable" distinguishing morning and afternoon trap checks, and include it as an additive effect on the logit of detection probability (i.e. a version of model M_t). As is standard in small mammal trapping studies, the traps were baited, here with rolled oats and chicken feed, and so it is reasonable to expect a positive or trap-happy behavioral response. This effect

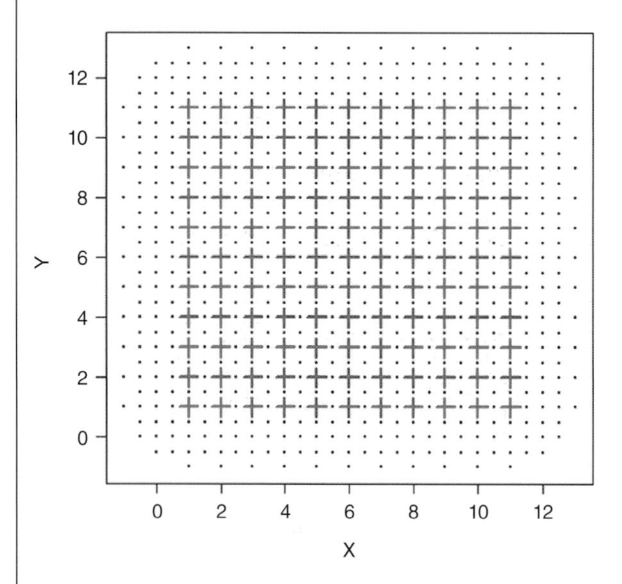

Figure B5.1.1 Trapping grid of 121 traps (red + signs) and the state-space (black dots) used in fitting the spatial capture-recapture (SCR) model.

was modeled using a dummy variable indicating whether, on a given occasion, an animal had been previously captured (M_b). Again, the behavior effect was included as an additive effect on the logit of detection probability. We considered the behavior effect alone and in combination with the time effect. We also fit two versions of model M_h: the logit-normal model and the two-component fixed mixture model. Model selection results based on AIC indicated a clear importance of both the behavioral response, which was strongly positive (i.e. trap happy), and a time effect with higher detection probability at morning trap checks.

We also fit several SCR models. These models contain an additional parameter σ (not related to the parameter of the same label in the logit-normal heterogeneity model) which relates encounter probability to the distance between individual activity center (a latent variable) and traps. Therefore, SCR models account for individual heterogeneity in the form of an explicit mechanism – spatial organization of traps relative to individual home ranges, as opposed to regarding it as a latent variable having no explicit interpretation – as is the case in the two versions of model M_h presented above. All SCR models require specification of an explicit state-space, defining the possible locations for activity centers. For the analysis here, we used a regular spacing of points near the trapping grid. We considered including either a behavioral response alone or in combination with the time effect. For the model fit here, we used a local behavioral response in which the probability of capture at a trap changes only for the trap of capture, but not other traps. It is not possible to formally compare the SCR models to ordinary capture-recapture models because the data used to fit each class of models is different. For ordinary capture-recapture models, the data are the 2d array of individual × occasion encounter histories, and for the SCR models the data are the 3d array of individual × trap × occasion encounter histories.

The key issue that motivates the importance of SCR models compared to ordinary capture-recapture models is that the interpretation of $\hat{N}$ under the ordinary capture-recapture models is imprecise. In other words, the description of any ordinary CR model does not involve any explicit statement about the relevant sampling area that N applies to. Conversely the SCR model uses the state-space which is part of the model in the sense that it defines the support of the individual random effect $\mathbf{s}_i$, and thus it interacts explicitly with the data and SCR model parameters as part of the formal estimation procedure. If you change S then N will change. As with the ordinary CR model, the best SCR model contains both a behavioral response and a time effect. We note that these are not directly comparable to those of the ordinary model because we fit a model in which the behavioral response is trap-specific, and the probability of encounter is also affected by distance. This model estimates about 6.75 individuals/ha which is more directly interpretable than is the estimate of $\hat{N} = 68$ under the similar ordinary CR model. If animals trapped on the grid only occupied the area of the trapping grid itself, 6.25 ha, we would have a density estimate of 68/6.25 ha or 10.88 individuals/ha. We can see that the spatial CR model has allowed us to account for the larger effective sample area, and without having to guess at its extent.

5.11 Future Directions

In this chapter we provided an overview of capture-recapture models for estimating population size. We covered the basic catalog of models introduced by Otis et al. (1978), including models M_0, M_b, M_t, and two treatments of heterogeneity models (model M_h), which allow unstructured individual variation in encounter or detection probability p. We also covered individual covariate models, in which one or more covariates thought to influence detection probability is measured on each individual that is captured (model M_x). Individual covariate models provide a natural transition to the relatively new class of models known as SCR or SECR. SCR models are similar to individual covariate models, where the individual covariate is the unobserved *activity center* of individuals. SCR models link observations to the activity center using a model for encounter probability that posits a decreasing function of distance between traps and activity centers. SCR models show promise for integrating many theories of spatial population ecology with individual encounter history data obtained by standard methods for the study of animal populations such as camera trapping and noninvasive genetic sampling (Royle et al. 2018). Last, we reviewed Jolly–Seber type models for estimation of abundance in open populations.

However, there are a number of topics not covered here, or only mentioned sparingly, which are active research areas. One topic of practical importance that needs more attention is the assessment of model fit for capture-recapture models that include individual-level effects. Model fit is an issue whether one adopts a likelihood or Bayesian approach to inference. Work is needed, both to develop fit statistics that are suitable under important assumption violations, and also studies to assess the power of those fit statistics.

Another methodological area of enormous practical relevance is that of modeling misidentification or misclassification of individuals in capture-recapture. Interest in this problem has been motivated largely by errors in genetic identification of individuals (Lukacs and Burnham 2005; Link et al. 2010) and, as a result of that specific context, stands to grow more in importance as genetic methods become less expensive and more widely used. However, the concept of uncertain identity is also relevant to camera trapping (Royle 2015; Augustine et al. 2018), and sampling based on multiple marking methods (McClintock 2015).

Methods that estimate population size or density in the absence of unique identification of individuals also seem promising. There are at least two distinct lines of work: one based on SCR, such that there is auxiliary information about where the encounters are made (Chandler and Royle 2013; Sollmann et al. 2013a,b), and one in which the spatial observations are assumed to be measurements on independent *replicate* populations (Dail and Madsen 2011). We covered SCR models in this chapter, but we think this topic will continue to expand as technology becomes cheaper and more widely deployed. A recent synthesis of SCR methods is Royle et al. (2014). Mark-resight models, which combine data on marked and unmarked individuals, are under active development and have great practical relevance to many studies in which capture of individuals is difficult (McClintock et al. 2009; Pledger et al. 2009; Sollmann et al. 2013b; Lyons et al. 2016).

Due to the explosive growth of new technologies such as camera trapping and DNA sampling, capture-recapture methods can be applied in situations that only a few years ago were impractical to study. Due to rapid and widespread adoption of these new technologies, capture-recapture methods are more practical and relevant than ever before. New technologies combined with new analytic methods stand to revolutionize the ways that animal populations are studied.

References

Alho, J.M. (1990). Logistic regression in capture-recapture models. *Biometrics* 46: 623–635.

Augustine, B., Royle, J.A., Kelly, M. et al. (2018). Spatial capture-recapture with partial identity: an application to camera traps. *Annals of Applied Statistics* 12: 67–95.

Basu, S. and Ebrahimi, N. (2001). Bayesian capture-recapture methods for error detection and estimation of population size: heterogeneity and dependence. *Biometrika* 88: 269–227.

Borchers, D. (2012). A non-technical overview of spatially explicit capture-recapture models. *Journal of Ornithology* 152: 435–444.

Borchers, D.L., Buckland, S.T., and Zucchini, W. (2002). *Estimating Animal Abundance: Closed Populations.* London: Springer-Verlag.

Borchers, D.L. and Efford, M.G. (2008). Spatially explicit maximum likelihood methods for capture-recapture studies. *Biometrics* 64: 377–385.

Boulanger, J. and McLellan, B. (2001). Closure violation in DNA-based mark-recapture estimation of grizzly bear populations. *Canadian Journal of Zoology* 79: 642–651.

Burnham, K.P. 1972. Estimation of population size in multiple capture-recapture studies when capture probabilities vary among animals. PhD Dissertation, Oregon State University.

Burnham, K.P. and Anderson, D.R. (2002). *Model Selection and Multi-Model Inference: A Practical Information-Theoretic Approach.* London: Springer-Verlag.

Burnham, K.P. and Overton, W.S. (1978). Estimation of the size of a closed population when capture probabilities vary among animals. *Biometrika* 65: 625–633.

Castledine, B.J. (1981). A Bayesian analysis of multiple-recapture sampling for a closed population. *Biometrika* 68: 197–210.

Chandler, R.B. and Royle, J.A. (2013). Spatially-explicit models for inference about density in unmarked or partially marked populations. *Annals of Applied Statistics* 7: 936–954.

Chandler, R.B., Royle, J.A., and King, D. (2011). Inference about density and temporary emigration in unmarked populations. *Ecology* 92: 1429–1435.

Chao, A. (1984). Nonparametric estimation of the number of classes in a population. *Scandinavian Journal of Statistics* 11: 265–270.

Chao, A. (1987). Estimating the population size for capture-recapture data with unequal catchability. *Biometrics* 43: 783–791.

Converse, S.J. and Royle, J.A. (2012). Dealing with incomplete and variable detectability in multi-year, multi-site monitoring of ecological populations. In: *Design and Analysis of Long-Term Ecological Monitoring Studies*, 560 p., vol. xxiv (eds. R.A. Gitzen, J.J. Millspaugh, A.B. Cooper and D.S. Licht), 426–442. Cambridge University Press.

Converse, S.J., White, G.C., and Block, W.M. (2006). Small mammal responses to thinning and wildfire in ponderosa pine-dominated forests of the southwestern United States. *Journal of Wildlife Management* 70: 1711–1722.

Cooch, E.G., White G.C. 2008. Program MARK: a gentle introduction.

Coull, B.A. and Agresti, A. (1999). The use of mixed logit models to reflect heterogeneity in capture-recapture studies. *Biometrics* 55: 294–301.

Dail, D. and Madsen, L. (2011). Models for estimating abundance from repeated counts of an open metapopulation. *Biometrics* 67: 577–587.

Dawson, D.K. and Efford, M.G. (2009). Bird population density estimated from acoustic signals. *Journal of Applied Ecology* 46: 1201–1209.

Dorazio, R.M. and Royle, J.A. (2003). Mixture models for estimating the size of a closed population when capture rates vary among individuals. *Biometrics* 59: 351–364.

Efford, M.G. (2004). Density estimation in live-trapping studies. *Oikos* 106: 598–610.

Efford, M.G. 2011. secr: Spatially explicit capture-recapture models. http://CRAN.R-project.org/package=secr

Fienberg, S.E., Johnson, M., and Junker, B. (1999). Classical multilevel and Bayesian approaches to population size estimation using multiple lists. *Journal of the Royal Statistical Society of London A* 163: 383–405.

Fiske, I.J. and Chandler, R.B. (2011). unmarked: an R package for fitting hierarchical models of wildlife occurrence and abundance. *Journal of Statistical Software* 43: 1–23.

Fuller, A.K., Sutherland, C.S., Royle, J.A., and Hare, M.P. (2016). Estimating population density and connectivity of American mink using spatial capture-recapture. *Ecological Applications* 26: 1125–1135.

Gardner, B., Reppucci, J., Lucherini, M., and Royle, J.A. (2010a). Spatially explicit inference for open populations: estimating demographic parameters from camera-trap studies. *Ecology* 91: 3376–3383.

Gardner, B., Royle, J.A., Wegan, M.T., and Rainbolt, R.E. (2010b). Estimating black bear density using DNA data from hair snares. *Journal of Wildlife Management* 74: 318–325.

Gelman, A., Carlin, J.B., Stern, H.S., and Rubin, D.B. (2004). *Bayesian Data Analysis*, 2e. Boca Raton, Florida, USA: CRC/Chapman and Hall.

Gelman, A., Meng, X.L., and Stern, H. (1996). Posterior predictive assessment of model fitness via realized discrepancies. *Statistica Sinica* 6: 733–759.

George, E.I. (1992). Capture-recapture estimation via Gibbs sampling. *Biometrika* 79: 677–683.

Gopalaswamy, A.M., Royle, J.A., Delampady, M. et al. (2012). Density estimation in tiger populations: combining information for strong inference. *Ecology* 93: 1741–1751.

Grueber, C.E., Nakagawa, S., Laws, R.J., and Jamieson, I.G. (2011). Multimodel inference in ecology and evolution: challenges and solutions. *Journal of Evolutionary Biology* 24: 699–711.

Hooten, M.B. and Hobbs, N.T. (2015). A guide to Bayesian model selection for ecologists. *Ecological Monographs* 85: 3–28.

Horvitz, D.G. and Thompson, D.J. (1952). A generalization of sampling without replacement from a finite universe. *Journal of the American Statistical Association* 47: 663–685.

Huggins, R.M. (1989). On the statistical analysis of capture experiments. *Biometrika* 76: 133–140.

Ivan, J., White, G.C., and Shenk, T.M. (2013). Using auxiliary telemetry information to estimate animal density from capture-recapture data. *Ecology* 94: 809–816.

Jolly, G.M. (1965). Explicit estimates from capture-recapture data with both death and dilution-stochastic model. *Biometrika* 52: 225–247.

Karanth, K.U. and Nichols, J.D. (1998). Estimation of tiger densities in India using photographic captures and recaptures. *Ecology* 79: 2852–2862.

Kellner, K. 2014. jagsUI: Run JAGS from R (an alternative user interface for rjags). R package version, 1(1).

Kendall, W.L., Pollock, K.H., and Brownie, C. (1999). A likelihood-based approach to capture-recapture estimation of demographic parameters under the robust design. *Biometrics* 51: 293–308.

Kéry, M. (2010). *Introduction to WinBUGS for Ecologists: A Bayesian Approach to Regression, ANOVA and Related Analyses*. London: Academic Press, Elsevier.

Kéry, M. and Royle, J.A. (2016). *Applied Hierarchical Modeling in Ecology: Analysis of distribution, abundance, and species richness in R and BUGS: Volume 1: Prelude and Static Models*. Academic Press.

Kéry, M. and Schaub, M. (2012). *Bayesian Population Analysis Using WinBugs*. London: Academic Press, Elsevier.

King, R. and Brooks, S.P. (2001). On the Bayesian analysis of population size. *Biometrika* 88: 317–336.

King, R., Morgan, B., Gimenez, O., and Brooks, S. (2010). *Bayesian Analysis for Population Ecology*. Boca Raton, Florida: Chapman and Hall/CRC.

Kuo, L. and Mallick, B. (1998). Variable selection for regression models. *Sankhya: The Indian Journal of Statistics Series B* 60: 65–81.

Link, W.A. (2003). Nonidentifiability of population size from capture-recapture data with heterogeneous detection probabilities. *Biometrics* 59: 1123–1130.

Link, W.A. and Barker, R.J. (2009). *Bayesian Inference: With Ecological Applications*. London: Academic Press.

Link, W.A., Yoshizaki, J., Bailey, L.L., and Pollock, K.H. (2010). Uncovering a latent multinomial: analysis of mark-recapture data with misidentification. *Biometrics* 66: 178–185.

Long, R.A., MacKay, P., Ray, J., and Zielinski, W. (eds.) (2008). *Noninvasive Survey Methods for Carnivores*. Washington, DC: Island Press.

Lukacs, P.M. and Burnham, K.P. (2005). Review of capture-recapture methods applicable to noninvasive genetic sampling. *Molecular Ecology* 14: 3909–3919.

Lunn, D.J., Thomas, A., Best, N., and Spiegelhalter, D. (2000). WinBUGS-a Bayesian modelling framework: concepts, structure, and extensibility. *Statistics and Computing* 10: 325–337.

Lyons, J.E., Kendall, W.L., Royle, J.A. et al. (2016). Population size and stopover duration estimation using mark-resight data and Bayesian analysis of a superpopulation model. *Biometrics* 72: 262–271.

MacKenzie, D.I. and Bailey, L.L. (2004). Assessing the fit of site-occupancy models. *Journal of Agricultural, Biological, and Environmental Statistics* 9: 300–318.

McClintock, B.T. (2015). multimark: an R package for analysis of capture-recapture data consisting of multiple “noninvasive” marks. *Ecology and Evolution* 5: 4920–4931.

McClintock, B.T., White, G.C., Antolin, M.F., and Tripp, D. W. (2009). Estimating abundance using mark-resight when sampling is with replacement or the number of marked individuals is unknown. *Biometrics* 65: 237–246.

Millar, R.B. (2009). Comparison of hierarchical Bayesian models for overdispersed count data using DIC and Bayes' factors. *Biometrics* 65: 962–969.

Norris, J.L. and Pollock, K.H. (1996). Nonparametric MLE under two closed capture-recapture models with heterogeneity. *Biometrics* 52: 639–649.

O'Connell, A.F., Nichols, J.D., and Karanth, U.K. (2010). *Camera Traps in Animal Ecology: Methods and Analyses*. London: Springer.

O'Hara, R.B. and Sillanpää, M.J. (2009). A review of Bayesian variable selection methods: what, how and which. *Bayesian Analysis* 4: 85–118.

Otis, D.L., Burnham, K.P., White, G.C., and Anderson, D.R. (1978). Statistical inference from capture data on closed animal populations. *Wildlife Monographs* 62: 3–135.

Pledger, S. (2000). Unified maximum likelihood estimates for closed capture-recapture models using mixtures. *Biometrics* 56: 434–442.

Pledger, S. (2004). Unified maximum likelihood estimates for closed capture–recapture models using mixtures. *Biometrics* 56: 434–442.

Pledger, S., Efford, M.G., Pollock, K.H. et al. (2009). Stopover duration analysis with departure probability dependent on unknown time since arrival. In: *Modeling Demographic Processes in Marked Populations*, 349–363. London: Springer.

Plummer, M. (2003). JAGS: A program for analysis of Bayesian graphical models using Gibbs sampling. In: *Proceedings of the 3rd international workshop on distributed statistical computing*, vol. 124, 125. Wien, Austria: Technische Universit at Wien.

Pollock, K.H. (1982). A capture-recapture design robust to unequal probability of capture. *Journal of Wildlife Management* 46: 757–757.

Royle, J.A. (2008). Hierarchical modeling of cluster size in wildlife surveys. *Journal of Agricultural, Biological, and Environmental Statistics* 13: 23–36.

Royle, J.A. (2009). Analysis of capture-recapture models with individual covariates using data augmentation. *Biometrics* 65: 267–274.

Royle, J.A. 2015. Spatial Capture-recapture with Partial Identity. arXiv preprint *arXiv*, 1503.06873.

Royle, J.A., Chandler, R.B., Sollmann, R., and Gardner, B. (2014). *Spatial Capture-Recapture*. Oxford: Academic Press, Elsevier.

Royle, J.A., Chandler, R.B., Sun, C.C., and Fuller, A.K. (2013). Integrating resource selection information with spatial capture-recapture. *Methods in Ecology and Evolution* 4: 520–530.

Royle, J.A. and Converse, S.J. (2014). Hierarchical spatial capture-recapture models: modeling population density from replicated capture-recapture experiments. *Methods in Ecology and Evolution* 5: 37–43.

Royle, J.A., Converse S.J., Link W.A. 2012. Data augmentation for hierarchical capture-recapture Models. arXiv preprint *arXiv*, 1211.5706

Royle, J.A. and Dorazio, R.M. (2008). *Hierarchical Modeling and Inference in Ecology: The Analysis of Data from Populations, Metapopulations and Communities*. London: Academic Press.

Royle, J.A. and Dorazio, R.M. (2012). Parameter-expanded data augmentation for Bayesian analysis of capture-recapture models. *Journal of Ornithology* 152: S521–S537.

Royle, J.A., Dorazio, R.M., and Link, W.A. (2007a). Analysis of multinomial models with unknown index using data augmentation. *Journal of Computational and Graphical Statistics* 16: 67–85.

Royle, J.A., Fuller, A.K., and Sutherland, C. (2016). Spatial capture–recapture models allowing Markovian transience or dispersal. *Population Ecology* 58: 53–62.

Royle, J.A., Fuller, A.K., and Sutherland, C. (2017). Unifying population and landscape ecology with spatial capture–recapture. *Ecography* 40: 1–12.

Royle, J.A., Kéry, M., Gautier, R., and Schmid, H. (2007b). Hierarchical spatial models of abundance and occurrence from imperfect survey data. *Ecological Monographs* 77: 465–481.

Sanathanan, L. (1972). Estimating the size of a multinomial population. *Annals of Mathematical Statistics* 43: 142–152.

Sandland, R.L. and Cormack, R.M. (1984). Statistical inference for Poisson and multinomial models for capture-recapture experiments. *Biometrika* 71: 27–33.

Schwarz, C.J. and Arnason, A.N. (1996). A general methodology for the analysis of capture-recapture experiments in open populations. *Biometrics* 52: 860–873.

Seber, G.A.F. (1965). A note on the multiple-recapture census. *Biometrika* 52: 249–259.

Seber, G.A.F. (1982). *The Estimation of Animal Abundance and Related Parameters*. New York: Macmillan Publishing Co.

Smith, P.J. (1991). Bayesian analyses for a multiple capture-recapture model. *Biometrika* 78: 399–407.

Sollmann, R., Gardner, B., Chandler, R.B. et al. (2013a). Using multiple data sources provides density estimates for endangered Florida panther. *Journal of Applied Ecology* 50: 961–968.

Sollmann, R., Gardner, B., Parsons, A.W. et al. (2013b). A spatial mark-resight model augmented with telemetry data. *Ecology* 93: 553–559.

Spiegelhalter, D.J., Best, N.G., Carlin, B.P., and Van Der Linde, A. (2002). Bayesian measures of model complexity and fit. *Journal of the Royal Statistical Society B* 64: 583–639.

Sturtz, S., Ligges, U., and Gelman, A.E. (2005). `R2WinBUGS`: a package for running Win-BUGS from R. *Journal of Statistical Software* 12: 1–16.

Sutherland, C., Fuller, A.K., and Royle, J.A. (2015). Modelling non-Euclidean movement and landscape connectivity in highly structured ecological networks. *Methods in Ecology and Evolution* 6: 169–177.

Sutherland, C., Royle, J.A., and Linden, D.W. (2019). oSCR: A spatial capture-recapture R package for inference about spatial ecological processes. *Ecography* 42: 1–11.

Tenan, S., O'Hara, R.B., Hendriks, I., and Tavecchia, G. (2014). Bayesian model selection: the steepest mountain to climb. *Ecological Modelling* 283: 62–69.

White, G.C. (1982). *Capture-Recapture and Removal Methods for Sampling Closed Populations*. Los Alamos National Laboratory.

White, G.C. and Burnham, K.P. (1999). Program `MARK`: survival estimation from populations of marked animals. *Bird Study* 46: 120–138.

White, G.C. and Shenk, T.M. (2001). Population estimation with radio-marked animals. In: *Radio Tracking and Animal Populations* (eds. J.J. Millspaugh and J.M. Marzluff), 329–350. San Diego, California, USA: Academic Press.

Williams, B.K., Nichols, J.D., and Conroy, M.J. (2002). *Analysis and Management of Animal Populations: Modeling, Estimation, and Decision Making*. San Diego, California, USA: Academic Press.

Yang, H.C. and Chao, A. (2005). Modeling animals' behavioral response by Markov chain models for capture–recapture experiments. *Biometrics* 61: 1010–1017.

6

Estimating Survival and Cause-specific Mortality from Continuous Time Observations

Dennis L. Murray and Guillaume Bastille-Rousseau

Department of Biology, Trent University, Peterborough, Ontario, Canada

Summary

Ecologists must understand the causes and demographic consequences of mortality to properly conserve and manage populations. Survival data collected in continuous time can yield robust survival estimates with high precision, but currently these methods are underused in ecology. Continuous time survival data can be derived from studies ranging from direct observation of stationary nests to radio telemetry of free-ranging animals; consistently, these designs involve intensive monitoring and high probability of detection of individual subjects. Continuous time survival rates are calculated using estimators that define time intervals either by constant risk periods (e.g. Mayfield, Heisey–Fuller) or mortality events (e.g. Kaplan–Meier, Nelson–Aalen); these estimators differ in their assumptions and suitability for specific study designs and datasets. Univariate survival rate comparison using simple nonparametric tests is ill-suited for ecological data because of common irregularities like few mortalities, staggered entry of subjects, and right censoring of survival timelines. Semi-parametric Cox proportional hazard (CPH) models offer robust insight into relative hazard while allowing researchers the flexibility to address study design complexities including multiple predictors, random effects, and time-dependent variables. Fully parametric survival models rarely provide improvement over a semi-parametric approach and require that underlying survival distributions are known, which is uncommon in ecology. An additional advantage of a continuous time study design is that precise timing of the mortality event is determined, potentially also allowing researchers to identify cause of death. Cause of death information is the basis for competing risks analysis, which extends the CPH approach to multiple mortality agents. Although infrequently used in ecology, competing risks analysis can be especially useful in conservation and management by revealing the relative importance of different risk types and whether they are additive or compensatory to other mortality sources. Knowledge gained from competing risks analysis can be especially valuable for appropriately targeting mitigation efforts. Newer approaches in survival analysis, including mixed-effects modeling and Bayesian methods, hold promise for refining inference from continuous time datasets. In sum, research in ecology will benefit from expanded collection and analysis of continuous time survival data, with new tracking technologies like camera-based monitoring and satellite-based radio telemetry being especially noteworthy for supporting novel analysis and insight. Ultimately, better integration of continuous time survival and competing risks analyses will contribute importantly to future advances in population ecology and conservation biology.

6.1 Introduction

Population ecologists have a longstanding interest in understanding the causes and consequences of mortality in organisms. Understanding when and where an animal or plant dies, what is the cause of death, and whether predisposing factors led to the mortality event, is of paramount relevance to research ranging from demographic analysis to conservation biology (Grosbois et al. 2008; Nussey et al. 2008). At its core, an individual's survival rate (or mortality rate, where survival = 1 − mortality) is central to estimates of fitness and thus is a necessary consideration when investigating evolutionary processes (Metcalf and Pavard 2007). Population-level survival and mortality estimates are fundamental to understanding population status and viability, community interactions, and ecosystem resilience (McCallum 2000; Morris and Doak 2004). Yet, population ecologists face substantive logistical challenges when studying survival and mortality in natural settings. Free-living organisms often occur at low density and have cryptic lifestyles, or else exhibit elusive or vagile behavior or prolonged periods of stasis or dormancy. Not only do such constraints make it difficult to implement a robust survival

Population Ecology in Practice, First Edition. Edited by Dennis L. Murray and Brett K. Sandercock.

Companion website: www.wiley.com/go/MurrayPopulationEcology

monitoring schedule, but they limit the ability to effectively survey mortality risk continuously through time. When survival monitoring is infrequent or has a low probability of detection, status and fate of individuals cannot be inferred directly, which challenges our ability to obtain unbiased survival estimates. In fact, the reliability and rigor of survival research is a general concern in population ecology (McCallum 2000; Williams et al. 2002; Murray 2006), leading to efforts to improve monitoring and estimation through both careful design and implementation of observational studies, and adoption of analytical procedures that provide robust inference while accounting for limitations in ecological datasets.

Contemporary methods in survival analysis repeatedly track uniquely marked individuals until they reach an endpoint like death or loss from the survey (Figure 6.1). Owing to the challenges of monitoring individual free-living organisms and documenting their fate, survival research in ecology can involve indirect and infrequent detection, leading to incomplete confirmation of the organism's status and discrete survey events that are separated by time gaps that can last days, months, or even years. In fact, ecologists commonly survey animals via live-capture, opportunistic field observation, or from noninvasive genetic methods, and then use *capture-mark-recapture* (CMR) statistics to estimate probability of survival (Chapter 7). Survival estimates based on CMR methods are usually approximations (i.e. "apparent survival," sensu White and Burnham 1999) because the sampling protocol is discrete, and mortality, emigration, and other fates usually are not known. Alternative models are being developed to more precisely estimate survival from *discrete time* data (Barbour et al. 2013; Schaub and Royle 2013; Chapter 7), but there is no substitute for monitoring individuals continuously through time and documenting their fate directly. Yet, this protocol requires specialized monitoring and statistical procedures based on *known-fate methods* (sensu White and Burnham 1999). *Continuous time* survival monitoring is logistically more demanding and not possible or

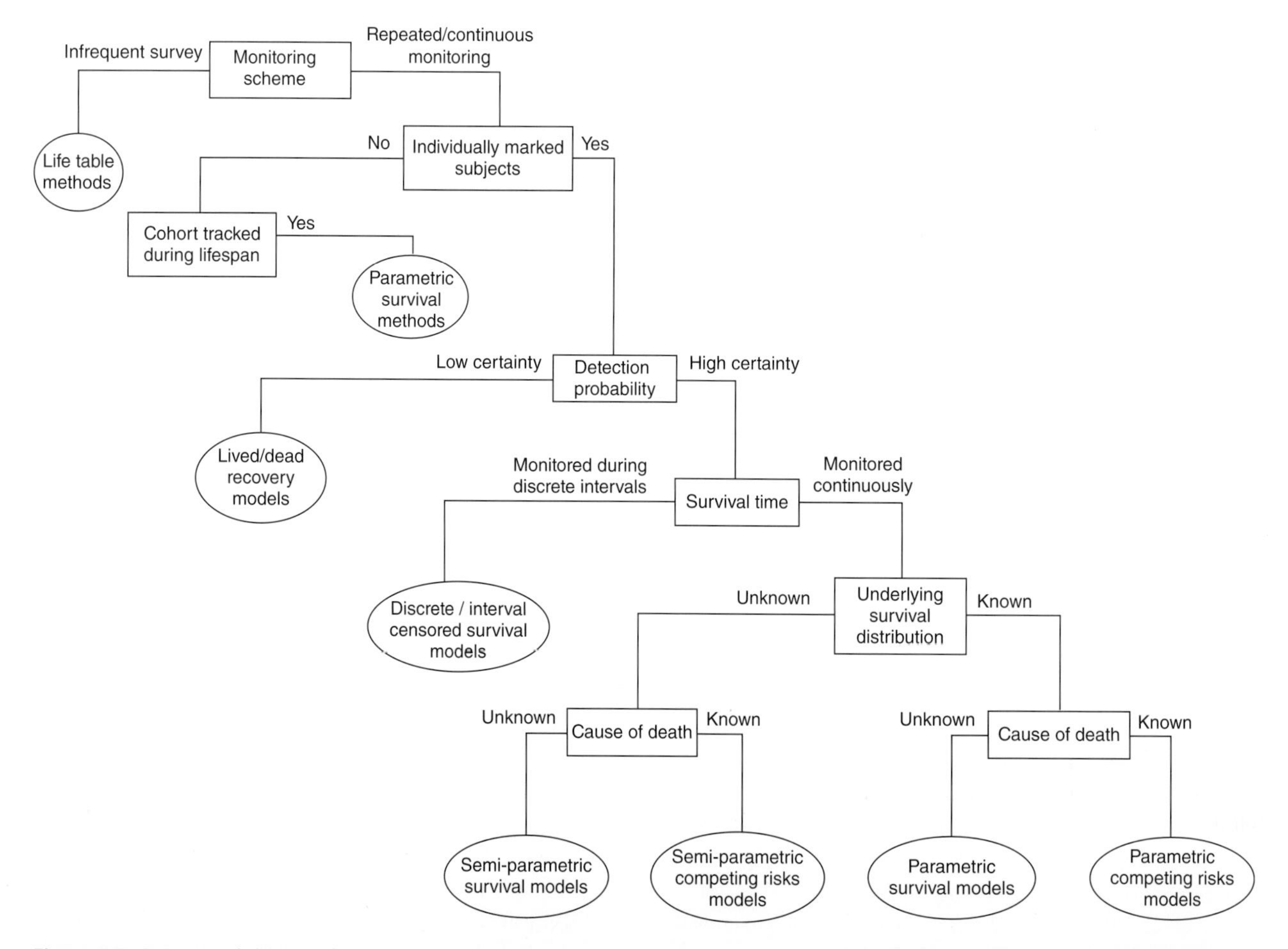

Figure 6.1 Conceptual diagram for selecting a continuous time survival analysis approach. Individuals must be distinguished and monitored repeatedly with high detection probability. Monitoring frequency should be sufficiently intense to develop a continuous survival timeline. Whether the underlying survival distribution and cause of death are known will determine whether fully parametric and competing risks methods can be used. *Source:* Adapted from Murray and Patterson (2006).

appropriate for all field studies, but when available, data from such studies can offer less bias and increased precision compared to their discrete time counterparts. Notably, discrete time and continuous time analytical methods yield similar survival estimates when monitoring and detection frequency converge, even though both approaches remain philosophically and computationally distinct (Efron 1988; Williams et al. 2002).

6.1.1 Assumption of No Handling, Marking or Monitoring Effects

Organisms can be monitored in a continuous or semi-continuous manner using a variety of approaches, and a fundamental assumption in survival research is that researcher activities and marking methods do not affect subjects. Stationary subjects like plants or nest sites can be individually tagged and assessed manually by observers or through deployment of field cameras; many vertebrates and some invertebrates can be tagged and monitored using very high frequency (VHF) transmitters that are detected remotely using radio telemetry; more recently, larger species are monitored using Global Positioning System (GPS) or satellite-based transmitters (Bridge et al. 2011; Mansfield et al. 2012; Kissling et al. 2014). These activities and devices allow researchers to potentially track the status of individuals continuously through time, and thereby facilitate precise death time estimation compared to more passive monitoring approaches that serve in discrete time survival research. Yet, procedures like repeated visitation of nest sites, or capture and handling of animals for radio transmitter deployment, or even the effects of radio transmitters themselves on animal behavior, condition, and fitness, are potentially impactful to subjects. In fact, it is reasonable to suggest that monitoring protocols and devices used in continuous time data collection tend to be more invasive, with more frequent monitoring and more obtrusive tags, than methods used for discrete time survival research. For example, we know that if predators follow human scent when searching for food, nest-site visitation by researchers may increase predation risk (Major 1990; Kurucz et al. 2015; but see Ibáñez-Álamo et al. 2012). We also know that stress associated with animal capture and handling can have marked impacts on post-capture survival probability (Gilbert et al. 2014; Chitwood et al. 2017; DelGiudice et al. 2018). Further, radio transmitters are sometimes deployed on individuals that are either too small or too sensitive to carry the tags (Paquette et al. 1997; Saraux et al. 2011). In fact, some research activities or tag types may have subtle effects that are not easily discerned or quantified using standard monitoring approaches (Hamel et al. 2004; Brooks et al. 2008; Ludynia et al. 2012; Vandenabeele et al. 2012). Although assessing the pros and cons of different monitoring and tagging methods is beyond the scope of our chapter, we remind that the potential effect of proposed monitoring activities and devices on subjects should be understood prior to initiating a survival study. Resources are available for evaluating the potential effects of handling and marking (Murray and Fuller 2000; Barron et al. 2010), and guidelines are available to help design field procedures having minimal researcher impacts (Hawkins 2004; Casper 2009). Thus, at the outset of a continuous time survival study researchers should consider carefully whether research activities are likely to affect subjects, how field procedures and tags can be adjusted to minimize potential negative effects, and if necessary, how such effects can be detected, quantified, and addressed via data processing or analytical adjustments.

6.1.2 Cause of Death Assessment

Continuous time detection of individuals provides additional benefits by allowing researchers to determine cause of death through contemporaneous recovery of carcasses and assessment of the death site. Plants and animals normally succumb to a variety of mortality agents, and knowing what agents are implicated is important for providing a range of options in survival analysis (Figure 6.1). *Cause of death* information is especially relevant when researchers aim to focus mitigation efforts to forestall population decline or promote recovery. Yet, there are obvious logistical challenges in determining cause of death in the field, including that carcasses are difficult to recover and post-mortem exams often provide equivocal results. Even survival studies prioritizing cause of death determination can fail to confirm the source of mortality for a sample of subjects, or else confound proximate causes of death or the factors directly causing mortality (e.g. predation, starvation) with the ultimate causes (e.g. sublethal infection, malnutrition). For example, even though predation is well-known as the primary proximate cause of death in snowshoe hares (*Lepus americanus*), an experimental manipulation of sublethal parasites in hares showed that ultimately, parasitism (Murray et al. 1997) and declining body condition (Murray 2002) contribute importantly to hare predation. Likewise, sublethal parasitism increases predation risk in Red Grouse (*Lagopus scotica*; Hudson et al. 1992) whereas poor body condition predisposes coral reef fish (*Pomacentrus amboinensis*) to predation (Hoey and McCormick 2004). Thus, additional information about predisposing factors is important for comprehensive cause of death assessment. Similarly, whether specific risk factors incur additive or compensatory mortality on populations remains a longstanding, albeit understudied, interest in ecology (Burnham and Anderson 1984; Boyce et al. 1999; Murray et al. 2010), and an understanding of proximate and ultimate causes of death is an important component of such an assessment. Accordingly, continuous time survival research can contribute to disentangling the roles of different causes of death and their demographic implications.

In this chapter, we review continuous time survival analysis in field-based ecology. Continuous time survival methods (i.e. "time-to-failure" or "time-to-event" models) are appropriate in research situations where survival monitoring is frequent with a high detection probability. Herein, we highlight important considerations and limitations when designing continuous time survival research (Winterstein et al. 2001; Williams et al. 2002; Murray 2006), and focus on the application of multiple-variable regression-based approaches which are especially well-designed for the challenges associated with observational field research. Our review extends to methods that account for cause-specific hazards and competing risks. *Competing risks analysis* is well-established in disciplines like epidemiology and industrial design (Crowder 2001; Pintillie 2006; Austin et al. 2016), but until recently was rarely used in ecology (Heisey and Patterson 2006). Owing to this oversight, and because collectively these methods are rarely covered in ecology undergraduate curricula and related statistical texts, review of this topic is important for students and researchers in population ecology.

6.1.3 Historical Origins of Survival Estimation

Human demographic analysis served as basis for contemporary methods in survival analysis, and an early demographer, John Graunt (1620–1674), accessed public records of births and deaths in Renaissance London to conduct "political arithmetick" in an effort to reveal demographic trends. Graunt pooled death records into 6–10 year (age) intervals to show changes in probability of survival and age-specific mortality (Figure 6.2). These records reveal a steady decline in survival probability as individuals age, such that by age 60 only 10% of the initial cohort remained alive and by age 80 only 1% had survived. This pattern translates to high mortality in age classes spanning 0–60 years and a lower risk in later ages, likely reflecting disproportionate loss of particularly frail individuals in the younger age classes.

In principle, we can evaluate these survival and mortality rates using either a cumulative function representing probability of survival considering all previous age classes (Figure 6.2a), or as age-specific mortality rates that reflect probability of death during a specific time interval (Figure 6.2b). Note that Figures 6.2a and b could have been expressed as the mortality function and the age-specific survival rate, respectively, because of the aforementioned "1 minus" property between survival and mortality.

Today, Graunt's analysis serves as an important foundation in statistical demography (Wainer and Velleman 2001; Egerton 2005); a similar example in ecology comes from Adolph Murie's study of Dall sheep (*Ovis dalli*) skulls collected over several years near Mt. McKinley, Alaska (Murie 1944). Murie estimated age of death from tooth wear and developed *life tables* to calculate mortality and lifespan in a manner similar to Graunt's approach with human public records (Box 6.1). Murie showed that Dall sheep have qualitatively similar survival patterns to those seen in humans, except that the decline in survival probability in the middle age classes is more restrained (Figure 6.2c). Age-specific variation results in lower age-specific mortality among middle-aged sheep compared to young and old individuals (Figure 6.2d).

Note that the parallels between the Graunt and Murie datasets extend beyond the basic shape of the survival functions. The life table approach used by both researchers constitutes a rather blunt survival analysis and differs from modern methods by being retrospective and reconstructive by using information from individuals who have already died rather than those whose risk was actively monitored (Figure 6.1). Use of age at death data is a *cross-sectional* assessment of a given population rather than a *longitudinal* study that follows survival of a cohort of individuals through real time. In fact, the human and sheep studies hinge on assumptions related to temporal consistency in both mortality risk and mortality reporting, and that the population age distribution and size are virtually stationary (Anderson et al. 1981); these assumptions are probably unrealistic and untestable for either dataset. More broadly, life table methods are especially problematic for long-term study of organisms living in highly variable environments where survival monitoring is opportunistic and probability of detection is low. For example, if emigrating sheep are subject to higher mortality risk but leave no evidence of their fate during field collections, they would be under-represented and the resulting survival estimate would be biased. Similarly, if smaller or more fragile lamb skeletons are less likely to be detected, juvenile survival may be overestimated (Gilbert et al. 2014). Fundamentally, life tables fail to track how individual mortality risk varies through time. In contrast, a well-designed continuous time survival study actively tracks individuals as they experience individualized and variable risk exposure. The specific approaches and limitations of life table methods in ecology, including as a means of survival analysis, are discussed elsewhere (Hastings 1997; McCallum 2000; Neal 2004). The remainder of our chapter focuses specifically on survival monitoring of individual subjects in continuous time.

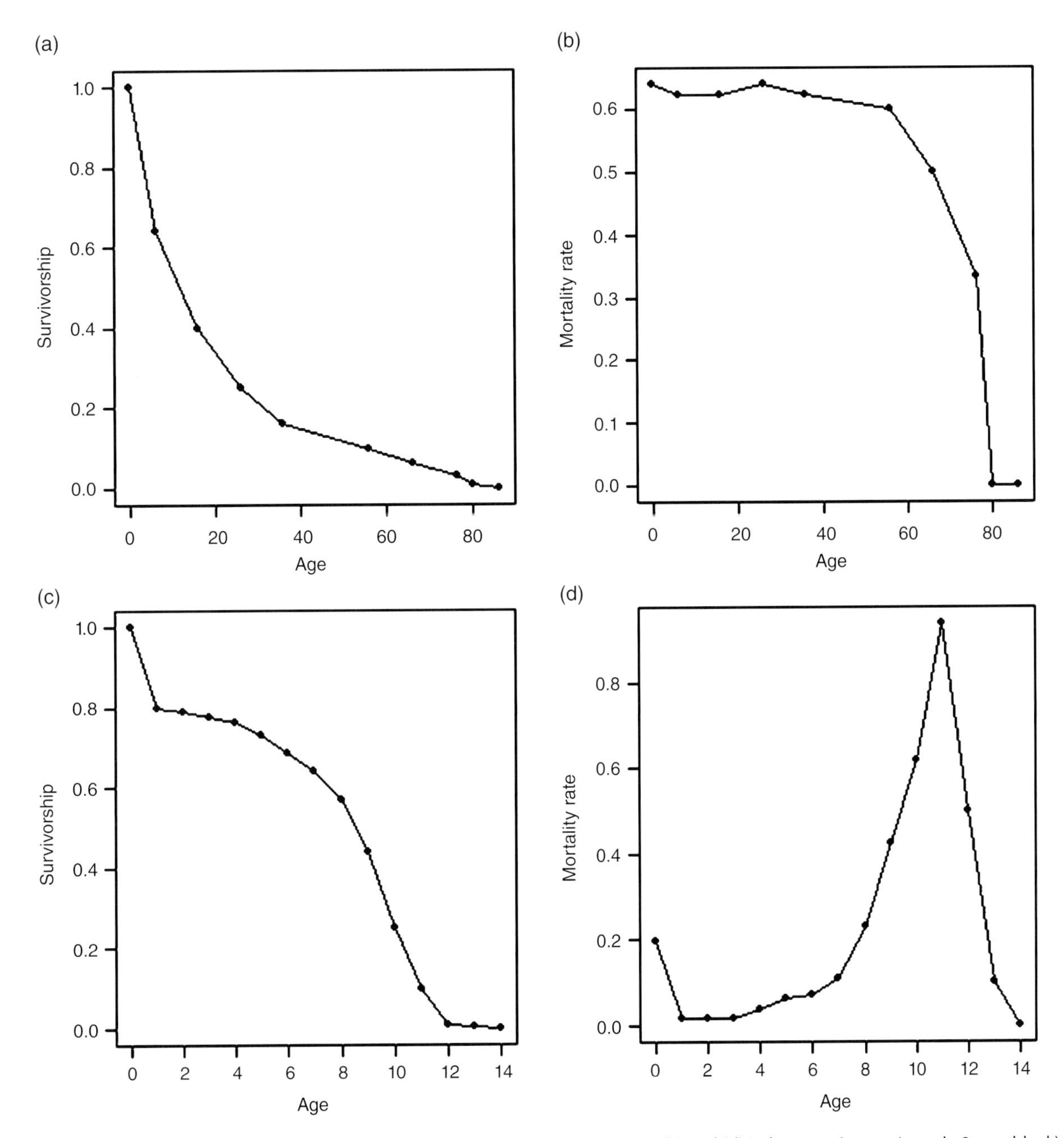

Figure 6.2 Cumulative survivorship (a) and (c) and age-specific mortality rate (b) and (d) in humans (approximately 8 year block) and Dall Sheep. Datasets for humans and sheep were from Graunt (1662) and Murie (1944), respectively.

6.2 Survival and Hazard Functions in Theory

To properly track risk through continuous time, we must establish representative survival timelines for individual subjects. Quantitative survival estimation measures the *cumulative* survival function $S(t)$ and the hazard function $h(t)$ of individuals. Assume that T is a non-negative random variable denoting time-to-failure (i.e. death), so that the survivor function represents the *cumulative probability of survival* of an individual at least until a specified time, t;

$$S(t) = \Pr(T \geq t). \tag{6.1}$$

The survivor function reflects the probability that there is no event prior to t, which is identified as the start of the monitoring period. Therefore, the survival rate equals 1 when $t = 0$ and decreases to zero as t approaches infinity (Figure 6.3a). Indeed, across a sufficiently long study duration all subjects will eventually succumb to mortality,

Box 6.1 Life Table Analysis

Murie (1944) collected skulls from Dall sheep found dead near Mt. McKinley in Alaska. Skulls were aged based on tooth wear, allowing estimation of the number of animals dying in each age class (x). We can then use a life table approach to reconstruct the demography of the population. From an initial sample of 608 sheep skulls, Murie calculated the number left alive (n_x) when expressing the starting population as 1000 individuals (i.e. conversion factor: 1000/608 = 1.645). Number of sheep dying in age class x is: $d_x = n_x - n_{x+1}$. Proportion of the total population surviving (l_x) is: $l_x = n_x/n_0$ and the mortality rate (q_x) is: $q_x = d_x/n_x$. The average number of individuals alive in each age class (L_x) is: $L_x = (n_x + n_{x-1})/2$. We can determine age-specific life expectancy (e_x) by calculating: $e_x = T_x/n_x$, where $T_x = \sum_x^\infty L_x$. The life table for the Dall sheep population is provided below, and proportion surviving and mortality rate relative to age are graphed in Figure 6.2c and d, respectively.

Age class (x)	Number alive (n_x)	Number dying (d_x)	Proportion surviving (l_x)	Mortality rate (q_x)	Avg. no. alive in age class (L_x)	T_x	Life expectancy (e_x)
0–1	1000	199	1.000	0.199	900.5	7053	7.0
1–2	801	12	0.801	0.015	795	6152.5	7.7
2–3	789	13	0.789	0.016	776.5	5357.5	6.8
3–4	776	12	0.776	0.015	770	4581	5.9
4–5	764	30	0.764	0.039	749	3811	5.0
5–6	734	46	0.734	0.063	711	3062	4.2
6–7	688	48	0.688	0.070	664	2351	3.4
7–8	640	69	0.640	0.108	605.5	1687	2.6
8–9	571	132	0.571	0.231	505	1081.5	1.9
9–10	439	187	0.439	0.426	345.5	576.5	1.3
10–11	252	156	0.252	0.619	174	231	0.9
11–12	96	90	0.096	0.937	51	57	0.6
12–13	6	3	0.006	0.500	4.5	6	1.0
13–14	3	3	0.003	1.000	1.5	1.5	0.5

although in a time-to-failure context, those who are lost from the study and thus succumb to an unknown fate only contribute to $S(t)$ while they are successfully detected. Variable T can be considered as either continuous or discrete, with the former tallying death over distinct, fixed-length time intervals and the latter tracking mortality across largely uninterrupted timelines. For example, if researchers monitor turtle nest survival by relocating nests on foot, daily visitation is probably necessary to achieve the uninterrupted timeline for a continuous time analysis; less frequent detection will result in discrete time intervals with uncertain timing of death events, potentially warranting a discrete time analysis.

As time progresses, the survival function based on cumulative probability of survival inevitably declines, as did survival functions for humans (Figure 6.2a) and Dall sheep (Figure 6.2c). The shape of the survival function reflects how mortality risk changes with time, with steeper declines indicating increased risk as the individual ages. Thus, we can consider the cumulative distribution function of the survival timeline, $F(t)$, which reflects the probability of dying by T and is the complement of the survivor function:

$$F(t) = 1 - S(t) = \Pr(T \leq t). \tag{6.2}$$

The trajectory of the cumulative distribution function under a range of scenarios is described in Figure 6.3b. It may be helpful to consider the cumulative distribution function as equivalent to a cumulative mortality function.

The *hazard function* $h(t)$ is the fundamental unit in the statistical analysis of survival and represents the instantaneous failure rate, or in other words, the instantaneous risk of death conditional upon the subject's survival to the beginning of the time interval (Cleves et al. 2010). Expressed in terms of probabilities, the hazard function is:

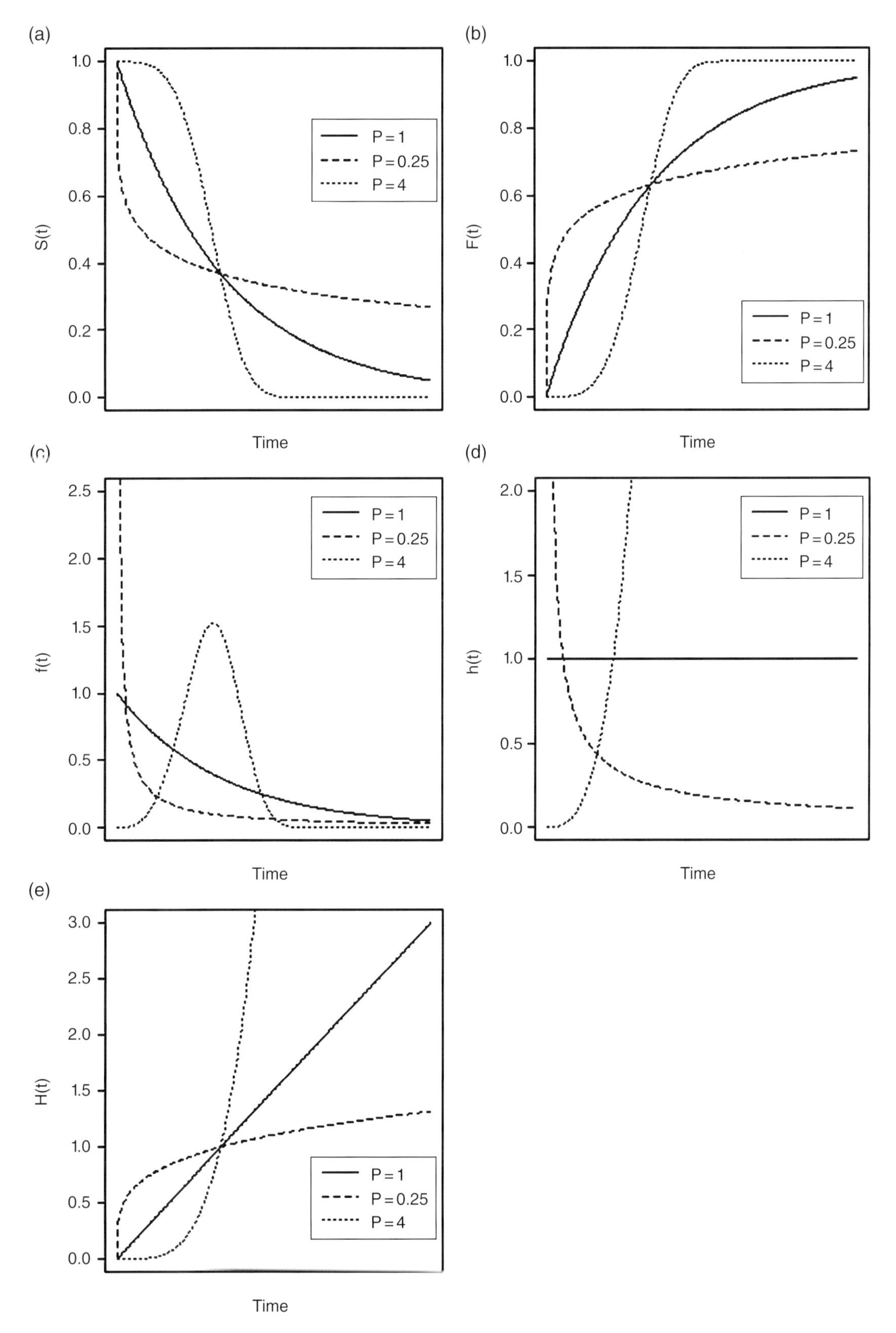

Figure 6.3 Theoretical functions relating cumulative survival (a), cumulative distribution (b), probability density (c), hazard (d), and cumulative hazard (e) in hypothetical scenarios. Parameter *P* refers to the dimensionless shape parameter defined by a Weibull distribution.

$$h(t) = \lim_{\Delta t \to 0} \frac{\Pr(t \le T < t\Delta t \mid T > 1)}{\Delta t}. \quad (6.3)$$

Equation (6.3) illustrates that the probability of survival past T is a function of the risk as time approaches zero (Figure 6.3c), meaning that the hazard rate represents the risk of death over an infinitesimally small time unit. Practically speaking, to achieve $\Delta t \to 0$, survival status should be monitored over short time intervals with a high probability of detection, although to some degree these criteria can be relaxed if hazard is very low. For instance, in the case of turtle nest monitoring it may be possible to relax daily monitoring if mortalities are especially infrequent or if the timing of death can be estimated with little error. Ultimately, continuous time survival analysis requires that subjects are monitored intensively so that timelines are largely uninterrupted, and timing of mortality is precisely known.

An important difference between the survival and hazard function is that while survival rates are bounded from zero to one and the survival function experiences a constant decline when plotted against time, hazard rates can range from zero (no risk) to infinity (certainty of death) and can increase, decrease, or follow irregular patterns depending on how risk changes with time. If mortality risk increases dramatically with age, the survival function declines rapidly while its corresponding hazard function increases; conversely, if risk decreases with age, the survival function declines gradually and the hazard declines as well (Figure 6.3c). The hazard function is linked to its survival function through the corresponding *probability density function*, $f(t)$, where:

$$h(t) = \frac{f(t)}{S(t)}. \quad (6.4)$$

Note that the notation used implies that the capitalized term (i.e. $F(t)$) represents the cumulative sum, whereas the lowercase (i.e. $f(t)$) is its first derivative [Figure 6.3]. To estimate the hazard function from the survival function, we note that the first derivative survival function, $S'(t)$, and probability density function, are linked [$f(t) = -S'(t)$]. Accordingly, we recognize that hazard and survival functions are related such that if hazard is constant through time, $h(t) = \lambda$, then $S(t) = e^{-\lambda t}$. We can then derive the survival function directly from the hazard function by:

$$S(t) = \exp\left[-\int_0^t h(u)du\right], \quad (6.5)$$

which also allows us to back-calculate to obtain the hazard function from the survival function:

$$h(t) = -\left[\frac{dS(t)/dt}{S(t)}\right]. \quad (6.6)$$

These algebraic exercises demonstrate that the survival, hazard, and related functions are all directly linked. To further illustrate this point, we can invoke an additional function, the *cumulative hazard function*, $H(t)$, which measures the accumulation of hazard at a given point in time and can be represented as:

$$H(t) = \int_0^t h(u)\,du. \quad (6.7)$$

$H(t)$ reflects the sum of hazards experienced by a subject and thus is a measure of risk accumulation. In other words, the derivative (slope) of the cumulative hazard function provides a time-specific estimate of hazard, and logically the accrual of this hazard can increase or remain stationary but never decrease with time (Figure 6.3e).

From the above calculations, we note that the cumulative hazard function, $H(t)$, is related to our other functions of interest, namely the hazard (6.7a), cumulative survival (6.7b), cumulative distribution (6.7c), and probability density functions (6.7d):

$$h(t) = \frac{d}{dt}H(t) \quad (6.8a)$$

$$S(t) = \exp\{-H(t)\} \quad (6.8b)$$

$$F(t) = 1 - \exp\{-H(t)\} \quad (6.8c)$$

$$f(t) = h(t)\exp\{-H(t)\}. \quad (6.8d)$$

At this juncture, we can consider more explicitly the linked expression of these various functions. As it turns out, the hazard function can follow a variety of functional forms, the most versatile being the *Weibull function*. The Weibull invokes a dimensionless shape parameter (p) that specifies how hazard varies through time. We can express the Weibull hazard function as:

$$h(t) = pt^{p-1} \quad (6.9a)$$

and the four corresponding related functions as:

$$S(t) = \exp(-t^p) \quad (6.9b)$$

$$F(t) = 1 - \exp(-t^p) \quad (6.9c)$$

$$f(t) = pt^{p-1}\exp(-t^p) \quad (6.9d)$$

$$H(t) = t^p \quad (6.9e)$$

Figure 6.3 illustrates several of these functions across a sample of p values.

6.3 Developing Continuous Time Survival Datasets

Continuous time survival monitoring gives rise to data that are comprised of consecutive non-negative observations from the same individual. As such, survival data do not conform to standard assumptions of independence and normality, and therefore require specialized dataset structure and analytical techniques. Further, continuous time survival analysis was originally developed for human medicine and industrial design, where subjects are tracked with relative ease, continuity, and certainty of detection compared to ecological studies. Concerns regarding the rigor of survival datasets and analyses in ecology highlight a need for careful data collection, dataset structuring, and data exploration, to avoid weak inference and biased results (Winterstein et al. 2001; Williams et al. 2002; Murray 2006).

6.3.1 Dataset Structure

We consider an example with a continuous time survival dataset for five individuals in a hypothetical population (Table 6.1). A survival dataset normally should reference when observations begin and end for each individual, and in standard survival analysis subject fate is recorded as a binary variable (death: "1"; non-death: "0"). Individuals failing to die during the study are *right-censored* and also coded as "0" (see ID no. 2 and 5, Table 6.1). As discussed below, censoring includes all non-mortality outcomes such as intentional withdrawal from the study, loss of contact, emigration, and study termination prior to a subject's death (Collett 2003; Murray 2006). *Left-censoring* refers to staggered entry and the timing that a new subject is recruited to the study.

Continuous time studies aim to eliminate interruptions in subject timelines, and Murray (2006) provides recommendations for establishing continuous time survival monitoring schedules. However, even with intensive monitoring longer gaps may arise when subjects are not surveyed or detected for extended periods. For

Table 6.1 Hypothetical continuous time survival dataset.

Id	Timein	Timeout	Days	Fate	Birthday	Mass	Sex	Ageclass	Year	COD1	COD2
Basic											
1	21jan12	22july13	548	1	15jan11	1250	1	1	2		
2	21jan11	27may12	492	0	2jan08	1100	0	3	1		
3	10feb12	12oct12	245	1	20jan10	1340	1	2	2		
4	21jan12	07mar13	411	1	8jan11	1190	1	1	2		
5	15may12	01dec12	200	0	10jan11	1350	0	1	2		
5	12feb13	11jun13	119	1	10jan11	1350	1	1	2		
Time-dependent											
1	21jan12	22july13	548	1	15jan11	1250	1	1	2		
2	21jan11	27may12	492	0	2jan08	1100	0	3	1		
3	10feb12	12oct12	245	1	20jan10	1340	1	2	2		
4	21jan12	07mar13	411	1	8jan11	1190	1	1	2		
5	15may12	01dec12	200	0	10jan11	1350	0	1	2		
5	12feb13	11jun13	119	1	10jan11	1350	1	2	3		
Competing risks											
1	21jan12	22july13	548	1	15jan11	1250	1	1	2	1	0
2	21jan11	27may12	492	0	2jan08	1100	0	3	1	0	0
3	10feb12	12oct12	245	1	20jan10	1340	1	2	2	1	0
4	21jan12	07mar13	411	1	8jan11	1190	1	1	2	0	1
5	15may12	01dec12	200	0	10jan11	1350	0	1	2	0	0
5	12feb13	11jun13	119	1	10jan11	1350	1	2	3	0	1

The dataset consists of five subjects (Id), each was recruited (Timein) and exited (Timeout) the study on a known date. The total number of days monitored is recorded for each individual (Days) as is the fate (1 = death, 0 = censor). Birthdate was known for each individual. When each subject was recruited to the study, its body mass (Mass: continuous variable), gender (Sex: binary variable), age category (Ageclass: categorical) and year of capture (Year) were recorded. In the Basic table, each row represents a continuous timeline and Subject no. 5 is distinguished by having two entries, reflecting a gap when the individual was not monitored. The Time-dependent dataset is extended to convert Ageclass and Year into time-dependent covariates that are adjusted at the beginning of each calendar year. The Competing risks dataset further extends the dataset to differentiate individuals according to their cause of death (COD1, COD2). Note that this data structure can be altered depending on the software package used and whether the study is designed to monitor a cohort with fixed recruitment timing (i.e. no staggered entry) or if all subjects are monitored to the time of death.

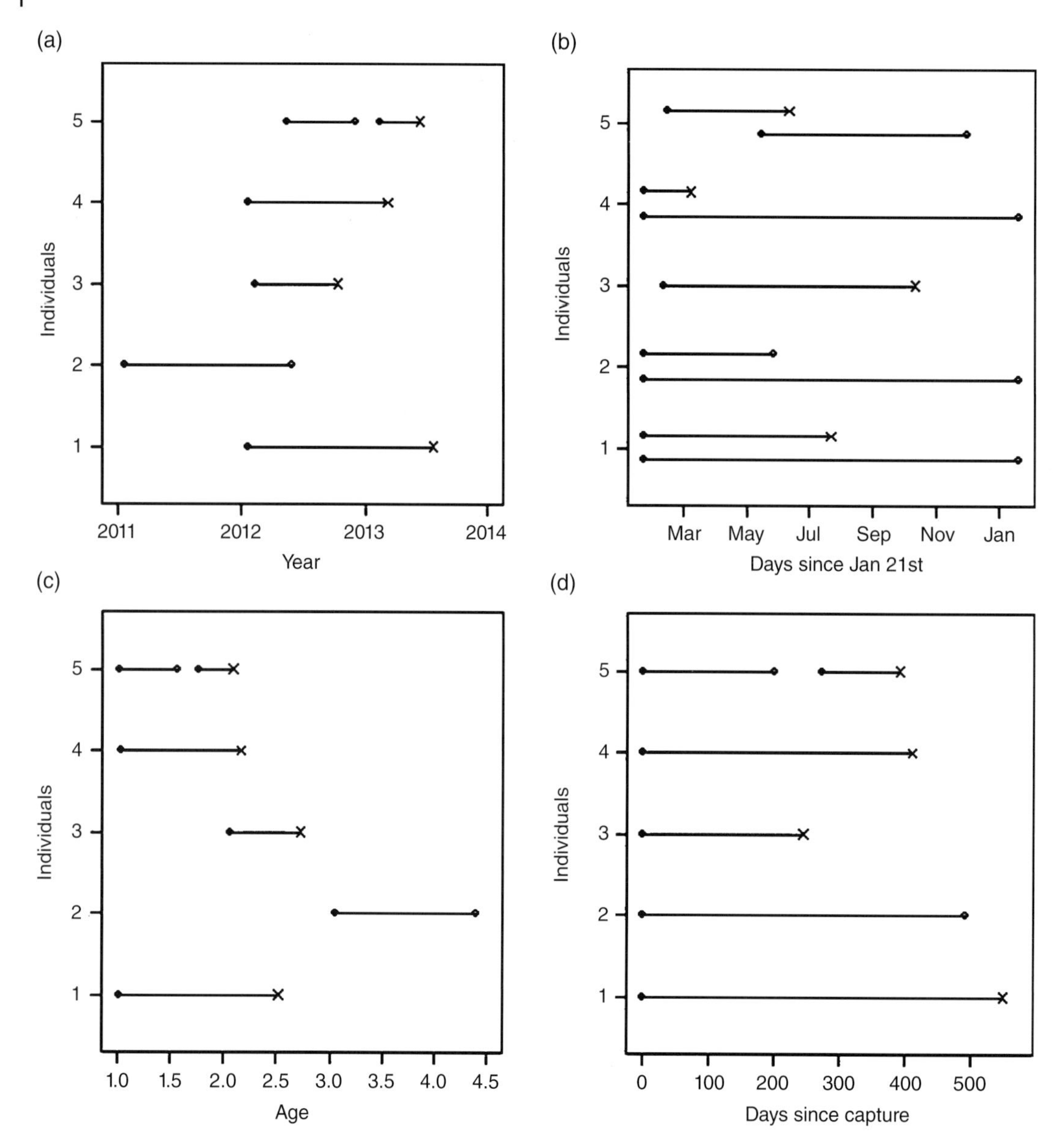

Figure 6.4 Timelines and fates of hypothetical individuals in a continuous time survival analysis. Each panel reflects a different time origin and structure for the same data. The five subjects correspond to those described in the basic dataset from Table 6.1, and are displayed according to study-related (a), recurrent (b), age (c), and time since capture (d) study designs.

example, in nest survival studies, periods of inclement weather may preclude daily visitation by researchers, whereas in radio telemetry research animals may temporarily leave the study area or else transmitters can fail and only be replaced at a later date. Normally, such gaps in an individual's timeline should be censored, which results in subject timelines being comprised of distinct uninterrupted segments punctuated by gaps with no data (see id no. 5, Table 6.1; Figure 6.4a). Yet, such gaps remain a point of contention as some authors suggest that missing observations can be imputed using standardized procedures, although the appropriate structure of such imputation models remains in doubt given that the dependent variable includes both event (i.e., death) and time (Van Buuren et al. 1999; White and Royston 2009). A critical point here is that, as a general rule, gaps in survival timelines should never be interpolated manually because this will bias the survival estimate in favour of subjects who re-enter the dataset and against those who fail to do so (Winterstein et al. 2001).

In theory, continuous time survival studies should detect mortality events more or less when they occur, leaving little uncertainty in the estimated death time. However, low detection success often adds uncertainty to the estimated timing of death, as may be the case when a subject is found dead after a gap in detection. How one corrects for the gap between the last live detection and the mortality event can be problematic because the common approach is to assume that mortality occurred at the *midpoint* of the monitoring gap (Winterstein et al. 2001; Murray 2006). In theory, this approach will underestimate variance and inject bias in the survival estimate (Lindsey and Ryan 1998; but see DeCesare et al. 2017). Likewise, assuming that several mortality events with uncertain dates occur at the midpoint of a monitoring gap creates *tied death times*, which is mathematically not possible in a continuous time context (Murray 2006). Some remedies for tied death times, including death time randomization or application of survival timeline likelihood functions, are implemented as defaults in some survival analysis software (Kalbfleisch and Prentice 2002; Collett 2003). Ultimately, while these adjustments may help address uncertainty in mortality timing, as a general rule, datasets that have extensive gaps in survival timelines and high uncertainty in death times may be best suited for approaches based on discrete time (Chapter 7) or interval censoring (Singer and Willett 1991).

6.3.2 Right-censoring

Right-censoring occurs when an individual's ultimate disposition is not known (see Id no. 2 and 5, Table 6.1). As a general guideline, right-censoring should be kept to a minimum and also be random or "noninformative" (Collett 2003). However, censoring rates can be a problem in ecological research because of the challenges associated with intensively tracking a representative sample of subjects under unpredictable field conditions to precisely estimate timing of the death event (Murray 2006). For example, field studies of long-lived organisms are almost always completed before all subjects have died, meaning that datasets inevitably include right-censored individuals. Likewise, if nests that are located in dense vegetation are less likely to be relocated at all times during the study, or if smaller individuals are prone to lose radio transmitters and thus succumb to an unknown fate, censoring will be biased. Field adjustments can be implemented to address censoring issues, including broadening relocation efforts to improve detection success or extending the study duration until most individuals have died. Sometimes, sources of censoring can be identified and reclassified using ancillary information obtained when the subject was last detected (Hays et al. 2007). Diagnostic tests for *informative censoring* involve assessing the significance of covariates potentially associated with censoring status, or plotting survival time of known deaths versus censored individuals (Oakes 2001; Collett 2003).

As an example of the problems that can arise with censoring, Smith et al. (2010) studied survival of recolonizing wolves (*Canis lupus*) for 22 years in the northwestern United States, where packs ranged over broad areas and were heavily persecuted by humans. A longstanding concern is that wolf survival estimates could be inflated if emigrating animals have a higher mortality risk or if radio-collared individuals are killed illegally and their transmitters are intentionally destroyed, leaving no evidence of their fate. To minimize overall censoring rates and the potential for informative censoring, Smith et al. (2010) intensified searches for missing animals when radio contact was lost and deployed new transmitters whenever animals were recaptured. Prior to their survival analysis, the authors confirmed that survival time was similar between censored and noncensored groups and that other potential confounding factors like proximity to human activity or to the edge of the monitoring area did not influence censoring patterns. Because censoring rates were comparable overall and only slightly higher for animals residing in remote locations, Smith et al. (2010) concluded that informative censoring was likely negligibly associated with emigration from the study area rather than from human persecution. In contrast, Liberg et al. (2011) found that high censoring rates among radio-collared wolves in Sweden were attributable to *cryptic poaching*, which took place when animals were killed intentionally by humans who then disposed of the transmitter and associated evidence of mortality. Consequently, an analysis that reclassified previously censored wolves as dead revealed an unsustainable mortality rate for the Swedish wolf population. This sensitivity analysis highlights both the potential role of informative censoring on bias in survival estimates, as well as the importance of intensive monitoring and data exploration for deciphering censoring patterns in survival datasets.

Once detected, informative censoring may be addressed either by establishing confidence intervals (CI) for survival times of censored animals or by conducting the survival analysis separately across censoring categories. It is also possible to develop survival models where censoring status is parameterized explicitly, although this latter option may involve making untestable assumptions about survival times and censoring patterns (Collett 2003). As a general observation, survival studies in ecology should pay much closer attention to censoring patterns, including reporting overall censoring rates, testing for potential censoring bias, and adopting appropriate corrective measures (Murray 2006).

6.3.3 Delayed Entry and Other Time Considerations

Survival estimates can be influenced by the structure of the *time scale* used in analysis (Fieberg and DelGiudice 2009), and continuous time survival datasets should be coded with time scales that are biologically sensible and that reflect study design considerations (Figure 6.4). This point is especially relevant to ecological studies, where mortality risk can be sensitive to the effects of age, seasonality, or calendar time. For example, if tree survival is repeatedly influenced by seasonal drought (Mueller et al. 2005), recurrent time (i.e. an annual cycle) may be the appropriate time scale for survival estimation; if mammals are subject to especially high mortality immediately postpartum (Gilbert et al. 2014), age-based tracking may be warranted (i.e. tracking starts at birth). In our basic dataset for a hypothetical organism (Table 6.1), calendar time was selected as the time scale because of the long duration of monitoring and lack of age-related or seasonal mortality pulses. Thus, researchers should consider the range of time scales that are possible (Figure 6.4), and select the best option for their study circumstances. Likewise, selecting the appropriate time unit for analysis (i.e. days, months, years) is important because it can influence the interpretation and utility of survival estimates and related model coefficients.

In many field studies, individuals are recruited through an extended period sometimes spanning months or years, leading to *delayed entry* (i.e. left-censoring or staggered entry) (Table 6.1; Figure 6.4a). Variation in timing of entry poses a number of challenges including that early mortalities can bias survival estimates if the initial sample size is small and some subjects die before all are recruited (Murray 2006; Fieberg and DelGiudice 2011). Sometimes, the survival dataset can be left-truncated to exclude early recruits, as was done by Smith et al. (2010) for a handful of wolves that were monitored during the 1980s. Here, the dataset was restricted to subjects monitored after 1994, which was justified because a small minority of animals were monitored during the early years of the study, survival estimates were comparable before versus after this date, and temporal variables did not distinguish early versus later recruits in an exploratory survival analysis (Smith et al. 2010). Occasionally, handling or marking protocols themselves may influence survival, and measuring changes in subject condition or behavior immediately post-capture can reveal the potential benefits of left truncating individual survival timelines prior to their full recovery (Dechen Quinn et al. 2012).

Left truncation also inevitably arises when survival monitoring is limited to select time periods. For neonatal ungulates, recruitment to a survival study typically happens within days after birth, when juveniles can be effectively captured and equipped with radio transmitters (Murray 2006). However, neonatal ungulates commonly experience high predation risk within hours after birth and this mortality pulse is usually missed when animals are recruited using traditional protocols and timelines (Gilbert et al. 2014). Thus, survival rates for juvenile ungulates are regularly overestimated, and this bias is specifically due to left truncation of subject timelines. Thus, researchers should conduct appropriate data exploration and report potential survival estimate bias from left truncation (Fieberg and DelGiudice 2011).

6.3.4 Sampling Heterogeneity

Across a cohort of subjects, individuals with *higher frailty* will die sooner, leaving those with lower risk to contribute disproportionately to mortality rate estimation. Likewise, if an initial group of subjects is recruited and tracked, over time the cohort of survivors will be entirely comprised of older individuals that may have higher or lower hazards. In either scenario, lack of replacement leads to progressive sampling heterogeneity, which biases the survival estimate and limits its generality (Zens and Peart 2003; Prichard et al. 2012). It follows that such problems will be aggravated by high mortality rates, prolonged study duration, or high initial heterogeneity among subjects. Thus, proper study design and subject recruitment are crucial for limiting progressive *sampling heterogeneity*, and important corrective measures include delayed recruitment of representative subjects to replace those that have died (Murray 2006). To detect whether a survival dataset is affected by sampling heterogeneity, baseline hazards can be plotted and checked for temporal variation (Vaupel and Yashin 1985). From an analytical perspective, sampling heterogeneity can be evaluated using study entry time as a covariate in survival models (Collett 2003; Murray 2006; Prichard et al. 2012), although this approach may cause unwanted correlation between the predictor and response variables.

Sometimes researchers intentionally recruit different types of subjects into a survival study, thereby making sampling heterogeneity an implicit aspect of study design. Smith et al. (2010) radio-collared wolves that were representative of the larger population as well those that were recently involved in livestock depredation. Baseline mortality risks were markedly higher for wolves that killed livestock, so Smith et al. (2010) necessarily treated each group separately in analyses and used only the representative sample in deriving survival estimates for the broader wolf population. Thus, researchers should be aware of how different sampling regimes can affect sampling heterogeneity and attendant survival estimates, and thereby assess the composition of the sampling population at different points during the study. When diagnostic tests reveal sampling heterogeneity and the potential for biased survival estimates, necessary adjustments may involve restricting the analysis to subjects that best reflect the population of interest.

6.3.5 Time-dependent Covariates

As individuals go through life, susceptibility to risks will vary with time; risks tend to increase or decrease according to intrinsic factors like age, body condition, or reproductive status, and to extrinsic factors like temperature, predator numbers, or season. Whenever possible, survival datasets should capture such variability by including time-dependent (or time-varying) covariates that account for temporal changes in risk owing to variability in predictor variables (Kleinbaum and Klein 2011). For example, if we want to document the effect of environmental conditions on survival of an endangered cactus at the northern edge of its geographical range, we should include daily temperature measurements in the survival dataset so we can relate daily temperature variation to plant survival.

Time-dependent or "time-varying" covariates are designed to reflect timing of the change in risk factors. Specifically, the dataset is constructed to reflect potential temporal change in the variable's impact on risk status of the individual. In Table 6.1, under the Basic classification, the five subjects are classified with a fixed value for the Age class and Year variables, meaning that these values represent conditions at the time that subjects were first recruited into the study. These values do not consider the time-varying nature of age or year on subject mortality risk. However, because subject no. 5 was monitored for an extended period during which it transitioned to an older age category and year of study, appropriate adjustments to these variables are needed to reflect temporal variability in risk. Under the Time-dependent classification (Table 6.1), Id no. 5 is recognized as being in a higher age category and subsequent year during its second monitoring period. Here, the Age class and Year variables were adjusted to represent the subject's current status with respect to those variables rather than conditions at the time of recruitment to the study, as was the case for the Basic classification.

Time-dependent variables can be incorporated into survival datasets by splitting subjects into different observation periods. As a rule of thumb, extrinsic variables are easier to present in a time-dependent context because they may not require intensive monitoring of subjects and related data on location, behavior, or condition. In the cactus example, coding for extreme weather conditions would be a simple matter of relating the recorded daily temperature, or daily minimum temperature, with the outcome of our survival monitoring for that time interval. The frequency of updating time-dependent variables, and the degree that subject timelines will need to be split to accommodate time-dependent variables, will depend on both the objectives of the study and the availability of ancillary information to include in the dataset. Intensive (i.e. daily) timeline splitting is required to document the effect of a specific temperature threshold on cactus survival, but it may be less crucial if we are more concerned with the cumulative effect of low temperature or with a factor that varies little over time. Another consideration is that time-dependent covariates can be coded either in real-time or following time delays to reflect lags between exposure to risk factors and expression of related changes in risk. For female wolves, reproductive status might be classified as a time-delayed time-dependent variable, because physiological effects of pregnancy and lactation on mortality risk may be most important especially in the months following parturition (Smith et al. 2010). Note that there are a variety of considerations and coding options that can guide the structure of time-dependent variables in a survival dataset (Kleinbaum and Klein 2011), and this aspect of dataset design requires careful evaluation of both data availability and study objectives.

6.4 Survival and Hazard Functions in Practice

6.4.1 Mayfield and Heisey–Fuller Survival Estimation

To understand how survival and hazard rates vary through time or relative to different factors, we first review the more common continuous time survival estimators. The Heisey–Fuller, Mayfield, and related estimators track survival over distinct time intervals where risk remains constant within an interval. The Kaplan–Meier and Nelson–Aalen estimators let mortality events rather than changes in risk determine the duration of time intervals. The *Heisey–Fuller estimator* (HF) was developed specifically for radio telemetry research where a group of individuals are monitored intensively for survival status and fate, whereas the Mayfield and related estimators originate from field studies involving continuous time monitoring of bird nest failure (Mayfield 1975; Williams et al. 2002). For illustrative purposes, we focus on the HF estimator, but the Mayfield estimator and related methods share a similar foundation and provide largely consistent results (Trent and Rongstad 1974; Heisey and Fuller 1985; Williams et al. 2002). We compute HF survival by:

$$\hat{S} = 1 - \frac{d}{r}, \tag{6.10}$$

where d is the number of deaths and r is the cumulative duration of monitoring. In nest survival or radio telemetry studies where individuals are monitored on a daily basis, parameter r is the number of individual days of exposure, and the term d/r is the *crude mortality rate*, where "crude" means that all causes of death are included. Confidence limits for the HF and related estimators can be calculated using methods outlined in Johnson (1979), Bart and Robson (1982), Heisey and Fuller (1985), and Powell (2007). Because in HF survival estimation risk is constant within a time interval, d/r should remain consistent from beginning to end of the interval. Accordingly, the HF is expanded to an interval spanning t days by: $S(t) = S^t$, and to i intervals by:

$$\hat{S} = \prod_{i=1}^{I} S_i. \tag{6.11}$$

For example, Wirsing et al. (2012) monitored mortality rates of nests of painted turtles (*Chrysemys picta*) and snapping turtles (*Chelydra serpentina*) by daily checks at nest sites for the duration of the nesting period (June–September). Most nest losses were due to predation by raccoons (*Procyon lotor*). They monitored 94 painted turtle and 198 snapping turtle nests, documented 54 and 166 nest mortality events, and accumulated 3561 and 3121 observation days, respectively. Using the HF, the daily survival rate was markedly higher for painted turtles [painted: $1 - (54/3561) = 0.98484$; snapping: $1 - (166/3121) = 0.94681$)]. When extrapolated to a roughly four-month (110-day) nesting period (which assumes a constant mortality risk through the four-month period), nest survival is almost 19% for painted turtles and roughly zero for snapping turtles (painted: $0.98484^{110} = 0.18631$; snapping: $0.94681^{110} = 0.00245$). As an aside, when working with daily survival rates it is helpful to retain multiple digits after the decimal because rounding error can alter rates when extrapolating to longer time intervals. For painted turtle nests, $0.98484^{110} = 0.186$; $0.9848^{110} = 0.185$; and $0.984^{110} = 0.189$.

The HF estimator is attractive because survival estimates are easily calculated from summary survival data on number of deaths and number of days of exposure. The model is robust to censoring, provided that censoring is random, and overall this estimator is especially useful for populations exposed to variable risk, where risk remains constant over shorter time periods that can be pooled into distinct time intervals. However, whether risk remains constant within an interval is critical to model performance, even though researchers rarely verify this assumption (Stanley 2004; Murray 2006). In fact, we show later that HF estimates for turtle nest survival are biased because risk was not constant through the nesting period. Further, the HF model can be biased if survival is monitored across irregular and infrequent intervals because of the imprecision in estimating timing of a mortality event; in such cases application of a likelihood-based alternative is advised (Hensler and Nichols 1981). Last, another important issue with the HF estimator is that while the rates themselves are easy to compute by hand or with a spreadsheet, the CI are less straightforward and to date researchers have done so mainly using a DOS-based program (`MICROMORT`, Heisey and Fuller 1985). To our knowledge, there is no accessible MS Windows-based replacement for calculating HF CI, meaning that on a long enough timeline, the HF estimator may become obsolete and its effective survival may drop to zero.

6.4.2 Kaplan–Meier Estimator

The *Kaplan–Meier estimator* (KM) is the most popular survival estimator and is widely encountered in both medical and ecological literature. KM estimation considers the timing of death events as the determinant of interval endpoints, so if there are k unique death times, the corresponding survival rate is the product of survival during each interval, i:

$$\hat{S} = \prod_{i=1}^{k} \left[1 - \frac{d_i}{v_i}\right], \tag{6.12}$$

where d_i is the number of deaths and v_i is the number of subjects at risk during i. KM variance is calculated as:

$$\text{var}(\hat{S}) = S^2 \left[\prod_{i=1}^{n} \left(\frac{d_i}{v_i(v_i - d_i)}\right)\right]. \tag{6.13}$$

A distinct feature of the KM estimator is that time interval endpoints are defined by actual mortality events and the focus of the estimator is on the individual subject rather than on the time interval; this is philosophically consistent with contemporary individual-based approaches to survival analysis, making the KM estimator a logical complement to standard analysis. The relationship between constant survival and interval endpoints follows a step-function, with Figure 6.5a showing high initial mortality during the early nesting period and lesser mortality later in the season, for both turtle species. Seasonal changes in mortality risk also emphasize why our earlier use of the HF estimator likely was not appropriate for determining turtle nest survival rates. The KM probability of cumulative survival for the 110-day nesting period is 0.357 (95% CI = 0.243, 0.524) for painted turtles and 0.165 (0.120, 0.226) for snapping turtles; these estimates differ substantially from the earlier estimates calculated using HF estimation (painted turtle: 0.186; snapping turtle: 0.002) and doubtless reveal bias in the original HF estimates (Table 6.2). However, the HF estimates could be improved to some extent by first tallying mortality risk within shorter time intervals when risk is more constant.

We can also determine nest mortality risk (hazard) by first estimating the cumulative hazard from the survival function ($H(t)$, see Eq. 6.8b) and then differentiating

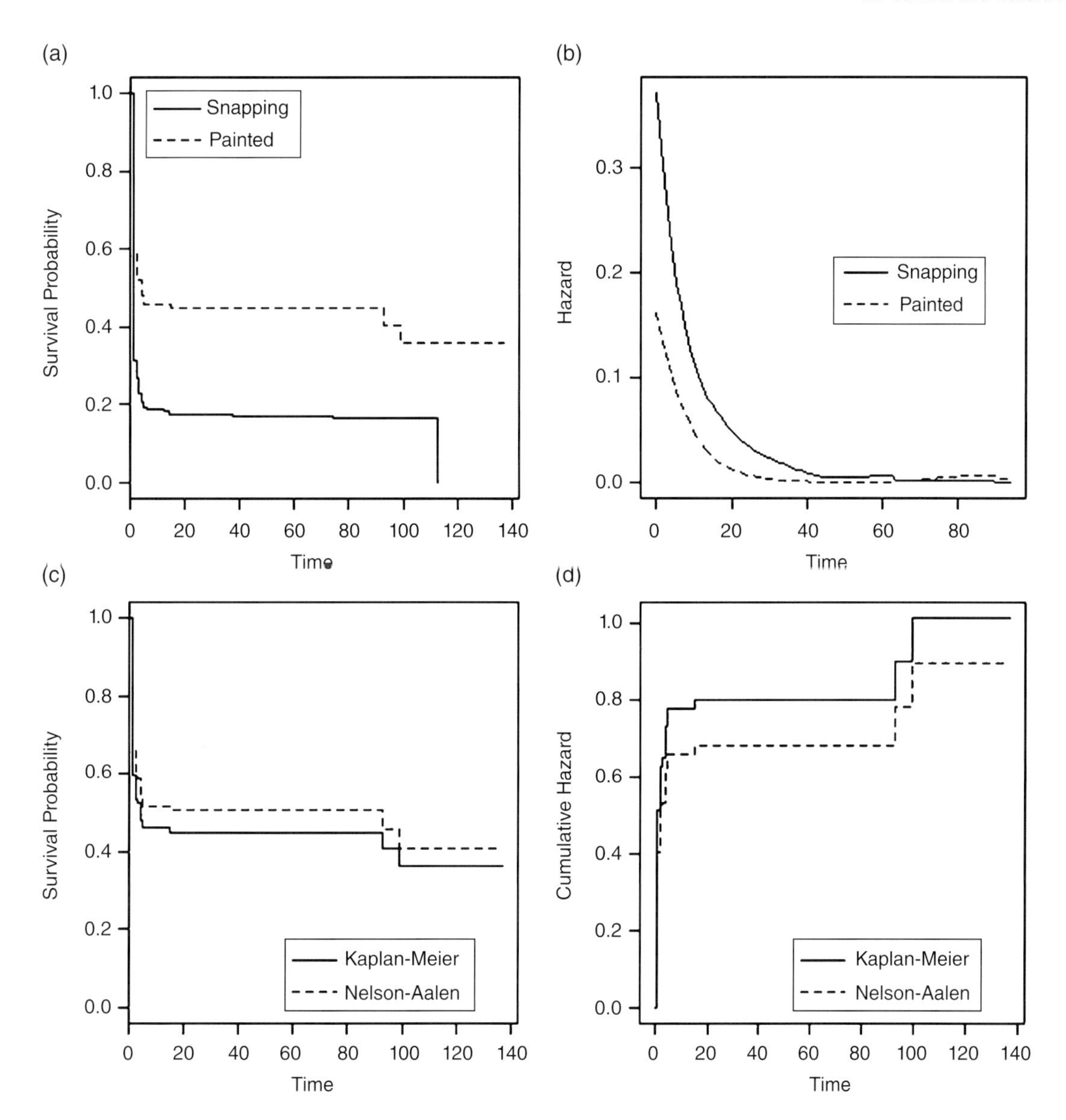

Figure 6.5 Kaplan–Meier survival functions (a), and hazard functions (b) for snapping turtle and painted turtle nests relative to time since clutch initiation. For snapping turtles only, panels (c) and (d) compare Kaplan–Meier and Nelson-Aalen survival (c), and Kaplan–Meier and Nelson–Aalen failure (d) rates. 95% CI are omitted for clarity. *Source:* The dataset was modified from Wirsing et al. (2012).

Table 6.2 Survival and failure rate estimates (±95% CI) during 110-day intervals for painted turtle and snapping turtle nests exposed to predation risk.

	Painted turtle	Snapping turtle
Heisey–Fuller survival	0.186 (0.119, 0.292)	0.002 (0.001, 0.006)
Kaplan–Meier survival	0.357 (0.244, 0.524)	0.165 (0.120, 0.226)
Nelson–Aalen survival	0.410 (0.288, 0.583)	0.274 (0.213, 0.354)
Kaplan Meier failure	1.030 (0.646, 1.411)	1.802 (1.487, 2.120)
Nelson–Aalen failure	0.892 (0.54, 1.245)	1.295 (1.038, 1.546)

Confidence intervals (CI) for the Heisey–Fuller estimator are obtained through Taylor series approximation (Heisey and Fuller 1985). Models were based on a modified dataset with a subset of variables and results differ from published estimates (Wirsing et al. 2012).

($h(t)$, see Eq. 6.8a). Practically speaking, this procedure is usually accomplished by using a nonparametric smoothing spline that is set to an appropriate *bandwidth* (Mueller and Wang 1994; Hess et al. 1999; Tsai et al. 1999). Bandwidth selection is largely arbitrary but the resulting graphed function should provide sufficient variation to reveal the most biologically meaningful patterns in hazards. For both turtle species, hazards decline dramatically with nest age, although there may be a slight increase in risk after 60 days (Figure 6.5b).

One drawback of the KM approach is that a new parameter is estimated with each death event, leading to lack of parsimony in studies having many deaths. Another issue is that in theory tied death times are not possible and require special treatment (Pollock et al. 1989). Since the KM estimator was initially developed for clinical research, the basic estimator is designed to deal with studies where subject recruitment occurs at the outset, but when staggered entry is present the *generalized Kaplan–Meier* model (GKM) makes adjustments for the conditional probability of late recruits (Pollock et al. 1989). For instance, the GKM would have been the appropriate estimator for survival analysis of wolves (Smith et al. 2010) or turtle nests (Wirsing et al. 2012), had calendar time rather than survival time been the temporal unit of interest. Yet, it is important to highlight that notwithstanding its advantages for late recruits, the GKM estimator is sensitive to small sample sizes, especially when mortalities occur early in the study before the full complement of subjects is under observation (Woodroofe 1985). Likewise, both the KM and GKM estimators give rise to absurd values in long-term studies when the entire initial group of subjects dies before subject recruitment is complete (Murray 2006). Thus, researchers should be judicious as to whether to include delayed entry of subjects in a survival analysis if KM or GKM survival estimation is a priority. Instead, left truncation may benefit datasets that have a small sample of early recruits. Currently, most statistical software packages do not account for staggered entry or GKM, so the default in almost all applications is the standard KM estimator.

In general, KM estimation may be superior over HF if risk varies widely with time (Figure 6.5a). In contrast, HF estimation may be more appropriate if mortality events occur in pulses or are particularly common or rare. Both HF and KM estimators are distinguished by how an interval endpoint is defined, and the methods are convergent when intervals are similarly identified (Pollock et al. 1989).

6.4.3 Nelson–Aalen Estimator

Owing to the limitations associated with both HF and KM estimation, there is a counting process analogue, the *Nelson–Aalen estimator* (NA), that can provide improved estimates, especially when sample sizes are small. The NA is often computed in the context of cumulative hazards. Recall that $H(t) = -\ln\{S(t)\}$; the cumulative hazards NA estimator is, through i time intervals:

$$\hat{H}(t) = \sum_{(j \mid t_i \leq t)} \frac{d_i}{v_i}, \tag{6.14}$$

where v_i is the number at risk at t_i and d_i is the number of deaths at t_i (Cleves et al. 2010). When dealing with small sample sizes, KM survival estimates and NA cumulative hazards tend to be superior to their counterparts, and as a general rule NA tends to overestimate survival whereas KM overestimates cumulative hazards (Table 6.2). Differences between the estimators become more pronounced as sample size diminishes (Figure 6.5c, d), but both estimators are convergent when sample sizes are large. Also, in contrast to the KM estimator, the NA estimator is robust to gaps when no subjects are at risk. Nevertheless, the NA estimator seems to be underused in ecological studies, as most researchers seem to accept the default KM, perhaps without fully appreciating its potential limitations. At a minimum, simulations with variable sample size and variation could be used to establish a general rule of thumb for when KM versus NA estimation should be used in survival estimation.

6.5 Statistical Analysis of Survival

6.5.1 Simple Hypothesis Tests

Sometimes, it may be convenient to initiate a survival analysis by comparing survival functions between populations or treatments without explicitly considering a suite of predictor variables. Such tests are structured around the null hypothesis of no differences between groups, and can provide preliminary insights and summary statistics related to the primary factors imposing risk. HF estimates can be compared using confidence limit overlap, using the z-statistic for two groups (Hensler and Nichols 1981; Bart and Robson 1982) or contingency tables for larger groups (Heisey and Fuller 1985; Sauer and Williams 1989). Survival data obtained from the KM estimator can be compared using a variety of nonparametric likelihood or rank tests (Box 6.2). Survival studies in ecology sometimes restrict their analysis to these simple tests, despite that they were originally designed for evaluating straightforward hypotheses from clinical trials having balanced designs, randomization, consistent distribution of mortality events, and limited delayed entry or censoring (Murray 2006). These conditions are unlikely to be upheld in a majority of ecological datasets, making these tests inappropriate for the multitude and complexity of factors affecting mortality risk under field conditions.

6.5.2 Cox Proportional Hazards

Risk factors in observational field studies are more appropriately analyzed using multivariate regression approaches. The *Cox proportional hazards model* (CPH) is the most widely used survival analysis method in the medical sciences and is firmly established in ecological research (Williams et al. 2002; Murray 2006). The model is based on the partial likelihood of the hazard, $h_i(t)$, where the ith individual is subject to a vector of covariates potentially influencing risk; $x_I = (x_{i1}, x_{i2},..., x_{ip})$. The corresponding CPH model is:

$$h_i(t) = h_0(t)\exp\left(\beta_1 x_{i1} + \beta_2 x_{i2} + \ldots + \beta_p x_{ip}\right), \quad (6.15)$$

where $h_0(t)$ is the *baseline hazard* (constant-only model) for an individual with covariate vector $x_I = (0,0\ldots0)$, and β is an unknown parameter (Murray 2006). The model is considered *semi-parametric* because the underlying shape of the hazard function remains unspecified, such that $h_i(t)$ and $h_j(t)$ differ only in that their ratio $[h_i(t)/h_j(t)]$ is proportional through time and differs by the exponential term related to the covariates. It follows that the absence of a specified hazard function makes the CPH model particularly versatile for the challenges of ecological survival research, where the functional form of baseline hazard is rarely known. CPH models generate *hazard ratios* (HR) or *model coefficients* (β) that are related by $\text{HR} = \exp(\beta)$, such that $\text{HR} > 1.0$ (or $\beta > 0$) indicates increased risk whereas $\text{HR} < 1.0$ (or $\beta < 0$) indicates reduced risk, relative to the baseline.

Wirsing et al.'s (2012) analysis of turtle nest survival considers a variety of risk factors potentially affecting nest predation. As an extension of our earlier survival estimate calculations (Table 6.2), multivariate CPH models (Table 6.3) reveal that for painted turtles, corridor proximity (Corridor: dummy variable, 1 = yes) is negatively associated with risk whereas date of nest laying (Datelaid: calendar date) is positively associated with risk. In other words, because Corridor has hazard <1.0, and 95% CI that do not overlap 1.0, we can infer that proximity to a corridor reduces predation risk. Datelaid has a hazard ratio > 1.0 (and 95% CI do not overlap 1.0), implying that nests laid later in the nesting season incur higher relative risk. We note that risk decreased by 81.4% ($1 - 0.186 \times 100$) when the nest was within a corridor, and risk increased by 4.3% ($|1 - 1.043| \times 100$) for each day that egg laying was delayed. Additionally, predation risk increased by 14% ($|1 - 1.014| \times 10 \times 100$) for every 10 m increase from vegetative cover (Veglaid variable was coded as continuous, in meters), but because the 95% CI for Veglaid overlapped 1.0, we infer that the effect of this variable was not statistically significant. For snapping turtle nests, risk increased by 22% ($|1 - 1.022| \times 10 \times 100$) for every 10 m increase in distance from water (Waterdist was coded as continuous, in meters), and with lesser (nonsignificant) influence from distance to vegetation (Vegdist was coded as continuous, in meters) and effect of disturbed habitat (Disturbed was

Table 6.3 Hazard ratios (HR) (±95% CI) from semi-parametric Cox proportional hazards (CPH) models, and hazard rates from parametric exponential and Weibull regression models, for nests of painted turtles and snapping turtles.

	Variable 1	Variable 2	Variable 3
Painted turtle	Corridor	Datelaid	Vegdist
Cox	0.186 (0.064, 0.542)	1.043 (1.014, 1.074)	0.997 (0.978, 1.016)
Exponential	0.129 (0.045, 0.369)	1.132 (1.088, 1.177)	1.013 (0.995, 1.031)
Weibull*	0.129 (0.056, 0.473)	1.062 (1.029, 1.096)	0.999 (0.980, 1.019)
**Weibull shape*	0.406 (0.326, 0.507)		
Snapping turtle	Disturbed	Vegdist	Waterdist
Cox	2.901 (0.832, 10.116)	0.997 (0.986, 1.010)	0.050 (0.023, 0.109)
Exponential	2.349 (0.735, 7.510)	1.000 (0.988, 1.012)	1.030 (1.014, 1.04)
Weibull*	2.988 (0.901, 9.920)	0.997 (0.985, 1.009)	1.025 (1.006, 1.043)
**Weibull shape*	0.509 (0.456, 0.568)		

Models for each species included only significant variables identified by Wirsing et al. (2012). Variables Corridor and Disturbed are dummy variables identifying whether the nest is on a predator travel corridor or in disturbed habitat, respectively (yes = 1), Datelaid is the Julian date when the nest was first laid, Vegdist and Waterdist is the distance to vegetation and open water, respectively. Models were based on a modified dataset with a subset of variables and results differ from published estimates (Wirsing et al. 2012).

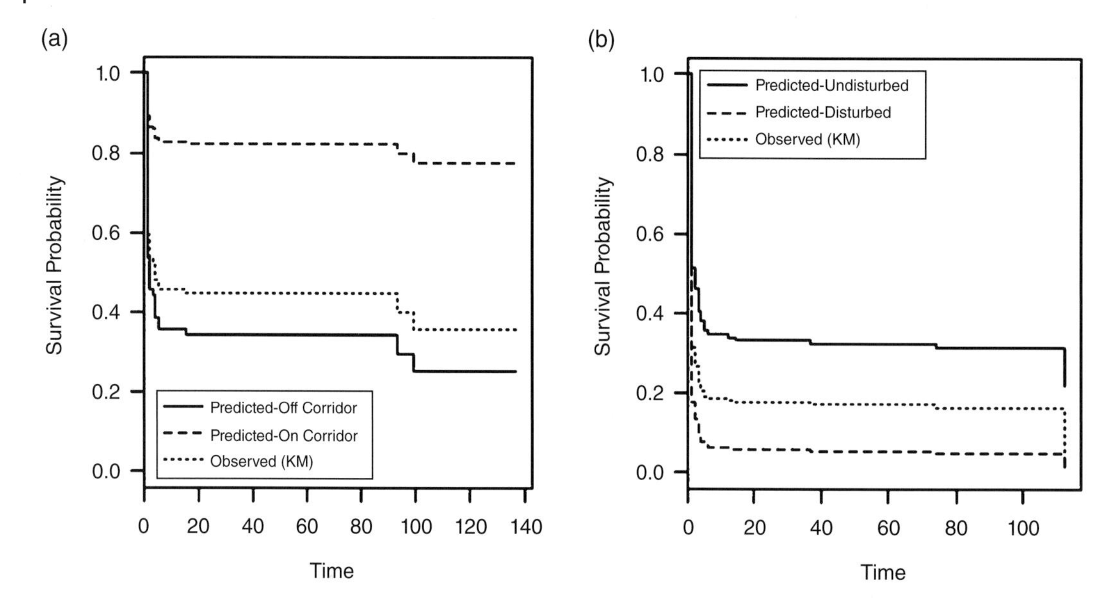

Figure 6.6 Survival probability for nests of snapping turtles (a) and painted turtles (b). Functions include Kaplan–Meier survival rates derived from observed data as well as those predicted by nest site location relative to habitat disturbance (a) and travel corridors (b) from a univariate Cox proportional hazards (CPH) model. *Source:* The dataset was modified from Wirsing et al. (2012).

coded as dummy variable, 1 = yes). In a similar study, Leighton et al. (2011) used CPH to determine sea turtle nest survival using a continuous time monitoring approach and determined that risk increased according to vegetative cover and time during the nesting season.

Once a hazard model is generated, it is possible to describe survival differences between groups by comparing HR (Table 6.3) or computing adjusted survival curves, which are computationally distinct from KM estimates because they factor in the possible effects of covariates on survival. Figure 6.6a and b provide survival curves for painted turtle and snapping turtle nests in light of predictor variables included in hazard models for each species. It is clear that there are notable differences in the estimates provided by the adjusted survival curves versus the KM estimates (Figure 6.5a) or hazards (Figure 6.5b) examined earlier.

CPH analysis requires continuous survival timelines but when ecological datasets are punctuated by *monitoring gaps* (i.e. low or variable detection probability), a counting process analogue of CPH can be used: the *Andersen–Gill model* (AG, Hosmer et al. 2008). We discussed counting processes previously in the context of the NA cumulative hazards estimator. Here, the counting process considers the survival timeline as an accumulation of conditional independent steps, which is a function of the total number of events (deaths) and the cumulative intensity process of the events up to t. Functionally, the counting process approach discretizes units of survival such that individual observations in time rather than continuous timelines underlie the analysis; the two approaches converge when there are no gaps in survival timelines (Murray 2006). For example, Johnson et al. (2004) used AG rather than CPH models to evaluate mortality risk among radio-collared grizzly bears (*Ursus arctos*) because the sample included many punctuated timelines and thereby favored a counting process approach. Similarly, Liebezeit and Kendall (2009) modeled the role of industrial activity on nest mortality in a variety of shorebirds and passerine birds; discontinuous nest survival timelines necessitated the application of AG rather than CPH models. Note, however, that researchers do not usually need to be concerned about the specific application of AG models in survival analysis because these are invoked as a default in most software programs when subject timelines are punctuated.

6.5.3 Proportionality of Hazards

The CPH approach is based on an assumption of *proportional hazards* across the model, meaning that there should be a linear relationship in how hazards vary according to predictor variables. The assumption can be verified using a variety of graphical approaches, and hazard proportionality between painted and snapping turtle nests can be assessed using linearity in plots of log(survival) vs. log(time) or -log(-log(survival)) vs. log (time) (Kleinbaum and Klein 2011). Our graphical assessment of the assumption using either method (Figure 6.7a and b) reveals largely parallel lines, and therefore proportional hazards. Note that here we could have used natural logs rather than $\log_{10}$ in our tests, and different software

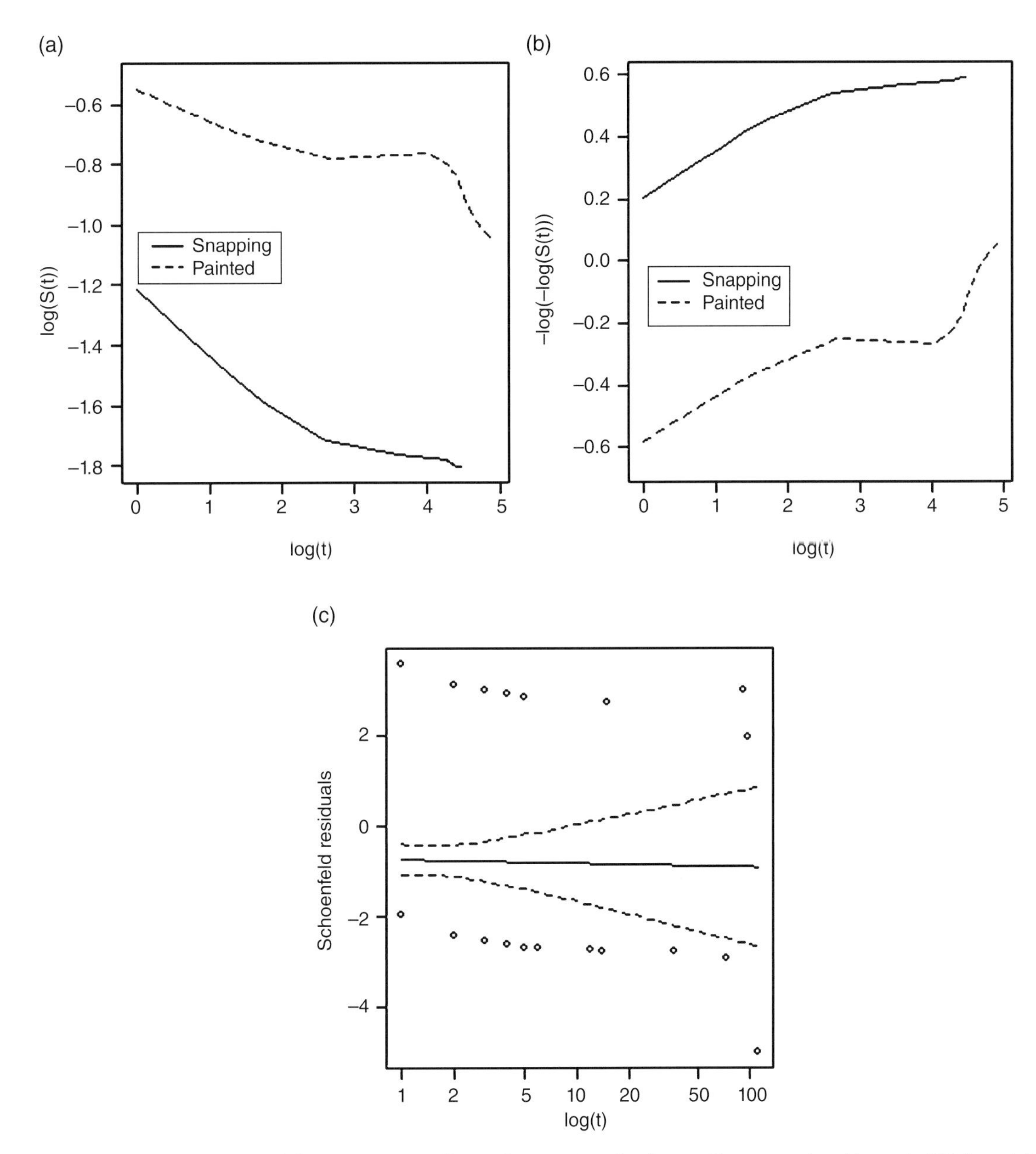

Figure 6.7 Graphical tests of the assumption of hazard proportionality from a Cox proportional hazards (CPH) model comparing nest survival of snapping and painted turtles. The most common tests are provided, including log(S(t)) (a), −log(−log(S(t)) (b), and Scaled Schoenfeld residuals (c). Dashed lines in panel c refer to 95% CI. *Source:* The dataset was modified from Wirsing et al. (2012).

programs use different transformations as defaults. Alternatively, we could plot the relationship between KM survival functions versus those from a CPH model with the predictor in question, and gauge parallelism between observed and expected functions (Kleinbaum and Klein 2011). Naturally, when predictor variables are continuous they should be reclassified into categories for proper graphical evaluation of the proportional hazards assumption.

Hazard proportionality can also be checked via analysis of residuals or model re-estimation (Therneau et al. 1990; Gail 1991). The primary tool for model checking in CPH analyses is the *Schoenfeld residuals*, which is based on the contribution of each covariate to the fit of the model specifically at each failure time. If residuals are distributed randomly, then there is no inherent structure and the proportional hazards assumption is met. For example, Schoenfeld's residuals from a simple univariate model comparing predation across turtle species are distributed randomly, with no inherent trend (Figure 6.7c), and with no statistically significant difference ($X^2 = 0.006$, $p = 0.94$). Thus, hazard proportionality is implied for

Box 6.2 Testing for Equality of Survival Curves

Simple tests are available for testing the hypothesis of no differences between two or more survival functions. These tests have limited utility in ecology but may be used when necessary, such as to test for differences between predictors that lack hazard proportionality. Let $t_1 < t_2 < \ldots < t_k$ represent the ranked failure times, with d_j being the number of deaths at t_j and n_j the number of subjects at risk prior to t_j; d_{ij} and n_{ij} are the same units for group i, $i = 1, \ldots, r$. The equality of survivor functions leads to the test:

$$H_0 = \lambda_1(t) = \lambda_2(t) = \ldots = \lambda_r(t), \tag{6.16}$$

where $\lambda(t)$ corresponds to the hazard function at t versus the alternate hypothesis that one or more $\lambda_i(t)$ is different for some t_j. If the null hypothesis is true, the expected number of failures in group i at t_j is $e_{ij} = n_{ij}d_j/n_j$ and the corresponding test statistic is:

$$\mathbf{u}' = \sum_{j=1}^{k} W(t_j)\left(d_{1j} - e_{1j}, \ldots, d_{rj} - e_{rj}\right). \tag{6.17}$$

Note that $W(t_j)$ is a weighting function that is zero when n_{ij} is zero. There are several different weighting functions that give rise to different test statistics, depending on data structure and limitations (StataCorp 2007).

Test Weight at each death (t_i)	Comments
Logrank 1	Best when hazards are proportional
Wilcoxon n_i	Best when hazards are not proportional and death and censoring patterns are similar
Tarone-Ware $\sqrt{n_i}$	Best when hazards are not proportional and test is less susceptible to death and censoring patterns
Peto-Peto-Prentice $\hat{S}(t_i)$	Best when hazards are not proportional and test is less susceptible to death and censoring patterns
Fleming-Harrington $\hat{S}(t_{i-1})^p\{1 - \hat{S}(t_{i-1})\}^q$	Test is flexible, if $p > q$, more weight to early deaths, if $p < q$ more weight to late deaths, if $p = q = 0$, test reduces to logrank test

the model under consideration. In cases where complex multivariate hazard models are considered, it is appropriate to check the fit of Schoenfeld residuals both for individual predictors as well as for the global model. Cleves et al. (2010) outline a variety of additional tests that can be used to further gauge hazard proportionality in CPH models.

If hazards are not proportional, a number of options are available, including developing separate models across the problematic variable and using nonparametric methods to compare survival rates. Better yet, a stratified CPH model can be fit, where different baseline hazards are assumed across the problematic variable. Specifically, stratification allows the form of the underlying hazard to vary across levels of the stratification variable, and thus enables model fitting without estimating the effect of the problematic covariate. However, predictors that are stratifiable usually must be binary or categorical, and the stratification process precludes direct assessment of the role of the predictor on risk, except through nonparametric tests (Box 6.2).

There are two other alternatives to stratification when faced with models failing to adhere to hazard proportionality. First, the time axis can be adjusted to accommodate shorter time periods that actually conform to the proportionality assumption. Second, it may be possible to use time-dependent variables to explicitly model nonproportionality or else develop models based on accelerated failure or additive hazards (Hosmer et al. 2008; Kleinbaum and Klein 2011). However, such methods are rarely invoked in ecological research, perhaps because the assumption of hazard proportionality itself is so rarely tested or because the complexity of such techniques tends to exceed the quality of many survival datasets (Murray 2006).

6.5.4 Extended CPH

In ecology, subject mortality risk is regularly influenced by *dynamic factors* such as time-dependent variation in weather, age, condition, or behavior. As described in Section 6.3.4, time-dependent covariates open a wealth of possibilities for evaluating the role of time-sensitive predictors on risk. Here we elaborate more formally on the extended CPH model, which includes one or more time-dependent variables. Recall that time-dependent variables can be either intrinsic or extrinsic and they are adjusted during the subject's timeline, as it progresses through distinct stages for the variable in question (see Id no. 5, Ageclass and Year variables, Time-dependent section, Table 6.1). We consider the expanded CPH model as:

$$h_i(t) = h_0(t)\exp\left[\left(\beta_1 x_{i1} + \beta_2 x_{i2} + \ldots + \beta_{p1} x_{ip1}\right) + \left(\alpha_1 x_{j1}(t) + \alpha_2 x_{j2}(t) + \ldots + \alpha_{p2} x_{jp2}(t)\right)\right]. \tag{6.18}$$

where x_i and x_j represent time-independent and time-dependent covariates, respectively, and β's and α's represent coefficients associated with time-independent and time-dependent variables, respectively. The extended CPH model assumes that the effect of a time-dependent variable on hazard at t is restricted to the value of the variable at t, and therefore does not consider its value either before or after the specified time (Kleinbaum and Klein 2011). Further, because the proportional hazards assumption is no longer satisfied for all variables under the extended CPH model, computation of the hazard ratio for the model now involves two sets of predictors, time-independent and time-dependent, and the hazard ratio becomes a function of time.

Extended CPH models are both flexible and robust to the complexities of field research, and thus should receive greater attention when risk is dynamic. We invoked time dependency in our earlier example of turtle nest survival by re-estimating rainfall and local turtle nest density on a daily basis (Wirsing et al. 2012). Similarly, Liebezeit and Kendall (2009) studied nest survival in birds and included temporal variability in risk by adjusting predictors through short time intervals. Smith et al. (2010) tallied wolf survival across three-month intervals and adjusted age, social status, dispersal status, and habitat-related predictors for each interval. In each case, time-dependent variables were assumed to be constant within time intervals and variable across intervals.

Note that the extended CPH model mimics the approach used for AG modeling, except that the CPH model structure is modified directly, whereas the AG model uses a counting process approach to accommodate time-dependent variables. In the aforementioned study of grizzly bear survival, dynamic variables representing age, road density, and habitat type were used in an AG modeling context to assess time-dependent mortality risk determinants (Johnson et al. 2004). To conclude this section, we submit that nominal additional investment in study design, data collection, or data structure and analysis may allow researchers to provide an important time-dependent context to studies that otherwise could yield weak or biased survival inference. A helpful first step would be to use simulations to clarify the optimal coding and time interval structure to best capture the dynamic effects of time-dependent variables on survival estimation.

6.5.5 Further Extensions

The CPH model has additional features that allow further refinements and improvements. Contemporary survival analysis hinges on the assumption that subjects are independent, but in reality individuals often share higher or lower frailty based on intrinsic or extrinsic similarities. Siblings may have related mortality risk depending on their genetic makeup, whereas group-living subjects may be exposed to particular *mortality agents* that affect the entire unit. In such cases, it is appropriate to generate robust standard errors that reflect shared mortality risk, a procedure that is analogous to adding a random variable to a regression model to explain correlation between clustered data (Kleinbaum and Klein 2011). Because the hazard in a *shared frailty model* is computed according to the assigned clusters, the analytical context reflects lack of independence among subjects, leading to increased precision in the estimated error. For example, in bird nest survival studies a frailty component can account for differences in predation risk between clusters of nests on the same plot (Liebezeit and Kendall 2009), whereas in wolf studies similar adjustments can improve the precision of estimated hazards for members of a common pack who share comparable exposure to natural and anthropogenic risks (Smith et al. 2010). It is also possible to use mixed-effects CPH models to explicitly estimate random coefficients to further refine comparisons between groups of subjects (Freitas et al. 2008), and such additions may be useful when dealing with unobserved heterogeneity that has strong effects on risk. However, to date such refinements and extensions of the CPH model have not gained strong traction in the ecological literature, possibly owing to sample size or study design limitations that constrain performance of more complex models.

6.5.6 Parametric Models

We consider that CPH models are well-suited for answering most survival questions in ecology, but in rare cases it may be appropriate to use fully *parametric models* to address uncertainty about the shape of the hazard function or to produce age-specific survival functions (Allison 1995; Collett 2003). For instance, Griffin et al. (2011) used parametric models to assess the role of climate and predators on the survival of neonatal elk (*Cervus elaphus*). Animals were radio-collared shortly after birth and researchers could infer birth date and age of individuals. Similarly, nest survival studies often infer the age of the nest based on the timing of its first discovery or nest and egg characteristics (Liebezeit and Kendall 2009; Wirsing et al. 2012). Accordingly, researchers can estimate the *baseline hazard* for the group of subjects according to age, and thereby use a parametric approach when estimating hazards. Yet, for most ecological studies the extent that a parametric approach will improve on inference that could be derived from a semi-parametric CPH analysis is not evident (Korschgen and Kenow 1996; Olson et al. 2014). For many field studies, researchers do not know either the underlying shape of the hazard function or the age distribution of subjects, meaning that parametric models are often simply not a suitable option. Thus, even when parametric models can be applied to ecological datasets they may not provide marked improvements in terms of hazard estimates, parameter precision, or model fit (Murray 2006).

In the case of predation on turtle nests (Wirsing et al. 2012), the date of nest deposition was known and therefore hazard functions could be estimated parametrically. To fit a parametric model it is necessary to select an underlying statistical distribution; the simplest of these is the *exponential distribution*, which assumes constant survival, followed by the *Weibull distribution*, which allows survival to monotonically increase or decrease (see Eq. 6.3). In the case of turtle nest predation (Wirsing et al. 2012), our previous CPH models fit to each species revealed a series of HR that are qualitatively different from the hazards for the same parameter estimated using the exponential model (Table 6.3). The difference was not surprising because we know from Figure 6.5 that hazard varies markedly during the nesting season and thus that the exponential distribution is an inappropriate fit. In fact, when we fit the more flexible Weibull model to the data, the hazard estimates for most predictors tend to be closer to the HR from CPH. Further, we note that the estimate for parameter p, the dimensionless shape unit defining the trend in the survival function, is below the value of one for either turtle species (Table 6.3). A shape unit <1.0 indicates an overall decline in hazard with time; we had qualitatively identified this trend earlier, after fitting hazard functions described in Figure 6.6b. To conclude, although fully parametric models may be warranted in specific instances when survival functions are known or if they serve as a basis for hypothesis testing, in reality there are few ecological applications where parametric approaches provide either improved parameter estimation or novel insights into the drivers of mortality risk.

6.6 Cause-specific Survival Analysis

6.6.1 The Case for Cause-specific Mortality Data

We began our chapter with a description of John Graunt's analysis of human mortality patterns in Renaissance London. Perhaps Graunt's most lasting contribution was the cataloging of causes of death befalling London residents during a time when sundry mortality agents prevailed across Europe. Graunt's original records report as many as 80 mortality agents, including the rather colorful "king's evil" (tubercular infection of lymph nodes), "hanged and made-away themselves," and "lunatique" (Graunt 1662). Yet, in retrospect one must question the accuracy and variety of Renaissance death records given the limited clinical and diagnostic tools available at that time. In contrast, contemporary ecologists are more restrained in their attribution of causes of death, with rarely more than four to six different proximate causes tending to be reported in a given study (Suzuki et al. 2003; Collins and Kays 2011; Tidemann and Nelson 2011).

Proper identification of cause of death is crucial when determining the relative importance of different risk factors for targeted mitigation. For example, if a population is declining due to predation, targeted management actions require an understanding of the prevalence of predation as a cause of death and which segments of the prey population are most vulnerable (e.g. juveniles, malnourished, or naïve individuals). Yet, there may be multiple predator species killing prey, but perhaps not all individuals are equally susceptible to risk from the same predators. Sympatric wolves and cougars (*Felis concolor*) kill different age and sex cohorts in natural populations of elk (*C. elaphus*, Husseman et al. 2003), as do lynx (*Lynx lynx*) and red fox (*Vulpes vulpes*) preying on roe deer (*Capreolus capreolus*, Melis et al. 2013). Note that it is possible that mortality agents target individuals that have higher frailty due to other factors, as is expected when predation is directed at sick, lame, or otherwise compromised individuals. Snowshoe hares and red squirrels (*Tamiasciurus hudsonicus*) are killed by a variety of predators. Body condition affects vulnerability of hares to risk of predation by specific predators, whereas for red squirrels, body condition does not influence predation risk (Wirsing et al. 2002). It follows that in such cases, predation is the proximate cause of death but malnutrition or other factors can be important ultimate causes. When both *proximate* and *ultimate* causes of death are involved in mortality, their respective role in population dynamics can become especially complex and difficult to disentangle.

The onus is on researchers to adopt field protocols that increase the likelihood of confirming proximate and ultimate causes of death. Frequent detection of subjects allows mortality events to be confirmed and diagnosed shortly after their occurrence, when evidence on the carcass itself and at the death site can be especially informative. With the advent of satellite-based telemetry, it is now possible to determine cause of death for migratory animals that range over expansive spatial scales (Hays 2014; Klaassen et al. 2014). But even intensive field research inevitably includes deaths that are attributed to unknown causes. The uncertainty reflects real-life challenges in conclusively determining cause of death from evidence at the death site, as well as difficulties associated with necropsy of incomplete or decomposed carcasses. Follow-up investigation and ancillary investigation may help discern the cause of death in instances where primary evidence is inconclusive. For example, Hays et al. (2003) inferred high rates of fishing-related mortality in leatherback turtles (*Dermochelys coriacea*) by focusing on unnatural changes in the location or movements of satellite-based radio transmitters. Alternatively, various diagnostic, genetic, and related tests are now available to complement traditional approaches in identifying cause of death, and thereby help reduce the proportion of mis-assigned or unknown mortality events (Onorato et al. 2006; Wengert et al. 2012; Mumma et al. 2014). In a detailed study of mortality of fishers (*Pekania pennanti*), researchers used gross necropsy, histology, toxicology, and molecular methods to distinguish among five proximate causes of death, several of which required multiple approaches for conclusive assessment (Wengert et al. 2013; Gabriel et al. 2015). A number of resources are available for diagnosis of cause of death for wildlife species (Friend et al. 1999; Mineau and Tucker 2002; Wengert et al. 2012; Alt and Eckert 2017).

It may also be possible to adopt field protocols to assess the potential for misdiagnosis of cause of death: in a study of radio-collared snowshoe hares, Murray et al. (1997) showed that deployment of hare carcasses in the field did not result in scavenging by predators, implying that predation rates in radio-collared hares were unlikely to be overestimated due to mistaking scavenging for predation. In contrast, deployment of carcasses of Ruffed Grouse (*Bonasa umbellus*) revealed relatively high rates of scavenging and carcass displacement (Bumann and Stauffer 2002). Regardless, cause of death assessment in field research should include prompt carcass retrieval, necropsy, and assessment of the death site, in conjunction with assessment of the timing of death. Ultimately, studies seeking to develop a robust assessment of cause-specific mortality rates and their demographic importance are compelled to redouble efforts to confirm the proximate and ultimate cause of death for study individuals.

6.6.2 Cause-specific Hazards and Mortality Rates

The quantitative analysis of cause of death allows us to assign specific risks to different mortality agents. To that end, we revisit our earlier expressions of survival and hazard and consider analogues for multiple risk types. Recalling that hazard is the instantaneous risk of death, then if death can occur from $p = 1,\ldots, q$ causes and T is the time to death from any cause, the *cause-specific hazard* for cause p at time t is:

$$h_p(t) = \lim_{\Delta t \to 0} \frac{\Pr(t \le T < t + \Delta t,\ death\,from\,p \mid T \ge t)}{\Delta t} \tag{6.19}$$

for T equal to the time of death from any cause (Cleves et al. 2010). Accordingly, we can consider that different causes of death are "competing" to define T, and therefore the total risk of any death type is $h(t) = \sum_j h_j(t)$, with the probability of death from cause p being $p(t)/h(t)$. If subjects die from either risk types p or q, the probability of death (cause-specific hazard) from p is $h_p(t)/\{h_p(t) + h_q(t)\}$ and the cause-specific hazard from q is one minus the above probability.

In reality, risk types p and q may not act independently, for example when particularly vulnerable individuals experience high risk from more than one mortality agent. This is an important point because the one-to-one correspondence between cause-specific hazard and the cumulative incidence of risk in standard survival analysis is lost in the case of cause-specific risk. Rather, we extend the survivor function to include multiple risks: $S(t) = \exp\{-H_p(t) - H_q(t)\}$, where $H_p(t)$ and $H_q(t)$ are the cumulative hazards from each risk type, respectively. The implications of this distinction for competing risks analysis are important, because the *cause-specific cumulative incidence* (i.e. $F_p(t)$) requires multiple risk types for proper calculation. If the term (1 – KM) estimates the failure function for standard survival data across all causes of death, the analogous term is not relevant to cause-specific failure because it assumes independence among risk types (Cleves et al. 2010; Andersen et al. 2012).The *cumulative incidence function* (CIF) is the appropriate unit in cause-specific mortality estimation, where the cumulative failure rate for risk type p ($F_p(t)$) now considers the association between $h_p(t)$ and related covariates. The relationship between $\text{CIF}_p(t)$ and cause-specific hazards can be summarized by: $p(t) = \int_0^t h_p(x)S(x)dx$, where x is a specified time. Practically speaking, it is more appropriate to consider a modified failure function rather than a modified survivor function when dealing with competing risks, and this is why researchers usually use CIF as the unit of interest (Cleves et al. 2010). However, the CIF can be biased low when there is staggered entry or a paucity of early recruits due to left truncation (Woodroofe 1985; Tsai 1988; but see Heisey and Patterson 2006). Alternatively, the HF estimator can be expanded to deal with multiple fates by considering that m_{ip} is the probability that an individual who is alive at the beginning of interval i dies as a result of mortality cause p (Heisey and Fuller 1985). The number of deaths in i from cause p (d_{ip}) is:

$$\hat{m}_{ip} = \frac{d_{ip}}{r_i}. \tag{6.20}$$

Collectively, the probability of death from cause p during interval i is the sum of the probability that the subject survives to a given day and then dies on the same day

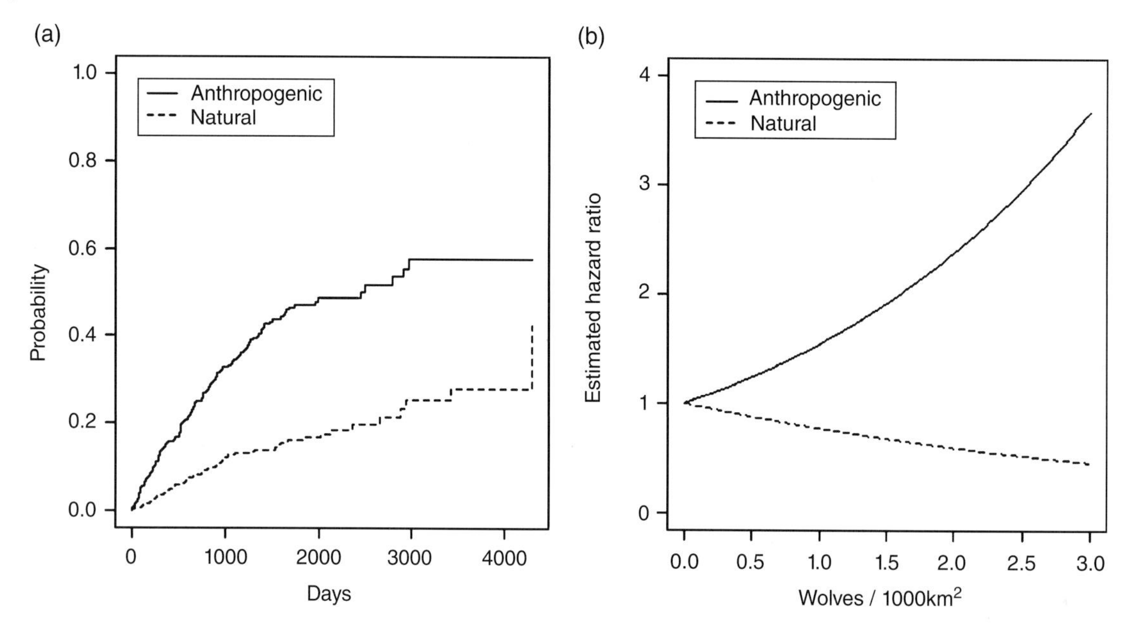

Figure 6.8 Cumulative incidence functions (CIF) (a) and estimated hazards (b) for wolves. Hazards are estimated from a competing risks model including wolf population density. The dataset was modified from Murray et al. (2010).

due to cause p, which can be calculated using the geometric progression:

$$\hat{M}_{ip} = m_{ip} + s_i m_{ip} + s_i^2 m_{ip} + \cdots + s_i^{(L_i - 1)} m_{ip} = \left(\frac{m_{ip}}{1 - s_i}\right)\left(1 - s_i^{L_i}\right). \tag{6.21}$$

It is important to recognize that the first quantity in the second expression above, $\left(\frac{m_{ip}}{1 - s_i}\right)$, is the relative risk of death from cause p in interval i, whereas the second quantity, $(1 - s_i^{L_i})$, is the total probability of death, all sources combined, during i (Heisey and Fuller 1985). More intuitively, if we accept that m represents a specific time unit (usually days), the expression reveals how on day 1 of interval i the subject experiences a mortality rate m_{ip}; on day 2 the mortality rate is a product of risk from m_{ip} and survival during the previous day (s_i), and so forth (Williams et al. 2002).

Our example dataset for a wolf population in western United States (Murray et al. 2010; Smith et al. 2010) included 170 anthropogenic mortalities and 63 natural mortalities, and CIFs were markedly higher for anthropogenic than natural risk (Figure 6.8a). Likewise, if we assume a single interval of constant hazard through the 22-year study, which is unlikely but useful to illustrate the point, 90-day cause-specific mortality rates are 0.029 (0.025, 0.035) for anthropogenic and 0.013 (0.010, 0.016) for natural causes where CIs were derived from Taylor series approximation (Heisey and Fuller 1985). Yet, both the CIF figure (Figure 6.8a) and cause-specific mortality estimates do little to inform about the potential inter-relation between different causes of death in the wolf population itself, and therefore should not be the final destination in our cause-specific mortality analysis.

6.6.3 Competing Risks Analysis

Competing or *proportional risks analysis* constitutes an extension of CPH for the case of cause-specific hazards; alternatively one could use fully parametric models to compare CIFs, although this latter approach is tenuous owing to the recognized limitations of parametric survival analysis in ecology. The most practical way to implement competing risks analysis is to use the *data augmentation* approach (Lunn and McNeil 1995), which takes advantage of the additivity of hazard functions by duplicating the dataset for each risk type and dummy coding within each duplicate to identify the appropriate risk. Accordingly, for each duplicate all mortalities from causes other than those outside the specific risk set are functionally right-censored (see Competing risks in Table 6.1). By stratifying according to risk type, we perform multiple regressions simultaneously, allowing us to deal with more than one risk type within the same analysis and to compare covariates that may relate differentially to each risk type.

There is an important note of caution when deciding on how to functionally conduct competing risks analysis. The number of different mortality agents affecting the subject population and the role of each risk on hazard should influence the number and identity of risk types included in the analysis. Notwithstanding the need to test hypotheses and predictions that have been established a priori (Chapter 2), exploratory analysis should be used to assess the composition of risk type structure in the dataset. If some causes of death are either uncommon or related through proximate-ultimate linkages, categorical outcomes should be pooled to achieve a more parsimonious risk set. Returning to the wolf mortality study, Murray et al. (2010) used a competing risks framework to compare factors promoting mortality from *natural causes* (disease, strife, senescence) versus *anthropogenic causes* (poaching, legal control, vehicle collision). To achieve this level of detail, animals were radio-collared and monitored regularly, and dead animals were recovered and necropsied shortly after mortality was detected. Some causes of death were relatively uncommon, and the objective of the study was primarily to assess the role of humans on wolf mortality risk. Thus, it was appropriate to consolidate the various causes of death into two main categories of risk: natural vs. anthropogenic causes. A subset of wolves (12%) could not be assigned to either of the two main risk types because they died of undetermined causes, but exploratory analysis of survival times and covariate influence confirmed that this group was comparable to the other two main risk types. Wolves dying of unknown causes probably constituted a representative sample of the population (i.e. included animals dying of anthropogenic and natural causes, in representative proportions) and thus could be included in the analysis as right censors. Note that inclusion of these individuals as right censors serves to illustrate two important reminders: (i) causes of death should be confirmed to the fullest extent possible when anticipating competing risks analysis to reduce the number of unknown mortalities; and (ii) all subjects, regardless of their fate, should be retained in the analysis (pending appropriate checks for bias) to ensure that informative censoring does not compromise parameter estimates.

The role of a variety of predictors was examined to identify determinants of anthropogenic and natural deaths in the wolf population (Table 6.4). Overall, animals that were dispersers (Disperser, 1 = yes) had qualitatively higher risk from both anthropogenic and natural causes of death compared to the nondispersing group, but the HRs were comparable between the two risk types, and in both cases CIs overlapped 1.0. However, risk of death for animals living in Montana (Montana, coded as 1 = yes) increased for anthropogenic causes but decreased for natural causes. Likewise, mortality risk from anthropogenic causes increased by 54.3% for each unit increase in wolf density ($|1 - 1.543| \times 100 = 54.3\%$), compared to a concomitant decline in risk from natural causes with increasing wolf density ($|1 - 0.772| \times 100 = 22.8\%$). The divergence in risk is especially evident when HR

Table 6.4 Cause-specific hazard ratios (HR) (±95% confidence intervals [CI]) for variables included in a multivariate competing risks model for wolves.

Variable	Anthropogenic	Natural
Disperser	1.684 (0.803, 3.531)	1.662 (0.850, 3.249)
Montana	5.421 (2.668, 11.02)	0.482 (0.233, 0.995)
Wolf density	1.543 (1.234, 1.930)	0.772 (0.545, 1.094)

Source: Data adapted from Murray et al. (2010).
The analysis was restricted to wolves that were representative of the population and thus excluded individuals captured following livestock depredation.

associated with each risk type are graphed against wolf density (Figure 6.5b). However, one must take care when comparing HR from competing risks analysis to their corresponding CIFs, owing to the possibility that causes of death are nonindependent (Andersen et al. 2012). Ultimately, because anthropogenic risks were higher, and natural risks were lower, in both Montana and across wolf density, we infer that the two risk types were not independent. In other words, anthropogenic mortality was to some extent compensatory to natural mortality (Figure 6.5b, see below). The complex relationships and contextual subtleties between causes of death would not have been revealed in the absence of a competing risks analysis isolating the effects of each risk type, specifically through inclusion of an interaction term with predictors (Heisey and Patterson 2006). As an outcome of this analysis, conservation efforts designed to manage wolf deaths from anthropogenic causes should focus primarily on the Montana population and relative to increasing wolf numbers. In a similar scenario, Robinson et al. (2015) used a CIF-based approach to compare cause-specific mortality rates of Amur tigers (*Panthera tigris*) to assess the role of a novel cause of death, canine distemper virus, on the tiger population.

6.6.4 Additive Versus Compensatory Mortality

Ecologists have long been preoccupied with understanding whether different risk types affect population dynamics, and in particular whether anthropogenic mortality, usually due to human harvest, is *additive* to other causes of death and thereby constrains population growth (Burnham and Anderson 1984; Boyce et al. 1999; Sandercock et al. 2011). Alternatively, harvest or other sources of mortality may play a *compensatory* role, meaning that their demographic influence is lessened by the fact that some mortality occurs irrespective of whether the cause of interest is implicated. For example, earlier we discussed instances when predators kill prey that are doomed to die from other causes; this type of mortality requires distinction between the proximate and ultimate cause of death and constitutes compensatory mortality because killed individuals would not otherwise contribute to population growth. At one extreme, compensation can be complete and all mortality replaces potential mortality from other causes; at the other extreme, mortality is fully additive. Studies often reveal *partially compensatory* effects (i.e. partially additive), depending on the magnitude and individual targets of the source of mortality. For example, power-line collisions are only partly compensatory to other sources of mortality in juvenile White Storks (*Ciconia ciconia*, Schaub and Lebreton 2004). This means that overall mortality rates in the stork population are higher due to such mortality. The status of a particular mortality agent in the context of the *additive-compensatory mortality continuum* will determine its influence on the population trajectory (Péron 2013). Interest in additive versus compensatory mortality has dominated longstanding discussions related to harvest management, population recovery, and species conservation (Burnham et al. 1984; Conroy and Krementz 1990; Pöysä et al. 2004; Cooley et al. 2009; Sandercock et al. 2011).

By allowing interaction terms between risk type and predictors, *competing risks models* offer a direct and robust individual-based tool in the assessment of additive versus compensatory mortality (Heisey and Patterson 2006; Murray et al. 2010). However, to date few studies have exploited this approach, preferring instead to focus on population-level changes in different risk types through space or time. Here, the focus is on the change in cause-specific mortality or HR and whether different risk types follow similar or opposite trends, which would indicate additive or compensatory effects, respectively. For example, in a large-scale harvest experiment, Sandercock et al. (2011) used a cause-specific hazards framework to show that the relative importance of hunting mortality in Willow Ptarmigan (*Lagopus lagopus*) often varied inversely with natural mortality, indicating that hunting is partially compensatory. Robinson et al. (2014) used changes in cause-specific mortality following a temporal shift in cougar harvest strategy to infer that mortality due to hunting of adult cats was an additive source of mortality. Griffin et al. (2011) showed that predation by grizzly bears was negatively related to survival of elk calves and therefore constituted additive mortality, whereas predation by other predator species had no discernible demographic significance and thus was compensatory. Bastille-Rousseau et al. (2016) found that predation on caribou calves (*Rangifer tarandus*) by invasive coyotes (*Canis latrans*) in Newfoundland was at least partly additive and therefore potentially contributing to caribou population decline.

Traditionally, the population-level approach to the additive–compensatory debate has involved regressing survival against cause-specific mortality, where rates are tallied for one or more populations, usually on an annual basis. Here, a negative slope implies additive mortality, with a steeper slope implying stronger additivity. For example, using annual cause-specific mortality rates for three separate wolf populations (Murray et al. 2010), we infer by the qualitative difference between slopes that anthropogenic mortality is more strongly additive in Montana and Idaho than in the Greater Yellowstone Area (Figure 6.9). Notably, the range of observed anthropogenic mortality rates is much greater in Montana, providing extra reassurance about the additive effects of anthropogenic mortality on that wolf population. The findings are consistent with other studies using a regression-based approach to infer additive effects of anthropogenic mortality on wolf populations (Creel and Rotella 2010; Sparkman et al. 2011).

When regression-based models are used to infer the demographic importance of a particular mortality agent, regression slopes are compared statistically to determine whether populations differ in the role of additive–compensatory factors (Zar 1999). One caution is that observations such as those in Figure 6.9 are serially autocorrelated when collected sequentially from the same population, leading to lack of independence. More importantly, survival and cause-specific mortality rates themselves are not independent, meaning that this approach may be biased toward showing additivity. To some extent, bias may be mitigated by calculating a corrected standard error of the slope, or perhaps by using a mixed-model design (Otis and White 2004; Schaub and Lebreton 2004). More recently, Servanty et al. (2010) advocated use of a *state-space approach* to properly assess correlations between causes of death, whereas Péron (2013) presented a metric for quantifying additivity based on the variance–covariance structure of the mortality rates themselves. Notwithstanding these developments, we must stress that population-level approaches relating cause-specific mortality to survival fail to take full advantage of continuous time information provided by tracking individual variability through space and time. In fact, Figure 6.8b presents a substantially more convincing case that anthropogenic mortality in wolves is partly compensatory than does Figure 6.9. Ultimately, competing risks analysis and supporting figures allow researchers to more fully dissect mechanisms underlying the additive–compensatory interplay than is possible with a population-level analysis. Thus, when combined with a study design based on a before-after-control-impact approach (Robinson et al. 2014) or replicated experimental treatments (Sandercock et al. 2011), competing risks models

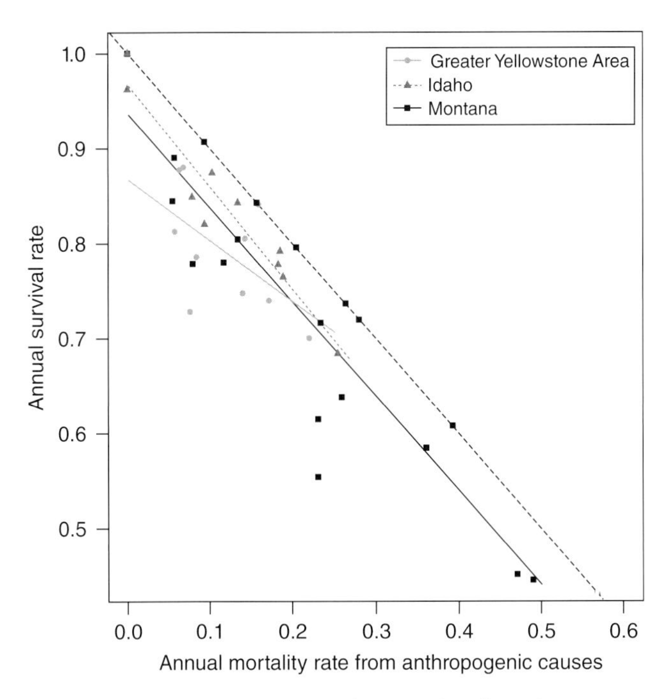

Figure 6.9 Tests for additive mortality in wolves from three populations, using regression of annual survival rate versus annual rate of anthropogenic mortality. Survival and mortality rates were calculated using Heisey–Fuller estimates. The dashed line has a slope = −1.0 which is expected with complete additivity. *Source:* The dataset was modified from Murray et al. (2010).

offer an especially powerful tool for rigorously disentangling additive versus compensatory effects of mortality.

6.7 Software Tools

Most advanced statistical software packages have built-in functions for the estimation of the survival curve with Kaplan–Meier models, and parametric and semi-parametric regression analyses. General computing environments or programming languages such as `MATLAB` and `PYTHON` also have functions or modules available for survival analysis. In `SAS`, the `lifetest` procedure is the principal function for calculating survival curves, instantaneous hazard curves, and associated tests. The `phreg` procedure is the main tool for parametric and semi-parametric regression analysis, with several possible specifications including the use of frailty terms and diagnostic tests. In `Stata`, the function `sts` is used for calculating survival functions and survival curves, whereas the functions `stcox` and `streg` can be used for semi-parametric and parametric regression, respectively. In program `R`, the `survival` package offers many functions related to survival analysis and is the primary reference. The `survival` package includes `survfit` to calculate the survival curve and `surv` to create survival objects that are used for semi-parametric (`coxph`) or parametric analyses (`survreg`). The `eha` package also offers different options for parametric and semi-parametric regression, notably using the functions `phreg` and `coxreg`. A distinction between `survreg` and `phreg` is that the former uses an accelerated-failure time model while the latter used a proportional hazards model that will give results analogous to `coxph` or `coxreg`. Other useful packages in `R` include: `muhaz` for the estimation of hazard functions, `coxme` for the addition of mixed-effects to CPH regression, and `cmprsk` for calculation of CIFs in a competing risk framework.

6.8 Online Exercises

In the exercises associated with our chapter, we use both simulated and real data to illustrate some of the main features covered herein. In Exercise 1, we use a simulated dataset to demonstrate preparations for a survival analysis, with particular attention given to the structure of the time variable. Next, we show how to conduct simple hypothesis tests for different covariates, and how to graph a variety of survival functions. In Exercises 2 and 3, we demonstrate how to develop more complex semi-parametric and parametric models to evaluate mortality risk determinants, using an example based on predation by raccoons on the nests of freshwater turtles. We test model assumptions, and then demonstrate how to identify the best out of a suite of candidate models. In Exercise 4, we demonstrate the steps for conducting a competing risks analysis, and generating corresponding graphical functions to compare hazards across alternative risk types for gray wolves exposed to both anthropogenic and natural mortality.

6.9 Future Directions

Understanding the factors affecting mortality risk and fate of organisms will remain a core interest in population ecology and conservation biology. Ecologists have largely abandoned cross-sectional studies of populations (i.e. life table analysis) in favor of longitudinal studies that track individuals through time as they encounter a variety of risks through life. However, despite a longstanding and well-established basis for investigating continuous time survival in observational research in other fields, several improvements are needed to take full advantage of the available opportunities in ecology. At its core, robust continuous time survival analysis relies on quality data consisting of intensive monitoring and high detection probability of individual subjects, prompt detection of mortality events, and application of contemporary diagnostic tools to improve cause of death determination. Although many research studies are already appropriately designed to meet many of these needs, for others even modest improvements in study design, data collection, and data treatment and analysis will dramatically improve inference. For example, we highlight the development of new technologies like remotely triggered trail cameras, bio-logging, or satellite-based telemetry that allow researchers to monitor individuals much more efficiently and with far greater detail than in the past (Naef-Danzer et al. 2005; Cagnacci et al. 2010). It is possible to use these devices to assign temporally or spatially explicit mortality risk to individuals (Smith et al. 2010; Loveridge et al. 2017). Doubtless, ongoing refinement and miniaturization of these new technologies means that the age of continuous time monitoring for all sorts of organisms is fast approaching, perhaps even for plants and small animals, which currently are under-represented in the continuous time survival literature.

Semi-parametric modeling approaches should meet the survival analysis needs of the vast majority of population ecologists. However, there remains a tendency to develop survival models without appropriately testing for model assumptions, in particular those related to proportional hazards, lack of progressive sample heterogeneity, and random censoring of subjects. These model assumptions are treated seriously in the human medicine and epidemiological literature (Grambsch and Therneau 1994; Leung et al. 1997; Zens and Peart 2003; Hosmer et al. 2008), and should receive similar attention in ecology. Additional concerns relate to sample-size requirements in survival analysis; observational field studies frequently include too few subjects, too few mortalities, or too many censors to provide robust statistical inference or to allow development of complex survival models (Murray 2006). Necessary steps in model diagnostics, including assessing model parsimony and overfitting, are often overlooked steps that relate directly to sample size and statistical power. Adjustments concerning these issues should include not only improved field protocols and refined statistical analysis, but also proper reporting of censoring rates as well as the outcome of exploratory analyses and model validation. Mixed-effects survival models, where random effects can be assigned and explicitly modeled for groups of individuals not under direct control of the observer, add an important dimension to traditional tools for analyzing survival in free-ranging organisms.

Our review is largely focused on survival analysis methods that are readily implemented using information-theoretic methods, including model selection and multimodel inference approaches (Chapter 2). Recent efforts adapted Bayesian methods to survival analysis (Ibrahim et al. 2005), specifically using Markov chain Monte Carlo and Gibbs sampling algorithms to appropriately weigh models according to their posterior probabilities. Bayesian approaches may be an informative solution to traditional survival analysis (Omerlu et al. 2009; Halstead et al. 2012) and thereby offer respite from otherwise thorny issues in standard survival analysis such as small sample size, missing values, censoring, nonproportionality of hazards, and even unknown causes of death. However, as with all Bayesian approaches, these methods rely on representative priors that can be readily translated from existing survival data; currently this may be a challenging requirement given the paucity of reliable and generalizable survival datasets for many taxa. Yet, development and acquisition of new survival data, combined with improved computational efficiency and user interface, will make Bayesian methods increasingly attractive for survival research in ecology.

One important extension of the suite of continuous time survival methods involves *multiple event analysis*, which is closely aligned to competing risks methods but considers rates of nonlethal reoccurring events. This approach takes advantage of the time-to-event origins of the methods discussed in this chapter, and applies them specifically to evaluate rate determinants for repeatable events. In a simple example, Merrill et al. (2010) used reoccurring events analyses (i.e. CPH and parametric analogues) to illustrate how environmental factors influence rates that wolves kill prey. Because wolves were equipped with satellite-based tracking units, monitoring and detection was quasi-continuous and allowed estimation of the number of prey kills per unit time and relative to habitat and prey density. Reoccurring events analysis holds possibilities for assessing rates of a variety of demographic characteristics such as breeding or dispersal or other events that may happen multiple times in an individual's lifetime. To date, reoccurring events in ecology have rarely been considered in an actual time-to-event context or for estimating demographic rates (Bastille-Rousseau et al. 2011; Whittington et al. 2011; McPhee et al. 2012). The logical extension here is the potential to adopt multistate models to evaluate the interplay between variable subject states and corresponding risks. For example, this approach could allow researchers to investigate how dispersal status increases mortality risk that is separate from natural nondispersal risk, while considering that dispersal status is reversible through time; these methods provide a sophisticated approach for assessing co-occurring and reoccurring risk types (Beyersmann et al. 2012; Devineau et al. 2014; Hightower and Harris 2017). However, we caution that these and other advanced methods are developed primarily for fields where sample sizes and study design tend not to limit modeling capabilities or to raise concerns over model complexity and model mis-specification. Thus, the extent that these methods overextend the capabilities of more modest ecological datasets is open for debate.

Ultimately, more careful application of the methods detailed in this chapter, specifically in the context of intensified fieldwork for developing robust continuous time datasets and the use of contemporary methods focusing on individual variability in risk through space and time, will serve well in raising the bar for ecological survival analysis. When combined with more extensive coverage of these topics in quantitative ecology courses and workshops, these advances should improve our capacity for robust survival analysis and thereby place us in a better position to address future questions and challenges related to population viability, sustainable harvest, or environmental impact mitigation.

References

Allison, P.D. (1995). *Survival Analysis Using the SAS System*. Cary, North Carolina, USA: SAS Institute, Inc.

Alt, K. and Eckert, M. (2017). *Predation ID Manual: Predator Kill and Scavenging Characteristics*. New York, NY: Skyhorse Publishing, Inc.

Andersen, P.K., Geskus, R.B., de Witte, T. et al. (2012). Competing risks in epidemiology: possibilities and pitfalls. *International Journal of Epidemiology* 41: 861–870.

Anderson, D.R., Wywialowski, A.P., and Burnham, K.P. (1981). Tests of the assumptions underlying life table methods for estimating parameters from cohort data. *Ecology* 62: 1121–1124.

Austin, P.C., Lee, D.S., and Fine, J.P. (2016). Introduction to the analysis of survival data in the presence of competing risks. *Circulation* 133: 601–609.

Barbour, A.B., Ponciano, J.M., and Lorenzen, K. (2013). Apparent survival estimation from continuous mark-recapture/resighting data. *Methods in Ecology and Evolution* 4: 846–853.

Barron, D.G., Brawn, J.D., and Weatherhead, P.J. (2010). Meta-analysis of transmitter effects on avian behavior and ecology. *Methods in Ecology and Evolution* 1: 180–187.

Bart, J. and Robson, D.S. (1982). Estimating survivorship when the subjects are visited periodically. *Ecology* 63: 1078–1090.

Bastille-Rousseau, G., Shaefer, J.A., Lewis, K.P. et al. (2016). Phase-dependent climate-predator interactions explain three decades of variation in neonatal caribou survival. *Journal of Animal Ecology* 85: 445–456.

Bastille-Rousseau, G., Fortin, D., Dussault, C. et al. (2011). Foraging strategies by omnivores: are black bears actively searching for ungulate neonates or are they simply opportunistic predators? *Ecography* 34: 588–596.

Beyersmann, J., Allignol, A., and Schumacher, M. (2012). *Competing Risks and Multistate Models with R*. New York: Springer.

Boyce, M.S., Sinclair, A.R.E., and White, G.C. (1999). Seasonal compensation of predation and harvesting. *Oikos* 87: 419–426.

Bridge, E.S., Thorup, K., Bowlin, M.S. et al. (2011). Technology on the move: recent and forthcoming innovations for tracking migratory birds. *BioScience* 61: 689–698.

Brooks, C., Bonyongo, C., and Harris, S. (2008). Effects of global positioning system collar weight on zebra behavior and location error. *Journal of Wildlife Management* 72: 527–534.

Bumann, G.B. and Stauffer, D.F. (2002). Scavenging of Ruffed Grouse in the Appalachians: influences and implications. *Wildlife Society Bulletin* 30: 853–860.

Burnham, K.P. and Anderson, D.R. (1984). Tests of compensatory vs. additive hypotheses of mortality in Mallards. *Ecology* 65: 105–112.

Burnham, K.P., White, G.C., and Anderson, D.R. (1984). Estimating the effect of hunting on annual survival rates of adult Mallards. *Journal of Wildlife Management* 48: 350–361

Cagnacci, F., Boitani, L., Powell, R.A. et al. (2010). Animal ecology meets GPS-based radiotelemetry: a perfect storm of opportunities and challenges. *Philosophical Transactions of the Royal Society, B: Biological Sciences* 365: 2157–2162.

Casper, R.M. (2009). Guidelines for the instrumentation of wild birds and mammals. *Animal Behaviour* 78: 1477–1483.

Chitwood, M.C., Lashley, M.A., DePerno, C.S. et al. (2017). Considerations on neonatal ungulate capture method: potential for bias in survival estimation and cause-specific mortality. *Wildlife Biology* 2017: wlb.00250.

Cleves, M., Gutierrez, R.G., Gould, W. et al. (2010). *An Introduction to Survival Analysis Using Stata*. College Station, TX: Sata Press.

Collett, D. (2003). *Modeling Survival Data in Medical Research*. New York, NY: Chapman and Hall.

Collins, C. and Kays, R. (2011). Causes of mortality in North American populations of large and medium-sized mammals. *Animal Conservation* 14: 474–483.

Conroy, M.J. and Krementz, D.G. (1990). A review of the evidence for the effects of hunting on American black duck populations. *Transactions of the North American Wildlife and Natural Resources Conference* 55: 511–517.

Cooley, H.S., Wielgus, R.B., Koehler, G.M. et al. (2009). Does hunting regulate cougar populations? A test of the compensatory mortality hypothesis. *Ecology* 90: 2913–2921.

Creel, S. and Rotella, J.J. (2010). Meta-analysis of relationships between human offtake, total mortality and population dynamics of gray wolves (*Canis lupus*). *PLoS One* 5: e12918.

Crowder, M.J. (2001). *Classical Competing Risks*. Chapman and Hall.

DeCesare, N.J., Hebblewhite, M., Lukacs, P.M. et al. (2017). Evaluating sources of censoring and truncation in telemetry-based survival data. *Journal of Wildlife Management* 80: 138–148.

Dechen Quinn, A.C., Williams, D.M., and Porter, W.F. (2012). Postcapture movement rates can inform data-censoring protocols for GPS-collared animals. *Journal of Mammalogy* 93: 456–463.

DelGiudice, G.D., Severud, W.J., Obermoller, T.R. et al. (2018). Gaining a deeper understanding of capture-induced abandonment of moose neonates. *Journal of Wildlife Management* 82: 287–298.

Devineau, O., Kendall, W.I., Doherty, P.F. et al. (2014). Increased flexibility for modeling telemetry and nest-survival data using the multistate framework. *Journal of Wildlife Management* 78: 224–230.

Efron, B. (1988). Logistic regression, survival analysis, and the Kaplan–Meier curve. *Journal of the American Statistical Association* 83: 414–425.

Egerton, F.N. (2005). A history of the ecological sciences, part 15: the precocious origins of human and animal demography and statistics in the 1600s. *Bulletin of the Ecological Society of America* 86: 31–38.

Fieberg, J. and DelGiudice, G.D. (2009). What time is it? Choice of time origin and scale in extended proportional hazards models. *Ecology* 90: 1687–1697.

Fieberg, J. and DelGiudice, G.D. (2011). Estimating age-specific hazards from wildlife telemetry data. *Environmental and Ecological Statistics* 18: 209–222.

Freitas, C., Kovacs, K.M., Lydersen, C. et al. (2008). A novel method for quantifying habitat selection and predicting habitat use. *Journal of Applied Ecology* 45: 1213–1220.

Friend, M., Franson, J.C., Ciganovich, E.A. et al. (eds.) (1999). *Field Manual of Wildlife Diseases: General Field Procedures and Diseases of Birds*. Biological Resources Division Information and Technology Report 1999–001. Washington, DC: U.S. Department of the Interior.

Gabriel, M.W., Woods, L.W., Wengert, G.M. et al. (2015). Patterns of natural and human-caused mortality factors of a rare forest carnivore, the fisher (*Pekania pennanti*) in California. *PLoS One* 10: e0140640.

Gail, M.H. (1991). A bibliography and comments on the use of statistical models in epidemiology in the 1980's. *Statistics in Medicine* 10: 1819–1885.

Gilbert, S.L., Lindberg, M.S., Hundertmark, K.J. et al. (2014). Dead before detection: addressing the effects of left truncation on survival estimation and ecological inference for neonates. *Methods in Ecology and Evolution* 5: 992–1001.

Grambsch, P.M. and Therneau, T.M. (1994). Proportional hazards tests and diagnostics based on weighted residuals. *Biometrika* 81: 515–526.

Graunt, J. (1662). *Natural and Political Observations Mentioned in a Following Index, and Made upon the Bills of Mortality*. London, UK: John Martin.

Griffin, K.A., Hebblewhite, M., Robinson, H.S. et al. (2011). Neonatal mortality of elk driven by climate, predator phenology and predator community composition. *Journal of Animal Ecology* 80: 1246–1257.

Grosbois, V., Giminez, O., Gaillard, J.-M. et al. (2008). Assessing the impact of climate variation on survival in vertebrate populations. *Biological Reviews* 83: 357–399.

Halstead, B.J., Wylie, J.D., Coates, P.S. et al. (2012). Bayesian shared frailty models for regional inference about wildlife survival. *Animal Conservation* 15: 117–124.

Hamel, N.J., Parrish, J.K., and Conquest, L.L. (2004). Effects of tagging on behavior, provisioning, and reproduction in the Common Murre (*Uria aalge*), a diving seabird. *Auk* 121: 1161–1171.

Hastings, A. (1997). *Population Biology: Concepts and Models*. New York, NY: Springer-Verlag.

Hawkins, P. (2004). Bio-logging and animal welfare: practical refinements. *Memoirs of the National Institute of Polar Research, Special Issue* 58: 58–68.

Hays, G.C. (2014). Tracking animals to their death. *Journal of Animal Ecology* 83: 5–6.

Hays, G.C., Bradshaw, C.J.A., James, M.C. et al. (2007). Why do Argos satellite tags deployed on marine animals stop transmitting? *Journal of Experimental Marine Biology and Ecology* 349: 52–60.

Hays, G.C., Broderick, A.C., Godley, B.J. et al. (2003). Satellite telemetry suggests high levels of fishing induced mortality for marine turtles. *Marine Ecology Progress Series* 262: 305–309.

Heisey, D.M. and Fuller, T.K. (1985). Evaluation of survival and cause-specific mortality rates using telemetry data. *Journal of Wildlife Management* 49: 668–674.

Heisey, D.M. and Patterson, B.R. (2006). A review of methods to estimate cause-specific mortality in presence of competing risks. *Journal of Wildlife Management* 70: 1544–1555.

Hensler, G.L. and Nichols, J.D. (1981). The Mayfield method of estimating nesting success: a model, estimators and simulation results. *Wilson Bulletin* 93: 42–53.

Hess, K.R., Serachipotol, D.M., and Brown, B.W. (1999). Hazard function estimators: a simulation study. *Statistics in Medicine* 18: 3075–3088.

Hightower, J.E. and Harris, J.E. (2017). Estimating fish mortality rates using telemetry and multistate models. *Fisheries* 42: 210–219.

Hoey, A.S. and McCormick, M.I. (2004). Selective predation for low body condition at the larval-juvenile transition of a coral reef fish. *Oecologia* 139: 23–29.

Hosmer, D.W., Lemeshow, S., and May, S. (2008). *Applied Survival Analysis: Regression Modeling of Time to Event Data*. New York, NY: Wiley.

Hudson, P.J., Dobson, A.P., and Newborn, D. (1992). Do parasites make prey vulnerable to predation? Red Grouse and parasites. *Journal of Animal Ecology* 61: 681–692.

Husseman, J.S., Murray, D.L., Power, G. et al. (2003). Assessing differential prey selection patterns between two sympatric large carnivores. *Oikos* 101: 591–601.

Ibáñez-Álamo, J.D., Sanllorente, O., and Soler, M. (2012). The impact of researcher disturbance on nest predation rates: a meta-analysis. *Ibis* 154: 5–14.

Ibrahim, J.G., Chen, M.-H., and Sinha, D. (2005). *Bayesian Survival Analysis*. New York, NY: Springer.

Johnson, D.H. (1979). Estimating nest success: the Mayfield method and an alternative. *Auk* 96: 651–661.

Johnson, C.J., Boyce, M.S., Schwartz, C.C. et al. (2004). Modeling survival: application of the Andersen–Gill model to Yellowstone grizzly bears. *Journal of Wildlife Management* 68: 966–978.

Kalbfleisch, J.D. and Prentice, R.L. (2002). *The Statistical Analysis of Failure Time Data*. Hoboken, New Jersey, USA: Wiley.

Kissling, W.D., Pattemore, D.E., and Hagen, M. (2014). Challenges and prospects in the telemetry of insects. *Biological Reviews* 89: 511–530.

Klaassen, R.H.G., Hake, M., Strandberg, R. et al. (2014). When and where does mortality occur in migratory birds? Direct evidence from long-term satellite tracking of raptors. *Journal of Animal Ecology* 83: 176–184.

Kleinbaum, D.G. and Klein, M. (2011). *Survival Analysis: A Self-Learning Text*. New York, NY: Springer.

Korschgen, C. and Kenow, K. (1996). Survival of radiomarked Canvasback ducklings in northwestern Minnesota. *Journal of Wildlife Management* 60: 120–132.

Kurucz, K., Batáry, P., Frank, K. et al. (2015). Effects of daily nest monitoring on predation rate – an artificial nest experiment. *North-Western Journal of Zoology* 11: 219–224.

Leighton, P., Horrocks, J., and Kramer, D.L. (2011). Predicting nest survival in sea turtles: when and where are eggs most vulnerable to predation? *Animal Conservation* 14: 186–195.

Leung, K.M., Elashoff, R.M., and Afifi, A.A. (1997). Censoring issues in survival analysis. *Annual Review of Public Health* 18: 83–104.

Liberg, O., Chapron, G., Wabakken, P. et al. (2011). Shoot, shovel, and shut up: cryptic poaching slows restoration of a large carnivore in Europe. *Proceedings of the Royal Society of London B* 279: 910–915.

Liebezeit, J. and Kendall, S. (2009). Influence of human development and predators on nest survival of tundra birds, Arctic Coastal Plain, Alaska. *Ecological Applications* 19: 1628–1644.

Lindsey, J.C. and Ryan, L.M. (1998). Tutorial in biostatistics-methods for interval-censored data. *Statistics in Medicine* 17: 219–238.

Loveridge, A.J., Valeix, M., Chapron, G. et al. (2017). Conservation of large predator populations: demographic and spatial responses of African lions to the intensity of trophy hunting. *Biological Conservation* 204: 247–254.

Ludynia, K., Dehnhard, N., Poisbleau, M. et al. (2012). Evaluating the impact of handling and logger attachment on foraging parameters and physiology in southern Rockhopper Penguins. *PLoS One* 7: e50429.

Lunn, M. and McNeil, D. (1995). Applying Cox regression to competing risks. *Biometrics* 51: 524–532.

Major, R.E. (1990). The effect of human observers on the intensity of nest predation. *Ibis* 132: 608–612.

Mansfield, K.L., Wyneken, J., Rittschof, D. et al. (2012). Satellite tag attachment methods for tracking neonate sea turtles. *Marine Ecology Progress Series* 457: 181–192.

Mayfield, H.F. (1975). Suggestions for calculating nest success. *Wilson Bulletin* 87: 456–466.

McCallum, H. (2000). *Population Parameters*. London: Blackwell.

McPhee, H.M., Webb, N.F., and Merrill, E.H. (2012). Time-to-kill: measuring attack rates in a heterogenous landscape with multiple prey types. *Oikos* 121: 711–720.

Melis, C., Nilsen, E.B., Panzacchi, M. et al. (2013). Roe deer face competing risks between predators along a gradient in abundance. *Ecosphere* 4: art111.

Merrill, E., Sand, H., Zimmerman, B. et al. (2010). Building a mechanistic understanding of predation with GPS-based movement data. *Philosophical Transactions of the Royal Society* 365: 2279–2288.

Metcalf, C.J. and Pavard, S. (2007). Why evolutionary biologists should be demographers. *Trends in Ecology and Evolution* 22: 205–212.

Mineau, P. and Tucker, K.R. (2002). Improving detection of pesticide poisoning in birds. *Journal of Wildlife Rehabilitation* 25: 4–13.

Morris, W.M. and Doak, D. (2004). *Quantitative Conservation Biology*. Sunderland, MA: Sinauer and Associates.

Mueller, H.G. and Wang, J.G. (1994). Hazard rates estimation under random censoring with varying kernels and bandwidths. *Biometrics* 50: 61–76.

Mueller, R.C., Scudder, C.M., Porter, M.E. et al. (2005). Differential tree mortality in response to severe drought: evidence for long-term vegetation shifts. *Journal of Ecology* 93: 1085–1093.

Mumma, M.A., Soulliere, C.E., Mahoney, S.P. et al. (2014). Enhanced understanding of predator-prey relationships using molecular methods to identify predator species, individual, and sex. *Molecular Ecology* 14: 100–108.

Murie, A. (1944). *The Wolves of Mount McKinley*. Seattle, WA: University of Washington Press.

Murray, D.L. (2002). Differential body condition and vulnerability to predation in snowshoe hares. *Journal of Animal Ecology* 71: 614–625.

Murray, D.L. (2006). On improving telemetry-based survival estimation. *Journal of Wildlife Management* 70: 1530–1543.

Murray, D.L., Cary, J.R., and Keith, L.B. (1997). Interactive effects of sublethal nematodes and nutritional status on snowshoe hare vulnerability to predation. *Journal of Animal Ecology* 66: 250–264.

Murray, D.L. and Fuller, M.R. (2000). A critical review of the effects of marking on the biology of vertebrates. In: *Research Techniques in Animal Ecology: Controversies and Consequences* (eds. L. Boitani and T.K. Fuller), 15–64. Columbia University Press.

Murray, D.L. and Patterson, B.R. (2006). Wildlife survival estimation: recent advances and future directions. *Journal of Wildlife Management* 70: 1499–1503.

Murray, D.L., Smith, D.W., Bangs, E.E. et al. (2010). Death from anthropogenic causes is partially compensatory in recovering wolf populations. *Biological Conservation* 143: 2514–2524.

Naef-Danzer, B., Früh, D., Stalder, M. et al. (2005). Miniaturization (0.2 g) and evaluation of attachment techniques of telemetry transmitters. *Journal of Experimental Biology* 208: 4063–4068.

Neal, D. (2004). *Introduction to Population Ecology*. Cambridge, UK: Cambridge University Press.

Nussey, D.H., Coulson, T., Festa-Bianchet, M. et al. (2008). Measuring senescence in wild animal populations: towards a longitudinal approach. *Functional Ecology* 22: 393–406.

Oakes, D. (2001). Biometrika centenary: survival analysis. *Biometrika* 88: 99–142.

Olson, K., Larsen, E., Mueller, T. et al. (2014). Survival probabilities of adult Mongolian gazelles. *Journal of Wildlife Management* 78: 35–41.

Omerlu, I.K., Ozdamar, K., and Ture, M. (2009). Comparison of Bayesian survival analysis and Cox regression analysis in simulated breast cancer datasets. *Expert Systems with Applications* (8): 11341–11346.

Onorato, D., White, C., Zager, P. et al. (2006). Detection of predator presence at elk mortality sites using mtDNA analysis of hair and scat samples. *Wildlife Society Bulletin* 34: 815–820.

Otis, D.L. and White, G.C. (2004). Evaluation of ultrastructure and random effects band recovery models for estimating relationships between survival and harvest rates in exploited populations. *Animal Biodiversity and Conservation* 27: 157–173.

Paquette, G.A., Devries, J.H., Emery, R.B. et al. (1997). Effects of transmitters on reproduction and survival of Mallards. *Journal of Wildlife Management* 61: 953–961.

Péron, G. (2013). Compensation and additivity of anthropogenic mortality: life history effects and review of methods. *Journal of Animal Ecology* 82: 408–417.

Pintillie, M. (2006). *Competing Risks: A Practical Perspective*. Wiley.

Pollock, K.H., Winterstein, S.R., Bunk, C.M. et al. (1989). Survival analysis in telemetry studies: the staggered entry approach. *Journal of Wildlife Management* 53: 7–15.

Powell, L.A. (2007). Approximating variance of demographic parameters using the delta method: a reference for avian biologists. *Condor* 109: 949–954.

Pöysä, H., Elmberg, J., Nummi, P. et al. (2004). Ecological basis of sustainable harvesting: is the prevailing paradigm of compensatory mortality still valid? *Oikos* 104: 612–615.

Prichard, A.K., Joly, K., and Dau, J. (2012). Quantifying telemetry collar bias when age is unknown: a simulation study with a long-lived ungulate. *Journal of Wildlife Management* 76: 1441–1449.

Robinson, H.S., Desimone, R., Hartway, C. et al. (2014). A test of the compensatory mortality hypothesis in mountain lions: a management experiment in West-Central Montana. *Journal of Wildlife Management* 78: 791–807.

Robinson, H.S., Goodrich, J.M., and Miquelle, D.G. (2015). Mortality of Amur tigers: the more things change, the more they stay the same. *Integrative Zoology* 10: 344–353.

Sandercock, B.K., Nielsen, E.B., Brøseth, H. et al. (2011). Is hunting mortality additive or compensatory to natural mortality? Effects of experimental harvest on the survival and cause-specific mortality of Willow Ptarmigan. *Journal of Animal Ecology* 80: 244–258.

Saraux, C., Le Bohec, C., Durant, J.M. et al. (2011). Reliability of flipper-banded penguins as indicators of climate change. *Nature* 471: 254–254.

Sauer, J.R. and Williams, B.K. (1989). Generalized procedures for testing hypotheses about survival or recovery rates. *Journal of Wildlife Management* 53: 137–142.

Schaub, M. and Lebreton, J.-D. (2004). Testing the additive versus the compensatory hypothesis of mortality from ring recovery data using a random effects model. *Animal Biodiversity and Conservation* 27: 73–85.

Schaub, M. and Royle, J.A. (2013). Estimating true instead of apparent survival using spatial Cormack–Jolly–Seber models. *Methods in Ecology and Evolution* 5: 1316–1326.

Servanty, S., Choquet, R., Baubet, E. et al. (2010). Assessing whether mortality is additive using marked animals: a Bayesian state–space modeling approach. *Ecology* 91: 1916–1923.

Singer, J.D. and Willett, J.B. (1991). Modeling the days of our lives: using survival analysis when designing and analyzing longitudinal studies of duration and the timing of events. *Psychological Bulletin* 110: 268–290.

Smith, D.W., Bangs, E.E., Oakleaf, J.O. et al. (2010). Survival of colonizing wolves in the northern Rocky Mountains of the United States, 1982–2004. *Journal of Wildlife Management* 74: 620–634.

Sparkman, A.M., Waits, L., and Murray, D.L. (2011). Social and demographic effects of anthropogenic mortality: a test of the compensatory mortality hypothesis in the red wolf. *PLoS One* 6: e20868.

Stanley, T.R. (2004). When should Mayfield model data be discarded? *Wilson Bulletin* 116: 267–269.

StataCorp (2007) Stata Statistical Software: Release 10. College Station, Texas, USA.

Suzuki, R.O., Kudoh, H., and Kachi, N. (2003). Spatial and temporal variations in mortality of the biennial plant, *Lysimachia rubida*: effects of intraspecific competition and environmental heterogeneity. *Journal of Ecology* 91: 114–125.

Therneau, T.M., Grambsch, P.M., and Fleming, T.R. (1990). Martingale-based residuals for survival models. *Biometrika* 77: 147–160.

Tidemann, C.R. and Nelson, J.E. (2011). Life expectancy, causes of death and movements of the grey-headed flying-fox (*Pteropus poliocephalus*) inferred from banding. *Acta Chiroptera* 13: 419–429.

Trent, T.T. and Rongstad, O.J. (1974). Home range and survival of cottontail rabbits in south-western Wisconsin. *Journal of Wildlife Management* 38: 459–472.

Tsai, K., Brownie, C., Nychka, D.W. et al. (1999). Smoothing hazard functions for telemetry survival data in wildlife studies. *Bird Study* 46 (supplement): S47–S54.

Tsai, W.-Y. (1988). Estimation of the survival function with increasing failure rate based on left truncated and right censored data. *Biometrika* 75: 319–324.

Van Buuren, S., Boshuizeni, H.C., and Knook, D.L. (1999). Multiple imputation of missing blood pressure covariates in survival analysis. *Statistics in Medicine* 18: 681–694.

Vandenabeele, S.P., Shepard, E.L., Grogan, A. et al. (2012). When three per cent may not be three per cent; device-equipped sea-birds experience variable flight constraints. *Marine Biology* 159: 1–14.

Vaupel, J.W. and Yashin, A. (1985). Heterogeneity's ruses: some surprising effects of selection on population dynamics. *American Statistician* 39: 176–185.

Wainer, H. and Velleman, P.A. (2001). Statistical graphics: mapping the pathways of science. *Annual Review of Psychology* 52: 305–335.

Wengert, G.M., Gabriel, M.W., and Clifford, D.L. (2012). Investigating cause-specific mortality and diseases in carnivores: tools and techniques. In: *Carnivore Ecology and Conservation: A Handbook of Techniques: A Handbook of Techniques* (eds. L. Boitani and R.A. Powell), 294–313. Oxford, UK: Oxford University Press.

Wengert, G.M., Gabriel, M.W., Foley, J.E. et al. (2013). Molecular techniques for identifying intraguild predators of fishers and other north American small carnivores. *Wildlife Society Bulletin* 37: 659–663.

White, G.C. and Burnham, K.P. (1999). Program MARK: survival estimation from populations of marked animals. *Bird Study* 46 (Suppl): S120–S139.

White, I.R. and Royston, P. (2009). Imputing missing covariate values for the Cox model. *Statistics in Medicine* 28: 1982–1998.

Whittington, J., Hebblewhite, M., DeCesare, N.J. et al. (2011). Caribou encounters with wolves increase near roads and trails: a time-to-event approach. *Journal of Applied Ecology* 48: 1535–1542.

Williams, B.K., Nichols, J.D., and Conroy, M.J. (2002). *Analysis and Management of Animal Populations*. New York, New York, USA: Academic.

Winterstein, S.R., Pollock, K.P., and Bunck, C.M. (2001). Analysis of survival data from radiotelemetry studies. In: *Radio Tracking and Animal Populations* (eds. J.J. Millspaugh and J.M. Marzluff), 351–380. New York, New York, USA: Academic.

Wirsing, A.J., Phillips, J., Obbard, M. et al. (2012). Incidental nest predation in freshwater turtles: inter- and intraspecific differences in vulnerability are explained by relative crypsis. *Oecologia* 168: 977–988.

Wirsing, A.J., Steury, T.D., and Murray, D.L. (2002). Relationship between body condition and vulnerability to predation in snowshoe hares and red squirrels. *Journal of Mammalogy* 83: 707–715.

Woodroofe, M. (1985). Estimating a distribution function with truncated data. *Annals of Statistics* 13: 163–177.

Zar, J.H. (1999). *Biostatistical Analysis*. Upper Saddle River, New Jersey, USA: Prentice Hall.

Zens, M.S. and Peart, D.R. (2003). Dealing with death data: individual hazards, mortality and bias. *Trends in Ecology and Evolution* 18: 366–373.

7

Mark-Recapture Models for Estimation of Demographic Parameters

Brett K. Sandercock

Department of Terrestrial Ecology, Norwegian Institute for Nature Research, Trondheim, Norway

Summary

Population ecologists require robust estimates of survival and other demographic parameters for understanding the ecological and evolutionary drivers of wildlife population dynamics. Live encounter data from individual animals marked with unique tags or natural marking patterns are a key source of information for many species. Imperfect detection and losses to emigration can be challenging issues when tracking mobile organisms under natural conditions. Mark-recapture models utilize encounter histories for marked individuals where consecutive sampling occasions are coded with detection or nondetection data. Alternative models can then be fit as fixed-effect models with maximum likelihood methods in a frequentist framework, or as hierarchical models with random effects in a Bayesian framework. The Cormack–Jolly–Seber (CJS) model conditions upon first capture and uses forward-time modeling to estimate apparent survival corrected for the probability of encounter. More complex models extend the basic CJS model to estimate additional parameters. Time-since-marking models estimate apparent survival corrected for losses due to transients, age effects, and other factors. Temporal symmetry and Jolly–Seber models combine forward- and reverse-time modeling to estimate recruitment and population change without the need to parameterize a full matrix model. Robust design models adopt a nested sampling approach and are useful for investigating the dynamics of temporary emigration due to regional movements, dormancy, and intermittent breeding. Multistate models extend the single-state CJS model to multiple categorical states and provide state-specific estimates of apparent survival and transition rates among sites, demographic classes, disease status, or other states. Extended multistate models allow for unobservable states or situations where state classifications may be uncertain. Other models combine live encounters from marked individuals with different types of auxiliary data. Mark-resighting models add counts of unmarked individuals to estimate total abundance. Joint models add data on dead recoveries or supplementary resightings to estimate true survival and site fidelity. Integrated population models combine live encounters with count and fecundity data to estimate immigration rates and population change. Mark-recapture models are powerful tools because they correct for imperfect detection and imperfect availability, they can control for transients, social structure, and other potential sources of heterogeneity, and they also allow joint analysis across independent datasets. Demographic parameters are estimated with less bias and greater precision, thereby providing a stronger foundation for addressing key questions in population biology, evolutionary ecology, and wildlife management.

7.1 Introduction

Estimation of demographic parameters is central to the population biology of wildlife species, with important applications for understanding the ecology of population dynamics, the evolution of life-history strategies, and for making management and conservation decisions. Survival is particularly difficult to estimate for wildlife populations because the timing and causes of mortality are usually unknown for free-living animals, and because imperfect detection is the rule rather than the exception in most field studies (Kellner and Swihart 2014). Marked animals are often difficult to detect due to the logistics of field effort, low densities, wide-ranging movements, or secretive behavior (Mazerolle 2015). Despite challenges for estimation, survival is often identified as the demographic parameter with the greatest impact on the *finite rate of population change* (λ, Doherty et al. 2004; Schorcht et al. 2009). Adult survival is often an important driver for long-lived vertebrates or declining populations, whereas juvenile survival can have greater impacts in short-lived species or growing populations (Oli and Dobson 2003; Stahl and Oli 2006; Vélez-Espino et al. 2006). The relative influence of survival rates and other demographic

Population Ecology in Practice, First Edition. Edited by Dennis L. Murray and Brett K. Sandercock.

Companion website: www.wiley.com/go/MurrayPopulationEcology

parameters is determined by both their mean value and variance, and demographic parameters that have a large effect on the rate of population change often have relatively low variance (Gaillard and Yoccoz 2003; Rotella et al. 2012; Péron et al. 2016). Estimation of the variance of survival rates is complicated because the maximum variance declines to zero when a probability approaches the boundary value of one (Morris and Doak 2004), and because temporal variation includes the *process variance* of biological interest, but also *sampling variance* which adds undesirable statistical noise (Gould and Nichols 1998; Ryu et al. 2016). The predicted impacts of conservation or management actions are determined by both the mean and the variance of demographic parameters in a population model. Hence, a central goal in population biology is to obtain parameter estimates that are *unbiased* with estimates that are close to the true value of a demographic parameter, but also have good *precision* with a low variance.

Estimation of survival rates and other demographic parameters for wildlife populations generally require one of four different types of data: age ratios from unmarked individuals, live encounters of marked individuals, dead recoveries of marked individuals, or intensive monitoring of animals marked with radio transmitters or other tags (Williams et al. 2002). *Age ratios* can be estimated from the standing age distribution of a population or by tracking cohorts through time, and may be the only demographic data available in a short-term study (Hernández-Matías et al. 2011). Calculation of survival from age distributions requires that the population has a stable age distribution, the rate of population change is either stable or stationary, and all age classes have an equal probability of encounter. All of these assumptions are likely to be violated in field studies of wildlife populations (Conn et al. 2005). *Dead recovery* data require that observers retrieve and report markers from animals that are harvested or found dead of natural causes. Dead recovery data can be a valuable source of information for harvested species, but are less useful for nongame species unless a large number of markers can be retrieved and reported (Robinson et al. 2009; Arnold et al. 2016). *Radio telemetry* data can be analyzed with time-to-event models to estimate survival and hazard rates (Zens and Peart 2003; Murray 2006, Chapter 6), and also provide insights into animal movements and space use (Chapters 13–14). The disadvantages of telemetry methods are mainly logistical considerations: transmitter size may limit battery life, attachment techniques should avoid impacts on survival, and financial costs of transmitters and tracking may limit sample size for a field study. Of the four sources of information, *live encounter* data are arguably the most widely used source of information for estimating survival and other demographic parameters for wild populations of animals and plants.

7.2 Live Encounter Data

Live encounter data include a variety of different types of information that can be collected for wildlife populations. In a mark–recapture study based on tagging, the field methods start with live capture and unique marking of individual animals. Standard techniques for physical marking vary among different groups of animals: numbered stickers for butterfly wings, injected tags for fish based on PIT (passive inductance transducers) or RFID (radio-frequency identification) technologies, toe-clipping or branding for amphibians and reptiles, neck collars, wing-tags or leg-bands on birds, and ear-tagging or tattoos for mammals (Silvy et al. 2012). In some species, natural marking patterns can be used to identify unique individuals without application of external tags: distinctive vocalizations of songbirds, spots of sharks and salamanders, notches in the tail flukes of whales, or coat patterns of wild cats (Vögeli et al. 2008; Bendik et al. 2013; Lee et al. 2014; McClintock 2015). Another noninvasive approach for tracking individuals is molecular genotyping based on DNA isolated from shed hair and feathers, or scat (Lukacs and Burnham 2005).

The basic assumptions of marking techniques are that handling and marking do not negatively affect animal survival or behavior, marks are read without error, and marks are retained for the duration of the field study. Once a uniquely marked individual has been released, observers monitor subsequent survival by attempting to find the same individual again. Assumptions of the detection process are that fates of marked individuals are independent, and that all marked individuals have the same probability of recapture. Live encounters of a marked individual or *detections* can include physical recaptures to read the mark, resighting of individuals with binoculars or a spotting scope, or registering the tag as the animal comes near a camera or other recording device. Detections can be recorded in several formats, including detection only, counts of the number of detection events, or information on the state of an individual. If an individual is successfully detected, the true state is usually unambiguous because the organism is observed to be alive or dead. Uncertainty arises for *nondetections* of marked individuals because they could be dead, emigrants from a study plot, alive but not available for encounter, or present but overlooked by the observer.

A mark-recapture study requires defining the *spatial* and *temporal* scale of the project relative to the movements and expected lifespan of the study organism (Lindberg 2012). Population studies often have one or more *study plots* of fixed size – a series of meadows for an alpine butterfly, a systematic trapping grid for rodents, a linear array of weirs for a stream fish, or a valley ecosystem for a resident predator. In most mark-recapture models, marked individuals are marked and encountered within the same network of study plots. In joint models, individuals are marked at a study plot of fixed size, but dead recovery data or resighting data can be taken from the extended range of a migratory population. Mark-recapture models that estimate number of immigrants or probability of recruitment usually require that the boundaries and size of the study area remain constant for the duration of the project. If a study area is enlarged midway through a project, newly captured individuals in the expanded zone cannot be distinguished from new immigrants.

Efforts to detect marked individuals are spread across two or more *occasions*. The time-step might be days for a short-lived insect, weeks or months for a rodent, or years for most vertebrate populations. Two sampling occasions permit estimation of return rates, but three to four occasions are the minimum needed to use mark-recapture models. Longer time series with six or more occasions are needed to model temporal covariates and to estimate the process variance of demographic parameters without the confounding effects of sampling variance. Mark-recapture models perform best if sampling is systematic with *equal intervals* between consecutive sampling occasions. However, systematic sampling may be impractical for remote field sites where logistics of site access are difficult, or for ectothermic animals which are only active during suitable environmental conditions. Most mark-recapture models can also accommodate *unequal intervals* among different sampling occasions. In *closed* models, the interval between occasions is relatively short, and population size is assumed to be constant and unchanging. Closed population models offer some advantages for estimation of abundance because encounter rates can be modeled more effectively, without a need to estimate survival or recruitment. In *open* models, the duration of the interval between sampling occasions is long enough to accommodate the dynamic processes, such that the number of marked individuals can increase due to gains from recruitment or immigration, or decrease due to losses from death or permanent emigration. Sampling should be instantaneous; gains and losses occur within long intervals but not during the short sampling occasions, although mark-recapture models can be robust to violations of this assumption (O'Brien et al. 2005).

Mark-recapture models are mainly used to estimate demographic parameters for marked individuals in populations, but can also be applied at different ecological levels. If encounter histories are coded for detections of species instead of individuals, the same set of mark-recapture models can be used to study community dynamics including species richness, persistence, colonization, and turnover (Dorazio et al. 2006; Zipkin et al. 2010). Noninvasive methods of detection may be preferable for threatened or secretive species where physical capture is undesirable or impossible. Detections or counts of animals from vocalizations, tracks, or other signs of animal activity (Pellet and Schmidt 2005; O'Connell et al. 2006; Richmond et al. 2012) can be used to estimate occupancy, abundance, and population dynamics (Dail and Madsen 2013; Chapters 3–4). Mark-recapture models for unmarked individuals can also be used to estimate survival and other dynamic rates for open populations (Zipkin et al. 2014). Here, I focus on mark-recapture models for estimating survival and other demographic parameters in field projects where at least some portion of the study population is individually marked or otherwise identifiable.

7.3 Encounter Histories and Model Selection

A starting point for any statistical analysis of live encounter data is to identify the mark-recapture model that best matches the sampling design of a field project, and will yield estimates of demographic parameters that are corrected for imperfect detection and other sources of heterogeneity (Horton and Letcher 2008; Lindberg 2012, Figure 7.1). The next step is to assemble encounter histories for the sample set of uniquely marked individuals. In a matrix of encounter histories, each row corresponds to a different individual and each column corresponds to a different sampling occasion. The cells of the resulting matrix are then coded with information from live encounter data (L), or some combination of live encounter and dead recovery information (LD). Live encounters in a single state are based on the L format and encounter histories are coded as: 1 = an observer detected a marked individual as a recapture or resighting, or 0 = an observer did not encounter the individual (Box 7.1). In a multistate model, detections are coded as categorical states: B = breeder, N = nonbreeder, and 0 = not detected. If the categorical states are the number of unmarked young attended by a marked parent, detections might be coded as digits: 3 = three young, 1 = one young, and 0 = no young detected. For joint models based on an LD format, codes would include: 10 = a live encounter, 01 = a dead recovery, 02 = a supplemental observation, or 00 = not detected by any method.

Once encounter histories are assembled, the next step is to select the intrinsic and extrinsic variables to be included in the starting global model. Selection of important covariates is conducted a priori to guard against the possibility of spurious or irrelevant results (Chapter 2). Covariates can be discrete or continuous, as well as static or dynamic. Variables that affect demographic

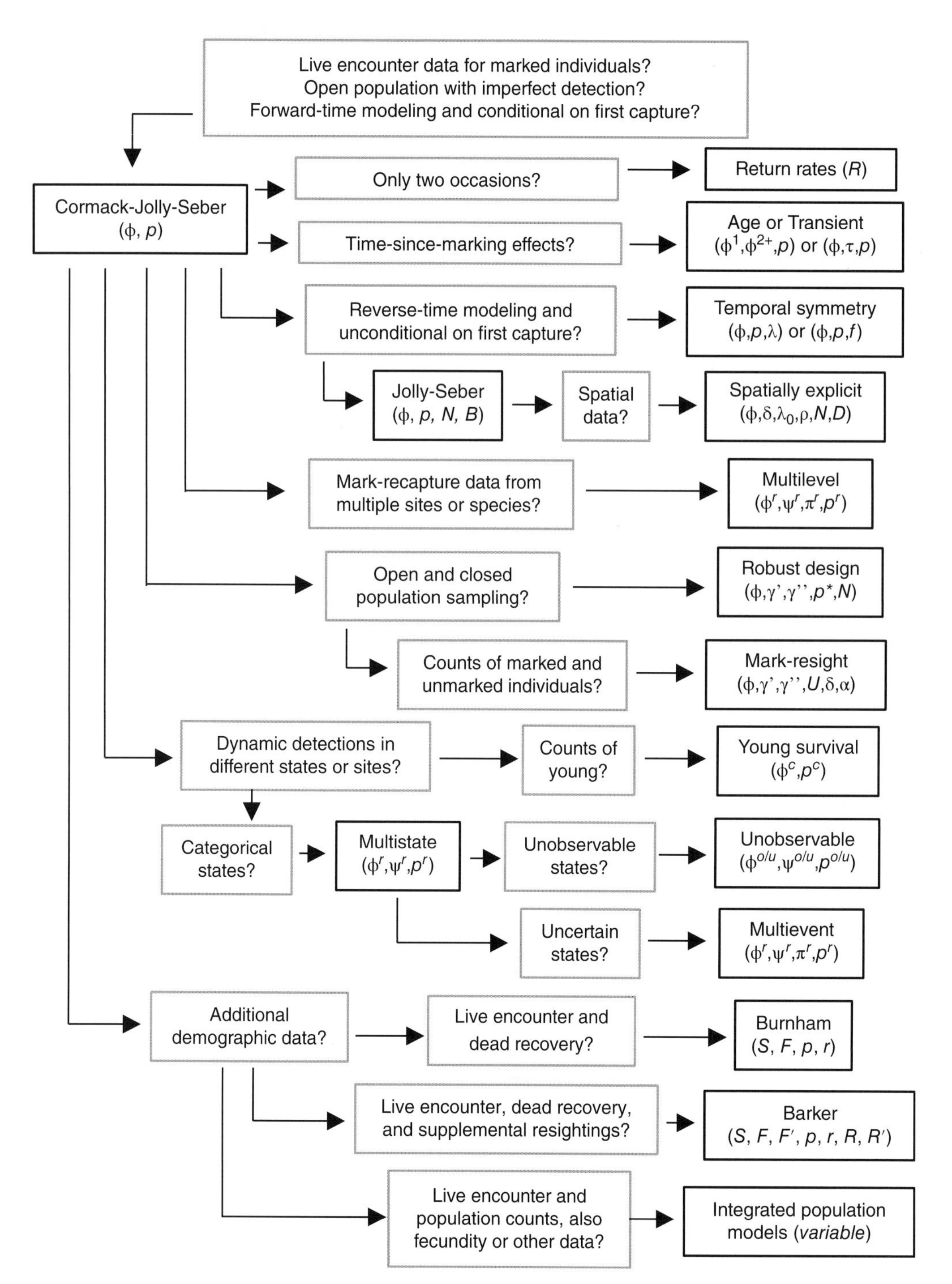

Figure 7.1 Conceptual diagram for open population models based on live encounter data for marked individuals. The *Cormack–Jolly–Seber model* is a fundamental model that estimates apparent survival (ϕ), corrected for the probability of encounter (p). If only two sampling occasions are available, *return rates* can be estimated as the product of the two parameters ($R = \phi\ p$). A large family of mark-recapture models are based on different extensions of the CJS model. If individuals are not detected after first capture, the *time-since marking model* estimates age-specific apparent survival (ϕ^n), whereas the *transient model* estimates the probability of transience (τ). *Temporal symmetry models* combine forward and reverse-time modeling to add estimates of realized population change (λ) or seniority (ζ). The *Jolly–Seber (JS) model* uses the complete encounter histories and estimates the population size of marked individuals (N), and number of immigrants (B). *Spatially explicit models* add spatial data on animal locations to also estimate recruitment (ρ) and density (D), adjusted for spatial detectability (δ) and encounters per individual (λ_0). *Multilevel models* use hierarchical approaches to model CJS data from multiple sites or species. *Robust design models* subdivide primary sampling occasions into secondary closed periods, and can be used to partition the probability of encounter into temporary emigration (γ) and true detection ($p*$). *Mark-resighting models* extend the robust design model by adding counts of unmarked individuals, and estimate transitions for observable (γ'') and unobservable individuals (γ'), abundance adjusted for the number of unmarked individuals (U), with corrections for individual heterogeneity (δ) and resighting (α). *Multistate models* code detections as different categorical states (r), and provide state-specific estimates of ϕ^r and p^r, along with the probability of changing states (ψ^r). If detections are coded as counts of unmarked young attending a marked parent, the *young survival model* gives age-specific estimates of ϕ^c and p^c. Multistate models for *unobservable states* are possible but usually require additional constraints to avoid parameter redundancy. *Multievent models* handle uncertainty in state classification, and include estimates the probability of assignment to a state (π). In the Burnham and Barker *joint models*, live encounter data are combined with dead recovery data and supplementary resighting data to decompose apparent survival into true survival (S) and movement (F, F'), corrected for the probabilities of recovery (r) and resighting (R and R'). *Integrated population models* combine mark-recapture data with counts of population size, fecundity, age ratios, or other types of demographic data to conduct an integrated demographic analysis.

Box 7.1 Encounter Histories and Parameter Index Matrices

The starting point for any capture-mark-recapture analysis is to assemble the detection–nondetection data for uniquely marked individuals into encounter histories. The rows of the file are the information for *individual* animals and the columns are the *sampling occasions*. The status of the animal is then recorded for each individual at each sampling occasion. In the case of the Cormack–Jolly–Seber (CJS) model for live encounter data, a single value is recorded for animal status at each occasion (LLLLLLL) with the two possible codes being detected alive (1) or not detected (0). Consider the following encounter history for one individual animal:

$$1101010 \tag{7.1}$$

The field study had seven sampling occasions, which produces an encounter history with seven columns. The values of one in the encounter history indicate that the animal was detected on occasions 1, 2, 4, and 6, whereas the zeros show it was not observed on occasions 3, 5, and 7. Given detections on occasions 2 and 6, the individual was definitely alive but was not detected on occasions 3 and 5. Inspecting the encounter histories can be a useful starting point for an analysis. If any encounter histories contain gaps with zeros nested within a string of ones, then the probability of encounter will be less than one. The fate of the animal on the last occasion is ambiguous: the individual could have been alive but not detected, or it might be dead.

The CJS model estimates two parameters, the probability of apparent survival (ϕ) and the probability of encounter (p). Depending on the data included in the encounter histories, the latter parameter can also be called the probability of capture or the probability of resighting. For a sample of n occasions, the CJS model calculates $n-1$ estimates of apparent survival and $n-1$ estimates of the probability of encounter.

$$\begin{array}{ccccccccccccc} t_1 & & t_2 & & t_3 & & t_4 & & t_5 & & t_6 & & t_7 \\ 1 & \longrightarrow & 1 & \longrightarrow & 0 & \longrightarrow & 1 & \longrightarrow & 0 & \longrightarrow & 1 & \longrightarrow & 0 \\ & \phi_1 & & \phi_2 & & \phi_3 & & \phi_4 & & \phi_5 & & \phi_6 & \\ & & p_1 & & p_2 & & p_3 & & p_4 & & p_5 & & p_6 \end{array} \tag{7.2}$$

Apparent survival is estimated for the *intervals* between consecutive occasions, whereas the estimates of encounter apply to the *sampling occasions*. In the CJS model, encounter rates are not estimated for the first occasion (t_1) because no animals were released before the start of the field study and none are available for recapture.

One possible summary of this model would be to collect all of the subscripts or *parameter index* numbers into separate vectors for each parameter.

$$\begin{array}{cccccccccccc} \phi & & & & & & p & & & & & \\ 1 & 2 & 3 & 4 & 5 & 6 & 1 & 2 & 3 & 4 & 5 & 6 \end{array} \tag{7.3}$$

In the above example, the index numbers correspond to different intervals and occasions. However, apparent survival and encounter are independent parameters and they might be renumbered consecutively to avoid confusion.

$$\begin{array}{cccccccccccc} \phi & & & & & & p & & & & & \\ 1 & 2 & 3 & 4 & 5 & 6 & 7 & 8 & 9 & 10 & 11 & 12 \end{array} \tag{7.4}$$

If the index numbers are different for each interval or occasion, then the parameter estimates from the model will be different among occasions or *time-dependent*.

One row of index numbers would be sufficient if a single batch of animals was marked and released on the first occasion of the field study. In an open population model, subsequent years of field work would include recapture or resighting effort, but also new capture events with marking and release of individuals on later occasions. If we add additional rows for animals first marked on the second and later occasions, our summary diagrams then become triangular *matrices*.

$$\begin{array}{cccccccccccc} \phi & & & & & & p & & & & & \\ 1 & 2 & 3 & 4 & 5 & 6 & 7 & 8 & 9 & 10 & 11 & 12 \\ & 2 & 3 & 4 & 5 & 6 & & 8 & 9 & 10 & 11 & 12 \\ & & 3 & 4 & 5 & 6 & & & 9 & 10 & 11 & 12 \\ & & & 4 & 5 & 6 & & & & 10 & 11 & 12 \\ & & & & 5 & 6 & & & & & 11 & 12 \\ & & & & & 6 & & & & & & 12 \end{array} \tag{7.5}$$

The triangular matrices are known as *parameter index matrices* or PIMs, and are one approach for describing the structure of a mark-recapture model and fitting alternative models in software packages such as Program `Mark`.

performance often include *group* effects such as sex, colony, or breeding status, *time* effects for environmental conditions that vary temporally, *time-since-marking* effects that change with time since first capture such as transience, age, or experience, and *individual covariates* such as body size and other morphometric traits (Grosbois et al. 2008; Fredericksen et al. 2014). Different sets of variables can then be combined in *additive* models with main effects only, or in *factorial* models with main effects and their interactions. As a rule of thumb, the maximum number of parameters in a starting model is usually capped at $n/10$, where n is the effective sample size, calculated as the number of captures and recaptures in a set of encounter histories based on live encounters or dead recoveries (Burnham and Anderson 2002, p. 245). The parameter count of a model (K) can be reduced by modeling parameters without any temporal variation, as a temporal *trend*, or as a *linear* or *nonlinear* function of explanatory covariates (Gimenez et al. 2006).

A common situation in mark-recapture analyses is that the detection or count data may be *overdispersed* due to lack of independence or heterogeneity among marked individuals, and the starting global model may not be a perfect fit to the encounter histories. A necessary first step is to use goodness-of-fit (GOF) procedures to calculate a *variance inflation factor* ($\hat{c}$) to correct for possible overdispersion caused by a lack of fit between the starting model and the encounter histories. The variance inflation factor approaches an asymptotic value of one if the starting model is a perfect fit with no overdispersion. If an estimation procedure returns a value of $\hat{c} < 1$, then the parameter is usually set to one. Values of $\hat{c} = 1$–3 are typical for most mark-recapture datasets, and higher values may indicate structural problems. A variety of procedures are available to estimate $\hat{c}$ for CJS and multistrata models, but not for more complex models, such as multievent models (Pradel et al. 2005; Choquet et al. 2009a; Kendall et al. 2013). Calibrated simulations have been used to test GOF for integrated population models, and might be an option for other mark-recapture models too (Besbeas and Morgan 2014). Last, if GOF tests are not available to test for overdispersion or to estimate $\hat{c}$, sensitivity analyses for model rankings can be conducted by varying $\hat{c}$ across a range of plausible values (e.g. $\hat{c} = 1$–3 by 0.5).

Mark-recapture analyses are often conducted in an information theory framework with the tools of model selection (Burnham et al. 2011; Lindberg et al. 2015, Chapter 2). If alternative models are fit with maximum likelihood estimation (MLE), model selection is based on different variants of Akaike's Information Criterion (AIC, AIC*c*, or QAIC*c*) or Bayes Information Criterion (BIC), which are tools for identification of parsimonious models (Burnham and Anderson 2004; Grueber et al. 2011, Chapter 2). Picking an information criteria depends on the goals of a study: AIC for exploratory analysis of model complexity in an observational study, and BIC for confirmatory analysis of the single-best model in a controlled experiment (Aho et al. 2014). Differences in information criterion values are used to calculate model weights (w_i), and ratios of model weights or Bayes factors can be used to calculate relative levels of support between different models. Caution should be taken when comparing models that differ by a single parameter because the minimum AIC model may retain an uninformative covariate (Arnold 2010). AIC and BIC are also less suitable for comparing hierarchical models with random effects, and tools for model selection in a Bayesian framework are an emerging field (Barker and Link 2015; Hooten and Hobbs 2015, Chapter 5).

Model testing starts with a global model, and proceeds by dropping variables to fit a series of reduced models which can be viewed as a suite of alternative hypotheses (Chapter 2). For a simple analysis, an *all-combinations* approach might be used to test candidate models that include all possible combinations of a limited set of explanatory variables. However, the number of possible models increases exponentially as additional explanatory variables or demographic parameters are added to the set of candidate models. In a *step-down* approach, terms are dropped from the global model in an iterative manner starting with the encounter rates or other nuisance parameters associated with the sampling process, proceeding to the demographic parameters of biological interest, and then with a final step that adds terms back to explore additional models close to the best model (Sandercock and Jaramillo 2002). In a *single-factor* approach, each parameter is modeled separately while retaining full model complexity for all other parameters in the model, and then highly ranked structures for each parameter type are combined in a composite model (Grosbois and Tavecchia 2003). The order that parameters are modeled may not be critical because model selection strategies seem to have little effect on bias or precision of parameter estimates (Doherty et al. 2012). Still, if computation times or issues with convergence are a consideration, the best approach may be to fit a concise set of 4–20 biologically plausible models and be less concerned with multimodel inference.

Once model testing has been completed, several procedures can be used to obtain parameter estimates. If a single model in the candidate set receives majority support ($w_i > 0.8$), then parameter estimates might be taken from that top-ranked model. If a subset of models each receive some support ($w_i > 0.3$), then *model-averaging* can be used where each set of parameter estimates is combined with the model weight, and then used to obtain a weighted average with an unconditional variance that also includes uncertainty due to model selection. An issue that arises is that demographic parameters or covariates may only appear in a subset of models in the candidate set (Grueber et al. 2011). Under the *natural average method*, parameters are averaged across the subset of models in which the parameter is present, after rescaling the model weights to sum to one. Under the *zero method*, missing parameters are set to zero in models where the parameters are absent, thereby shrinking predictors toward zero if models without the effect have strong support. The *relative importance* of a parameter is sometimes calculated by summing the weights across models where the parameter appears, but simulations have shown that sums of weights can be unreliable for assessing the importance of predictor variables (Galipaud et al. 2014). Model averaging can also lead to flawed parameter estimates if the predictor variables are correlated, but can be addressed with tests for multicollinearity and by standardizing predictor variables for their covariance structure (Cade 2015).

7.4 Return Rates

Return rates (R) are a good starting point for understanding the value of mark-recapture models. If a sample of individuals is marked on an occasion at a sampling site, the return rate is the proportion of marked individuals encountered the following year, or during some block of future years. Return rates are challenging to interpret because the probability of capturing an individual animal in two consecutive years is a function of four independent probabilities: *true survival* (S), *site fidelity* (F), *site propensity* (δ), and *true detection* ($p*$, Box 7.2). Return rates are a minimum estimate of true survival because they are the product of four probabilities ($R = S \times F \times \delta \times p^*$). If return rates are high ($R > 0.9$), then true survival and the other three probabilities must also be high (≥ 0.9). Difficulties arise in the interpretation of low or moderate return rates ($0.2 < R < 0.6$), and with comparisons of return rates among different groups or years. If any of the last three parameters are <1, return rates are negatively biased as an estimate of true survival. A low return rate of 0.4 could be the result of poor survival, weak site fidelity, or a

Box 7.2 Definitions of Demographic Parameters in Cormack–Jolly–Seber Models

Return rates have been widely used in population biology as an index of survival. If a group of individuals are marked one year at a sampling site, the "return rate" (R) can be calculated as the proportion of marked individuals encountered the following year, or during some block of future years. Return rates are difficult to interpret because they are the product of four independent probabilities ($R = S \times F \times \delta \times p^*$):

True survival (S): the probability that an individual survives between two sampling occasions. The complement of survival includes losses to mortality ($1 - S$).

Site fidelity (F): the probability that an individual returns to the same sampling area, conditional upon true survival. The complement of site fidelity includes losses to permanent emigration ($1 - F$).

Site propensity (δ): the probability that an individual is available for encounter in the same sampling area the next occasion, conditional upon survival and site fidelity. The complement of site propensity includes losses to temporary emigration ($\gamma = 1 - \delta$).

True detection (p^*): the probability that an observer detects the individual under field conditions, conditional upon survival, site fidelity, and availability for encounter. The complement of true detection is imperfect detection ($1 - p^*$).

Cormack–Jolly–Seber (CJS) models are an improvement over return rates because they use encounter histories of marked individuals to estimate two independent probabilities:

Apparent survival ($\phi = SF$): the probability that an individual survives between two sampling occasions and returns to the sampling area. The complement of apparent survival includes losses to mortality or permanent emigration.

Encounter rate ($p = \delta\, p^*$): the probability that an individual is detected under field conditions given that is available for encounter in the sampling area. The complement of the probability of encounter includes individuals not available for detection and individuals that were present in the sampling area but not detected.

Extended mark-recapture models can be used to decompose the demographic parameters even further. Joint models combine live encounter with dead recovery data and can be used to decompose apparent survival into true survival and site fidelity. Conversely, robust design models can be used to separate temporary emigration from the probability of true detection. An integrated model that combines live encounter data, supplementary observations, and a robust design framework can estimate all four parameters separately (Kendall et al. 2013).

variety of other combinations of the four parameters. Given the expected problems of bias and interpretation, return rates should be avoided if possible. On the other hand, return rates may be the only estimator available if the duration of a project is 2–3 years. In that case, return rates may be used as a crude index of true survival, but assumptions must be made about the remaining three probabilities. If site fidelity, site propensity, and true detection are assumed to be equivalent among groups or years, then differences in return rates might be due to variation in true survival. Similarly, temporal trends in return rates may be compared among different locations if the distributions of the other probabilities are assumed to be stationary over time.

7.5 Cormack–Jolly–Seber Models

The *Cormack–Jolly–Seber* (CJS) model is a fundamental model for analysis of live encounter data, and many other models for live encounter data are extensions of the CJS model (Figure 7.1). The CJS model conditions upon first capture and uses forward-time modeling to estimate two demographic parameters for a population of marked individuals: *apparent survival* (ϕ) corrected for the *probability of encounter* (p, Box 7.2). Apparent survival is less biased as an estimator of true survival compared to return rates because it is the product of two instead of four parameters ($\phi = S \times F$). Like return rates, if apparent survival is high, then both S and F must be high. CJS models can be used to estimate true survival for sessile organisms if movements are limited and site fidelity is effectively one ($\phi = S$ if $F \approx 1$). Conversely, CJS models can also be used to estimate site fidelity if true survival rates are expected to be high during a short-term study ($\phi = F$ if $S \approx 1$). The probability of encounter ($p = \delta \times p^*$) is usually regarded as a nuisance parameter because it corrects for imperfect detection (p^*), but p can also be of biological interest if the dynamics of temporary emigration lead to low site propensity ($\gamma = 1 - \delta$).

CJS and other mark-recapture models require live encounter data from at least three sampling occasions because internal gaps in the encounter histories are used to estimate the probability of encounter (e.g. `101`, Box 7.1). Individuals that are newly marked on the last occasion of a field study do not contribute to parameter estimates in a CJS model (e.g. `001`), and the terminal field season of a field project is best spent trying to recapture or resight marked individuals. CJS models can be fit to encounter histories with a multinomial likelihood using MLE in a frequentist framework (Lebreton et al. 1992), or with a state-space likelihood and Markov chain Monte Carlo (MCMC) methods in a Bayesian framework (Gimenez et al. 2007; Royle 2008). Alternative models with different structures can be defined with model statements, parameter index matrices (PIMs), or design matrices (DM) in the different software tools (Box 7.3).

One issue that arises with the CJS model and many other mark-recapture models is that the intrinsic model structure may prevent estimation of a desired parameter. In a fixed-effects CJS model with time-dependence in both ϕ and p, it is not possible to estimate p for the last sampling occasion because no capture data are available from occasions following the end of the project. If ϕ and p cannot be estimated separately for the last interval in a time series, then the product of the last two transition rates is estimated as a single parameter (ϕp). The terminal parameters of ϕ or p are *nonidentifiable* because they cannot be estimated separately, and the model is then considered *parameter redundant* because the likelihood is expressed as a function of fewer parameters than the original count (Gimenez et al. 2003; Hubbard et al. 2014). The terminal product of ϕp can be decomposed if some constraint is applied to the CJS model: by modeling ϕ or p as constant over time or as a random effect in a hierarchical model. Nevertheless, n years of study may yield $n - 2$ estimates of apparent survival, and each additional year adds another annual estimate. Longer time series are needed to model apparent survival as a function of annual covariates, to estimate components of variance, and to estimate additional demographic parameters in more complex models.

7.6 The Challenge of Emigration

Capture, marking, and monitoring of individually marked animals for a population study are labor-intensive activities. Logistics and cost often constrain field projects to sampling on a fixed-area study plot that encompasses a random or representative sample of individual territories or home ranges. Marked individuals may be detected with intensive area-search sampling (Efford 2011; Royle et al. 2011), or with grids or webs of live traps, hair snares, or camera stations (O'Connell et al. 2006; Monterroso et al. 2014). If a fixed-area study plot is located in contiguous habitat, the plot boundaries may not be a barrier to movements. Movements can lead to *permanent emigration* where marked individuals disperse away from the study plot and never return, whereas during *temporary emigration*, the dispersing individuals may leave or be unavailable for encounter for some period but then return again to the study plot in the future.

Box 7.3 Alternative Approaches to Fitting Candidate Models in `RMark` and Program `Mark`.

Fitting mark-recapture models with formulae in `RMark`, or with parameter index matrices (PIM) and design matrices (DM) in Program `Mark`. Model effects are constructed with model statements in `RMark`. With PIMs, shared or different index numbers are used to pool or separate estimates for different groups or times. Each row is a cohort that was first captured on different sampling occasions. In DMs, dummy variables are used to code separate effects for the intercept, groups, time, or interactions. The example is based on six transitions for seven sampling occasions, and two groups for sex.

RMark	PIMs	DM
Constant model `formula = ~1`	$\begin{array}{rrrrrr} 1 & 1 & 1 & 1 & 1 & 1 \\ & 1 & 1 & 1 & 1 & 1 \\ & & 1 & 1 & 1 & 1 \\ & & & 1 & 1 & 1 \\ & & & & 1 & 1 \\ & & & & & 1 \end{array}$	$\begin{bmatrix} 1 \\ 1 \\ 1 \\ 1 \\ 1 \\ 1 \end{bmatrix}$
Group model `formula = ~sex`	$\begin{array}{rrrrrr} 1 & 1 & 1 & 1 & 1 & 1 \\ & 1 & 1 & 1 & 1 & 1 \\ & & 1 & 1 & 1 & 1 \\ & & & 1 & 1 & 1 \\ & & & & 1 & 1 \\ & & & & & 1 \\ 2 & 2 & 2 & 2 & 2 & 2 \\ & 2 & 2 & 2 & 2 & 2 \\ & & 2 & 2 & 2 & 2 \\ & & & 2 & 2 & 2 \\ & & & & 2 & 2 \\ & & & & & 2 \end{array}$	$\begin{bmatrix} 1 & 1 \\ 1 & 1 \\ 1 & 1 \\ 1 & 1 \\ 1 & 1 \\ 1 & 1 \\ 1 & 0 \\ 1 & 0 \\ 1 & 0 \\ 1 & 0 \\ 1 & 0 \\ 1 & 0 \end{bmatrix}$
Time model `formula = ~time`	$\begin{array}{rrrrrr} 1 & 2 & 3 & 4 & 5 & 6 \\ & 2 & 3 & 4 & 5 & 6 \\ & & 3 & 4 & 5 & 6 \\ & & & 4 & 5 & 6 \\ & & & & 5 & 6 \\ & & & & & 6 \end{array}$	$\begin{bmatrix} 1 & 1 & 0 & 0 & 0 & 0 \\ 1 & 0 & 1 & 0 & 0 & 0 \\ 1 & 0 & 0 & 1 & 0 & 0 \\ 1 & 0 & 0 & 0 & 1 & 0 \\ 1 & 0 & 0 & 0 & 0 & 1 \\ 1 & 0 & 0 & 0 & 0 & 0 \end{bmatrix}$
Factorial model `formula = ~sex * time`	$\begin{array}{rrrrrr} 1 & 2 & 3 & 4 & 5 & 6 \\ & 2 & 3 & 4 & 5 & 6 \\ & & 3 & 4 & 5 & 6 \\ & & & 4 & 5 & 6 \\ & & & & 5 & 6 \\ & & & & & 6 \\ 7 & 8 & 9 & 10 & 11 & 12 \\ & 8 & 9 & 10 & 11 & 12 \\ & & 9 & 10 & 11 & 12 \\ & & & 10 & 11 & 12 \\ & & & & 11 & 12 \\ & & & & & 12 \end{array}$	$\begin{bmatrix} 1 & 1 & 1 & 0 & 0 & 0 & 0 & 1 & 0 & 0 & 0 & 0 \\ 1 & 1 & 0 & 1 & 0 & 0 & 0 & 0 & 1 & 0 & 0 & 0 \\ 1 & 1 & 0 & 0 & 1 & 0 & 0 & 0 & 0 & 1 & 0 & 0 \\ 1 & 1 & 0 & 0 & 0 & 1 & 0 & 0 & 0 & 0 & 1 & 0 \\ 1 & 1 & 0 & 0 & 0 & 0 & 1 & 0 & 0 & 0 & 0 & 1 \\ 1 & 1 & 0 & 0 & 0 & 0 & 0 & 0 & 0 & 0 & 0 & 0 \\ 1 & 0 & 1 & 0 & 0 & 0 & 0 & 0 & 0 & 0 & 0 & 0 \\ 1 & 0 & 0 & 1 & 0 & 0 & 0 & 0 & 0 & 0 & 0 & 0 \\ 1 & 0 & 0 & 0 & 1 & 0 & 0 & 0 & 0 & 0 & 0 & 0 \\ 1 & 0 & 0 & 0 & 0 & 1 & 0 & 0 & 0 & 0 & 0 & 0 \\ 1 & 0 & 0 & 0 & 0 & 0 & 1 & 0 & 0 & 0 & 0 & 0 \\ 1 & 0 & 0 & 0 & 0 & 0 & 0 & 0 & 0 & 0 & 0 & 0 \end{bmatrix}$

Emigration can lead to bias in the demographic parameters estimated with CJS models. If emigration is temporary and random, only the probability of encounter is biased, whereas nonrandom temporary emigration leads to bias in both apparent survival and the probability of encounter (Schaub et al. 2004a). If emigration is permanent, apparent survival is biased low as an estimate of true survival because losses to mortality cannot be distinguished from losses to permanent emigration (Ergon and Gardner 2014; Schaub and Royle 2014). Emigration limits the utility of CJS models for estimating true survival for any group of animals with low site fidelity, such as juvenile age classes with strong natal dispersal or species that use ephemeral habitats (Paradis et al. 1998; Zimmerman et al. 2007; Roche et al. 2012). If long-distance dispersers cannot be detected, measures of dispersal distance are truncated by the longest axis of a fixed-area study plot, and dispersal distance will also be underestimated (Cunningham 1986; Koenig et al. 1996).

A variety of mark-recapture models and study designs have been proposed for estimation of *true survival* instead of apparent survival. In some cases, the size of the study plot remains fixed and the different types of emigration are handled with model structure or auxiliary data. If dispersing individuals are transients or migratory individuals that permanently emigrate from a study plot, *time-since-marking* or *transient* models can be used to estimate apparent survival that is corrected for losses of individuals that are never encountered after first capture. If dispersing individuals are temporary emigrants that are likely to return, *robust design* models can be used to estimate apparent survival corrected for losses to temporary emigration (Horton and Letcher 2008). Marked individuals with home ranges close to the boundary of a study plot may be more likely to emigrate, and modeling ϕ or p as a function of distance to plot edge may improve parameter estimates from CJS models (Boulanger and McLellan 2001; but see Marshall et al. 2004). Losses to emigration may be less important in linear habitats such as streams or coastal beaches. In terrestrial habitats, circular or square study plots that minimize perimeter-area ratios may help to reduce losses to emigration. Estimation of survival can also be improved if a subset of the marked population is marked with radio transmitters or transponders to track movements (Powell et al. 2000). *Multistate* or *robust design* models can be used for joint analysis of encounter data for marked individuals with or without radio transmitters to model survival corrected for imperfect detection and emigration from a fixed-area study plot (Devineau et al. 2010; Horton et al. 2011; Bird et al. 2014).

Movements of marked individuals within a study plot are another source of data that can be used to estimate true instead of apparent survival. One approach is to make post hoc corrections to adjust apparent survival for losses due to local movements. *Area-ratio methods* correct for bias by comparing observed movement distances vs. random pairs of locations within the study plot to estimate a probability of detection for different dispersal distances (Zeng and Brown 1987; Baker et al. 1995). Area-ratio methods are flexible and can accommodate study plots of diverse sizes and shapes, or various dispersal distributions. Unfortunately, area-ratio methods can perform poorly for survival estimation if adjustments for rare long-distance dispersal events cause survival to be overestimated (Cooper et al. 2008). Still, Taylor et al. (2015) showed that post hoc corrections for local breeding dispersal led to less-biased estimates of apparent survival for arctic-breeding sandpipers. *Spatially explicit* models include information on locations of marked animals in analyses of demographic parameters (Borchers 2012). Most attention has been on use of closed population models for estimation of abundance (Efford 2004; Royle and Young 2008), but new applications for open populations have also been developed (Gardner et al. 2010, Chapter 5). Last, *spatial CJS models* (sCJS) allow joint analysis of capture and spatial data, and may yield estimates of true survival if local dispersal movements are relatively short compared to the spatial scale of a fixed-area study plot (Schaub and Royle 2014; Weiser et al. 2018).

Losses to emigration can also be reduced by expanding the size of a fixed-area study plots. One study design is to conduct captures and marking in a core study area but search for marked individuals in an expanded *buffer zone* (Marshall et al. 2004). Inclusion of additional detections from an expanded buffer zone can increase return rates and estimates of apparent survival (Reed and Oring 1993; Cilimburg et al. 2002; McKim-Louder et al. 2013). The challenge with buffer zones is that the total search area increases exponentially with potential dispersal distance and quickly becomes unmanageable. Zimmerman et al. (2007) modeled the effects of plot size on bias in CJS models and found that apparent survival of juvenile owls increased linearly among plots ranging in area from approximately 80 to 1400 km^2, suggesting unbiased estimates of survival would require even larger plots.

Another possible sampling design is to use a *distributed network* of study plots that can potentially detect both short and long-distance dispersal movements. Inclusion of spatial data on animal captures allows survival and emigration to be jointly estimated from movements both within and among fixed-area study plots (Schaub and von Hirschheydt 2009; Gilroy et al. 2012; Ergon and Gardner 2014; Lagrange et al. 2014). A network approach has been successfully used for birds of conservation concern where multiple research groups are working with the same species, populations are small with a high proportion of marked individuals, and patches of suitable habitat are discrete (Cooper et al. 2008; Gilroy et al. 2012; Roche et al. 2012). Distributed networks of study plots are also a feature of national programs for constant effort banding of landbirds in Europe and North America (Saracco et al. 2008; Robinson et al. 2009). Repeated captures and information from neighboring stations can be combined in a hierarchical model to investigate spatial variation in survival and residency on the study plots (Saracco et al. 2012). Moreover, correction factors for the difference between apparent and true survival can be calculated as the difference between estimates of the finite rate of population change based on demographic rates versus population counts (Ryu et al. 2016). At the broadest possible spatial scale, *joint models* can be used to decompose apparent survival into true survival and site fidelity if the live encounter data come from a fixed-area study plot, but other sources of information such as dead recovery or supplementary resighting data are taken from a larger geographic area such as a regional flyway (Bacheler et al. 2009; Bowerman and Budy 2012; Kendall et al. 2013; Lok et al. 2013).

7.7 Extending the CJS Model

The CJS model provides the foundation for most mark-recapture models for live encounter data. *Time-since-marking* and *transient* models are also CJS models, but use a different model structure to estimate apparent survival separately for different intervals after first capture. *Temporal symmetry* and *Jolly–Seber* models are based on the same set of encounter histories as the CJS model but do not condition upon first capture, and include both forward- and reverse-time modeling. Other mark-recapture models for open populations extend the CJS model by including various kinds of *auxiliary data* in the encounter histories. Use of auxiliary data usually requires additional sampling but allows apparent survival and other demographic parameters to be estimated with less bias and greater precision (Kendall et al. 2006, 2013; Schaub and Abadi 2011).

Multilevel models use data from multiple populations of one species or multiple species at a single site in a random effects framework, thereby expanding CJS models based on a single population of one species to multiple higher levels. *Spatially explicit* models incorporate spatial data with the locations where marked individuals are encountered, allowing joint analysis of movement and capture-recapture data. The auxiliary data in a *robust design* model come from dividing the primary occasions of a CJS model into shorter secondary periods when the population is assumed to be closed to gains or losses. CJS models are based on live encounters of marked individuals, but *mark-resight* models incorporate auxiliary data with counts of unmarked individuals. The CJS model is a single-state model that is coded with detections and non-detections, and the *young survival* model extends the detections to include counts of young. Similarly, *multi-state* models include categorical information about the state of an individual when it is detected, such as information about site, breeding status, physiological condition, or different methods of detection. Multistate models with *unobservable states* handle the special situation where individuals in some states cannot be encountered, whereas *multievent* models address situations where an individual cannot be assigned to a state with certainty. *Joint* models are based on CJS models but also allow inclusion of auxiliary data on dead recoveries or supplemental resightings that are recorded during the intervals between sampling occasions. Last, *integrated population* models are a flexible approach for development of customized models that allow joint modeling of mark-recapture data with population counts, and other demographic data on the components of fecundity, age ratios in harvest bags, or known fate survival data.

7.8 Time-since-marking and Transient Models

A common feature of live encounter data is that a subset of marked individuals are never detected again after first capture, with encounter histories characterized by detection at only one occasion (e.g. `1000`, `0100`, Box 7.1). In this case, apparent survival of newly marked individuals will be lower in the interval after first capture (ϕ^1) than returning individuals detected in subsequent intervals (ϕ^{2+}). In a sample of animals marked as young, apparent survival may be lower after first capture if juveniles have lower true survival or site fidelity than adults ($\phi^1 < \phi^{2+}$). The same difference can occur in a sample of animals marked as adults, if capture and handling negatively affect survival or site fidelity, if a sample includes transients, or if capture rates are heterogeneous. If any of these effects are present, ϕ^{2+} from a time-since-marking model may be less biased as an estimate of true survival than ϕ from a standard CJS model where ϕ^1 and ϕ^{2+} are pooled (Johnston et al. 1997; Korfanta et al. 2012).

Time-since-marking and transient models are CJS models that control for losses to mortality or permanent emigration that occur during the interval after the first capture occasion. The encounter histories are coded in the same manner as for standard CJS models, but the model structure is more complex to estimate additional parameters. The model structure is set up so that ϕ^1 and ϕ^{2+} are estimated separately, along with the encounter rate (p). The structure can be termed a *time-since-marking* model if a sample includes individuals of unknown age, or an *age* model if the sample is known-aged individuals. A *transient* model can be applied in situations where losses after first capture are likely due to inclusion of transients captured as unmarked individuals (Pradel et al. 1997). The probability that an unmarked individual is a transient at time t (τ_t) is then estimated as a derived parameter: $\hat{\tau}_t = 1 - \hat{\phi}_t^1/\hat{\phi}_t^{2+}$. The proportion of transients in the population ($\hat{T}_t$) can then be estimated as $\hat{T}_t = \hat{\tau}_t[N_t/(N_t + m_t)]$, where N_t and m_t are the numbers of newly marked and recaptured individuals at time t (Jessopp et al. 2004). Multistate versions of the transient model include τ_t as a parameter in the likelihood (Schaub et al. 2004b), allowing greater flexibility for modeling transient dynamics as a response to temporal covariates.

A common pattern in time-since-marking models is that apparent survival rates are often ranked: juveniles after first capture < adults after first capture < adults in subsequent intervals (Sandercock 2006). Unfortunately, the ecological causes are often difficult to distinguish because age, handling effects, and capture heterogeneity are all expected to produce the same general pattern of $\phi^1 < \phi^{2+}$. Thus, a prudent approach before starting any CJS analysis is to examine the relative frequency of individuals that were encountered on one versus multiple sampling occasions. If a set of encounter histories is dominated by individuals encountered on only one sampling occasion, time-since-marking effects should probably be included in the set of candidate models. A minimum of four occasions are needed to estimate the extra parameters of the time-since-marking or transient models, and the terminal parameters of each age class may be nonidentifiable in models with full time dependence (Hubbard et al. 2014).

7.9 Temporal Symmetry Models

CJS models condition upon first capture and proceed forward in time, whereas *temporal symmetry* models analyze the encounter histories with both forward- and reverse-time modeling (Pradel 1996; Nichols and Hines 2002; Nichols 2016). Temporal symmetry models do not require additional sampling and can be based on the same set of encounter histories as standard CJS models. With addition of reverse-time modeling, temporal symmetry models use the complete encounter histories and can be considered an alternative parameterization of a Jolly–Seber (JS) model (see below, Cooch and White 2018). Forward-time modeling yields estimates of apparent survival (ϕ) and encounter rates (p). Modeling of the same encounter histories from the last capture backwards gives a seniority probability (ζ) as a reverse-time analogue of ϕ, which is defined as the probability that an individual did not enter the population between the previous and current occasion. Forward-time modeling assumes that capture probabilities are homogeneous among marked animals. In the temporal symmetry and JS models, reverse-time modeling extends the assumption of equal catchability to marked and unmarked individuals.

The temporal symmetry model has three alternative parameterizations. The *seniority* or ζ-parameterization (ϕ_t, p_t, ζ_t) yields separate estimates of ϕ and ζ corrected for p. Alternatively, the *lambda* or λ-parameterization is perhaps the most useful model (ϕ_t, p_t, λ_t), where ϕ and ζ are combined to estimate the finite rate of population change $\hat{\lambda}_t = \hat{\phi}_t/\hat{\zeta}_{t+1}$. Here, λ can be modeled as a function of environmental covariates, and the variance of λ can be used to estimate risk of extinction (Nichols and Hines 2002). In the *recruitment* or *f*-parameterization of the temporal symmetry model (ϕ_t, p_t, f_t), ϕ and ζ are again combined but instead to estimate the per capita rate of recruitment $\hat{f}_t = \hat{\phi}_t(1 - \hat{\zeta}_{t+1})/\hat{\zeta}_{t+1}$. Bayesian versions of the temporal symmetry model allow the parameters to be modeled as random effects (Saracco et al. 2008; Tenan et al. 2014).

Temporal symmetry models allow estimation of λ from live encounter data alone, without the need to determine abundance from population censuses or estimate demographic rates to parameterize a matrix model (Sandercock and Beissinger 2002; Pradel and Henry 2007; Currey et al. 2011; Schorr 2012). In temporal symmetry models, λ is a realized estimate of population change that includes gains from recruitment and is based on the demographic classes included in the encounter histories. Seniority parameters are comparable to elasticity values from a matrix model in identifying sensitivity of $\hat{\lambda}$: survival has a greater effect on $\hat{\lambda}$ if $\hat{\zeta} > 0.5$, whereas recruitment has a greater effect on $\hat{\lambda}$ if $\zeta < 0.5$ (Nichols et al. 2000; Korfanta et al. 2012).

Temporal symmetry models require longer time series than CJS models to estimate ζ as an additional parameter. Time dependence in ϕ and p creates inestimable terms in the first and last intervals, and n occasions yields $n - 3$ estimates of $\hat{\lambda}$ (Dreitz et al. 2002). Modeling can be tricky because model fit for p affects parameter estimates for ϕ and ζ, and vice versa. Modeling p as a trend or a function of a covariate may induce a pattern in λ or the other model parameters (Tenan et al. 2014). Estimates of λ from temporal symmetry models are also sensitive to changes in sampling area. An increase in the size of a fixed-area study plot could increase the numbers of newly marked individuals which would affect estimates of recruitment into the population. Last, numerical simulations suggest that $\hat{\lambda}$ is robust to the effects of individual heterogeneity in capture, but behavioral responses to trapping and failure to account for losses at capture can bias estimates (Hines and Nichols 2002; Marescot et al. 2011).

7.10 Jolly–Seber Model

The JS model is an open population model that estimates four demographic parameters from the encounter histories: the probability of *apparent survival* ($\phi = S \times F$), the probability of *encounter* ($p = \delta \times p^*$), *abundance* or population size (N), and *net recruitment* or the number of new individuals entering the population (B, Pollock et al. 1990). The JS model is similar to a CJS model but is an unconstrained model, and the apparent survival and encounter parameters are not equivalent between the two classes of models. CJS models condition upon first capture and assume that marked individuals have the same probability of capture. Like a temporal symmetry model, the JS model uses the complete encounter histories including the leading zeros before first capture. The JS model extends the assumption of equal probability of capture to both marked and unmarked individuals, which allows the complete encounter histories to be used in the estimation of N and B. Thus, ϕ and p from a CJS model apply to marked individuals only, but to both marked and unmarked individuals in a JS model. The difference between the parameters in the CJS and JS models is subtle but has practical implications for study design. In a CJS model, different field methods can be used for marking and recapture, and individuals marked on the last occasion can be discarded because they do not contribute to the parameter estimates. If capture and marking are costly, the final season of a field project might be devoted to resighting efforts only. In a JS model, the same field methods should be used for marking of unmarked individuals and detections of marked individuals to meet the assumption of equal catchability. Individuals should still be captured for marking on the last occasion because leading zeros in the encounter histories will also contribute to parameter estimation in the analysis.

Several parameterizations of the JS model are available but differ in number of sources of data, how recruitment is modeled, and whether estimation is conditional upon individuals detected in the study area (Madon et al. 2011; Cooch and White 2018). Criteria for choosing a particular model include whether or not losses at capture are present, and what combination of demographic parameters are desired as model output, including abundance, net recruitment, per capita recruitment (f), and the finite rate of population change (λ). JS models can be useful for modeling patterns of recruitment into a population, and for separating gains from in situ recruitment vs. immigration (Collier et al. 2013). Unfortunately, individual heterogeneity and other issues often lead to bias in estimates of N and B from JS models (Chapter 5). If estimation of abundance is the goal of a mark-recapture analysis, mark-resight or closed population models may be better methods with additional options for control of problems due to heterogeneity of capture (Abadi et al. 2013).

7.11 Multilevel Models

Mark-recapture data collected from natural populations are often structured at multiple levels: individuals in a population with different probabilities of survival or capture, different populations of the same species, or related species in the same guild or community. Here, replication among sites or species provides a type of auxiliary data that can be used for parameter estimation. Under a frequentist framework, groups are either pooled in a constant model or treated as a fixed effect with a separate parameter estimates for each population or species. An assumption of homogeneity is required if groups are pooled, whereas fixed-effect models can be problematic if data are sparse for a subset of groups. Multilevel models offer a useful compromise because groups retain their individual identity but information is shared among groups during the estimation process (Cam 2012). A Bayesian framework allows groups to be modeled as random effects and at different ecological scales: individual heterogeneity, multipopulation, or multispecies.

If individual identity is treated as a random effect, heterogeneity in survival can be used to investigate patterns of senescence or other questions with *frailty* models (Marzolin et al. 2011; Cam et al. 2013). Individual heterogeneity in catchability can affect bias and precision of estimates of apparent survival and abundance, but may be handled with *finite-mixture* models or as a random effect in *multilevel* models (Cubaynes et al. 2012; Abadi et al. 2013; Péron et al. 2016). Multipopulation studies can treat time as a random effect to investigate spatial patterns of synchrony (Grosbois et al. 2009), or site as a random effect to calculate overall estimates of survival from multiple populations of a single species (Papadatou et al. 2011; Jansen et al. 2014; Weiser et al. 2018). Last, multilevel models can be used to treat species as a random effect to investigate patterns of synchrony in apparent survival among species in a community (Lahoz-Monfort et al. 2011; Papadatou et al. 2012), or to compare different communities (Lloyd et al. 2014).

7.12 Spatially Explicit Models

Spatially explicit capture-recapture models provide a robust methodology to combine live encounter data with location of capture information for joint estimation of demographic parameters (Efford 2011; Borchers 2012). Encounters can be based on standard methods for live capture or resighting, or alternatively, camera traps or other devices that record proximity of marked individuals to a sampling location. Location data for encounters can be recorded with different sampling designs, including coordinates of a site within a trap grid, angle, and distance from a point count station or line transects, or as a location within the boundaries of a fixed-area study plot. A general feature of spatially explicit models is that the spatial scale of a trap grid or study plot should be larger than the home range of the study organism.

Closed population models estimate abundance (N) corrected for the probability of capture (p). In *spatially explicit* models for estimation of abundance, detections of individuals at known sampling sites provide information on the probability of detection at the center of a home range (g_0). Animal distributions can be viewed as a point process model where range centers are assumed to be distributed at some density (D). The locations of individual activity centers are usually unknown but the mean distances between recaptures on different occasions can provide information on the scale of spatial movements (σ). For example, Gardner et al. (2010) generalized a JS model to include spatial data for a grid of camera traps, and estimated abundance of activity centers (N_t), density per unit area (D_t), and the per capita recruitment rate (ρ), adjusted for detectability (δ), and the expected number of encounters per individual at a given trap station (λ_0). Spatially explicit models can be fit in either a MLE or Bayesian framework, or with other methods (Efford 2004). Another challenge for spatially explicit models is that estimates of N are required for the likelihood but the true abundance is unknown. This issue can be addressed by *data augmentation* where the data set is supplemented with a large number of encounter histories for individuals that were never captured and have all zeros. These additional encounter histories are used to calculate recruitment, and thereby estimate the proportion of additional individuals that were present but never captured.

Spatially explicit models for open populations allow inclusion of spatial data of locations in the individual encounter histories. Schaub and Royle (2014) developed a sCJS model that includes a kernel for dispersal movements, which allows estimation of apparent survival corrected for local movements within a fixed-area study plot. Raabe et al. (2014) extended the CJS model to include spatial data for detections of stream fish in a linear array of weirs. A spatial version of a robust design model has also been proposed by Ergon and Gardner (2014). Spatial CJS models can be effective at correcting for bias due to permanent or temporary emigration (Weiser et al. 2018), but simulations suggest that parameter estimates may be sensitive to the statistical distributions used to model dispersal movements (Ergon and Gardner 2014). Estimates of dispersal kernels from mark-recapture data suggest that use of Gaussian kernels can lead to parameter bias (Fujiwara et al. 2006). Nevertheless, spatially explicit models for open populations have great potential for improving demographic analyses because auxiliary data on animal locations are often recorded in population studies but the information is not typically included in mark-recapture analyses.

7.13 Robust Design Models

Robust design models differ from CJS models in that additional sampling is required at the outset of a field project, and primary sampling periods are subdivided into shorter secondary sampling occasions. Robust design models assume that the population is open to gains and losses among *primary* sampling periods but closed to change within *secondary* sampling occasions (Kendall et al. 1997; Fujiwara and Caswell 2002; Schaub et al. 2004a). Like CJS models, encounter histories are coded: 1 = detected and 0 = not encountered. Robust design models control for both imperfect detection and imperfect availability by modeling a *superpopulation* with marked individuals moving in and out of the sampling area where they can be encountered. Survival rates are assumed to be the same for individuals both inside and outside of the sampling area. Standard robust design models can perform well if the assumption of closure is not met (Kendall 1999). Open robust design models relax the assumption of geographic closure to allow staggered entry and exit of individuals from a breeding population (Schwarz and Stobo 1997; Kendall and Bjorkland 2001), or the assumption of demographic closure to allow mortality between the secondary sampling occasions (Bailey et al. 2004a).

In a CJS model, the probability of encounter (p) is the product of site propensity (σ) and the true probability of detection (p^*). If true detection is close to unity, then the encounter rates are effectively an estimate of site propensity ($p = \sigma$ if $p^* = 1$). In a robust design model, the open population part of the model gives the same p as a CJS model, whereas the closed captures provide estimates of $p*$ and abundance (N), or alternatively, the number of individuals never captured (f_0). Closed population models were first used to estimate abundance for a single closed period and the original notation used a capital M to denote different model structures (Otis et al. 1978). A variety of closed population models can be used to model p^* during the secondary occasions of a robust design model, including a null model (p or M_0), time dependence (p_t or M_t), behavioral effects (p vs. c, or M_b), heterogeneity of capture with finite-mixture models (p_π sensu M_h), or different combinations of the effects. Robust design models usually estimate the probability of *temporary emigration* as the complement of site propensity ($\gamma = 1 - \delta$, but see Kendall et al. 2013).

A variety of alternative models can be used to model the probability of temporary emigration. If movements are random and not affected by previous events, temporary emigration at occasion t is calculated as: $\hat{\gamma}_t = 1 - \hat{p}_t / \hat{p}_t^*$. If movements are nonrandom, the probability of temporary emigration is modeled as a function of an individual's status on the previous occasion. Temporary emigration is estimated separately for absent individuals that remained unavailable for capture (γ_t') versus individuals that were present but dispersed away from the sampling area (γ_t'', Kendall et al. 1997). Individuals may be unavailable for capture either because they have dispersed away the sampling area, such as marine animals that have left breeding sites to be at sea, or because marked individuals are hidden in refugia where they cannot be sampled, such as fossorial animals in belowground burrows. The parameter γ_t' has been termed "immigration," but the probability that an absent individual re-enters the sampling area is calculated as the complement: $\delta' = 1 - \gamma'$. Alternative models for movement among the different segments of a superpopulation include no temporary emigration ($\gamma' = \gamma'' = 0$), random emigration ($\gamma' = \gamma''$), nonrandom emigration ($\gamma' \neq \gamma''$), or even-flow models with balance between immigration and emigration ($1 - \gamma' = \gamma''$, Cooch and White 2018). Depending on model structure, robust design models quickly become complex with estimation of multiple parameters for the transitions between primary periods in the open part of the model (ϕ, γ',γ''), and also for the secondary sampling occasions in the closed part of the model (p, c, N or f_0). Estimation of these additional parameters may require long-term datasets with large samples of marked individuals.

Robust design models have two advantages compared to the CJS model. One issue for both models is parameter estimation during the last interval if demographic parameters are time-dependent. Under random temporary emigration, robust design provides estimates of ϕ and p for all intervals in a time series because p^* is available for all occasions. If temporary emigration is nonrandom, then the last movement parameters γ_t' and γ_t'' are still confounded with survival (Peñaloza et al. 2014). As in the CJS model, constraints would have to be applied to the robust design model to estimate survival for the last interval. Robust design models can yield estimates of ϕ and p with less bias and greater precision than CJS models if temporary emigration is present. Temporary emigration causes heterogeneity in encounter rates because individuals that are present have a nonzero probability of encounter, whereas temporary emigrants are unavailable for capture and probability of encounter is zero. Under random temporary emigration, estimates of ϕ and p from CJS models are relatively unbiased but precision is reduced. Under nonrandom temporary emigration, estimates of ϕ and p from standard CJS models can be strongly biased if $\gamma_t' \neq \gamma_t''$ (Kendall et al. 1997, Peñaloza et al. 2014).

The demographic processes that lead to temporary emigration remain poorly understood in ecology. Empirical applications of robust design models include investigations of regional movements and partial migration (Dinsmore et al. 2003; Jahn et al. 2010; Cantor et al. 2012), dormancy or use of refugia where animals are not available for capture (Schaub and Vaterlaus-Schlegel 2001; Bailey et al. 2004b), as well as variation in breeding propensity among animals at reproductive sites (Kendall and Nichols 1995; Kendall and Bjorkland 2001; Sedinger et al. 2001; Schmidt et al. 2002). Breeding propensity can be imperfect ($\sigma < 1$) because of delayed maturity among juveniles or subadults, intermittent breeding among adults, or early reproductive failure that causes emigration before detection. The closed portion of a robust design model also provides estimates of abundance, and can be used to estimate population gains from in situ recruitment vs. immigration (Nichols and Pollock 1990). However, estimates of abundance from the closed part of the model are restricted to the number of marked individuals available for capture in the fixed-area study plot (N or f_0). Thus, estimates of abundance from robust design models may be an underestimate of total population size unless a high proportion of the population has been marked, or in the special case where individuals have been identified by natural marks (Cantor et al. 2012; Lee et al. 2014).

7.14 Mark-resight Models

Mark-resight models take their name from early field studies where marked individuals were released on one occasion and then resighted on multiple occasions.

A single release event might be more cost-effective if capture and marking are expensive procedures, and less invasive if capture events pose a risk to sensitive species of wildlife. "Mark-resight" is a confusing term because the CJS and joint models can also be based on encounter histories with resightings of marked animals. The fundamental difference among different types of models is that counts of *unmarked individuals* are included as inputs for the likelihood function in mark-resight models. Mark-resight models can also allow *batch marking*, where all captured individuals are marked with the same tag as a group, instead of marking each individual with a unique tag. Even if individual tags were used, mark-resight models can include partial information for individuals that were seen as marked but were not individually identified. Mark-resight models improve upon closed population and robust design models by allowing estimation of the total abundance for the marked and unmarked individuals that are observable in the fixed-area study plot (N), and also the nonobservable individuals that are part of a larger superpopulation (N^*). A hybrid model that combines mark-resighting and closed population models has been used to estimate abundance of bobcats (*Lynx rufus*, Alonso et al. 2015). Similarly, a combined mark-resighting and robust design model has been used to estimate survival and total abundance for open populations (McClintock and White 2009).

Three estimators for mark-resight data include the logit-normal estimator (LNE), the immigration–emigration logit-normal estimator (IELNE), and the zero-truncated Poisson log-normal estimator ([Z]PNE, McClintock and White 2012). The LNE and IELNE estimators can be used with batch marks instead of individually marked animals, but have the restrictive requirements that the total number of marked individuals must be known, and sampling with replacements is not allowed. Number of marks is rarely known in field studies of open populations, with the possible exception of situations where all animals are marked immediately before release, or if all marked individuals receive radio collars with mortality switches. Double-counting of individuals may be difficult to avoid if sampling is partitioned into multiple occasions and animals are mobile. Still, if the model assumptions can be met, the IELNE estimator produces estimates of the total superpopulation because geographic closure is not required.

The robust design version of the (Z)PNE estimator is perhaps the most useful model for open populations. The (Z)PNE estimator requires individually identifiable marks and geographic closure, but is a better fit to modeling of open populations because it does not require the number of marks to be known and allows sampling with replacements within secondary occasions. Ideally, individual identity should be obtained on most encounters of marked individuals in the field (>90%). If partial sightings are common, extended mark-resight models allow for individual heterogeneity in detection and control for uncertainty in individual identification (McClintock et al. 2014). Encounter histories for marked individuals do not condition upon distinct secondary occasions and instead tally the total number of individual sightings per primary occasion (Cooch and White 2018). If a marked individual is not detected on a primary occasion, separate codes are used if the animal is known to be alive (+0) or of unknown fate (–0). Additional input includes counts of unmarked animals, marked individuals sighted but not identified, and number of marked individuals in the population (if known). Like the robust design model, the (Z)PNE estimator provides estimates of ϕ, γ', and γ'', but also provides estimates of the number of unmarked individuals (U) and total population size ($N = U + n$) that are corrected for variation in resighting rates (α, σ, and λ). The (Z)PNE estimator has been used to estimate demographic parameters for sharks (Lee et al. 2014), migratory sandpipers (Lyons et al. 2016), and river dolphins (Ryan et al. 2011).

7.15 Young Survival Model

The *young survival* model extends the CJS model by coding detections as counts of group size (Lukacs et al. 2004; Cooch and White 2018). Like the CJS model, the young survival model estimates apparent survival (ϕ) corrected for probability of encounter (p). The young survival model was developed to monitor groups where one or more individuals are individually marked but the remainder of the group is unmarked. Group size must be known at the start of the encounter history, but if group size is one, then the young survival model is equivalent to a CJS model. Survival rates of the different unmarked individuals within each group are assumed to be equivalent. Groups should be independent, and the model cannot be used in field situations where fission or fusion of social groups leads to changes in numbers that are not due to mortality.

The young survival model has been to model chick survival for precocial birds where an individually marked parent attends a brood of unmarked young (Dreitz 2009; Brudney et al. 2013). Detection is often imperfect because broods are mobile and young can escape observers by hiding or moving away. However, hatchlings may be too small to be marked, tags might affect risk of mortality, or individual tags may be too expensive. Thus, an advantage of the young survival model is that chick survival can still be estimated even if the young are not individually marked. If the dependent young are unable to survive without parental care, then the model will estimate true survival because all losses result in mortality. A high maximum group size leads to many possible transitions between groups of different size, which can reduce the precision of parameter estimates (Cooch and White 2018). It may not be possible to use the model if the maximum group size is large but the sample size of family groups is relatively small. Still, the young survival model can be useful for estimating juvenile survival for species where litter or brood sizes are small (≤3–4 young) and detection of mobile young is imperfect (Kendall et al. 2003; Ryan et al. 2011; Tarwater et al. 2011).

7.16 Multistate Models

Multistate models provide a particularly flexible modeling approach that have been widely used in population ecology (Sandercock 2006; White et al. 2006; Lebreton et al. 2009), and can be set up as mark-recapture (Boxes 7.2 and 7.3) or multievent models (Box 7.4). The term multistate refers to the coding of detections as dynamic categorical states that potentially change between consecutive occasions. The Arnason–Schwarz (AS) model is a multistate version of a CJS model because both models are conditional on first capture, but multistate versions of the JS model can also be unconditional (Brownie et al. 1993; Lebreton et al. 2009; Pledger et al. 2013). Many types of survival models, including the CJS, transient, robust design, dead recovery, and time-to-event models can also be set up as multistate models (Lebreton et al. 1999; Schaub et al. 2004a, 2004b; Gauthier and Lebreton 2008; Lebreton et al. 2009; Devineau et al. 2014).

A CJS model is a single-state model because the encounter histories are coded as 1 = detected, or 0 = not encountered. In a multistate model with two states (r), detections might be coded as A = state A versus B = state B, and 0 = not encountered. The estimates of apparent survival from a CJS model (ϕ) can then be decomposed into *state-specific estimates* of apparent survival and changing states ($\phi = \phi^r \psi^r$). A multistate model with two states will have a minimum of six parameters: state-specific estimates of apparent survival (ϕ^A and ϕ^B) and probability of encounter (p^A and p^B), as well as the transitional *probability of changing states*, such as from state A to B (ψ^{A-B}) or from B to A (ψ^{B-A}). The total number of parameters increases rapidly with addition of more states ($K = r^2 + r$), with 12 parameters for $r = 3$ states, 30 parameters for $r = 5$ states, and so forth. Standard assumptions of multistate models are that apparent survival does not depend on a previous state, all individuals make the transition at the end of the interval, individuals do not temporarily emigrate to states where they cannot be detected, and that observers can correctly assign individuals to states. The different

Box 7.4 Fitting Multistate Models with Elementary Matrices

Multievent models are defined by elemental matrices that specify three sets of probabilities: initial states (**Π**), transitions (**Φ**), and conditional events (**B**). Initial states are a vector but **Φ** and **B** can each be one or more matrices in a Markov chain. Multievent models differ from standard mark-recapture models because the dead state (*D*) is explicitly included as a terminal state even if dead individuals are never observed. The usual conventions are that the dead state is the last column in **Π** and **Φ**, whereas not detected is the first column in **B**. Intermediate states may not be the same as initial states, and elementary matrices need not be square but each matrix row sums to one and is row-stochastic.

A standard multistate model might include two observable states for breeders (*B*) and nonbreeders (*N*). Initial states (**Π**) for the *B*, *N* and *D* states would be:

$$\mathbf{\Pi}_t = \begin{bmatrix} \pi^B & 1-\pi^B & 0 \end{bmatrix}, \tag{7.6}$$

where π^B and $1 - \pi^B$ are the probabilities that an individual is a breeder or nonbreeder at first encounter, and 0 indicates that none of the individuals are dead at the start. Two transition matrices (**Φ**) summarize the state-specific rates for the *B*, *N*, and *D* states:

$$\mathbf{\Phi}_t^{r(\phi)} = \begin{bmatrix} \phi^B & 0 & 1-\phi^B \\ 0 & \phi^N & 1-\phi^N \\ 0 & 0 & 1 \end{bmatrix} \text{ and } \mathbf{\Phi}_t^{r(\psi)} = \begin{bmatrix} \psi^{BB} & 1-\psi^{BB} & 0 \\ 1-\psi^{NN} & \psi^{NN} & 0 \\ 0 & 0 & 1 \end{bmatrix}, \tag{7.7}$$

where ϕ^r is the state-specific probability of apparent survival and ψ^r is the probability of changing states. The top rows indicate that an individual must survive, remain in the same state, or change states. The right column absorbs individuals that die or permanently emigrate, whereas the bottom row indicates that lost individuals never return to the population. The columns of the conditional events matrix (**B**) summarize the state-specific probabilities of encounter for three possible events: not detected, detected as a nonbreeder, or detected as a breeder:

$$\mathbf{B}_t^r = \begin{bmatrix} 1-p^N & p^N & 0 \\ 1-p^B & 0 & p^B \\ 1 & 0 & 0 \end{bmatrix}. \tag{7.8}$$

assumptions are relaxed in modified multistate models that model transitions as a function of states previously occupied (Rouan et al. 2009), subdivide events during transitions (Grosbois and Tavecchia 2003), include temporary emigrants as an unobservable state (Kendall 2004; Schaub et al. 2004a), or account for uncertainty in state assignment (Pradel 2005).

Dynamic categorical states have been used to investigate four basic types of information: location, age or stage classes, disease or physiological state, and alternative sources of encounter data. *Location data* are based on detections of marked individuals at discrete geographic sites such as ecoregions, islands, breeding colonies, or core vs. peripheral habitat patches. In these cases, state-specific estimates of apparent survival might be used to identify the habitat strata with the best demographic performance (Serrano et al. 2005; Low et al. 2010). Movement rates can be modeled as a function of distance, colony size, or habitat conditions to better understand the effects of connectivity in spatially structured populations (Brown et al. 2003; Breton et al. 2006; Roche et al. 2012). Movement rates can also be estimated for a core study plot with a contiguous buffer zone (Powell et al. 2000; Devineau et al. 2010; Horton et al. 2011), or between staying in a site or moving elsewhere in a network of sites (Lagrange et al. 2014). If a network of study sites represents the entire known distribution for a metapopulation, apparent survival may approach true survival because losses will be due to mortality and not permanent emigration.

Multistate models can also be used to model demographic variation among different *age or stage classes*. Demographic classes are important sources of heterogeneity that can be treated separately in a multistate model but are pooled in a CJS model. In the case of breeding status, encounter histories might be coded as B = breeder, N = nonbreeder, and 0 = not detected (Sandercock et al. 2000). With sufficient data, strata might be extended to multiple classes: juveniles, nonbreeders, inexperienced breeders, or experienced breeders (Schaub et al. 2011; Rotella et al. 2012), or to a combination of breeding status and site (Anderson et al. 2012). Alternatively, stage classes might be based on social dominance (subordinate vs. dominant, Cohas et al. 2007), reproductive output (unsuccessful vs. successful, Schaub and von Hirschheydt 2009), migratory status (resident vs. migrant, Grayson et al. 2011), or sociality (solitary vs. groups, Genton et al. 2015). Multistate models based on stage classes yield state-specific estimates of ϕ and p, and ψ becomes the probability of maturation or changing stage class. If age or experience accrue over time, the transitional probabilities of returning to a younger or inexperienced state can be fixed to zero. Multistate models based on breeding status have been used to test for the cost of reproduction, as a life-history tradeoff between reproduction and survival (Nichols et al. 1994; Nichols and Kendall 1995). Observational methods are not the best approach to test for tradeoffs because breeders or experienced individuals may have high reproductive success, high survival, and strong site fidelity (Sandercock et al. 2000; Sanz-Aguilar et al. 2008; Schaub and von Hirschheydt 2009). Multistate models can also be used to test for demographic responses to manipulated treatments in an experimental context, but relatively few studies have used this approach (Doligez et al. 2002; Lyet et al. 2009).

Multistate models are a useful tool for investigating *disease dynamics* in wild populations (Cooch et al. 2012). Here, encounter histories might be coded as U = uninfected, I = infected, and 0 = not detected. Determination of disease state can be determined by external appearance, immunoassays, or molecular tests of different tissues. Depending on disease latency and test sensitivity, the U-state can be a heterogeneous mixture of uninfected individuals and individuals that are infected but still asymptomatic (Conn and Cooch 2009). If disease negatively impacts survival or activity, infected individuals may have lower state-specific estimates of apparent survival ($\phi^I < \phi^U$) or encounter ($p^I < p^U$, Faustino et al. 2004). Alternatively, encounter rates might be higher for infected individuals if behavioral changes increase detection by observers (Senar and Conroy 2004). Here, the transition rates provide estimates of the probability of infection (ψ^{U-I}) and potential recovery (ψ^{I-U}), and asymmetry between the infection and recovery rates can provide an index of disease virulence ($\psi^{U-I} > \psi^{I-U}$). Multistate models allow the transitions to be modeled as a function of environmental covariates, such as the relationship between infection rates and prevalence of a disease in a population (Lachish et al. 2007; Ozgul et al. 2009).

Multistate models have also been used to investigate the effects of *body mass or size* on survival. Body mass and size are usually continuous variables which must be converted to states based on quartiles or standardized scores (i.e. z-values, Letcher and Horton 2008; Boulanger et al. 2013; Monticelli et al. 2014). Ordinal covariates such as rank scores of body condition are often pooled to a few states too (Miller et al. 2003). Multistate models based on size or mass classes yield state-specific estimates of ϕ and p, and larger, mobile individuals might be predicted to have higher survival or a lower probability of encounter. The transition ψ becomes the probability of growing or regressing into a different stage class. Multistate models were a useful method for modeling mass and size because early mark-recapture models could not handle continuous covariates that were also dynamic (Nichols and Kendall 1995). However, pooling is potentially problematic because information may be lost when continuous variables are collapsed into categorical bins. Improved methods for including individual covariates in CJS models now allow ϕ and p to be modeled as a function of covariates that are both continuous and dynamic (Cooch and White 2018).

A final application for multistate models is for combining multiple sources of information such as different marker types, live encounter with dead recovery data, or both (Kendall et al. 2006; Besnard et al. 2007; Juillet et al. 2011). Different marker types might include a standard mark (band, tag, or molecular identity) vs. an additional auxiliary mark such as a neck collar (Reed et al. 2005) or a radio tag (Powell et al. 2000; Boulanger et al. 2004). Here, the encounter histories might be coded as M = detected with a standard mark, A = detected with an auxiliary mark, or 0 = not detected. If neck collars or radio tags negatively impact animal performance, apparent survival or encounter rates may be lower for individuals with auxiliary marks ($\phi^A < \phi^M$). Additionally, the transition rates can provide an estimate of marker loss (ψ^{A-M}). In the case of combining live encounter and dead recovery information, encounter histories might be coded as A = alive, D = newly dead, or 0 = not detected (Lebreton et al. 1999; Gauthier and Lebreton 2008; Devineau et al. 2014). A third possible state of dead for multiple years is unobservable and not necessary to include. To investigate competing risks of mortality, codes for dead recoveries can be expanded to multiple sites (Kendall et al. 2006), or to different causes of mortality such as predation, harvest, or collisions (Schaub and Pradel 2004). The transition rates of the multistate model then become estimates of cause-specific mortality rates for the different sources of mortality (ψ^{A-D}) and encounter rates are calculated separately for the live encounters (p^A) and cause-specific dead recoveries (p^D or r, Bischof et al. 2009). For example, Besnard et al. (2010) used multistate models to show that monthly risks of harvest mortality were up to three times higher than natural mortality in hunted populations of Gray Partridge (*Perdix perdix*).

The main limitation of multistate models is that the number of state-specific parameters and transitions increases exponentially with the number of states, especially in *memory* models where the parameters are also dependent on previous states. In practice, most multistate models are limited to ~2–4 categorical states, although networks with multiple sites can still be modeled with two states for staying at a site ("here") or dispersing ("elsewhere," Lagrange et al. 2014). Model complexity can also be reduced by fixing parameters to zero for any impossible transitions that cannot occur, such as regression of adults to juveniles (ψ^{A-J}) or resurrection from dead to alive (ψ^{D-A}). Complex multistate models can have problems with parameter redundancy as an *intrinsic* function of model structure, or due to *extrinsic* problems with sparse data if the encounter histories do not include all possible transitions (Fujiwara and Caswell 2002; Gimenez et al. 2003; Bailey et al. 2010; McCrea et al. 2012). Complex multistrata models can also be difficult to fit with MLE methods due to problems with convergence and multiple solutions for the likelihood function (Lebreton and Pradel 2002; Pradel et al. 2008). Problems with model fitting can be addressed by using different sets of random values as starting values, or by using *simulated annealing* as an optimization algorithm that uses random jumps to sample the likelihood function (White et al. 2006).

7.17 Multistate Models with Unobservable States

One challenge for multistate models is when one or more states are *unobservable* (Kendall and Nichols 2002; Kendall 2004; Schaub et al. 2004a). Unobservable states are a feature of herbaceous plants where belowground rhizomes or corms may be dormant (Shefferson et al. 2003; Kéry et al. 2005), and among long-lived vertebrates where nonbreeders are unobservable due to delayed maturity among juveniles or intermittent breeding among adults (Fujiwara and Caswell 2002; Kendall et al. 2009; Stauffer et al. 2013). Two states might be coded as O = observable and U = unobservable, and more complex models may have multiple observable or unobservable states (Converse et al. 2009; Bailey et al. 2010). By definition, $p^O > 0$ but $p^U = 0$, and the state U does not appear in the encounter histories. The multistate model has the potential to give state-specific estimates of apparent survival (S^O and S^U), but the two parameters may not be identifiable unless constrained to be equal (Henle and Gruber 2018). The transition rates are equivalent to a robust design model with non-random movements where ψ^{U-O} is immigration (or $1 - \gamma'$) and ψ^{O-U} is temporary emigration (or γ'', White et al. 2006). If $\psi^{U-O} = 0$, then emigration is not temporary but is permanent. Conceptually, multistate models with unobservable states can be used to model temporary emigration as *absent* (ψ^{U-U} and $\psi^{O-U} = 0$), *random* ($\psi^{U-U} = \psi^{O-U}$), *nonrandom* ($\psi^{U-U} \neq \psi^{O-U}$), or balanced in a model with *even-flow* dynamics ($\psi^{U-O} = \psi^{O-U}$). One potential advantage is that multistrata models can still be used even if a field study lacks the secondary sampling periods needed for a robust design model (Schaub et al. 2004a). Unfortunately, unobservable states can cause problems with intrinsic parameter redundancy, such that not all parameters may be identifiable in multistate models (Hunter and Caswell 2009; Bailey et al. 2010; Cole 2012), and may affect bias and precision for the parameter estimates (Henle and Gruber 2018).

Issues of parameter redundancy can be resolved by setting constraints on the structure of multistate models (Kendall and Nichols 2002; Kendall 2004). Three alternatives include setting survival to be equal for the different states ($S_t^O = S_t^U$), setting transitional probabilities to be time-constant ($\psi_c^{U-O} \neq \psi_c^{O-U}$), or by placing constraints on a subset of the transitions, for example if breeders become obligate nonbreeders when interbirth intervals are greater than a year ($\psi^{O-O} \equiv 0$, Fujiwara and Caswell 2002; Reed et al. 2003; Monk et al. 2011). Alternatively, constraints on p may be preferred if the probability of encounter has less biological interest. Multistate models for herbaceous plants have included three states: dormant (D), vegetative (V), and flowering plants (F). Dormant life-stages are unobservable ($p^D \equiv 0$), and the dormant state is not included in the encounter histories. Encounter rates for aboveground life-stages of a sessile plant can be arbitrarily set to one ($p^V = p^F \equiv 1$, Kéry et al. 2005), or can be estimated with closed population models (p^V and $p^F > 0.97$, Shefferson et al. 2003). Encounter rates of mobile animals are unlikely to be close to unity, and different methods are required.

One approach for investigating dynamics of unobservable states is a combined *multistate open robust design* model for open populations (MSORD) with a multistate framework for estimating state-specific demographic parameters coupled with a robust design component for estimating encounter rates. Encounters on secondary occasions can include repeated recaptures in the *robust gateway* model (Bailey et al. 2004a; Church et al. 2007), or resightings only in the *less-invasive robust design* (Kendall et al. 2009). Problems with parameter redundancy can still arise with these models, and survival of individuals in the observed and unobserved states is usually constrained to be equal (Kendall and Bjorkland 2001; Bailey et al. 2009). Despite the increased complexity of the combined MSORD models, empirical applications have provided insights into the sex-specific costs of breeding (Muths et al. 2013; Stauffer et al. 2013; Rendón et al. 2014), reproductive strategies of failed and successful breeders (Converse et al. 2009), and the key environmental drivers that affect these life-history transitions (Church et al. 2007; Grayson et al. 2011).

7.18 Multievent Models with Uncertain States

Another challenge for multistate models arises when individuals cannot be assigned to categorical states with confidence. Uncertain states differ from unobservable states because *uncertain states* can still be detected, whereas unobservable states are not available for capture (δ or $p^* < 1$). Uncertain states can arise for two non-exclusive reasons (Conn and Cooch 2009). Under *misclassification errors*, marked individuals may be assigned to an incorrect state, such as a genotype or species (Lukacs and Burnham 2005; Runge et al. 2007). Under *partial observability*, marked individuals are detected but states remain ambiguous because of difficulties in determining sex (Nichols et al. 2004), disease status (Conn and Cooch 2009), breeding status (Sanz-Aguilar et al. 2011), or association with dependent offspring (Kendall et al. 2003; Taylor and Boor 2012). Early attempts to deal with uncertain states included adding an additional unknown state to the encounter histories (Wood et al. 1998; Conroy et al. 1999; Faustino et al. 2004), or using auxiliary data to develop state assignment matrices that could be used as model constraints (Fujiwara and Caswell 2002; Lebreton and Pradel 2002). Censoring of unknown observations or including unknowns as an additional state are undesirable solutions because both procedures affect the bias and precision of parameter estimates from multistate models (Faustino et al. 2004; Nichols et al. 2004; Conn and Cooch 2009).

Like unobservable states, uncertain states can also be handled with a combined MSORD. In mammals, large-bodied females can be relatively easy to detect but it be harder to determine if a female is attending dependent offspring. In a population study of manatees (*Trichechus manatus*), Kendall et al. (2003) sampled marked females twice per year and recorded three states: female with calf (C), lone female with no calf (N), or undetected (0). Females had imperfect detection ($p = 0.48$), but if a female was detected, then the probability of detecting her calf was also imperfect ($\delta = 0.72$). Estimates of the transitional probability of becoming a breeder were biased low in a standard multistate model ($\hat{\psi}^{N-B} = 0.31$) compared to a model controlling for uncertain states ($\hat{\psi}^{N-B} = 0.61$). Nichols et al. (2004) used a similar model to address uncertainty in sex determination. One feature of these models is that probability of classification is often hierarchical; a female with a calf cannot be misclassified as a lone female, and unsexed juveniles mature to become known sex adults, but cannot regress to an unknown state.

Uncertain states can also be handled with *multievent* models, which belong to the family of *hidden Markov models* (HMM, Pradel 2005, 2009; Choquet et al. 2009b; Conn and Cooch 2009). The models are *hidden* because the latent state dynamics are only partially observable and *Markov* because individuals move independently among a finite set of states in an ordered series

Box 7.5 Fitting Multievent Models with Elementary Matrices

Multievent models can also handle situations where the state assignments are *uncertain* in a multistate model. Desprez et al. (2014) used a multievent model to investigate costs of first-time breeding in an animal population with three latent states: prebreeders (*P*), new breeders (*N*), and experienced breeders (*E*). Dead (*D*) was also included as a state in the elementary matrices. In the field, individuals were observed in four possible states: not detected, detected as a prebreeder, detected in an unknown state, or detected as an adult (new or experienced combined). All individuals were first marked as prebreeders and the initial states ($\mathbf{\Pi}$) for the *P*, *N*, *E*, and *D* states were then:

$$\mathbf{\Pi}_t = \begin{bmatrix} 1 & 0 & 0 & 0 \end{bmatrix}. \tag{7.9}$$

The transition matrices ($\mathbf{\Phi}$) summarized the state specific rates for the *P*, *N*, *E*, and *D* states:

$$\mathbf{\Phi}_t^{r(\phi)} = \begin{bmatrix} \phi^P & 0 & 0 & 1-\phi^P \\ 0 & \phi^N & 0 & 1-\phi^N \\ 0 & 0 & \phi^E & 1-\phi^E \\ 0 & 0 & 0 & 1 \end{bmatrix} \text{ and } \mathbf{\Phi}_t^{r(\psi)} = \begin{bmatrix} 1-\psi^{PN} & \psi^{PN} & 0 & 0 \\ 0 & 0 & 1 & 0 \\ 0 & 0 & 1 & 0 \\ 0 & 0 & 0 & 1 \end{bmatrix}, \tag{7.10}$$

where ϕ^r and ψ^r are the state-specific probabilities of apparent survival and the probability of changing states. Prebreeders had an imperfect probability of becoming a new breeder (ψ^{PN}) whereas new or experienced breeders automatically transitioned to becoming experienced breeders.

The model had two conditional events matrices (**B**). The columns of the first matrix summarized probabilities of encounter (p^r) for the four latent states as three possible events: not detected, detected as a prebreeder, or detected as an adult. The columns of a second matrix then summarized the probability of classification (δ^r) for each of the three events as four possible observation codes: not detected, detected and classified as a prebreeder, detected in an uncertain state, or detected and classified as an adult:

$$\mathbf{B}_t^{r(p)} = \begin{bmatrix} 1-p^P & p^P & 0 \\ 1-p^A & 0 & p^A \\ 1-p^A & 0 & p^A \\ 1 & 0 & 0 \end{bmatrix} \text{ and } \mathbf{B}_t^{r(\delta)} = \begin{bmatrix} 1 & 0 & 0 & 0 \\ 0 & \delta^P & 1-\delta^P & 0 \\ 0 & 0 & 1-\delta^A & \delta^A \end{bmatrix}. \tag{7.11}$$

of events. Multievent models are defined by elemental matrices that specify three sets of probabilities: initial states ($\mathbf{\Pi}$), transitions ($\mathbf{\Phi}$), and conditional events (**B**, Box 7.5). Like the challenges for multistate models with unobservable states, multievent models face problems of parameter redundancy and the possibility of multiple optima. Multievent models provide a flexible framework for developing custom models to investigate capture heterogenity (Crespin et al. 2008; Péron et al. 2010), memory models for movements (Rouan et al. 2009), or patterns of breeding propensity and mate fidelity (Cubaynes et al. 2011; Sanz-Aguilar et al. 2011; Culina et al. 2013; Desprez et al. 2014).

7.19 Joint Models

Joint models combine live encounter data collected at sampling occasions with auxiliary data on dead recoveries or supplemental resightings taken from intervals between occasions (Burnham 1993; Barker 1997, 1999). Live encounters are usually taken from a fixed-area study plot at set sampling occasions, whereas dead recoveries and supplemental resightings are collated opportunistically from other sites and at any time of year. In Program `Mark`, the encounter histories are coded in a LD format where each type of information is recorded separately for each occasion (Cooch and White 2018). Thus, the code 11 is released alive but recovered dead in the following interval, 12 is released alive and resighted in the following interval, 10 is a live encounter only, 01 is a dead recovery only (not allowed on the first occasion), 02 is an observation only, and 00 is not detected. A key advantage of using information from dead recoveries and supplemental observations is that losses to permanent emigration are reduced if the auxiliary data are taken from a large geographic area (Horton and Letcher 2008). Many mark-recapture models require systematic sampling at discrete occasions with regular intervals to meet the assumption of instantaneous sampling. A second advantage is that auxiliary data can be collected at any time, which may be a better sampling design for field projects where marked individuals are sampled opportunistically at irregular or continuous intervals instead of discrete occasions (Ruiz-Gutiérrez et al. 2012; Barbour et al. 2013; Kendall et al. 2013).

The Burnham model combines live encounter and dead recovery data to estimate four parameters: true survival (*S*), site fidelity (*F*), probability of encounter for live individuals (*p*), and the probability of reporting (*r*) for dead individuals. Similarly, the Barker model combines live encounters, dead recoveries, and auxiliary observations to estimate S, p, and r, and four additional parameters: the probabilities of resighting for individuals that survive or die during an interval (R, R'), and probabilities that an individual in the population is at risk of capture or not (F, F'). Joint models can also be parameterized as multistate (Lebreton et al. 1999; Kendall et al. 2006) or multievent models (Kendall et al. 2013). If dead recoveries are not available, the reporting parameters can be fixed to zero ($r \equiv 0$) and the Barker model can be based on live encounters and auxiliary observations alone (Collins and Doherty 2006; LeDee et al. 2010). If most data are live encounters with relatively few auxiliary observations, then site fidelity and resighting parameters might be modeled as constants without time dependence (Ruiz-Gutiérrez et al. 2012). Extended versions include models that allow for continuous covariates (Bonner 2013), control for marker loss (Conn et al. 2004), and include a robust design framework (Lindberg et al. 2001; Barker et al. 2004; Kendall et al. 2013).

The main challenge for use of joint models is that multiple sources of information must be available for individually marked animals. Burnham and Barker models have been widely used with harvested species where live and dead individuals are routinely encountered (Doherty et al. 2002; Blums et al. 2005; Sandercock 2006), but applications to nongame species are also possible (LeDee et al. 2010; Lok et al. 2013; Cohen et al. 2014). Joint models effectively decompose apparent survival into the component probabilities, thereby providing separate estimates of true survival (S) and site fidelity (F). Thus, joint models have also allowed for variation in the two demographic parameters to be modeled as a function of demographic classes, environmental conditions, and other explanatory factors.

7.20 Integrated Population Models

Integrated population models provide a powerful modeling approach for joint analysis of live encounters with counts or abundance from population surveys, along with other available sources of demographic data, such as productivity data from nest monitoring, dead recoveries or age ratios from harvest monitoring, or known fate survival models (Gauthier et al. 2007; Schaub and Abadi 2011; Zipkin and Saunders 2018). In a traditional population model, each dataset might be analyzed separately with different statistical models, and the estimates of fecundity, survival, and recruitment would then be combined in a matrix population model (Chapter 8). Integrated population models can be more efficient because they allow simultaneous analysis of all available information, explicitly handle uncertainty, and can be used to estimate immigration and other demographic parameters that are not part of the input datasets (Abadi et al. 2010a; Schaub et al. 2013; Chapter 9). Integrated models are also flexible enough to cope with messy features of long-term monitoring programs such as gap years of missing data, datasets that vary in temporal overlap or duration, as well as predictions into the future.

Integrated population models are developed in three steps: defining an age- or stage-structured population model that links population size to demographic rates, defining separate likelihood functions for each available dataset, and creating a joint likelihood for an integrated model as the product of all component likelihoods (Schaub and Abadi 2011). The population model is usually a projection matrix where the matrix elements are comprised of lower-level demographic parameters including fecundity, apparent survival, and immigration. Individual likelihoods can be simple for parameters measured without error such as direct counts of number of young. To control for imperfect detection in survey counts or apparent survival, state-space formulations split the likelihood for each dataset into separate components for the latent state dynamics and the observation process. If input datasets are independent, the joint likelihood for the integrated model is then calculated as the product of the component likelihoods. If input datasets are not independent, simulations suggest parameter estimates may still be robust (Abadi et al. 2010b) or could be tackled with spatially explicit integrated population models (Chandler and Clark 2014). The joint likelihood can be illustrated conceptually as a directed acyclic graph (DAG), with nodes and arcs similar to the life-cycle diagram of a matrix population model (Schaub and Abadi 2011; Chapter 9).

7.21 Frequentist vs. Bayesian Methods

Mark-recapture models can be fit to encounter histories with different modeling approaches, including MLE in a *frequentist* or *information theory framework*, or MCMC methods in a *Bayesian framework*. The methods have different advantages and disadvantages but often converge to identical results given the same set of encounter histories and mark-recapture model. Maximum-likelihood models are usually limited to fixed-effects models but have the advantage that less computation time is needed and a suite of alternative models can be fit to a dataset relatively quickly. Fitting mark-recapture models in a Bayesian framework can be challenging but offers several advantages. Hierarchical models in a Bayesian framework are a good framework for modeling explanatory factors as random instead of fixed effects, and may perform better with sparse data (Gimenez et al. 2007; Calvert et al. 2009). Random-effect models can be used to investigate individual heterogeneity (Cam et al. 2013), calculate the process variance of a demographic parameter (Rotella et al. 2012), examine the functional relationship between two demographic parameters without the confounding effects of sampling variance (Link and Barker 2005; Sedinger et al. 2010), or jointly analyze data from multiple sites or species (Papadatou et al. 2011, 2012; Jansen et al. 2014).

Another advantage of Bayesian models is greater flexibility in developing customized models that may be a better fit to the sampling design of a field study, such as cases where states are unobservable or uncertain (Pradel 2005), or integrated population models that combine live encounter data with population counts and other demographic information (Gauthier et al. 2007; Schaub and Abadi 2011; Zipkin and Saunders 2018). Early integrated population models were analyzed with MLE methods based on complex integrals and process equations that had to be approximated by normal distributions (Gauthier et al. 2007). Integrated population models are now usually fit with MCMC in a Bayesian framework (Schaub and Abadi 2011, Chapter 9). The Bayesian framework requires definition of prior distributions for model parameters but allows use of binomial distributions for probabilities and Poisson or negative binomial distributions for count data, and may be more efficient for sparse datasets (Véran and Lebreton 2008; Schaub et al. 2012).

MLE and Bayesian methods differ in model selection procedures. In MLE methods, model selection procedures can be based on AIC (or AIC*c* or QAIC*c*) or BIC. In Bayesian models, model selection can be based on the Deviance Information Criterion (DIC) where a model with a low DIC value is a better approximation of the underlying biological processes than other models with higher values (Barnett et al. 2010). Unfortunately, model selection based on DIC sometimes works poorly with Bayesian models because of challenges in counting the effective number of parameters in hierarchical models with random effects (Millar 2009; Barker and Link 2015). Hooten and Hobbs (2015) reviewed alternatives for Bayesian model selection, and recommend the Watanabe-Akaike Information Criterion (WAIC) as an alternative for model selection with hierarchical models, Bayes factors for conducting Bayesian model averaging, and model-based methods such as stochastic search variable selection for an integrated approach to model fitting and selection.

In a MLE analysis, parameter estimates can be taken from the minimum AIC model as the most parsimonious model that contains the variables of interest, or by model averaging across the candidate models via multimodel inference. One potential disadvantage of MLE methods is that the 95% confidence intervals (CI) of the parameter estimates are based upon asymptotic assumptions which may be not be appropriate for small datasets. If a parameter is not included in the likelihood, it cannot be modeled but must be calculated as a derived parameter once a model has been fit. For example, abundance is treated as a derived parameter in some closed population models, and is calculated as the number of uniquely marked individuals plus an estimate of the number never caught: $\hat{N} = M_{t+1} + \hat{f}_0$. Derived parameters can also include real parameters that are projected to a different time period or combined in a function. For example, expected longevity might be extrapolated from apparent survival: $\hat{L} = -1/\ln\hat{\phi}$. In an MLE framework, confidence intervals for derived parameters must be calculated analytically with the delta method, or numerically with bootstrapping methods (Powell 2007; Cooch and White 2018).

In a Bayesian model, estimates of variance and 95% credible intervals (CRI) for parameter estimates are taken directly from the posterior distributions after the model has converged. The CRI from the posterior distribution are exact for any arbitrary sample, which may be an advantage if sample sizes are small (Gardner et al. 2010). Moreover, posterior distributions can be calculated for any real or derived parameter, and use of the delta method or bootstrapping are not required. One drawback for Bayesian models is that computational times to reach convergence can be considerably longer than MLE methods, potentially restricting the number of alternative models that can be tested. Implementation of Bayesian models also requires programming expertise with specialist software, but available textbooks with sample code provide a useful starting point (Kéry and Schaub 2012; Royle et al. 2014).

7.22 Software Tools

A growing number of software tools are available for analyses of mark-recapture data for marked individuals with imperfect detection (Mazerolle 2015). Two widely used software packages with a wide range of alternative models and extensive documentation include Programs `MARK` (White and Burnham 1999; Cooch and White 2018), and `E-SURGE` (Choquet et al. 2009a; Choquet and Gimenez 2012). The `U-CARE` software provides tools for assessing GOF tests for CJS, multistate, and other models (Choquet et al. 2009a). The `R` software environment offers a suite of specialized packages for mark-recapture analyses including: the `R2ucare` package for GOF tests (Gimenez et al. 2017), the `RMark` package as an interface to `Mark` (Cooch and White 2018: Appendix C), the marked package for fitting basic mark-recapture models with MLE or Bayesian MCMC methods (Laake et al. 2013), the secr package for spatially explicit capture-recapture models (Efford 2017), and the multimark package for modeling encounters of individuals identified by natural marks (McClintock 2015). The unmarked package offers a range of hierarchical models for unmarked individuals, including distance sampling, occupancy models, and count-based models (Fiske and Chandler 2011). Bayesian mark-recapture models can be run from an R environment with the R2WinBUGS package as an interface to WinBUGS or OpenBUGS, or with the rjags or R2jags packages as an interface to JAGS (Kéry and Schaub 2012; Royle et al. 2014). Many of these software tools are open-source programs available as free downloads, and have online support from dedicated communities, including the phidot forum, or the unmarked forum at Google Groups.

7.23 Online Exercises

Creating the encounter histories for marked individuals is a necessary first step for any mark-recapture analysis. Exercise 1 is an example of an R script for a songbird dataset that shows the steps for converting a vertical file with captures on different occasions into a horizontal encounter history with ones and zeros for detection and nondetection events. Another common input format for mark-recapture analyses is the m-array. Exercise 2 is an R script that can be used to summarize encounter histories into a standard m-array for input to different software packages. Last, Exercise 3 provides an example of a basic CJS analysis for a classic dataset on European Dippers (*Cinclus cinclus*) using the RMark software package as an interface to Program MARK.

7.24 Future Directions

Many questions in population biology, evolutionary ecology, and wildlife management require robust estimates of demographic parameters and their variance. Mark-recapture analyses based on live encounter data are often cost effective, and provide estimates of apparent survival and other parameters that are unbiased with good precision. Fundamental models such as the CJS, Jolly–Seber, and multistate models will remain important as stand-alone tools and as building blocks for more complex models. Mark-recapture methods are an area of active research among quantitative ecologists and major advances continue to be made in five areas. One area has been development of new models that relax the assumptions of standard mark-recapture models, such as multistate models that allow for unobservable or uncertain states. A second area has been integration of live encounter data with other sources of auxiliary data, such as movement data in spatial CJS models, dead recovery data in joint models, or population counts and fecundity data in integrated population models. Third, tests for model fit and corrections for overdispersion are standard procedures for fundamental models like the CJS and multistate models. GOF tests have not yet been developed for multistate models with unobservable states, integrated population models, and other complex models but remain an area of active research. Fourth, continuing development of new software tools with comprehensive documentation has led to widespread adoption of mark-recapture methods, including hierarchical models in a Bayesian framework which provide flexibility for model design. Last, open data and open source software are quickly becoming the new standards for ecological research. Archiving of long-term datasets will facilitate retrospective analyses of existing datasets as new statistical tools become available, and will provide the necessary baseline for understanding future patterns of ecological change. Documentation of software code will allow the next generation of ecologists to continue to use mark-recapture models to tackle the most challenging questions in population biology.

References

Abadi, F., Botha, A., and Altwegg, R. (2013). Revisiting the effect of capture heterogeneity on survival estimates in capture-mark-recapture studies: does it matter? *PLoS One* 8: e62636.

Abadi, F., Gimenez, O., Arlettaz, R. et al. (2010a). An assessment of integrated population models: bias, accuracy, and violation of the assumption of independence. *Ecology* 91: 7–14.

Abadi, F., Gimenez, O., Ullrich, B. et al. (2010b). Estimation of immigration rate using integrated population models. *Journal of Applied Ecology* 47: 393–400.

Aho, K., Derryberry, D., and Peterson, T. (2014). Model selection for ecologists: the worldviews of AIC and BIC. *Ecology* 95: 631–636.

Alonso, R.S., McClintock, B.T., Lyren, L.M. et al. (2015). Mark-recapture and mark-resight methods for estimating abundance with remote cameras: a carnivore case study. *PLoS One* 10: e0123032.

Anderson, K.E., Fujiwara, M., and Rothstein, S.I. (2012). Demography and dispersal of juvenile and adult Brown-headed Cowbirds (*Molothrus ater*) in the eastern Sierra Nevada, California, estimated using multistate models. *Auk* 129: 307–318.

Arnold, T.W. (2010). Uninformative parameters and model selection using Akaike's information criterion. *Journal of Wildlife Management* 74: 1175–1178.

Arnold, T.W., De Sobrino, C.N., and Specht, H.M. (2016). Annual survival rates of migratory shore and upland game birds. *Wildlife Society Bulletin* 40: 470–476.

Bacheler, N.M., Buckel, J.A., Hightower, J.E. et al. (2009). A combined telemetry-tag return approach to estimate fishing and natural mortality rates of an estuarine fish. *Canadian Journal of Fisheries and Aquatic Sciences* 66: 1230–1244.

Bailey, L.L., Converse, S.J., and Kendall, W.L. (2010). Bias, precision, and parameter redundancy in complex multistate models with unobservable states. *Ecology* 91: 1598–1604.

Bailey, L.L., Kendall, W.L., Church, D.R. et al. (2004a). Estimating survival and breeding propensity for pond-breeding amphibians: a modified robust design. *Ecology* 85: 2456–2466.

Bailey, L.L., Kendall, W.L., and Church, D.R. (2009). Exploring extensions to multi-state models with multiple unobservable states. In: *Modeling Demographic Processes in Marked Populations* (eds. D.L. Thomson, E.G. Cooch and M.J. Conroy), 693–709. New York, NY: Springer.

Bailey, L.L., Simons, T.R., and Pollock, K.H. (2004b). Estimating detection probability parameters for *Plethodon* salamanders using the robust capture-recapture design. *Journal of Wildlife Management* 68: 1–13.

Baker, M., Nur, N., and Geupel, G.R. (1995). Correcting biased estimates of dispersal and survival due to limited study area: theory and an application using Wrentits. *Condor* 97: 663–674.

Barbour, A.B., Ponciana, J.M., and Lorenzen, K. (2013). Apparent survival estimation from continuous mark-recapture/resighting data. *Methods in Ecology and Evolution* 4: 846–853.

Barker, R.J. (1997). Joint modeling of live-recapture, tag-resight, and tag-recovery data. *Biometrics* 53: 666–677.

Barker, R.J. (1999). Joint analysis of mark-recapture, resighting and ring-recovery data with age-dependence and marking-effect. *Bird Study* 46 (Suppl): S82–S91.

Barker, R.J., Burnham, K.P., and White, G.C. (2004). Encounter history modeling of joint mark-recapture, tag-resighting and tag-recovery data under temporary emigration. *Statistica Sinica* 14: 1037–1055.

Barker, R.J. and Link, W.A. (2015). Truth, models, model sets, AIC, and multimodel inference: a Bayesian perspective. *Journal of Wildlife Management* 79: 730–738.

Barnett, D.G., Koper, N., Dobson, A.J. et al. (2010). Using information criteria to select the correct variance-covariance structure for longitudinal data in ecology. *Methods in Ecology and Evolution* 1: 15–24.

Bendik, N.F., Morrison, T.A., Gluesenkamp, A.G. et al. (2013). Computer-assisted photo identification outperforms visible implant elastomers in an endangered salamander, *Euryacea tonkawae*. *PLoS One* 8: E59424.

Besbeas, P. and Morgan, B.J.T. (2014). Goodness-of-fit of integrated population models using calibrated simulation. *Methods in Ecology and Evolution* 5: 1373–1382.

Besnard, A., Novoa, C., and Gimenez, O. (2010). Hunting impact on the population dynamics of Pyrenean Grey Partridge *Perdix perdix hispaniensis*. *Wildlife Biology* 16: 135–143.

Besnard, A., Piry, S., Berthier, K. et al. (2007). Modeling survival and mark loss in molting animals: recapture, dead recoveries, and exuvia recoveries. *Ecology* 88: 289–295.

Bird, T., Lyon, J., Nicol, S. et al. (2014). Estimating population size in the presence of temporary migration using a joint analysis of telemetry and capture-recapture data. *Methods in Ecology and Evolution* 5: 615–625.

Bischof, R., Swenson, J.E., Yoccuz, N.G. et al. (2009). The magnitude and selectivity of natural and multiple anthropogenic mortality causes in hunted brown bears. *Journal of Animal Ecology* 78: 656–665.

Blums, P., Nichols, J.D., Hines, J.E. et al. (2005). Individual quality, survival variation and patterns of phenotypic selection on body condition and timing of nesting in birds. *Oecologia* 143: 365–376.

Bonner, S.J. (2013). Implementing the trinomial mark-recapture-recovery model in Program `Mark`. *Methods in Ecology and Evolution* 4: 95–98.

Borchers, D. (2012). A non-technical overview of spatially explicit capture-recapture models. *Journal of Ornithology* 152 (Suppl): S435–S444.

Boulanger, J., Cattet, M., Nielsen, S.E. et al. (2013). Use of multi-state models to explore relationships between changes in body condition, habitat and survival of grizzly bears *Ursus arctos horribilis*. *Wildlife Biology* 19: 274–288.

Boulanger, J. and McLellan, B. (2001). Closure violation in DNA-based mark-recapture estimation of grizzly bear populations. *Canadian Journal of Zoology* 79: 642–651.

Boulanger, J., McLellan, B.N., Woods, J.G. et al. (2004). Sampling design and bias in DNA-based capture-mark-recapture population and density estimates of grizzly bears. *Journal of Wildlife Management* 68: 457–469.

Bowerman, T. and Budy, P. (2012). Incorporating movement patterns to improve survival estimates for juvenile bull trout. *North American Journal of Fisheries Management* 32: 1123–1136.

Breton, A.R., Diamond, A.W., and Kress, S.K. (2006). Encounter, survival, and movement probabilities from an Atlantic Puffin (*Fratercula arctica*) metapopulation. *Ecological Monographs* 76: 133–149.

Brown, C.R., Covas, R., Anderson, M.D. et al. (2003). Multistate estimates of survival and movement in relation to colony size in the Sociable Weaver. *Behavioral Ecology* 14: 463–471.

Brownie, C., Hines, J.E., Nichols, J.D. et al. (1993). Capture-recapture studies for multiple strata including non-Markovian transitions. *Biometrics* 49: 1173–1187.

Brudney, L.J., Arnold, T.W., Saunders, S.P. et al. (2013). Survival of Piping Plover (*Charadrius melodus*) chicks in the Great Plains region. *Auk* 130: 150–160.

Burnham, K.P. (1993). A theory for the combined analysis of ringed recovery and recapture data in marked individuals. In: *Marked Individuals in the Study of Bird Populations* (eds. J.-D. Lebreton and P.M. North), 199–213. Basel, Switzerland: Birkhauser Verlag.

Burnham, K.P. and Anderson, D.R. (2002). *Model selection and multimodel inference: a practical information-theoretic approach.* New York, USA: Springer Science.

Burnham, K.P. and Anderson, D.R. (2004). Multimodel inference – understanding AIC and BIC in model selection. *Sociological Methods and Research* 33: 261–304.

Burnham, K.P., Anderson, D.R., and Huyvaert, K.P. (2011). AIC model selection and multimodel inference in behavioral ecology: some background, observations, and comparisons. *Behavioral Ecology and Sociobiology* 65: 23–35.

Cade, B.S. (2015). Model averaging and muddled multimodel inferences. *Ecology* 96: 2370–2382.

Calvert, A.M., Bonner, S.J., Jonsen, I.D. et al. (2009). A hierarchical Bayesian approach to multi-state mark-recapture: simulations and applications. *Journal of Applied Ecology* 46: 610–620.

Cam, E. (2012). "Each site has its own survival probability, but information is borrowed across sites to tell us about survival in each site": random effects models as means of borrowing strength in survival studies of wild vertebrates. *Animal Conservation* 15: 129–132.

Cam, E., Gimenez, O., Alpizar-Jara, R. et al. (2013). Looking for a needle in a haystack: inference about individual components in a heterogeneous population. *Oikos* 122: 739–753.

Cantor, M., Wedekin, L.L., Daura-Jorge, F.G. et al. (2012). Assessing population parameters and trends of Guiana dolphins (*Sotalia guianensis*): an eight-year mark-recapture study. *Marine Mammal Science* 28: 63–83.

Chandler, R.B. and Clark, J.D. (2014). Spatially explicit integrated population models. *Methods in Ecology and Evolution* 5: 1351–1360.

Choquet, R. and Gimenez, O. (2012). Towards built-in capture-recapture mixed models in program `E-SURGE`. *Journal of Ornithology* 152: 625–639.

Choquet, R., Lebreton, J.D., Gimenez, O. et al. (2009a). `U-CARE`: utilities for performing goodness of fit tests and manipulating capture recapture data. *Ecography* 32: 1071–1074.

Choquet, R., Rouan, L., and Pradel, R. (2009b). Program `E-SURGE`: a software application for fitting multievent models. In: *Modeling Demographic Processes in Marked Populations* (eds. D.L. Thomson, E.G. Cooch and M.J. Conroy), 845–865. New York, NY: Springer.

Church, D.R., Bailey, L.L., Wilbur, H.M. et al. (2007). Iteroparity in the variable environment of the salamander *Ambystoma tigrinum*. *Ecology* 88: 891–903.

Cilimburg, A.B., Lindberg, M.S., Tewksbury, J.J. et al. (2002). Effects of dispersal on survival probability of adult Yellow Warblers (*Dendroica petechia*). *Auk* 119: 778–789.

Cohas, A., Bonenfant, C., Gaillard, J.M. et al. (2007). Are extra-pair young better than within-pair young? A comparison of survival and dominance in alpine marmot. *Journal of Animal Ecology* 76: 771–781.

Cohen, E.B., Hostetler, J.A., Royle, J.A. et al. (2014). Estimating migratory connectivity of birds when re-encounter probabilities are heterogeneous. *Ecology and Evolution* 4: 1659–1670.

Cole, D.J. (2012). Determining parameter redundancy of multi-state mark-recapture models for sea birds. *Journal of Ornithology* 152 (Suppl 2): S305–S315.

Collier, B.A., Kremer, S.R., Mason, C.D. et al. (2013). Immigration and recruitment in an urban White-winged Dove breeding colony. *Journal of Fish and Wildlife Management* 4: 33–40.

Collins, C.T. and Doherty, P.F. Jr. (2006). Survival estimates for Royal Terns in southern California. *Journal of Field Ornithology* 77: 310–314.

Conn, P.B. and Cooch, E.G. (2009). Multistate capture-recapture analysis under imperfect state observation: an application to disease models. *Journal of Applied Ecology* 46: 486–492.

Conn, P.B., Doherty, P.F. Jr., and Nichols, J.D. (2005). Comparative demography of New World populations of thrushes (*Turdus*): comment. *Ecology* 86: 2536–2541.

Conn, P.B., Kendall, W.L., and Samuel, M.D. (2004). A general model for the analysis of mark-resight, mark-recapture, and band-recovery data under tag loss. *Biometrics* 60: 900–909.

Conroy, M.J., Senar, J.C., Hines, J.E. et al. (1999). Development and application of a mark-recapture model incorporating predicted sex and transitory behaviour. *Bird Study* 46 (Suppl): S62–S73.

Converse, S.J., Kendall, W.L., Doherty, J.R. et al. (2009). Multistate models for estimation of survival and reproduction in the Grey-headed Albatross (*Thalassarche chrysostoma*). *Auk* 126: 77–88.

Cooch, E.G., White, G.C. (2018) Program `MARK`: A gentle introduction, 18th Edition. http://www.phidot.org/software/mark/docs/book

Cooch, E.G., Conn, P.B., Ellner, S.P. et al. (2012). Disease dynamics in wild populations: modeling and estimation: a review. *Journal of Ornithology* 152 (Suppl 2): S485–S509.

Cooper, C.B., Daniels, S.J., and Walters, J.R. (2008). Can we improve estimates of juvenile dispersal distance and survival? *Ecology* 89: 3349–3361.

Crespin, L., Choquet, R., Lima, M. et al. (2008). Is heterogeneity of catchability in capture-recapture studies a mere sampling artifact or a biologically relevant feature of the population? *Population Ecology* 50: 247–256.

Cubaynes, S., Doherty, P.F. Jr., Schreiber, E.A. et al. (2011). To breed or not to breed: a seabird's response to extreme climatic events. *Biology Letters* 7: 303–306.

Cubaynes, S., Kavergne, C., Marboutin, E. et al. (2012). Assessing individual heterogeneity using model selection criteria: how many mixture components in capture-recapture models? *Methods in Ecology and Evolution* 3: 564–573.

Culina, A., Lachish, S., Pradel, R. et al. (2013). A multievent approach to estimating pair fidelity and heterogeneity in state transitions. *Ecology and Evolution* 3: 4326–4338.

Cunningham, M.A. (1986). Dispersal in White-crowned Sparrows: a computer simulation of the effect of study-area size on estimates of local recruitment. *Auk* 103: 79–85.

Currey, R.J.C., Dawson, S.M., Schneider, K. et al. (2011). Inferring causal factors for a declining population of bottlenose dolphins via temporal symmetry capture-recapture modeling. *Marine Mammal Science* 27: 554–566.

Dail, D. and Madsen, L. (2013). Estimating open population site occupancy from presence–absence data lacking the robust design. *Biometrics* 69: 146–156.

Desprez, M., Harcourt, R., Hindell, M.A. et al. (2014). Age-specific cost of first reproduction in female southern elephant seals. *Biology Letters* 10: e20140264.

Devineau, O., Kendall, W.L., Doherty, P.F. Jr. et al. (2014). Increased flexibility for modeling telemetry and nest-survival data using the multistate framework. *Journal of Wildlife Management* 78: 224–230.

Devineau, O., Shenk, T.M., White, G.C. et al. (2010). Evaluating the Canada lynx reintroduction programme in Colorado: patterns in mortality. *Journal of Applied Ecology* 47: 524–531.

Dinsmore, S.J., White, G.C., and Knopf, F.L. (2003). Annual survival and population estimates of Mountain Plovers in southern Phillips County, Montana. *Ecological Applications* 13: 1013–1026.

Doherty, P.F. Jr., Nichols, J.D., Tautin, J. et al. (2002). Sources of variation in breeding-ground fidelity of Mallards (*Anas platyrhynchos*). *Behavioral Ecology* 13: 543–550.

Doherty, P.F. Jr., Schreiber, E.A., Nichols, J.D. et al. (2004). Testing life history predictions in a long-lived seabird: a population matrix approach with improved parameter estimation. *Oikos* 105: 606–618.

Doherty, P.F., White, G.C., and Burnham, K.P. (2012). Comparison of model building and selection strategies. *Journal of Ornithology* 152 (Suppl 2): S317–S323.

Doligez, B., Clobert, J., Pettifor, R.A. et al. (2002). Costs of reproduction: assessing responses to brood size manipulation on life-history and behavioral traits using multi-state capture-recapture models. *Journal of Applied Statistics* 29: 407–423.

Dorazio, R.M., Royle, J.A., Söderstron, B. et al. (2006). Estimating species richness and accumulation by modeling species occurrence and detectability. *Ecology* 87: 842–854.

Dreitz, V.J. (2009). Parental behaviour of a precocial species: implications for juvenile survival. *Journal of Applied Ecology* 46: 870–878.

Dreitz, V.J., Nichols, J.D., Hines, J.E. et al. (2002). The use of resighting data to estimate the rate of population growth of the Snail Kite in Florida. *Journal of Applied Statistics* 29: 609–623.

Efford, M. (2004). Density estimation in live-trapping studies. *Oikos* 106: 598–610.

Efford, M.G. (2011). Estimation of population density by spatially explicit capture-recapture analysis of data from area searches. *Ecology* 92: 2202–2207.

Efford, M.G. (2017). `secr`: Spatially explicit capture-recapture models. R package version 3.1.3. CRAN.R-project.org/package=secr

Ergon, T. and Gardner, B. (2014). Separating mortality and emigration: modelling space use, dispersal and survival with robust-design spatial capture-recapture data. *Methods in Ecology and Evolution* 5: 1327–1336.

Faustino, C.R., Jennelle, C.S., Connolly, V. et al. (2004). *Mycoplasma gallisepticum* infection dynamics in a House Finch population: seasonal variation in survival, encounter and transmission rate. *Journal of Animal Ecology* 73: 651–669.

Fiske, I. and Chandler, R. (2011). `unmarked`: an `R` package for fitting hierarchical models of wildlife occurrence and abundance. *Journal of Statistical Software* 43: 1–23.

Fredericksen, M., Lebreton, J.-D., Pradel, R. et al. (2014). Identifying links between vital rates and environment: a toolbox for the applied ecologist. *Journal of Applied Ecology* 51: 71–81.

Fujiwara, M., Anderson, K.E., Neubert, M.G. et al. (2006). On the estimation of dispersal kernels from individual mark-recapture data. *Environmental and Ecological Statistics* 13: 183–197.

Fujiwara, M. and Caswell, H. (2002). A general approach to temporary emigration in mark-recapture analysis. *Ecology* 83: 3266–3275.

Gaillard, J.M. and Yoccoz, N.G. (2003). Temporal variation in survival of mammals: a case of environmental canalization? *Ecology* 84: 3294–3306.

Galipaud, M., Gillingham, M.A.F., David, M. et al. (2014). Ecologists overestimate the importance of predictor variables in model averaging: a plea for cautious interpretations. *Methods in Ecology and Evolution* 5: 983–991.

Gardner, B., Reppucci, J., Lucherini, M. et al. (2010). Spatially explict inference for open populations: estimating demographic parameters from camera-trap studies. *Ecology* 91: 3376–3383.

Gauthier, G., Besbeas, P., Lebreton, J.-D. et al. (2007). Population growth in Snow Geese: a modeling approach integrating demographic and survey information. *Ecology* 88: 1420–1429.

Gauthier, G. and Lebreton, J.-D. (2008). Analysis of band-recovery data in a multistate capture-recapture framework. *Canadian Journal of Statistics* 36: 59–73.

Genton, C., Pierre, A., Cristescu, R. et al. (2015). How Ebola impacts social dynamics in gorillas: a multistate modeling approach. *Journal of Animal Ecology* 84: 166–176.

Gilroy, J.J., Virzi, T., Boulton, R.L. et al. (2012). A new approach to the "apparent survival" problem: estimating true survival rates from mark-recapture studies. *Ecology* 93: 1509–1516.

Gimenez, O., Choquet, R., and Lebreton, J.-D. (2003). Parameter redundancy in multistate capture-recapture models. *Biometrical Journal* 45: 704–722.

Gimenez, O., Covas, R., Brown, C.R. et al. (2006). Nonparametric estimation of natural selection on a quantitative trait using mark-recapture data. *Evolution* 60: 460–466.

Gimenez, O, Lebreton, J-D, Choquet, R et al. (2017) `R2ucare`: Goodness-of-fit tests for capture-recapture models. R package version 1.0.0. CRAN.R-project.org/package=R2ucare

Gimenez, O., Rossi, V., Choquet, R. et al. (2007). State-space modelling of data on marked individuals. *Ecological Modelling* 206: 431–438.

Gould, W.R. and Nichols, J.D. (1998). Estimation of temporal variability of survival in animal populations. *Ecology* 79: 2531–2538.

Grayson, K.L., Bailey, L.L., and Wilbur, H.M. (2011). Life history benefits of residency in a partially migrating pond-breeding amphibian. *Ecology* 92: 1239–1246.

Grosbois, V., Gimenez, O., Gaillard, J.-M. et al. (2008). Assessing the impact of climate variation on survival in vertebrate populations. *Biological Reviews* 83: 357–399.

Grosbois, V., Harris, M.P., Anker-Nilssen, T. et al. (2009). Modeling survival at multi-population scales using mark-recapture data. *Ecology* 90: 2922–2932.

Grosbois, V. and Tavecchia, G. (2003). Modeling dispersal with capture-recapture data: disentangling decisions of leaving and settlement. *Ecology* 84: 1225–1236.

Grueber, C.E., Nakagawa, S., Laws, R.J. et al. (2011). Multimodel inference in ecology and evolution: challenges and solutions. *Journal of Evolutionary Biology* 24: 699–711.

Henle, K. and Gruber, B. (2018). Performance of multistate mark-recapture models for temporary emigration in the presence of survival costs. *Methods in Ecology and Evolution* 9: 657–667.

Hernández-Matías, A., Real, J., and Pradel, R. (2011). Quick methods for evaluating survival of age-characterizable long-lived territorial birds. *Journal of Wildlife Management* 75: 856–866.

Hines, J.E. and Nichols, J.D. (2002). Investigations of potential bias in the estimation of λ using Pradel's (1996) model for capture-recapture data. *Journal of Applied Statistics* 29: 573–587.

Hooten, M.B. and Hobbs, N.T. (2015). A guide to Bayesian model selection for ecologists. *Ecological Monographs* 85: 3–28.

Horton, G.E. and Letcher, B.H. (2008). Movement patterns and study area boundaries: influences on survival estimation in capture-mark-recapture studies. *Oikos* 117: 1131–1142.

Horton, G.E., Letcher, B.H., and Kendall, W.L. (2011). A multistate capture-recapture modeling strategy to separate true survival from permanent emigration for a passive integrated transponder tagged population of stream fish. *Transactions of the American Fisheries Society* 140: 320–333.

Hubbard, B.A., Cole, D.J., and Morgan, B.J.T. (2014). Parameter redundancy in capture-recapture-recovery models. *Statistical Methodology* 17: 17–29.

Hunter, C.M. and Caswell, H. (2009). Rank and redundancy of multistate mark-recapture models for seabird populations with unobservable states. In: *Modeling Demographic Processes in Marked Populations* (eds. D.L. Thomson, E.G. Cooch and M.J. Conroy), 797–825. New York, NY: Springer.

Jahn, A.E., Levey, D.J., Hostetler, J.A. et al. (2010). Determinants of partial bird migration in the Amazon Basin. *Journal of Animal Ecology* 79: 983–992.

Jansen, D.Y.M., Abadi, F., Harebottle, D. et al. (2014). Does seasonality drive spatial patterns in demography? Variation in survival in African Reed Warblers *Acrocephalus baeticatus* across southern Africa does not reflect global patterns. *Ecology and Evolution* 47: 889–898.

Jessopp, M.J., Forcada, J., Reid, K. et al. (2004). Winter dispersal of leopard seals (*Hydruga leptonyx*): environmental factors influencing demographics and seasonal abundance. *Journal of Zoology* 263: 251–258.

Johnston, J.P., Peach, W.J., Gregory, R.D. et al. (1997). Survival rates of tropical and temperate passerines: a Trinidadian perspective. *American Naturalist* 150: 771–789.

Juillet, C., Choquet, R., Gauthier, G. et al. (2011). A capture–recapture model with double-marking, live and dead recoveries, and heterogeneity of reporting due to auxiliary mark loss. *Journal of Agricultural, Biological, and Environmental Statistics* 16: 88–104.

Kellner, K.F. and Swihart, R.K. (2014). Accounting for imperfect detection in ecology: a quantitative review. *PLoS One* 9: e111436.

Kendall, W.L. (1999). Robustness of closed capture-recapture methods to violations of the closure assumption. *Ecology* 80: 2517–2525.

Kendall, W.L. (2004). Coping with unobservable and mis-classified states in capture-recapture studies. *Animal Biodiversity and Conservation* 27: 97–107.

Kendall, W.L., Barker, R.J., White, G.C. et al. (2013). Combining dead recovery, auxiliary observations and robust design data to estimate demographic parameters from marked individuals. *Methods in Ecology and Evolution* 4: 828–835.

Kendall, W.L. and Bjorkland, R. (2001). Using open robust design models to estimate temporary emigration from capture-recapture data. *Biometrics* 57: 1113–1122.

Kendall, W.L., Conn, P.B., and Hines, J.E. (2006). Combining multistate capture-recapture data with tag recoveries to estimate demographic parameters. *Ecology* 87: 169–177.

Kendall, W.L., Converse, S.J., Doherty, P.F. Jr. et al. (2009). Sampling design considerations for demographic studies: a case of colonial seabirds. *Ecological Applications* 19: 55–68.

Kendall, W.L., Hines, J.E., and Nichols, J.D. (2003). Adjusting multistate capture-recapture models for misclassification bias: manatee breeding proportions. *Ecology* 84: 1058–1066.

Kendall, W.L. and Nichols, J.D. (1995). On the use of secondary capture-recapture samples to estimate temporary emigration and breeding proportions. *Journal of Applied Statistics* 22: 751–762.

Kendall, W.L. and Nichols, J.D. (2002). Estimating state-transition probabilities for unobservable states using capture-recapture/resighting data. *Ecology* 83: 3276–3284.

Kendall, W.L., Nichols, J.D., and Hines, J.E. (1997). Estimating temporary emigration using capture-recapture data with Pollock's robust design. *Ecology* 78: 563–578.

Kéry, M., Gregg, K.B., and Schaub, M. (2005). Demographic estimation methods for plants with unobservable life-states. *Oikos* 108: 307–320.

Kéry, M. and Schaub, M. (2012). *Bayesian Population Analysis Using* `WinBUGS`*: A Hierarchical Perspective*. Oxford, UK: Academic Press.

Koenig, W.D., Van Vuren, D., and Hooge, P.N. (1996). Detectability, philopatry, and the distribution of dispersal distances in vertebrates. *Trends in Ecology and Evolution* 11: 514–517.

Korfanta, N.M., Newmark, W.D., and Kauffman, M.J. (2012). Long-term demographic consequences of habitat fragmentation to a tropical understory bird community. *Ecology* 93: 2548–2559.

Laake, J.L., Johnson, D.S., and Conn, P.B. (2013). `marked`: an `R` package for maximum likelihood and Markov chain Monte Carlo analysis of capture–recapture data. *Methods in Ecology and Evolution* 4: 885–890.

Lachish, S., Jones, M., and McCallum, H. (2007). The impact of disease on the survival and population growth rate of the Tasmanian Devil. *Journal of Animal Ecology* 76: 926–936.

Lagrange, P., Pradel, R., Bélisle, M. et al. (2014). Estimating dispersal among numerous sites using capture-recapture data. *Ecology* 95: 2316–2323.

Lahoz-Monfort, J.J., Morgan, B.J.T., Harris, M.P. et al. (2011). A capture-recapture model for exploring multi-species synchrony in survival. *Methods in Ecology and Evolution* 2: 116–124.

Lebreton, J.-D., Almeras, T., and Pradel, R. (1999). Competing events, mixtures of information and multistratum recapture models. *Bird Study* 46 (Suppl): S39–S46.

Lebreton, J.-D., Burnham, K.P., Clobert, J. et al. (1992). Modeling survival and testing biological hypotheses using marked animals: a unified approach with case studies. *Ecological Monographs* 62: 67–118.

Lebreton, J.-D., Nichols, J.D., Barker, R.J. et al. (2009). Modeling individual animal histories with multistate capture-recapture models. *Advances in Ecological Research* 41: 87–173.

Lebreton, J.-D. and Pradel, R. (2002). Multistate recapture models: modelling incomplete individual histories. *Journal of Applied Statistics* 29: 353–369.

LeDee, O.E., Arnold, T.W., Roche, E.A. et al. (2010). Use of breeding and nonbreeding encounters to estimate survival and breeding-site fidelity of the Piping Plover at the Great Lakes. *Condor* 112: 637–643.

Lee, K.A., Huveneers, C., Gimenez, O. et al. (2014). To catch or to sight? A comparison of demographic parameter estimates obtained from mark-recapture and mark-resight models. *Biodiversity and Conservation* 23: 2781–2800.

Letcher, B.H. and Horton, G.E. (2008). Seasonal survival in size-dependent survival of juvenile Atlantic salmon (*Salmo salar*): performance of multistate capture-mark-recapture models. *Canadian Journal of Fisheries and Aquatic Sciences* 65: 1649–1666.

Lindberg, M.S. (2012). A review of designs for capture-mark-recapture studies in discrete time. *Journal of Ornithology* 152 (Suppl 2): S355–S370.

Lindberg, M.S., Kendall, W.L., Hines, J.E. et al. (2001). Combining band recovery data and Pollock's robust design to model temporary and permanent emigration. *Biometrics* 57: 273–281.

Lindberg, M.S., Schmidt, J.H., and Walker, J. (2015). History of multimodel inference via model selection in wildlife science. *Journal of Wildlife Management* 79: 704–707.

Link, W.A. and Barker, R.J. (2005). Modeling association among demographic parameters in analysis of open population capture-recapture data. *Biometrics* 61: 46–51.

Lloyd, P., Abadi, F., Altwegg, R. et al. (2014). South temperate birds have higher apparent adult survival than tropical birds in Africa. *Journal of Avian Biology* 45: 493–500.

Lok, T., Overdijk, O., Tinbergen, J.M. et al. (2013). Seasonal variation in density dependence in age-specific survival of a long-distance migrant. *Ecology* 94: 2358–2369.

Low, M., Arlt, D., Eggers, S. et al. (2010). Habitat-specific differences in adult survival rates and its links to parental workload and on-nest predation. *Journal of Animal Ecology* 79: 214–224.

Lukacs, P.M. and Burnham, K.P. (2005). Review of capture-recapture methods applicable to noninvasive genetic sampling. *Molecular Ecology* 14: 3909–3919.

Lukacs, P.M., Dreitz, V.J., Knopf, F.L. et al. (2004). Estimating survival probabilities of unmarked dependent young when detection is imperfect. *Condor* 106: 926–931.

Lyet, A., Cheylan, M., Prodon, R. et al. (2009). Prescribed fire and conservation of a threatened mountain grassland specialist: a capture-recapture study on the Orsini's viper in the French Alps. *Animal Conservation* 12: 238–248.

Lyons, J.E., Kendall, W.L., Royle, J.A. et al. (2016). Population size and stopover duration estimation using mark-resight data and Bayesian analysis of a superpopulation model. *Biometrics* 72: 262–271.

Madon, B., Gimenez, O., McArdle, B. et al. (2011). A new method for estimating animal abundance with two sources of data in capture-recapture studies. *Methods in Ecology and Evolution* 2: 390–400.

Marescot, L., Pradel, R., Duchamp, C. et al. (2011). Capture-recapture population growth rate as a robust tool against detection heterogeneity for population management. *Ecological Applications* 21: 2898–2907.

Marshall, M.R., Diefenbach, D., Wood, L.A. et al. (2004). Annual survival estimation of migratory songbirds confounded by incomplete breeding site-fidelity: study designs that may help. *Animal Biodiversity and Conservation* 27: 59–72.

Marzolin, G., Charmantier, A., and Gimenez, O. (2011). Frailty in state-space models: application to actuarial senescence in the Dipper. *Ecology* 92: 562–567.

Mazerolle, M.J. (2015). Estimating detectability and biological parameters of interest with the use of the R environment. *Journal of Herpetology* 49: 541–559.

McClintock, B.T. (2015). multimark: an R package for analysis of capture–recapture data consisting of multiple "noninvasive" marks. *Ecology and Evolution* 5: 4920–4931.

McClintock, B.T., Hill, J.M., Fritz, L. et al. (2014). Mark-resight abundance estimation under incomplete identification of marked individuals. *Methods in Ecology and Evolution* 5: 1294–1304.

McClintock, B.T. and White, G.C. (2009). A less field-intensive robust design for estimating demographic parameters with mark-resight data. *Ecology* 90: 313–320.

McClintock, B.T. and White, G.C. (2012). From NOREMARK to MARK: software for estimating demographic parameters using mark-resight methodology. *Journal of Ornithology* 152 (Suppl 2): S641–S650.

McCrea, R.S., Morgan, B.J.T., and Bregnaballe, T. (2012). Model comparison and assessment for multi-state capture-recapture-recovery models. *Journal of Ornithology* 152 (Suppl 2): S293–S303.

McKim-Louder, M., Hoover, J.P., Benson, T.J. et al. (2013). Juvenile survival in a neotropical migratory songbird is lower than expected. *PLoS One* 8: e56059.

Millar, R.B. (2009). Comparison of hierarchical Bayesian models for overdispersed count data using DIC and Bayes' factors. *Biometrics* 65: 962–969.

Miller, M.W., Aradis, A., and Landucci, G. (2003). Effects of fat reserves on annual apparent survival of Blackbirds *Turdus merula*. *Journal of Animal Ecology* 72: 127–132.

Monk, M.H., Berkson, J., and Rivalan, P. (2011). Estimating demographic parameters for loggerhead turtles using mark-recapture data and a multistate model. *Population Ecology* 53: 165–174.

Monterroso, P., Rich, L.N., Serronha, A. et al. (2014). Efficiency of hair snares and camera traps to survey mesocarnivore populations. *European Journal of Wildlife Research* 60: 279–289.

Monticelli, D., Araujo, P.M., Hines, J.E. et al. (2014). Assessing the role of body mass and sex on apparent adult survival in polygynous passerines: a case study of Cetti's Warblers in Central Portugal. *Journal of Avian Biology* 45: 75–84.

Morris, W.F. and Doak, D.F. (2004). Buffering of life histories against environmental stochasticity: accounting for a spurious correlation between the variabilities of vital rates and their contributions to fitness. *American Naturalist* 163: 579–590.

Murray, D.L. (2006). On improving telemetry-based survival estimation. *Journal of Wildlife Management* 70: 1530–1543.

Muths, E., Scherer, R.D., and Bosch, J. (2013). Evidence for plasticity in the frequency of skipped breeding opportunities in common toads. *Population Ecology* 55: 535–544.

Nichols, J.D. (2016). And the first one now will later be last: time-reversal in Cormack–Jolly–Seber models. *Statistical Science* 31: 175–190.

Nichols, J.D. and Hines, J.E. (2002). Approaches for the direct estimation of lambda, and demographic

contributions to lambda, using capture-recapture data. *Journal of Applied Statistics* 29: 539–568.
Nichols, J.D., Hines, J.E., Lebreton, J.-D. et al. (2000). Estimation of contributions to population growth: a reverse-time capture-recapture approach. *Ecology* 81: 3362–3376.
Nichols, J.D., Hines, J.E., Pollock, K.H. et al. (1994). Estimating breeding proportions and testing hypotheses about costs of reproduction with capture-recapture data. *Ecology* 75: 2052–2065.
Nichols, J.D. and Kendall, W.L. (1995). The use of multi-state capture-recapture models to address questions in evolutionary ecology. *Journal of Applied Statistics* 22: 835–846.
Nichols, J.D., Kendall, W.L., Hines, J.E. et al. (2004). Estimation of sex-specific survival from capture-recapture data when sex is not always known. *Ecology* 85: 3192–3201.
Nichols, J.D. and Pollock, K.H. (1990). Estimation of recruitment from immigration versus in situ reproduction using Pollock's robust design. *Ecology* 71: 21–26.
O'Brien, S., Robert, B., and Tiandry, H. (2005). Consequences of violating the recapture duration assumption of mark–recapture models: a test using simulated and empirical data from an endangered tortoise population. *Journal of Applied Ecology* 42: 1096–1104.
O'Connell, A.F. Jr., Talancy, N.W., Bailey, L.L. et al. (2006). Estimating site occupancy and detection probability parameters for meso- and large mammals in a coastal ecosystem. *Journal of Wildlife Management* 70: 1625–1633.
Oli, M.K. and Dobson, F.S. (2003). The relative importance of life-history variables to population growth rate in mammals: Cole's prediction revisited. *American Naturalist* 161: 422–440.
Otis, D.L., Burnham, K.P., White, G.C. et al. (1978). Statistical inference from capture data on closed animal populations. *Ecological Monographs* 62: 3–135.
Ozgul, A., Oli, M.K., Bolker, B.M. et al. (2009). Upper respiratory tract disease, force of infection, and effects on survival of gopher tortoises. *Ecological Applications* 19: 786–798.
Papadatou, E., Ibáñez, C., Pradel, R. et al. (2011). Assessing survival in a multi-population system: a case study on bat populations. *Oecologia* 165: 925–933.
Papadatou, E., Pradel, R., Schaub, M. et al. (2012). Comparing survival among species with imperfect detection using multilevel analysis of mark-recapture data: a case study on bats. *Ecography* 35: 153–161.
Paradis, E., Baillie, S.R., Sutherland, W.J. et al. (1998). Patterns of natal and breeding dispersal in birds. *Journal of Animal Ecology* 67: 518–536.
Pellet, J. and Schmidt, B.R. (2005). Monitoring distributions using call surveys: estimating site occupancy, detection probabilities and inferring absence. *Biological Conservation* 123: 27–35.
Peñaloza, C.L., Kendall, W.L., and Langtimm, C.A. (2014). Reducing bias in survival under nonrandom temporary emigration. *Ecological Applications* 24: 1155–1166.
Péron, G., Crochet, P.-A., Choquet, R. et al. (2010). Capture–recapture models with heterogeneity to study survival senescence in the wild. *Oikos* 119: 524–532.
Péron, G., Gaillard, J.-M., and Barbraud, C. (2016). Evidence of reduced heterogeneity in adult survival of long-lived species. *Evolution* 70: 2909–2914.
Pledger, S., Baker, E., and Scribner, K. (2013). Breeding return times and abundance in capture-recapture models. *Biometrics* 69: 991–1001.
Pollock, K.H., Nichols, J.D., Brownie, C. et al. (1990). Statistical inference for capture-recapture experiments. *Wildlife Monographs* 107: 1–97.
Powell, L.A. (2007). Approximating variance of demographic parameters using the delta method: a reference for avian biologists. *Condor* 109: 949–954.
Powell, L.A., Conroy, M.J., Hines, J.E. et al. (2000). Simultaneous use of mark-recapture and radiotelemetry to estimate survival, movement, and capture rates. *Journal of Wildlife Management* 64: 302–313.
Pradel, R. (1996). Utilization of capture-mark-recapture for the study of recruitment and population growth rate. *Biometrics* 52: 703–709.
Pradel, R. (2005). `Multievent`: an extension of multistate capture-recapture models to uncertain states. *Biometrics* 61: 442–447.
Pradel, R. (2009). The stakes of capture-recapture models with state uncertainty. In: *Modeling Demographic Processes in Marked Populations* (eds. D.L. Thomson, E.G. Cooch and M.J. Conroy), 781–795. New York, NY: Springer.
Pradel, R., Gimenez, O., and Lebreton, J.-D. (2005). Principles and interest of GOF tests for multistate capture-recapture models. *Animal Biodiversity and Conservation* 28: 189–204.
Pradel, R. and Henry, P.-Y. (2007). Potential contributions of capture-recapture to the estimation of population growth rate in restoration projects. *Ecoscience* 14: 432–439.
Pradel, R., Hines, J.E., Lebreton, J.-D. et al. (1997). Capture-recapture survival models taking account of transients. *Biometrics* 53: 60–72.
Pradel, R., Maurin-Bernier, L., Gimenez, O. et al. (2008). Estimation of sex-specific survival with uncertainty in sex assessment. *Canadian Journal of Statistics* 36: 29–42.
Raabe, J.K., Gardner, B., and Hightower, J.E. (2014). A spatial capture-recapture model to estimate fish survival and location from linear continuous monitoring arrays. *Canadian Journal of Fisheries and Aquatic Sciences* 71: 120–130.
Reed, E.T., Gauthier, G., Pradel, R. et al. (2003). Age and environmental conditions affect recruitment in Greater Snow Geese. *Ecology* 84: 219–230.
Reed, E.T., Gauthier, G., and Pradel, R. (2005). Effects of neck bands on reproduction and survival of female

Greater Snow Geese. *Journal of Wildlife Management* 69: 91–100.

Reed, J.M. and Oring, L.W. (1993). Philopatry, site fidelity, dispersal, and survival of Spotted Sandpipers. *Auk* 110: 541–551.

Rendón, M.A., Garrido, A., Rendón-Martos, M. et al. (2014). Assessing sex-related chick provisioning in Greater Flamingo *Phoenicopterus roseus* parents using capture–recapture models. *Journal of Animal Ecology* 83: 479–490.

Richmond, O.M.V., Tecklin, J., and Beissinger, S.R. (2012). Impact of cattle grazing on the occupancy of a cryptic, threatened rail. *Ecological Applications* 22: 1655–1664.

Robinson, R.A., Julliard, R., and Saracco, J.F. (2009). Constant effort: studying avian population processes using standardized ringing. *Ringing and Migration* 24: 199–204.

Roche, E.A., Gratto-Trevor, C.L., Goosen, J.P. et al. (2012). Flooding affects dispersal decisions in Piping Plovers (*Charadrius melodus*) in Prairie Canada. *Auk* 129: 296–306.

Rotella, J.J., Lin, W.A., Chambert, T. et al. (2012). Evaluating the demographic buffering hypothesis with vital rates estimated for Weddell seals from 30 years of mark-recapture data. *Journal of Animal Ecology* 81: 162–173.

Rouan, L., Choquet, R., and Pradel, R. (2009). A general framework for modeling memory in capture-recapture data. *Journal of Agricultural, Biological, and Environmental Statistics* 14: 338–355.

Royle, J.A. (2008). Modeling individual effects in the Cormack–Jolly–Seber model: a state-space formulation. *Biometrics* 64: 364–370.

Royle, J.A., Chandler, R.B., Sollmann, R. et al. (2014). *Spatial Capture-Recapture*. Oxford, UK: Academic Press.

Royle, J.A., Kéry, M., and Guélat, J. (2011). Spatial capture-recapture models for search-encounter data. *Methods in Ecology and Evolution* 2: 602–611.

Royle, J.A. and Young, K.V. (2008). A hierarchical model for spatial capture-recapture data. *Ecology* 89: 2281–2289.

Ruiz-Gutiérrez, V., Doherty, P.F. Jr., Santana, C.E. et al. (2012). Survival of resident neotropical birds: considerations for sampling and analysis based on 20 years of bird-banding efforts in Mexico. *Auk* 129: 500–509.

Runge, J.P., Hines, J.E., and Nichols, J.D. (2007). Estimating species-specific survival and movement when species identification is uncertain. *Ecology* 88: 282–288.

Ryan, G.E., Dove, V., Trujillo, F. et al. (2011). Irrawaddy dolphin demography in the Mekong River: an application of mark-resight models. *Ecosphere* 25: art58.

Ryu, H.Y., Shoemaker, K.T., Kneip, É. et al. (2016). Developing population models with data from marked individuals. *Biological Conservation* 197: 190–199.

Sandercock, B.K. (2006). Estimation of demographic parameters from live encounter data: a summary review. *Journal of Wildlife Management* 70: 1504–1520.

Sandercock, B.K. and Beissinger, S.R. (2002). Estimating rates of population change for a neotropical parrot with ratio, mark-recapture and matrix methods. *Journal of Applied Statistics* 29: 589–607.

Sandercock, B.K., Beissinger, S.R., Stoleson, S.H. et al. (2000). Survival rates of a neotropical parrot: implications for latitudinal comparisons of avian demography. *Ecology* 81: 1351–1370.

Sandercock, B.K. and Jaramillo, A. (2002). Annual survival rates of wintering sparrows: assessing demographic consequences of migration. *Auk* 119: 149–165.

Sanz-Aguilar, A., Tavecchia, G., Genovart, M. et al. (2011). Studying the reproductive skipping behavior in long-lived birds by adding nest inspection to individual-based data. *Ecological Applications* 21: 555–564.

Sanz-Aguilar, A., Tavecchia, G., Pradel, R. et al. (2008). The cost of reproduction and experience-dependent vital rates in a small petrel. *Ecology* 89: 3195–3203.

Saracco, J.F., Desante, D.F., and Kaschube, D.R. (2008). Assessing landbird monitoring programs and demographic causes of population trends. *Journal of Wildlife Management* 72: 1665–1673.

Saracco, J.F., Royle, J.A., DeSante, D.F. et al. (2012). Spatial modeling of survival and residency and application to the monitoring avian productivity and survivorship program. *Journal of Ornithology* 152 (Suppl): S469–S476.

Schaub, M. and Abadi, F. (2011). Integrated population models: a novel analysis framework for deeper insights into population dynamics. *Journal of Ornithology* 152 (Suppl 1): S227–S237.

Schaub, M., Gimenez, O., Schmidt, B.R. et al. (2004a). Estimating survival and temporary emigration in the multistate capture-recapture framework. *Ecology* 85: 2107–2113.

Schaub, M., Jakober, H., and Stauber, W. (2011). Demographic response to environmental variation in breeding stopover and non-breeding areas in a migratory passerine. *Oecologia* 167: 445–459.

Schaub, M., Jakober, H., and Stauber, W. (2013). Strong contribution of immigration to local population regulation: evidence from a migratory passerine. *Ecology* 94: 1828–1838.

Schaub, M., Liechti, F., and Jenner, L. (2004b). Departure of migrating European Robins, *Erithacus rubecula*, from a stopover site in relation to wind and rain. *Animal Behaviour* 67: 229–237.

Schaub, M. and Pradel, R. (2004). Assessing the relative importance of different sources of mortality from recoveries of marked animals. *Ecology* 85: 930–938.

Schaub, M., Reichlin, T.S., Abadi, F. et al. (2012). The demographic drivers of local population dynamics in two rare migratory birds. *Oecologia* 168: 97–108.

Schaub, M. and Royle, J.A. (2014). Estimating true instead of apparent survival using spatial Cormack–Jolly–Seber models. *Methods in Ecology and Evolution* 5: 1316–1326.

Schaub, M. and Vaterlaus-Schlegel, C. (2001). Annual and seasonal variation of survival rates in the garden dormouse (*Eliomys quercinus*). *Journal of Zoology* 255: 89–96.

Schaub, M. and von Hirschheydt, J. (2009). Effect of current reproduction on apparent survival, breeding dispersal, and future reproduction in Barn Swallows assessed by multistate capture-recapture models. *Journal of Animal Ecology* 78: 625–635.

Schmidt, B.R., Schaub, M., and Anholt, B.R. (2002). Why you should use capture-recapture methods when estimating survival and breeding probabilities: on bias, temporary emigration, overdispersion and common toads. *Amphibia-Reptilia* 23: 375–388.

Schorcht, W., Bontadina, F., and Schaub, M. (2009). Variation of adult survival drives population dynamics in a migrating forest bat. *Journal of Animal Ecology* 78: 1182–1190.

Schorr, R.A. (2012). Using a temporal symmmetry model to assess population change and recruitment in the Preble's meadow jumping mouse (*Zapus hudsonicus preblei*). *Journal of Mammalogy* 93: 1273–1282.

Schwarz, C.J. and Stobo, W.T. (1997). Estimating temporary migration using the robust design. *Biometrics* 53: 178–194.

Sedinger, J.S., Lindberg, M.S., and Chelgren, N.D. (2001). Age-specific breeding probability in Black Brant: effects of population density. *Journal of Animal Ecology* 70: 798–807.

Sedinger, J.S., White, G.C., Espinosa, S. et al. (2010). Assessing compensatory versus additive harvest mortality: an example using Greater Sage-Grouse. *Journal of Wildlife Management* 74: 326–332.

Senar, J.C. and Conroy, M.J. (2004). Multi-state analysis of the impacts of avian pox on a population of Serins (*Serinus serinus*): the importance of estimating recapture rates. *Animal Biodiversity and Conservation* 27: 133–146.

Serrano, D., Oro, D., Ursúa, E. et al. (2005). Colony size determines adult survival and dispersal preferences: Allee effects in a colonial bird. *American Naturalist* 166: E22–E31.

Shefferson, R.P., Proper, J., Beissinger, S.R. et al. (2003). Life history trade-offs in a rare orchid: the costs of flowering, dormancy, and sprouting. *Ecology* 84: 1199–1206.

Silvy, N.J., Lopez, R.R., and Peterson, M.J. (2012). Techniques for marking wildlife. In: *The Wildlife Techniques Manual* (ed. N.J. Silvy), 230–257. Baltimore, MD: John Hopkins University Press.

Stahl, J.T. and Oli, M.K. (2006). Relative importance of avian life-history variables to population growth rate. *Ecological Modelling* 198: 23–39.

Stauffer, G.E., Rotella, J.J., and Garrott, R.A. (2013). Variability in temporary emigration rates of individually marked female Weddell seals prior to first reproduction. *Oecologia* 172: 129–140.

Tarwater, C.E., Ricklefs, R.E., Maddox, J.D. et al. (2011). Pre-reproductive survival in a tropical bird and its implications for avian life histories. *Ecology* 92: 1271–1281.

Taylor, C.M., Lank, D.B., and Sandercock, B.K. (2015). Using local dispersal data to reduce bias in apparent survival and mate fidelity. *Condor* 117: 598–608.

Taylor, R.K. and Boor, G.K.H. (2012). Beyond the robust design: accounting for changing, uncertain states and sparse biased detection in a multistate mark-recapture model. *Ecological Modelling* 243: 73–80.

Tenan, S., Pradel, R., Tavecchia, G. et al. (2014). Hierarchical model of population growth rate from individual capture-recapture data. *Methods in Ecology and Evolution* 5: 606–614.

Vélez-Espino, L.A., Fox, M.G., and McLaughlin, R.L. (2006). Characterization of elasticity patterns of North America freshwater fishes. *Canadian Journal of Fisheries and Aquatic Sciences* 63: 2050–2066.

Véran, S. and Lebreton, J.-D. (2008). The potential of integrated modelling in conservation biology: a case study of the Black-footed Albatross (*Phoebastria nigripes*). *Canadian Journal of Statistics* 36: 85–98.

Vögeli, M., Laiolo, P., Serrano, D. et al. (2008). Who are we sampling? Apparent survival differs between methods in a secretive species. *Oikos* 117: 1816–1823.

Weiser, E.L., Lanctot, R.B., Brown, S.C. et al. (2018). Environmental and ecological conditions at arctic breeding sites have limited effects on true survival rates of adult shorebirds. *Auk* 135: 29–43.

White, G.C. and Burnham, K.P. (1999). Program MARK: survival estimation from populations of marked animals. *Bird Study* 46: S120–S139.

White, G.C., Kendall, W.L., and Barker, R.J. (2006). Multistate survival models and their extensions in Program MARK. *Journal of Wildlife Management* 70: 1521–1529.

Williams, B.K., Nichols, J.D., and Conroy, M.J. (2002). *Analysis and Management of Animal Populations: Modeling, Estimation, and Decision Making*. San Diego, California: Academic Press.

Wood, K.V., Nichols, J.D., Percival, H.F. et al. (1998). Size-sex variation in survival rates and abundance of pig frogs *Rana grylio*, in northern Florida wetlands. *Journal of Herpetology* 32: 527–535.

Zeng, Z. and Brown, J.H. (1987). A method for distinguishing dispersal from death in mark-recapture studies. *Journal of Mammalogy* 68: 656–665.

Zens, M.S. and Peart, D.R. (2003). Dealing with death data: individual hazards, mortality and bias. *Trends in Ecology and Evolution* 18: 366–373.

Zimmerman, G.S., Gutiérrez, R.J., and LaHaye, W.S. (2007). Finite study areas and vital rates: sampling effects on estimates of Spotted Owl survival and population trends. *Journal of Applied Ecology* 44: 963–971.

Zipkin, E.F. and Saunders, S.P. (2018). Synthesizing multiple data types for biological conservation using integrated population models. *Biological Conservation* 217: 240–250.

Zipkin, E.F., Royle, J.A., Dawson, D.K. et al. (2010). Multi-species occurrence models to evaluate the effects of conservation and management actions. *Biological Conservation* 143: 479–484.

Zipkin, E.F., Sillett, T.S., Grant, E.H.C. et al. (2014). Inferences about population dynamics from count data using multistate models: a comparison to capture-recapture approaches. *Ecology and Evolution* 4: 417–426.

Part III

Population Models

8

Projecting Populations

Stéphane Legendre

Institut de Biologie de l'Ecole Normale Supérieure (IBENS), Centre National de la Recherche Scientifique (CNRS), Paris, France

Summary

Population dynamics models allow us to predict, explain, or remedy population trends given the biology of the species and the characteristics of its environment. In the approach proposed here, the key concept is the life cycle graph, a representation of the life history of a given species, parameterized by the demographic parameters, the fecundity and survival rates. I describe how deterministic systems in discrete time are built from the life cycle graph, and how demographic descriptors (growth rate, generation time, sensitivities) can be computed. Regulation by population density and the spatial component (metapopulation) are introduced. Stochastic models are then constructed from the deterministic models, to account for fluctuations in the environment and for demographic stochasticity, that is, random population drift. Though the chapter contains many mathematical formulas, it mostly relies on biological intuition.

8.1 Introduction

The aim of this chapter is to provide a broad overview of population dynamics methods that are useful for practical applications. These methods rely on solid theoretical grounds, and this necessitates the use of mathematical concepts and equations. The reader should not be put off by the presence of mathematical formulas, and can skip some of them (in particular Boxes 8.3–8.5) and rely on the intuitive meaning given in the text. Experience shows that biological intuition is enough to develop pertinent population models without being mathematically inclined. I however believe that mathematical formulas remain the most concise and universal way of transmitting information.

The key concept of the chapter is the *life cycle graph*, introduced in Section 8.2. It is a simple concept which has far-reaching consequences. Once mastered, it goes well along with biological sense, but its construction and interpretation are crucial first steps. Then, matrix algebra and the theory of stochastic processes are convenient tools to analyze the properties of a given life cycle.

Population dynamics is the study of population trajectories or variation in population sizes over time, which is necessary for predicting, explaining, or managing population trends. In the framework of population dynamics, a *population* is defined as a set of individuals of the same genotype, living in the same environment. Accordingly, all individuals in a population are considered identical to an average individual. The average individual is described by the life cycle graph, a schematic description of the stages traversed by an organism during its life. The life cycle graph, whose parameters are determined from individual data (survival rates, fecundities, life span), is constructed according to the biology of the species and the question of interest.

From the life cycle graph, *matrix population models* and variants thereof are built, allowing to project the population in *discrete time* $t = 0, 1, 2, \ldots$ Discrete time is well suited because many organisms reproduce seasonally. Also, demographic parameters, survival and fecundity rates, are measured in the field or in the laboratory at regular dates, in relation with the periodical schedule of the species.

Matrix population models provide pertinent demographic descriptors: growth rate, generation time, and the most sensitive parameters, those that contribute most to the dynamics. These models can also account for environmental characteristics such as limitation of resources leading to population regulation or environmental fluctuations, as well as other specific features such as spatial structure or harvesting, thereby leading to demographic predictors like the probability of extinction. Matrix models form a multipurpose toolbox for the management and conservation of species, and can also address evolutionary issues.

Population Ecology in Practice, First Edition. Edited by Dennis L. Murray and Brett K. Sandercock.

Companion website: www.wiley.com/go/MurrayPopulationEcology

8.2 The Life Cycle Graph

8.2.1 Description

All organisms traverse different stages during their life, typically immature stages where they develop followed by mature stages where they reproduce. This scheme is represented by the life cycle graph, a directed graph whose *nodes* describe the stages, and whose *arcs* describe transitions from one stage to the next. Demographic parameter values (vital rates) are associated with the *transitions*, representing the contribution of a stage to the next (Figure 8.1). The life cycle graph can be seen as a simplification and quantification of the developmental process and of the contribution of one generation to the next. The same duration is associated with each transition. It will be the time step of the discrete time dynamical system built from the life cycle graph.

To project a population, first construct the life cycle graph of the species. All population dynamics models are based, explicitly or not, on such a representation. The life cycle graph accounts for the biology of the species, depends on the available individual data, and is constructed to address the question of interest.

The life cycle is in general assumed to be *female-based*: the population is considered to be constituted of females only. It means that there are always enough males to fertilize the females, and that the male life cycle does not differ from the female life cycle. These two assumptions are not always met, and will be relaxed when considering *two-sex models* (Section 8.8.2).

Figure 8.1b represents the life cycle graph of a short-lived species such as a small songbird or a small lizard. This example will be used throughout the text and referred as the *passerine model* (Legendre et al. 1999). In this example, the population census occurs shortly before reproduction (*prebreeding census*). There are two age classes: individuals aged one year (subadults), and individuals aged two years or more (adults). Subadults reproduce with fecundity f_1 and become adults with subadult survival rate s_1. Adults reproduce with fecundity f_2 and stay in the adult class according to the survival rate s_2, as expressed by the self-loop in the adult stage. Because of the prebreeding census, a newborn will be censused the following year, when it is (almost) one year old: it survives with juvenile survival rate s_0 before entering the one-year-old age class. This is why the juvenile survival rate s_0 multiplies the fecundities in the reproductive transitions. As the life cycle is female-based, only females giving birth to females are considered: the primary female sex ratio σ (the proportion of females at birth) multiplies the fecundities. In many species the primary sex ratio is balanced: $\sigma = 0.5$.

In a *postbreeding census*, the population census occurs shortly after reproduction, and the first stage consists of zero-year-old individuals. In the passerine model, there are now three age classes instead of two. The age-0 individuals are descendants of the juveniles, subadults, and adults censused the previous year, which survived (for almost one year) at rates s_0, s_1, and s_2 respectively (Figure 8.1c). To summarize, in the prebreeding census the juvenile survival rate s_0 multiplies the fecundities, whereas in the postbreeding census there is a supplementary stage (age-0 individuals), and age-specific survival rates multiply the fecundities (Figure 8.1b). Both representations lead to the same demographic descriptors (Section 8.3.2).

8.2.2 Construction

The life cycle graph is constructed in three steps:

1) Determine the largest *time step* compatible with the biology of the species and the observations. Typically, the time step might be a year for birds and large mammals, a month for small mammals, or a day for insects. Recall that all arcs in the life cycle graph correspond to the same time step.
2) Determine the pertinent stages. For many organisms, the convenient stages are *age classes*, with the identification of the age at first reproduction. However, several organisms are better classified according to other criteria, like size. *Size classes* are often preferred for plants whose development is much more plastic than animals, because growth can be delayed until sufficient light exposure is met, and because observations are more conveniently performed in terms of size rather than age. Size-classified models (Figure 8.1f) are also used for animal species like turtles, fish, and reptiles (Morris and Doak 2002; Crouse et al. 1987). In Figure 8.1f, g_i is the probability to grow from a size class to the next.

   In general, any structure can be used for the stages, provided that they are biologically relevant. In fact, life cycles are very diverse across plant and animal taxa (Jones et al. 2014). In many cases, it is convenient to incorporate in an age- or size-classified model stages that account for biological specificities or for the question of interest. For example, the life cycle of the wolf *Canis lupus* (Figure 8.1c2, Chapron et al. 2003) is not strictly age-classified. We have pups (0–6 months), juveniles (6–18 months), subadults (18–30 months), adults (30+ months), and two more stages that account for characteristics of this social species: dispersers (dispersing juveniles must wait one year before looking for a mate), and pack leaders. In stage-classified models, a same duration is still associated with each transition. However, individuals in a stage are of different ages, and individuals of the same age

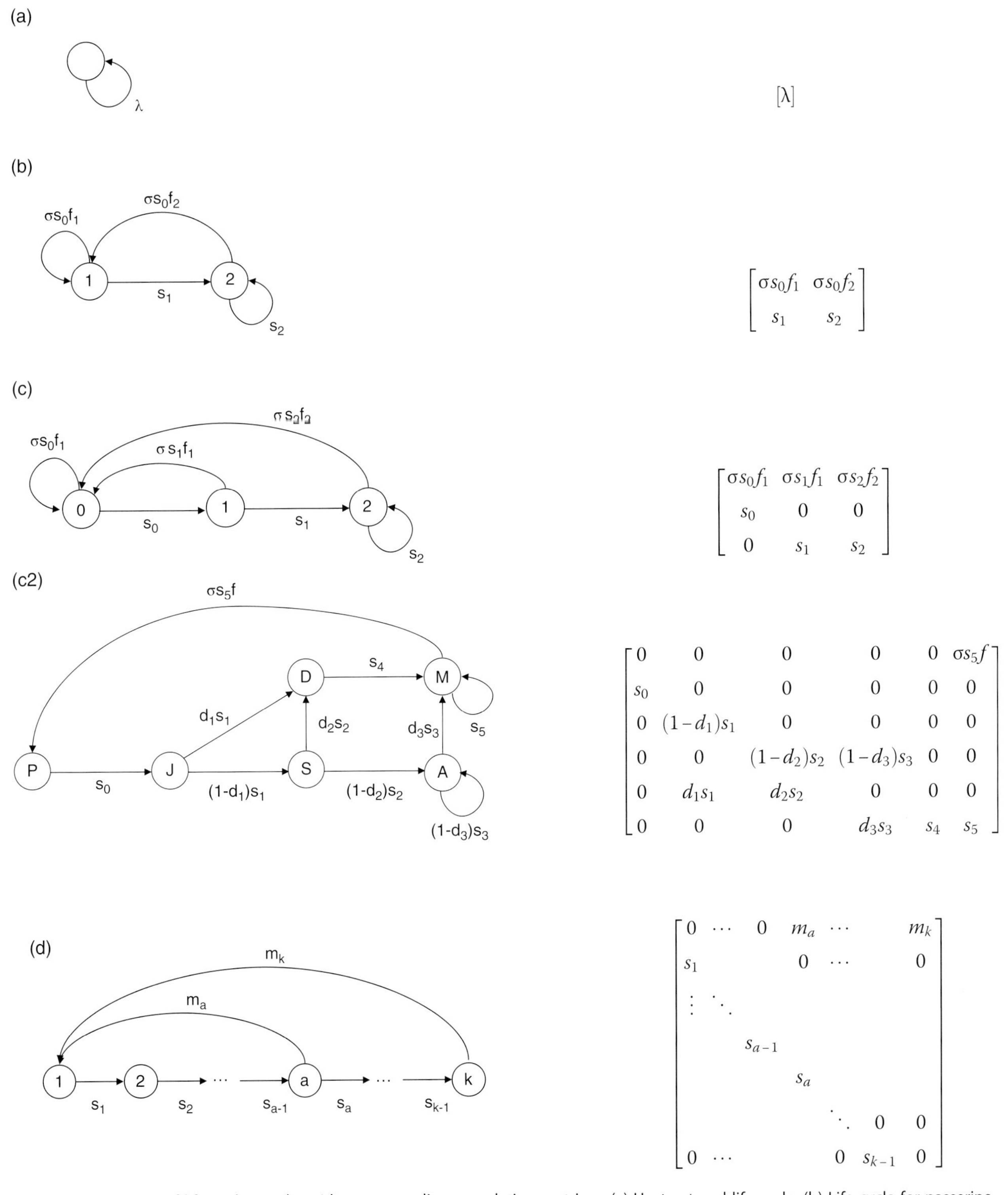

Figure 8.1 A library of life cycle graphs with corresponding population matrices. (a) Unstructured life cycle. (b) Life cycle for passerine (prebreeding census). (c) Life cycle for passerine (postbreeding census). (c2) Life cycle for the wolf *Canis lupus*. Stages: P = pups, J = juveniles, S = subadults, A = adults, D = dispersers, M = dominant. (d) Leslie age-classified life cycle (s_i survival rate, m_i fertility rate, a age at first reproduction). (e) Age-classified life cycle (Leslie model with self-loop on the last age class). (f) Size-classified life cycle (s_i survival rate, g_i probability to grow, m_i fertility rate, a age at first reproduction). (g) Age-classified life cycle with postreproductive stage (reducible matrix). (h) Semelparous life cycle (periodic matrix). (i) Life cycle with gap in reproduction (periodic matrix). (j) Life cycle with stages for immigrant pool (I) and emigrant pool (E). (k) Life cycles of two sites, A and B, connected by migrations. (l) Two-sex life cycle with underlying mating function μ. (m) Individual heterogeneity: two developmental pathways according to function J.

(e)

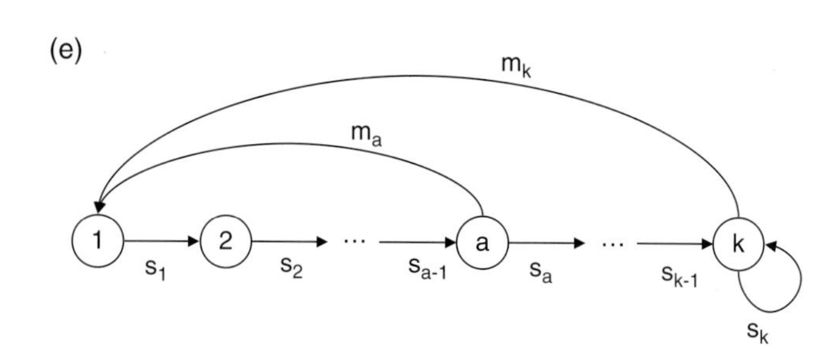

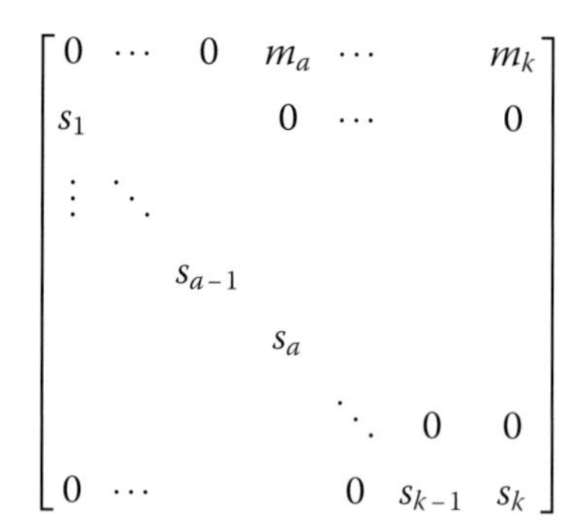

(f)

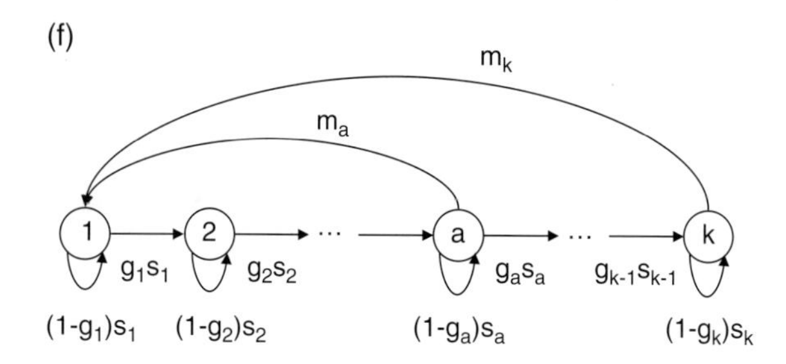

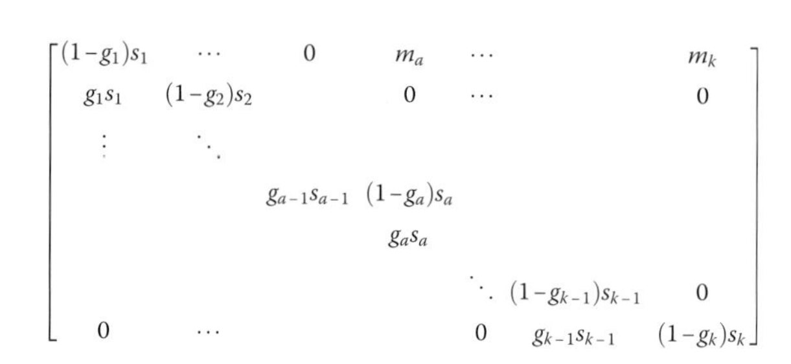

(g)

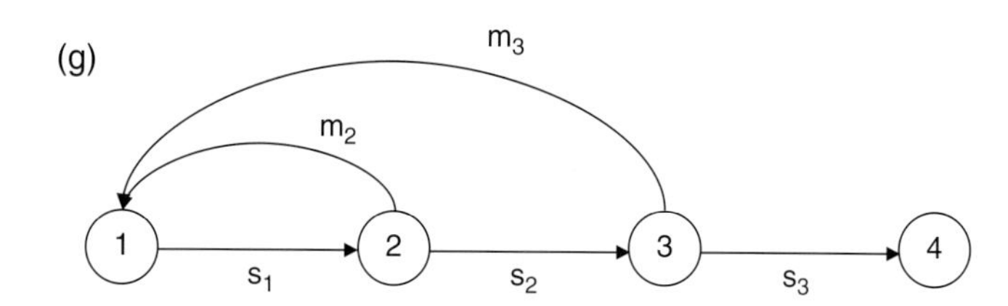

(h)

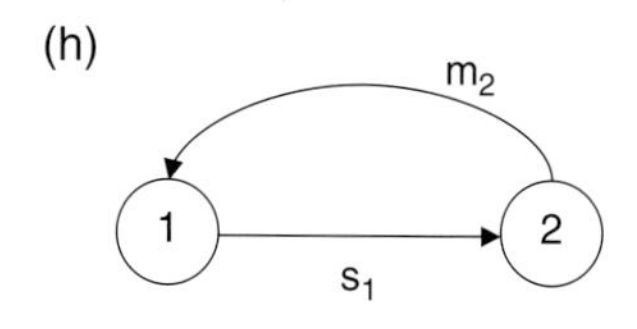

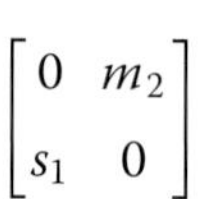

(i)

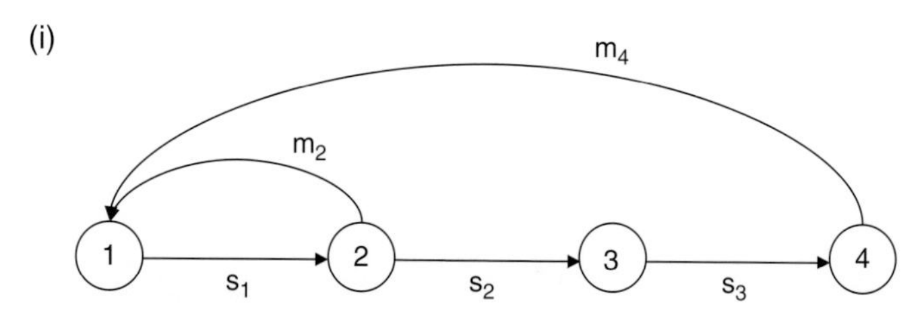

$$\begin{bmatrix} 0 & m_2 & 0 & m_4 \\ s_1 & 0 & 0 & 0 \\ 0 & s_2 & 0 & 0 \\ 0 & 0 & s_3 & 0 \end{bmatrix}$$

Figure 8.1 (Continued)

(j)

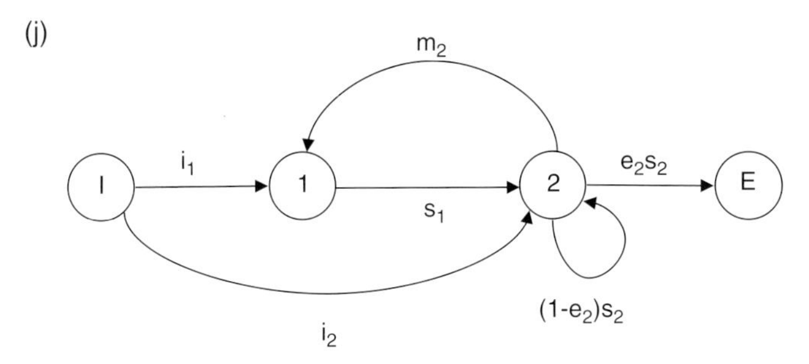

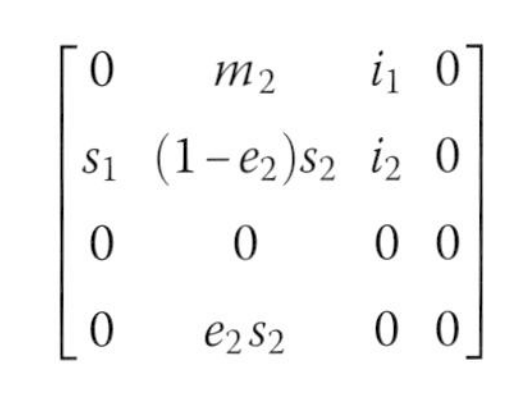

$$\begin{bmatrix} 0 & m_2 & i_1 & 0 \\ s_1 & (1-e_2)s_2 & i_2 & 0 \\ 0 & 0 & 0 & 0 \\ 0 & e_2 s_2 & 0 & 0 \end{bmatrix}$$

(k)

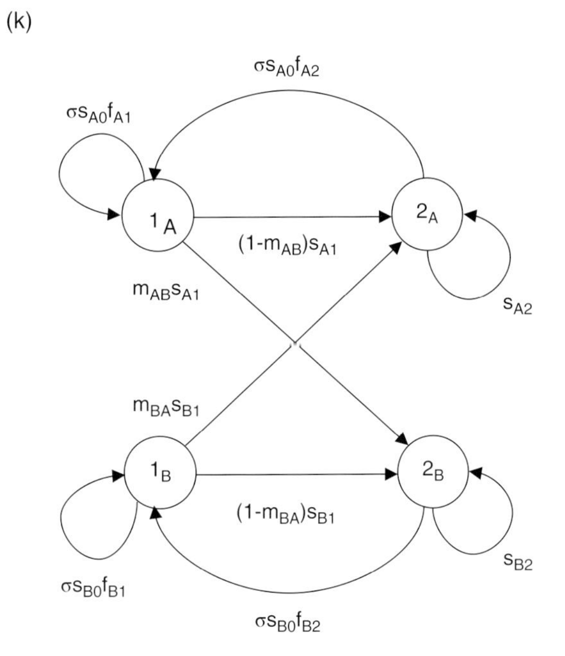

$$\begin{bmatrix} \sigma s_{A0} f_{A1} & \sigma s_{A0} f_{A2} & 0 & 0 \\ (1-m_{AB})s_{A1} & s_{A2} & m_{BA}s_{B1} & 0 \\ 0 & 0 & \sigma s_{B0} f_{B1} & \sigma s_{B0} f_{B2} \\ m_{AB}s_{A1} & 0 & (1-m_{BA})s_{B1} & s_{B2} \end{bmatrix}$$

(l)

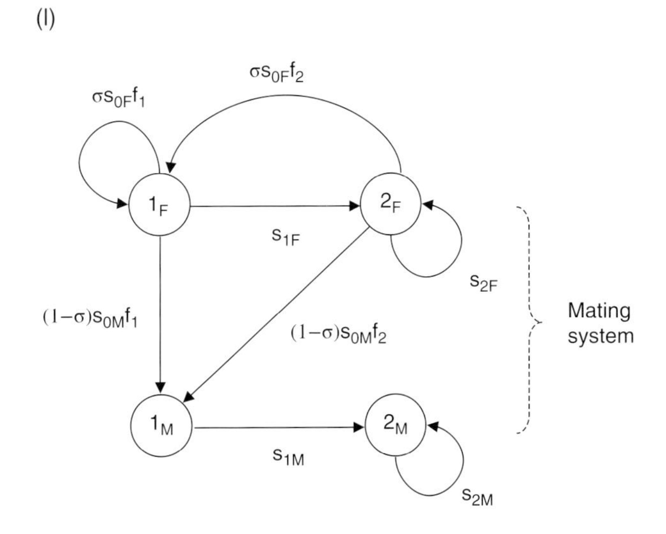

$$\begin{bmatrix} \sigma s_{0F} f_1 \dfrac{\mu}{f} & \sigma s_{0F} f_2 \dfrac{\mu}{f} & 0 & 0 \\ s_{1F} & s_{2F} & 0 & 0 \\ (1-\sigma) s_{0M} f_1 \dfrac{\mu}{f} & (1-\sigma) s_{0M} f_2 \dfrac{\mu}{f} & 0 & 0 \\ 0 & 0 & s_{1M} & s_{2M} \end{bmatrix}$$

(m)

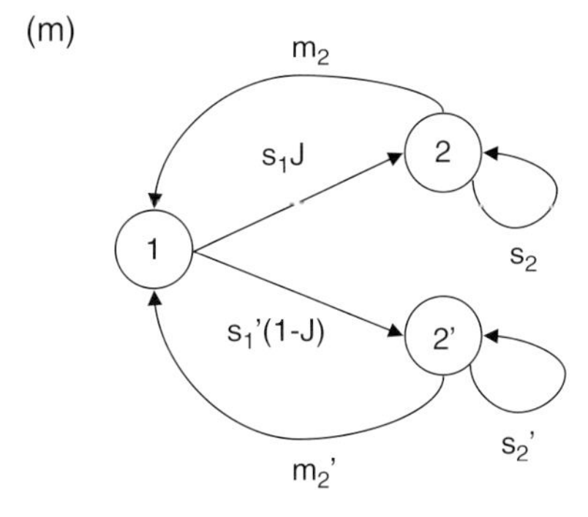

$$\begin{bmatrix} 0 & m_2 & m_2' \\ s_1 J & s_2 & 0 \\ s_1'(1-J) & 0 & s_2' \end{bmatrix}$$

Figure 8.1 (Continued)

may be in different stages, which can make the results more difficult to intuit.

3) Determine the *demographic parameters* associated with the transitions of the life cycle from individual data. For wild populations, survival rates are usually estimated using known fate models (Chapter 6), or with capture-mark-recapture methods (Lebreton et al. 1992; Sandercock 2006; Chapter 7). Uninformed parameters may be left as variables that can be explored using different scenarios, or filled with guess values taken from the literature for a related species. It will be seen that some demographic parameters have little influence on population dynamics (sensitivities, Section 8.3.3), so that accurate estimation is not crucial.

8.3 Matrix Models

8.3.1 The Projection Equation

The simplest life cycle graph has a single class (Figure 8.1a). In this *unstructured model*, population size n grows by a constant multiplicative factor from one time step to the next:

$$n(t+1) = \lambda n(t). \tag{8.1}$$

The *population growth rate* (λ) describes the (average) contribution of each individual alive at time t to population size at time $t+1$. The contribution involves survival and fecundity: $\lambda = s + f$. We have *nonoverlapping generations* when, in a time step, individuals give birth to f offspring and then die ($s = 0$). In this case, a time step corresponds to a generation, and the *generation time* is $T = 1$. Otherwise, generations overlap.

Population size is given by a geometric series in Eq. (8.1), so that

$$n(t) = \lambda^t n(0), \tag{8.2}$$

where $n(0)$ is initial population size. From Eq. (8.2), we deduce that population size grows exponentially to infinity when $\lambda > 1$, and decreases exponentially to 0 when $\lambda < 1$.

Let us consider a general life cycle graph with k stages (e.g. Figure 8.1b where $k = 2$, or Figure 8.1d–f). A square matrix $\mathbf{A} = (a_{ij})$ of size $k \times k$ is associated with the life cycle graph: the entry a_{ij} of $\mathbf{A}$ describes the contribution of an individual from stage j to stage i during a time step when there is an arc from j to i, and is 0 otherwise. Thus a nonzero entry a_{ij} of $\mathbf{A}$ corresponds to the arc $j \to i$. Its value is the demographic parameter associated with the transition $j \to i$. Let $n_i(t)$ denote the number of individuals in stage i at time t, then summing contributions,

$$n_i(t+1) = \sum_{j=1}^{k} a_{ij} n_j(t). \tag{8.3}$$

Introducing the column population vector $\mathbf{n}(t) = (n_1(t), \cdots, n_k(t))$, we recognize in Eq. (8.3) the product of the matrix $\mathbf{A}$ by the vector $\mathbf{n}(t)$. Hence, the population vector is updated from a time step to the next according to

$$\mathbf{n}(t+1) = \mathbf{A}\mathbf{n}(t), \tag{8.4}$$

which gives

$$\mathbf{n}(t) = \mathbf{A}^t \mathbf{n}(0). \tag{8.5}$$

A convention in matrix algebra is that boldface type is used to denote matrices and vectors.

Let us describe the procedure for the passerine model. Let $n_1(t)$ be the number of one-year old individuals at time t, $n_2(t)$ the number of individuals aged two years or more at time t. Inspection of the life cycle graph (Figure 8.1b) tells us that population sizes in the stages are updated from one time step to the next according to

$$\begin{aligned} n_1(t+1) &= \sigma s_0 f_1 n_1(t) + \sigma s_0 f_2 n_2(t), \\ n_2(t+1) &= s_1 n_1(t) + s_2 n_2(t). \end{aligned} \tag{8.6}$$

These equations can be written in matrix form:

$$\begin{bmatrix} n_1 \\ n_2 \end{bmatrix}_{t+1} = \begin{bmatrix} \sigma s_0 f_1 & \sigma s_0 f_2 \\ s_1 & s_2 \end{bmatrix} \begin{bmatrix} n_1 \\ n_2 \end{bmatrix}_t. \tag{8.7}$$

For an age-classified life cycle with k stages and pre-breeding census, the associated $k \times k$ matrix is shown in Figure 8.1e, where a is age at first reproduction and it is assumed that reproductive individuals have constant survival s_k and constant fecundity f_k from age k. When all individuals die at age k, $s_k = 0$, we obtain the Leslie matrix (Figure 8.1d).

The total population size at time t is $n(t) = n_1(t) + \cdots + n_k(t)$. The population structure at time t is defined as the column vector of proportions of individuals in the stages:

$$\mathbf{w}(t) = \left(\frac{n_1(t)}{n(t)}, \ldots, \frac{n_k(t)}{n(t)} \right). \tag{8.8}$$

In fact, the dynamical properties of the discrete time system (Eq. 8.4) depend entirely on algebraic properties of the matrix. The main result is that the population grows (or decreases) at an *exponential rate*. More precisely, the population first goes through a transient regime of damped oscillations, followed by an asymptotic regime of exponential growth where

$$n(t+1) \approx \lambda n(t) \quad \text{for large } t. \tag{8.9}$$

Hence, the expected behavior of the structured model (Eq. 8.4) is close to that of the unstructured model (Eq. 8.1). The growth rate λ is the dominant eigenvalue of the matrix (Boxes 8.1 and 8.4). The asymptotic regime is reached concomitantly with the stabilization of the population structure:

$$\mathbf{w}(t) \to \mathbf{w}.$$

Box 8.1 The Life Cycle Graph and Dynamical Behavior

The existence of the dominant eigenvalue $\lambda > 0$ (a unique real eigenvalue of largest modulus), and the associated left and right positive eigenvectors **v** and **w**, is guaranteed under conditions pertaining to the structure of the life cycle graph, its stages and arcs, and not to the values associated with the arcs (the demographic parameters). The discrete dynamical system described by Eq. (8.4) is equivalent to performing paths of length t in the life cycle graph, traversing cycles from newborn stages via survival arcs and back to newborn stages via reproductive arcs. The condition of existence of the dominant eigenvalue λ is that the life cycle graph should be *irreducible* and *aperiodic*:

A) Irreducibility means the existence of an oriented path from any stage to any other. Reducibility occurs, for example, when postreproductive stages are included (Figure 8.1g; see also the case of source-sink metapopulations, Section 8.7). Such stages do not belong to reproductive cycles since newborn stages cannot be reached from them. They do not contribute to growth and can be omitted. In case of reducibility, there may nevertheless exist a dominant eigenvalue, in which case the main results still hold, but this is not guaranteed, and the vectors **v** and **w** usually have some zero entries.

B) Aperiodicity means that the lengths of cycles in the life cycle graph have no common divisor >1. Periodicity occurs, for example, with the life cycle of Figure 8.1h, which produces survival-reproduction cycles of length 2, 4, 6, …, when the life cycle is traversed 1, 2, 3… times: the common divisor is 2. The Leslie matrix is aperiodic (Figure 8.1d, assuming $m_i > 0$, but see also Figure 8.1i) because the cycle lengths have 1 as common divisor, hence are relatively prime. In case of periodicity, the population trajectories present periodic oscillations that are not damped, though the growth is still exponential on average. Aperiodicity is guaranteed when there is a self-loop (a cycle of length 1) in the life cycle graph, i.e. a nonzero entry on the diagonal of the matrix, a condition that is often met (e.g. Figure 8.1e, f).

Conditions (A) and (B) are equivalent to the matrix **A** being primitive: there exists a positive integer q such that $\mathbf{A}^q$ has all entries positive. To summarize, the generic behavior (primitive matrix) is the rule, with exceptions to be aware of: reducibility and periodicity, the latter being rare.

Here, the *stable stage distribution* $\mathbf{w} = (w_i)$, normalized $\sum w_i = 1$, is the right eigenvector associated with the eigenvalue λ. The proportion of individuals in stage i tends to w_i geometrically fast. The transient fluctuations depend on the initial population structure $\mathbf{w}(0)$. In fact, if $\mathbf{w}(0)$ equals $\mathbf{w}$, the asymptotic regime is attained instantly. The transient behavior is easily understood for an age-classified model. For example, if the initial condition is 100 individuals in the first (newborn) stage, and 0 individuals in the other stages, these 100 individuals traverse the pre-reproductive stages step by step (population size decreases); the surviving ones then reach the reproductive stages, from which the newborn stage is alimented again (population size bursts). We thus have oscillatory behavior until the number of individuals homogenize across stages to eventually meet the stable stage distribution.

Equation (8.2) generalizes into

$$\mathbf{n}(t) \approx C\,\lambda^t \mathbf{n}(0) \quad \text{for large } t. \tag{8.10}$$

Here, the constant term $C = \mathbf{vw}(0)$ is the dot product of the reproductive value, the left eigenvector associated with the eigenvalue λ, a row vector $\mathbf{v} = (v_i)$, normalized $\sum v_i = 1$, and the initial population structure $\mathbf{w}(0)$. The interpretation of the *reproductive value* can be understood from Eq. (8.10). The population grows exponentially at rate λ by Eq. (8.9), independently of the initial condition $\mathbf{n}(0)$. However, according to Eq. (8.10), population size depends on the initial structure $\mathbf{w}(0)$ and on the reproductive value $\mathbf{v}$: the stages i with large reproductive values v_i contribute most to population size (but not to the growth rate), weighted by the initial structure, as expressed by the dot product. For the Leslie matrix (Figure 8.1d), the typical pattern is that v_i increases up to age at first reproduction, because individuals have an increasing potential to contribute to future population size, and decreases thereafter.

The *transient oscillatory behavior* of the model should not be overlooked (Koons et al. 2007; Gamelon et al. 2014). Indeed, even if the population is at the demographic equilibrium, where the population structure is stationary, perturbations from the environment may alter the structure and drive the population into transient fluctuations.

Taking logs on each side of Eq. (8.10) provides an estimate of the growth rate:

$$\lambda \approx \exp\left(\frac{\ln(n(t)) - \ln(n(0))}{t}\right). \tag{8.11}$$

According to Eq. (8.11), if you observe a population at date t_0 and later on at date t_1 and measure population sizes $n_{obs}(t_0)$ and $n_{obs}(t_1)$, then the growth can be estimated by

$$\hat{\lambda} = \exp\left(\frac{\ln(n_{obs}(t_1)) - \ln(n_{obs}(t_0))}{t_1 - t_0}\right). \quad (8.12)$$

8.3.2 Demographic Descriptors

We have already encountered the growth rate λ which tells us if the demographic parameters of the population lead to increase ($\lambda > 1$) or decrease ($\lambda < 1$). The stable stage distribution **w** and the reproductive value **v** describe stage-specific contributions to the dynamics. For example, the population structure of introduced individuals should be chosen close to the stable stage distribution to avoid initial population fluctuations which could put the program at risk. If individuals in a single stage are introduced, those with the largest reproductive value should be preferred, since they contribute more to future population size.

In the age-classified model, let us consider an individual that survives up to age i and produces f_i offspring. The traversal of a survival-reproduction cycle in the life cycle graph can be associated with the cumulated rate

$$\phi_i = s_0 s_1 \cdots s_{i-1} f_i. \quad (8.13)$$

The *net reproductive rate*

$$R_0 = \sum_i \phi_i \quad (8.14)$$

represents the per individual contribution to the renewal of the population. The conditions $R_0 > 1$ and $\lambda > 1$ are equivalent.

The dominant eigenvalue λ is the largest root of the *characteristic equation* $\sum_i \phi_i X^{-i} = 1$, so that the terms

$$p_i = \phi_i \lambda^{-i} \quad (8.15)$$

satisfy $\sum p_i = 1$, hence constitute a probability distribution. A measure of the generation time is then defined as the mean of the distribution:

$$T = \sum_i i p_i. \quad (8.16)$$

By construction, T is the *mean age of the mothers*, assuming the population at the stable age distribution. *Entropy*, the Shannon diversity index of the p_is,

$$S = -\sum_i p_i \ln(p_i), \quad (8.17)$$

is a measure of the complexity of the life cycle: it is linearly related to the logarithm of body size, metabolic rate, and maximal life span (Demetrius et al. 2009). A semelparous life cycle has an entropy of zero (Figure 8.1h).

For the passerine model (Figure 8.1b, model file `pass_0.ulm`), the demographic parameter values are $s_0 = 0.2$, $s_1 = 0.35$, $s_2 = 0.5$ for survival, and $f_1 = f_2 = 7$ for fecundity, with primary sex ratio $\sigma = 0.5$. We obtain the growth rate $\lambda = 1.1050$ and generation time $T = 1.67$. For the wolf, survival rates are not known precisely in this population: various scenarios can be considered, from pessimistic to optimistic, leading to growth rates ranging from $\lambda = 0.93$ to $\lambda = 1.16$ (Figure 8.1c2, model file `wolf_0.ulm`).

Demographic descriptors can also be computed for models that are not age-classified. The analysis of the cycles subtending the life cycle graph allows us to compute the characteristic equation (of which λ is the largest root), and expressions for the eigenvectors **w** and **v**, and the net reproductive rate R_0 in terms of the demographic parameters (Caswell 2001). For size-classified models (Figure 8.1f), it can be convenient to translate the size currency into the more intuitive time currency. Simple formulas exist in this case, allowing, for example, to compute the *average residence time* in a stage (Barot et al. 2002). The general case for any nonnegative matrix that has relevant biological meaning is more complicated (Cochran and Ellner 1992). However, generation time has a simple expression (Bienvenu and Legendre 2015).

8.3.3 Sensitivities

When a parameter of the life cycle varies due to some condition, will the population still grow, or will it decrease? What change in a parameter will ensure the restoration of a declining population? These questions can be addressed using sensitivity analysis. The impact of small changes in the parameter x on the growth rate is measured by the *sensitivity* of λ to changes in x:

$$s_\lambda(x) = \frac{\partial \lambda}{\partial x}. \quad (8.18)$$

The meaning of the partial derivative is that if x changes by an amount Δx, then λ changes by an amount $\Delta\lambda = s_\lambda(x)\Delta x$. The sensitivity s_{ij} of λ to changes in the matrix entry a_{ij} is computed from the left and right eigenvectors by the formula

$$s_{ij} = \frac{v_i w_j}{\mathbf{vw}} \quad (8.19)$$

where the denominator is a dot product. From the s_{ij} values, the sensitivity to changes in a lower level parameter x is obtained using the chain rule:

$$s_\lambda(x) = \sum_{i,j} \frac{\partial \lambda}{\partial a_{ij}} \frac{\partial a_{ij}}{\partial x} = \sum_{i,j} s_{ij} \frac{\partial a_{ij}}{\partial x}. \quad (8.20)$$

Sensitivities of other demographic descriptors (e.g. T) or of the population vector entries can also be computed.

A related quantity is the *elasticity*, which measures proportional changes in a lower-level parameter (x):

$$e_\lambda(x) = \frac{x}{\lambda}\frac{\partial\lambda}{\partial x}. \quad (8.21)$$

If x changes by α% then λ changes by β% with $\beta = \alpha e_\lambda(x)$. The elasticities e_{ij} of λ to changes in the matrix entries sum to 1. Zero entries in the population matrix usually have nonzero sensitivities, reflecting the impact a small change in these entries would have on λ, even if it is not biologically meaningful, but their elasticities are zero, as seen from the definition. Sensitivities and elasticities are generally positive, but may be negative. Indeed, as the growth rate obviously increases with an increase in survival or fecundity, these parameters have positive sensitivity. But if survival is written $s = 1 - m$, where m is mortality (Chapter 6), the parameter m has negative sensitivity. Sensitivities and elasticities provide similar information, that of a perturbation analysis: sensitivities reflect (linearly) the effects of additive perturbations; elasticities reflect (nonlinearly) the effects of proportional perturbations.

Sensitivities or elasticities allow us to determine the relative contribution of the demographic parameters to population growth (Box 8.2). In consequence, the most sensitive parameters should be determined with the greatest accuracy. Such parameters should be targeted to restore the growth of a population. Moreover, sensitivities quantify the change in specific parameters to reach a desired goal. Conversely, parameters with low sensitivities require less accuracy, as they contribute little to population dynamics.

Changes in matrix entries can be described by random variables A_{ij} with expectations given by the constant reference matrix $\mathbf{A}$, $E(A_{ij}) = a_{ij}$. The dominant eigenvalue of the corresponding random matrix is a random variable Λ such that $E(\Lambda) = \lambda$ (capital letters are used for random variables). Using the methods of Box 8.3, its variance is approximated using the sensitivities of $\mathbf{A}$:

$$\mathrm{Var}(\Lambda) \approx \sum_{i,j,k,l} s_{ij} s_{kl} \mathrm{Cov}(A_{ij}, A_{kl}). \quad (8.28)$$

Box 8.2 Short-lived Versus Long-lived Species

The elasticity of λ to changes in a parameter c multiplying all reproductive transitions of the life cycle – the entries involving fecundities – is given by the inverse of the generation time (Houllier and Lebreton 1986; Bienvenu and Legendre 2015):

$$e_\lambda(c) = \frac{1}{T}. \quad (8.22)$$

This is the case of the primary sex ratio σ, or the juvenile survival s_0 in the age-classified model with prebreeding census (e.g. Figure 8.1b). As the elasticities sum to 1, the elasticity of λ to changes in a parameter d multiplying all nonreproductive transitions is given by

$$e_\lambda(d) = 1 - \frac{1}{T}. \quad (8.23)$$

The formula provides a rationale for the classification of species along the so-called fast-slow continuum. It leads to a useful dichotomy between short-lived and long-lived species. Typically, short-lived species (small T) with a fast life history (short life spans, large progeny sets, low adult survival) invest in reproduction, whereas long-lived species (large T) with a slow life history (long life spans, small progeny sets, high adult survival) invest in survival. In general, the most sensitive parameters of short-lived species are the fecundities and the juvenile survival rate. The most sensitive parameters of long-lived species are the adult survival rates.

The generation time also reflects the tempo of biochemical reactions within the organism, measured by the metabolic rate P, and related to body mass W by the Kleiber allometric relation (Brown et al. 2004):

$$P \propto W^{3/4}. \quad (8.24)$$

Hence, the metabolic rate per unit mass,

$$\frac{P}{W} \propto W^{-1/4}, \quad (8.25)$$

decreases with increasing body mass. This is consistent with short-lived species having small body size and long-lived species, large body size. In a related way, it is observed that the rate of increase $r = \ln(\lambda)$ of natural populations verifies

$$r \approx \frac{\ln(R_0)}{T} \quad (8.26)$$

where R_0 is the net-reproductive rate, leading to the allometric relation:

$$r \propto T^{-1}. \quad (8.27)$$

Box 8.3 Environmental Stochasticity

Denoting $\lambda(\mathbf{M})$ the function associating to a matrix **M** its dominant eigenvalue, a Taylor expansion to the first order gives

$$\lambda(\mathbf{M}+\Delta\mathbf{M}) \approx \lambda(\mathbf{M}) + \sum_{i,j} \frac{\partial\lambda(\mathbf{M})}{\partial m_{ij}} (\Delta\mathbf{M})_{ij} \tag{8.29}$$

where we recognize the sensitivities $s_{ij} = \frac{\partial\lambda}{\partial m_{ij}}$ under the sum. To account for environmental noise, we write

$$\mathbf{A} = \mathrm{A} + \mathbf{E} \text{ with } \mathrm{E}(\mathbf{A}) = \mathrm{A} \text{ and } \mathrm{E}(\mathbf{E}) = 0 \tag{8.30}$$

where the matrix $\mathbf{E} = (E_{ij})$ represents a small deviation $\Delta\mathbf{A}$ from matrix **A**. Using the Taylor expansion above, the growth rate $\Lambda = \lambda(\mathbf{A}) = \lambda(\mathbf{A}+\mathbf{E})$ is a random variable such that

$$\Lambda \approx \lambda + \sum_{i,j} s_{ij} E_{ij}. \tag{8.31}$$

We have $\mathrm{E}(\Lambda) = \lambda$ and

$$\mathrm{Var}(\Lambda) \approx \sum_{i,j,k,l} s_{ij} s_{kl} \mathrm{Cov}(E_{ij}, E_{kl}) = \sum_{i,j,k,l} s_{ij} s_{kl} \mathrm{Cov}(A_{ij}, A_{kl}). \tag{8.32}$$

Tuljapurkar's formula (Tuljapurkar 1990) quantifies the impact of environmental noise on population growth depending on characteristics of the environmental process:

$$r_e \approx r - \frac{\sigma_e^2}{2\lambda^2} + \frac{\theta}{\lambda^2}. \tag{8.33}$$

Here, $r = \ln(\lambda)$ is the rate of increase in absence of stochasticity. The stochastic rate of increase r_e decreases linearly with the environmental variance σ_e^2. The environmental variance is identified with the variance of Λ from variations in the matrix, as given by Eq. (8.32). It depends on changes in the entries of the mean matrix, quantified by the sensitivities, and on the covariance across matrix entries. Positive covariance is more detrimental than no covariance, whereas negative covariance is less. The alteration is (quadratically) less important with increasing λ. The term θ quantifies the contribution of temporal autocorrelation to stochastic growth. When there is no autocorrelation (iid environment), $\theta = 0$, and

$$r_e \approx r - \frac{\sigma_e^2}{2\lambda^2}. \tag{8.34}$$

When environmental variance is defined on the logarithmic scale as $\omega_e^2 = \mathrm{Var}(\ln(\Lambda))$, we can write Eq. (8.32) with logarithms and use $\frac{\partial r}{\partial a_{ij}} = \frac{1}{\lambda}\frac{\partial\lambda}{\partial a_{ij}}$. Then Eq. (8.34) becomes.

$$r_e \approx r - \frac{1}{2}\omega_e^2, \text{ with } \omega_e^2 = \frac{\sigma_e^2}{\lambda^2}. \tag{8.35}$$

If the matrix entries vary independently:

$$\mathrm{Var}(\Lambda) \approx \sum_{i,j} s_{ij}^2 \mathrm{Var}(A_{ij}). \tag{8.36}$$

Similarly, if some lower level parameters x_i vary independently, the variance in growth rate is

$$\sigma_\lambda^2 \approx \sum_i \left(\frac{\partial\lambda}{\partial x_i}\right)^2 \sigma_{x_i}^2. \tag{8.37}$$

The formula allows us to compute confidence intervals on λ, given confidence intervals on measured parameter values. The distribution of Λ, which can be estimated using resampling methods (`bootstrap`; Caswell 2001), is generally skewed, but can often be approximated by a normal distribution. Under this assumption, the growth rate belongs to the interval

$$\lambda \pm 1.96\sigma_\lambda \text{ with a probability of } 95\%,$$

and more generally,

$$\lambda \pm z_\alpha\sigma_\lambda \text{ with a probability of } \alpha\%,$$

where z_α is the upper $\alpha/2$ percentage point of the standard normal distribution.

For the passerine model (file `pass_0.ulm`), the sensitivities of the growth rate to changes in the demographic parameters are $s_\lambda(s_0) = 3.309$, $s_\lambda(s_1) = 0.693$, $s_\lambda(s_2) = 0.401$ for survival, $s(f_1) = 0.060$, $s_\lambda(f_2) = 0.035$ for fecundity. Juvenile survival s_0 is the most sensitive parameter in this short-lived species. It is checked for the elasticities, $e_\lambda(s_0) = e_\lambda(\sigma) = 1/T = 0.599$ (Box 8.2). We can also infer that if s_0 is known within $\sigma_{s_0} = 0.05$ then $0.78 \le \lambda \le 1.43$ with 95% confidence.

For the long-lived wolf (model file `wolf_0.ulm`), under the intermediate scenario ($\lambda = 1.085$), the parameter with the largest elasticity is the survival rate of the pack leaders, $e_\lambda(s_5) = 0.46$. The elasticity of fecundity is lower: $e_\lambda(f) = 0.14$.

8.4 Accounting for the Environment

The constant matrix model allows us to project the population if the conditions under which the demographic parameters have been measured were to be maintained.

The life cycle graph represents the genotype and part of the phenotype of an average individual. The part of the phenotype that is not accounted for in the constant matrix model depends on the influence of the biotic and abiotic environment on the life cycle. Three main phenomena are involved: (i) *density dependence* coming from the regulation of the demographic parameters by the limitation of resources; (ii) *environmental stochasticity* as the impact of variation in the environment on the demographic parameters, considered as random processes; and (iii) *spatial structure*, where subpopulations are connected by dispersal.

8.5 Density Dependence

All populations have the potential to grow exponentially, which is what the constant matrix model shows (Section 8.3.1). However, exponential growth cannot be sustained forever because living organisms depend on finite resources. Density dependence – the fact that population growth has to decrease with increasing population density – comes from *intra-specific competition* for resources (nutriment, space, mate), but may also come from *inter-specific competition* for shared resources, or even from predation (more prey sustain more predators, in turn decreasing the growth of the prey). Here, we restrict ourselves to the point of view of the population, and density dependence is basically described by a unique parameter, the carrying capacity K.

8.5.1 Density-dependent Scalar Models

A continuous time formulation of the unstructured model (8.1) is

$$\frac{1}{n}\frac{dn}{dt} = r. \tag{8.38}$$

The increase of the population is $\frac{dn}{dt}$, and the per individual rate of increase, $\frac{1}{n}\frac{dn}{dt}$, is the constant r. The differential equation is integrated as

$$n(t) = e^{rt}n(0). \tag{8.39}$$

By identification with Eq. (8.2), we obtain a relation between the rate of increase r and the growth rate:

$$\lambda = \exp(r). \tag{8.40}$$

The increase versus decrease criterion, $\lambda > 1$ versus $\lambda < 1$, translates into $r > 0$ versus $r < 0$.

To account for density dependence, one assumes that the rate of increase is maximal, equal to r, when population size is 0, decreases linearly with increasing population size, and is < 0 when population size is above the carrying capacity K, the largest number of individuals that the environment can accommodate. Hence,

$$\frac{1}{n}\frac{dn}{dt} = r\left(1 - \frac{n}{K}\right). \tag{8.41}$$

The integration of the differential equation leads to the so-called logistic equation. Here, in our discrete time context, using Eq. (8.40), we may write the density-dependent growth rate as

$$\lambda_K(n) = \exp\left(r\left(1 - \frac{n}{K}\right)\right), \tag{8.42}$$

which leads to the Ricker equation:

$$n(t+1) = \lambda\exp\left(-\frac{r}{K}n(t)\right)n(t). \tag{8.43}$$

(I do not claim that this equation is equivalent to the logistic equation.)

More generally, to account for density dependence, we make the growth rate decrease with increasing population size via some function f:

$$n(t+1) = \lambda f(n(t))n(t). \tag{8.44}$$

8.5.2 Density-dependent Matrix Models

In a density-dependent matrix model, some of the matrix entries are regulated by population sizes in the stages. The projection equation is

$$\mathbf{n}(t+1) = \mathbf{A}(\mathbf{n}(t))\ \mathbf{n}(t). \tag{8.45}$$

For example, the demographic parameter x regulated by the Ricker function is

$$x_{reg} = \exp(-(\alpha_1 n_1 + \cdots + \alpha_k n_k))x \tag{8.46}$$

where the coefficients $\alpha_1, \ldots, \alpha_k$, some of which may be 0, express the relative contribution of the stages to resource consumption. Contrarily to the unstructured model, the carrying capacity K is not apparent in this formulation. The coefficients α_i must be adjusted in order to match a pre-defined carrying capacity.

The Ricker function $f(n) = \exp(-\alpha n)$ is overcompensatory in the sense that a population overshooting the carrying capacity will be penalized by a large decrease. Other regulation function can be used, for example, the compensatory function $f(n) = \frac{1}{1+\beta n}$, for which decrease will compensate the increase.

Under density dependence, the long-term growth rate of the regulated population is 1 on average. The expected dynamical behavior is that the population increases up to the carrying capacity and then stabilizes (single point equilibrium). The trajectory then presents an S-shaped (sigmoid) pattern. However, complex dynamics can occur, and more easily when the growth rate λ (in absence of regulation) is large. Indeed, when λ is large, the population will overshoot the carrying capacity more importantly, creating a stronger feedback regulation. For not-too-large λ, damped oscillations result, and population size equilibrates back to the carrying capacity. For large λ, oscillations may not damp because of strong growth as soon as density-dependent regulation is relaxed. This can produce periodic, quasi-periodic, or chaotic dynamics.

Density dependence can also be incorporated without altering the matrix, but considering the carrying capacity K as a ceiling. If the model predicts $n(t+1) > K$, then we set $n(t+1) = K$: the population sizes in the stages are reduced in some proportion (for example, proportionately to the stable stage distribution) so that they sum up to K.

It is often assumed that small populations are far from the carrying capacity, so that density dependence need not be modeled. This is, however, not always the case (Mugabo et al. 2013). For example, observed probabilities of extinction in small populations of spiders could not be recovered without incorporating density dependence (Schoener et al. 2003). It was also shown in this study that population structure cannot be disregarded at small population sizes.

8.5.3 Parameterizing Density Dependence

Using the observed population sizes $n(t)$ along time, the model of population regulation can be fit to the data using linear or nonlinear regression of the growth rate at time t: $\lambda_t = n(t+1)/n(t)$ against $n(t)$. The procedure gives the best function f such that $\lambda_t = \lambda f(n(t))$ (Morris and Doak 2002). For the Ricker model, $f(n) = \exp(-\alpha n)$.

8.5.4 Density-dependent Sensitivities

It is possible to compute the sensitivity or elasticity of the equilibrium population size n_{eq}, or of an average of population size over some time period in case of complex dynamics, to changes in a parameter x (Grant and Benton 2000, 2003; Caswell et al. 2004).

8.6 Environmental Stochasticity

Environmental stochasticity refers to the impact of the environment on the population due to biotic interactions (e.g. competing species, predators) or abiotic factors (e.g. temperature), considered as a random process. Models for the environment range from *environmental noise*, repeatedly deviating the vital rates from their average values, to rare *catastrophic events* significantly altering these values. The demographic parameters affected by the environment are now considered as random variables, defined according to the model chosen for the environment. Population size is a random variable, N, and the population matrix a random matrix $\mathbf{A}$. Under generic conditions of *ergodicity*, which essentially means that past events are progressively forgotten by the process so that initial conditions are not determinant for the long-term dynamics, population size N admits an asymptotic distribution whose mean and variance bring information on the influence of the environment on the dynamics.

8.6.1 Models of the Environment

I) In the simplest form of a stochastic environment, the values along time of a parameter impacted by the environment are drawn from a fixed probability distribution, with mean the reference value of the parameter. Each value is drawn independently of the previous values with no temporal autocorrelation. Hence, the random variables X_t describing the variations of the parameter x over time t are *independent and identically distributed* (iid). For example, the impact of the environment on fecundity f is modeled by drawing the stochastic fecundities along time according to the normal distribution:

$$f_e \equiv \mathrm{N}^+ (f, \sigma_f). \qquad (8.47)$$

Here, the + sign indicates that only nonnegative values are kept (as fecundity cannot be negative), and σ_f is the standard deviation, measuring the impact of environmental stochasticity on f. A similar approach is taken for survival rates, where the random values are constrained in the interval [0,1]. A beta distribution can be used in this case.

II) A *Markovian environment* accounts for *temporal autocorrelation* where the outcome of an event at a given date depends on previous outcomes. In the iid environment, no memory is kept of previous outcomes of the random parameters, as when casting a die. However, environmental variables often exhibit positive autocorrelation. For example, temperatures in May in one year tend to follow temperatures in May the previous year. To give an example of the Markovian environment, assume that there are good years (G) and bad years (B), and that the parameter X has value x_G in G-years and value x_B in B-years (with $x_G > x_B$). Assume that G-years and B-years are equiprobable. We parameterize positive autocorrelation in the environment (i.e. G-years are more likely to be followed by G-years) by α, $0 \leq \alpha \leq 0.5$, and use a two-states Markov chain. The states occur with the same frequency, 0.5. Switching states occurs with probability α, and remaining in the same state occurs with probability $1 - \alpha$, as described by the Markov matrix:

$$\begin{bmatrix} 1-\alpha & \alpha \\ \alpha & 1-\alpha \end{bmatrix}. \tag{8.48}$$

Hence, a low value of α means a strong autocorrelation. As α increases, the strength of autocorrelation decreases, and $\alpha = 0.5$ corresponds to no autocorrelation. For $\alpha > 0.5$, we have negative autocorrelation (switching states is more likely). More generally an autocorrelated environment can be modeled by a finite state Markov chain and a population matrix associated with each state (Cohen 1977; Caswell and Kaye 2001; Tuljapurkar and Haridas 2006). At each time step, an environmental state is drawn using the Markov chain, and the corresponding matrix is used in the projection equation.

A simple way to implement temporal autocorrelation is to use an autoregressive process of order 1 to update a parameter X with mean μ and variance σ^2:

$$X(t+1) = \alpha(X(t) - \mu) + B(t), \tag{8.49}$$

where $B(t)$ is iid with mean μ and variance $\sigma^2(1 - \alpha^2)$. Autocorrelation is parameterized by α, $-1 \leq \alpha \leq 1$, with positive autocorrelation for $\alpha > 0$ (see model file `pass_ea.ulm`).

III) *Catastrophic events* can be modeled by an occurrence frequency and a given impact. For example, when a catastrophe occurs, some parameters are reduced to 50% of their reference value, or total population size is reduced to a given proportion. Catastrophes may also present temporal autocorrelation. Population dynamics under catastrophic regimes (large deviations) are less known theoretically, but can be easily simulated (Lande 1993).

At a given time step, different demographic parameters affected by the environment may covary due to joint effects of environmental conditions, or to life-history tradeoffs. For example, a random reduction in survival is associated with a reduction in fecundity (positive covariance), or, less likely, with an increase in fecundity (negative covariance).

8.6.2 Stochastic Dynamics

Under environmental stochasticity, the population matrix is stochastic, depending on t, so that the deterministic model for constant conditions (Eq. 8.4) is now written with a stochastic population vector as:

$$\mathbf{N}(t+1) = \mathbf{A}_t\mathbf{N}(t). \tag{8.50}$$

A *population trajectory* is a realization of this random process. Underlying the process is the constant matrix $\mathbf{A}$, with $\mathrm{E}(\mathbf{A}) = \mathbf{A}$ in case of environmental noise (environments I and II above).

Population growth is a multiplicative process, here affected at each time step by random events. It can be shown by the multiplicative version of the central limit theorem that the distribution of total population size, $N(t)$, is lognormal meaning that $\ln(N(t))$ is normally distributed. Hence, it is convenient to use the logarithmic scale. Eq. (8.11) shows that, for a constant matrix $\mathbf{A}$:

$$\lambda = \lim_{t\to\infty} n(t)^{\frac{1}{t}}, \text{giving}\, r = \ln(\lambda) = \lim_{t\to\infty} \frac{1}{t}\ln(n(t)). \tag{8.51}$$

The formula indicates that the *stochastic rate of increase* can be defined as

$$r_e = \lim_{t\to\infty} \frac{1}{t}\mathrm{E}[\ln(N(t))], \tag{8.52}$$

and the *environmental variance* can be calculated as $\omega_e{}^2 = \lim_{t\to\infty} \frac{1}{t}\mathrm{Var}[\ln(N(t))]$. It can be demonstrated that these are indeed the relevant descriptors (Caswell 2001). The rate r_e is in general less than the rate $r = \ln(\lambda)$ in absence of environmental stochasticity, the difference increasing with variance in the environment (Box 8.3). The logarithm of population size, $\ln(N(t))$, is normally distributed with mean $r_e t$ and variance $\omega_e{}^2 t$. Thus, the mean of $\ln(N)$ increases or decreases linearly with time, depending on $r_e > 0$ or $r_e < 0$, and the distribution of $\ln(N)$ spreads over time, with linearly increasing variance (Morris and Doak 2002). Consequently, the mean of population size $N(t)$ increases or decreases exponentially with time, and its variance increases exponentially.

To summarize, the stochastic matrix model keeps on average the exponential behavior of the underlying constant model, at rate $\lambda_e = \exp(r_e) < \lambda$, but random fluctuations create opportunity for extinction. Stochasticity may entail negative growth, hence certain extinction in the long run even when the underlying constant model has positive growth. In this latter case, the mean values of the demographic parameters entail positive growth, but fluctuations around these values may produce negative growth on average: the threshold depends on the intensity of the fluctuations, namely on the environmental variance (Box 8.3).

If the *probability of extinction* is plotted against time, the typical pattern is an S-shaped curve. The curve starts from 0 at $t = 0$, then increases to eventually plateau at the

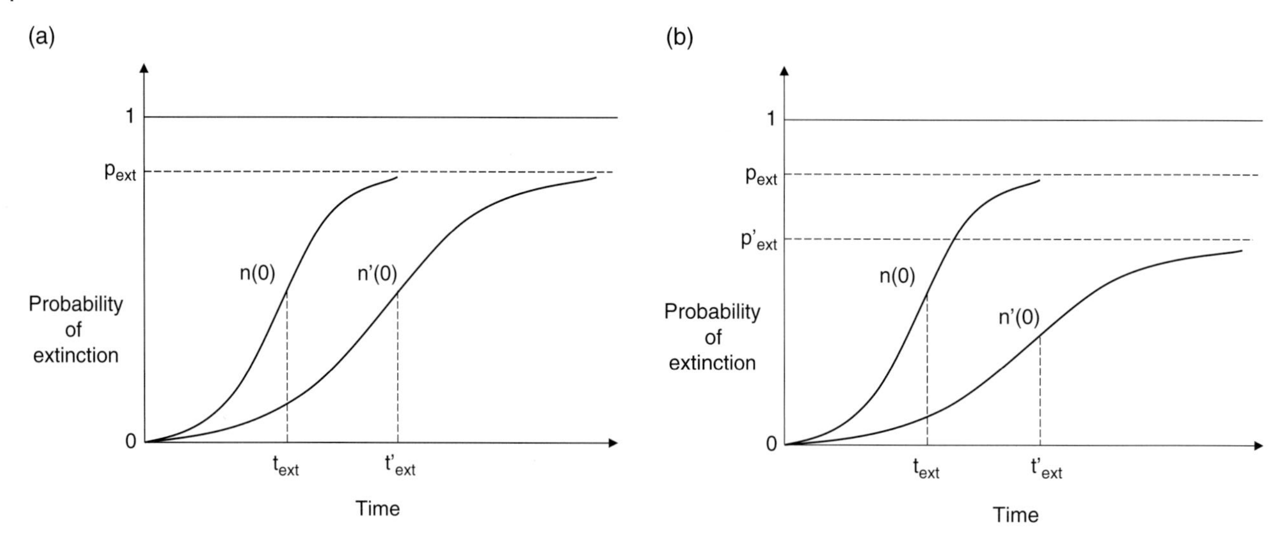

Figure 8.2 Schematic representation of the probability of extinction along time: p_{ext} is the ultimate probability of extinction, t_{ext} the mean time to extinction, and $n(0) < n'(0)$ is the initial population size. (a) Environmental stochasticity: the ultimate probability of extinction is independent of initial population size, the mean time to extinction increases with increasing initial population size. (b) Demographic stochasticity: the probability of extinction increases with decreasing population size, the mean time to extinction increases with increasing population size.

ultimate probability of extinction for large t. The portion of the curve where the probability of extinction increases sharply corresponds to the mean time to extinction. An important feature of environmental stochasticity is that the ultimate probability of extinction where the extinction curve plateaus does not depend on the initial population size $n(0)$. However, the mean time to extinction does, the extinction curve shifting to the left with decreasing $n(0)$ so that the ultimate probability of extinction is reached sooner (Figure 8.2a).

From a computational point of view, stochastic processes are best studied using Monte Carlo simulation (Box 8.6). A large number m of trajectories (e.g. $m = 1000$) are drawn up to some time horizon, and statistics are performed on the set of trajectories. The trajectories that go below a predefined threshold h at time τ (i.e. $n(\tau) < h$) are declared extinct, with associated extinction time τ. Usually the threshold $h = 1$ is used. Larger values of h (e.g. $h = 50$) define a *quasi-extinction threshold*. The probability of extinction at time t, $p_{ext}(t)$, is estimated as the proportion of the m trajectories going extinct at time $\tau \leq t$. The extinction time is estimated as the mean value of the extinction times of the extinct trajectories. Quasi-extinction thresholds and probabilities provide estimates of *minimum viable population sizes*, a mainstay of *Population Viability Analysis* (PVA, Boyce 1992). Denoting $n^{(i)}(t)$ the i-th trajectory of the Monte Carlo simulation, the expected population size along time, $\mathrm{E}(N(t))$, is estimated by the average trajectory

$$\bar{n}(t) = \frac{1}{m}\sum_{i=1}^{m} n^{(i)}(t). \tag{8.53}$$

The mean population size over non-extinct trajectories, $\mathrm{E}(N(t)|N(t) > h)$, can also be computed. The *stochastic rate of increase* is estimated using:

$$r_e = \frac{1}{m}\sum_{i=1}^{m} \frac{\ln\left(n^{(i)}(t)\right) - \ln\left(n(0)\right)}{t}. \tag{8.54}$$

Underlying this computation is the meaning of ergodicity: the long-term time average of a quantity over a single trajectory can be estimated using the short-term average over a large number of trajectories.

For the passerine model, let us consider that juvenile survival s_0 is subjected to random fluctuations under an iid model of the environment. We use a beta distribution, here parameterized according to mean and standard deviation: $s_{0e} \equiv \mathrm{Beta}(s_0, \sigma_{s_0})$, with the reference value $s_0 = 0.2$ and $\sigma_{s_0} = 0.15$. Monte Carlo simulation up to time 50 is run over 10,000 trajectories with 20 individuals as initial population size, 10 subadults and 10 adults, close to the stable age distribution. At time 50, the stochastic growth rate is $\lambda_e = 1.008$, the probability of extinction (threshold $h = 1$) is $p_{ext}(50) = 0.25$, the mean time to extinction is $t_{ext}(50) = 7$ years, the mean time to extinction over non-extinct trajectories is $t_{ext}^*(50) = 28$ years. We observe that small fluctuations in the most sensitive parameter s_0 reduce the growth rate by 10% (from $\lambda = 1.1050$ to $\lambda_e = 1.008$) and produce a large probability of extinction with important disparity across trajectories. The probability of quasi-extinction with threshold $h = 10$ at time 50 is $q_{ext}(50) = 0.73$ (model file `pass_e.ulm`).

Box 8.4 Computing the Growth Rate

The algorithm uses the fact that $\frac{n(t+1)}{n(t)} \to \lambda$ is geometrically fast when iterating Eq. (8.4) (power method). In the following, **w**, **w**′ are vectors for the population structure, **v**, **v**′ are vectors for the reproductive value, and **A** is the $k \times k$ projection matrix, assumed primitive (Box 8.1). The norm of a vector **u** is defined as $|\mathbf{u}| = \sum_i u_i$. The algorithm stops when successive values of λ are within $\varepsilon = 10^{-9}$ (or when t is above some time horizon $t_{\max} = 10\,000$, in which case it fails).

$\mathbf{w} \leftarrow [1/k, \cdots, 1/k]$ initial value of population structure
$\mathbf{v} \leftarrow [1/k, \cdots, 1/k]$ initial value of reproductive value

$\lambda \leftarrow 1$

$t \leftarrow 0$

repeat
- $t \leftarrow t+1$ increment time
- $\mathbf{w}' \leftarrow \mathbf{Aw}$ update population structure
- $norm \leftarrow |\mathbf{w}'|$ compute norm
- $\mathbf{w} \leftarrow \mathbf{w}'/norm$ rescale population structure
- $\lambda_1 \leftarrow \lambda$ memorize previous value of λ
- $\lambda \leftarrow norm$ update λ
- $\mathbf{v}' \leftarrow \mathbf{vA}$ update reproductive value
- $norm \leftarrow |\mathbf{v}'|$ compute norm
- $\mathbf{v} \leftarrow \mathbf{v}'/norm$ rescale reproductive value

until $|\lambda - \lambda_1| < \varepsilon$ or $t > t_{\max}$.

At the end of the algorithm, λ contains the dominant eigenvalue, **w** the stable population structure, and **v** the reproductive value.

Box 8.5 Environmental Variance and Demographic Variance

Engen et al. (1998) apply the law of total variance to the stochastic contribution W of a female individual to population size at the next time step, given the environment Z:

$$\mathrm{Var}(W) = \mathrm{E}(\mathrm{Var}(W|Z)) + \mathrm{Var}(\mathrm{E}(W|Z)). \tag{8.55}$$

This can be interpreted as

$$\mathrm{Var}(W) = \sigma_d^2 + \sigma_e^2, \tag{8.56}$$

where demographic variance $\sigma_d^2 = \mathrm{E}(\mathrm{Var}(W|Z))$ is the mean through time of the variance of individual contributions to population size from a time step to the next, and environmental variance $\sigma_e^2 = \mathrm{Var}(\mathrm{E}(W|Z))$ is the variance through time of the mean individual contribution to population size from a time step to the next.

The stochastic population matrix can be decomposed $\mathbf{A} = \mathbf{A} + \mathbf{E} + \mathbf{D}$ according to two sources of stochasticity, the environment represented by **E** and demographic stochasticity represented by **D**. It is assumed that there is no covariance between **E** and **D**, and that they have 0 expectation. The methods of Box 8.3 lead to (Engen et al. 2005):

$$\sigma_e^2 = \mathrm{Var}(\mathbf{E}|N) \approx \sum_{i,j,k,l} s_{ij} s_{kl} \mathrm{Cov}(E_{ij}, E_{kl}), \tag{8.57}$$

$$\sigma_d^2 = \mathrm{Var}(\mathbf{D}|N) \approx \sum_{i,j,k,l} s_{ij} s_{kl} \mathrm{Cov}(D_{ij}, D_{kl}). \tag{8.58}$$

The total variance in the stochastic growth rate Λ is then

$$\mathrm{Var}(\Lambda|N) = \sigma_e^2 + \frac{\sigma_d^2}{N}. \tag{8.59}$$

The formula highlights the $1/N$ dependency of the variance in Λ with respect to demographic stochasticity. The corresponding variance in population size is

$$\mathrm{Var}(N(t+1)|N(t) = N) = \sigma_e^2 N^2 + \sigma_d^2 N. \tag{8.60}$$

Hence, although environmental variance is usually smaller than demographic variance, the environmental term dominates for large populations. Demographic stochasticity can in fact be neglected when $N \gg \frac{\sigma_d^2}{\sigma_e^2}$ (Lande et al. 2003).

The stochastic rate of increase (environmental plus demographic stochastitity) can be written:

$$s(N) \approx r - \frac{\sigma_e^2}{2\lambda^2} - \frac{\sigma_d^2}{2\lambda^2 N}, \tag{8.61}$$

where the negative impact of the environment agrees with Eq. (8.34). The environmental and demographic variances can be estimated from individual data, provided that the dataset is large enough (Engen et al. 2005). The preceding results also hold when demographic stochasticity is replaced by demographic heterogeneity (Vindenes et al. 2008).

8.6.3 Parameterizing Environmental Stochasticity

The value of the variance σ_X^2 of a random parameter X can be estimated from the observed time series of the parameter, and put in relation with known environmental conditions on the site, such as a time series of temperatures. Environmental variance σ_e^2 can also be recovered from individual data using the methods of Engen et al. (2005) (Box 8.5).

More generally, environmental stochasticity can be introduced to test the robustness of the species and estimate its probability of extinction, without precise reference to the environment. The strength of environmental noise can be parameterized from low to high to produce various scenarios. A rule of thumb is that $\sigma_X \approx 0.3\mathrm{E}(X)$ in natural populations (Mills et al. 1996).

8.7 Spatial Structure

Populations are in general not closed, but subject to immigration and emigration. A *metapopulation* is a set of populations of the same species living on separate sites connected by dispersal from one site to another (Morris and Doak 2002). In a metapopulation, a local population going extinct may start again through recolonization from neighboring sites, ensuring the persistence of the metapopulation (rescue effect). The metapopulation framework can, to a certain extent, be studied using matrix models. Demographic descriptors generalizing those of Section 8.5 can be computed for constant age-classified multisite matrix models (Lebreton 1996). A simple example is given in Figure 8.1j, where an age-classified model is connected from a stage I representing the pool of immigrants, and to a stage E representing the pool of emigrants. To account for space, matrix population models can also be incorporated into *diffusion models*, for example, to study expanding waves of invading species (Neubert and Caswell 2000). They could also underlie the local rules of cellular automata.

We use the passerine model to show counterintuitive features of metapopulation demography (model file `pass_ab.ulm`). Figure 8.1k presents the life cycle graph of two passerine populations living on sites A and B, with juveniles dispersing from A to B at rate m_{AB} and from B to A at rate m_{BA}.

We first assume that the two populations have the same reference demographic parameters so that their growth rates are identical: $\lambda_A = \lambda_B = 1.1050$. With the dispersal rates $m_{AB} = 0.2$, $m_{BA} = 0.2$, the growth rate of the metapopulation is $\lambda = 1.0978$, slightly lower than in absence of dispersal. The A-entries and the B-entries of the metapopulation age structure and reproductive value vector are identical and positive. When the dispersal rates are $m_{AB} = 0.2$, $m_{BA} = 0$ (no dispersal from B to A, the matrix is reducible, see Box 8.1), the growth rate of the metapopulation is $\lambda = \lambda_A = 1.1050$. In this case, the A-entries of the metapopulation structure are 0 because the A-population is depopulated; the entries of the metapopulation reproductive value are all positive, those of the B-population being larger than those of the A-population.

Box 8.6 Monte Carlo Simulation

The Monte Carlo simulation casts m trajectories up to some time horizon t_{max}. We compute the average population size $\bar{n}(t)$ along time, the stochastic rate of increase r_e, and the probability of quasi-extinction p_{ext} with threshold h at time t_{max}.

In the following, $\mathbf{n}$ is the population vector, $|\mathbf{n}| = \sum_i n_i$ is population size, $\mathbf{A}$ is a trajectory-specific realization of the stochastic projection matrix $\mathbf{A}$, and $\bar{\mathbf{n}}$ is a vector of size at least t_{max} for computing the average population size along time.

```
for t ← 1 to t_max do n̄[t] ← 0   initialize average trajectory
ext ← 0   initialize number of extinctions
r_e ← 0   initialize stochastic rate of increase
for i ← 1 to m do   loop over trajectories
    SetSeed(i)   set random generator seed to trajectory-specific value
    n ← n_0   initialize population vector
    n_0 ← |n|   memorize initial population size
    for t ← 1 to t_max do   loop over time
        n ← An   update population vector
        n ← |n|   compute population size
        if n < h then ext ← ext + 1   quasi-extinction
    endloop
    r_e ← r_e + (ln(n) − ln(n_0))/t_max   update stochastic rate of increase
endloop
for t ← 1 to t_max do n̄[t] ← n̄[t]/m   compute average trajectory
p_ext ← ext/m   probability of extinction
r_e ← r_e/m   stochastic rate of increase
```

Note that a trajectory declared quasi-extinct is nevertheless computed up to t_{max}, hence contributes to the average trajectory $\bar{\mathbf{n}}$, and to the stochastic rate r_e.

We now assume that the B-population is decreasing, with reduced fecundities (f_{B1} = 5, f_{B2} = 6, leading to $\lambda_B = 0.9582$). When the dispersal rates are $m_{AB} = m_{BA} = 0.2$, the metapopulation growth rate is altered, $\lambda = 1.0651$, the A-entries of the metapopulation structure are larger than the B-entries, and a similar pattern holds for reproductive values. When the dispersal rates are $m_{AB} = 0.2$, $m_{BA} = 0$, we are in the case of a source-sink metapopulation. The metapopulation growth rate is further altered, $\lambda = 1.0539$, the A-entries of the metapopulation structure are larger than the B-entries, and the B-entries of the metapopulation reproductive value are 0 because the B-population does not contribute to growth. When the dispersal rates are $m_{AB} = 0$, $m_{BA} = 0.2$, juveniles disperse from the decreasing population to the increasing one. The metapopulation growth rate is $\lambda = \lambda_A = 1.1050$, the B-entries of the metapopulation structure are 0, and the A-entries of the metapopulation reproductive value are larger than the B-entries.

8.8 Demographic Stochasticity

Demographic stochasticity is the chance realization of the transitions in the life cycle graph by individuals. Demographic stochasticity is an intrinsic feature of the demographic process, independent of the environment. Moreover, it is unavoidable. As we shall see, the effects of demographic stochasticity are only sensible when population size is small, but the contribution to the extinction risk can be important for small populations.

8.8.1 Branching Processes

We still assume the individuals are identical, but account for heterogeneity in their fates; that is, different realizations of their vital rates. *Demographic heterogeneity*, where individuals are not assumed to be identical, accounts for different realizations of heterogeneous life cycles (Section 8.9).

In the life cycle graph, when n individuals survive from one stage to the next, say with rate s, some of them may either survive (with probability s) or die (with probability $1 - s$). The number n' of survivors is the sum of n draws of the Bernoulli (head and tail) distribution with parameter s, or equivalently, a single draw according to the binomial distribution:

$$n' \equiv \mathrm{Binom}(n,s). \tag{8.62}$$

Every 0/1 transition (sex determination, dispersal status) can be treated in the same way. Similarly, when n individuals reproduce with fecundity rate f, the number n' of offspring is the sum of n draws of the Poisson distribution with parameter f, which we denote as

$$n' \equiv \mathrm{Poisson}(n,f). \tag{8.63}$$

Integer distributions other than the Poisson distribution can be used for reproduction.

The above described randomization of the life cycle events amounts to constructing a branching process on the relations associated with the life cycle graph (Eq. 8.3). For the passerine model, the relations of the branching process built from the matrix relations are:

$$n_1(t+1) = \mathrm{Binom}(\mathrm{Poisson}(n_1(t),f_1) + \mathrm{Poisson}(n_2(t),f_2),\sigma s_0), \tag{8.64}$$

$$n_2(t+1) = \mathrm{Binom}(n_1(t),s_1) + \mathrm{Binom}(n_2(t),s_2). \tag{8.65}$$

In the framework of demographic stochasticity, population sizes are integer numbers. The population is extinct as soon as its size is 0. Even when $\lambda > 1$, the probability of extinction is nonzero. Indeed, there is a nonzero probability that all individuals in a stage do not survive or do not reproduce. Such an event is rare when population size is large. Hence, demographic stochasticity mostly plays at small population sizes. This moreover suggests that at large population size, with or without demographic stochasticity, dynamical behaviors will be close. When incorporating demographic stochasticity, the average behavior of the random process is that of the constant underlying model, with exponential growth or decrease at rate λ and average population structure $\mathbf{w}$. The behavior differs for the probability of extinction. In the constant model, the probability of extinction is 1 when $\lambda < 1$, and 0 when $\lambda \geq 1$. Under demographic stochasticity, the probability of extinction is 1 when $\lambda \leq 1$, and has a definite nonzero value when $\lambda > 1$.

Denoting $q_i(t)$ the probability of extinction at time t when the initial population consists in a single individual in stage i, the probability of extinction at time t when the initial population consist in $n_i(0)$ individuals in stage i is

$$p(t) = q_i(t)^{n_i(0)}. \tag{8.66}$$

If $q_i(t) = 1$, certain extinction occurs. If $q_i(t) < 1$, the above relation shows that the extinction risk from demographic stochasticity decreases exponentially with increasing population size. The overall probability of extinction at time t is

$$p(t) = q_1(t)^{n_1(0)} \cdots q_k(t)^{n_k(0)}, \tag{8.67}$$

depending on initial population size and structure. The ultimate probability of extinction is

$$p = \lim_{t \to \infty} p(t). \tag{8.68}$$

When $\lambda \leq 1$ in the underlying constant matrix model, we have certain extinction under demographic stochasticity (in finite time for $\lambda < 1$, and in infinite expected time for $\lambda = 1$). The probability of extinction along time presents an S-shaped pattern, as for environmental stochasticity. However, under demographic stochasticity, for $\lambda > 1$ the curve plateaus at a value which depends on initial population size, whereas this is not the case for environmental stochasticity (Figure 8.2b).

8.8.2 Two-sex Models

When population size is small (say less than 50–100 individuals), each individual behavior matters, and demographic stochasticity can be an important factor of the extinction risk. Individual fitness may decrease at low population density due to the deterioration of social bonds or cooperative behavior, a phenomenon known as the *Allee effect* (Courchamp et al. 1999, 2008). An important driver of the Allee effect comes from the mating system which relies on behavior (sexual selection, Andersson 1994), but it is also sensitive to demographic stochasticity acting on the number of males and females (Bessa-Gomes et al. 2004).

Consider a population of males and females with monogamous pair formation and primary 1 : 1 sex ratio where the proportion of females at birth is 0.5. Computation shows that, because of demographic stochasticity alone, the probability of an individual being mated is 0.9 when there are 100 individuals, but drops to 0.75 when there are 10 individuals. The probability drops even further if females are choosy due to sexual selection, and to 0.55 when population size is 10 (Møller and Legendre 2001). For small populations with sexual reproduction, the male portion of the population cannot be ignored, even if the male and female life cycles are identical. The population is doomed to extinction when there are either no males or no females left, a situation which can occur because of demographic stochasticity. Hence, for small populations, two-sex models are recommended. For example, observed probabilities of extinction in the polygynous bighorn sheep *Ovis canadensis* could not be reconstructed using a model based on females only (Legendre 2004).

Two-sex models incorporate the life cycle of males, which may differ from the life cycle of females, notably in the case of sexual dimorphism. In the two-sex life cycle (Figure 8.11), the male and female parts are coupled by the pair formation process. The mating process is modeled using a function giving the number of matings at a given time. For example, the monogamous mating function

$$\mu(f,m) = \min(f,m) \tag{8.69}$$

counts the maximum number of monogamous pairs that can be formed given the number of reproductive males (m) and females (f). The polygynous mating function:

$$\mu(f,m) = \min(f,hm) \tag{8.70}$$

gives the number of matings when each male mates on average with h females, h being harem size. The *breeding sex ratio* is the proportion of reproductive females in the reproductive population at a given time:

$$\rho = \frac{f}{m+f}. \tag{8.71}$$

We have $\dfrac{\min(f,hm)}{f+m} = \min\left(\dfrac{f}{f+m}, \dfrac{hm}{f+m}\right) = \min(\rho, h(1-\rho))$, so that the optimal breeding sex ratio is obtained when $\rho = h(1-\rho)$, or

$$\rho_{opt} = \frac{h}{h+1}. \tag{8.72}$$

For monogamy, $\rho_{opt} = 0.5$, as expected. Care must be taken to appropriately design the mating function (Bessa-Gomes et al. 2010).

The mating function introduces a nonlinearity in the underlying constant model which becomes *frequency dependent*, depending on the relative proportions of each sex (Caswell and Weeks 1986). The generic behavior is nevertheless that of exponential growth after transient fluctuations. However, in the two-sex model, transient fluctuations come from both the convergence toward the stable stage distribution and toward the stable sex structure, where the breeding sex ratio becomes constant. The transient fluctuations in sex structure superimposed on those in stage structure suggest that the two-sex model under demographic stochasticity leads to a larger extinction risk than the corresponding female-based model (although fluctuations could cancel one another in some instances). This is indeed the case, and the effect is more pronounced for short-lived species (Legendre et al. 1999), in which parameters associated with reproductive transitions are more sensitive (Box 8.2). Moreover, the growth rate λ_μ of the two-sex model with mating function μ verifies $\lambda_\mu \leq \lambda$. The difference between the two-sex growth rate λ_μ and the female-based growth rate λ depends on how far the realized breeding sex ratio ρ is from the optimal one ρ_{opt} (Legendre 2004). A two-sex model for passerines with monogamous pair formation (Figure 8.11), and incorporating demographic stochasticity, is given in model file `pass_2d.ulm`.

8.9 Demographic Heterogeneity

Demographic heterogeneity accounts for the fact that individuals in a population are not identical. Individuals can differ genetically, by their ontogenic trajectories, and by the plastic adaptation of their phenotypes to heterogeneous environmental pressures, possibly involving epigenetics. In addition, social status and behavior may vary over time. Figure 8.1m displays a simple example: newborn individuals j in stage 1 can mature according to two pathways depending on a given developmental function $J(j)$.

Within demographic heterogeneity, demographic stochasticity is at play as the chance realization of heterogeneous demographic parameters, or heterogeneous life cycle trajectories as in Figure 8.1m. Thus, demographic stochasticity is a component of demographic heterogeneity, but demographic heterogeneity also has an intrinsic component which can be deterministic or random, and is termed *individual heterogeneity*. Exploration of demographic heterogeneity is relatively recent, and the terminology is not yet standardized.

Demographic heterogeneity can be modeled by *individual based models* (IBM, Chapter 9). Insight is nevertheless gained from mathematical models (Box 8.5). A notable feature is that, under demographic heterogeneity, the relation between individual variation and population variation is complex: demographic variance may be either larger or smaller as compared to the corresponding homogeneous case (Kendall and Fox 2002, 2003; Vindenes et al. 2008). Individual heterogeneity is more likely to play a role in long-lived species than in short-lived species, where there is less correlation from one generation to the next.

Another insight is that more accurate models can be built when using total reproductive value instead of total population size (Engen et al. 2009). The *total reproductive value* at time t is calculated as $V(t) = \mathbf{vn}(t) = \sum_i v_i n_i(t)$, where the reproductive value $\mathbf{v}$ is normalized such that $\mathbf{vw} = 1$. The metric is less sensitive than total population size N to fluctuations in population structure $\mathbf{w}(t)$ over time, and nevertheless verifies a similar relationship (Box 8.5):

$$\mathrm{Var}(V(t+1)|V(t)=V) \approx \sigma_e{}^2 V^2 + \sigma_d{}^2 V. \qquad (8.73)$$

The total reproductive value is the sum of individual reproductive values, which can be computed from individual data. *Environmental variance* $\sigma_e{}^2$ and *demographic variance* $\sigma_d{}^2$ can then be estimated. Fluctuations in population growth are affected by heterogeneity in individual reproductive values rather than heterogeneity in survival and reproduction.

A general framework is that, on top of the life cycle, individuals go along with their lives through a finite number of states with probabilities of switching between states given by a Markov matrix. The diversity of the individual trajectories or the successive states taken by an individual during its life reflects the demographic heterogeneity in the population.

Tuljapurkar et al. (2009) exemplify the case using capture-mark-recapture data with states corresponding to offspring number. For example, if the states labeled 1, 2, 3, correspond to offspring number 0, 1–3, 4+, a possible individual trajectory is $(1, 2, 2, 1, 3, \ldots)$. These authors quantify the degree of heterogeneity in the population (called *dynamic heterogeneity*) using the entropy of the Markov matrix.

Caswell (2009) built a Markov matrix $\mathbf{u}$ from the stage-classified life cycle graph by removing reproductive arcs and incorporating an absorbing state corresponding to death. The states of this Markov chain are all transient, and probabilities to switch between states are given by the Markov matrix:

$$\mathbf{z} = \sum_{i=0}^{\infty} \mathbf{u}^i = (\mathbf{1} - \mathbf{u})^{-1}. \qquad (8.74)$$

Sensitivities of demographic descriptors can be computed from the matrix $\mathbf{z}$ to quantify demographic heterogeneity (called *individual stochasticity*).

Physiologically structured population models (PSPM, González-Suárez et al. 2011) work in a continuous-time continuous-state framework. The models rely on the bioenergetic mechanisms conditioning the vital rates, and allow us to finely track dynamics of cohorts, that is, sets of individuals born at the same time step, and to detect heterogeneous phenotypes in a population (Claessen et al. 2000).

8.9.1 Integral Projection Models

Some traits, like size, are best represented by continuous variables. Variations over time may follow smooth functions which can be parameterized using regression techniques. *Integral projection models* (IPM, Easterling et al. 2000; Ellner and Rees 2006, 2007; Briggs et al. 2010; Merow et al. 2014) use *continuous states* instead of the discrete stages or age-classes of matrix models. IPMs are good where sample sizes are small, or if discrete age- or size-classes are difficult to identify. They should sometimes be preferred to matrix models (Ramula et al. 2009). Projection equations analogous to Eq. (8.3) are obtained by summing over the space Ω of continuous states:

$$n(y, t+1) = \int_{\Omega} k(y, x) n(x, t) dx. \qquad (8.75)$$

Here $n(x, t)$ is the distribution at time t of the number of individuals bearing the value (state) x of the trait. The kernel $k(y, x)$, which is an analogue of the projection matrix, can in practice be replaced by a large finite matrix. Integral projection models can also be used to account for demographic heterogeneity (Vindenes et al. 2011; Vindenes and Langangen 2015).

8.10 Software Tools

Matrix population analyses can be conducted with the `popbio` and `IPMpack` packages in Program `R`, or with general purpose programs for mathematical analyses such as `Matlab` or `Mathematica`. For this purpose of the online exercises, I demonstrate how matrix population modeling can be conducted with the `ULM` computer program (Legendre and Clobert 1995; Ferrière et al. 1996; Legendre 2008).

8.11 Online Exercises

The online exercises include example projection matrices for songbirds and wolves. Seven exercises illustrate models based on constant deterministic matrices, models with environmental stochasticity (with and without autocorrelation), a two-site model, and a two-sex model with demographic stochasticity.

8.12 Future Directions

I conclude by reviewing the main steps to construct a population dynamics model in a user-specific case. The principal advice is to keep the model as simple as possible, using the most basic assumptions, and analyzing different effects first separately, then in conjunction.

1) Construct the life cycle graph (Section 8.2) from what is known about the biology of the species of interest (life history, demographic parameter values), and according to the questions the biologist wants to answer concerning his case study. For unknown demographic rates, use data from the literature on a related species, or leave the entry as a parameter and explore several values (e.g. as in model file `wolf_0.ulm`). Here are examples of specific situations: Harvesting, poaching can be modeled by reducing appropriately the survival rates of the classes that are affected; Immigration, emigration, introductions can be dealt with by adding classes to the life cycle of the closed population (Figure 8.1j, Sarrazin and Legendre 2000). For uncertainty in demographic parameter values, or observation errors, see Morris and Doak (2002).
2) Build the associated matrix population model (Section 8.3.1) and compute the main demographic descriptors (growth rate, stable age distribution, reproductive value, generation time; Section 8.3.2).
3) Always perform a sensitivity analysis (Section 8.3.3). It provides insights into the demographic parameters that matter the most in the organism's life cycle (Box 8.2). Sensitivity analysis helps to determine the dynamical behavior when more complex dynamics are introduced, such as density dependence, environmental stochasticity, and demographic stochasticity. Life-table Response Experiments (LTRE) are not discussed in this chapter, but can also be useful for decomposing the contributions of different demographic rates and their covariances to the overall variance of the population growth rate (Caswell 2001).
4) An alternative to matrix models is the use of IPMs for which there exists a convenient `R` package (`IPMpack`, Metcalf et al. 2013).
5) For small populations, say less than 100 individuals, it is recommended to introduce demographic stochasticity, and build two-sex models with a mating function. Two-sex models are necessary for the PVA of dimorphic species with mating systems that feature strong sexual selection (Section 8.8.2).
6) For metapopulations, spatial modeling comes into play (Section 8.7). Matrix models can be used as a first step toward more elaborate models of dispersion.
7) In many cases, an analysis based on a deterministic matrix model without variation in environmental conditions is enough to get a good idea of the demography of the studied population (Section 8.3). One must, however, recall that the constant matrix model mainly reveals potential short-term trends under a constant environment.
8) Density dependence requires the demographic data to be parameterized by a function, acting on the right classes and with the right parameters (Section 8.5). This is rarely met, but density dependence can be modeled for exploratory purposes with guessed parameter values.
9) Environmental stochasticity requires information on demographic parameter values over time, for example, in good and bad years (Section 8.6). Like density dependence, environmental stochasticity can be introduced for exploration, for example, to test the resilience of the species and get an estimate of its probability of extinction in an environment that is more realistic than the constant conditions assumed by the deterministic model (Section 8.3).

References

Andersson, M. (1994). *Sexual Selection*. Princeton, NJ, USA: Princeton University Press.

Barot, S., Gignoux, J., and Legendre, S. (2002). Stage-classified matrix models and age estimations. *Oikos* 96: 56–61.

Bessa-Gomes, C., Legendre, S., and Clobert, J. (2004). Allee effects, mating systems and the extinction risk in populations with two sexes. *Ecology Letters* 7: 802–812.

Bessa-Gomes, C., Legendre, S., and Clobert, J. (2010). Discrete two-sex models of population dynamics: on modelling the mating function. *Acta Oecologica* 36: 439–445.

Bienvenu, F. and Legendre, S. (2015). A new approach to the generation time in matrix population models. *American Naturalist* 185: 834–843.

Boyce, M.S. (1992). Population viability analysis. *Annual Review of Ecology and Systematics* 23: 481–506.

Briggs, J., Dabbs, K., Riser-Espinoza, D. et al. (2010). Structured population dynamics and calculus: an introduction to integral modeling. *Mathematics Magazine* 83: 243–257.

Brown, J.H., Gillooly, J.F., Allen, A.P. et al. (2004). Toward a metabolic theory of ecology. *Ecology* 85: 1771–1789.

Caswell, H. (2001). *Matrix Population Models – Construction, Analysis, and Interpretation*, 2e. Sunderland, Massachusetts: Sinauer Associates.

Caswell, H. (2009). Stage, age and individual stochasticity in demography. *Oikos* 118: 1763–1782.

Caswell, H. and Kaye, T.N. (2001). Stochastic demography and conservation of an endangered perennial plant *(Lomatium bradshawii)* in a dynamic fire regime. *Advances in Ecological Research* 32: 1–51.

Caswell, H., Takada, T., and Hunter, C.M. (2004). Sensitivity analysis of equilibrium in density-dependent matrix population models. *Ecology Letters* 7: 380–387.

Caswell, H. and Weeks, D.E. (1986). Two-sex models: chaos, extinction, and other dynamic consequences of sex. *American Naturalist* 128: 707–735.

Chapron, G., Legendre, S., Ferrière, R. et al. (2003). Conservation and control strategies for the wolf (*Canis lupus*) in western Europe based on demographic models. *Comptes Rendus Biologies* 326: 575–587.

Claessen, D., de Roos, A.M., and Persson, L. (2000). Dwarfs and giants: cannibalism and competition in size-structured populations. *American Naturalist* 155: 219–237.

Cochran, M.E. and Ellner, S. (1992). Simple methods for calculating age-based life history parameters for stage-structured populations. *Ecological Monographs* 62: 345–364.

Cohen, J.E. (1977). Ergodicity of age structure in populations with Markovian vital rates, III: finite-state moments and growth rate; an illustration. *Advances in Applied Probability* 9: 462–475.

Courchamp, F., Berec, L., and Gascoigne, J. (2008). *Allee Effects in Ecology and Conservation*. Oxford: Oxford University Press.

Courchamp, F., Clutton-Brock, T., and Grenfell, B. (1999). Inverse density dependence and the Allee effect. *Trends in Ecology and Evolution* 14: 405–410.

Crouse, D.T., Crowder, L.B., and Caswell, H. (1987). A stage-based population model for loggerhead sea turtles and implications for conservation. *Ecology* 68: 1412–1423.

Demetrius, L., Legendre, S., and Harremöes, P. (2009). Evolutionary entropy: a predictor of body size, metabolic rate and maximal life span. *Bulletin of Mathematical Biology* 71: 800–818.

Easterling, M.R., Ellner, S.P., and Dixon, P.M. (2000). Size-specific sensitivity: applying a new structured population model. *Ecology* 81: 694–708.

Ellner, S.P. and Rees, M. (2006). Integral projection models for species with complex demography. *American Naturalist* 167: 410–428.

Ellner, S.P. and Rees, M. (2007). Stochastic stable population growth in integral projection models: theory and application. *Journal of Mathematical Biology* 54: 227–256.

Engen, S., Bakke, Ø., and Islam, A. (1998). Demographic and environmental stochasticity – concepts and definitions. *Biometrics* 54: 840–846.

Engen, S., Lande, R., Sæther, B.-E., and Dobson, F.S. (2009). Reproductive value and the stochastic demography of age-structured populations. *American Naturalist* 174: 795–804.

Engen, S., Lande, R., Sæther, B.-E., and Weimerskirch, H. (2005). Extinction in relation to demographic and environmental stochasticity in age-structured models. *Mathematical Biosciences* 195: 210–227.

Ferrière, R., Sarrazin, F., Legendre, S., and Baron, J.-P. (1996). Matrix population models applied to viability analysis and conservation: theory and practice with ULM software. *Acta Oecologica* 17: 629–656.

Gamelon, M., Gimenez, O., Baubet, E. et al. (2014). Influence of life-history tactics on transient dynamics: a comparative analysis across mammalian populations. *American Naturalist* 184: 673–683.

González-Suárez, M., Le Galliard, J.-F., and Claessen, D. (2011). Population and life-history consequences of within-cohort individual variation. *American Naturalist* 178: 525–537.

Grant, A. and Benton, T.G. (2000). Elasticity analysis for density dependent populations in stochastic environments. *Ecology* 81: 680–693.

Grant, A. and Benton, T.G. (2003). Density-dependent populations require density-dependent elasticity analysis: an illustration using the LPA model of *Tribolium*. *Journal of Animal Ecology* 72: 94–105.

Houllier, F. and Lebreton, J.-D. (1986). A renewal equation approach to the dynamics of stage grouped populations. *Mathematical Biosciences* 79: 185–197.

Jones, O.R., Scheuerlein, A., Salguero-Gomez, R. et al. (2014). Diversity of ageing across the tree of life. *Nature* 505: 169–173.

Kendall, B.E. and Fox, G. (2002). Variation among individuals and reduced demographic stochasticity. *Conservation Biology* 16: 109–116.

Kendall, B.E. and Fox, G. (2003). Unstructured individual variation and demographic stochasticity. *Conservation Biology* 17: 1170–1172.

Koons, D.N., Holmes, R.R., and Grand, J.B. (2007). Population inertia and its sensitivity to changes in vital rates and population structure. *Ecology* 88: 2857–2867.

Lande, R. (1993). Risks of population extinction from demographic and environmental stochasticity and random catastrophes. *American Naturalist* 142: 911–927.

Lande, R., Engen, S., and Sæther, B.E. (2003). *Stochastic Population Dynamics in Ecology and Conservation.* Oxford University Press.

Lebreton, J.-D. (1996). Demographic models for subdivided populations: the renewal equation approach. *Theoretical Population Biology* 49: 291–313.

Lebreton, J.D., Burnham, K.P., Clobert, J., and Anderson, D. R. (1992). Modeling survival and testing biological hypotheses using marked animals – a unified approach with case-studies. *Ecological Monographs* 62: 67–118.

Legendre, S. (2004). Influence of age structure and mating system on population viability. In: *Evolutionary Conservation Biology* (eds. R. Ferrière, U. Dieckmann and D. Couvet), 41–58. Cambridge University Press.

Legendre S 2008. `ULM` computer program, version 4.5. http://www.biologie.ens.fr/~legendre/ulm/ulm.html.

Legendre, S. and Clobert, J. (1995). ULM, a software for conservation and evolutionary biologists. *Journal of Applied Statistics* 22: 817–834.

Legendre, S., Clobert, J., Møller, A.P., and Sorci, G. (1999). Demographic stochasticity and social mating system in the process of extinction of small populations: the case of passerines introduced to New Zealand. *American Naturalist* 153: 449–463.

Merow, C., Dahlgren, J., Metcalf, C.J.E. et al. (2014). A user's guide to advances in demography with integral projection models. *Methods in Ecology and Evolution* 5: 99–110.

Metcalf, C.J.E., McMahon, S.M., Salguero-Gómez, R., and Jongejans, E. (2013). `IPMpack`: an `R` package for integral projection models. *Methods in Ecology and Evolution* 4: 195–200.

Mills, L.S., Hyes, S.G., Baldwin, C. et al. (1996). Factors leading to different viability predictions for a grizzly bear data set. *Conservation Biology* 10: 863–873.

Møller, A.P. and Legendre, S. (2001). Allee effect, sexual selection and demographic stochasticity. *Oikos* 92: 27–34.

Morris, W.F. and Doak, D.F. (2002). *Quantitative Conservation Biology: Theory and Practice of Population Viability Analysis.* Sinauer Associates.

Mugabo, M., Perret, S., Legendre, S., and Le Galliard, J.-F. (2013). Density-dependent life history and the dynamics of small populations. *Journal of Animal Ecology* 82: 1227–1239.

Neubert, M.G. and Caswell, H. (2000). Demography and dispersal: calculation and sensitivity analysis of invasion speed for structured populations. *Ecology* 81: 1613–1628.

Ramula, S., Rees, M., and Buckley, Y.M. (2009). Integral projection models perform better for small demographic data sets than matrix population models: a case study of two perennial herbs. *Journal of Applied Ecology* 46: 1048–1053.

Sandercock, B.K. (2006). Estimation of demographic parameters from live-encounter data: a summary review. *Journal of Wildlife Management* 70: 1504–1520.

Sarrazin, F. and Legendre, S. (2000). Demographic approach to releasing adults versus young in reintroductions. *Conservation Biology* 14: 488–500.

Schoener, T.W., Clobert, J., Legendre, S., and Spiller, D.A. (2003). Life-history models of extinction: a test with island spiders. *American Naturalist* 162: 558–573.

Tuljapurkar, S. (1990). *Population Dynamics in Variable Environments.* New York: Springer Verlag.

Tuljapurkar, S. and Haridas, C.V. (2006). Temporal autocorrelation and stochastic population growth. *Ecology Letters* 9: 327–337.

Tuljapurkar, S., Steiner, U.K., and Orzack, S.H. (2009). Dynamic heterogeneity in life histories. *Ecology Letters* 12: 93–106.

Vindenes, Y., Engen, S., and Sæther, B.-E. (2008). Individual heterogeneity in vital parameters and demographic stochasticity. *American Naturalist* 171: 455–467.

Vindenes, Y., Engen, S., and Sæther, B.-E. (2011). Integral projection models for finite populations in a stochastic environment. *Ecology* 92: 1146–1156.

Vindenes, Y. and Langangen, Ø. (2015). Individual heterogeneity in life histories and eco-evolutionary dynamics. *Ecology Letters* 18: 417–432.

9

Combining Counts of Unmarked Individuals and Demographic Data Using Integrated Population Models

Michael Schaub

Swiss Ornithological Institute, Sempach, Switzerland

Summary

Integrated population models are powerful models that can be used to jointly analyze population counts and data that are specific on one or more demographic rates. Joint analysis of all available datasets has the advantage that demographic parameters for which no explicit data are available can often be estimated and that the precision of parameter estimates is improved. Both advantages are a direct consequence of the more complete extraction of the information in the data. Population count data, which are required for all integrated population models as defined here, contain information about all demographic processes operating in the study population. A key part of an integrated population model is a state-transition model which links age- or stage-specific population sizes with demographic rates. Thus, integrated population models combine different models such as capture-recapture models, regression models, and matrix projection models, and can be viewed as a unifying framework for population analyses. I demonstrate applications of these models for temporally variable environments, to model density dependence and illustrate their use for population viability analyses.

9.1 Introduction

Central aims in population ecology are the understanding of reasons of population changes and the ability to predict the future behavior of populations (Sibly and Hone 2002). Population size changes because individuals die, emigrate to other populations, produce recruits, or the population receives immigrants from other populations. At the level of the population, these events are summarized by four demographic parameters: survival, recruitment, emigration, and immigration. Since population growth is a function of these demographic parameters, knowledge about the demographic parameters and their link to population growth are necessary to investigate the reasons for population change and sets the basis for predicting population size in the future. Various sampling designs and associated statistical models have been developed to estimate demographic parameters and population size (Chapters 3–7). Classically, each of the sampled datasets is analyzed separately to obtain estimates of demographic parameters or population size, and inference about population dynamics is typically obtained based the application of projection matrix models parameterized with estimates of demographic parameters (Chapter 8). Here, I present a method for combining different data sources into a single population model: integrated population models.

Integrated models in general can be viewed as a joint analysis of multiple datasets, that is, several datasets are analyzed simultaneously with a single statistical model. Inference is based on the joint likelihood which is usually composed of the product of the likelihoods of the single datasets. A key feature is that one or several parameters are shared among two or more likelihoods of the different component datasets. Examples in the context of population analyses include the joint analysis of capture-recapture and mark-recovery data (Burnham 1993; Lebreton et al. 1995), of capture-recapture and carcass inspection data (Goodman 2004), of age ratios and mark-recovery data (Zimmerman et al. 2010), of age-at-harvest with mark-recovery data (Conn et al. 2008), or the combination of telemetry and camera-trapping data (Sollmann et al. 2013). The integration of information stemming from different sources into a single integrated model is naturally achieved with hierarchical models (Royle and Dorazio 2008; Schaub and Kéry 2012). Here I define *integrated population models* as models that specify a joint likelihood for population counts data which are informative about all demographic rates and

Population Ecology in Practice, First Edition. Edited by Dennis L. Murray and Brett K. Sandercock.

Companion website: www.wiley.com/go/MurrayPopulationEcology

about population changes, and for data which are informative about only one or a few demographic rates.

Population analyses that combine information on population size and demographic data have been conducted for a long time, however, these different datasets have usually been analyzed separately (Jenouvrier et al. 2003; Schaub et al. 2004). Typically, the demographic datasets were analyzed with a specific model for estimation of each set of demographic parameters. Parameter estimates were then combined to parameterize a matrix projection model to either estimate population growth rate or to predict the likely future dynamics of a population. Comparisons with the trajectory of the observed population size, characterized either by the size or the population growth rate, allowed one to tell whether the population model contained all relevant demographic parameters (Jenouvrier et al. 2003), or whether observation errors were substantial. Correlations between the observed population growth rate and the demographic parameters (Peach et al. 1999; Freeman and Crick 2003; Robinson et al. 2004; Freeman et al. 2007) or retrospective population analyses based on *life-table response experiments* may suggest the most important drivers of population growth (Caswell 2001). These modeling approaches have been widely used for wildlife management and conservation, but have a number of drawbacks. First, they are inefficient, because they do not use all the available information. Population count data include information about demography, but the data are not exploited for parameter estimation. Second, it is difficult to properly account for the uncertainty in the demographic parameters and the population growth rate (but see McGowan et al. 2011). Last, unless all relevant demographic parameters are included, population growth rates derived from projection matrices are biased low (Caswell 2001). For example, the growth rate of a Whiskered Tern (*Chlidonias hybrida*) population calculated from census data was 1.29 while the asymptotic population growth rate estimated from a matrix projection model that was parameterized with apparent survival and fecundity was only 1.02 (Ledwon et al. 2014). The discrepancy in growth rates was attributed to immigration which was not included in the matrix projection model.

Recently developed integrated population models hold promise to overcome these drawbacks (Besbeas et al. 2002, 2005; Brooks et al. 2004). The main difference from conventional *matrix projection models* is that in an integrated population model, all available datasets are analyzed *simultaneously*, that is, the single data likelihoods are used to construct one joint likelihood upon which inference is based. Joint analysis offers several important advantages. All uncertainty emerging from the fact that the data stem from a random sample of individuals in the population of interest are accounted for. Proper error propagation is particularly important for population viability analyses. Second, *immigration* and other demographic parameters for which no explicit demographic data are available can often be estimated in the integrated analysis (Besbeas et al. 2002; Abadi et al. 2010a; Lahoz-Monfort et al. 2014). Last, demographic parameters can be estimated with greater *precision* (Besbeas et al. 2002; Tavecchia et al. 2009; Abadi et al. 2010a). The latter two points are a direct consequence of the more complete exploitation of the available information in the joint analysis: population growth is a function of demography, hence the population size data contain information about all demographic processes in the population, and this information is explicitly exploited with an integrated population model.

Here, I use the acronym IPM to denote *integrated population models*. This acronym is also used to describe *integral projection models* (Chapter 8) but the two models have different structure and must not be confused. In my chapter, I present examples of how an IPM is constructed. I then introduce important generalizations of the basic model with temporal random effects and density dependence, and show how the model is naturally used for population viability analyses.

9.2 Construction of Integrated Population Models

To construct an IPM, I find it useful to distinguish between three basic steps: (i) the development of a *population model* that links demographic rates with population size; (ii) the construction of *separate likelihoods* for each available dataset; and (iii) the construction of a *joint likelihood* for making inferences (Figure 9.1). I describe each of these three steps in detail with the help of an example.

Starting by assuming that our focus is a small bird species with the life history typical for many songbirds, that is, all individuals start to reproduce at an age of one year and the species has a monogamous mating system. Three different datasets are available: annual counts of the number of breeding pairs, capture-recapture data based on live resightings, and fledgling counts. We further assume that the data are sampled in an area that is large enough that the majority of dispersal occurs within the study area, so that emigration and immigration can be neglected. I will show later how some of these assumptions can be relaxed by extending the model.

9.2.1 Development of a Population Model

In the first step, we need to define a population model that links demography and population growth, and such a model is typically a matrix projection model (Caswell 2001, Chapter 8). For our example of a short-lived

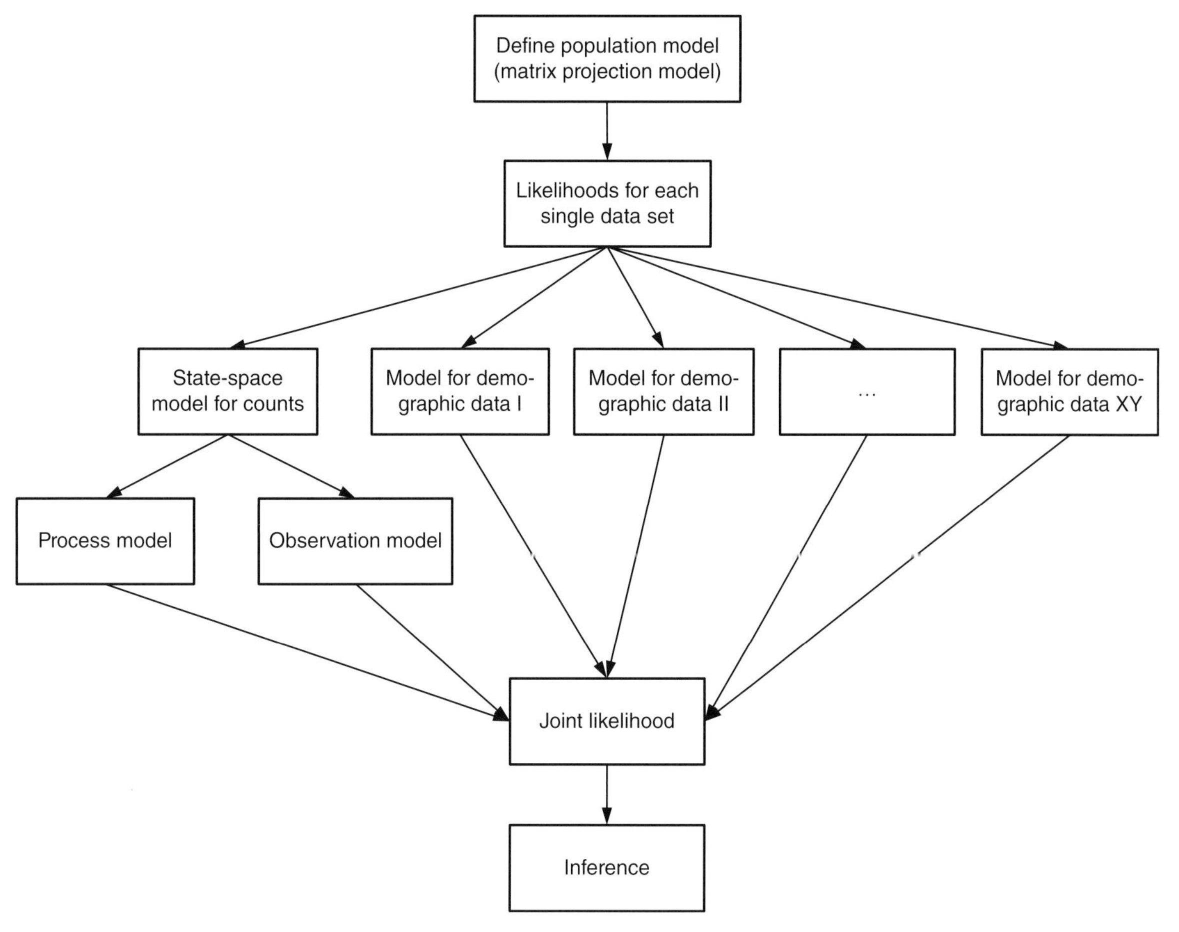

Figure 9.1 Workflow diagram for the construction of an integrated population model.

songbird, an age-structured model with two age classes is used to describe population dynamics. We consider a female-based model with a prebreeding census and thus describe the dynamics of the number of one-year-old females (N_1) and of the number of females that are at least two years old (N_{2+}) as a function of the demographic rates:

$$\begin{aligned} N_{1,t+1} &= N_{1,t}\frac{f_t}{2}s_{j,t} + N_{2+,t}\frac{f_t}{2}s_{j,t}. \\ N_{2+,t+1} &= N_{1,t}s_{a,t} + N_{2+,t}s_{a,t} \end{aligned} \tag{9.1}$$

Here, $s_{j,t}$ is the probability that a fledgling alive in year t survives until the breeding season in year $t + 1$ (juvenile survival probability), $s_{a,t}$ is the probability that an adult alive in year t survives until year $t + 1$ (adult survival probability), and f_t is the number of fledglings that a female is producing in year t (fecundity). Our definition of fecundity implicitly means that all females are reproducing in each year or that f_t includes a component for the portion of the females that skip reproduction in year t. We assume an even sex ratio among fledglings and divide total fecundity by 2 to ensure that the model is female based. It is equivalent to write this model in matrix notation:

$$\begin{bmatrix} N_{1,t+1} \\ N_{2+,t+1} \end{bmatrix} = \begin{bmatrix} \frac{f_t}{2}s_{j,t} & \frac{f_t}{2}s_{j,t} \\ s_{a,t} & s_{a,t} \end{bmatrix} \begin{bmatrix} N_{1,t} \\ N_{2+,t} \end{bmatrix}. \tag{9.2}$$

The population is growing exponentially under this model and the growth rate is calculated as $\lambda_t = \frac{f_t}{2}s_{j,t} + s_{a,t}$.

Written in this way, the model includes *environmental stochasticity*, because all parameters have a time index, meaning that they could vary from one year to the next. Yet, we also want to include *demographic stochasticity*, which is always present and becomes important when the population size is small (Lande 2002). Therefore, we write the model by using appropriate distributions. The number of one-year-old females must be an integer, larger or equal than 0, which has no upper boundary in principle and its expected value is $\frac{f_t}{2}s_{j,t}(N_{1,t} + N_{2+,t})$. We can use the Poisson distribution and write:

$$N_{1,t+1} \sim Poisson\left(\frac{f_t}{2}s_{j,t}(N_{1,t} + N_{2+,t})\right). \tag{9.3}$$

For the number of females that are at least two years old, we use the binomial distribution, as it produces integers, is bounded between 0 and the binomial total, and has expected values of $s_{a,t}(N_{1,t}+N_{2+,t})$, thus:

$$N_{2+,t+1} \sim Binomial((N_{1,t}+N_{2+,t}), s_{a,t}). \qquad (9.4)$$

This population model describes one possible link between population size and demography. As always in a statistical analysis, the inference under the IPM will be based on the implicit assumption that the population model is true or at least sufficiently close to truth; the estimated parameter uncertainty does not reflect structural uncertainty about the population model. Therefore, model construction is a crucial step. Great flexibility is possible in defining alternative models; one may choose a different number of age classes, assume different fecundity of first-time and experienced breeders, include males in a two-sex model as well, or use different distributions for the parameters. The structure of the model typically depends on the research question, prior knowledge of the species, and the available data.

9.2.2 Construction of the Likelihood for Different Datasets

The second step consists of the construction of the likelihoods for all the available datasets. Selection of the likelihoods requires a good knowledge of a range of statistical models to estimate demographic parameters, such as state-space models (De Valpine and Hastings 2002; Newman et al. 2014), capture-recapture (Lebreton et al. 1992) and multistate capture-recapture models (Lebreton et al. 2009, Chapter 7), or generalized linear models (McCullagh and Nelder 1989), to name just the most frequently used models. In the following, I assume a reasonable working knowledge of these models and therefore do not explain them in detail. Previous chapters of this book (Chapters 6–7) and Kéry and Schaub (2012) present these and other models in detail.

In the IPM as defined here, count data, or an index of the number of individuals, must always be included, therefore there is always a model that links these count data to demographic rates. *State-space models* are perfectly suited for this purpose because they describe a partially observed state that develops over time as a first-order Markov process (De Valpine and Hastings 2002; Buckland et al. 2004; Newman et al. 2014). A state-space model in the context of an IPM consists of the state process equations that describe the dynamics of the number of individuals, stratified by age or stage class, as a function of the age-/stage-specific number of individuals one time step before, and the demographic rates. The state in the first year cannot be described as a function of the previous size, therefore a specific model that describes the initial state of the population in the first year is also needed. The state-space model also consists of observation equations that link the observations with the true state of the population. Generally, the *state-process equations* are exactly equivalent to the population model defined above (Eqs. 9.3 and 9.4), whereas the *observation model* should reflect the sampling design of the population count data. I assume here that a single survey to count the number of breeding pairs was conducted each year and that the counts are subject to imperfect detection and double counting, and that the two sources of uncertainty cancel each other out, on average. Therefore, we model the counts (C_t) as:

$$C_t \sim Normal(N_{1,t}+N_{2,t}, \sigma^2), \qquad (9.5)$$

where σ^2 is the residual error that contains lack of fit and observation errors. In our example, we assume that we cannot determine the age of the individuals, so only the sum of the two age classes is included. It is possible to specify other observation models that may be better descriptions of the observation process (Section 9.3.4).

The likelihood of the state-space model (L_{SS}) is the product of the different likelihoods, thus

$$L_{SS}(\boldsymbol{N},\boldsymbol{s},\boldsymbol{f},\sigma^2|\boldsymbol{C}) = L_i(\boldsymbol{N}_1) \times L_S(\boldsymbol{N},\boldsymbol{s},\boldsymbol{f}) \times L_O(\boldsymbol{N},\boldsymbol{\sigma}^2|\boldsymbol{C}), \qquad (9.6)$$

where L_i is the likelihood of the initial population size at the first occasion, L_s is the likelihood of the state process, and L_o is the likelihood of the observation process. Note that I adopt the usual notation here of denoting vectors in bold face. The state-space model includes all the parameters to be estimated, and it is possible in principle to base parameter estimation directly on this model. However, most parameters will not be separately estimable using this model since there are many possible combinations of $\boldsymbol{s}$ and $\boldsymbol{f}$ resulting in the same dynamics of $\boldsymbol{N}$ (Box 9.1). Therefore, additional information on at least some parameters is needed to render the model/parameters identifiable. One solution is the inclusion of additional datasets that are informative about one or several parameters. In a Bayesian context, the additionally required information could also be included by the specification of informative priors (Thomas et al. 2005; King et al. 2010). Conceptually, the two approaches are closely related because we could also use the additional datasets to construct informative priors for the state-space model.

The second dataset is the capture-recapture data that we analyze with a Cormack–Jolly–Seber model (Lebreton et al. 1992, Chapter 7), which provides separate estimates of the probabilities of apparent survival ($\boldsymbol{\phi}$) and of recapture ($\boldsymbol{p}$). I assume here that permanent emigration is negligible and therefore true instead of apparent

Box 9.1 Identifiability of Parameters

When fitted in a Bayesian manner, IPMs may yield estimates of plenty of parameters, but there is no guarantee that all parameters in a fitted model are indeed identified and informed by the data rather than the priors. There are two kinds of nonidentifiability: *intrinsic* and *extrinsic*. A model has intrinsically identified parameters if the same likelihood for the data cannot be obtained by a smaller number of parameters, while *parameter-redundant* models with at least one unidentified parameter can be expressed in terms of a smaller number of parameters (Catchpole and Morgan 1997). Extrinsic nonidentifiability refers to the situation where a parameter should be identifiable due to the structure of the model, but cannot be identified because information in a particular dataset is missing. Such problems can arise for instance when data for certain years are lacking. In a Bayesian analysis, there is strictly no such thing as nonidentifiability, because the posterior is a combination of the prior and the likelihood, and is therefore always defined, even if the likelihood is completely noninformative about a certain parameter (provided that the prior is proper, Gelman et al. 2004). However, if the information in the data is low for a particular parameter (extrinsic nonidentifiability) and/or if the likelihood surface is completely flat for a parameter (intrinsic nonidentifiability) the posterior will simply reflect the prior. Clearly, in a Bayesian analysis as well, we would normally want to know whether our posterior is informed by the observed data and is identifiable in a classical sense, or simply by the priors or the structure of the model. Therefore, a prior sensitivity analysis is of interest. Another

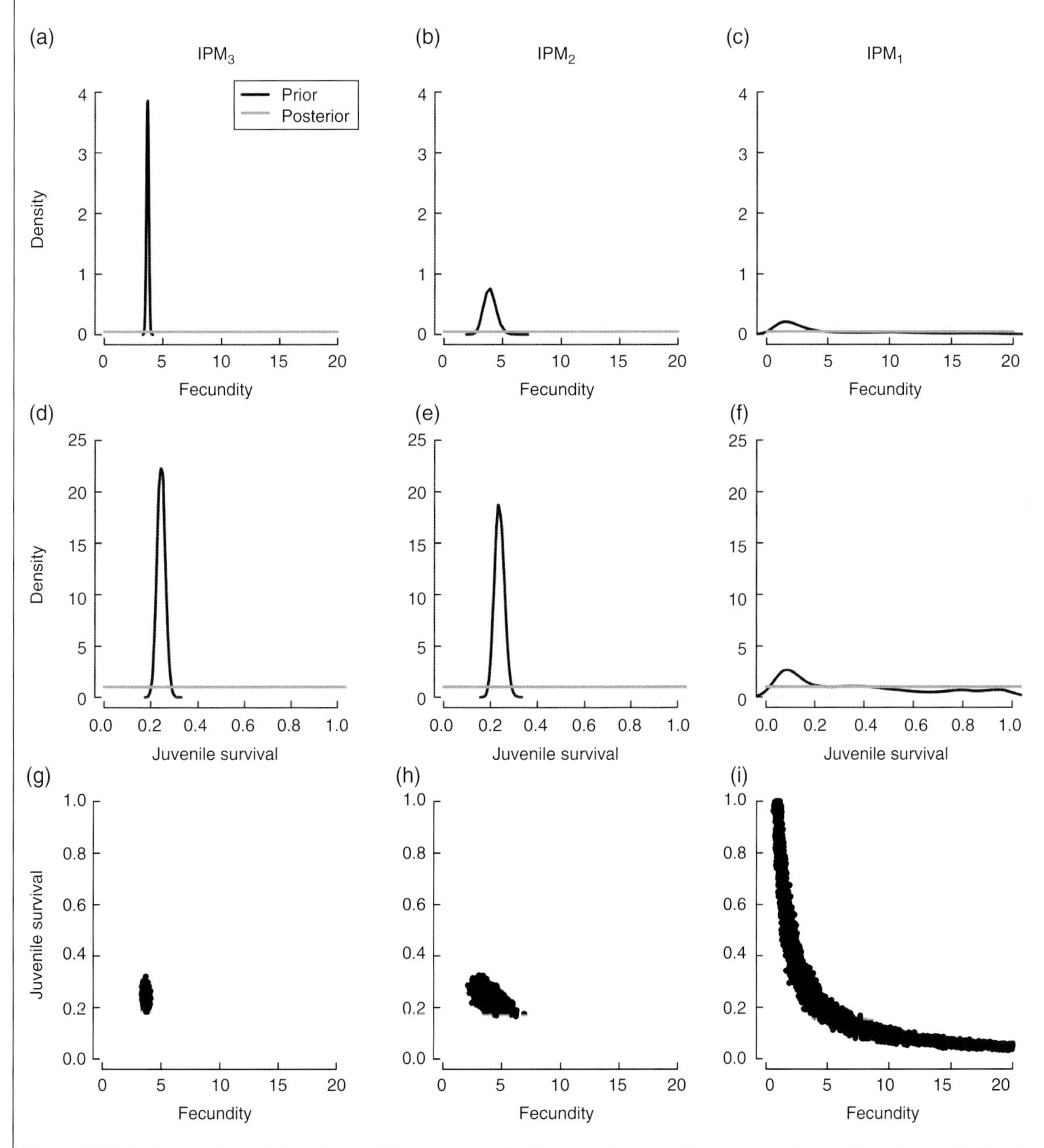

Figure B9.1.1 Comparison of the prior and the posterior distributions for mean fecundity and mean juvenile survival and the correlation between mean fecundity and mean juvenile survival for an integrated population model with all three data types (counts, capture-recapture data of juveniles and adults, and data on fecundity) (panels a, d, g), for a model without data on fecundity (panels b, e, h) and for a model without data on fecundity and capture-recapture data of juveniles (panels c, f, i).

approach to assess parameter identifiability is the comparison between the prior and the posterior distribution (Gimenez et al. 2009). If flat priors are specified and if the overlap between prior and posterior is large, the parameter is weakly identifiable only. Last, a high sampling correlation between parameters of interest is also an indication that the set of parameters is not separately identifiable.

As an example we would like to know whether fecundity and juvenile survival can be estimated in an integrated population model if no data on fecundity are available (IPM_2), or in a model that only contains counts and capture-recapture data of adults (IPM_1), but no explicit data about fecundity and juvenile survival. For comparison, the model with all datasets (IPM_3) is also considered. From Figure B9.1.1 (panels a and d) it becomes clear that fecundity and juvenile survival can be estimated from IPM_3, as the overlap between their posteriors and priors is small. The sampling correlation between fecundity and juvenile survival is small (Figure B9.1.1, panel g, $r = -0.15$). When there are no explicit data about fecundity (IPM_2), fecundity and juvenile survival can still be estimated as their posteriors do not overlap largely with their priors (Figure B9.1.1, panels b and e). However, the posteriors are less peaked and have a larger spread than the posterior from the IPM_3 indicating that the information about fecundity is less. The sampling correlation between fecundity and juvenile survival becomes stronger (Figure B9.1.1, panel h, $r = -0.57$), but there is still no reason for much concern. If only counts and capture-recapture data of adults are available (IPM_1), there is a strong overlap between posteriors and priors of fecundity and juvenile survival (Figure B9.1.1, panels c and f). Thus, fecundity and juvenile survival are not separately estimable in the model. The sampling correlation becomes strongly negative (Figure B9.1.1, panel i, $r = -0.91$), which is an indication that the two parameters are linked and that possibly only a function of the two parameters could be identified.

survival is estimated ($\boldsymbol{\phi} = \boldsymbol{s}$). CJS models are introduced in details in Chapter 7; for a Bayesian approach see Chapter 7 in Kéry and Schaub (2012). The data for individual capture histories can be summarized in the m-array format ($\boldsymbol{m}$) (Burnham et al. 1987). Given the number of released individuals ($\boldsymbol{R}$) at each occasion, each row in the m-array is modeled using a multinomial likelihood,

$$\boldsymbol{m} \sim Multinomial(\boldsymbol{R}, \boldsymbol{\Pi}), \tag{9.7}$$

where $\boldsymbol{\Pi}$ is a function of the underlying parameters $\boldsymbol{s}$ and $\boldsymbol{p}$ (recapture probability). Symbolically the likelihood of the CJS model is $L_{CJS}(\boldsymbol{s},\boldsymbol{p}|\boldsymbol{m})$. The analysis of capture-recapture data with the multinomial likelihood runs much more quickly than with the likelihood of the individual encounter histories (Kéry and Schaub 2012).

Last, the third dataset consists of the total number of fledglings ($\boldsymbol{J}$, productivity data) that were produced annually from a number of surveyed broods ($\boldsymbol{B}$). We use a Poisson regression model:

$$\boldsymbol{J} \sim Poisson(\boldsymbol{f} \times \boldsymbol{B}). \tag{9.8}$$

Symbolically the associated likelihood can be written as $L_F(\boldsymbol{f}|\boldsymbol{B},\boldsymbol{J})$.

9.2.3 The Joint Likelihood

Once all the puzzle pieces needed for an IPM are defined, in a third step, we combine the likelihoods for the different datasets and analyze the resulting integrated model. Under the assumption of independence, the *joint likelihood* of our integrated model (IPM_3) is formed by the multiplication of each individual data likelihood, and thus:

$$\begin{aligned} &L_{IPM}(\boldsymbol{N},\boldsymbol{s},\boldsymbol{f},\boldsymbol{p},\boldsymbol{\sigma}^2|\boldsymbol{C},\boldsymbol{m},\boldsymbol{J},\boldsymbol{B}) = L_{SS}(\boldsymbol{N},\boldsymbol{s},\boldsymbol{f},\boldsymbol{\sigma}^2|\boldsymbol{C}) \times L_{CJS}(\boldsymbol{s},\boldsymbol{p}|\boldsymbol{m}) \times L_F(\boldsymbol{f}|\boldsymbol{J},\boldsymbol{B}) \\ &= L_i(\boldsymbol{N}_1) \times L_S(\boldsymbol{N},\boldsymbol{s},\boldsymbol{f}) \times L_O(\boldsymbol{N},\boldsymbol{\sigma}^2|\boldsymbol{C}) \times L_{CJS}(\boldsymbol{s},\boldsymbol{p}|\boldsymbol{m}) \times L_F(\boldsymbol{f}|\boldsymbol{J},\boldsymbol{B}). \end{aligned} \tag{9.9}$$

A graphical representation of this model which highlights the flux of information is provided in Figure 9.2. The *assumption of independence* is crucial. One rather strict view is that independence means that different datasets must be composed of completely different individuals. Clearly, the independence assumption is respected under such a sampling protocol, but the integrated population model is based on the assumption that the dynamics and the demography of the population segments from which the different datasets stem are identical. In practice, the different datasets are often sampled from a single population and therefore some of the monitored individuals may appear in different datasets. A simulation study mimicking exactly the model and data structure of the example in this chapter has shown that the violation of the assumption of independence hardly affects the accuracy of the estimated parameters (Abadi et al. 2010a). However, in a different context where several populations as well as exchanges of individuals between the populations were modeled, the violation of the independence assumption had a stronger

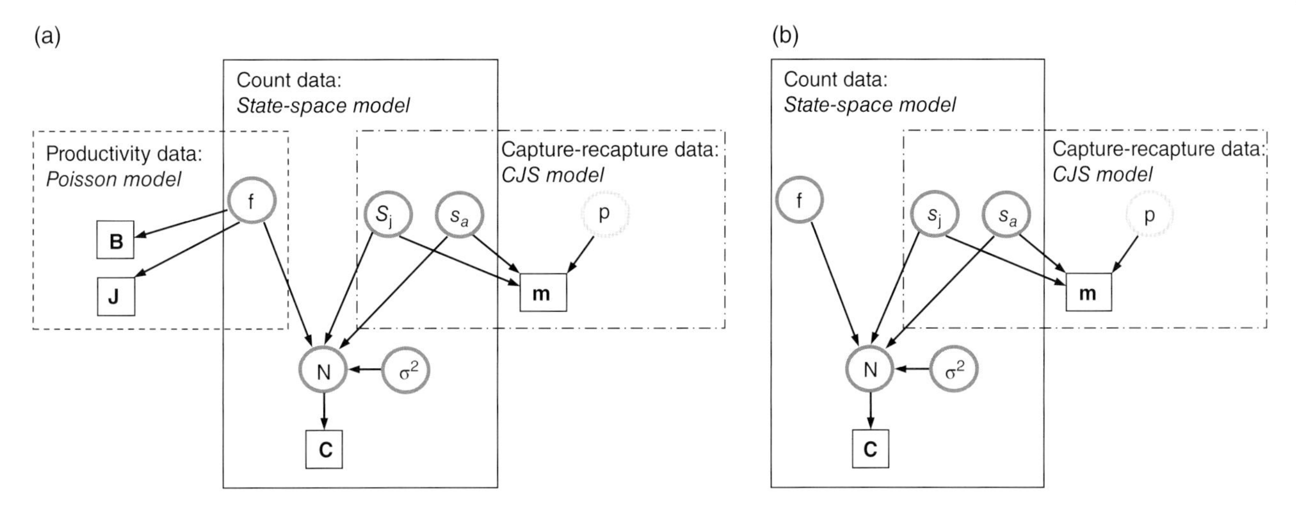

Figure 9.2 Graphical representation of different integrated population models. Graphs are similar to directed acyclic graphs (DAG) without the priors. Data are symbolized with small squares, estimated parameters with circles. Large squares show the individual submodels and the arrows the flux of information. Circles appearing in two submodels indicate that they are informed by two data sources. (a) integrated population model with count, productivity, and capture-recapture data (IPM_3, Eq. 9.9). (b) integrated population model with count and capture-recapture data (IPM_2, Eq. 9.10). For the notation of parameters and data see the text.

impact on parameter accuracy (Besbeas et al. 2009). Whether or not the violation of the independence assumption has an effect on the accuracy of the parameter estimates seems to depend on the information content that the different datasets share. If the degree of sharing is low as in our example, where the information about fecundity and survival in the count data is small compared to the information in the productivity and capture-recapture data, violation of this assumption has a negligible impact. Simulation is a good way to study whether there is a problem with a particular dataset. Ideally, a single data set should not be used twice. For example, the independence assumption is strongly violated if capture-recapture data are used first to estimate survival and then again to provide an index of population size for use in the state-space model. Thus, under a less restrictive, but probably often reasonable view of independence, the different datasets should be obtained by different sampling protocols, but not necessarily contain separate samples of individuals.

The joint likelihood of an integrated population model can be analyzed by *maximum likelihood* (Besbeas et al. 2002; Besbeas and Freeman 2006; Tavecchia et al. 2009) or by *Bayesian inference* (Brooks et al. 2004; Schaub et al. 2007; King et al. 2010). Maximum likelihood usually requires additional assumptions of normality and linearity, and requires use of the Kalman filter (Besbeas et al. 2002; Besbeas et al. 2003; Besbeas et al. 2005; Gauthier et al. 2007; Besbeas and Morgan 2012b). Advantages of maximum likelihood include faster computation and the ability to use likelihood ratio tests or Akaike's Information Criterion (AIC) for model selection (Chapter 2). On the other hand, Bayesian inference is more flexible and nonlinear relationships due to demographic stochasticity or density dependence are easily dealt with, but computation time is typically much longer and model selection is less straightforward (Hooten and Hobbs 2015). The Bayesian approach also requires the formulation of *prior distributions* for all parameters that are estimated. The *posterior distribution* is obtained by the combination of the joint likelihood and the prior distributions via Bayes theorem and inference is obtained from the posterior by simulation, typically with Markov chain Monte Carlo (MCMC) methods (Ntzoufras 2009; Lunn et al. 2013). Often prior distributions are chosen in such a way that they are not informative about the parameters in question. Nevertheless, the possibility to include external information in a formal way is an asset of the Bayesian analysis of IPMs and will often result in additional estimable parameters and increased precision. The examples in my chapter applied the Bayesian framework and used vague priors.

9.2.4 Fitting an Integrated Population Model

As an example, I use simulated data from 19 years (Online Exercise 9.1). To simulate the data, I set the mean juvenile survival probability as $s_j = 0.26$, the mean adult survival probability as $s_a = 0.5$, and the mean fecundity as $f = 4$. Each of these parameters was allowed to vary over time according to a normal distribution on a logit scale (temporal variance of juvenile survival: 0.3; temporal variance of adult survival: 0.15), or on the $\log_{10}$ scale (temporal variance of fecundity: 0.1). The recapture probabilities of the

Table 9.1 Posterior means (SD in parentheses) of the mean demographic parameters and their temporal variability (σ^2) obtained from five different integrated population models.

	Fecundity (f)		Juvenile survival (s_j)		Adult survival (s_a)	
Model	Mean	σ^2	Mean	σ^2	Mean	σ^2
IPM_3	3.709 (0.101)	-	0.240 (0.017)	-	0.540 (0.025)	-
IPM_2	3.805 (0.586)	-	0.238 (0.022)	-	0.538 (0.027)	-
IPM_1	2.941 (2.627)	-	0.459 (0.250)	-	0.545 (0.030)	-
IPM_{3R}	3.744 (0.327)	0.122 (0.056)	0.237 (0.022)	0.099 (0.121)	0.542 (0.031)	0.085 (0.136)
IPM_{2R}	3.670 (0.624)	0.128 (0.246)	0.240 (0.025)	0.088 (0.105)	0.545 (0.031)	0.077 (0.109)

IPM_3: includes counts, capture-recapture data, and productivity data, all parameters are constant over time; IPM_2: includes counts and capture-recapture data, all parameters are constant over time; IPM_1: includes counts and capture-recapture data of adults, all parameters are constant over time; IPM_{3R}: includes counts, capture-recapture data, and productivity data, all parameters with random time effects; IPM_{2R}: includes counts and capture-recapture data, all parameters with random time effects.

capture-recapture data were set to p =0.6, and I assumed that, on average, 80% of the reproducing individuals were counted and that the success of 70% of all broods was recorded. Thus, detection was relatively high, but not perfect for any dataset. The demographic rates varied from year to year, but I first fit simple integrated population models where the demographic rates are assumed to be constant over time (see Online Exercise 9.1 for sample code). The estimates of the demographic parameters are provided in Table 9.1. The population trajectory is smoothed compared to the counts, that is the variation around the trajectory is random noise due to observation errors (Figure 9.3). We obtain estimates of the total population size (i.e. $N_t = N_{1,t} + N_{2+,t}$), as well as estimates of the number of individuals in each age class that are defined in the population model (i.e. $N_{1,t}, N_{2+,t}$). The Markovian structure of the model makes this possible, even though the ages of the counted individuals are unknown. In the current example, these estimates are not of particular interest, but this may be interesting in other applications of IPMs (Koons et al. 2017).

Now assume that no fecundity data are available. The model only needs a few adaptions; basically, the likelihood of the productivity data has to be removed. The joint likelihood for a model based on two datasets (IPM_2) is therefore

$$L_{IPM2}(\boldsymbol{N},\boldsymbol{s},\boldsymbol{f},\boldsymbol{p},\boldsymbol{\sigma}^2|\boldsymbol{C},\boldsymbol{m}) = L_{SS}(\boldsymbol{N},\boldsymbol{s},\boldsymbol{f},\boldsymbol{\sigma}^2|\boldsymbol{C}) \times L_{CJS}(\boldsymbol{s},\boldsymbol{p}|\boldsymbol{m}). \quad (9.10)$$

Fecundity can still be estimated because information about it is included in the counts and this information is extracted (Figure 9.2). Compared to the model where productivity data are available, the fecundity parameter is estimated with a lower precision, but the mean is close to that from the model with all data (Table 9.1). A lower precision of fecundity was expected, as fecundity is estimated only from the count data which contain less

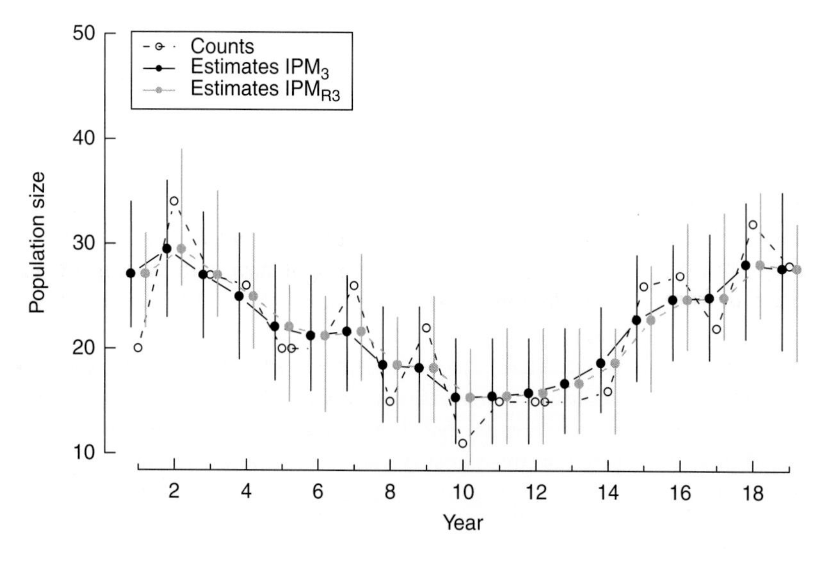

Figure 9.3 Posterior means and 95% credible intervals (vertical lines) of total population size ($N_1 + N_2$) from integrated population models that jointly analyzed counts, capture-recapture and productivity data with constant demographic rates (IPM_3) and with demographic rates with random time effects (IPM_{3R}). The black dots show the population counts. See exercise solutions (9.1. and 9.2) for code.

information about fecundity than the productivity data themselves.

If we extend the model further and assume that we have collected only counts and capture-recapture data of adults because we never visited or searched for nests, fecundity and juvenile survival could not be estimated separately (Box 9.1). The credible intervals then cover almost the complete range of the prior (Table 9.1) and these two parameters are strongly correlated. However, the predicted population sizes would still be identifiable, albeit with a lower precision. The inclusion of prior information becomes more important and possibly the only option when more than one parameter for which no data were sampled are included and estimated with IPMs (Thomas et al. 2005; King et al. 2010; Matthiopoulos et al. 2014).

9.3 Model Extensions

The basic models described up to now make a number of restrictive assumptions that we may wish to relax. Most importantly, we assumed that the demographic parameters are constant over time and the absence of any density-dependent feedback. In the following, I first show how environmental stochasticity can be included in an IPM, and then second, how density dependence can be modeled. The models also assumed geographically closed populations and I show how IPMs can be applied to open populations by explicit estimation of immigration. Last, I comment on different observation models for the count data.

9.3.1 Environmental Stochasticity

The basic models that were introduced so far assumed that the demographic rates were constant over time, but this assumption was not met for the simulated dataset and will often be inadequate for empirical data. Therefore, we now adapt the IPM to allow demographic rates to vary from one year to the next by including *environmental stochasticity*. To model temporal variability, we have the choice to consider the year as either a fixed or a random effect. A *fixed effect* for a year means that the estimate for one year's effect is completely independent from the estimates of the other years' effects. By contrast, if the year is treated as a *random effect*, an estimate from one year is not independent from the estimates of the other years. Rather, the year-specific estimates are assumed to be draws from a normal or other distribution where the mean and variance are estimated. By treating the year as a random effect, we also obtain estimates of each single year, and compared to the fixed-effects estimates these random-effects estimates are pulled toward the overall mean. The degree of this so-called *shrinkage* depends on the precision of the annual estimates, which is a desired property (Burnham and White 2002). In addition, random effects modeling is required if the model is to be used to make predictions of future population size. Therefore, I show here only this option. For juvenile survival we use the following formulations:

$$\begin{aligned} \text{logit}\left(s_{j,t}\right) &= \boldsymbol{\mu}_{s_j} + \boldsymbol{\varepsilon}_{s_j,t} \\ \boldsymbol{\varepsilon}_{s_j,t} &\sim Normal\left(0, \sigma^2_{s_j}\right), \end{aligned} \tag{9.11}$$

where $\boldsymbol{\mu}_{s_j}$ is the mean juvenile survival on the logit scale, $\boldsymbol{\varepsilon}_{s_j,t}$ are the random year effects, and $\boldsymbol{\sigma}^2_{s_j}$ is the temporal variability of juvenile survival on the logit scale, or the degree of how strongly juvenile survival varies over time. The same model can also be written as $\text{logit}\left(s_{j,t}\right) \sim Normal\left(\boldsymbol{\mu}_{s_j}, \boldsymbol{\sigma}^2_{\varepsilon_j}\right)$. The logit link function is necessary to ensure that all estimates of $s_{j,t}$ are bounded by the interval between 0 and 1. For adult survival, we use an analogous formulation. For fecundity, we use the log-link function to ensure that all estimates of f_t are positive:

$$\begin{aligned} \log(f_t) &= \mu_f + \varepsilon_{f,t} \\ \varepsilon_{f,t} &\sim Normal\left(0, \sigma^2_f\right). \end{aligned} \tag{9.12}$$

Here $\boldsymbol{\mu}_f$ is the mean fecundity on the log scale, $\boldsymbol{\varepsilon}_{f,t}$ are the random year effects, and $\boldsymbol{\sigma}^2_f$ is the temporal variability of fecundity. It is also necessary to specify prior distributions for the means ($\boldsymbol{\mu}$) and the variances ($\boldsymbol{\sigma}^2$).

The estimated means of all demographic parameters are similar to the ones obtained from the constant model, but the precision is slightly reduced (Table 9.1). *Loss of precision* is an expected behavior as the random effects formulation has more quantities that need to be estimated. The population trajectories of the two models are hardly distinguishable (Figure 9.3).

Random temporal variation of fecundity can even be estimated when only population count and capture-recapture data are available. The model structure is then similar to IPM_2 (Eq. 9.10), but has random temporal effects for the demographic parameters. The resulting estimates have lower precision, as expected (Table 9.1), but the point estimates are close to those of the other models.

Modeling the demographic parameters with random temporal variation offers the possibility to include environmental variables that might have had an impact on the demographic parameters. For example, if it is assumed that the number of frost days had an effect on juvenile survival and the annual number of frost days is stored in vector ($\boldsymbol{X}$), the model 9.12 could be extended to

$$\begin{aligned} \text{logit}\left(s_{j,t}\right) &= \boldsymbol{\mu}_{s_j} + \boldsymbol{\beta} X_t + \boldsymbol{\varepsilon}_{s_j,t} \\ \boldsymbol{\varepsilon}_{s_j,t} &\sim Normal\left(0, \boldsymbol{\sigma}^2_{s_j*}\right). \end{aligned} \tag{9.13}$$

β is the estimated relationship of juvenile survival with the number of frost days and $\sigma^2_{s_j*}$ is the environmental variability of juvenile survival that is not explained by the variation of the number of frost days. As a formal test of whether frost days had a "significant" effect on juvenile survival, we can check whether the 95% credible interval of β contains zero. It is also informative to calculate the amount of temporal variability in juvenile survival that is explained by the variation of frost days as $\left(\sigma^2_{s_j} - \sigma^2_{s_j*}\right) / \sigma^2_{s_j}$ (Grosbois et al. 2008).

A further possible extension is to relax the assumption that different demographic rates vary independently over time, and to assume that they are correlated. *Positive correlations* in demographic rates can be induced by shared effects of environment conditions. For example, harsh conditions during winter may not only affect juvenile survival, but also the survival of adults. The implementation of correlated effects can be performed with a multivariate Normal distribution (see Chapter 7 in Kéry and Schaub 2012 for details and Schaub et al. 2013).

9.3.2 Direct Density Dependence

Populations do not grow exponentially over long time periods and therefore there must be a regulatory negative feedback that limits their growth (Turchin 2001). The main biological mechanism underlying population regulation is competition among individuals for limited resources. *Competition* induces a demographic response such as a decrease in productivity or survival that eventually results in a reduction of the population growth rate. Including *density dependence* in population analyses is important for a sound understanding of population dynamics. However, the estimation of density dependence is not an easy task. At the level of a population, density dependence is typically assessed with regression-like models that relate population growth with population size – a negative relationship is then an indication of the presence of density dependence (Dennis and Taper 1994). However, observation errors and the fact that the growth rate and population size are not independent will result in negative bias of the estimator of density dependence (Freckleton et al. 2006; Lebreton 2009; Lebreton and Gimenez 2013). Bias is an undesired result since density dependence is then detected too often (Knape and De Valpine 2012). Statistical models that account for observation errors in population size such as state-space models are preferred, but it is still difficult to obtain sound estimates due to weak identifiability even in simple models (Knape 2008).

Detecting density dependence at a demographic level can also be done using regression-like approaches where a demographic parameter is modeled as a function of population size. If the population size contains observation errors that are not accounted for, the power to detect density dependence is reduced (Lebreton and Gimenez 2013). However, the risk of *not* being able to detect density dependence is perhaps less problematic than the risk to detect it when in fact it is absent. Therefore, dealing with density dependence at the demographic level is generally easier than dealing with it at the population level.

IPMs provide a promising framework to study density dependence. The advantages of adopting integrated models for the study of density dependence are twofold: first, population sizes (or indices) are estimated and therefore no longer affected by observation errors. Consequently, the risk to spuriously detect density dependence at the population level is reduced whereas the power to detect density dependence at the demographic level is increased. Second, joint models allow the study of density dependence at the population *and* at the demographic level. We therefore obtain both a *phenomenological description* (Is the population regulated by density?) along with a more *mechanistic understanding* (Which demographic process is inducing the density dependence we observe at the population level?). Abadi et al. (2012) have developed an IPM with which density dependence can be studied and I will present this next.

In principle, including density dependence is easy and straightforward. All that is needed in addition to the IPM introduced so far is the specification of a model that relates the demographic parameter with the estimated population size at a previous point in time. The simplest such relationship is a Ricker-type of model that includes a linear function on the appropriate scale, and we adopt this here. It is necessary to also include *temporal random variation*, otherwise we would unrealistically assume that the entire temporal variation of a demographic rate is due to changes in population size. To model density dependence in juvenile survival in the case where the estimated population size is $\hat{N}_t = \hat{N}_{1,t} + \hat{N}_{2+,t}$, we can use the following relationship:

$$\text{logit}(s_{j,t}) = \mu_{s_j} + \beta \hat{N}_t + \varepsilon_{s_j,t}$$
$$\varepsilon_{s_j,t} \sim Normal\left(0, \sigma^2_{s_j}\right). \quad (9.14)$$

μ_{s_j} is the mean juvenile survival on the logit scale (if $\hat{N}_t$ is centred to zero), β is the estimated strength of density dependence, $\varepsilon_{s_j,t}$ the random year effects, and $\sigma^2_{s_j}$ is the residual temporal variation of juvenile survival. The estimate of β is negative (−0.045), but the credible interval included zero (−0.111, 0.001), which is not surprising

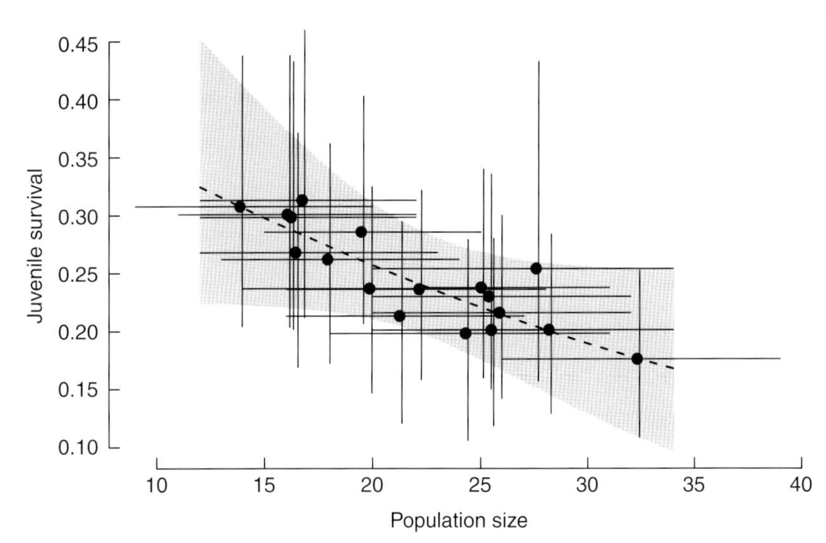

Figure 9.4 Relationship between estimated juvenile survival probabilities and population size. The dots are the posterior mean and the horizontal and vertical lines show the limits of the 95% credible intervals. The broken line shows the predicted regression line $(\text{logit}(s_{j,t})^{-1} = \mu_{s_j} + \beta \hat{N}_t)$ and the light gray area indicates its 95% credible interval. All estimates are obtained from an integrated population model with density-dependent juvenile survival that jointly analyzing counts, capture-recapture, and productivity data.

given that the data were simulated without density dependence. However, the probability that β is negative is 0.973, thus there is strong evidence for density dependence on juvenile survival in our example. Figure 9.4 shows the annual estimates of juvenile survival with the estimated total population size. Note that I used the same data as before which were simulated without density dependence. The fact that density dependence is found here is therefore a chance event. If data simulation and data analysis is repeated many times, we would expect to find no density dependence, on average.

There is a close similarity between Eqs. (9.13) and (9.14); the impact of the population size on juvenile survival is assessed in the same way as that of any other explanatory variable. A difference is, however, that $\hat{N}_t$ is an estimated quantity (Eq. 9.14) while conventional explanatory variables are considered to be free of measurement error (Eq. 9.13). Use of MCMC methods fully accommodates the uncertainty in $\hat{N}_t$ in the analysis and accomplishes error propagation in a fully adequate way.

The demographic parameter and population size as well as their relationship (i.e. $\boldsymbol{\beta}$) are estimated within the same model – there is no two-step approach in which the demographic rate and population size are estimated first, and then a second model estimates their relationship. Direct estimation is possible with Bayesian methods, but is prohibitively complex with classical methods (Jamieson and Brooks 2004; Besbeas and Morgan 2012a).

As density dependence at the population level is the result of density dependence at a demographic level, it is not necessary to specify density dependence at the population level in the IPM. The assessment of whether the specified density relationships at the demographic level were strong enough to impose a response at the population level should be conducted outside the IPM. A possibility is to use again a Ricker type of model as described in Dennis and Taper (1994). Thus, we may fit the following model:

$$\log\left(\frac{\hat{N}_{t+1}}{\hat{N}_t}\right) = r_0 + \boldsymbol{\beta}_p \hat{N}_t + \boldsymbol{\varepsilon}_t, \quad \boldsymbol{\varepsilon}_t \sim Normal(0, \boldsymbol{\sigma}^2) \qquad (9.15)$$

where $\boldsymbol{\beta}_p$ is the strength of density dependence at the population level, r_0 is the population growth rate when $N = 0$, $\boldsymbol{\varepsilon}_t$ the random year effects, and $\boldsymbol{\sigma}^2$ is the residual temporal variance of population size that is not due to density dependence (environmental stochasticity). The population sizes are estimates from the IPM and are therefore not affected by observation errors. In this case, density dependence should correctly be estimated (Lebreton and Gimenez 2013). These calculations are done for each MCMC sample allowing one to obtain posterior distributions for r_0 and $\boldsymbol{\beta}_p$. For our dataset, the mean of $\boldsymbol{\beta}_p$ is −0.021 (95% credible interval: −0.043 to −0.009) and the probability that $\boldsymbol{\beta}_p$ is negative is >0.99. Thus, there is strong evidence that density dependence in juvenile survival resulted in a negative feedback at the population level.

In the example above, it was assumed that only juvenile survival was affected by density dependence, but it is of course possible to include density dependence in more than one or even in all demographic rates. Alternative relationships between a demographic rate and population size could also be considered. These include smoothing functions such as splines (Gimenez et al. 2006), threshold models (Besbeas and Morgan 2012a), or delayed density dependence (Paradis et al. 2002). As a caveat, it must be mentioned that Markov chains in IPMs with density dependence generally do not mix well and therefore long runs are often necessary to achieve convergence.

9.3.3 Open Population Models and Other Extensions

We have assumed so far that the focal population is geographically closed. For vagile species and studies at small geographic scales, this assumption is unreasonable because emigration and immigration can be strong. If the same data types are available, but stem from a small-scale, local study, the capture-recapture data will estimate apparent and not true survival. *Apparent survival* is the combined probability of true survival *and* site fidelity to remain in the study area (Chapter 7). Losses to *permanent emigration* are therefore already included in the complement of the estimate of apparent survival, even though it is not explicitly estimated. *Immigration* is occurring as well, but usually there are no data available allowing the estimation of immigration. *Reverse-time models* allow the estimation of total recruitment to the population (Pradel 1996; Chapter 7), and data sampling for *robust design models* even allows for obtaining separate estimates of local recruitment from immigrations (Nichols and Pollock 1990; Chapter 7). With slight modifications to the IPM it is easily possible to estimate immigration (Abadi et al. 2010b; Schaub and Fletcher 2015), or movement probabilities among populations (McCrea et al. 2010), and thus to study demography and dynamics of geographically open populations (Borysiewicz et al. 2009; McCrea et al. 2010; Péron et al. 2010; Schaub et al. 2012; Brown and Collopy 2013; Altwegg et al. 2014; Szostek et al. 2014; Tempel et al. 2014; Duarte et al. 2016).

IPMs are flexible and therefore there are almost no limits for further extensions or adaptations to a specific situation. Extensions can include the structure of the population model itself and/or the types of data and the associated likelihoods that are considered. For example, we could model the production of fledglings and their survival until they become one year old separately (Schaub et al. 2013). We need to include the number of female fledglings (N_F) as an additional stage in the population model whose number is generated with a Poisson process $\left(N_{F,t} \sim Poisson\left(\frac{f_t}{2}(N_{1,t}+N_{2+,t})\right)\right)$. The survival of the fledglings is then generated with a Binomial process as $N_{1,t+1} \sim Binomial\left(N_{F,t}, s_{j,t}\right)$. These formulations would have the advantage that the variability of the two different demographic processes is fully accounted for. Other examples include the adaptation of the structure of the population model to include seasonal dynamics (Buckland et al. 2004) or populations of two species whose synchrony is assessed (Péron and Koons 2012). Other types of data that have been included besides the classical capture-recapture and productivity data are dead recoveries of marked individuals (Brooks et al. 2004; Baillie et al. 2009; Reynolds et al. 2009), telemetry data (Johnson et al. 2010; Schaub et al. 2010), age ratios of unmarked dead individuals (Hoyle and Maunder 2004; Schaub et al. 2010; Fieberg et al. 2010), nest records (Robinson et al. 2012), harvesting data (Lee et al. 2015), or occupancy data (Chandler and Clark 2014). In a review, Schaub and Abadi (2011) summarize the types of models and data that have been used in studies of birds and mammals with integrated population models.

9.3.4 Alternative Observation Models

The *observation process* in the state-space model links the population counts to the underlying true population size. Therefore, the observation process plays an important role in the interpretation of the estimated population sizes. The estimated residual error (σ^2, Eq. 9.5) is often loosely referred to as *observation error*, but it is in fact composed of errors due to observation and errors due to lack of fit of the state-process model. If the count data (C_t) are collected in such a way that they are correct on average with the possibility that some individuals are not detected (*false negative errors*) while others are double-counted (*false positive errors*), the state-space model provides estimates of the true population sizes N_t. This case arises because the average detection probability or the probability that an individual is included in the count is 1 in such a situation. Often it is reasonable to assume that the relative error rather than the absolute error is constant. For example, the accuracy of the counts may be ±10% regardless of population size. Consequently, the variance of the counts changes with variable population size counts. By using the lognormal distribution instead of the normal distribution for the observation model ($C_t \sim logNormal(N_{1,t}+N_{2+,t}, \sigma^2)$), we can account for increasing variance with increasing population size.

When population counts are performed often only one type of error is relevant: not all individuals are detected and thus only false-negative errors occur. In this case, the state-space model estimates expected counts, i.e. $N_t p_t$ where p_t is the detection probability. The application of the state-space model is still useful as it helps to get rid of the random sampling variation, which is the variation that is induced due to the binomial sampling nature of the counts. The expected counts ($N_t p_t$) change about in parallel to the true population size N_t if the detection probability is either constant or varies randomly over time. In cases where detection probability varies nonrandomly, for example by increasing over time, the expected counts and the true population size will no longer be parallel and inference about population dynamics from the IPM will be biased (Chapter 5 in Kéry and Schaub 2012). To avoid this scenario, one could try to keep the detection probability constant over time or put so much effort that the detection probability is close to 1, which might be possible in intensive, small-scale studies. The best solution is, however, to adapt the sampling protocol in such a way that the detection probability can be estimated. Useful sampling protocols include point counts (Royle 2004), distance sampling (Buckland et al. 2001), or double-observer surveys (Nichols et al. 2000). In all of these sampling situations unmarked animals are counted and the models allow the estimation of abundance corrected for imperfect detection. The specific model to estimate p should be included in the IPM, meaning that the likelihood should be written explicitly and become part of the joint likelihood.

Sometimes it is possible to distinguish age classes in the field and then count the number of individuals in each. Age-specific counts are valuable because changes of age ratios over time contain information about demography (Link et al. 2003). The inclusion of age-specific counts requires an adaptation of the observation equations. For example, had it been possible in our analysis to distinguish first-year and after first-year individuals in the counts ($C_{1,t}$, $C_{2+,t}$), we would simply have specified two observation equations ($C_{1,t} \sim Normal(N_{1,t}, \sigma_1^2)$ and $C_{2+,t} \sim Normal(N_{2+,t}, \sigma_{2+}^2)$). The residual errors may or may not have been assumed to be the same. Such additional demographic information results in more precise parameter estimates (Tavecchia et al. 2009).

9.4 Inference About Population Dynamics

Results from IPMs are usually estimates of demographic rates and of stage-dependent population sizes. The estimates may be interesting on their own, but usually we want to make an inference about population dynamics. Here, I show how results from an IPM can be used in a *retrospective analysis* to infer demographic drivers of past population changes, and then how an IPM must be adapted to perform a *prospective analysis* as a population viability analysis.

9.4.1 Retrospective Population Analyses

A fundamental aim in many population studies is the understanding of demographic reasons of population change. The tools to perform such retrospective population analyses include life-table response experiments (Horvitz et al. 1997; Caswell 2001; Caswell 2010; Koons et al. 2016), life-stage simulation analyses (Wisdom et al. 2000) and simple correlation analyses (Robinson et al. 2004). All these approaches require estimates of demographic rates and the latter also of population growth rates. Since these estimates are results from an IPM, no changes in the IPM are needed to perform any of these retrospective analyses.

Robinson et al. (2014) investigated demographic drivers of several British bird species. The researchers first fitted IPMs and then decomposed the variation of the population growth rates into contributions of demographic rates using a life-table response experiment. Owing to the Bayesian mode of analyses, posterior distributions of these demographic contributions could easily be computed and thus uncertainty due to data sampling was expressed. The quantification of uncertainty in demographic contributions is an important step forward for sound inference; most past retrospective population analyses did not consider uncertainty. Van Oosten et al. (2014) used life-table response experiments to understand differential dynamics of three populations of Wheatears (*Oenanthe oenanthe*). Estimates of the demographic rates stem from IPMs fitted to each of the three populations. The authors found that differential immigration contributed strongly to the different growth rates of these three populations.

In another study, Schaub et al. (2013) investigated the dynamics of a Red-backed Shrike (*Lanius collurio*) population, where numbers ranged between 35 and 74 breeding pairs during a 36-year period. Schaub et al. (2013) fit an IPM that included immigration and temporal random effects for all demographic rates. The estimated demographic rates were then correlated with the estimated population growth rate to assess how strongly a demographic parameter contributed to population change (Figure 9.5). Because all of the demographic estimates were uncertain, the authors calculated correlation coefficients for each MCMC draw, i.e. they computed their posterior distributions. From Figure 9.5 it is obvious that all demographic parameters – juvenile and adult apparent survival, fecundity, and immigration – were positively related to the population growth rate. The strongest correlation was found for the immigration rates, suggesting that this local population was substantially driven by immigration.

9.4.2 Population Viability Analyses

The purpose of a *population viability analysis* is to gauge the likely future trajectory of a population (Morris and Doak 2002). Typical objectives of a population viability analysis are the assessment of the *extinction risk* of a population under study, the evaluation of different possible

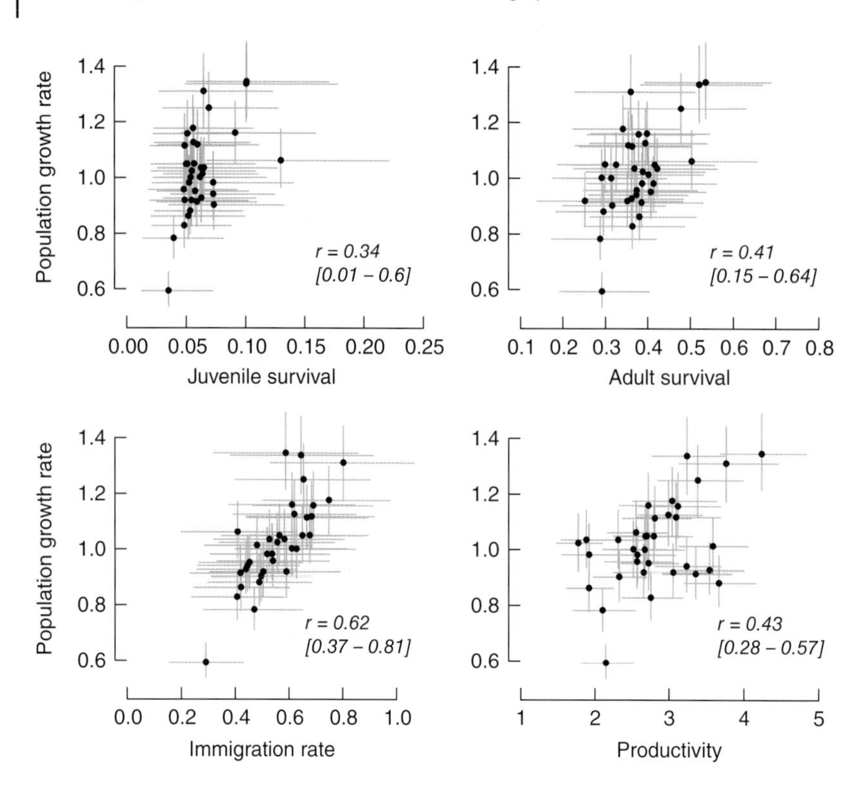

Figure 9.5 Correlations between annual population growth rates and demographic parameters of Red-backed Shrike (*Lanius collurio*) females from a population in southern Germany. *Source:* From Schaub et al. (2013), adapted. The vertical and horizontal lines show the limits of the 95% credible intervals. The posterior means along with the 95% credible intervals (parentheses) of the correlation coefficients (*r*) are also given.

management options, or the development of *sustainable harvest strategies* (Punt and Hilborn 1997). All objectives require a prediction of population size into the future.

Under the assumption that the environment does not change in future and given estimates about demographic rates and population size, it is straightforward to predict the expected population sizes in future years, but it is more difficult to assess their uncertainty. Yet, *measures of uncertainty* are important for inference. The future population size is a function of the past population size and the demographic rates. Since both are estimated with uncertainty, the uncertainty of the future population size is a complex function of the uncertainty of model components. In a frequentist framework, prediction uncertainty can be estimated by numerical simulation (Morris and Doak 2002; Lande et al. 2003). In a Bayesian framework, the error propagation occurs "automatically," and therefore the prediction of the future population size with a measure of uncertainty is easy to obtain (Wade 2002). The Bayesian framework also offers the possibility to make probability statements about past or future population sizes, and about population trends or extinction risks that are not possible in the frequentist framework or only with difficulties and/or restrictive assumptions (Wade 2000).

The prediction of future population sizes or future demographic rates can be viewed as a *missing data* problem in the Bayesian framework (chapter 11 in Kéry and Schaub 2012). Computation is relatively easy: it only requires extending the loop for the quantities that need to be predicted (population sizes and demographic rates) into the future for a period of years. For illustration, assume that we want to predict the likely population development over the next five years in our example study, to estimate the probability that the number of females drops to fewer than 5 in five years (quasi-extinction probability) and to estimate the probability that the population will be smaller in five years than it was in the last year of the study. Figure 9.6 shows the past and the future development of the population. The population is likely to increase in the future and the uncertainty about the predicted population sizes is increasing the further ahead we make a prediction. An increase in uncertainty for predictions or other extrapolations is a general result of population viability analyses and therefore it often makes little sense to project too far into the future (Fieberg and Ellner 2000). The uncertainty increases faster and with greater uncertainty among demographic rates, population sizes, and the more stochastic elements that are built into the model. In our case, the probability that the population size five years from now is less than 5 females (which is our extinction threshold) can simply be computed as the fraction of the MCMC samples of N_{24} that are less than 5. The *probability of quasi-extinction* is estimated to be 0.043 – thus it is unlikely that the population would go extinct within the next five years unless the conditions experienced by the population during the past 19 years change.

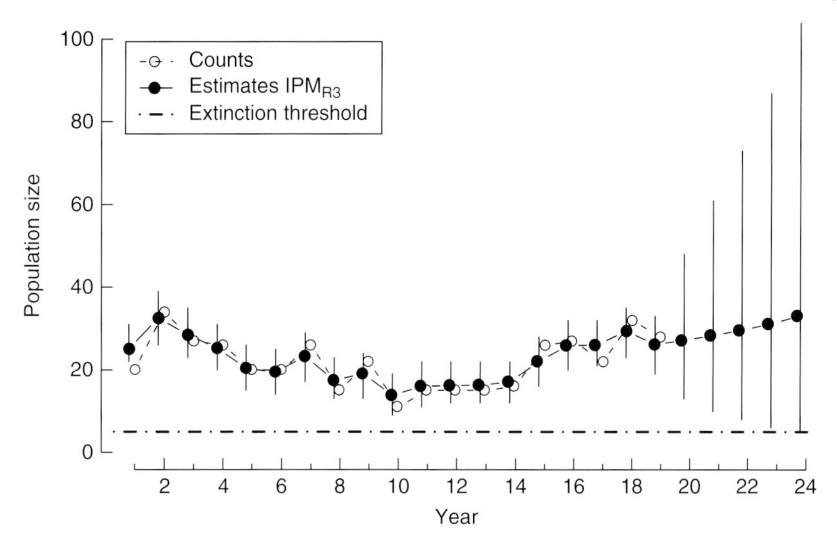

Figure 9.6 Posterior means and 95% credible intervals (vertical lines) of total population size ($N_1 + N_2$) from an integrated population model with demographic rates with random time effects (IPM_{3R}), and the predicted population sizes in the next five years. The black dots show the counts. The extinction threshold is indicated by a broken horizontal line. Only a small part of the 95% credible interval is below the broken line indicating that the extinction probability in five years is low.

Second, we wanted to know the probability that the population is smaller in five years than the population in the last study year. Compared to before, the population size in five years is now not compared to a fixed value, but to an estimate with associated uncertainty. The calculation of the probability is, however, analogous. Now, we compare each MCMC sample of N_{19} with that of N_{24} and tally up the number of these comparisons where N_{24} is smaller than N_{19}. The fraction of comparisons in which $N_{24} < N_{19}$ is our estimate of the probability that the population size in five years is smaller than the actual population size. In our example this is 0.52, hence, it is about as likely that the terminal population size will be smaller as it will be larger. The result may seem surprising, given that the population size is predicted to increase on average. However, there is a positive correlation between N_{19} and N_{24}: if a draw in the MCMC chain predicted a small N_{19}, then N_{24} tended to be smaller than average, too. This sampling correlation is taken into account in the MCMC-based estimate of the probability.

Predictions are also possible under scenarios of environmental changes or of possible management options. Knowledge or educated guesses about how the mean or the temporal variability of demographic rates change as a function of environmental changes or management actions are required. Such functions can be built into the IPM. Under the Bayesian framework it is straightforward to compute the probability that one management option will lead to a larger population size in the future compared to another. See Online Exercise 9.3 for such an example.

Although population viability analyses are straightforward to conduct using Bayesian IPMs, only a handful of applications have been published. Maunder (2004) predicted the future size of a fish population, Oppel et al. (2014) estimated the extinction probability of a population of a rare songbird, and Tenan et al. (2012) investigated the potential effect of the reduction of poison-related mortality in a raptor population. I expect that we will see many more applications of IPMs in the context of population viability assessment in the future.

It is evident that many tools that are available for the analysis of matrix projection models can also be used in connection with IPMs (Caswell 2001; Chapter 8). Measures of uncertainty reflecting measurement errors of demographic rates and population size can in this way be obtained easily.

9.5 Missing Data

We can easily handle *missing data* with IPMs. We have already seen two examples of missing data problems: the count and survival model without productivity data (IPM_2, Eq. 9.10), and in the population viability analysis there were no data available for future years. Nevertheless, the demographic parameters or the population sizes can still be estimated when data are missing, because the available count and demographic information as well as the model structure are informing them. In the former case information about fecundity in the counts is extracted, while in the latter case the Markovian nature of the population model along with the random-year assumptions allows one to infer future population size. Since IPMs can deal with severe cases of missing data, it is perhaps not so surprising that the models can deal also with less severe cases, for instance when population counts are missing for some years. It may also happen that the different datasets do not cover exactly the same time periods. Incomplete overlap is no problem either; one then specifies the model for the time period with count data and considers periods with missing demographic data as missing data. It might also be possible to specify an IPM for the complete duration where any data are available. It is certainly advantageous if the different datasets overlap partially. If this is not the case, an integrated analysis may still be possible, but requires an additional assumption that the distribution of the demographic parameters is stationary over the entire period of time. Clearly, the Bayesian MCMC-based analysis of IPMs confers great flexibility to handle missing data, but increasing numbers of missing values always come at the price of lower precision and additional assumptions. Moreover, there are limitations about how many datasets can be missing (Box 9.1).

9.6 Goodness-of-fit and Model Selection

Goodness-of-fit testing of any model is important to reduce the risk of obtaining biased parameter estimates, poor uncertainty assessments, or otherwise inadequate inference. Recently, Besbeas and Morgan (2014) suggested the use of *calibrated simulation* to assess the fit of an IPM. The principle of their suggestion is to simulate replicate data under the model using the parameter estimates, and to compare these replicate data with the observed data using some discrepancy measures. If the replicate and observed data are similar, it can be concluded that the model fits. Because several datasets are analyzed in an IPM, tests are conducted for all of them. Ensuring that the different submodels fit the data is of particular importance when a parameter is estimated for which no explicit data have been sampled, such as for IPM_2 (Eq. 9.10). For example, if survival had increased and fecundity remained constant over time, and IPM_2 is fitted with constant survival and time-dependent fecundity, we would get estimates of fecundity that show an increasing trend over time.

Model selection is straightforward for IPMs when they are analyzed in the frequentist framework, because established methods such as information-theoretic approaches (e.g. AIC) or likelihood ratio tests can be used (Chapter 2). When the Bayesian framework is applied model selection is more challenging, but a recent, excellent review shows many possibilities (Hooten and Hobbs 2015). In principle, Bayesian model selection criteria such as the *deviance information criteria* (DIC, Spiegelhalter et al. 2002) could be applied (Schaub et al. 2007), but there is controversy about whether the DIC is valid when applied to certain classes of hierarchical models (Celeux et al. 2006), especially to mixture models with discrete latent states. Davis et al. (2014) used the *Watanabe-Akaike information criterion* (WAIC) for model selection of complex IPMs. *Reversible jump MCMC* (RJMCMC) is a good option for model selection (King et al. 2010), but usually requires development of custom MCMC code (RJMCMC is implemented in `BUGS` for simple cases) and hence is currently probably outside the reach of most nonstatisticians.

9.7 Software Tools

IPMs can be analyzed in either the frequentist or the Bayesian framework, and consequently the available software differs between the two approaches. Owing to its flexibility, there is so far no canned program with which integrated population models can be fitted. Therefore, skills in coding are necessary to fit an IPM and several computing platforms can be used. For the frequentist framework `MATLAB` (Besbeas et al. 2002; Gauthier et al. 2007), `C++` (Baillie et al. 2009), and `ADMB` (Maunder 2004; Maunder and Punt 2013) are computing platforms for the implementation of the IPMs. For the Bayesian framework `Python` (Fonnesbeck and Conroy 2004), but most often `BUGS` and `JAGS` (Kéry and Schaub 2012), have been used. Writing code for IPMs requires knowledge of a variety of models and experience in programming. The `BUGS` language used in the software packages `WinBUGS`, `OpenBUGS` (Lunn et al. 2000), and `JAGS` (Plummer 2003) is particularly well suited to define IPMs in a manner that is accessible not only to statisticians, but also to many ecologists. The reason for this is that a complicated joint likelihood does not need to be formulated explicitly as a single term, but rather as sequence of simpler, conditionally independent, "local" relationships (Lunn et al. 2013). The relationships are typically fairly simple to write down and put the analysis of complex IPMs in the reach of many quantitative biologists.

The IPMs developed for this chapter were fit with the program `JAGS` run from `R` (R Development Core Team, 2004) using package `jagsUI` (Kellner 2015). Code for some models is provided in the web exercises and I also refer to chapter 11 in Kéry and Schaub (2012) for comments on the code. `JAGS` and its sister programs `WinBUGS` and `OpenBUGS` use MCMC simulation to obtain samples from the posterior distributions. As in all Bayesian analyses that use MCMC, output should be checked carefully to ensure the MCMC chains have reached stationary distributions. Moreover, the impact of the priors on the posteriors should be evaluated by prior sensitivity analyses. Several introductory books on Bayesian modeling provide background on these and related issues (Ntzoufras 2009; Kéry 2010; Lunn et al. 2013).

9.8 Online Exercises

The online exercises provide examples of an integrated population model fitted to counts, capture-recapture data, and data on productivity. The models are based on `JAGS` code and are run from `R` using package `jagsUI`. Exercise 1 uses the data in the main text and is modified for the case where no data on productivity are available. Exercise 2 extends the model based on constant demographic rates to one with random temporal effects. Last, exercise 3 adapts the model to consider the possible outcomes from alternative management scenarios as part of a population viability analysis.

9.9 Future Directions

IPMs can naturally be developed as *hierarchical models*. Hierarchical models constitute a class of statistical models that describe an observed response as a nested sequence of random variables (Royle and Dorazio 2008). The models express a complex joint likelihood as a product of simpler, conditionally independent probabilities and are therefore particularly suited to properly accommodate process variability and uncertainty at multiple levels of a stochastic process. Thus, it is unsurprising that hierarchical models are considered by some to be unifying framework for inference and prediction in ecology (Clark and Björnstad 2004; Cressie et al. 2009).

A key feature of IPMs is that demographic information coming from different data sources is combined into a single model in a coherent way. Synthesis is achieved by the analysis of a joint likelihood that is the product of the likelihoods of the different datasets. The combination of information is powerful and has several important advantages. Generally, it enables one to get more precise parameter estimates and to estimate parameters for which no explicit data have been collected. Combination of information is not restricted to IPMs, but is widely applicable. Recently, an increasing number of studies has been published that integrate information from various sources to obtain better inference (Cornulier et al. 2011; Sollmann et al. 2013; Halstead et al. 2012; Papadatou et al. 2012) and this trend will likely continue, and probably at an increased rate (Schaub and Kéry 2012).

Although already applied for some time, IPMs are still relatively novel and further development will certainly happen. For most current IPMs, inference from the joint likelihood is based on the assumption that the combined datasets are independent. In practice, this assumption may often be violated and depending on the context this violation may affect parameter accuracy (Besbeas et al. 2009; Abadi et al. 2010a). The most elegant solution to this potential problem would be development of a joint likelihood that takes the nonindependence of the single datasets into account, but this is not obvious to achieve. Chandler and Clark (2014) developed a model that integrates capture-recapture and occupancy data. The independence assumption was relaxed because both models were conditional on the same spatial process of population dynamics. The development of goodness-of-fit assessment for IPM has just begun (Besbeas and Morgan 2014) and it is desirable to achieve a better understanding of the performance of the proposed or alternative goodness-of-fit assessments.

Recently developed open *N-mixture models* allow interesting perspectives for modeling population dynamics at large spatial scales: using spatially and temporally replicated counts of unmarked individuals for population abundance, population growth rate, and demographic parameters such as survival and recruitment can be estimated (Dail and Madsen 2011). Chandler and King (2011) used this model to estimate habitat-specific abundance, apparent survival, and recruitment of Golden-winged Warblers (*Vermivora chrysoptera*) in Costa Rica. Zipkin et al. (2014) extended this model such that it can deal with stage-structured counts such as size classes, and applied it to a population of Dusky Salamanders (*Desmognathus fuscus*) to estimate stage-specific survival, immigration, and recruitment. The N-mixture models are appealing because they can be applied to monitoring data, and thus inference about demography is possible even if only unmarked individuals are counted. However, parameter estimates are often quite imprecise and violations of model assumptions can result in large bias (Hostetler and Chandler 2015; Bellier et al. 2016). To improve parameter accuracy, approaches where independent information on demographic parameters via integrated modeling or prior knowledge is included are promising. A challenge might be the differential spatial scale at which data are often sampled. Nevertheless, it is likely that we will see future development of integrated models combining spatially and temporally replicated counts with demographic data.

Demographic analyses are essential for conservation, management, and harvesting of populations (Morris and Doak 2002; Conroy and Carroll 2009; Mills 2013). Demographic data must extend over several years for making sound inference and are generally expensive to sample. Therefore, it is important that existing demographic data are analyzed with the most efficient methods. Existing data can often not be analyzed properly with conventional matrix projection models, because of a small number of individuals or because of a lack of specific data on productivity or other parameters. IPMs may go a long way toward mitigating such small-sample problems. By making efficient use of existing data they may help to obtain the best possible inference even from limited data (Schaub et al. 2007). Moreover, especially when fitted using Bayesian MCMC methods, they can be specified extremely flexibly and adapted to virtually any particular study provided that a minimal field protocol has been respected and that a minimum amount of observed data are sampled.

References

Abadi, F., Gimenez, O., Arlettaz, R., and Schaub, M. (2010a). An assessment of integrated population models: bias, accuracy, and violation of the assumption of independence. *Ecology* 91: 7–14.

Abadi, F., Gimenez, O., Jakober, H. et al. (2012). Estimating the strength of density dependence in the presence of observation errors using integrated population models. *Ecological Modelling* 242: 1–9.

Abadi, F., Gimenez, O., Ullrich, B. et al. (2010b). Estimation of immigration rate using integrated population modeling. *Journal of Applied Ecology* 47: 393–400.

Altwegg, R., Jenkins, A., and Abadi, F. (2014). Nestboxes and immigration drive the growth of an urban Peregrine Falcon *Falco peregrinus* population. *Ibis* 156: 107–115.

Baillie, S.R., Brooks, S.P., King, R., and Thomas, L. (2009). Using a state-space model of the British Song Thrush *Turdus pilomenos* population to diagnose the causes of a population decline. In: *Modeling Demographic Processes in Marked Populations* (eds. D.L. Thomson, E.G. Cooch and M.J. Conroy), 541–561. New York: Springer.

Bellier, E., Kéry, M., and Schaub, M. (2016). Simulation-based assessment of dynamic N-mixture models in the presence of density dependence and environmental stochasticity. *Methods in Ecology and Evolution* 7: 1029–1040.

Besbeas, P., Borysiewicz, R.S., and Morgan, B.J.T. (2009). Completing the ecological jigsaw. In: *Modeling Demographic Processes in Marked Populations* (eds. D.L. Thomson, E.G. Cooch and M.J. Conroy), 513–539. New York: Springer.

Besbeas, P. and Freeman, S.N. (2006). Methods for joint inference from panel survey and demographic data. *Ecology* 87: 1138–1145.

Besbeas, P., Freeman, S.N., and BJT, M. (2005). The potential of integrated population modelling. *Australian and New Zealand Journal of Statistics* 47: 35–48.

Besbeas, P., Freeman, S.N., BJT, M., and Catchpole, E.A. (2002). Integrating mark-recapture-recovery and census data to estimate animal abundance and demographic parameters. *Biometrics* 58: 540–547.

Besbeas, P., Lebreton, J.D., and Morgan, B.J.T. (2003). The efficient integration of abundance and demographic data. *Applied Statistics* 52: 95–102.

Besbeas, P. and Morgan, BJT. (2012a). A threshold model for heron productivity. *Journal of Agricultural, Biological, and Environmental Statistics* 17: 128–141.

Besbeas, P. and Morgan, B.J.T. (2012b). Kalman filter initialisation for integrated population modelling. *Applied Statistics* 61: 151–162.

Besbeas, P. and Morgan, B.J.T. (2014). Goodness of fit of integrated population models using calibrated simulation. *Methods in Ecology and Evolution* 5: 1373–1382.

Borysiewicz, R.S., Morgan, B.J.T., Hénaux, V. et al. (2009). An integrated analysis of multisite recruitment, mark-recapture-recovery and multisite census data. In: *Modeling Demographic Processes in Marked Populations* (eds. D.L. Thomson, E.G. Cooch and M.J. Conroy), 579–591. New York: Springer.

Brooks, S.P., King, R., and Morgan, B.J.T. (2004). A Bayesian approach to combining animal abundance and demographic data. *Animal Biodiversity and Conservation* 27: 515–529.

Brown, J.L. and Collopy, M.W. (2013). Immigration stabilizes a population of threatened cavity-nesting raptors despite possibility of nest box imprinting. *Journal of Avian Biology* 44: 141–148.

Buckland, S.T., Anderson, D.R., Burnham, K.P. et al. (2001). *Introduction to Distance Sampling*. Oxford: Oxford University Press.

Buckland, S.T., Newman, K.B., Thomas, L., and Koesters, N.B. (2004). State-space models for the dynamics of wild animal populations. *Ecological Modelling* 171: 157–175.

Burnham, K.P. (1993). A theory for combined analysis of ring recovery and recapture data. In: *Marked Individuals in the Study of Bird Populations* (ed. J.D. Lebreton), 199–213. Basel: Birkhäuser.

Burnham, K.P., Anderson, D.R., White, G.C. et al. (1987). Design and analysis methods for fish survival experiments based on release-recapture. *American Fisheries Society Monograph* 5: 1–437.

Burnham, K.P. and White, G.C. (2002). Evaluation of some random effects methodology applicable to bird ringing data. *Journal of Applied Statistics* 29: 245–264.

Caswell, H. (2001). *Matrix Population Models. Construction, Analysis, and Interpretation*. Sunderland, Massachusetts: Sinauer Associates.

Caswell, H. (2010). Life table response experiment analysis of the stochastic growth rate. *Journal of Ecology* 98: 324–333.

Catchpole, E.A. and Morgan, B.J.T. (1997). Detecting parameter redundancy. *Biometrika* 84: 187–196.

Celeux, G., Forbes, F., Robert, C.P., and Titterington, D.M. (2006). Deviance information citeria for missing data models. *Bayesian Analysis* 1: 651–674.

Chandler, R.B. and Clark, J.D. (2014). Spatially explicit integrated population models. *Methods in Ecology and Evolution* 5: 1351–1360.

Chandler, R.B. and King, D.I. (2011). Golden-winged Warbler habitat selection and habitat quality in Costa Rica: an application of hierarchical models for open populations. *Journal of Applied Ecology* 48: 1038–1047.

Clark, J.S. and Björnstad, O.N. (2004). Population time series: process variability, observation errors, missing values, lags, and hidden states. *Ecology* 85: 3140–3150.

Conn, P.B., Diefenbach, D.R., Laake, J.L. et al. (2008). Bayesian analysis of wildlife age-at-harvest data. *Biometrics* 64: 1170–1177.

Conroy, M.J. and Carroll, J.P. (2009). *Quantitative Conservation of Vertebrates*. Oxford: Wiley-Blackwell.

Cornulier, T., Robinson, R.A., Elston, D.A. et al. (2011). Bayesian reconstitution of environmental change from disparate historical records: hedgerow loss and farmland bird declines. *Methods in Ecology and Evolution* 2: 86–94.

Cressie, N., Calder, C.A., Clark, J.S. et al. (2009). Accounting for uncertainty in ecological analysis: the strengths and limitations of hierarchical statistical modeling. *Ecological Applications* 19: 553–570.

Dail, D. and Madsen, L. (2011). Models for estimating abundance from repeated counts of an open metapopulation. *Biometrics* 67: 577–587.

Davis, A.J., Hooten, M.B., Phillips, M.L., and Doherty, P.F. (2014). An integrated modeling approach to estimating Gunnison Sage-Grouse population dynamics: combining index and demographic data. *Ecology and Evolution* 4: 4247–4257.

De Valpine, P. and Hastings, A. (2002). Fitting population models incorporating process noise and observation error. *Ecological Monographs* 72: 57–76.

Dennis, B. and Taper, M.L. (1994). Density dependence in time series observations of natural populations: estimation and testing. *Ecological Monographs* 64: 205–224.

Duarte, A., Weckerly, F.W., Schaub, M., and Hatfield, J.S. (2016). Estimating Golden-cheeked Warbler immigration: implications for the spatial scale of conservation. *Animal Conservation* 19: 65–74.

Fieberg, J. and Ellner, S.P. (2000). When is it meaningful to estimate an extinction probability? *Ecology* 81: 2040–2047.

Fieberg, J.R., Shertzer, K.W., Conn, P.B. et al. (2010). Integrated population modeling of black bears in Minnesota: implications for monitoring and management. *PLoS One* 5: e12114.

Fonnesbeck, C.J. and Conroy, M.J. (2004). Application of integrated Bayesian modeling and Markov chain Monte Carlo methods to the conservation of a harvested species. *Animal Biodiversity and Conservation* 27: 267–281.

Freckleton, R.P., Watkinson, A.R., Green, R.E., and Sutherland, B.J. (2006). Census error and the detection of density dependence. *Journal of Animal Ecology* 75: 837–851.

Freeman, S.N. and Crick, H.Q.P. (2003). The decline of the Spotted Flycatcher *Muscicapa striata* in the UK: an integrated population model. *Ibis* 145: 400–412.

Freeman, S.N., Robinson, R.A., Clark, J.A. et al. (2007). Changing demography and population decline in the Common Starling *Sturnus vulgaris*: a multisite approach to integrated population monitoring. *Ibis* 149: 587–596.

Gauthier, G., Besbeas, P., Lebreton, J.D., and BJT, M. (2007). Population growth in Snow Geese: a modeling approach integrating demographic and survey information. *Ecology* 88: 1420–1429.

Gelman, A., Carlin, J.P., Stern, H.S., and Rubin, D.B. (2004). *Bayesian Data Analysis*. Boca Raton: CRC/ Chapman & Hall.

Gimenez, O., Crainiceanu, C., Barbraud, C. et al. (2006). Semiparametric regression in capture-recapture modeling. *Biometrics* 62: 691–698.

Gimenez, O., Morgan, B.J.T., and Brooks, S.P. (2009). Weak identifiability in models for mark-recapture-recovery data. In: *Modeling Demographic Processes in Marked Populations* (eds. D.L. Thomson, E.G. Cooch and M.J. Conroy), 1055–1067. New York: Springer.

Goodman, D. (2004). Methods for joint inference from multiple data sources for improved estimates of population size and survival rates. *Marine Mammal Science* 20: 401–423.

Grosbois, V., Gimenez, O., Gaillard, J.M. et al. (2008). Assessing the impact of climate variation on survival in vertebrate populations. *Biological Reviews* 83: 357–399.

Halstead, B.J., Wylie, G.D., Coates, P.S. et al. (2012). Bayesian shared frailty models for regional inference about wildlife survival. *Animal Conservation* 15: 117–124.

Hooten, M.B. and Hobbs, N.T. (2015). A guide to Bayesian model selection for ecologists. *Ecological Monographs* 85: 3–28.

Horvitz, C., Schemske, D.W., and Caswell, H. (1997). The relative "importance" of life-history stages to population growth: prospective and retrospective analyses. In: *Structured Population Models in Marine, Terrestrial, and Freshwater Systems* (ed. S. Tuljapurkar), 247–271. New York: Chapman & Hall.

Hostetler, J.A. and Chandler, R.B. (2015). Improved state-space models for inference about spatial and temporal variation in abundance from count data. *Ecology* 96: 1713–1323.

Hoyle, S.D. and Maunder, M.N. (2004). A Bayesian integrated population dynamics model to analyze data for protected species. *Animal Biodiversity and Conservation* 27: 247–266.

Jamieson, L.E. and Brooks, S.P. (2004). Density dependence in North American ducks. *Animal Biodiversity and Conservation* 27: 113–128.

Jenouvrier, S., Barbraud, C., and Weimerskirch, H. (2003). Effects of climate variability on the temporal population dynamics of Southern Fulmars. *Journal of Animal Ecology* 72: 576–587.

Johnson, H.E., Mills, L.S., Wehausen, J.D., and Stephenson, T.R. (2010). Combining ground count, telemetry, and mark-resight data to infer population dynamics in an endangered species. *Journal of Applied Ecology* 47: 1083–1093.

Kellner, K.F. (2015) Package `jagsUI`. A Wrapper Around '`rjags`' to Streamline '`JAGS`' Analyses. http://cran.r-project.org/web/packages/jagsUI/index.html (accessed 2 July 2015).

Kéry, M. (2010). *Introduction to* `WinBUGS` *for Ecologists. – A Bayesian Approach to Regression, ANOVA, Mixed Models and Related Analyses*. Burlington: Academic Press.

Kéry, M. and Schaub, M. (2012). *Bayesian Population Analysis Using* `WinBUGS`. *A Hierarchical Perspective*. Waltham: Academic Press.

King, R., Morgan, B.J.T., Gimenez, O., and Brooks, S.P. (2010). *Bayesian Analysis for Population Ecology*. Boca Raton: Chapmann & Hall.

Knape, J. (2008). Estimability of density dependence in models of time series data. *Ecology* 89: 2994–3000.

Knape, J. and De Valpine, P. (2012). Are patterns of density dependence in the global population dynamics database driven by uncertainty about population abundance? *Ecology Letters* 15: 17–23.

Koons, D.N., Arnold, T.W., and Schaub, M. (2017). Understanding the demographic drivers of realized population growth rates. *Ecological Applications* 27: 2107–2115.

Koons, D.N., Iles, D.T., Schaub, M., and Caswell, H. (2016). A life-history perspective on the demographic drivers of structured population dynamics in changing environments. *Ecology Letters* 19: 1023–1031.

Lahoz-Monfort, J.J., Harris, M.P., BJT, M. et al. (2014). Exploring the consequences of reducing survey effort for detecting individual and temporal variability in survival. *Journal of Applied Ecology* 51: 534–543.

Lande, R. (2002). Incorporating stochasticity in population viability analysis. In: *Population Viability Analysis* (eds. S. R. Beissinger and D.R. McCullough), 18–40. Chicago: The University of Chicago Press.

Lande, R., Engen, S., and Saether, B.E. (2003). *Stochastic Population Dynamics in Ecology and Conservation*. Oxford: Oxford University Press.

Lebreton, J.D. (2009). Assessing density dependence: where are we left? In: *Modeling Demographic Processes in Marked Populations* (eds. D.L. Thomson, E.G. Cooch and M.J. Conroy), 19–42. New York: Springer.

Lebreton, J.D., Burnham, K.P., Clobert, J., and Anderson, D.R. (1992). Modeling survival and testing biological hypotheses using marked animals: a unified approach with case studies. *Ecological Monographs* 62: 67–118.

Lebreton, J.D. and Gimenez, O. (2013). Detecting and estimating density dependence in wildlife populations. *Journal of Wildlife Management* 77: 12–23.

Lebreton, J.D., Morgan, B.J.T., Pradel, R., and Freeman, S.N. (1995). A simultaneous survival rate analysis of dead recovery and live recapture data. *Biometrics* 51: 1418–1428.

Lebreton, J.D., Nichols, J.D., Barker, R.J. et al. (2009). Modeling individual animal histories with multistate capture-recapture models. *Advances in Ecological Research* 41: 87–173.

Ledwon, M., Betleja, J., Stawarczyk, T., and Neubauer, G. (2014). The Whiskered Tern *Chlidonias hybrida* expansion in Poland: the role of immigration. *Journal of Ornithology* 155: 459–470.

Lee, A.M., Björkvoll, E., Hansen, B.B. et al. (2015). An integrated population model for a long-lived ungulate: more efficient data use with Bayesian methods. *Oikos* 124: 806–816.

Link, W.A., Royle, J.A., and Hatfield, J.S. (2003). Demographic analysis from summaries of an age-structured population. *Biometrics* 59: 778–785.

Lunn, D., Jackson, C., Best, N. et al. (2013). *The* `BUGS` *Book. A Practical Introduction to Bayesian Analysis*. Boca Raton: CRC Press.

Lunn, D.J., Thomas, A., Best, N., and Spiegelhalter, D. (2000). `WinBUGS` – a Bayesian modelling framework: concepts, structure, and extensibility. *Statistics and Computing* 10: 325–337.

Matthiopoulos, J., Cordes, L., Mackey, B. et al. (2014). State-space modelling reveals proximate causes of harbour seal population declines. *Oecologia* 174: 151–162.

Maunder, M.N. (2004). Population viability analysis based on combining Bayesian, integrated, and hierarchical analyses. *Acta Oecologica* 26: 85–94.

Maunder, M.N. and Punt, A.E. (2013). A review of integrated analysis in fisheries stock assessment. *Fisheries Research* 142: 61–74.

McCrea, R.S., Morgan, B.J.T., Gimenez, O. et al. (2010). Multi-site integrated population modelling. *Journal of Agricultural, Biological, and Environmental Statistics* 15: 539–561.

McCullagh, P. and Nelder, J.A. (1989). *Generalized Linear Models*. London: Chapman & Hall.

McGowan, C.P., Runge, M.C., and Larson, M.A. (2011). Incorporating parameteric uncertainty into population viability analysis models. *Biological Conservation* 144: 1400–1408.

Mills, L.S. (2013). *Conservation of Wildlife Populations. Demography, Genetics and Management*. West Sussex: Wiley-Blackwell.

Morris, W.F. and Doak, D.F. (2002). *Quantitative Conservation Biology*. Sunderland, Massachusetts: Sinauer.

Newman, K.B., Buckland, S.T., Morgan, B.J.T. et al. (2014). *Modelling Population Dynamics. Model Formulation, Fitting and Assessment Using State-Space Methods.* New York: Springer.

Nichols, J.D., Hines, J.E., Sauer, J.R. et al. (2000). A double-observer approach for estimating detection probability and abundance from point counts. *Auk* 117: 393–408.

Nichols, J.D. and Pollock, K.H. (1990). Estimation of recruitment from immigration versus in situ reproduction using Pollock's robust design. *Ecology* 71: 21–26.

Ntzoufras, I. (2009). *Bayesian Modeling Using* WinBUGS. Hoboken, New Jersey: Wiley.

van Oosten, H.H., van Turnhout, C., Hallmann, C.A. et al. (2014). Site-specific dynamics in remnant populations of Northern Wheatears *Oenanthe oenanthe* in the Netherlands. *Ibis* 157: 91–102.

Oppel, S., Hilton, G., Ratcliffe, N. et al. (2014). Assessing population viability while accounting for demographic and environmental uncertainty. *Ecology* 95: 1809–1818.

Papadatou, E., Pradel, R., Schaub, M. et al. (2012). Comparing survival among species with imperfect detection using multilevel analysis of mark-recapture data: a case study on bats. *Ecography* 35: 153–161.

Paradis, E., Baillie, S.R., Sutherland, W.J., and Gregory, R.D. (2002). Exploring density dependent relationships in demographic parameters in populations of birds at large spatial scale. *Oikos* 97: 293–307.

Peach, W.J., Siriwardena, G.M., and Gregory, R.D. (1999). Long-term changes in over-winter survival rates explain the decline of Reed Buntings *Emberiza schoeniclus* in Britain. *Journal of Applied Ecology* 36: 798–811.

Péron, G. and Koons, D.N. (2012). Integrated modeling of communities: parasitism, competition, and demographic synchrony in sympatric species. *Ecology* 93: 2456–2464.

Péron, G., Crochet, P.-A., Doherty, P.F., and Lebreton, J.D. (2010). Studying dispersal at the landscape scale: efficient combination of population surveys and capture-recapture data. *Ecology* 91: 3365–3375.

Plummer, M. (2003). JAGS: a program for analysis of Bayesian graphical models using Gibbs sampling. In: *Proceedings of the 3rd International Workshop on Distributed Statistical Computing (DSC 2003)* (eds. K. Hornik, F. Leisch and A. Zeileis), 1–10. Vienna: Technische Universität, Wien.

Pradel, R. (1996). Utilization of capture-mark-recapture for the study of recruitment and population growth rate. *Biometrics* 52: 703–709.

Punt, A.E. and Hilborn, R. (1997). Fisheries stock assessment and decision analysis: the Bayesian approach. *Reviews in Fish Biology and Fisheries* 7: 35–63.

R Development Core Team (2004). *R: A Language and Environment for Statistical Computing.* Vienna: R Foundation for Statistical Computing.

Reynolds, T.J., King, R., Harwood, J. et al. (2009). Integrated data analysis in the presence of emigration and mark loss. *Journal of Agricultural, Biological, and Environmental Statistics* 14: 411–431.

Robinson, R.A., Baillie, S.R., and King, R. (2012). Population processes in European Blackbirds *Turdus merula*: a state-space approach. *Journal of Ornithology* 152: 419–433.

Robinson, R.A., Green, R.E., Baillie, S.R. et al. (2004). Demographic mechanisms of the population decline of the Song Thrush *Turdus philomelos* in Britain. *Journal of Animal Ecology* 73: 670–682.

Robinson, R.A., Morrison, C.A., and Baillie, S.R. (2014). Integrating demographic data: towards a framework for monitoring wildlife populations at large spatial scales. *Methods in Ecology and Evolution* 5: 1361–1372.

Royle, J.A. (2004). Generalized estimators of avian abundance from count survey data. *Animal Biodiversity and Conservation* 27: 375–386.

Royle, J.A. and Dorazio, R.M. (2008). *Hierarchical Modeling and Inference in Ecology. The Analysis of Data from Populations, Metapopulations and Communities.* New York: Academic Press.

Schaub, M. and Abadi, F. (2011). Integrated population models: a novel analysis framework for deeper insights into population dynamics. *Journal of Ornithology* 152: S227–S237.

Schaub, M., Aebischer, A., Gimenez, O. et al. (2010). Massive immigration balances high human induced mortality in a stable Eagle Owl population. *Biological Conservation* 143: 1911–1918.

Schaub, M. and Fletcher, D. (2015). Estimating immigration using a Bayesian integrated population model: choice of parametization and priors. *Environmental and Ecological Statistics* 22: 535–549.

Schaub, M., Gimenez, O., Sierro, A., and Arlettaz, R. (2007). Use of integrated modeling to enhance estimates of population dynamics obtained from limited data. *Conservation Biology* 21: 945–955.

Schaub, M., Jakober, H., and Stauber, W. (2013). Strong contribution of immigration to local population regulation: evidence from a migratory passerine. *Ecology* 94: 1828–1838.

Schaub, M. and Kéry, M. (2012). Combining information in hierarchical models improves inferences in population ecology and demographic population analyses. *Animal Conservation* 15: 125–126.

Schaub, M., Pradel, R., and Lebreton, J.D. (2004). Is the reintroduced White Stork (*Ciconia ciconia*) population in Switzerland self-sustainable? *Biological Conservation* 119: 105–114.

Schaub, M., Reichlin, T.S., Abadi, F. et al. (2012). The demographic drivers of local population dynamics in two rare migratory birds. *Oecologia* 168: 97–108.

Sibly, R.M. and Hone, J. (2002). Population growth rate and its determinants: an overview. *Philosophical Transactions of the Royal Society of London B* 357: 1153–1170.

Sollmann, R., Gardner, B., Parsons, A.W. et al. (2013). A spatial mark-resight model augmented with telemetry data. *Ecology* 94: 553–559.

Spiegelhalter, D.J., Best, N.G., Carlin, B.P., and van der Linde, A. (2002). Bayesian measure of model complexity and fit. *Journal of The Royal Statistical Society Series B* 64: 583–639.

Szostek, K.L., Schaub, M., and Becker, P.H. (2014). Immigrants are attracted by local pre-breeders and recruits in a seabird colony. *Journal of Animal Ecology* 83: 1015–1024.

Tavecchia, G., Besbeas, P., Coulson, T. et al. (2009). Estimating population size and hidden demographic parameters with state-space modeling. *American Naturalist* 173: 722–733.

Tempel, D.J., Peery, M.Z., and Gutierrez, R.J. (2014). Using integrated population models to improve conservation monitoring: California Spotted Owls as a case study. *Ecological Modelling* 289: 86–95.

Tenan, S., Adrover, J., Navarro, A.M. et al. (2012). Demographic consequences of poison-related mortality in a threatened bird of prey. *PLoS One* 7: e49187.

Thomas, L., Buckland, S.T., Newman, K.B., and Harwood, J. (2005). A unified framework for modelling wildlife population dynamics. *Australian and New Zealand Journal of Statistics* 47: 19–34.

Turchin, P. (2001). Does population ecology have general laws? *Oikos* 94: 17–26.

Wade, P.R. (2000). Bayesian methods in conservation biology. *Conservation Biology* 14: 1308–1316.

Wade, P.R. (2002). Bayesian population viability analysis. In: *Population Viability Analysis* (eds. S.R. Beissinger and D. R. McCullough), 213–238. Chicago: The University of Chicago Press.

Wisdom, M.J., Mills, L.S., and Doak, D.F. (2000). Life stage simulation analysis: estimating vital-rate effects on population growth for conservation. *Ecology* 81: 628–641.

Zimmerman, G.S., Link, W.A., Conroy, M.J. et al. (2010). Estimating migratory game-bird productivity by integrating age ratio and banding data. *Wildlife Research* 37: 612–622.

Zipkin, E.F., Thorson, J.T., See, K. et al. (2014). Modeling structured population dynamics using data from unmarked individuals. *Ecology* 95: 22–29.

10

Individual and Agent-based Models in Population Ecology and Conservation Biology

Eloy Revilla

Department of Conservation Biology, Estación Biológica de Doñana CSIC, Sevilla, Spain

Summary

Individual-based or agent-based models are a type of stochastic simulation model in which explicit agents or individuals interact with each other and the environment to generate system dynamics. The use of these models is linked to questions dealing with complex systems and is more akin to a research program than a method in itself, borrowing techniques from many different disciplines. First, the general aim and the questions to be addressed with the model, including the a priori expectations, must be explicit. The second step includes building the conceptual model based on the aim and the empirical and theoretical knowledge available. The conceptual model is then implemented in a core model which should be able to perform a single simulation run. The core model includes the definition of individuals and their traits, the functional relationships, the environment and its properties, the temporal and spatial domains, resolutions and boundary conditions, and model scheduling. A single-model run should produce an output that allows for an early evaluation of model consistency and that can be analyzed later on. At this stage, the conceptual model and the core model should be carefully documented. Finally, analyzing the model may require several steps, including model debugging at run time and an evaluation of the consistency of model behavior at the relevant parameterizations and at extreme values; the evaluation of structural uncertainty and sensitivity analyses, including uncertainty analyses; the use of model selection techniques, if there are alternative model specifications; and model validation and calibration, which consists of estimating model parameters by systematically comparing empirical and simulated data. Ultimately, the successful use of these models is highly dependent on having a clear aim and a good conceptual model. Given the complexity of the questions these models can address and the large flexibility that is allowed in analyzing them, this chapter is just a brief introduction to their construction and use.

10.1 Individual and Agent-based Models

Individual-based models (IBMs) belong to a broad class of stochastic simulation models in which the individuals (or more generally agents) of a population are explicit and identifiable, interacting under a set of rules within a given environment (DeAngelis and Mooij 2005; Grimm and Railsback 2005). Each individual is characterized by specific properties and state variables such as sex, age, reproductive status, body condition, and the coordinates defining its spatial location or its genetic make-up. IBMs may range from very simple to extremely complex implementations. Nevertheless, the conceptual simplicity is one of the reasons why IBMs are becoming so pervasive in disciplines dealing with complex systems, such as astrophysics, cell biology, the social sciences, or ecology (Grimm et al. 2005; Gilbert 2008). Complex systems are characterized by *emergent properties* generated by the interaction between its components and the environment. Typically, the behavior of those emergent properties is affected by stabilizing negative feedbacks and/or destabilizing positive feedbacks, as occurs with density-dependent processes or with Allee effects. Conceptually, it is easy to grasp what IBMs are, as it is to build them if we have an intermediate command of a programming language. The difficult part is using these models in a way that is useful for our purposes and then communicating the methods and results to third parties in a clear and logical way. In this chapter I will try to help you in doing so.

Populations are just collections of *different* individuals. The uniqueness of individuals affects their realized fitness thus contributing in different amounts to the dynamics of the population to which they belong. Fortunately, the heterogeneity of individuals can be categorized into

Population Ecology in Practice, First Edition. Edited by Dennis L. Murray and Brett K. Sandercock.

Companion website: www.wiley.com/go/MurrayPopulationEcology

several main types that summarize the most relevant sources of heterogeneity in fitness, such as demographic classes, phenotypes, or genotypes. In population ecology, we can take advantage of this structuring by averaging reproduction, survival, and movement parameters within each of these groups and then describe or project population dynamics using those estimates (Chapter 8). Nevertheless, class-specific demographic parameters vary through time and for individuals in different spatial locations, normally as a consequence of changes in relevant environmental variables.

Populations belong to the most challenging type of complex systems: *adaptive systems* where the responses of individuals can change (Grimm and Railsback 2006). Apart from *evolutionary responses*, which may occur within a small number of generations making them relevant for population dynamics (DeAngelis and Mooij 2005), individuals can show behavioral and other *phenotypic responses* including memory, maternal effects, or the effect of previous conditions within the domain of each individual. Thus, individuals have the capacity to adapt their responses to environmental conditions in unexpected ways, making demographic functional responses dynamic (Kuparinen and Merila 2007; Doak and Morris 2010). Methods dealing with complexity are especially useful for questions dealing with real populations. Nowadays, the major challenge of population ecology lies in having some forecasting capacity for populations composed of heterogeneous and adaptive individuals living in an environment which is also heterogeneous and dynamic in time and space.

10.1.1 What an IBM Is and What it Is Not

The typical implementation of an IBM comes in the form of a computer program that executes, in a dynamic way, the processes describing the interactions between a set of individuals and their environment, generating relevant emergent properties at the population level, such as trajectories of population size in time, age, stage or sex distributions, or distributions of density in space. Therefore, IBMs are simply a way to generate simulated data using stochastic numerical simulations. In itself, an IBM is not a method of analysis based on some statistical paradigm and therefore it departs from most of the methods described in other chapters of this book. To be of any use, the simulated data needs to be summarized by analyzing it in a similar fashion to that of field data, using everything we have learned so far, from how to generate and test sensible hypotheses, to estimating demographic parameters or analyzing time series and spatial structure. Therefore, the use of IBMs requires some skills in coding and an advanced research plan, including an adequate initial design for a clearly stated question, testing the general behavior of the model against empirical data or theoretical expectations, and finally conducting some simulation experiments in which we systematically evaluate alternative scenarios to make some useful predictions.

Building an IBM requires software coding, either implicitly or explicitly. Nevertheless, coding is by no means the limiting factor when building an IBM. The main challenge is making explicit the question and designing a sensible and logical procedure to address it. Above all, using IBMs is an excellent way to make explicit our knowledge and assumptions in order to generate new hypotheses and predictions. It is therefore clear that IBMs are most relevant when aiming at complex questions for which other approaches are limited. To be able to do so we need a priori knowledge about how the system might work as well as information to be able to parameterize the model, even if using scenarios with hypothetical parameterizations (DeAngelis and Mooij 2005; Grimm and Railsback 2005).

10.1.2 When to Use an Individual-based Model

The use of IBMs has increased significantly in the last few decades, and so has the diversity of research questions covered (Grimm 1999). Models are often used to investigate complex questions, such as those having highly discordant *spatiotemporal scales* for different processes and patterns (generally local interactions generating data patterns at large scales), or feedbacks and conditional parameter values affecting functional responses and strong impacts of spatial environmental heterogeneity on individual traits and responses. In many cases, the use of IBMs links population ecology to other disciplines, such as genetics, landscape ecology, behavioral ecology, ecotoxicology, and economics. Typical studies range from population viability analysis (PVA) of small populations for which demographic stochasticity is important (Chapters 8 and 9), to management questions including the evaluation of different harvest regimes (Wiegand et al. 1998; Whitman et al. 2004), and questions dealing with population genetics, such as genetic structure or effective population size, and their relationship with demography and population viability (Storz et al. 2002; Bruggeman et al. 2010; Perez-Figueroa et al. 2012). Authors often explore the role that different mechanisms can play at the population level under different environmental conditions, including physiological processes, such as individual energetics, growth and biomass dynamics, or their interaction with diseases (Willis 2007; Buckley 2008; Boyles and Willis 2010), as well as behavioral mechanisms, such as the link between individual behavioral responses and their impact on demographic parameters, the role of group living and sociality or spatial ecology, and individual movements, including dispersal and how it impacts population dynamics (Stephens et al. 2002; Kramer-Schadt et al. 2004; Revilla et al. 2004; Goss-Custard et al. 2006; Rands et al. 2006; Revilla and Wiegand 2008; Tablado and Revilla 2012). Last, IBMs can also be used to address complex multispecific questions, such as predator–prey interactions, community dynamics, or epidemiology of diseases (Rushton et al. 2000; Schmitz 2000; Wilkinson et al. 2004; Carlo and Morales 2008; Ramsey and Efford 2010).

10.1.3 Criticisms on the Use of IBMs: Advantages or Disadvantages

When first used, IBMs were heavily criticized along four main lines of thought. First, these models were described as too complex and therefore data-hungry and prone to overfitting and error propagation problems. This critique has been based on a simplifying generalization and on some erroneous analyses (Beissinger and Westphal 1998; Mooij and DeAngelis 1999). If properly designed, calibrated, and analyzed, IBMs are no more prone to those problems than any other applicable method (Wiegand et al. 2004b). The generalization about overcomplexity is unfair since it is by definition not part of IBMs, but rather a consequence of addressing complex questions. Additionally, it confuses the definition of complexity used for statistical inference in statistics probability theory, defined by the number of parameters of a statistical model, with structural complexity under algorithmic theory. The situation leads to an axiomatic application of Occam's razor, which should be applied to empirically or theoretically supported process descriptions when those descriptions are similarly supported by data. Only then should the model with fewer parameters be favored. The usefulness of a model is not given by the number of parameters, but rather its ability to address a question.

It is often assumed that the lower the number of parameters of a model the more generalizable the results, forgetting that the assumptions are also part of the model, and that to be able to make generalizations to other systems (not to say to make predictions) the set of assumptions must be sensible and comparable among systems. Structural realism is an important advantage of IBMs, especially in relation to model assumptions and even if model parameterization is not fully resolved or specified (Wiegand et al. 2004b; Ajelli et al. 2010). For example, the structural complexity of IBMs allows for the direct inclusion of demographic stochasticity with no need to parameterize it.

The remaining three criticisms are that IBMs are difficult to analyze, difficult to communicate, and, finally, the results are difficult to generalize to make inferences on the functioning of other systems (Bolker et al. 2003). These points are relevant and represent the main challenge of using IBMs. The poor implementation of some early models, for some of which it appears the aim was to build the model itself, combined with poor documentation, made the models too obscure and difficult to follow, not to mention replicate (Müller et al. 2014). The only way to minimize those problems consists in using a research program aiming to understand how a complex system works (individual-based ecology, sensu Grimm and Railsback 2005). In doing so we should take advantage of the flexibility of IBMs, including the possibility of linking them to other methods, the capacity to make use of many sources of data with varying quality, including ancillary data, or the capacity to introduce difficult structures, such as covariation between model parameters, in a natural way. Last, an important advantage of using IBMs is that if properly built, they force us to make explicit all the relevant knowledge on a population, including how different processes interact, and the capacity to generate predictions that are testable in the field.

10.2 Building the Core Model

10.2.1 Design Phase: The Question and the Conceptual Model

The first step in building an IBM is to identify and make explicit the general aim of the model. In the early days of IBMs, it was not uncommon to find examples of models that were described with no further aim, consequently generating a lot of criticism. IBMs, as any other model, should be built to address a specific question. The general aim should be developed in the form of *specific questions* that can be directly linked to a priori predictions as well as data, both empirical and simulated. The theoretical and empirical context must be set, together with the general simplifying assumptions that are made a priori, such as no role for space or evolutionary processes.

The second step in the design phase consists in developing a *conceptual model* in which we summarize the knowledge in relation to the question to be addressed. At this stage it is quite useful to perform an in-depth review of the state of the art of the question, which should be made available to readers, either as part of the final manuscript or as a stand-alone publication. The conceptual model should make explicit the processes at the level of individuals that are known to affect some of their fitness traits such as age-mediated survival, the environmental factors modulating fitness such as higher mortality at low temperatures, and the available parameter estimates including their central value, variability, and uncertainty. It is also important, especially if we are dealing with a question related to a specific species and population, that we clearly differentiate the information coming from (i) general theory (including empirically derived heuristic patterns); (ii) from species with a similar life history and ecology; (iii) from the same species in other populations; and (iv) from the focal population itself. These distinctions will help us later on when defining model uncertainty, parameterizations, and the alternative scenarios.

From the design phase, we should have a summary list with the working plan and the required pieces, including: (i) the individual traits (both those directly and indirectly linked with fitness); (ii) processes and their parameters directly modifying individual traits (including rules and equations); (iii) environmental processes and their parameters (indirectly affecting individual traits, their rules, and equations); (iv) a well-planned schedule for how all those processes occur and integrate along the iterations of the model, that is along the individuals in the population or through time; and, finally, (v) the emergent properties directly linked to the questions at hand (Figure 10.1).

The design phase is critical and our final success will depend on doing a good job at this stage (Figure 10.1). It is also the most difficult part of the entire process, requiring some experience to master. The good news is that there is no single correct way to do it, and that we have a lot of freedom to follow our own preferences and style. A good starting point is to consult relevant papers using IBMs (Section 10.1.2) and see how different authors deal with stating and breaking the general aim into questions and predictions, and how they explain and justify their conceptual model. Building an IBM is about creating a *conceptual model* with an explicit and dynamic representation of the available knowledge on the relevant processes and their parameters affecting some variables of interest as the *emergent properties* linked to the questions and predictions. We will have a chance to eventually be successful only if we have a clear question and a good conceptual model.

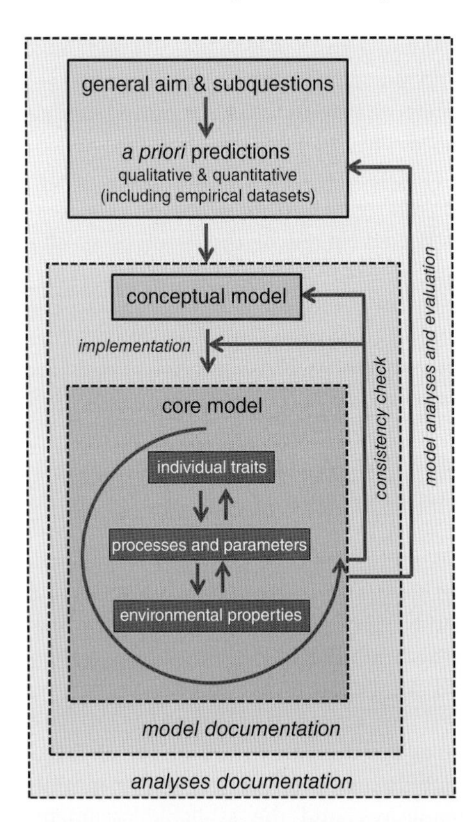

Figure 10.1 Simplified scheme of the modeling cycle for model design, including the modifications that often need to be introduced during consistency checking and analyses, both in the conceptual model and its implementation in the core model and even in the way we develop the question and predictions at hand.

10.2.2 Implementation of the Core Model

The next step is the implementation of the conceptual model in a core model that by iteration of the processes generates some type of dynamics in a single simulation run. Normally, the core model is implemented using a programming language. The best language is the one you already know, or the one mastered by a colleague who can provide logistical support. There are so many potential choices that here we can only offer a brief field guide to help you in deciding (Box 10.1), and make general recommendations that are useful across platforms and languages. There is no single best approach since different systems and languages have both advantages and disadvantages. Running simulations will require a modeling environment that allows for an efficient characterization of individuals and the proper integration across scales. Additionally, it is convenient that the system allows for debugging while coding and while running simulations, which will help in detecting errors and in the evaluation of model consistency (Figure 10.1). Last, the selected platform should allow for fast simulation runs to be time-efficient in the analyses (Box 10.1).

10.2.3 Individuals and Their Traits

The population is a collection of individuals, but before creating any individual, we have to define their attributes by describing the traits and properties characterizing them, as defined in our conceptual model. For example, if we need to distinguish their sex, age, and reproductive status, we will need to define those three identifiers. Even if two individuals have the same values for all traits, they must be unique and it should be possible to distinguish and find them within the population. Individual traits can be constant throughout their lifetime, for example their genetic makeup, or – depending on the taxa – their sex; or dynamic, if they change during the life of the individual, such as age or reproductive status (Box 10.2). The questions to be addressed with the model will help us in

Box 10.1 Programs and Software: A Field Guide to Some Individual-based Coding Environments

We can use three types of approaches: using software that allows for scripting using interpreted languages, general multipurpose programming languages that allow for object oriented programming, or specialized development environments created specifically to build agent-based models.

Approaches Useful for Building Demonstrator Models

We can create IBMs using software which allows for scripting, as is possible in some spreadsheets such as Gnumeric, LibreOffice Calc, or proprietary `MS Excel`, noting that you need some knowledge of `Visual Basic for Applications`, `Python`, or any other supported scripting language to program the macros (e.g. Macal and North 2010). These implementations are useful as demonstrators for learning concepts and teaching or for implementing structurally simple IBMs for which the analyses are not complex. We can also build IBMs in environments that are efficient in making generalized scalar operations such as in vectorial or array programming languages, such as `R` or `Matlab`, or even in more eclectic languages such as `Wolfram` (running in proprietary `Mathematica`).

`R` is a software platform that allows for the efficient manipulation and analysis of relatively small datasets. It is so flexible that we can also build IBMs with it. However, doing so is only reasonable for learning purposes or when dealing with simple IBMs with few parameters and individuals. `R` uses array programming, operating with all the data simultaneously, making the processing of large datasets inefficient. Therefore, it is slow and resource hungry in dealing with the data we create when, for example, running a sensitivity analysis across many-dimensional spaces. It is also an interpreted language, i.e. does not compile the commands we write into machine code, making simulations much slower than other alternatives.

General Purpose Development Environments

This group refers to compiling object-oriented programming languages that allow programming totally ad hoc models. Normally, the source code is written within a computer program called *compiler* that transforms the source language into a machine compatible language that can be executed by the computer. This approach is more efficient than interpreted languages, allowing for much faster simulations. Creating individuals is straightforward using *objects* or *classes*. After compilation we can obtain a range of possibilities, from a self-contained executable file to a sophisticated application with a detailed Graphical User Interface (GUI, normally created by using *forms*) that may allow for interaction with the user during the initialization (e.g. for parameterization), a graphical inspection of model behavior during run time and also the exploration of the results. We can cite `C++`, `Python` (to some extent), or `Java` as general languages, with different derivations of `Fortran` and `Object Pascal` being very popular in academic and scientific applications. All of them have many compilers available. If you have some experience programming this would probably be your best way to proceed.

To run the model we have several alternatives, very much dependent on the language we are using and the environment (compiler and operating system). The most basic is a batch-like mode in which, after asking for execution (e.g. by clicking in the .exe file created by the compiler after a successful compilation), all the code is executed at once with no further intervention on our part. In most modern programming languages we interact with a compiler that includes prewritten components (library-like) that can be used and reused, allowing for fast model construction and deployment. Forms are the most basic of such components when running the program. They create a window that allows for interaction between the user and the model at run time. Many other components can be used, including buttons to be inserted in the form which execute some code when we click on them. Forms and other components with which we interact are part of the GUI of our model. If, for example, the pseudocode in Box 10.3 was written in a compiler allowing for forms, we could add to it a button which on a click would run the subroutine for population dynamics.

Specialized Development Environments

These are just implementations built using general programming languages but that offer through an Application Programming Interface (API) access to precoded libraries that can simplify the initial work of making explicit the conceptual model (Railsback et al. 2006). Using a specific environment would save you a lot of time if you have no experience programming. Running the model in specialized development environments is straightforward, just follow the program instructions. Specific environments for building IBMs have their own detailed documentation and many examples to build upon. A nonexhaustive list would include:

`ALMaSS`, Animal, Landscape and Man Simulation System. A complex, highly specific model, with detailed implementations built for different species (e.g. voles, skylarks). The model is spatially explicit, including individual movement behavior, a landscape model that can be dynamic, and a weather simulator. Open source project written in `C++`. Topping et al. (2003). http://ccpforge.cse.rl.ac.uk/gf/project/almass.

GAMA. A highly flexible system that allows for the development of complex, spatially explicit models of potentially very large populations. The conceptual model is coded in GAML language, which is a derivative of XML. Allows for calling R and SQL code using several DBMS (database management systems). The user interface is based on the Eclipse platform (which is itself mostly written in Java). Grignard et al. (2013). https://github.com/gama-platform.

Repast. A set of open-source platforms to perform agent-based modeling and simulations, including spatially explicit models. Different implementations, including either Java or C++ coding systems. Allows for fast simulations and large and very complex models to be built. Very complete and with many tools available. Macal and North (2009). http://repast.sourceforge.net.

Mason. Multiagent simulation of neighborhoods. It is a discrete event agent based simulation platform implemented in Java (requires experience with this language). It is fast, flexible, and portable across machines, with good capacity to run in batch mode with no visualization. Luke et al. (2005). http://cs.gmu.edu/~eclab/projects/mason.

NetLogo. An intuitive and easy to use system to develop simple grid-based models. Recommended for people with no programming experience. Based on a language derived from Logo (but built in Java), with many primitives (built-in commands). Includes a collection library with many ecological model examples. Well suited for educational purposes, but simulations are very slow (does not compile into binary). Can be linked and called from R using Rnetlogo. Wilensky (1999). http://ccl.northwestern.edu/netlogo.

Swarm. It was the first platform developed for agent-based simulation modeling. Initially designed in Objective-C, currently runs in Java. Well organized and stable. www.swarm.org.

Box 10.2 The Population: Creating the Individuals

There are two general ways to define and create individuals in general purpose development environments. The methods used in specific development environments can match these or be more graphical.

Lists of Objects

It consists of using a *list* to generate a collection of objects where the list refers to the population, and a *class* template of objects is used to represent agents or individuals (Box 10.3). Within the object oriented programming paradigm, classes are created to serve as templates to define objects, which in our case will refer to individuals and the properties or variables characterizing them. They can be seen as, data structures. Additionally, in all languages, classes can have methods associated with them. In principle, we can create our template for individuals without needing methods, using simplified class versions, if available (e.g. record in Pascal, or struct in C++). Once we have created (declared in programming jargon) the data structure for our individuals, we need to declare and create a list to manage a collection of pointers, each of which will be used to link each individual we create. In such a manner we will be able to locate and distinguish individuals even if they have the same trait values. The list can be seen as a container that facilitates the management of individuals, allowing for adding, removing (and destroying), searching, sorting, and counting among other useful methods. In summary, we simply have to create the population (*list*) and add the number of individuals (*objects*) we need, each of them with their own set of descriptors as specified in by the conceptual model. Running many simulations can lead to problems of memory usage and allocation in the computer, depending on the environment, language, and compiler. To avoid this situation we need to do the housekeeping of managing memory when destroying individuals (or any other class) and when dealing with subroutines (for example, freeing resources such as virtual memory).

Dynamic Arrays

The second method consists of using dynamic arrays (arrays are simply vectors or matrices in programming jargon). Obviously, they also represent a data structure in which each cell has a single value. In dynamic arrays we can keep the number of dimensions variable in run-time. Therefore, by keeping constant the dimensions characterizing the traits, and variable one dimension representing the number of individuals, we can describe a population. It is easy to understand how they work by analogy with a table in a database: the columns describing trait variables will be a fixed dimension, each of which represents a trait, and each of the rows will be an individual. This second dimension will be dynamic, i.e., with a variable size because we should be able to create and delete items. Dynamic arrays also come with useful methods associated with the management of the items they contain.

Box 10.3 Pseudocode Algorithm Describing a Basic IBM

We consider a population with N individuals and with reproduction and survival as demographic processes. We follow the list-class approach to create the population. The model represents an exponential growth system (for example to evaluate a reintroduction in the short term or a population collapse). The explicit parameters of this model are N_0 initial population size; P_R reproduction probability; P_S survival probability; *max_age* maximum age; *t* number of time steps simulated. Note that there are other implicit parameters such as litter size, a constant that works as a model assumption. We move along all individuals of the population using conditional loops (such as Do While- or For-loops, which are sections of code that are repeated as long as a condition is met); note that we can call one subroutine from another (as for survival called from population dynamics subroutine).

```
//Declaring a container for our population, named "Population"
1: list Population

//Declaring the data structure for individuals (their traits)
2: class Individual
        Sex: string
        Age: integer

//Initializing a population of size N0;
3: procedure Initialize
4: create Population
5: with Population do
6: for 1 to N0
7:      create individual
8:      individual.sex = random
        (female/male)
9:      individual.age = random
        (maximum_age)
10:     add individual
11: endfor

//subroutine for reproduction with a breeding probability PR
12: procedure Reproduction
13: with Population do
14: N = Population size // assign current population size to variable N
15: for i = 1 to N do
16:     individual = [i]
17:     if individual.sex = f then
18:         if random<PR then
19:         begin
20:             create individual
21:             individual.sex = random
                (f/m)
22:             individual.age = 0
23:             add individual
24:         end
25: endfor

//subroutine for survival with a survival probability PS
26: procedure Survival
27: with Population do
28: N = Population size
29: for i = 1 to N do
30:     individual = [i]
31:     if individual.age > max_age then
        delete individual else
32:         if random>PS then delete
            individual else
33:         individual.age = individual.
                age+1
34: endfor

//subroutine for population dynamics; this is the procedure we call to run the model
35: procedure Dynamics
36: N0 = #
37: t = #
38: PR = #
39: PS = #
40: max_age = #
41: Initialize
42: for time = 1 to t do
43:     Reproduction
44:     Survival
45:     N = Population size
46:     plot time vs N
47:     save results
48: endfor
```

defining the initial population, which needs to be created from scratch at the beginning of a simulation. This population will have a given number of individuals, each with its own traits. As such, we create a population with a specific distribution of, for example, sexes, ages, and statuses. Obviously, the initial condition imposed by this population will have a profound impact on model dynamics: the dynamics generated by an initial population of 10 or 500 individuals will be quite different. Therefore, the design and justification of the initial conditions should be thought out carefully and its impact analyzed.

10.2.4 Functional Relationships

Individuals should interact in such a way that their fitness traits are affected. In classical population ecology we broadly distinguish between processes dealing with survival, reproduction, and movement. Conceptualizing survival and reproduction as processes removing or adding individuals from the population is straightforward (Box 10.3). Movement is more complicated as it is a process mediating the addition or removal of individuals by migration. We can distinguish three types of processes directly affecting individuals: (i) adding individuals by recruitment or immigration; (ii) removing individuals by mortality or emigration; and (iii) processes modifying individual traits, including responses to environmental conditions, behavioral responses, and automatic modifiers of traits, such as age-related performance. The processes may range from simple rules, for example, if the maximum age is reached the individual must die deterministically, to complex sets of conditional equations, such as a function calculating the probability of breeding as a function of local density and a set of environmental variables only if age and body condition allow for it. The possibilities are incredibly broad, but fortunately, we have a conceptual model at hand to identify what processes are potentially relevant.

Implementing functional relationships is normally done by programming subroutines, which is nothing more than a packed sequence of instructions that is executed whenever we call for it. Subroutines take different names in different software languages but work in a similar way (e.g. functions, procedures, methods). Functional relationships are implemented by modifying variables (Box 10.4) with mathematical, logical, and other types of operators as well as functions (for example, to obtain the absolute value of a floating number or to truncate it). In the case of complex equations we can make use of precoded libraries which are subroutines in themselves that can simplify the task. A key characteristic of subroutines in IBMs is that many of them need to go through the entire population, individual by individual, in order to perform the required calculations. For example, to apply an annual mortality rate we need to go through all individuals, one by one, and stochastically check if they have survived to the following year (Box 10.3).

10.2.5 The Environment and Its Relevant Properties

The environment represents the set of variables that act as direct or indirect modifiers of the traits of individuals. For example, the probability of reproduction of a female depends on its age, the actual density, and the amount of rain in that year, with some specific parameters estimated with field data. Age is an individual property with its own dynamics whereas density and rain are external variables for each individual. In this case, we need to calculate and keep track of population size and then calculate density during each simulation (Box 10.4). Note here that density dependence is probably one of the simplest impacts that the environment may have on the traits of each focal individual. The same applies to rain, which, depending on our needs, may be a predefined set of values or have its own dynamics depending on additional processes. In the case of predefined values, rainfall might be a constant included in a one-dimensional array of integer values, indexed from the first to the last year of data. Fixed environmental properties are included in the model as variables, with or without associated variability (Box 10.4). In the case of dynamic environmental properties, we need to include the processes describing the dynamics in specific subroutines as we do with other processes, including any rules, functional relationships, and their parameters. Environmental properties, which are also part of the initial condition, will have to be set up when starting the simulation.

10.2.6 Time and Space: Domains, Resolutions, Boundary Conditions, and Scheduling

A critical element is how time and space are dealt with. Both are defined in all conceptual models, either *implicitly* or *explicitly*. In explicit definitions, we need to keep track of them, either in continuous or discrete ways. If time or space are not explicit, we still need to acknowledge them by clarifying the assumptions made on their reference domains. A *domain* is just the range of allowed values. Even in nonspatial models we have a spatial domain in the form of an assumption. Therefore, the first step is defining the temporal and spatial domains. Time is explicit in most cases (but not all), whereas both spatially implicit and explicit IBMs are common. For example, if we define the temporal domain of our model as 10 years for evaluation of a short-term reintroduction effort, we know that a simulation can run at most for that amount of time; or, if the spatial domain is 100 x 500 km, that is the area in which our population occurs.

Within its domain, time can be represented by one or more temporal resolutions as required by the processes affecting individuals and the environment. The study of the interaction between processes at highly discordant temporal resolutions is essential for understanding the dynamics of complex systems (Grimm et al. 2005). In the above example, the 10 years can run in steps of one day or one year depending on the relevant processes. For example, in the case of univoltine species, reproduction can occur only once a year and therefore reproduction would require steps of one year. On the other hand, if we need to evaluate the role of the mortality imposed by short-term cold spells, we may think of a finer temporal

Box 10.4 Parameters, Arguments, and Pseudorandom Numbers

Parameters and Arguments

With model parameters we refer to values that are relevant in our conceptual model and that need to be considered either by themselves or as part of the functional relationships. Their value can be constant in any parameterization (e.g. maximum life expectancy) or can change between parameterizations. Additionally, model parameters can be sampled from a distribution to represent not only the means but the variability of their estimates. Arguments are information that we track at run time. They are normally needed by subroutines or commands, for example, population size at a given time of a simulation, which may be required in itself as output or to calculate density. They are sometimes referred as summary statistics (Hartig et al. 2011).

Parameters and arguments are stored as variables which are identified by a symbolic name (N for the argument population size or P_S for the parameter defining survival probability in Box 10.3). Variables can be local or global depending on their scope. Typically, we tend to use local variables when dealing with information required only within a subroutine (e.g. the variable describing the counter of a loop) and global ones when needed throughout the model. Depending on the language that we are using, variables may need to be explicitly declared, initialized, and emptied before reuse and the type of information they can store needs to be defined a priori (for example, a string or an integer value). One important distinction is between variables that can hold a single value and arrays that can have multiple ordered values in one or more dimensions (i.e., vectors and matrices).

Variability and Pseudorandom Numbers

Some (or most) of the parameters used to parameterize a model have some associated variability in relation to both uncertainty in the empirical estimates and natural variability, typically in time, space, or associated with interindividual variability. These sources of stochasticity need to be dealt with, first in the conceptual model by identifying and justifying which of them are relevant, and then when defining the parameterizations that will be used for sensitivity and further analyses.

In order to obtain a stochastic value from a known distribution, we use standard procedures that generate pseudorandom numbers and that are available in all compilers. These procedures need to be initialized with a seed number. If we always use the same seed, we will obtain the same sequence of numbers, which is helpful in detecting errors in the code. Typically, when running simulations, we use different seeds coming from a highly variable source (such as the clock of the computer, with the help of the relevant function), thus making the sequence more unpredictable (be aware that some of the algorithms can be poor, with relatively short return rates).

Pseudorandom number generators produce numbers from a given distribution, usually a uniform distribution between 0 and 1. Unless the probability density distribution that we need is already implemented in the compiler, as often occurs with the normal distribution (with a given mean and variance that we need to specify), we can use the pseudorandom numbers obtained from the uniform distribution to randomly sample any other probability density distribution or discrete probability histogram, with a bit of thought and simple math: by rejection sampling or using the inversion method (inverse transform sampling), in which we use the cumulative distribution function of the known probability distribution.

Often we may have erratic errors occurring at low rates. To locate where they occur in the code, it helps to switch off the randomization process used to generate pseudorandom numbers. In that way, the error will always occur at the same point of the simulation, allowing you to locate the problem. We can use breakpoints in the code just before the error happens and then run the code line by line from within the compiler.

resolution. Time is normally introduced as a conditional loop in which there is a counter that keeps track of the current time step (see subroutine for population dynamics in Box 10.3). If we have several temporal resolutions, we can nest several conditional loops in a way that allows accounting for time as a clock does. For example, if we need days for survival and years for reproduction, we will code two nested loops, one counting years and another, within the previous one, counting days. Once the day loop runs for 365 days we start it again and the yearly loop moves to the next year.

In spatially explicit models, we can proceed from simple to complex descriptions of space (Box 10.5). Typically, we need to use explicit space when movement is a relevant process and therefore it needs to be implemented in subroutines, with rules or equations describing when, how, and where individuals move. Modeling of movement can be handled by changing the values of the traits describing the coordinates of individual location. The subroutines tend to have fine-scale temporal resolutions to allow for individual movement decisions. All the rules and equations should be clearly specified and justified in the conceptual model (Nathan et al. 2008). Associated with individual movement decisions is the concept of *boundary conditions*. What happens if individuals move to the edge of the spatial domain? Individuals can

Box 10.5 Space Representations

We can use two simple approaches to define space by using either a continuous or a discrete space.

Continuous Space

In this case, the location of each individual within the spatial domain is defined using a Cartesian or polar representation. This approach is typical of applications in which individuals move independently of an environment or at most their movement is affected by only a few spatial references that we can track with their coordinates, such as the location of other individuals or the location of a nest. The location of each individual is kept as individual traits (its coordinates) that change when it moves, whereas the spatial resolution is given by the resolution of the numeric values used (e.g. integer or floating types). Nevertheless, it is perfectly possible to use more complex vectorial map representations, which will require a bit more thinking and recalling the trigonometry we learned in secondary school.

Discrete Space

This approach is used in cases with more complex spatially explicit environmental properties, such as several levels of habitat quality affecting survival or movement. In that case we can represent a map as an array of one, two, or three dimensions (more akin to a raster GIS landscape map), depending on the required dimensionality: one for landscapes, such as rivers, that can be represented linearly; two for x and y landscapes, and three if we need *x*, *y*, and *z* coordinates such as in the ocean, or if using a dynamic landscape (*x*, *y*, and *t*). In this array, each dimension is indexed between 0 and a maximum value (as defined by the domain), with the index representing the spatial location (coordinates) and the value at that location some relevant environmental property (for example, 1 for presence of a nest, 0 for absence; or different values representing different habitat qualities). The discrete space represented by the array has a typical resolution (e.g. 10×10 m or 5×5 km) which is not explicit in itself. A good way to visualize this is to think about the typical bidimensional map represented as a grid or a raster map with x and y coordinates and a stored value within each grid cell. Grid cells can be square or take other shapes (hexagonal grids; Liu et al. 1995; Letcher et al. 1998). Very often the resolution of the map is also used to define the coordinates of the position of individuals, thus using only one spatial resolution in the model. If we do not use the same resolution we have to deal with the scaling between the two, one for individuals and one for the map, with some rules (such as rounding or truncation, behavior at the border of grid cells, etc.). For most applications grid-based approaches may be sufficient, whereas for very large domains it can be computationally demanding.

basically do two things: either be *reflected back* into the domain, which would be the case for movements within closed populations bounded by a fenced area, an island, or an oversized spatial domain, or *emigrate* by leaving the domain. If we implement emigration, we may need to implement immigration as well. In some cases, it is sufficient using a balanced emigration–immigration function by moving individuals back into the domain at the other end of the dimension they left in a torus-like fashion. In any case, the best answer depends entirely on the system and the question at hand.

Last, a critical concept we must think about carefully is that of *scheduling*, or how processes having different resolutions are nested and, for those with the same resolution, how they are ordered. Even in simple models, sometimes it is not easy answering questions such as what or who should be first, as is the case for survival and reproduction, in a model with only one temporal resolution (see for example the model in Box 10.3 and think about the effect of calling survival first instead of reproduction). In models with an implicit time, as occurs with some short-term IBMs dealing with individual decision-making, or within a temporal resolution, we still need to define the order of interaction between individuals, that is their cuing or implicit timing of interindividual interactions. Different schedules affect model behavior and results. Again, the conceptual model is critical here as well as the explicit listing of how many temporal and spatial resolutions we have for each of the processes involved (Berec 2002). Once you have a schedule, it also helps to plot a diagram describing it (Figure 10.2).

10.2.7 Single Model Run, Data Input, Model Output

The core model can be used to run single simulations. As such, it is not of much use apart from demonstration or educational purposes with regard to our conceptual model. Most compilers allow for a process called *debugging*, which permits detecting the existence of programming errors, often locating the place where the code is flawed. Therefore, this debugging compilation will

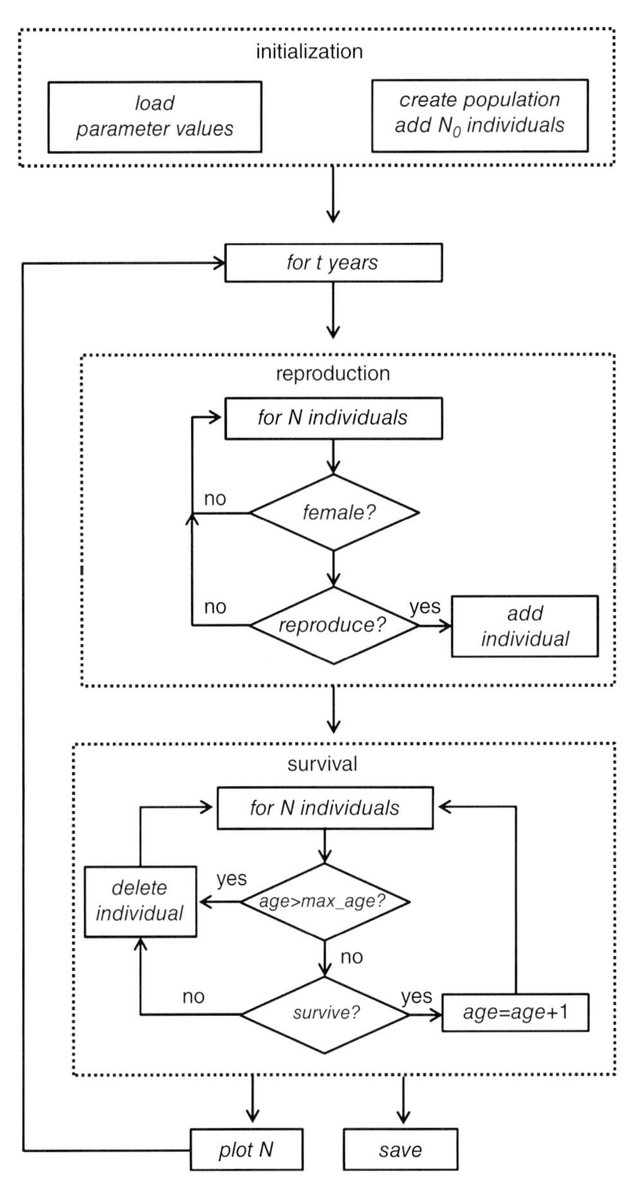

Figure 10.2 Schematic flow chart depicting the scheduling of a time step for the model described in Box 10.3. Time resolution is one year and space is implicit.

probably be the first manner of execution that we face, to our despair, but it is needed to obtain a clean and consistent core model. Nevertheless, debugging does not solve the inconsistencies that we introduced in the conceptual model or in the questions (Figure 10.1).

To run the model, we need to parameterize it by setting *initial values* to all model parameters (Box 10.6) and defining the *initial conditions* with the starting population size and structure and the environmental setting. After running a simulation, we need to obtain some *output* describing model behavior and predictions (Box 10.6). Remember that in the conceptual model we had identified simulated data directly linked to specific questions and their a priori predictions. IBMs are stochastic models and, therefore, the output variables will yield different results in different simulation runs with the exact same parameterization and set of initial conditions. To estimate the probability distributions of each of the output data, a number of simulation runs must be repeated with each parameterization. A reasonable rule of the thumb is enough runs to obtain stabilized estimates of the mean and standard deviation of the output variables.

10.3 Protocols for Model Documentation

At this stage, we have a general aim that breaks into a set of specific questions and their potential responses based on a priori expectations, a conceptual model describing the system and the potentially relevant processes involved (and their parameters), and a description of how those processes drive the interactions between individuals, between those and the environment, and the environmental dynamics itself, generating the dynamics of the population. We have implemented the conceptual model into a simulation model in what I have called the *core dynamic model*. At this stage, it is crucial to document what we did so far before the model gets too complex. During the process of building the model, we probably needed to modify some parts and details of the conceptual model to accommodate the explicit way we built it and why we did so (Figure 10.1). Once we start analyzing the model, we will probably need to revise both the conceptual and the core dynamic models again. A process of continuous refinement is normal and it is not a problem in itself. Nevertheless, and as complexity grows, we have to document what we have, even if it needs be modified later on.

Traditionally, model documentation has ranged from simple verbal descriptions to detailed descriptions and justifications, including *pseudocode* or even the full code of the model. Model documentation should run together with model building, as it forces us to go through a process of thinking about how we are designing things and how all the components integrate. The documentation should include both model justification and a detailed description of its processes. For that reason, the refined version of the conceptual model, after the revision when constructing the model, should be the main part of the documentation.

Some general guidelines can help properly inform our work. We need to be as clear as possible about the general aim and the specific questions to be addressed, including the a priori predictions and the list of model behaviors and variables dynamically predicted by the model that will be used in the analyses. If using field or theoretical

Box 10.6 Data In, Data Out

There are three ways to parameterize a model. The simplest is by typing assigning statements in the code. For example, we can define the variable storing the maximum age that an individual could reach equals 10 years (*max_age* = 10 in Box 10.3). This can be done with all the required information. Nevertheless, this approach is normally used with parameters that will not change in between simulations (such as constants).

If our model has a GUI, we can add components to it on which we can specify parameter values. There are many types of components, such as text, combo, or drop-down list boxes, all of which have a default value that can be changed again in the form once the code is executed. Those values can easily be assigned to the relevant parameters. This method is useful to explore model behavior.

The most efficient way for the analyses is using standalone files in which we specify all the parameterization(s) at once. The easiest is using text files with information delimited in some way (e.g. comma, space, or tab separated values) to allow for easy identification of the values. Once the file is open and read, we can use a series of assigning statements to initialize all the variables. All this can be programmed in a subroutine which will be run early in the model to load all the parameters. Other types of files that can be used are data tables belonging to a database. This is a bit more complex since we would need to install the required ODBC (Open Database Connectivity) drivers for the specific database engine (e.g. MySQL, PostgreSQL, or DB2) and some libraries in our compiler.

Retrieving output data is done in a similar way to input data: plotting graphical output in the GUI, saving it in text files, or using a database engine from within the model. For example, we can add a graph component to plot the trajectory of population size (Figure 10.3). Retrieving graphical output is very useful in the initial phases of model evaluation and analysis, whereas saving data in files is the standard for in-depth analyses. Keeping the output data together in the same files with the model parameters used (and the constants) is always a good recommendation to avoid future confusion.

data to compare with the predictions of your model, be as clear as possible about the methods used and the quality of those data sources. Make explicit all rules, equations, and schedules included in each of the processes, with the help of graphs and other schemes, if needed (Figure 10.2). Use mathematical notation to declare equations and also rules, such as conditional probability or Boolean algebra notation. List model parameters, including constants, in association with the submodels they are implicated in, their description, and the available estimates with both their variability and the associated uncertainty, explaining and justifying the field and statistical methods used and/or the data sources. Make explicit all scales, domains, resolutions, and how they integrate in each of the processes. Explain carefully how stochasticity is handled, including parameter sampling, randomization, and any other decision that may affect the interpretation of the results such as data rounding or truncation. Last, consider seriously publishing a final version of your code, either in the form of annotated pseudocode

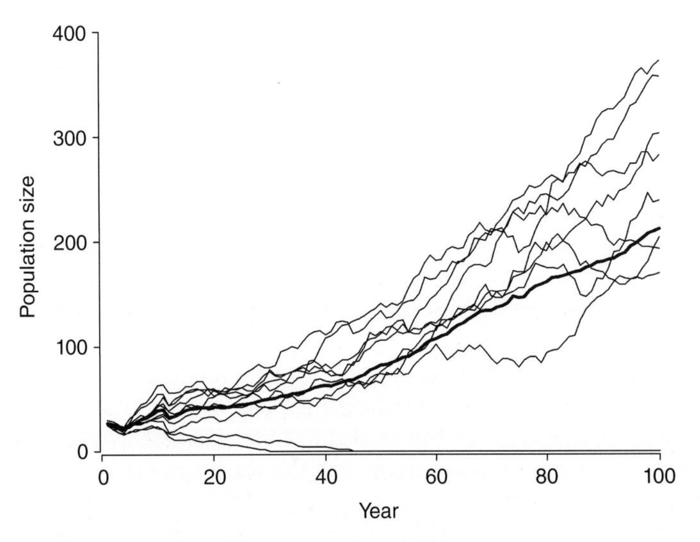

Figure 10.3 Graphical output for population size simulated with the model given in Box 10.3 and parameterized with N_0 = 30; P_R = 0.6; P_S = 0.9; *max_age* = 10; *t* = 100. The plot corresponds to 10 simulated population trajectories and their average (bold line). With this parameterization we observe two extinctions and the effect of the initial condition lasting for the first 15 years.

(Box 10.3), the code of your core model, or all code produced for both the core model and the analyses, with separate versions to help in understanding what was done.

There have been several attempts to make explicit a list of minimum requirements to document IBMs in the form of *model documentation protocols* (Mooij and Boersma 1996). The most popular is the *Overview, Design Concepts and Details* (ODD) protocol presented by Grimm et al. (2006), which has been updated and expanded by Grimm et al. (2010) and by Topping et al. (2010) who created the `ODdox` version for `C++` code annotation and documentation. The result is a set of documents providing a heavily annotated and hyperlinked version of the ODD protocol linking model description to the source code. The ODD protocol or any other alternative can be used as a guideline to check that we considered and described properly all the components of a model. The ODD protocol is a good way to organize and present information, but other alternatives maybe be more consistent with the aims and level of complexity of your model (Müller et al. 2014).

10.3.1 The Overview, Design Concepts, and Details Protocol

The ODD protocol aims to offer a standard that provides an ordered sequence of information that allows readers to follow the logic and details of any IBM (Grimm et al. 2006, 2010). It first starts with general information in the Overview section (Table 10.1), described by three elements: the *purpose* of the model, the *state variables* and *spatiotemporal scales*, and finally a short overview of the *processes* and *scheduling*. The next section, the *design concepts*, describes the strategic design of the model. The current version includes a list of 11 elements, ranging from emergence and adaptation to collectives or stochasticity. The list of elements is a bit arbitrary and it is not in a particularly relevant order. Go through them and build an ad hoc list by selecting the ones relevant for you. The final section goes into an explanation of the model in detail, including the *initialization*, the *input data*, and finally, a detailed description of all *processes*. All sections and subsections of the ODD are articulated as groups of questions (Table 10.1). The final result is a document in which relevant details of the model are described. Nevertheless, following the guidelines of the ODD does not ensure that the explanations make sense, especially if your conceptual model is not consistent and well thought out. In the process of building your conceptual model you can use the ODD questions to check what you might be skipping.

Grimm et al. (2010) assume that a single protocol can suit all potential model implementations and that the ODD protocol should be strictly followed. However, the question of whether a single protocol can be applied to a variety of implementations built to address different questions remains unresolved. My view is a bit more unorthodox because depending on the aims, we can find alternative ways to efficiently communicate our work. For example, in my view the clarity of the documentation of a model improves by clearly separating what belongs to the description of the core model from the description of the analyses. Details include different model parameterizations and initial conditions that are typically associated with specific analyses. In doing so, it is easier to understand the different steps, especially if the parameterization and initial conditions differ between analyses. Additionally, separating those two parts simplifies the distinction between what we consider as supported knowledge and the part that we will investigate in detail both in relation to model structure and parameterization.

10.4 Model Analysis and Inference

Analyzing a model is about understanding its behavior and its emergent dynamic properties under different conditions. The analysis of complex models is not a simple task. At this stage, the ecologist will use all of her knowledge about experimental design and statistical analyses, including the methods explained in this book. There is no single best way to analyze an IBM, with different approaches ideally yielding similar conclusions. Nevertheless, I offer some general guidelines to simplify the challenge. It is often difficult to distinguish between the phases of model building and model analysis because during the analyses we may be forced to redefine once again the initial conceptual model and the code, in another iteration of the modeling cycle (Figure 10.1; Grimm and Railsback 2005). Normally we will follow a step-by-step program of analysis. I distinguish between four main steps. First, we need to go through a process of model debugging and consistency checking, followed by an evaluation of the consistency of model structure and a sensitivity analysis. Next come the steps of model selection, validation, and calibration. Last, you should try to answer the questions that motivated the model within the inference constraints imposed by the previous results (Figure 10.4).

10.4.1 Model Debugging and Checking the Consistency of Model Behavior

Before going into your questions of interest, you should perform a thorough evaluation of model performance to detect errors arising from model design or implementation and determine if the behavior of the model makes sense. In this a priori checking you will detect many small problematic details and bugs that once removed will

Table 10.1 The Overview, Design concepts and Details (ODD) protocol for documentation of individual-based models (IBM).

Elements	Questions
Overview	*Context and general information*
1 Purpose	What is the purpose of the model?
2 Entities, state variables, and scales	What entities (e.g., individuals, collectives) are in the model? By what state variables (attributes and traits) are these entities characterized? What are the temporal and spatial resolutions and domains of the model?
3 Process overview and scheduling	Who (entity) does what, and in what order? When are state variables updated? How is time modeled, as discrete steps or as a continuum over which both continuous processes and discrete events can occur?
Design	*Strategic considerations*
4 Design concepts	
4.1 basic principles	Which theories, hypotheses, assumptions, or modeling approaches are behind a model's design? How were they taken into account? Are they used in submodels or at the system level? Will the model provide insights into the basic principles themselves?
4.2 emergence	What model results are expected to vary in complex and perhaps unpredictable ways when particular characteristics of individuals or their environment change? Are there other results that are more tightly imposed by model rules and hence less dependent on interactions?
4.3 adaptation	What adaptive traits do the individuals have? What rules do they have for making decisions or changing behavior in response to changes in themselves or their environment? Do these traits explicitly seek to increase some measure of individual success regarding its objectives, or, instead, cause individuals to reproduce previously observed behaviors?
4.4 objectives	If adaptive traits explicitly act to increase some measure of individual fitness, what exactly is that objective and how is it measured? When individuals make decisions by ranking alternatives, what criteria do they use?
4.5 learning	Do individuals change their adaptive traits over time as a consequence of experience? If so, how?
4.6 prediction	How do individuals predict the future conditions (either environmental or internal) they will experience? What internal models do they use to estimate future conditions or the consequences of their decisions? What tacit or hidden predictions are implied in these internal model assumptions?
4.7 sensing	What internal and environmental state variables (including those of other individuals) are individuals assumed to sense and consider in their decisions? Are there mechanisms by which individuals obtain information, or are they assumed to know these variables?
4.8 interaction	What kinds of interactions among agents are assumed? Are there direct interactions in which individuals encounter and affect others, or are interactions indirect? If the interactions involve communication, how is it represented?
4.9 stochasticity	What processes are modeled as random or partly random? Is stochasticity used to reproduce variability in processes for which the actual causes of the variability are unknown or not relevant? Is it used to model events or behaviors with a specified probability?
4.10 collectives	Are there social networks? If so, are their structures imposed (a priori additional entity) or emergent? Are collectives affecting, or have been affected by the individuals?
4.11 observation	What data are collected from the simulations for testing, understanding, and analyzing the model? How and when are they collected?
Details	*Detailed technical description*
5 Initialization	What is the initial state of the model at the beginning of a simulation run? Is initialization always the same, or is it allowed to vary among simulations? Are the initial values chosen arbitrarily or based on data?
6 Input data	Does the model use input from external sources such as data files or other models to represent processes that change over time?
7 Submodels	What, in detail, are the processes listed in point 3? How were they designed, parameterized, and tested? What are their parameters, dimensions, and reference values?

Source: Adapted from Grimm et al. (2006, 2010).

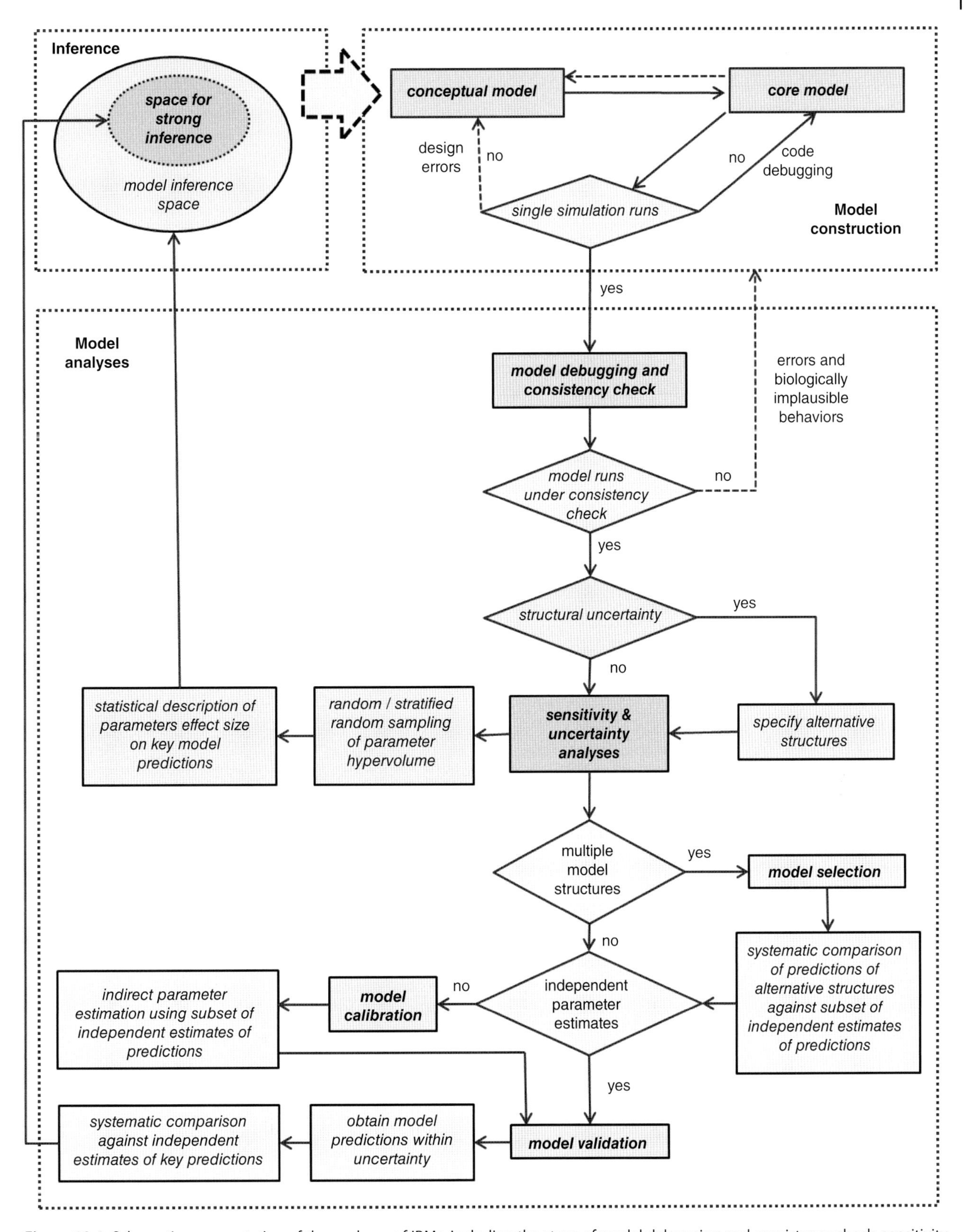

Figure 10.4 Schematic representation of the analyses of IBMs, including the steps of model debugging and consistency check, sensitivity, and uncertainty analyses and model selection, calibration, and validation. Key model predictions refer to the questions related with the aims for which the model was built. In the end, the initial questions should be answered within the inference constraints imposed by the results. Ideally, the results should help in improving the conceptual model.

improve model consistency, saving a lot of time later on. Note that while writing the code of your model you were already debugging it at compilation time: any error appearing during compilation should have been corrected already (Box 10.4). Now we search for errors during execution time. The model should be able to run simulations with no errors during a single simulation run, using a standard parameterization with the mean value and variability for all parameter estimates.

The next step consists of forcing model behavior with different combinations of parameters set at extreme values, such as very low or high survival rates. Testing boundary conditions will force working with many zeros and with large numbers (including many individuals), thus causing errors to appear. It is a good idea to repeat this step by step, going through the different processes before making overall extreme parameterizations of the model. Tests may generate problems by making forbidden or undefined calculations, such as floating point divisions by zero and other exceptions that the code does not handle properly. Many of the errors will be associated with exception handling, which depending on the language and compiler will be easy to solve. Other important sources of errors will be associated with logical failures in scheduling and the way we introduce stochasticity into parameter values.

Simultaneous to model debugging during execution, it is important to look for biologically implausible behaviors, especially when working at extreme parameterizations. Before concluding that an interesting or unexpected behavior is a new finding, we must consider the possibility that it is associated with something incorrect in model specification or coding. The dynamics of the model should be consistent with the general expectations of the conceptual model. It is a good idea to use graphical output to check the relevant output in run time, as well as saving simulated data together with parameters and tracking other data not directly related with the model aims and emergent properties, such as realized reproductive and mortality rates. All this information will serve as a log file, helping to determine whether an unexpected model behavior is due to a problem with design or programming, or if it is a new emergent result. Be sure to update the documentation of the model to describe the changes made in the conceptual or core models.

10.4.2 Model Structural Uncertainty and Sensitivity Analyses

The next step in analyzing an IBM should deal with setting the context in which to interpret the results: what are the limits for the inference? This step has two complementary sides, one related to *model structural consistency*, as defined by the processes and how they are integrated, and the other to the *parameterization* of those processes (Figure 10.4). Thinking about the structural uncertainty of a model consists of specifying alternative definitions of the processes that we have implemented, such as using additive or multiplicative processes, or different functions such as power or exponential laws. It is important when we do not have a good empirical description or theoretical justification for the choices. For example, imagine that based on empirical data we implemented a function in which survival is affected by temperature, but there is no data on which function is best and how it needs to be integrated with other factors such as density. If the main reason to build your model is addressing questions regarding the impact of temperature variation on some relevant population traits, it will be a good idea to think of alternative ways to implement the processes, such as an additive or multiplicative interaction with density. The idea is to create two or more alternative model structures that will be subject to sensitivity analyses. Further analyses will be repeated for each of the alternatives and the results compared for consistency under a model selection framework. Sensitivity analyses will help to gain confidence on how the specification of the model may affect inference. Structural uncertainty should be evaluated for processes that have some level of uncertainty and for which we expect, a priori, a relevant role on model behavior (Figure 10.4).

In *sensitivity analyses*, we quantify how changes in the values of model parameters affect the value of the key output variables. This is achieved by repeatedly running the model with different parameterizations and measuring how the relevant outcomes respond. Depending on the aim of the analyses, we can differentiate between two different types: sensitivity analysis in a strict sense versus uncertainty analysis. In sensitivity analyses we define the range of values to be explored using biologically realistic values for each model parameter that we want to explore. For example, the boundary conditions might set the parameter hypervolume, using parameters between the minimum and maximum values reported in the literature. In this way, we can explore the potential behaviors of the system under plausible conditions. Conversely, in *uncertainty analysis* we sample only within the existing uncertainty around each of the parameter estimates to determine the variability of the response of the model in relation to the available information. Typically for some parameters we do not have accurate estimates from the literature or from empirical data, for instance, the probabilistic parameters used in stochastic rules, and this uncertainty needs to be taken into account to avoid overinterpreting the results.

Sensitivity analysis is generally considered a key component of the quality evaluation of any model, for understanding the model itself and providing the context in which the rest of the results will be interpreted. For example, if the model aims to evaluate a conceptual hypothesis then the actual parameterization is not so relevant, whereas model behavior in a range of plausible conditions is. On the other hand, uncertainty analysis is particularly useful in indicating which parameters are candidates for additional research to narrow the degree of uncertainty in model results, and is a key component of models built for making predictions based on empirically estimated parameters. Something that is often overlooked in sensitivity analyses is the possibility of including how parameters interact by including *covariation* in parameter values. A final recommendation is avoiding sensitivity analyses using the central estimate of parameter values and an arbitrary small amount of variation (typically ±5 to 10%). The range of values to be used should be well justified.

In sensitivity or uncertainty analyses two general approaches are used depending on whether all parameters are considered simultaneously or not. In *local* and *one-at-a-time analyses* we sample the range of values of just one parameter while keeping all the others constant at their central estimate and then measuring to what extent the output of the model is affected. One-at-a-time approaches perform poorly when dealing with complex models such as IBMs and should in general be avoided (Saltelli and Annoni 2010; but see Beaudouin et al. 2008). In *global* or *multivariate* sensitivity analyses we explore all the parameters simultaneously, repeatedly sampling the n-dimensional parameter hypervolume.

The sensitivity analysis will require a substantial amount of coding only for this purpose. Therefore, making a specific version of the model for this is a good idea. By coding loops, one for each parameter and with as many steps as values needed for each of them, you can run a global analysis at once even if you have a lot of parameters to sample. There are several ways to sample the *parameter hypervolume*, from simply randomly choosing parameter values (very inefficient) to a complete factorial sampling design, which may be reasonable for a reduced number of parameters. The alternative approaches become computationally challenging for relatively small models. Just 10 parameters with 5 values each running with 100 simulation replicates to estimate the variability of the output requires 10^7 simulation runs.

Latin hypercube sampling is an efficient approach for addressing the issue of a large number of simulation runs (Iman and Helton 2006). Briefly, this technique is a stratified sampling method commonly used to reduce the number of simulation runs necessary for sampling the parameter hypervolume. Each parameter is sampled using an even sampling method and then randomly combined sets containing all parameters are used to run the model. For each parameter the range of possible values is divided into nonoverlapping intervals of equal probability size (Box 10.4). One value from each interval is chosen at random and this process is repeated for each parameter until we obtain a parameterization set. The key is that for every parameter each interval must be sampled only once until all intervals of all parameters have been used once. Then the process starts again. If the model is complex, it may be necessary to use a refined version of the Latin hypercube sampling that reduces the dimensionality of the problem by carefully analyzing some relevant processes before going into a simplified global analysis.

In the end, we obtain a dataset including the parameter values used and one or more relevant model predictions directly related with the questions (such as overall population size, density, growth rates, extinction probability, mean time to extinction, or sex ratio). All this information needs to be summarized in order to obtain a picture of the differential role of the parameters and their associated uncertainty. The most basic way to do this is simply by using a *partial rank correlation analysis* (Segovia-Juarez et al. 2004). A more inclusive approach is to run *generalized regressions* between model predictions with the average of the replicates for each parameterization as a dependent variable and model parameters as independent predictors (McCarthy et al. 1995). The resulting equations approximate the functions that relate the parameters of the simulation model to predictions in a simple way, while the standardized coefficients of the regression can be used to describe the sensitivity of model predictions to each of the input parameters (Revilla et al. 2004; Revilla and Wiegand 2008). The generalized version of this approach is referred to as *Gaussian process analysis*, in which the behavior of the simulation model with regard to each of its predictions is approximated by a Gaussian statistical model in which the predictors are the parameters of the simulation model (Dancik et al. 2010). Remember that you need to report effect sizes and confidence intervals to give readers an idea of the magnitude and relative importance of each parameter effect. P values do not make sense here since the input parameters are known to generate the output, while the unlimited power provided by large simulated sample sizes makes their interpretation irrelevant.

Last, we need to warn you against using sensitivity or elasticity analyses to make strong inferences about the actual factors driving the dynamics of a real population. These analyses do not necessarily tell you much about which parameters should be managed in the field. It specifies what each of the parameters does and the strength of the effect, so avoid making any definitive conclusion on what might be going on unless you have some empirical indication that the parameters identified as important in

Table 10.2 Some issues to consider when comparing empirical and simulated data for model selection, validation, and calibration.

Data	
Key data	Empirical data directly related with the questions to be answered with the model.
Ancillary or secondary data	Empirical data not directly related with the questions. It contains information useful in model selection and calibration. Often corresponds to data at discordant spatiotemporal scales.
Estimates	Key and secondary data can be quantitative, including point estimates and their uncertainty and variability, or qualitative, such as trends
Summary statistics	Aggregation of data into new simplified yet informative statistics (for example calculating a growth rate from a raw series of census data). This is often done to simplify the comparison between data and predictions.
Single vs multiple	The amount of data can vary from a single key variable to multiple key variables and secondary data.
Predictions	
Symmetry	We need to calculate as model output the same key and secondary predictions as with the empirical data.
Single parameterization	For a given parameterization we generate a frequency distribution of model predictions by repeating a number of simulations with the parameterization.
Output formats	Predictions can be obtained as graphical outputs to visualize the results and saved into files. It is convenient saving the parameterization within the output files.
Multiple parameterizations	Often we need to repeat the process for multiple parameterizations obtained by moving across the parameter space.
Comparisons	
The logic	Systematically compare data and predictions to estimate the likelihood of reproducing the observed data with a given parameterization and model structure.
Types of comparisons	Rejection filtering by using pattern oriented modeling or informal likelihoods
	Direct calculation of the likelihood by running a sufficiently large number of simulations
	Informal likelihoods (e.g. sum of squared differences between data and predictions)
	Nonparametric likelihood approximations (e.g. kernel density estimation)
	Parametric likelihood approximations (e.g. Z scores)
	Approximate Bayesian computation
Methods to define parameterizations	Systematic search of the parameter space when the number of parameters is low
	Latin hypercube sampling for more complex models
	Markov chain Monte Carlo strategies: Metropolis–Hastings and Gibbs sampling algorithms and their variants.
	Sequential Monte Carlo approaches, also known as particle filters or bootstrap filters
	Numerical optimization methods such as genetic algorithms, simulated annealing, simplex algorithm, or support vector machine algorithms

the sensitivity analyses are the ones that need to be managed. For example, the fact that adult survival is the parameter with the highest sensitivity or elasticity does not guarantee that the population is declining due to low adult survival; it could be entirely due to a lack of recruitment.

10.4.3 Model Selection, Validation, and Calibration

A bit trickier is comparing the outcome or outcomes of the model against a specific dataset. The comparison is usually made for different reasons, such as model selection, model validation, and model calibration (Figure 10.4, Table 10.2). If we are dealing with uncertainty in model structure, we will have alternative process specifications which can be assessed by their capacity to reproduce the observed data. In the case of *validation*, we typically have estimates of model parameters with their variability and uncertainty, which are then validated by evaluating their capacity to replicate an empirical dataset or some empirically observed behaviors, setting a credibility standard for that model structure, parameterization, and question (Figure 10.4). *Calibration* is a kind of model parameterization in which we estimate parameters from observed field data on model predictions by filtering out the parameterizations that do not match the data, by Gaussian process approximation or any other likelihood approximation (Hartig et al. 2011). It is important to note that we leave model parameterization for the analysis-inference and not for model building since this step is important in understanding how the model behaves. Parameterization is often first about defining and then reducing the dimensionality of the model before making any strong inference such as management recommendations. Model parameterization by calibration (or inverse modeling) may use no a priori information on the actual parameters, or may use the available information as priors under a Bayesian calibration framework (Hartig et al. 2011). In *mechanistic modeling*, we assume that we can use information about the processes and how they integrate from other populations, whereas the parameters are just different realizations that we may observe. In model calibration, we can simultaneously perform the parameterization and the uncertainty analysis.

This step requires the systematic comparison of empirical and simulated data to decide which of the tested parameterization sets or model structures reproduce the empirical data in a reasonable way by calculating the probability of reproducing the field data with a given model structure and parameterization. Typically, we run simulations until we obtain a distribution of the frequencies of the simulated observations that the model structure and parameterization can generate and from them calculate the probability of observing the field values. The comparison between the observed and simulated data can be straightforward, as the difference or the sum of squared distances between the observed values and those obtained from the simulated data, or more efficient if we make the comparison only once against the summary statistics of the simulated frequency distribution (mean and variance). Conceptually, we can generalize all the alternative approaches as a kind of point-wise likelihood approximation of the goodness-of-fit of our model to the data (Hartig et al. 2011). As such, we need to calculate the likelihood of observing the empirical data for each model structure and parameterization. The final goal is finding the structure and parameterization that maximizes that likelihood, thus obtaining a parameterization of the model with field data on model predictions, obtaining an estimate of the uncertainty by knowing how many alternative parameterizations match our threshold of fit, or simply helping us to select the model structure that is best supported by the available data (Figure 10.4). Hartig et al. (2011) review the different methods under a useful likelihood-based inference conceptual framework. The methods range from those that explicitly approximate the likelihood, such as approximate Bayesian computation, simulated (synthetic) pseudo-likelihoods, or indirect inference, to those that allow calibrating the model without explicitly approximating the likelihood, such as pattern-oriented modeling or informal likelihoods (Beaumont 2010; Hartig et al. 2011). The value of these methods is that the structural realism in the definition of processes at the right scales allows for inverse parameter estimation (Wood 2010; Hartig et al. 2011, 2014).

One of the classic ways to calculate the likelihood of obtaining the observed data given a model structure and parameterization makes use of central limit theorem, which allows us to calculate the probability of obtaining an empirical measurement from the summary statistics of the distribution of model outcomes for a given parameterization, if the simulated distribution can be approximated with a normal distribution as a parametric likelihood approximation (following the notation of Hartig et al. 2011). For each model prediction, we calculate a *match score*, for example, a Z score using the mean and the standard deviation of the simulated replicates (Revilla et al. 2004); while by setting different threshold probabilities for acceptance we can simultaneously evaluate multiple model predictions using a *multicriteria approach*, such as Pareto optimality assessment (Reynolds and Ford 1999). Alternatively, we can use a Bayesian framework to calculate the posterior distribution and proceed in a similar manner (Beaumont 2010; Hartig et al. 2014). If the simulated frequency distribution generated by the model does not conform to a normal distribution, which typically occurs when using highly aggregated data that generate multimodal distributions, then we may instead use a *kernel density estimator* to obtain a nonparametric estimation of the probability density function of the simulated distribution and subsequently calculate the probability of observing the empirical data from it (Tian et al. 2007). There are cases in which the variability in the observed data is high due to measurement error, but the predictions of the model for the same type of data shows lower variability. In these cases it is advisable to add a tractable error term (parametric or nonparametric) on the side of the observed data to account for noise (Hartig et al. 2011). If we are evaluating alternative model structures, and therefore cannot be sure of the origin of the mismatch between observed and simulated data (structure, parameterization, or stochasticity), it is advisable to use simpler measures of *mismatch*, such as the sum of squared distances between the observed and simulated data (informal likelihoods; Hartig et al. 2011) or some kind of ad hoc rejection filtering under the pattern-oriented approach (Grimm et al. 2005).

Pattern-oriented modeling, also termed rejection or performance filtering (Grimm et al. 2005; Webb et al. 2010; Hartig et al. 2011), can be applied to models of dynamical systems. It is probably the most liberal approach with regard to model selection, validation, and calibration, because it can also be used when the data to be adjusted (both the empirical and/or the simulated data) have complex distributions such as multimodal or multidimensional, or when the quality of the empirical data is poor or simply unknown. The method consists of defining criteria that allow classifying whether model structures or parameterizations match the observed data within a given explicit threshold, instead of calculating the actual likelihood of obtaining the observed value or a close enough value. The criteria used to define the thresholds can be diverse or even ad hoc, and may include some of the indexes of adjustment discussed above, such as a mean squared difference or a Z score threshold. Additionally, we can use the error of the field data estimates to define the criteria. It allows using multiple ancillary data which in isolation do not contain much information, but that in combination can provide a robust approximation to constrain model behavior within the limits of the available information (Wiegand et al. 2004b).

Potentially, the number of variables that may be included in the empirical dataset to be directly used in the comparison with simulated output can be large. Often we aggregate the available information in some way to obtain a simplified set of data that can be compared with the simulated output. These variables are referred to in the literature as patterns, state variables, output variables, or simply summary statistics (Hartig et al. 2011). The difficulty lies in deciding, which of the many alternatives are statistically sufficient given the purpose of the model. The statistics need to convey information on the relevant properties of model dynamics. A good recommendation is to choose variables that operate at different spatial or temporal scales and hierarchical levels, including variables describing stationary and nonstationary dynamics (Grimm et al. 2005; Wiegand et al. 2004b; Wood 2010). Nevertheless, the question behind your model should be the key when you decide which data are relevant, obviously, within the limits imposed by the available empirical information.

All the methods discussed above require searching the potential parameter space in order to find the model structure or parameterizations best supported by data using some kind of *numerical approximation* (Bolker 2008). In models with a reduced dimensionality, we can use a Latin hypercube sampling strategy. In more complex models, say above 20 parameters, depending on the availability of computing power, the programming language, and how efficiently the model was coded, we will need a more efficient sampling strategy, such as *Markov chain Monte Carlo* strategies, including the Metropolis–Hastings and the Gibbs sampling algorithms, which start with an initial parameterization obtained from the parameter space, from which we generate a new parameterization by randomly moving a small amount within the parameter space. Then the likelihood, or similar, of the two consecutive parameterizations is compared, retaining the best one from which a new parameterization is obtained. There are lots of variants aiming to increase the speed, for example by reducing the correlation between consecutive parameterizations, and to avoid getting stacked in local likelihood maxima by going downhill with some probability. Another alternative is using sequential Monte Carlo approaches in which, starting with a set of parameterizations obtained from the whole parameter space, we calculate the point-wise likelihood and then weight each of them, for example, by their normalized importance weight, according to their estimates. From this initial set, we obtain a new set of parameterizations with probabilities according to their weights and repeat the process until some convergence criteria are met, such as that all parameterizations within the set are within a given likelihood threshold. Last, we can consider using a numerical optimization algorithm when dealing with multiple data to be fitted under a pattern-oriented approach (Table 10.2). Hartig et al. (2011) provide pseudocode algorithms for some of these numerical sampling methods. Applying these methods is most efficiently done by programming the routines within the coding environment. The methods in themselves are not complicated (though the specific jargon is) but require extensive coding. Remember making a specific version of the model for the purpose of validation and calibration. A potentially less efficient alternative is generating the simulated datasets and then using some of the algorithm implementations available within `R`.

10.4.4 Answering your Questions

At this stage, and after all the work is done, we should have a clear idea of the questions to answer. The potential uses of IBMs are broad and flexible, as are other stochastic simulation models, making it difficult to summarize their uses (see examples in Section 10.1.2). The first and most basic use consists of reviewing and integrating the available knowledge on a system. This step is completed by building the conceptual model and its implementation in a core model plus the sensitivity analysis over the biologically plausible parameter space and a validation of the model with independent data. We must provide all the available information, making clear what is supported by knowledge and data and what are the assumptions and hypotheses which should be investigated further. Following the initial step, the typical use of IBMs consists of gaining new knowledge about how a system usually works, often evaluating the predictions of theoretical models and empirical generalizations for population regulation, movement, density dependence, or interspecific interactions such as predation or diseases. Last, practical applications represent a broad field of use, including population viability analyses, the evaluation of alternative management scenarios for conservation, population control or exploitation, the evaluation of strategies to control diseases, or measuring the impact of infrastructures on connectivity among spatially structured populations, just to mention a few.

All these uses have in common the description of model behavior under different scenarios. A scenario is defined by a model structure, an initial condition, and a parameterization, which also includes the space definitions used in spatially explicit models, normally as maps. For the scenario, we obtain frequency distributions of the relevant model outcomes by running multiple stochastic simulations. The simplest approach is just a qualitative or quantitative description of those outcomes, for example, by plotting the results in figures. It is much more common that we need to compare the results of one scenario against other scenarios, empirical data, or theoretical expectations in a qualitative and/or quantitative way, as discussed in the previous section. Comparing the output of the model for alternative scenarios is more or less straightforward, especially if what we need is the relative evaluation against a desired standard. For example, we may need to evaluate alternative hunting strategies to estimate maximum yield, to reduce interannual variability in population size, or to minimize extinction risk. We can also use statistical descriptions to compare the distributions of outcomes for the different scenarios. The comparison of multiple scenarios, such as management alternatives, needs to be carefully thought out under the standard framework of experimental design in a *virtual ecologist approach* (Zurell et al. 2010).

Last, one important issue to consider when designing experiments is *model hysteresis* or the dependence of model behavior on both its current and past states. The initial conditions or a perturbation often impose a *transient-state phase* after which the system may reach a *steady state* with stationary stochastic dynamics, which occurs when the dynamic properties of the model do not change over time, with the frequency distributions of model outcomes remaining stable. Depending on the aims, we may need to focus on the nonstationary dynamics, for example, when studying the impact of an event or perturbation, such as the success rate of different reintroduction scenarios varying in the number of animals released (Kramer-Schadt et al. 2005), or a PVA affected by the initial conditions imposed by an empirical estimate of population size and structure (Wiegand et al. 1998). We can also focus on the steady-state phase, as we do when calculating the intrinsic mean *time to extinction* in a PVA (Grimm and Wissel 2004), or on both transient and steady phases, for example, when investigating the impact of different management activities starting with an observed initial population size (Wiegand et al. 2004a).

10.5 Software Tools

Individual and agent-based models can be constructed in a variety of different programming languages. For the online exercises associated with this chapter, I chose a generic language and a relatively simple open-source compiler platform to implement the three exercises. I provide pseudocode in `Object Pascal` language using the `Free Pascal` compiler in Program `Lazarus`. `Lazarus` is cross-platform, and as such it can be used in Linux, Windows, or Mac OS environments. It is an integrated development environment (IDE) that allows for rapid application development. I selected a language based on `Pascal` because it is intuitive and as a result the learning curve is not as steep as with other languages such as `Java` or `C++`. Therefore, following the logic of the examples will be easier for naïve programmers. Although `Pascal` is an old programming language, it has a long tradition of use in academic circles.

10.6 Online Exercises

Three online exercises illustrate the basics of individual based models. Exercise 1 is an exponential model that represents an iteroparous organism with a single reproductive event each year. Exercise 2 extends the exponential model by adding the effects of density dependence. Last, Exercise 3 extends the density-dependent model from a single population to a spatially explicit model with three different types of habitat: nonhabitat, sink habitat, and source habitat.

10.7 Future Directions

My chapter is a bit different from the others. More than discussing a specific method with a lot of examples, it deals with a research approach that can be implemented in many alternative ways to address a potentially broad range of questions. As such, the approach borrows methods from many disciplines, including not only ecology, but also statistics, complex systems, and algorithmic theories and software engineering. I did not intend to present a thorough review of the literature in regard to examples of IBM implementations and applications. Instead, I aimed to provide an overview of the whole process, from the beginning to the end of the research program, focusing on those parts that might be more challenging for newcomers and hopefully providing some useful guidelines. Using IBMs is by no means easy. The challenge remains in having a good conceptual model and clear questions early on. Analyzing the model requires some experience in order not to be overwhelmed or lost in irrelevant detail. As with using any other approach that relies on programming, the learning curve may be steep, but it should lead somewhere, and knowing where to go is on the side of the user. Remember that, by itself, building a model is not the question to answer.

I provide some example models in the online materials. The examples are built merely to illustrate one of the many different ways you may choose to start coding an IBM. The examples are intended to help you to feel more comfortable with how IBMs are built. The examples are not core models, just out-of-the-box toy models for you to play with, modify, corrupt, and modify again, and in this manner learn a bit more about the logic behind this research approach. With the help of this chapter and the methods presented in the rest of the book, you should be able to address your research questions.

References

Ajelli, M., Gonçalves, B., Balcan, D. et al. (2010). Comparing large-scale computational approaches to epidemic modeling: agent-based versus structured metapopulation models. *BMC Infectious Diseases* 10: 190.

Beaudouin, R., Monod, G., and Ginot, V. (2008). Selecting parameters for calibration via sensitivity analysis: an individual-based model of mosquitofish population dynamics. *Ecological Modelling* 218: 29–48.

Beaumont, M.A. (2010). Approximate Bayesian computation in evolution and ecology. *Annual Review of Ecology, Evolution, and Systematics* 41: 379–406.

Beissinger, S.R. and Westphal, M.I. (1998). On the use of demographic models of population viability in endangered species management. *Journal of Wildlife Management* 16: 821–841.

Berec, L. (2002). Techniques of spatially explicit individual-based models: construction, simulation, and mean-field analysis. *Ecological Modelling* 150: 55–81.

Bolker, B. (2008). *Ecological Models and Data in R.* Princeton: Princeton University Press.

Bolker, B., Holyoak, M., Křivan, V. et al. (2003). Connecting theoretical and empirical studies of trait-mediated interactions. *Ecology* 84: 1101–1114.

Boyles, J.G. and Willis, C.K.R. (2010). Could localized warm areas inside cold caves reduce mortality of hibernating bats affected by white-nose syndrome? *Frontiers in Ecology and the Environment* 8: 92–98.

Bruggeman, D., Wiegand, T., and Fernandez, N. (2010). The relative effects of habitat loss and fragmentation on population genetic variation in the Red-cockaded Woodpecker (*Picoides borealis*). *Molecular Ecology* 19: 3679–3691.

Buckley, L.B. (2008). Linking traits to energetics and population dynamics to predict lizard ranges in changing environments. *American Naturalist* 171: E1–E19.

Carlo, T.A. and Morales, J.M. (2008). Inequalities in fruit-removal and seed dispersal: consequences of bird behaviour, neighbourhood density and landscape aggregation. *Journal of Ecology* 96: 609–618.

Dancik, G.M., Jones, D.E., and Dorman, K.S. (2010). Parameter estimation and sensitivity analysis in an agent-based model of *Leishmania major* infection. *Journal of Theoretical Biology* 262: 398–412.

DeAngelis, D.L. and Mooij, W.M. (2005). Individual-based modeling of ecological and evolutionary processes. *Annual Review of Ecology Evolution and Systematics* 36: 147–168.

Doak, D.F. and Morris, W.F. (2010). Demographic compensation and tipping points in climate-induced range shifts. *Nature* 467: 959–962.

Gilbert, N. (2008). *Agent-based Models.* Sage No. 153.

Goss-Custard, J., Burton, N., Clark, N. et al. (2006). Test of a behavior-based individual-based model: response of shorebird mortality to habitat loss. *Ecological Applications* 16: 2215–2222.

Grignard, A., Taillandier, P., Gaudou, B. et al. (2013). GAMA 1.6: advancing the art of complex agent-based modeling and simulation. *The 16th International Conference on Principles and Practices in Multi-Agent Systems (PRIMA)* 8291: 242–258.

Grimm, V. (1999). Ten years of individual-based modelling in ecology: what have we learned and what could we learn in the future? *Ecological Modelling* 115: 129–148.

Grimm, V., Berger, U., Bastiansen, F. et al. (2006). A standard protocol for describing individual-based and agent-based models. *Ecological Modelling* 198: 115–126.

Grimm, V., Berger, U., DeAngelis, D.L. et al. (2010). The ODD protocol: a review and first update. *Ecological Modelling* 221: 2760–2768.

Grimm, V. and Railsback, S.F. (2005). *Individual-based Modeling and Ecology*, Princeton Series in Theoretical and Computational Biology. Princeton: Princeton Press.

Grimm, V. and Railsback, S.F. (2006). Agent-based models in ecology: patterns and alternative theories of adaptive behaviour. In: Billari, F.C., Fent, T., Prskawetz, A., Scheffran, J. (eds) *Agent-Based Computational Modelling. Contributions to Economics*. Physica-Verlag HD.

Grimm, V., Revilla, E., Berger, U. et al. (2005). Pattern-oriented modeling of agent-based complex systems: lessons from ecology. *Science* 310: 987–991.

Grimm, V. and Wissel, C. (2004). The intrinsic mean time to extinction: a unifying approach to analyzing persistence and viability of populations. *Oikos* 105: 501–511.

Hartig, F., Calabrese, J.M., Reineking, B. et al. (2011). Statistical inference for stochastic simulation models – theory and application. *Ecology Letters* 14: 816–827.

Hartig, F., Dislich, C., Wiegand, T. et al. (2014). Technical note: approximate Bayesian parameterization of a process-based tropical forest model. *Biogeosciences* 11: 1261–1272.

Iman, R.L. and Helton, J.C. (2006). An investigation of uncertainty and sensitivity analysis techniques for computer models. *Risk Analysis* 8: 71–90.

Kramer-Schadt, S., Revilla, E., and Wiegand, T. (2005). Lynx reintroductions in fragmented landscapes of Germany: projects with a future or misunderstood wildlife conservation? *Biological Conservation* 125: 169–182.

Kramer-Schadt, S., Revilla, E., Wiegand, T. et al. (2004). Fragmented landscapes, road mortality and patch connectivity: modelling influences on the dispersal of Eurasian lynx. *Journal of Applied Ecology* 41: 711–723.

Kuparinen, A. and Merila, J. (2007). Detecting and managing fisheries-induced evolution. *Trends in Ecology and Evolution* 22: 652–659.

Letcher, B.H., Priddy, J.A., Walters, J.R. et al. (1998). An individual-based, spatially explicit simulation model of the population dynamics of the endangered Red-cockaded Woodpecker, *Picoides borealis*. *Biological Conservation* 86: 1–14.

Liu, J., Dunning, J.B. Jr., and Pulliam, H.R. (1995). Potential effects of a forest management plan on Bachman's Sparrows (*Aimophila aestivalis*): linking a spatially explicit model with GIS. *Conservation Biology* 9: 62–75.

Luke, S., Cioffi-Revilla, C., Panait, L. et al. (2005). MASON: a multi-agent simulation environment. *Simulation: Transactions of the Society for Modeling and Simulation International* 82: 517–527.

Macal, C.M. and North, M.J. (2009). Agent-based modelling and simulation. In: *Proceedings of the 2009 Winter Simulation Conference* (eds. M.D. Rossetti, R.R. Hill, B. Johansson, et al.), Institute of Electrical and Electronics Engineers IEEE, 86–98.

Macal, C.M. and North, M.J. (2010). Tutorial on agent-based modelling and simulation. *Journal of Simulation* 4: 151–162.

McCarthy, M.A., Burgman, M.A., and Ferson, S. (1995). Sensitivity analyses for models of population viability. *Biological Conservation* 73: 93–100.

Mooij, W.M. and Boersma, M. (1996). An object-oriented simulation framework for individual-based simulations (OSIRIS): *Daphnia* population dynamics as an example. *Ecological Modelling* 93: 139–153.

Mooij, W.M. and DeAngelis, D.L. (1999). Error propagation in spatially explicit population models: a reassessment. *Conservation Biology* 13: 930–933.

Müller, B., Balbi, S., Buchmann, C.M. et al. (2014). Standardised and transparent model descriptions for agent-based models – current status and ways ahead. *Environmental Modelling and Software* 55: 156–163.

Nathan, R., Getz, W.M., Revilla, E. et al. (2008). A movement ecology paradigm for unifying organismal movement research. *Proceedings of the National Academy of Sciences USA* 105: 19052–19059.

Perez-Figueroa, A., Wallen, R., Antao, T. et al. (2012). Conserving genomic variability in large mammals: effect of population fluctuations and variance in male reproductive success on variability in Yellowstone bison. *Biological Conservation* 150: 159–166.

Railsback, S.F., Lytinen, S.L., and Jackson, S.K. (2006). Agent-based simulation platforms: review and development recommendations. *Simulation* 82: 609–623.

Ramsey, D.S.L. and Efford, M.G. (2010). Management of bovine tuberculosis in brushtail possums in New Zealand: predictions from a spatially explicit, individual-based model. *Journal of Applied Ecology* 47: 911–919.

Rands, S.A., Pettifor, R.A., Rowcliffe, J.M. et al. (2006). Social foraging and dominance relationships: the effects of socially mediated interference. *Behavioral Ecology and Sociobiology* 60: 572–581.

Revilla, E. and Wiegand, T. (2008). Individual movement behavior, matrix heterogeneity, and the dynamics of spatially structured populations. *Proceedings of the National Academy of Sciences USA* 105: 19120–19125.

Revilla, E., Wiegand, T., Palomares, F. et al. (2004). Effects of matrix heterogeneity on animal dispersal: from individual behavior to metapopulation-level parameters. *American Naturalist* 164: E130–E153.

Reynolds, J.H. and Ford, D. (1999). Multi-criteria assessment of ecological process models. *Ecology* 80: 538–553.

Rushton, S.P., Lurz, P.W.W., Gurnell, J. et al. (2000). Modelling the spatial dynamics of parapoxvirus disease in red and grey squirrels: a possible cause of the decline in the red squirrel in the UK? *Journal of Applied Ecology* 37: 997–1012.

Saltelli, A. and Annoni, P. (2010). How to avoid a perfunctory sensitivity analysis. *Environmental Modelling and Software* 25: 1508–1517.

Schmitz, O.J. (2000). Combining field experiments and individual-based modeling to identify the dynamically relevant organizational scale for a field system. *Oikos* 89: 471–484.

Segovia-Juarez, J.L., Ganguli, S., and Kirschner, D. (2004). Identifying control mechanisms of granuloma formation during *M. tuberculosis* infection using an agent-based model. *Journal of Theoretical Biology* 231: 357–376.

Stephens, P.A., Frey-Roos, F., Arnold, W. et al. (2002). Model complexity and population predictions. The alpine marmot as a case study. *Journal of Animal Ecology* 71: 343–361.

Storz, J., Ramakrishnan, U., and Alberts, S. (2002). Genetic effective size of a wild primate population: influence of current and historical demography. *Evolution* 56: 817–829.

Tablado, Z. and Revilla, E. (2012). Contrasting effects of climate change on rabbit populations through reproduction. *PLoS One* 7: e48988.

Tian, T., Xu, S., Gao, J. et al. (2007). Simulated maximum likelihood method for estimating kinetic rates in gene expression. *Bioinformatics* 23: 84–91.

Topping, C.J., Hansen, T.S., Jensen, T.S. et al. (2003). ALMaSS, an agent-based model for animals in temperate European landscapes. *Ecological Modelling* 167: 65–82.

Topping, C.J., Hoye, T.T., and Olesen, C.R. (2010). Opening the black box: development, testing and documentation of a mechanistically rich agent-based model. *Ecological Modelling* 221: 245–255.

Webb, C.T., Hoeting, J.A., Ames, G.M. et al. (2010). A structured and dynamic framework to advance traits-based theory and prediction in ecology. *Ecology Letters* 13: 267–283.

Whitman, K., Starfield, A.M., Quadling, H.S. et al. (2004). Sustainable trophy hunting of African lions. *Nature* 428: 175–178.

Wiegand, T., Knauer, F., Kaczensky, P. et al. (2004a). Expansion of brown bears (*Ursus arctos*) into the eastern Alps: a spatially explicit population model. *Biodiversity and Conservation* 13: 79–114.

Wiegand, T., Naves, J., Stephan, T. et al. (1998). Assessing the risk of extinction for the brown bear (*Ursus arctos*) in the Cordillera Cantabrica, Spain. *Ecological Monographs* 68: 539–570.

Wiegand, T., Revilla, E., and Knauer, F. (2004b). Dealing with uncertainty in spatially explicit population models. *Biodiversity and Conservation* 13: 53–78.

Wilensky U. (1999) NetLogo. http://ccl.northwestern.edu/netlogo. Center for Connected Learning and Computer-Based Modeling, Northwestern University. Evanston, IL.

Wilkinson, D., Smith, C.G., Delahay, R.J. et al. (2004). A model of bovine tuberculosis in the badger *Meles meles*: an evaluation of different vaccination strategies. *Journal of Applied Ecology* 41: 492–501.

Willis, J. (2007). Could whales have maintained a high abundance of krill? *Evolutionary Ecology Research* 9: 651–662.

Wood, S.N. (2010). Statistical inference for noisy nonlinear ecological dynamic systems. *Nature* 466: 1102–1104.

Zurell, D., Berger, U., Cabral, J.S. et al. (2010). The virtual ecologist approach: simulating data and observers. *Oikos* 119: 622–635.

Part IV

Population Genetics and Spatial Ecology

11

Genetic Insights into Population Ecology

Jeffrey R. Row[1] *and Stephen C. Lougheed*[2]

[1] *School of Environment, Resources and Sustainability, University of Waterloo, Waterloo, Ontario, Canada*
[2] *Department of Biology, Queen's University, Kingston, Ontario, Canada*

Summary

Although population ecology is fundamentally concerned with understanding the factors that influence population dynamics across space and time, collecting the required empirical data can be time consuming and logistically challenging. Quantifying population dynamics can be particularly daunting for rare or cryptic species where individuals are often difficult to observe, diminishing the utility of traditional population ecology techniques. Using genetic data in population ecology may alleviate some of these issues, but the wide variety of statistical techniques and types of molecular data make it challenging for non-specialists to easily grasp the array of questions that can be addressed and the types of data required. In this chapter we introduce the genetic markers that are most commonly deployed in population genetics and subsequently provide an outline of the genetic terminology and approaches that will be most useful in population ecology. We focus on methodology used to delineate genetic population structure, estimate population parameters such as population size, growth rate, and demographic history, and quantify dispersal and gene flow among defined populations. In addition to the insights that molecular data may provide for population ecology, a more complete integration of population ecology and population genetics will lead to more biologically realistic genetic models and a more comprehensive understanding of the genetics of demographic independence. Further, combining genetic and demographic data with landscape and environmental information provides a clearer understanding of how local demographic characteristics (e.g. population density, growth rates, and carrying capacity) and species interactions influence and interact with dispersal to shape landscape connectivity and source-sink dynamics.

11.1 Introduction

At its core, population ecology is concerned with understanding the abiotic and biotic factors that influence population abundance and trajectories across space and time (Rockwood 2006). As evidenced in many of the chapters in this book, collecting this information can be onerous, often requiring extensive field studies over multiple years. Quantifying population dynamics is particularly challenging for cryptic or rare species, where it can be difficult to determine the presence of individuals, let alone obtain the multiple recaptures required for application of capture-recapture techniques for population estimation (Chapter 5) or collecting enough occurrence records for occupancy or abundance models (Chapter 3). Genetic data can alleviate many of the issues associated with obtaining reliable population estimates, defining spatial structure, and estimating dispersal and gene flow among populations (Manel et al. 2005; Broquet and Petit 2009; Luikart et al. 2010). Some of the benefits of genetic data can be realized by simply using genetic techniques to identify individuals from hair or scat collected from snags or ground surveys (Taberlet and Luikart 1999; Waits 2004; Waits and Paetkau 2005) and using these data in capture-recapture analyses (Lukacs and Burnham 2005). These methods are not only less invasive than live trapping or capturing of individuals, but can also reduce the time and resources required for data collection.

Genetic sampling can reveal much more about a population than just numbers of individuals. Information such as sex (Aasen and Medrano 1990; Griffiths et al. 1996, 1998), age class (Barrett and Richardson 2011), genetic diversity of individuals or populations, and even the probability of an individual being a resident or migrant (Pritchard et al. 2000; Piry et al. 2004) can be determined using noninvasive genetic sampling. To this end, De Barba et al. (2010) advocated for "genetic" monitoring of populations and illustrated the benefits of this approach by monitoring a reintroduced Eurasian brown bear (*Ursus arctos arctos*) population entirely through noninvasive

Population Ecology in Practice, First Edition. Edited by Dennis L. Murray and Brett K. Sandercock.

Companion website: www.wiley.com/go/MurrayPopulationEcology

methods. Through the identification of individuals, De Barba et al. (2010) established that the population size had increased from 9 to 27 over a 10-year time period. Using the **genotypes** of individuals, however, they were also able to determine that all individuals were descendants of introduced individuals (i.e. no immigration) and that genetic diversity was decreasing due to inbreeding. This was likely due to the fact that one dominant male had sired most of the young, which the researchers were able to infer using parentage analysis based on DNA markers. Given the negative consequences of inbreeding (Madsen et al. 1999; De Barba et al. 2010), these results point to a much more informed and less positive assessment of the bear population than would have been possible without insights from genetic techniques.

Although it is usually beneficial for population ecologists to integrate molecular tools into their research program, the number and types of molecular analyses available can be overwhelming, particularly to the nonspecialist. In this chapter, we introduce the terminology and approaches used to: (i) delineate genetic population structure; (ii) estimate population parameters, such as population size and growth rate, as well as infer demographic history; and (iii) quantify dispersal and gene flow among populations. It is worth noting that there has been a move over the last decade to incorporate historical and contemporary landscape variables into population genetic models (Manel et al. 2003; Storfer et al. 2007). This is an important advance as it allows for the development of more biologically realistic models and can extend the level of inference beyond simply *how populations are distributed and the effects of various distribution patterns,* to also include *why populations are distributed in a particular manner,* and *how landscape changes will impact population connectivity.* Although a detailed consideration of the specific landscape genetics methodologies is beyond the scope of this chapter, many are relevant and thus we draw on research in this field and provide examples illustrating how landscape variables can and should be included. Further, reflecting our background in vertebrate evolutionary genetics and conservation, most of our empirical examples come from the animal literature, although we recognize the rich body of plant population genetics research (Brown et al. 1990; Wright and Gaut 2005).

11.2 Types of Genetic Markers

The methods used to isolate and identify genetic markers have dramatically increased over the last 20 years, reducing the costs and time associated with targeting specific genomic regions in population genetic studies (Freeland et al. 2011). Each class of marker potentially conveys different information about the individuals or populations under study and thus, it is important to have a general idea of the basic properties of the different markers available. For example, in sexually reproducing species, one can target genomic regions that are inherited from both parents (biparental) or only from either the mother or the father (uniparental). Uniparentally inherited haploid markers have approximately one-quarter the **effective population size** of biparentally inherited diploid markers and largely track the history of the parent from which they were inherited, which has important implications for the interpretation of genetic patterns. Another important consideration when choosing or analyzing genetic markers is mutation rate, which largely dictates the expected variability among individuals and populations, but also the temporal period over which inferences can be drawn. For example, under the simple scenario where two populations began to diverge in the recent past, markers with high mutation rates are more likely to accumulate differences and thus reveal the demographic signature of the population split. Considering all these factors together, a suite of different types of markers will likely prove useful to population ecologists, their choice depending on the research question.

11.2.1 Mitochondrial DNA

Uniparentally inherited markers are typically assayed using direct sequencing of the mitochondrial (animals or plants) or chloroplast (plants) genomes (Birky 1995), or from sequencing portions of sex chromosomes (Selelstad and Hebert 1994), resulting in individual **haplotypes**. Mitochondrial DNA (mtDNA) is typically maternally inherited, and since the 1980s sequence data have been the most common markers for animal species delineations and establishing large-scale *phylogeographic* patterns of population structure, gene flow, and demographic history (Pakendorf and Stoneking 2005; Zink and Barrowclough 2008; Galtier et al. 2009) of animals. This widespread use of mtDNA sequence data has been largely due to lack of recombination, high substitution rates relative to nuclear counterparts, high copy number making isolation relatively easy, and the aforementioned small effective population size, which means mtDNA markers may more closely track demographic changes. All of these factors combined make it relatively cost- and time-efficient to generate large mtDNA datasets that are more easily interpreted than many nuclear DNA datasets (nDNA) (Zink and Barrowclough 2008). However, due to questions regarding the assumptions of strict maternal inheritance in vertebrates and selective neutrality, and because all mtDNA markers are inherited together (i.e. not independent loci), basing demographic inference and population structure solely on mtDNA patterns is not recommended and decreasing in practice (Galtier et al. 2009; Lougheed et al. 2013). A mitochondrial assessment of the range-wide genetic structure of foxsnakes, for example, found no genetic differences between Western and Eastern foxsnakes (*Pantheris* spp., Crother et al. 2011). However, a more recent test with microsatellites found strong genetic differentiation between the species, and even within Eastern foxsnakes, with this differentiation corresponding with geographic disjunctions (Row et al. 2011). Given the endangered status of Eastern foxsnakes in Canada, which houses 70% of the range of this species, a genetic assessment based on mtDNA alone could have had strong negative management implications.

11.2.2 Nuclear Introns

Due to the relatively low substitution rate of nDNA compared to mtDNA, the direct sequencing of **introns** is often used to elucidate evolutionary relationships among taxa and/or deeper phylogeographic patterns. When nuclear introns are used in concert with mtDNA markers, the results can often be illuminating given differences in temporal resolution and mode of inheritance (Eytan and Hellberg 2010). The difficulties of deducing haplotypes for diploid markers, which are required for many types of analysis (Stephens et al. 2001), and the challenges in accumulating the large amount of data necessary for robust inference (because introns typically show low variability) mean that nuclear introns are unlikely to be particularly relevant to population ecologists.

11.2.3 Microsatellites

In part due to their high mutation rate, genotyping short tandem repeats (STR) or microsatellites emerged as one the most popular choices for estimating recent patterns of divergence, gene flow, and current population size (Selkoe and Toonen 2006). Microsatellites are tandem repeats of a core DNA motif, typically composed of two to six nucleotides that are biparentally inherited and generally assumed to be selectively neutral. The number of repeats defines distinct alleles, with between 5 and 40 repeats being common (Selkoe and Toonen 2006). Microsatellites are assumed to mutate usually in a stepwise fashion (i.e. gain or loss of single repeat due to DNA polymerase slippage). There is no need to obtain the DNA sequence of every individual as the number of repeats can be determined by calculating the size of the amplified fragment against standards of known size using capillary electrophoresis and fluorescent labels (Koumi et al. 2004). Such automated genotyping is typically cheaper and more time efficient than direct DNA sequencing, especially when multiple microsatellite loci are assayed in the same **polymerase chain reaction** (PCR) – called multiplexing. Recent sequencing advances have also made the identification of microsatellite markers much more cost- and time-effective (Perry and Rowe 2011). Regardless of these advances, testing microsatellites for variation among populations, determining optimal PCR conditions, and scoring microsatellites for nonmodel organisms can still be time-consuming and the number of markers per study for wild organisms is still typically less than 20 or 30 (Selkoe and Toonen 2006).

11.2.4 Single Nucleotide Polymorphisms

Single nucleotide polymorphisms (SNP) have been gaining in popularity largely due to an increased ability to identify and genotype large numbers of markers from across the genome (Seeb et al. 2011). As with microsatellites, they are typically derived from nDNA, and thus are biparentally inherited and used in similar types of analysis, which are based on allele frequencies. Due to the lower mutation across most of the nuclear genome, SNPs generally have much less variability and are on average less informative than microsatellites for population structure (4–12 times less informative; Rosenberg et al. 2003; Liu et al. 2005); however, SNPs have reduced scoring error rates, higher reproducibility, and can be highly informative of population structure when large numbers are used (Rosenberg et al. 2003; Liu et al. 2005). Finally, the theoretical foundations of population genetics are largely based on simple di-allelic systems like SNPs (see: Hedrick 2009), and thus our understanding of how to analyze such data is more advanced than for highly variable markers like microsatellites.

11.2.5 Next-generation Sequencing

Next-generation sequencing (NGS – also called high throughput or massively parallel sequencing) is a term used to cover an array of technologies that allow us to quickly and relatively inexpensively obtain millions of base pairs of DNA or RNA sequences – from a single individual to suites of individuals across populations of a focal species (Shendure and Ji 2008). Current NGS platforms include Illumina, Roche 454, Ion Torrent, and SOLiD. The field is moving quickly and the term next-generation sequencing is considered by some to be passé – indeed we are now speaking of third-generation sequencing technologies (van Dijk et al. 2014), including novel techniques like single-molecule real-time sequencing (SMRT), and single-molecule nanopore DNA sequencing (Madoui et al. 2015). Regardless, it is now technically feasible and cost effective to obtain genome-wide panels of thousands or tens of thousands of DNA markers (often SNPs but also contiguous stretches of DNA sequence) from individuals from multiple populations (Malenfant et al. 2015).

All of the markers discussed so far are selectively neutral and most approaches that will be of interest to population ecologists and described throughout this chapter assume this to be the case; in other words, that patterns of variation reflect the relative influences of **gene flow** and **genetic drift** only. With NGS approaches, we can move beyond simple quantitative surveys of large numbers of DNA markers intended to estimate these traditional population genetic parameters. For example, so-called genome scan techniques look for "outlier" loci – and those SNPs that exhibit greater divergence than background, which may be the signature of selection (Nosil et al. 2008). We can also sequence the transcriptome of individuals, the set of all RNA transcripts produced by a genome, rather than the DNA itself to test hypotheses on differences in gene expression in varying environments or divergence in developmental pathways among distinct populations or species (Wang et al. 2009). These genetic markers of adaptive significance may be of interest to population ecologists who wish to identify genes or genomic regions responsible for individual variation in biologically important traits, such as the growth rate of individuals (Hemmer-Hansen et al. 2011). Even with NGS, however, the methods to link known, complex adaptive traits to a specific genomic region (or more likely regions), remain fairly lab intensive and thus most approaches that we discuss in this chapter utilize neutral genetic markers.

11.3 Quantifying Population Structure with Individual-based Analyses

The origins of population genetics can be tied back to the seminal works of several prominent biologists and mathematicians. In particular, Ronald Fisher, J.B.S. Haldane, and Seawell Wright are largely credited with developing the mathematical underpinnings of population genetics by combining their understanding of Mendelian inheritance, statistics, and sampling theory with ideas on selection in captive and wild populations (Fisher 1930; Wright 1931; Haldane 1932; Provine 1971). Their ideas were a major contribution to the Modern Evolutionary Synthesis (Huxley 1942), which underpins much of modern evolutionary biology. The work also laid the foundations of classical population genetics, which largely compares changes in allele frequencies within and among a priori defined populations – it relies on comparisons to expectation under **Hardy–Weinberg Equilibrium** (HWE). HWE describes the expected genotypic proportions in an ideal population at equilibrium, the characteristics of which include random mating, no selection, and negligible mutation. Given a single **genetic locus** example at HWE where there are two alleles, A_1 and A_2 with frequencies p and q, the expected respective genotypic proportions of genotypes A_1A_1, A_1A_2, and A_2A_2 in subsequent generations can be calculated using:

$$p^2 + 2pq + q^2 \qquad (11.1)$$

HWE is treated as a "null" hypothesis where significant departures from expectation can be used to infer particular processes. For example, lower than predicted heterozygote frequencies across surveyed loci imply inbreeding; departures for a single locus might imply the action of selection or the presence of a nonamplifying "null" allele for PCR-based genotyping (Hedrick 2009).

In practice, traditional population-based analyses using HWE are of most utility for species where there are obvious population delineations or breeding aggregations, such as aquatic species inhabiting a pond, or a forest-dwelling species occupying a forest patch in a heavily fragmented landscape. In many cases, however, individuals are distributed more continuously or obvious population boundaries are not apparent. Nonetheless cryptic divisions may exist, which can range from subtle restrictions in dispersal (Rueness et al. 2003; Sacks et al. 2004), through admixture of historical genetic lineages that confound attempts to understand contemporary genetic patterns (Austin et al. 2002; Gibbs et al. 2006), to evolutionarily-independent cryptic species that have not diverged morphologically (Elmer et al. 2007). These divisions have strong implications for quantifying and interpreting the dynamics of populations and if not identified, can introduce bias and lead to spurious interpretations.

Three major classes of analysis that can be used to quantify major patterns of genetic population structure in a given dataset: Bayesian clustering, multivariate, and spatial autocorrelation. These broadly defined categories are all primarily individual-based methods, where the researcher is interested in identifying the patterns of genetic structure without prior delineation of populations. Although not always possible, the most appropriate sampling strategy for individual-based approaches is systematic sampling across a landscape and on either side of putative population boundaries or impediments to dispersal. The specific sampling regime with random, uniform, or hierarchical protocols will largely depend on the research question, resources, and the species being studied (Storfer et al. 2007).

It is important to first define "population," and discussion in both population ecology and population genetics has concerned what might be most appropriate (Wells and Richmond 1995; Waples and Gaggiotti 2006; Palsbøll et al. 2007). Most researchers seem to agree that population designations require some **demographic independence** and not a simple rejection of a lack of **panmixia** or statistical lack of genetic structure, but how to define demographic independence is still unclear. Many of the analyses below are designed to simply test for panmixia, and thus in most cases, to identify true "populations" in a demographic sense will require studies linking genetic and demographic patterns. We refer to the genetic structure identified through the various analyses as genetic groups or genetic clusters, with the understanding that more information is required to determine their status as "true" populations in any population ecological sense.

11.3.1 Bayesian Clustering

The essence of model-based Bayesian (see Chapter 2 for introduction into Bayesian analysis) clustering algorithms is that individuals can probabilistically be assigned to genetic clusters, based solely on their individual multilocus genotypes. Since the seminal paper by Pritchard et al. (2000), Bayesian clustering has been used in a broad range of topics in ecology and evolution, including quantifying species interactions (Wang et al. 2006; Rudge et al. 2009; Kobmoo et al. 2010), identifying hybrid zones and their dynamics (Pierpaoli et al. 2003; Mavárez et al. 2006; Tung et al. 2008), and understanding the mechanisms underlying adaptation (Kitano et al. 2008). Most commonly and perhaps most relevant to population ecologists, this approach is used to determine both the number and spatial extent of genetic clusters within focal species. Once identified, clusters can be used as population groupings in subsequent population-based analysis (Baums et al. 2005; Hoelzel et al. 2007; Janssens et al. 2007) or mapped onto the landscape to identify anthropogenic (Riley et al. 2006; Row et al. 2010) or natural landscape (Fouquet et al. 2012) features that restrict dispersal and gene flow.

The simplest clustering algorithms generally do not assume an underlying mutational model and thus can be used with any class of Mendelian-inherited codominant marker where allele frequencies from independent loci can be derived (e.g. SNPs, microsatellites). Most clustering algorithms assume that each individual can be placed into its own unique genetic cluster, within which, individuals are in HWE and **linkage equilibrium** (LE), and attempt to find the groupings that best satisfy these assumptions using Bayesian statistical methods (Pritchard et al. 2000; Guillot et al. 2005a; Chen et al. 2007). Under this framework the most likely number of genetic clusters (K) and a probability of membership (q) for each individual to a particular genetic cluster (Pritchard et al. 2000) can be determined (Box 11.1). In their original paper, Pritchard et al. (2000) also suggested a modification of this simple model which incorporates admixture between genetic clusters and thus calculates the proportion of each individual's genome derived from each cluster rather than a probability of individual membership. In the absence of strong geographic barriers or low dispersal rates an admixture model is likely more biologically realistic and has extended the used of Bayesian clustering to hybridization studies, where by definition admixture occurs. For example, Stewart et al. (2016) used individual-based Bayesian cluster analysis to examine the outcome of secondary contact between lineages of a temperate frog species, spring peeper (*Pseudacris crucifer*), finding asymmetrical hybridization between lineages.

Early Bayesian clustering models used only the genotype to assign individuals to genetic clusters and for the most part did not assume any spatial structure or incorporate geographical information. This may be appropriate in some scenarios, where barriers to dispersal are not spatially constrained (e.g. host specificity), however, in most cases individuals closer in space are more likely to belong to the same genetic cluster. Thus, Bayesian clustering algorithms have been developed to incorporate such spatial information using three main approaches that are described well in Guillot et al. (2009). In the simplest approach, individuals collected from the same sampling location are assumed to be more likely in the same genetic cluster. More complex approaches require coordinates of individual sample locations and define proximity using **Voronoi tessellation**, which deconstructs the landscape into a number of spatial polygons with no overlap or gaps. Tessellation is typically conducted using either free Voronoi tessellation, which builds a defined number of polygons independent of sampling locations (Guillot et al. 2005a, b) or constrained tessellation where polygons are built using the sampling locations (Chen et al. 2007; Corander et al. 2008b). In both approaches, the tessellated landscape is used as prior information in the clustering analysis under the assumption that geographically proximate individuals have a higher probability of belonging to the same cluster. Guillot et al. (2009) provide a detailed description of the merits of each approach. Most importantly, methods using free Voronoi tessellation can extrapolate genetic structure to areas that have not been sampled, which may be a positive or negative feature depending on the real amount of spatial structure in a dataset. Overall, the benefits of including spatial information can be substantial. Guillot et al. (2005) used spatial clustering analysis to re-analyze a DNA microsatellite dataset of wolverines (*Gulo gulo*) and diagnosed an additional genetic cluster and five

Box 11.1 Bayesian Clustering Analysis of Foxsnake Populations Using `Structure`

Row et al. (2010) examined the genetic population structure of eastern foxsnakes (*Patherophis gloydi*) across a fragmented region in southwestern Ontario, Canada, and found significant genetic structure. Here we use Bayesian clustering, as implemented in Program `Structure` (ver. 2.3.3), on a subset of data (N = 203 individuals) to determine number and extent of genetic clusters across four sample locations (Figure 11.2). Given a high degree of natural and anthropogenic fragmentation, it was possible that each sampling location would comprise a separate genetic cluster. Thus, we determined the number genetic clusters (K) by running 20 replicates for K = 1–5 (100 replicates total), with the maximum value of K simply being one more than the total number of locales. For each replicate we ran 100 000 MCMC iterations with 50 000 iterations discarded after an initial burn-in, as prescribed by the program authors. Preliminary runs suggested longer burn-ins, and run lengths did not change the results. We expected some gene flow between the clusters, and used admixture analysis with correlated allele frequencies (Falush et al. 2003).

We considered four alternative statistics for estimating the number of genetic clusters (Figure B11.1.1). Evanno et al. (2005) suggested that a peak in ΔK, in our case at K = 3, is the best determinant of the likely number of true genetic clusters in simulated studies. When a clear peak is not observed, an asymptote of L[K] (a) is also commonly used as evidence for K and the barcharts below are derived for a range of K values. Because we did have a clear peak, we determined individual admixture coefficients by conducting another 80 simulations in `Structure` for K = 3, and averaged the top 5% of the lowest L[K]. A barchart of the Q-matrix representing admixture coefficients from the genetic clusters where each individual is represented by a single bar, and each cluster with a different color is a typical way to display the results of Bayesian clustering (Figure B11.1.2).

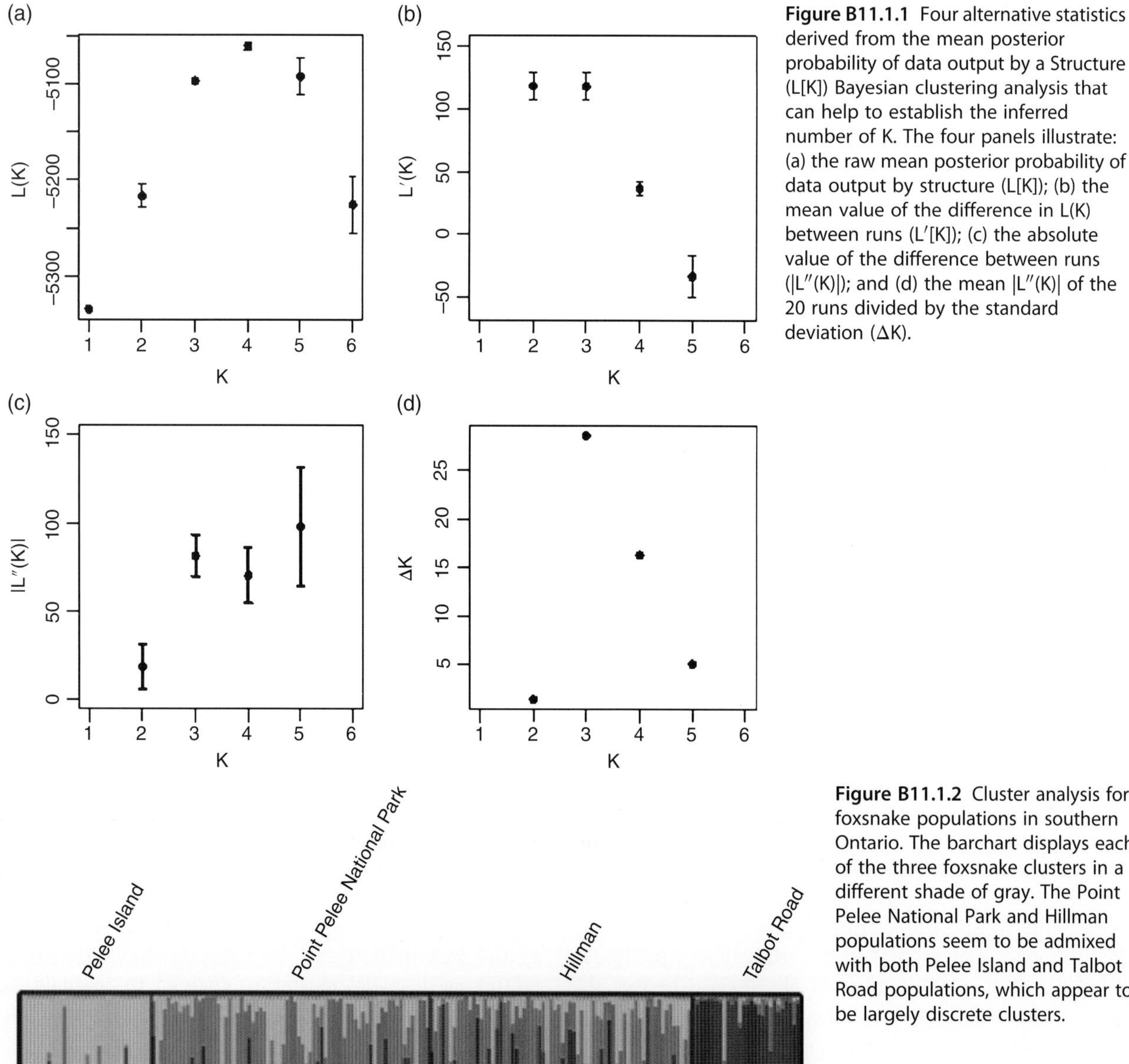

Figure B11.1.1 Four alternative statistics derived from the mean posterior probability of data output by a Structure (L[K]) Bayesian clustering analysis that can help to establish the inferred number of K. The four panels illustrate: (a) the raw mean posterior probability of data output by structure (L[K]); (b) the mean value of the difference in L(K) between runs (L′[K]); (c) the absolute value of the difference between runs (|L″(K)|); and (d) the mean |L″(K)| of the 20 runs divided by the standard deviation (ΔK).

Figure B11.1.2 Cluster analysis for foxsnake populations in southern Ontario. The barchart displays each of the three foxsnake clusters in a different shade of gray. The Point Pelee National Park and Hillman populations seem to be admixed with both Pelee Island and Talbot Road populations, which appear to be largely discrete clusters.

first-generation migrants that were not identified in previous nonspatial analysis (Cegelski et al. 2003).

11.3.2 Multivariate Analysis of Genetic Data Through Ordinations

Multivariate ordination approaches, and in particular principal component analysis (PCA), have proved effective for analyzing population genetic structure and have several advantages over Bayesian clustering (Patterson et al. 2006; Jombart et al. 2008; Reich et al. 2008). Because such ordination methods simply summarize major trends in variation among individual genotypes, population-based allele frequencies or matrices of genetic distances (using principal coordinates analysis [PCoA]), they are free from key genetic assumptions (HWE, LE) and thus can accommodate multiple types of genetic structure. This includes **isolation-by-distance** and **genetic clines** (Jombart et al. 2009), which Bayesian clustering analyses struggle with given their assumption of discrete clusters in equilibrium (Frantz et al. 2009; Schwartz and Mckelvey 2009). Another advantage of ordination methods is that they do not require **Markov chain Monte Carlo** (MCMC) algorithms to solve complex equations and thus are less computer-intensive than Bayesian clustering and can accommodate large genetic datasets (Ma and Amos 2012); this advantage will become more important with the large amounts of genomic data being generated through NGS (Shendure and Ji 2008).

Ordination methods summarize genetic variation (Jombart et al. 2009) and the most widely used is PCA (Paschou et al. 2007; Montarry et al. 2010; van Heerwaarden et al. 2010), which attempts to identify a series of composite multivariate axes (called principal axes or factors) that best summarize major trends in allelic variation. Multivariate space can be composed of individual genotypes or population-based allele frequencies, but in both cases, derived axes or factors are defined by the contribution of each allele (allele loading) and their eigenvalue (explained variation), with the locations of individuals or populations defined by their scores along each axis (Box 11.2).

A major challenge of PCA is to establish statistical and biological meaning of derived axes (Patterson et al. 2006; Reich et al. 2008; Jombart et al. 2009). This is perhaps best illustrated by the controversies surrounding Cavalli-Sforza et al. (1994), who attributed continent-wide genetic clines in human populations detected using PCA analysis to ancient migration patterns. While some of the assertions presented by Cavalli-Sforza et al. (1994) have been corroborated with other lines of evidence, some researchers have suggested that these patterns can also arise through other processes like simple isolation-by-distance (Novembre and Stephens 2008; Reich et al. 2008). This example certainly points to the importance of combining multiple lines of evidence when interpreting genetic data, but also the importance of taking a statistical approach for ordination analysis (Patterson et al. 2006). In particular, many have advocated determining the number of PCA axis to retain for further analysis by comparing the distribution of eigenvectors to a Tracy–Widom (TW) distribution (Johnstone 2001; Patterson et al. 2006). The resulting significant axes can be visualized to assess population structure, and clustering algorithms such as Ward's clustering (Ward 1963) can be used on distances derived from the retained axes to identify population structure (Box 11.2; van Heerwaarden et al. 2010, 2011).

In many cases we would advocate examining population structure using multiple types of analyses to provide confidence in the assertions regarding the underlying patterns. For example, consider the populations of eastern foxsnakes (*Pantherophis gloydi*) described in Box 11.1. We find that the eigenvalues of the first four axes show significant deviation from the TW distribution, suggesting that these best capture overall patterns of genetic structure. Plotting the first and second principal axes, we find a clear separation between the same three clusters identified with Bayesian clustering (Figure 11.1a). Plots of the third and fourth axes, however, show no additional power to separate among populations, despite significant deviation from the TW distribution (Figure 11.1b). These axes were marginally significant ($P = 0.03$ for both) and some have suggested that this method tends to

Box 11.2 Multivariate Analysis of Genetic Structure for Simulated Populations

We used Program `Easypop` (ver. 2.01) to simulate four genetic datasets (20 loci mutating according to a stepwise mutation model): (i) a single panmictic population; (ii) a hierarchical population structure where two populations (P1 and P2) are connected by high gene flow (0.004 probability of an individual migrating from one population to another each generation with no breeding restrictions) with a third population (P3) with less gene flow between it and the other two (0.002 probability of migration); (iii) three populations with equivalent high amounts of migration (0.004 probability of migration); and (iv) three populations with half the amount of migration (0.002 probability of migration) as in scenario c. All simulations were run for 5000 generations with fixed population sizes

Table B11.2.1 Eigen analysis of genetic population structure for four simulated datasets.

Eigenvalue	Sim i	Sim ii	Sim iii	Sim iv
1	0.62	**<0.001**	**<0.001**	**<0.001**
2	0.76	**0.003**	**<0.001**	**<0.001**
3	0.86	0.45	0.37	0.26
4	0.79	0.32	0.73	0.23
5	0.70	0.69	0.53	0.20
6	0.62	0.67	0.94	0.72
7	0.51	0.67	0.84	0.60
8	0.49	0.50	0.79	0.51
9	0.73	0.66	0.86	0.53
10	0.65	0.90	0.85	0.89

of 500 individuals (equal sex ratios), 30 of which were sampled for genetic analysis. When genetic structure was present in the dataset (ii–iv) the first, and usually second, eigenvalues were disproportionately larger than the others, but this was not the case for the panmictic population (i). Indeed, we find that only when there is genetic structure in the dataset and not panmixia do the eigenvalues show significant deviations from a Tracy–Widom (TW) distribution (bolded values) (Table B11.2.1). Model results can be visualized by plotting the first two PCA axes to summarize the genetic variation of each dataset illustrated the genetic structure (or lack thereof) in each of the four simulated scenarios (Figure B11.2.1).

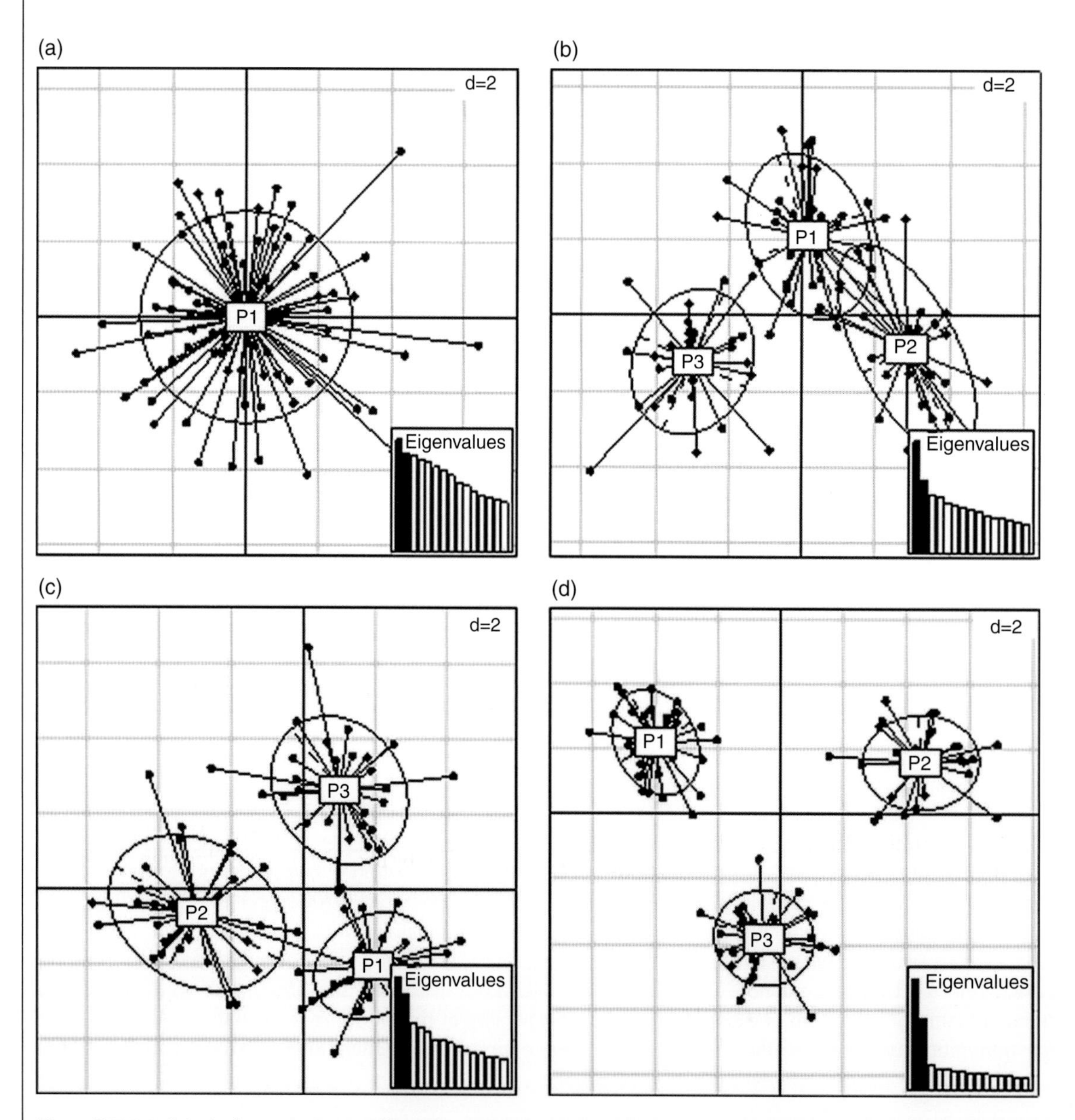

Figure B11.2.1 Principal components analysis to summarize genetic variation in simulated datasets. Shown are 95% confidence ellipses around the centroid of each simulated population with vectors attaching individuals to the centroid of their population.

(a)

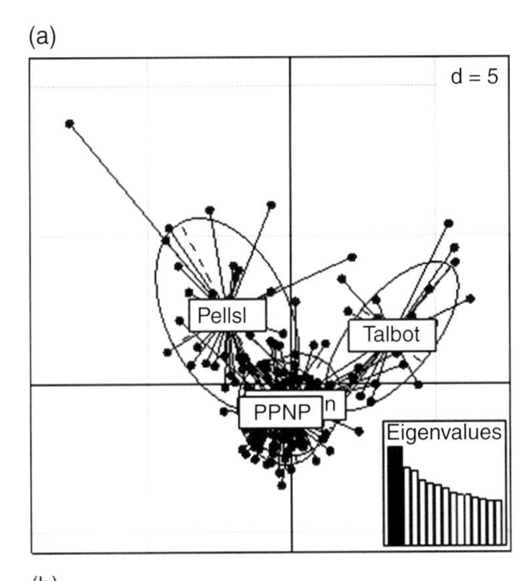

(b)

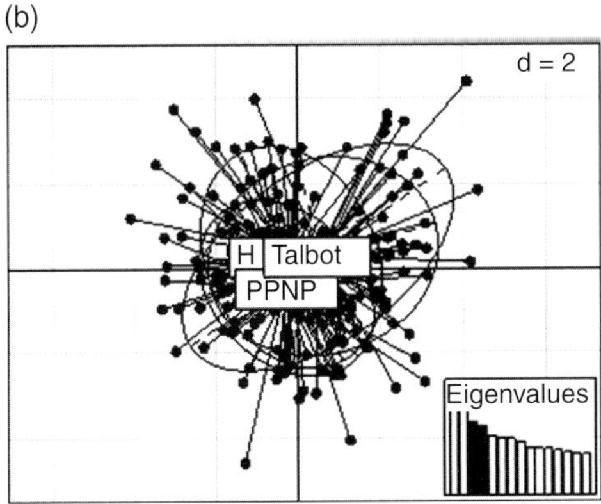

Figure 11.1 Biplots of the principal axes of a principal components analysis (PCA) on foxsnake microsatellite genotypes distributed across southwestern Ontario (Figure 11.2), with insets of eigenvalues (displayed axis indicated by black bars). The first four eigenvalues significantly deviated from a Tracy–Widom (TW) distribution.

overestimate the number of significant axes (Lee et al. 2011), as is likely in this example.

Recently, the incorporation of spatial and landscape data in ordination analysis has improved the ability of these methods to detect fine-scale population structure and establish associations between genetic and landscape variation. Jombart et al. (2008) introduced spatial principal components analysis (sPCA), a modification of PCA that simultaneously maximizes the genetic variance and spatial autocorrelation of the principal axes (i.e. maximizes the product of variance and spatial autocorrelation measured using Moran's I). Before conducting sPCA analysis, Jombart et al. (2008) advocated for, and derived a test to, detect the presence of spatially correlated genetic structure by comparing individual allele frequencies to Moran's Eigenvectors Maps (MEM; Dray et al. 2006). MEMs are derived from a connection network of sampled individuals, and when spatially correlated genetic structure is present, the correlation (Pearson R^2 values) between individual allele frequencies with MEMs should be higher than R^2 values calculated for randomly permuted allele frequencies (Jombart et al. 2008).

Using genetic simulations, Jombart et al. (2008) found sPCA to perform better than PCA in revealing simple (e.g. clines, discrete patches) and complex genetic structure, such as mixture of clines and patches mixed with randomly distributed individuals. These results have been corroborated by many empirical studies that have used sPCA and the method has been utilized to identify various spatial genetic patterns, including genetic clines (Vandewoestijne and Van Dyck 2010; Koen et al. 2011) and discrete populations (DiLeo et al. 2010; Chiappero et al. 2011). Other spatial analysis such as PCA of Neighbor Matrices can incorporate landscape information to explicitly test for associations between allele frequencies and the distribution of habitat variables (Dray et al. 2006; Legendre and Fortin 2010). This can be particularly relevant for quantifying associations between particular alleles and environmental variables and be useful in identifying genes of adaptive importance (Manel et al. 2010; Bothwell et al. 2013).

11.3.3 Spatial Autocorrelation Analysis

Fine-scale spatial genetic structure not revealed by ordinations or Bayesian analyses may be present within continuously distributed populations or genetic clusters, and relate to the distribution of individuals or patterns of dispersal. Individual patterns of spatial genetic structure can arise through selection and historical processes, such as range fragmentation or expansion (Epperson 2003), but for populations at **mutation-drift equilibrium** the predominant patterns likely reflect limited dispersal, with neighboring individuals exhibiting higher relatedness than individuals at greater distances apart (i.e. individual-based isolation-by-distance; Rousset 2000). Spatial autocorrelation analysis quantifies the correlation between pairs of individuals within increasing geographical distance classes (d) and is frequently used for continuously distributed populations (Hardy and Vekemans 1999; Smouse and Peakall 1999).

Dewey and Heywood (1988) suggested using Moran's I on individuals by first coding individual genotypes based on allele frequencies (q; 1.0 – homozygotes, 0.5 – heterozygotes, and 0.0 – not present) and calculating Moran's I for each allele in each distance class (d). In the formula (Sokal and Oden 1978)

$$I(d) = \frac{n\sum_i \sum_j w_{ij}(d) Z_i Z_j}{\left[\sum_i \sum_j w_{ij}(d)\right]\left[\sum_i Z_i^2\right]} \tag{11.2}$$

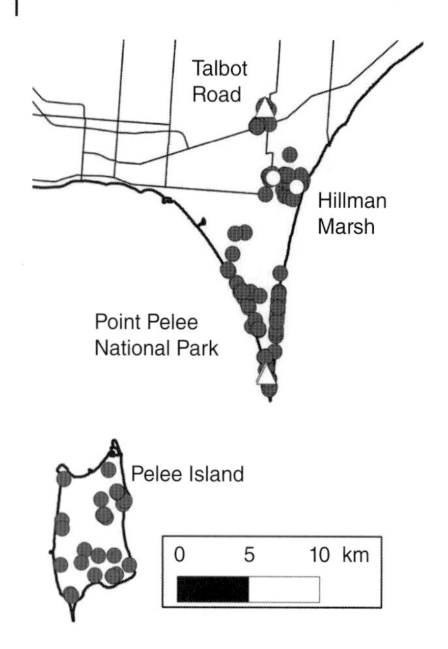

Figure 11.2 Sampling locations of eastern foxsnakes used in Bayesian clustering (Box 11.1) and subsequent assignment tests to identify migrants. Assignment tests identified migrants or their descendants ($G = 2$) among the three clusters. We set $k = 3$ and ran an admixture analysis for 100 000 MCMC iteration (50 000 burn-in) and set the probability of an individual being a migrant (v) at 0.1. Significant migrants out of Pelee Island (white triangles) and the Talbot (white circles) population are shown.

n is the number of individuals, $Z_i = q_i - \bar{q}$ and $w_{ij}(d)$ is a weighting scheme, typically constrained to equal 1 if separated by a distance within d, and 0 if outside that distance class. Averaging over alleles and loci gives an estimate of the correlation of allele frequencies between individuals, within each distance class.

Calculating Moran's I across any set of geographic distance classes and permuting individuals among categories to derive a null expectation for no spatial structure is a common approach to determine significance across distance classes (Hardy and Vekemans 2002). As an example, we determined the extent of significant, positive correlation (Moran's I) for genotypes of Canada lynx (*Lynx canadensis*) distributed across Manitoba to Quebec (Figure 11.3a). Canada lynx have a wide distribution across northern Canada and very little genetic population structure, with most of mainland North America belonging to a single genetic cluster (Row et al. 2012). Here, using individuals genotyped at 14 DNA microsatellite loci, we show positive spatial autocorrelation for individuals separated by less than 400 km (Figure 11.3b). This fits well with our current knowledge of movement patterns of Canada lynx, which regularly disperse distances greater that 250 km and up to 1000 km (Poole 1997). Although we have not established specific dispersal parameters (e.g. range and mean of dispersal distance), the extent of positive spatial genetic structure gives us an idea of the dispersal potential of this species without having to conduct an extensive radio-telemetric study, as in Poole (1997).

Several genetic distance and relatedness coefficients can be used as individual-based estimators of genetic differentiation in spatial autocorrelation analysis. Many studies have compared and estimated the properties of each (e.g. error, precision – see Hardy 2003; Vekemans and Hardy 2004; Rousset 2008) and related spatial genetic patterns to actual dispersal parameters and genetic neighborhood size (Vekemans and Hardy 2004; Epperson 2005; Wagner et al. 2005). The choice of genetic distance or relatedness estimator, and the number and size of distance classes, can have a large effect on the estimated parameters for spatial autocorrelation. In general, the number of pairwise comparisons and the percentage of individuals present at least once in the distance class, and the overall distance in relation to expected dispersal distance or movement, can be used to select appropriate distance classes (Hardy and Vekemans 2002; Double et al. 2005). Systematically varying the distance classes in the analyses can also give some measure of the sensitivity of the desired statistic to the choice of a particular set of distance classes (Troupin et al. 2006).

Smouse and Peakall (1999) developed a multivariate correlation coefficient (r) for estimating individual spatial genetic structure, which is commonly used in the literature. Their approach is similar to Moran's I and is also calculated across distance classes, but is a "true" correlation coefficient between individuals in the sense that it is bounded by −1 and 1, with 0 reflecting no correlation. Another possible advantage of this approach is that instead of calculating and averaging separate coefficients for each locus, a multivariate distance is used in the calculation, potentially increasing the statistical power (Smouse and Peakall 1999; Peakall et al. 2003).

Most empirical studies having used spatial autocorrelation analysis typically compare the observed patterns between groups in regions with different habitat compositions, such as disturbed versus undisturbed sites, to gain insight into dispersal potential under varying environmental conditions (Schmuki et al. 2006; Jolivet et al. 2010; Wang et al. 2012). Similar comparisons can be made between species or different reproductive classes in the same region or habitat to examine within and among species variation. For example, Paquette et al. (2010) used spatial autocorrelation analysis for a population of radiated tortoises (*Astrochelys radiata*) and found that males were less genetically similar than females at short distance classes, providing evidence for male-biased dispersal in this species. This result was not apparent when population-based approaches were used, and given

(a)

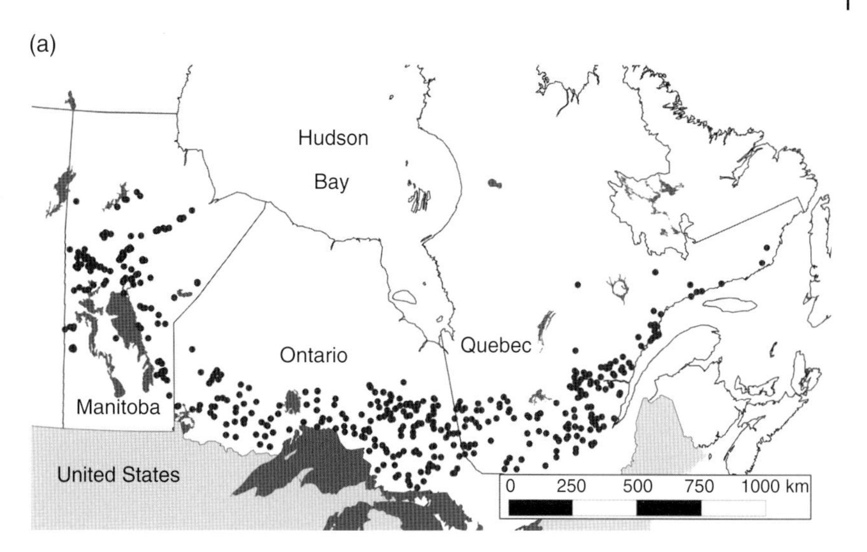

(b)

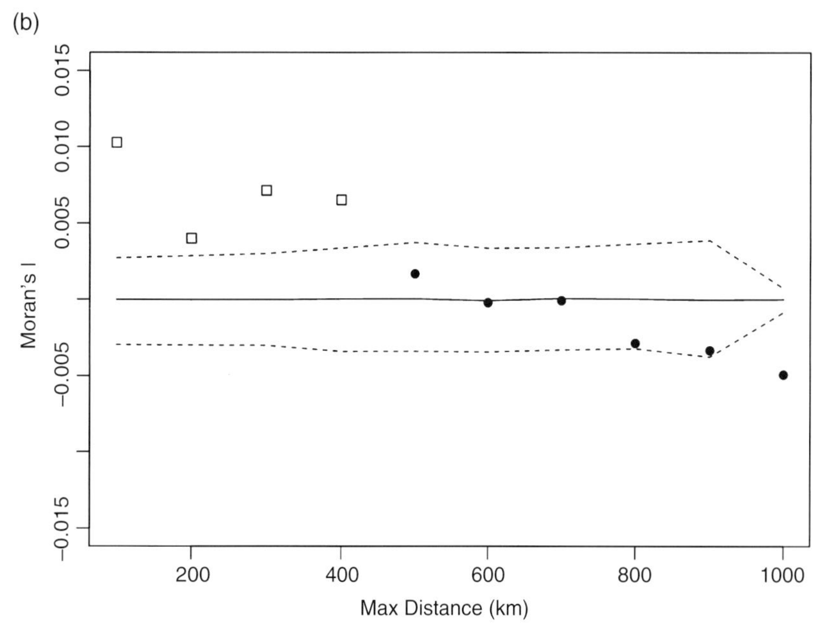

Figure 11.3 Results from Spatial Autocorrelation analysis of individual Canada lynx (*Lynx canadensis*) genotypes distributed from Manitoba to Quebec in Canada. (a) Distribution of samples, and (b) degree of autocorrelation (Moran's I) between individuals within increasing 100 km distance classes are shown. Dotted line represents correlation between genotypes randomly permuted across distances classes, and open squares and closed circles delineate distance classes where observed autocorrelation is significantly above or within the permuted range, respectively.

that the species is critically endangered it would have been difficult to obtain through capture-recapture records.

11.3.4 Population-level Considerations

In many cases, individual-based analyses would be a logical starting point for evaluations of population structure as they provide a nonarbitrary population definition and suggest boundaries to dispersal on the landscape. Many molecular methods, however, consider groups of individuals and estimate group-based statistics, such as effective population size or gene flow between patches, and thus populations must be defined. A typical approach might be hierarchical: to use individual-based methods to determine the number and extent of genetic clusters, which are then used as "populations" in the population-based analyses (Bergl and Vigilant 2007). For example, we diagnosed three genetic clusters for eastern foxsnakes (Box 11.1) and proceed to use these as sampling units in population-based analysis to determine effective population sizes and directionality of gene flow (Box 11.4).

11.4 Estimating Population Size and Trends

A fundamental property of organisms in nature is population size. While population size is a key parameter for both ecologists and geneticists, and important for management and predicting extinction risk of species

Box 11.3 Introduction to Coalescent Theory

Coalescent theory dates back to the 1970s (Kingman 2000), but was more formally described by Kingman (1982a, 1982b) who mathematically derived the *n*-coalescent as the time to the most recent common ancestor (MRCA) of a contemporary sample of genes, under an idealized Wright–Fisher population model (Fisher 1930; Wright 1931). This formulation allowed a shift from classical population genetics, which describe the properties of entire populations, to considering only the direct ancestors of the sampled genes through their gene genealogies. In his original derivation, Kingman (1982a, 1982b) determined that the probability of a coalescent event for two genes in the prior generation is $1/(2N)$, where $2N$ is the haploid population size (N is the diploid population size) and thus tying the coalescent process to the effective size of the population. Consider an example of 10 genes in a small population of 10 individuals (Figure B11.3.1).

When two genes choose the same ancestor, a coalescent event occurs, and this process continues until all genes coalesce into a single common ancestor, the MRCA. In a panmictic, stable population, the rate of coalescent events depends on the number of lineages and population size, thus allowing for demographic inference based on the genealogies of sampled genes.

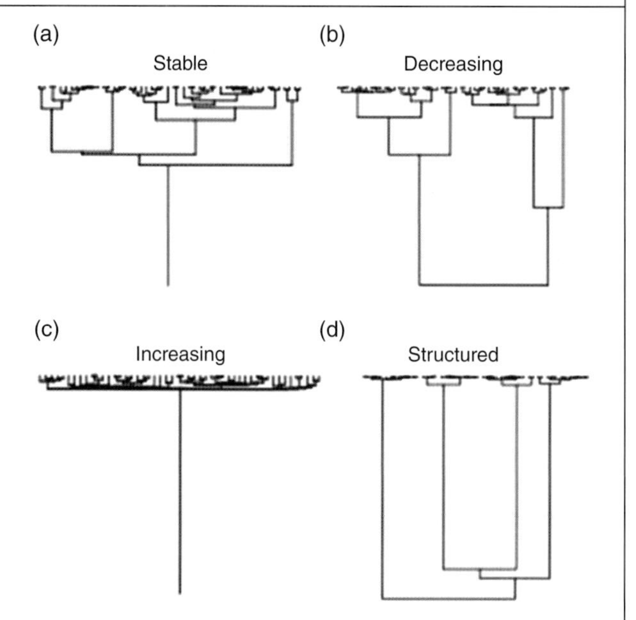

Figure B11.3.2 Visualizing the individual gene genealogies for 50 individuals generated using `Simcoal` (ver. 2.0, Laval and Excoffier 2004) under four different scenarios: (a) a stable population; (b) an exponentially decreasing population; (c) an exponentially increasing population; and (d) four populations with no gene flow occurring among them.

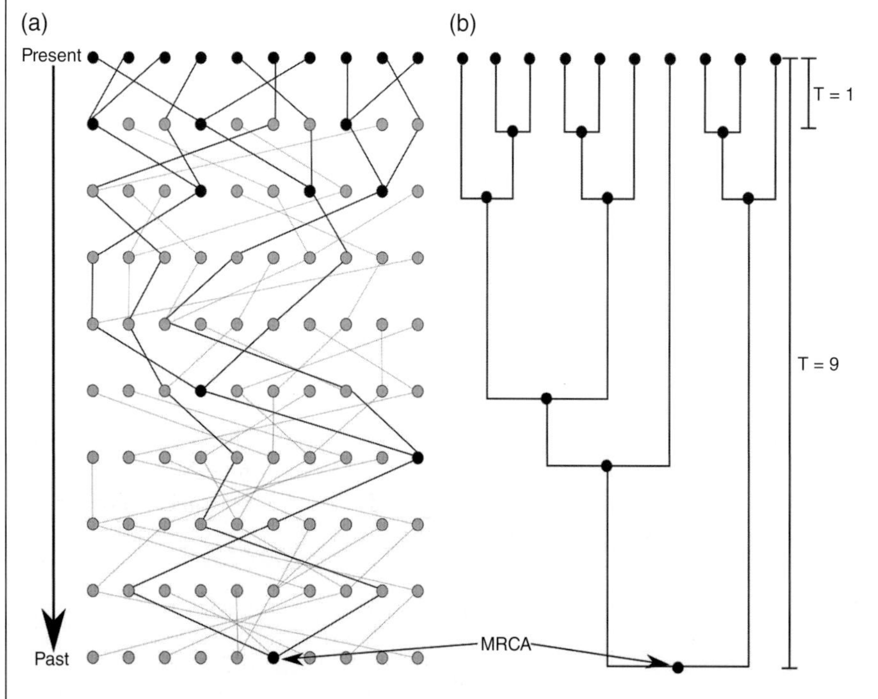

Figure B11.3.1 Example of coalescent events based on the complete genealogy of 10 sampled genes in a haploid population of fixed size (10 individuals) where the immediate ancestors for each gene are randomly selected from the proceeding generation. *Source:* Adapted from Rosenberg and Nordborg (2002).

Many studies have expanded on the basic coalescent to derive mathematical models that describe this process under a variety of demographic scenarios, such as increasing or decreasing population size (Slatkin and Hudson 1991; Griffiths and Tavaré 1994) or structured populations with gene flow (Notohara 1990; Beerli and Felsenstein 1999), which can all drastically affect shape of individual gene genealogies (Figure B11.3.2).

Using these models, genealogical samplers have been developed that estimate demographic parameters such as effective population size, gene flow, and population change parameters (e.g. degree of increase or decrease in size), for a given set of sampled genes (Table B11.3.1). Many of these models have been instrumental in extending population genetic analysis to infer demographic history for modern populations. Demographic inferences for coalescent analysis have ranged across time from the very recent, such as establishing the origins of viral strains (Pybus and Rambaut 2009) or paths of invasive species (Taylor and Keller 2007), to the demographic history of populations thousands of years in the past (Row et al. 2011; Dudaniec et al. 2012; Fouquet et al. 2012).

Table B11.3.1 Examples of coalescent samplers used to estimate population demography and gene flow between populations.

Program	Description	Main Assumptions	Platforms	References
`Migrate`	Estimate asymmetric gene flow and effective population size for interacting populations	Constant population sizes and migration rates	Win, MacOS, Linux	Beerli and Felsenstein (1999), Beerli (2006)
`Lamarc`	Estimate asymmetric gene flow and effective population size for interacting populations which may be increasing or decreasing	Constant migration rates; exponential increasing or decreasing populations	Win, MacOS, Linux	Kuhner (2006)
`IM`, `IMa2`	Estimate divergence time and gene flow between two populations that have split at some population in the past;	Two populations split with or without gene flow in the recent past	Win, Linux	Nielsen and Wakeley (2001), Hey (2010)

All programs can incorporate DNA sequence data and multi-allelic markers and assume a particular mutation model that can be defined.

of conservation concern, definitions differ among disciplines (Luikart et al. 2010). In population genetics, as originally conceived by R.A. Fisher (1930) and S. Wright (1931, 1978) effective population size (N_e) is broadly defined as the number of individuals of an idealized population – one with random mating (with each parent having the same expected number of progeny), and assuming no immigration or emigration, equal sex ratio, no selection, negligible mutation – that exhibits the same behavior in terms of inbreeding or genetic drift as the natural population under consideration (Fisher 1930; Wright 1931). The strict definition may seem to be overly restrictive but the intent is not to mirror real biological systems, but rather to have a "common currency" through which we can (i) compare populations and species with different life histories; and (ii) understand the genetic consequences of variation in population sizes. Beyond this general definition there are different conceptions of effective size that reflect different properties of genetic populations: (i) *inbreeding effective population size* emphasizing identity of descent of two randomly chosen alleles (Crow 1954; Crow and Denniston 1988); (ii) *variance effective population size* that reflects variance in gene frequencies due to drift (Crow 1954; Crow and Denniston 1988); (iii) *eigenvalue effective population size* that is based on a matrix of probabilities of changes in allelic states and is calculated as the maximum nonunit eigenvalue of this *transition matrix* (Ewens 1982); and (iv) *coalescent effective population size* based on tracking the history of individual genes using coalescence theory (Box 11.3; Sjödin et al. 2005). Below we discuss some methods for estimating contemporary effective population size and inferring temporal demographic patterns, although a detailed discussion of all of versions of N_e is beyond the purview of this chapter. Suffice to say that versions of N_e can vary in the means by which they are estimated, in the values that they assume for the same population (e.g. evolutionary equilibrium; Chesser et al. 1993; Wang 1997a, 1997b), and the degree to which they incorporate contemporary versus historical demographic changes (Schwartz et al. 1998; Wang 2005; Crandall et al. 2006).

Box 11.4 Effective Population Sizes and Gene Flow for Populations of Foxsnakes

Previous Bayesian clustering (Box 11.1) suggested three genetic clusters (Pelee Island, Point Pelee National Park [PPNP]-Hillman, Talbot) for eastern foxsnakes (*P. gloydi*) populations sampled in southwestern Ontario. Here, we compare capture-recapture analysis, ABC (`ONeSAMP`; Tallmon et al. 2008) and coalescent analysis (`migrate-n`, ver. 3.2.2; Beerli and Felsenstein 1999; Beerli 2006) to estimate census (N_c) and effective population size (N_e) for one local population (Hillman), and to estimate effective population size and gene flow among three identified genetic clusters.

We monitored the population at Hillman Marsh Conservation Area from 2006 to 2010 and each new individual captured was marked with a passive integrated transponder or PIT tag. We derived a census population size using a Jolly–Seber open population model (Cormack 2012) in `Rcapture` (Baillargeon and Rivest 2007). We estimated N_c for adults (>700 mm snout-to-vent length) to be 119.47 individuals (standard error 23.97). Although open population models are known to give biased estimates of abundance if capture heterogeneity is present, our sampling protocol did not fit a closed population model as the preferred model for estimation of abundance. To estimate N_e for this local population, we ran the ABC analysis for 50 000 iterations, and set minimum and maximum population sizes to 2 and 500, respectively. This analysis suggested an effective population size of $N_e = 34.2$ individuals (95% CI: 26.9–53.7), and thus a N_e/N_c ratio of 0.28, which is equal to the average value for small populations of conservation concern reported in Palstra and Ruzzante (2008).

We also used the ABC analysis to estimate the effective population size for each of the three clusters with the same parameters as above. We found the PPNP-Hillman cluster to have the largest N_e (mean: 43.3; 95% CI: 33.4–67.2), which was significantly larger (nonoverlapping confidence intervals) than the Talbot value (mean: 19.6; 95% CI: 15.5–28.54), but not the Pelee Island estimate (mean: 31.8; 95% CI: 25.1–49.41).

For these same samples, we estimated N_e and directional rates of migration using a coalescent analysis as implemented in `Migrate` (ver 3.2.20). We ran 10 independent replicates, each having a 2.0×10^7 burn-in and 4.0×10^8 iterations. We assessed convergence by evaluating the effective population size and the posterior distribution of migration and population size parameters and confirmed the results with multiple runs. The analysis estimated Θ ($4N_e\mu$) and a mutation-scaled effective immigration rate M (m/μ) for each of the genetic clusters (1 – Pelee Island; 2 – PPNP-Hillman; 3 – Talbot). To obtain population sizes we divided Θ by 4×0.0005, which is the average mutation rate for microsatellites (Estoup and Angers 1998) and obtained the absolute number of immigrants by Θ X M. Again, the coalescent analysis suggests the PPNP-Hillman cluster has the largest effective population size (Table B11.4.1).

Effective population sizes in general were larger for the coalescent analysis, except for the Talbot cluster, which had a much smaller size. The Pelee Island and Talbot clusters also have much higher effective immigration rates with most of the migrants coming from the PPNP-Hillman cluster. These higher rates are largely due to the fact that these populations had much smaller effective population sizes and thus smaller amounts of migration have a larger effect. In terms of actual migrants there was a larger number of individuals migrating from the Talbot and Pelee Island clusters into the PPNP-Hillman cluster, which would be consistent with the Bayesian clustering and assignment results.

Table B11.4.1 Estimation of effective population size (N_e) with coalescent analysis.

Parameter	Mode	95% CI	N_e	Number of Immigrants
Θ_1	0.23	2.28–4.20	115	NA
Θ_2	3.46	0.00–0.59	1730	NA
Θ_3	0.01	0.00–0.37	5	NA
$M_{1->2}$	3.00	NA	NA	10.38
$M_{1->3}$	0.07	NA	NA	<0.01
$M_{2->1}$	13.93	NA	NA	3.20
$M_{3->1}$	3.27	NA	NA	0.75
$M_{2->3}$	15.53	NA	NA	0.15
$M_{3->2}$	3.40	NA	NA	11.76

For population ecology, census population size (N_c) is the more obviously important parameter. Census population size has been variously estimated as (i) the total number of individuals, including juveniles, and breeding and nonbreeding adults; (ii) the total number of adults; or (iii) the number of breeding individuals only (Nunney and Elam 1994). Nunney and Elam (1994) suggest that N_c should be the number of adults where adulthood is "… defined by the likelihood of breeding." Authors should be unequivocal as to which definition they adhere to as it has important implications for how we understand the ratio of effective to census population size. A number of studies have compared the N_e/N_c ratio across taxa with N_e generally found to be smaller (Frankham 1995; Palstra and Ruzzante 2008), although the ratio of N_e/N_c varies enormously among species with divergent life history attributes, from 10^{-6} in oysters to almost 1.0 for humans (Frankham 1995). Luikart et al. (2010) caution that we know little about the temporal stability of this ratio for specific populations, the consistency of the N_e/N_c ratio among regions across species ranges, or the taxonomic generality of the N_e/N_c ratio among species sharing life history attributes. We are thus far from a general understanding of the relation between the N_e and N_c.

11.4.1 Estimating Census Population Size

Methods for estimating both N_c and N_e from genetic data, details on assumptions, and computational methodology have been reviewed by Charlesworth (2009) and Luikart et al. (2010) (Table 11.1). Here, we provide an overview of methods and issues but defer to these other works where appropriate. Traditional mark-recapture methods require initial capture and physical marks such as bands, passive integrated transponder tags, branding, or some other reasonably durable mark, including toe clips for amphibians and lizards, scale clips for reptiles, and fin clips in fish. After marking individuals, they are subsequently recaptured with N_c estimated from the proportion of marked versus unmarked recaptures relative to the number marked initially (Seber 1973; Chapter 7). For secretive, cryptic, fossorial, or otherwise challenging-to-study species such traditional marking approaches may be unfeasible or yield so few data that population size estimates are inaccurate. Genetic capture-recapture methods may resolve many of the logistical challenges using noninvasive DNA sampling of items like scat, hair, feathers, shed skins, or urine (Waits and Paetkau 2005; Beja-Pereira et al. 2009). Genetic capture-recapture methods rely on unique multilocus genotypes typically using high-resolution markers, such as microsatellites or SNPs, that can distinguish individuals. Of great value is the possibility of using samples collected over a short time span – what amounts to a single temporal sample – to estimate N_c with genotypes matched across samples treated as recaptures (Luikart et al. 2010). Values of N_c can be estimated using **rarefaction**, where a plot of the accumulation of unique genotypes approaches an asymptotic value of population size (Kohn et al. 1999; Frantz and Roper 2006), although new Bayesian or likelihood methods have largely displaced rarefaction in recent studies (Miller et al. 2005; Lukacs et al. 2007). These genetic approaches to estimating population size can lead to unique insights as in Hájková et al. (2008) where the population size estimate for the Eurasian otter (*Lutra lutra*) at one of their sites was 76 (95% CI = 49–96), twice the estimate of 38 derived using a method based on tracks in the snow.

Noninvasive DNA sampling and subsequent genotyping can also be used for traditional two- or multisample capture-recapture estimation of N_c (Chapter 7), treating genotypes like any other of the permanent marks indicated above. Unlike more traditional methods of capture-recapture models, there can be error in individual identification arising from **allelic dropout** or genotyping errors. However, this has become less of an issue with improved genotyping and genomics techniques irrespective of potentially degraded DNA samples (Luikart et al. 2010) and can be incorporated into the population model (Lukacs and Burnham 2005).

11.4.2 Estimating Contemporary Effective Population Size with One Sample Methods

There are two general approaches to estimating present-day or short-term N_e from genetic data: one-sample, and multisample or temporal methods. One-sample methods use contemporary signatures of population size that are reflected in key genetic parameters to estimate present-day N_e: **linkage disequilibrium** (LD) (Hill 1981), **heterozygote excess** (Pudovkin et al. 1996; Luikart and Cornuet 1999; Balloux 2004), relatedness or parentage (Wang 2009), or difference in allelic diversity between juvenile and adult cohorts (Hedgecock et al. 2007).

The LD method uses the nonrandom association between alleles at different loci to infer N_e (Hill 1981). LD can be caused by overlapping generations, selection, dispersal, or mixing of two genetically distinct populations or the **Wahlund effect** (Wahlund 1928), and genetic drift (Wang 2005). However, the LD method assumes that, in geographically isolated populations, **linkage** between neutral DNA loci will be influenced by drift alone and thus the amount of disequilibrium will be proportional to N_e (Hill 1981). A basic problem with this approach, and all one-sample approaches, is that it necessarily integrates and reflects historical rather than contemporary N_e. For example, the LD method better

Table 11.1 Programs for estimating census (N_c) and effective (N_e) population size, and diagnosing population bottlenecks using molecular data.

Program	Description	Main Assumptions	Markers	Platforms	Source
Estimating Census Population Size					
Capwire	Estimates census population size from genetic sampling.	Individuals must be identified with certainty.	MULT	R (all platforms)	Miller et al. (2005), Pennell et al. (2012)
Estimating Effective Population Size					
ONeSAMP	Estimates effective population size from a single sample using approximate Bayesian computation (ABC).	Population follows a Wright-Fisher model	STR	Web-based (all platforms)	Tallmon et al. (2008)
LDNe	Estimates N_e using linkage disequilibrium.	Mating is random or there is lifetime monogamy. Other assumptions consistent with a Wright-Fisher model.	MULT	Win	Waples and Do (2008)
Nb_HetEx	Offers two methods to estimate N_c: Using an excess heterozygosity and a temporal method (Waples 1989b).	Small number of breeding (<30); large number of sample progeny (>200). Assumes that loci are independent.	MULT	Win	Zhdanova and(2008)
NeEstimator	Estimates N_e using allele frequency data. Encodes 3 internal methods (two one-sample method, and one temporal method using moment-based F-statistics), and three third-party programs described below: TM3, MLNE and Mcleeps.	Various depending on the method employed.	MULT, mtDNA	Win	Ovenden et al. (2007)
TM3	Estimates N_e with a coalescent Bayesian-based inference.	Two samples at distinct times from a single closed (i.e. no immigration) population.	MULT	Win	Berthier et al. (2002)
MLNE	Estimates N_e and m (the migration rate) from temporal changes in allele frequencies using maximum likelihood.	Assumes an infinitely large source population from which immigrants flow into the focal population.	MULT	Win, Source code available.	Wang (2001); Wang and Whitlock (2003)
Mcleeps	Estimates N_e using temporal changes in allele frequencies from individuals sampled in different generations.	Assumes underlying Wright–Fisher model.	MULT	MacOS, Source code available.	Anderson et al. (2000)
CoNe	Employs a coalescent-based likelihood approach to estimate N_e using data from two temporal samples	Closed population with no selection or immigration and negligible mutation. Allele frequency changes arise from drift alone.	MULT	(all platforms)	Anderson (2005)
TMVP	As in TM3 but allows for more than 2 temporal samples.	Closed population with no selection or immigration and negligible mutation. Allele frequency changes arise from drift alone.	MULT	Win	Beaumont (2003)
TempoFs	F-statistic moment method based on multiple temporal genetic samples.	Closed population with no selection or immigration and negligible mutation. Allele frequency changes arise from drift alone.	MULT	Win, Linux	Jorde and Ryman (2007)
Diagnosing Population Bottlenecks and Size changes					
Bottleneck	Identifies the molecular signal of recent population bottlenecks using 3 different approaches.	Only identifies extreme and sustained bottlenecks.	MULT	Win	Cornuet and Luikart (1996)
M_P_Val.exe critical_M.exe	M_P_Val.exe calculates a statistic, M (ratio of the number of alleles to range in allele size) and simulates an equilibrium null distribution for comparison.	Mutations follow a version of stepwise mutation with a proportion p_s adding or deleting one repeat, and a proportion ($1-p_s$) with larger mutations.	STR	Win, MacOS	Garza and Williamson (2001)

Table 11.1 (Continued)

Program	Description	Main Assumptions	Markers	Platforms	Source
	`Critical_M.exe` uses sample size, the number of loci, and three parameter values for the two-phase mutation model and calculate a critical value, M_c.				
MSVAR	Estimates a current effective population size (N_0) that has changed from a historical population size (N_1) beginning at some point (ta) in the past. Can incorporate linear or exponential population size changes.	Assumes a demographic history of a single population undergoing expansion or decline following a coalescent model.	STR	Win, Web-based (all platforms)	Beaumont (1999)

Temporal methods typically assume negligible mutation over the time span considered. Typically they also require a known duration between samples measured in number of generations.
MULT: multi-allelic markers (e.g. SNPs or microsatellites); STR: short tandem repeats or microsatellites; mtDNA: mitochondrial DNA sequence.

estimates N_e when loci are tightly linked, but disequilibrium between such loci reflects the signature of deeper demographic events and integrates information on founder events, changes in population size, or dispersal (Rudge et al. 2009). Waples (1991) indicated that the LD method has greater power when N_e is small because the signal of LD will be proportionally large. Concomitantly, large sample sizes of >90 individuals may be required for reasonable estimates of N_e (Luikart et al. 2010) because if sample size is less than effective size there is marked downward bias on the estimate of N_e (England et al. 2006), which represents a challenge for the study of rare or cryptic species, although corrections for sample size biases have been developed (Waples 2006).

The heterozygote excess method depends on the observation that, in small populations of dioecious organisms, allele frequencies in females and males will differ because of binomial sampling error, with N_e or more accurately inbreeding N_e being a function of this disparity. Similarly, in small populations we expect an excess of heterozygosity among progeny due to nonrandom mating (i.e. departures from HWE – see Balloux 2004), which can be used to estimate the effective number of breeders (N_b) (Zhdanova and Pudovkin 2008). The **sibship** method for estimating effective population size proposed by Wang (2009) evaluates relatedness among offspring taken at random from a population and uses the frequency of inferred half- and full-siblings to infer N_e. Wang et al. (2010) developed a one-sample parentage assignment method for organisms with overlapping generations that requires a random sample from a focal population, and for each individual a multi-locus genotype and sex and age. Hedgecock et al. (2007) developed a rarefaction method which was used for European oysters, *Ostrea edulis,* where they simulated the effective number of parents required to produce allelic diversity in a juvenile cohort relative to their adult samples.

Tallmon et al. (2008) introduced a promising approach that may provide better estimates of contemporary N_e as it uses approximate Bayesian computation (ABC) (Beaumont et al. 2002). Their approach uses more of the information contained within the genetic data, estimating eight summary statistics, each of which have been shown to relate to effective population size, and deploying upper and lower bounds on N_e for simulated populations, each with an effective size drawn from a uniform distribution of random numbers between the lower and upper N_e priors. Initial genetic variation within each replicate simulated population is defined by theta, which is defined by $4N_e{*}\mu$ where N_e is the historical effective population size and μ is the mutation rate of the considered loci. Each simulated population is allowed to reproduce following a **Wright–Fisher model** for between two and eight generations (number determined by a random draw of a uniform distribution bounded by these values) before sampling an identical number of individuals and loci as contained in the observed dataset. The best value of N_e is the one where summary statistics of the simulated population are closest to that of the target population. We use this approach to compare estimates of contemporary N_e to historical N_e and census population sizes for isolated populations of eastern foxsnakes (Box 11.4).

11.4.3 Estimating Contemporary Effective Population Size with Temporal Sampling

Temporal methods require estimates of allele frequencies from two or more time periods (Waples 1989a), and thus necessarily require that populations have been sampled in the past. Generally, this class of estimators is based on the observation that the magnitude of change in allele frequencies over time brought about by genetic drift is inversely related to effective population size. Luikart et al. (2010) provide a detailed list and discussion of temporal methods that are based on (i) declines in heterozygosity (Harris 1989; Miller and Waits 2003); (ii) F (inbreeding) statistics (Nei and Tajima 1981; Jorde and Ryman 2007); (iii) a pseudo-maximum likelihood approach (Wang 2001); (iv) descent-based MCMC estimators (Anderson 2005); and (v) a coalescent Bayesian perspective (Beaumont 2003). Temporal methods, as well as most one-sample methods, typically assume nonoverlapping generations, no age structure, no migration, and negligible mutation. Luikart et al. (2010) systematically evaluated these assumptions, and the means to mitigate their impacts or methods most robust to violations of them. Ultimately conclusions regarding population size for use in management benefit from clear articulation of assumptions of any method deployed, a good understanding of the ecology of focal species, and comparison across estimators.

A need remains for detailed comparisons of the many one-sample and temporal methods of estimating N_e, but some nice empirical work has been done. Skrbinšek et al. (2012) compared four newer methods for estimating N_e in brown bears (*Ursus arctos*). They genotyped 510 bears of varying ages for 20 DNA microsatellites and organized them into age cohorts. Using the unbiased LD estimator of Waples (2006), ABC as per Tallmon et al. (2008), the sibship method of Wang (2009), and the parentage assignment method of Wang et al. (2010), they found that the four methods produced comparable results. Thus, they concluded that one-sample methods might be very useful for long-term monitoring of species of conservation concern. For example, the harmonic mean N_e across cohorts was 276 (95% CI: 183–350) for the parentage assignment method, and was similar to the median long-term N_e estimate of 305 (95% CI: 241–526) from their ABC analysis. Hoehn et al. (2012) also quantified N_e using four different methods for the Australian reticulated velvet gecko (*Oedura reticulata*). They concluded that single-sample estimators produced comparable values to temporal methods for estimating N_e, although the two temporal estimates exhibited substantial variation in confidence intervals with concerns regarding prior information incorporated in estimates.

11.4.4 Diagnosing Recent Population Bottlenecks

Due to seasonality and other environmental factors such as disease and drought, populations are known to fluctuate in size over time with obvious ramifications for demography and genetic diversity. **Population bottlenecks** are marked reductions in effective population size and can have profound direct and indirect consequences for population persistence (Frankham 2005). Smaller populations have greater probabilities of local extinction due to both intrinsic demographic factors such as demographic stochasticity, and extrinsic environmental variability (e.g. droughts, severe storms, changes in habitat; May 1973). Small populations are also more vulnerable to genetic stochasticity defined as the loss of genetic diversity through random genetic drift and inbreeding (Shaffer 1981) and may experience reduced mean fitness (Kalinowski and Waples 2002). Indeed when population sizes change over time, small populations may have a disproportionate impact on genetic diversity (Hedrick 2011). Because of the effect of size changes on genetic diversity, long-term effective population size should be estimated as the harmonic mean of N_e calculated at different time periods (Sjödin et al. 2005).

Genetic markers can be used to diagnose recent population bottlenecks because the events leave a distinct, albeit transient genetic signature. Thus, even in a population for which we have no current demographic insights, we can make inferences into some recent demographic trends. Populations that have undergone a bottleneck are expected to exhibit both reduced heterozygosity and allelic richness (Nei 1975), but we predict fewer alleles than expected from the observed heterozygosity, assuming that the population is at *mutation-drift equilibrium* (Maruyama and Fuerst 1985). The difference in allelic richness compared to that predicted from heterozygosity, and our ability to detect the genetic signature of a bottleneck, depend on time elapsed since the bottleneck began, the difference in N_e before and after the bottleneck, the number of genetic markers assayed, and the mutation rate of the markers (Maruyama and Fuerst 1985; Cornuet and Luikart 1996). Cornuet and Luikart (1996) and Luikart et al. (1998) developed three simple statistical tests to detect whether a population has a significant number of loci with a heterozygote excess as a signature of bottlenecks: (i) A Sign test on the difference between observed and expected heterozygosity across all loci; (ii) A standardized differences test, which compares the standardized difference between observed and expected heterozygosity to zero; and (iii) A Wilcoxon sign-rank test, which is a nonparametric paired method. In a different approach to the same question, Garza and Williamson (2001) showed that the mean ratio of the number of alleles to the range in allele size across assayed loci, denoted as M, also can be used to test for the signature of bottlenecks up to 100 generations after the initial population reduction. Significance is assessed using simulations and calculation of a critical value, M_c, under mutation-drift equilibrium and assuming a particular model of mutation (Garza and Williamson 2001). Population size changes can be similarly identified using coalescence theory (Box 11.3) and MCMC, to evaluate either population declines or expansion across longer periods (Beaumont 1999).

Bottleneck tests have not proved particularly effective at identifying known or strongly suspected population declines, primarily because of limited sampling or violations of some of the key assumptions (e.g. mutational models; Peery et al. 2012). Another caveat to this discussion is the aforementioned disparity in the relation between census and effective population size across populations and taxa, which means that changes in the N_c may not be immediately evident in the N_e (Frankham 1995).

11.5 Estimating Dispersal and Gene Flow

Since the conception of population genetics many summary statistics have been developed to estimate genetic differentiation among populations using allele frequencies. The most widely used estimator is F_{ST}, which was originally described by Sewall Wright in his derivation of F-statistics (Wright 1943, 1965) for two alleles. F_{ST} varies from zero to one and can be defined as the reduction in genetic diversity caused by population subdivision relative to the total population. When multiallelic markers became available, the derivation from a biallelic system would not work and many other fixation indices were derived that could describe differentiation at more than 2 alleles. The most popular of these is G_{ST}:

$$G_{ST} = \left(\frac{H_T - H_S}{H_T}\right) \tag{11.3}$$

where H_T is the total gene diversity (expected heterozygosity for diploid species) and H_S is the within population expected heterozygosity (Nei 1973). Some have raised concerns about this derivation because the index no longer ranges between zero and one, but has a maximum of $1 - H_s$ as H_t will always be larger than H_s. Thus, maximum G_{ST} is not independent of diversity, and when diversity is high, G_{ST} can have low values even when populations do not share any common alleles (Jost 2008; Meirmans and Hedrick 2011). This issue has led to several proposed solutions for more standardized measures of differentiation. For example, using the maximum possible G_{ST}, $G_{ST(max)}$, described as a function of H_S and number of populations (k):

$$G_{ST(\max)} = \frac{(k-1)(1-H_S)}{k-1+H_S}. \tag{11.4}$$

Hedrick (2005) proposed G'_{ST}, which standardizes G_{ST} and forces an upper limit of 1:

$$G'_{ST(\max)} = \frac{G_{ST}}{G_{ST(\max)}}. \tag{11.5}$$

Jost (2008) similarly recognized the problem of a lack of standardization in G_{ST}, but also suggested that expected heterozygosity is altogether unsuitable as it does not scale linearly with genetic diversity (i.e. asymptotes at 1). For example, given that expected heterozygosity is calculated as $1 - \sum_{i=1}^{k} p^2$ for k alleles, a change in two equally frequent alleles (p = 0.5 for each allele) to 20 alleles (p = 0.05 for each allele) results in a change of H_s from 0.5 to 0.95 and not the 10-fold increase that would be intuitive. Thus, he proposed a new measure of differentiation based on a "true diversity" quantifying the effective number alleles in a population. True diversity can be calculated from heterozygosity as $1/(1 - H_s)$ and this metric scales linearly with diversity – in the example above this would result in a change of 2–20 effective alleles. Using true diversity, Jost (2008) proposed the differentiation statistic D which can be expressed using heterozygosities and the total number of populations (k):

$$D = \left(\frac{k}{k-1}\right)\left(\frac{H_T - H_S}{H_s}\right). \tag{11.6}$$

This statistic is a purer measure of differentiation between populations and gives the proportion of each subpopulation's alleles that are unique to that subpopulation. In practice, we have found that many of these measures of genetic differentiation are highly correlated, but nevertheless it is wise to use multiple measures. Large differences in estimators can largely be explained (i.e. high diversity) and can lead to unique insights and assist with determining the appropriate statistic for a given question.

While most measures of genetic differentiation should correlate with gene flow (Neigel 2002) and there are obvious advantages of standardized estimates of genetic differentiation among populations, they are generally ineffective at quantifying the actual amount of movement occurring under most biological scenarios (Charlesworth 1998; Whitlock and McCauley 1999; Jost 2008). Given that actual dispersal rates will be of greater interest to population ecologists than any measure of genetic differentiation, we focus on the approaches that more directly estimate this parameter.

The terms "dispersal" and "gene flow" are often used interchangeably, but there is a critical distinction between the two. Dispersal is the movement of individuals, or their gametes, from one population to another. If that movement results in successful transfer of genes because matings or fertilizations produce offspring, then effective dispersal or gene flow has occurred. Dispersal does not result in gene flow if the dispersing individual does not reproduce. A variety of newer methods can estimate either recent dispersal patterns or more sustained levels of gene flow through time. Different approaches have even led some to compare recent and historical gene flow and make inferences about the impacts of human habitat alterations on population connectivity (Howes et al. 2008; Burbrink 2010; Chiucchi and Gibbs 2010). We discuss approaches to estimate (i) recent dispersal or gene flow events; or (ii) sustained levels of gene flow over long periods of time. We then introduce the recent application of network analysis to study genetic connectivity, which may be a particularly useful approach given its relevance to metapopulation dynamics and the similarity in analytical methodology between both approaches. All of these approaches require population grouping, which will largely depend on the biology of the species under study.

11.5.1 Estimating Dispersal and Recent Gene Flow

Within a Bayesian clustering analysis, the posterior probability of membership or admixture coefficients can be used to identify possible immigrants within a population. In our example, several individuals that were captured on the mainland (PPNP) have admixture coefficients that suggest island origins (Box 11.1). Instead of simply basing the identification of migrants on qualitative assessments of admixture coefficients, Pritchard et al. (2000) suggested a formal test to determine the probability that an individual or its ancestors are migrants. In theory, their approach can provide evidence of migrant ancestry up to G generations in the past, but in practice the power to detect ancestry beyond a few generations (G = 2) is low (Pritchard et al. 2000). Their model assumes that most individuals originate from their sampled population, but incorporates a small probability that an individual is an immigrant or has an immigrant ancestor. Because of these assumptions, the method is most informative when the number of clusters matches the number of sampling locations, with few migrant individuals. For our foxsnake populations it was therefore necessary to reclassify the dataset to designate individuals from geographically proximate mainland locales, Hillman and PPNP, as coming from the same location because they could not be separated in our original Bayesian clustering analysis (Box 11.1). Subsequently, we used an assignment test to identify individuals that were migrants or had migrant ancestry (G = 2) among the three locations. Using this approach we diagnosed four migrants in the PPNP-Hillman cluster; two of the southern migrants were from Pelee Island and two migrants in the north were from the Talbot population (Figure 11.3). A lack of migrants going from PPNP-Hillman to either Talbot or Pelee Island pointed to an apparent asymmetry in movement, suggesting that PPNP-Hillman may be acting as a sink. This fits with the results from our Bayesian clustering analysis (Box 11.1), which implies some admixture within the PPNP-Hillman cluster but little within the Pelee Island or Talbot clusters.

In addition to testing for the probability on an individual belonging to a particular cluster in a Bayesian analysis (Rannala and Mountain 1997; Pritchard et al. 2000), there are a variety of other approaches to identify recent migrants using genotypic data. One method simply calculates individual pairwise genetic distance (Takezaki and Nei 1996), and identifies migrants as individuals within a priori defined populations that are more closely related to individuals from a population different from the one in which they were sampled (Cornuet et al. 1999). Likelihood-based approaches, which maximize the likelihood of an individual belonging to a particular population, are also available (Paetkau et al. 1995). In a comparison of all three methods, Cornuet et al. 1999 found the Bayesian and likelihood methods provided much greater power to assign unknown individuals to their population of origin than the distance-based method. The power of such assignment tests increases with the sample size for each population and when all populations exchanging migrants have been sampled (Paetkau et al. 2004).

In many cases, it will be more useful to determine a recent migration rate as opposed to identifying individual migrants. Likely for this reason, the method introduced by Wilson and Rannala (2003) has been particularly well used (Vignieri 2005; Johnson et al. 2007). Because the underlying model in this approach incorporates and estimates an inbreeding coefficient, it does not require populations to be in HWE. The model does assume low to moderate migration (less than 30% of the population composed of migrants), and that migration has been constant and genetic drift negligible over the past three generations where the migration rates are being estimated. In an independent sensitivity analysis, Faubet et al. (2007) found that, if differentiation was not too low (F_{ST} > 0.05) and model assumptions were not violated, the approach provided relatively accurate migration estimates.

11.5.2 Estimating Sustained Levels of Gene Flow

In contrast to assignment methods, which attempt to identify individual migrants in extant populations or establish recent migration rates from allele frequencies, coalescent approaches (Box 11.3) estimate population parameters such as effective population size or gene flow, over much longer periods of time. There are various considerations before choosing the most appropriate coalescent model for estimating demographic parameters (Table B11.3.1). In general, coalescent approaches search the parameter space of an assumed demographic model to derive population parameter estimates and associated probability distributions, given the observed data (i.e. *gene genealogies* of sampled genes). It is important to understand the underlying demographic or evolutionary models and accompanying assumptions to determine whether they match with the biological underpinnings of collected data, the ecology of the focal species, and the research questions being asked.

Because of the mathematical complexity of even simple coalescent models, parameter estimates cannot typically be calculated directly and the methods used to derive parameter estimates vary. Parameters are typically estimated using MCMC algorithms, with parameter searching guided by either a likelihood function or a Bayesian approach (Metropolis et al. 1953; Hastings 1970). Beerli (2006) provides a nice overview and comparison of likelihood and Bayesian approaches using the same migration model. For both methods, it is important to assess convergence of the model parameter estimates and if done properly, they should yield similar results (Beerli 2006). In part, this can be done by assessing the consistency of multiple runs with different starting parameters and prior probabilities in the case of Bayesian analysis. We provide an example of a coalescent-based approach for estimating gene flow between populations of eastern foxsnakes (Box 11.4).

In addition to providing the basis for model-based approaches, coalescent theory can be used to simulate generate genealogies under an assumed demographic and mutational model (Hudson 2002; Laval and Excoffier 2004; Carvajal-Rodríguez 2008). Using these simulations, expected distributions of genetic summary statistics for the simulated data can be calculated and compared to observed data to determine the likelihood of, or power to detect, a given demographic scenario (Thalmann et al. 2007; Rutledge et al. 2012). To this end, some have advocated for ABC to be used to make statistical comparisons between generated and observed summary statistics and to compare coalescent simulations generated under competing demographic models (Beaumont et al. 2002). This approach has many advantages over the more traditional model-based approaches due to the flexibility of the demographic models that can be specified, the possibility of incorporating prior information, and the ability to compare competing evolutionary and demographic models (Bertorelle et al. 2010; Csilléry et al. 2010). Typically, model comparisons have been employed to test among models with different invasion scenarios (Pascual et al. 2007), population status as stable, increasing or decreasing populations (François et al. 2008; Row et al. 2011), or types of population subdivision (Peter et al. 2010). If the demographic history of sampled populations is known or has been selected through a model comparison approach, the posterior distributions can also be used to derive parameter estimates of splitting times and/or effective population sizes. Potential limitations to ABC include a lack of power in model comparisons (Robert et al. 2011), and care must be taken to choose appropriate genetic summary statistics and conduct proper model checking and validation (Bertorelle et al. 2010). Eliminating the summary statistics altogether, and instead using full allelic distributions, may also address these deficiencies (Sousa et al. 2009).

In both ABC and model-based approaches, a variety of types of genetic markers can be incorporated into the analysis. In fact, both approaches can incorporate more than one type of marker in the same analysis, which may increase the power to establish demographic history across spatial and temporal scales (Peters et al. 2008; Wegmann and Excoffier 2010). Note that because coalescence results from random processes, there can be large variation among individual genealogies and thus it is important to sample a number of unlinked genes to quantify variation and derive a confidence interval for a given result (Rosenberg and Nordborg 2002). This has led some to debate the merits of using more individuals versus obtaining more data from other unlinked genetic loci, with most researchers suggesting that increasing the number of loci at the expense of increasing the number of sampled individuals leads to more accurate parameter estimates (Felsenstein 2006).

11.5.3 Network Analysis of Genetic Connectivity

Most of the population approaches we have discussed so far largely involve pairwise comparisons between populations, but since Dyer and Nason's (2004) key paper introduced population graphs, the use of network-based approaches in population genetics has increased. Network methodology relies on graph theory, which is widely applied in many disciplines (e.g. sociology, transportation, biology, information systems, and physics) and to address a variety of research problems. Thus, the statistical and mathematical underpinnings of graph theory are well developed. In general, networks are composed of nodes or vertices that are connected by a set of arcs or edges. In population genetics, nodes generally represent populations with edges characterizing some form of genetic connectivity. Through systematic pruning or the removal of edges, one can gain insight into the genetic structure, connectivity of populations, and the importance of individual nodes.

In their original introduction of networks to population genetics, Dyer and Nason (2004) proposed "population graphs" where pruning is done to a saturated graph, where all populations are connected, in a manner which minimizes the genetic covariance structure. In their method, individuals are first defined by their multilocus genotype within multivariate space, and pairwise distances between populations (nodes) are derived from the centroids of individuals within multivariate space that comprise each population. Following Magwene (2001), Dyer and Nason (2004) transformed a population distance matrix (d_{ij}) to a standardized precision matrix (r_{ij}) where off diagonal elements represent the partial correlation coefficients. Values that do not significantly deviate from zero represent connections that are conditionally independent such that their removal (i.e. pruning) will not affect the overall genetic covariance structure. To date, this method of pruning genetic networks has been utilized in landscape genetics to identify barriers to dispersal (Giordano et al. 2007) or local factors affecting migration rates (e.g. snow depth; Garroway et al. 2008). However, there are several other methods to build (e.g. minimum spanning network, Gabriel graph; Dale and Fortin 2010) or prune (e.g. percolation threshold; Stauffer and Aharony 1994; Rozenfeld et al. 2008) genetic networks that may be equally useful for displaying and summarizing genetic patterns.

The overall network, as well as node-specific characteristics, can also provide valuable insights into population structure and dispersal patterns. For example, nodes with many edges linking them to other nodes are disproportionately important for the maintaining overall connectivity of the metapopulation; the *degree* of a node connotes the total number of nodes to which it is joined by edges. Assessing the degree values in a pruned network of genetic distances for populations across the range of a clonal sea grass, Rozenfeld et al. (2008) found a small number of populations with high degree values suggesting that these populations were acting as important centres of gene flow and were essential to maintaining connectivity across the entire system. In contrast, Garroway et al. (2008) found a well-connected genetic network for fishers (*Martes pennanti*) across northern Ontario with comparable degree values across all populations. This pointed to a high level of resiliency in fishers to the loss of any particular node (population) on the network.

Networks can also be used to test predictions or to make comparisons among regions or within and among species. For example, tests can be carried out to determine if the number of connections in graphs pruned using genetic covariance significantly deviate from the saturated graph or to test for the effects of a particular habitat or landscape boundaries. Giordano et al. (2007) used this approach and after pruning a network with conditional independence found significantly fewer edges among than within high- and low-altitude populations of long-toed salamanders (*Ambystoma macrodactylum*), suggesting restricted gene flow between populations at different elevations in Montana. Fortuna et al. (2009) compared networks for four Mediterranean woody plant species across the same fragmented landscape in southern Spain. The importance of any particular habitat patch varied drastically among species, implying difficulties for devising a multispecies conservation plan. Certainly the ability to include local (at-site) and regional (between-site) characteristics and examine their combined effects on metapopulation dynamics, make network-based approaches attractive for combining population ecology and genetics.

11.6 Software Tools

Throughout this chapter we outlined many different individual- and population-based approaches that were developed to (i) quantify population structure using individual genotypes; (ii) estimate census and effective population sizes and their trends; and (iii) estimate gene flow and dispersal between populations. Many of these approaches have accompanying programs that can be used for analyses of genetic data. Although the availability of these programs increases the ability of nonexperts to conduct complex analyses, their sheer number can be overwhelming. Here we outline a few of the major platforms for conducting some of most common analyses and summarize this information in three tables.

11.6.1 Individual-based Analysis

Bayesian clustering approaches are a popular first approach for population genetic analyses, and various programs have been developed (Table 11.2). Although the objectives of the analyses underlying these programs are similar, many of the assumptions are different and they can accommodate varying types of data. The largest distinction among the major programs is whether they incorporate geographic locations into the clustering algorithm, such as `Tess` (Chen et al. 2007) or `Geneland` (Guillot et al. 2005), or only utilize genotypes with or without a population identifier (`Structure`, `Geneclass2`). Incorporating information on sample provenance can increase the spatial resolution (Guillot et al. 2005b), but it often does not make sense for well-spaced population level data, or clustered sampling. Further, when utilizing a population identifier, both `Structure`

Table 11.2 Bayesian clustering and assignment test programs used to identify genetic clusters and estimate dispersal.

Program	Description	Assumptions	Spatial	Platform	Source
`Structure`	Identifies genetic structure using Bayesian clustering of individual genotypes or identifies migrants using population analysis.	HWE and LE within clusters; can allow admixture between clusters; recent versions can accommodate some linkage	Population group	Win, MacOS, Linux	Pritchard et al. (2000)
`Geneland`	Identifies genetic structure using Bayesian clustering of individual genotypes; new versions can incorporate phenotypic information.	HWE and LE within clusters; can incorporate null alleles; can allow admixture between clusters	Individual locations	R (all platforms)	Guillot et al. (2005b)
`Baps`	Identifies genetic structure using Bayesian clustering of individual genotypes or identifies migrants and admixed individuals using population analysis.	HWE within clusters; can allow admixture between clusters	Population group	Win, MacOS, Linux	Corander et al. (2004, 2008a, b)
`Tess`	Identifies genetic structure using Bayesian clustering of individual genotypes.	HWE within clusters; can allow admixture between clusters	Individual locations	Win, MacOS,	Chen et al. (2007)
`Geneclass2`	Assigns unknown individuals to a priori defined populations and identifies first generation migrants using three methods.	Different for each of the three methods	Population group	Win, Linux	Piry et al. (2004)
`BayesAss`	Uses Bayesian inference to identify recent migrants (>3 generations) and estimate directional migration rates.	LE but allows for inbreeding within populations (i.e. deviations from HWE); Proportion of immigrants <30%	Population group	Win, MacOS, Linux	Wilson and Rannala (2003)
`BIMr`	Uses Bayesian inference to identify recent migrants and estimate directional migration rates, and relates patterns to environmental variables.	LE but allows for inbreeding within populations (i.e. deviations from HWE)	Population group	Win, MacOS, Linux	Faubet and Gaggiotti (2008)

All programs can utilize unlinked multi-allelic, such as microsatellite and SNP markers with no assumed mutation model.
HWE: Hardy–Weinberg Equilibrium; LE: Linkage equilibrium.

and `Geneclass2` allow for specific tests to identify immigrants in sampled populations. The program `Geneclass2` (Piry et al. 2004) is more complete in this aspect and implements three different methods to identify first-generation migrants in a sample of populations: (i) a method utilizing an array of individual pairwise genetic distances (Cornuet et al. 1999); (ii) a method that uses Bayesian criteria and is similar to those used in the migrant test of `Structure` (Rannala and Mountain 1997); and (iii) implements a likelihood-based approach which maximizes the likelihood of an individual belonging to a particular population (Paetkau et al. 1995).

Other programs that incorporate individual level data, include both `Genalex 6` (Peakall and Smouse 2006) and the `R` package `adegenet` (Jombart 2008), which can be used for PCA and other multivariate analysis. In particular, `adegenet` is a versatile and expanding `R` package that can conduct spatial PCA analysis and incorporate various types of data, including genomic level SNPs. Although `Genalex 6` (Peakall and Smouse 2006) is based in `Excel`, it incorporates an array of analyses including a spatial autocorrelation analysis across defined distance classes (Peakall et al. 2003). The program `SPAGeDi` (Hardy and Vekemans 2002) can similarly conduct spatial autocorrelation analysis using a number of different individual-based genetic distance measures and tests for deviations from random for any set of geographic distance classes, using permutation tests. For example, we used `SPAGeDi` to determine the extent of significant, positive correlation (Moran's *I*) for Canada lynx (*Lynx canadensis*) genotypes distributed across Manitoba to Quebec (Figure 11.3).

11.6.2 Population-based Population Size

The software for analyzing census and effective population sizes is similarly vast and we have summarized many of the available programs in Table 11.1. For capture-recapture, a newer `R` package `capwire` implements the likelihood methods of Miller et al. (2005) to estimate N_c (Pennell et al. 2013). This approach allows researchers to: (i) obtain maximum likelihood estimates of census population size; (ii) select between two capture rate models (Equal Capture Model versus Two-Innate Rates

Model) using a log likelihood ratio test; and (iii) estimate 95% confidence intervals of the population estimate using parametric bootstraps.

Effective population size estimators can largely be classed into single-sample and multitemporal sample estimators (Table 11.1). A good starting point is the program NeEstimator (Do et al. 2014), as it implements three single-sample methods as well as a temporal estimator. Because this program incorporates multiple approaches, comparisons of effective population sizes using different estimators can be easily accomplished. Currently, ONeSAMP (Tallmon et al. 2008) is the only program available that utilizes ABC, but this program was recently suggested to overestimate effective population sizes for small populations and has much longer run times than many comparable programs (Gilbert and Whitlock 2015). This led the authors to recommend against using it and instead suggested the LD method available in NeEstimator (Do et al. 2014). If temporal data are available, MLNe was suggested as the best option (Wang 2001; Wang and Whitlock 2003). Because this method and program also estimates migration from a source population, it was found to be less sensitive to higher immigration.

There are also some programs that allow, and explicitly test for changes in population size. Two options for use with microsatellites are the program MSVAR (Beaumont 1999) and more recently the R package VarEff (Nikolic and Chevalet 2014). Because VarEff relies on an approximation of the exact likelihood, it is much faster than MSVAR and thus the effect of mutational model and priors can be tested more efficiently.

11.6.3 Dispersal and Gene Flow

Multipopulation programs that estimate gene flow between populations typically utilize coalescent approaches that search the parameter space of an assumed genetic model to derive population parameter estimates given the observed gene genealogies of sampled genes. These models can range from migration between stable populations at mutation-drift equilibrium (Migrate, Beerli and Felsenstein 1999, 2001; IM, Nielsen and Wakeley 2001), to more complex models that incorporate and estimate population size changes through time (Lamarc, Kuhner 2006). Given the rise in popularity in ABC, several programs that incorporate coalescent simulations into an ABC framework have been recently developed. These programs range from standalone user-friendly programs (DIYABC, Cornuet et al. 2010) to collections of packages and scripts that assist with running simulations and comparing the output of simulations and observed data (ABCtoolbox, Wegmann et al. 2010). Along these lines, the R packages abc (Csilléry et al. 2012) and easyABC (Jabot et al. 2015) offer R solutions to running ABC analyses, but rely on external simulation software to conduct the demographic simulations. Given that all these programs allow for the comparison and parameterization of customized models, more biologically realistic models can be derived and parameters such as effective population size and gene flow can be estimated. However, as always there is a trade-off, as increasing complexity and number of parameters will invariably lead to longer run times and concomitantly diminished power. Moreover, complexity may also suffer from problems of failure to converge.

11.7 Online Exercises

The online exercises for our chapter illustrate both individual and population-based genetic analyses, for example, data from different populations of foxsnakes (*Pantheris* spp.) and Canada lynx (*Lynx canadensis*). Exercise 1 uses Bayesian models in the Geneland package of Program R to examine spatial clustering and genetic structure among foxsnakes that were genotyped with microsatellite markers. Exercise 2 is an example of spatial clustering of simulated genotypes across a climatic cline. Exercises 3 and 4 use sPCA to identify spatially explicit genetic structure with the tools of the adegenet package. Exercises 5 and 6 use Bayesian models in the VarEff package to estimate effective population size and population trends based on coalescent analysis.

11.8 Future Directions

In this chapter, we have highlighted the genetic techniques that can reveal key insights likely to be of most interest to population ecologists. Despite this perhaps asymmetrical focus on the influence of genetics in ecology, both theoretical and empirical population genetics would also benefit from incorporating more of the biological complexity embedded within population ecology. Certainly, there have been recent suggestions that some of the simplistic demographic scenarios such as idealized Fisher–Wright populations and stable population sizes may compromise our ability to make truly insightful biological inferences using population genetic analyses (Neuhauser et al. 2003; Lambert 2010). Genetic simulations have revealed the relevance of population stochasticity and density-dependent dispersal on the effects of genetic drift and the evolution of spatial genetic structure, independent of mean population characteristics (Kaitala et al. 2006; Björklund et al. 2010).

Beyond simply incorporating more biological realism into population genetics, there are other benefits to be gained from merging demographic and genetic inferences in empirical research. Perhaps the value of a more integrated approach is best illustrated by the utility of both demographic and genetic information for accurately estimating the viability of populations (Reed et al. 2002; Traill et al. 2010) or even their geographic definition (Wells and Richmond 1995; Palsbøll et al. 2007; Lambert 2010), usually involving some form of demographic independence. Theory suggests that dispersal rates below 10% lead to demographic independence (Hastings 1993), but there are few empirical data to support this assertion (Waples and Gaggiotti 2006; Lowe and Allendorf 2010). Given that many life history attributes and population size itself all likely influence the impact of dispersal on local demographics, the effect of a 10% dispersal rate is likely to vary widely among species and even populations. Only through demographic and genetic monitoring over multiple years will we begin to understand the genetics of demographic independence. With increased awareness of the power of molecular insights, more ecologists are collecting genetic material, and shared databases have emerged for both genetic (Benson et al. 2005) and demographic information (NERC Centre for Population Biology, IC 2010), so these data are likely to become increasingly available in the coming years. Over the shorter term, spatially explicit demographic simulations that explore the relationship between complex demography and allele frequencies under different life history and demographic scenarios would help shed light on benefits of a more complete synthesis of these fields.

A recent merging of landscape ecology and population genetics approaches has led to the new field of landscape genetics (Storfer et al. 2007; Holderegger and Wagner 2008; Parisod and Holderegger 2012). Much of the research in this field has identified key landscape and other environmental attributes that promote or impede dispersal among populations for a variety of species. Given the importance of landscape connectivity in population persistence and viability (Brooker and Brooker 2002; Bonte et al. 2004; Stevens and Baguette 2008), this information is invaluable to conservation initiatives. Some recent research, however, has suggested that overall population connectivity may be similarly impacted by local factors such as the density of interacting species (Manier and Arnold 2006) or fine-scale environmental characteristics (Murphy et al. 2010; Veysey et al. 2011). Some of this importance is undoubtedly tied to the attractiveness of site characteristics for dispersing individuals, but the demographic characteristics of a given region, such as the production of an excess or deficiency of individuals, likely have large effects on dispersal and overall population connectivity (Dias 1996; Figueira 2009). Only through quantifying local demographic characteristics and species interactions, and how they influence and interact with dispersal, can we truly understand the complexity underlying landscape connectivity and source-sink dynamics.

Although a complete amalgamation of population genetics and population ecology is not likely, there are many overlapping objectives and both disciplines have much to benefit from a more combined approach. NGS will vastly increase the ease with which large genetic datasets can be generated and, combined with statistical and computational advances, this will augment the power of genetic analysis for demographic inference, and allow for more complex and demographically realistic genetic models. As this occurs, the lines between many aspects of population ecology and population genetics will likely blur, leading to a more synthetic understanding of the factors influencing population dynamics and shaping patterns of genetic diversity.

Glossary

Allelic dropout The absence of one or more alleles in an individual genotype due to technical issues that include failure to amplify because of degraded DNA template or because of mutations in the DNA priming site.

Demographic independence Two populations are demographically independent when the demographic characteristics of one does not influence the other. As migration between populations increases, demographic independence decreases.

Effective population size (N_e) The size of an ideal population in which one would see the same rate of evolution via genetic drift as in the real, natural population under study. An ideal population is one comprised of *N* diploid, hermaphroditic (and self-compatible), and random mating individuals with nonoverlapping generations. In such ideal populations, each individual has the same probability of contributing to the subsequent generation. Other assumptions include no selection and negligible mutation.

Gene flow Simply the movement of genes between populations. This involves the migration of individuals from one population to another and subsequent successful mating, or the movement of gametes (e.g. plant pollen) between populations and subsequent successful fertilization.

Gene genealogy Representation of ancestry and descent or of evolutionary affinities for a particular gene (usually represented as a tree or network).

Genetic cline A geographic gradient in the frequency of alleles or mean trait value. Clines may arise due to secondary contact of previously geographically isolated populations, or due to selection or restricted dispersal across a gradient in environmental features (called a primary contact zone).

Genetic drift A process driven by random sampling of gametes each generation that can produce changes in allele frequencies over time. Genetic drift is more pronounced in small versus large populations. All else being equal, the probability of fixation of any given allele in a population by drift alone, given sufficient time, is its frequency at any considered time.

Genetic locus The specific location or position of a gene or other DNA sequence (e.g. microsatellite) on a chromosome. Loci for a group of individuals can have many different variants, referred to as alleles, and can be used to derive a genotype.

Genotype Often tacitly assumed to be the complete suite of genes across all loci for a particular individual. Alternatively, it is the allelic composition of a locus or set of loci under study.

Haplotype Alternate forms of a particular DNA sequence or gene. Often used in reference to variants of genes of haploid genomes of organelles (e.g. the mitochondrion in vertebrates).

Hardy–Weinberg Equilibrium (HWE) The expected genotypic proportions in an ideal population (characteristics of which include random mating, no selection, and negligible mutation). The HWE may be treated as a "null" hypothesis, significant statistical departures from which may indicate the action of such processes as selection or inbreeding. Expected genotypic frequencies are predicted from a binomial expansion of the allele frequencies. For example, for two alleles A_1 and A_2 with frequencies p and q, the expected respective genotypic proportions of A_1A_1, A_1A_2 and A_2A_2 are $(p + q)^2 = p^2 + 2pq + q^2$.

Heterozygote excess A higher proportion of observed heterozygotes (i.e. two different alleles on different homologous chromosomes at the same locus) than expected under HWE.

Intron Regions of DNA within a gene that are transcribed, but do not code for protein. Introns are spliced out to produce a mature RNA transcript comprised only of coding exon sequences.

Isolation-by-distance A pattern where populations that are closer geographically, are less genetically differentiated, than populations further apart (Wright 1943; Malécot 1948). More recently this pattern has been extended to include the differentiation or relatedness among individuals (Rousset 2000).

Linkage The physical association of two or more loci on a chromosome such that they do not assort independently during meiosis and gamete formation.

Linkage disequilibrium (LD) The nonrandom association between alleles at two or more genetic loci. Disequilibrium can be caused by selection, genetic drift, or mixing of two genetically differentiated populations. The persistence or decay of linkage disequilibrium is related to the physical genetic linkage between considered loci. Note that the term is a bit misleading as LD can occur between loci that are unlinked, and physically linked loci can achieve LE.

Linkage equilib rium (LE) The genotype at one locus is independent of the genotypes present at other loci.

Markov chain Monte Carlo Algorithms used to explore probability space of defined models. All the algorithms are based on the Markov chain, which is a random process where each new state of a variable is based on its current state only, and not any previous values.

Mutation-drift equilibrium A situation where the loss of genetic diversity within populations due to genetic drift is balanced by its gain through new mutations. Populations which have undergone large, recent changes in size or which are expanding in geographic range are likely not at mutation-drift equilibrium.

Next-Generation Sequencing (NGS) Post-Sanger (Sanger et al. 2007) DNA sequencing techniques, which allow for the massively parallel sequencing of many short DNA sequences from a single sample. NGS sequencing can produce numbers of DNA sequences (reads) in the tens or hundreds of millions.

Panmixia Any population in which breeding is at random (i.e. uninfluenced by the genotypes of breeding individuals or spatial structure).

Phylogeography The study of the geographic distribution of genetic variation emphasizing the evolutionary (genealogical) relationships among populations.

Polymerase chain reaction Molecular technique used to amplify a targeted region of DNA from a few copies to several thousands or millions of copies. It may involve a single target sequence or multiple sequences (termed a multiplex reaction).

Population bottleneck A severe, temporary reduction in effective population size.

Rarefaction (curve) A curve describing the relationship between genotyped genetic samples and the number of unique genotypes (different individuals). In population genetics, typically used with noninvasive genetic sampling to estimate census population size (N_c).

Sibship Either a group of individuals related by descent from a common ancestor or a group of siblings. The relatedness among a sample of individuals using genotypic data can be used in the estimation of effective population size.

Voronoi tessellation deconstructs a landscape into a number of spatial polygons with no overlap or gaps. The polygons are typically built using either free Voronoi tessellation, which builds a defined number of polygons independent of sampling locations or constrained tessellation where polygons are built using the sampling locations.

Wahlund effect Named after Sven Wahlund (Wahlund 1928). Indicates departures from Hardy–Weinberg expectation (e.g. heterozygote deficiencies) because of mixing of two or more temporally or spatially genetically distinct populations.

Wright–Fisher model A theoretical ideal population of finite and constant size, with random mating, nonoverlapping generations, and selective neutrality for the markers considered.

References

Aasen, E. and Medrano, J.F. (1990). Amplification of the Zfy and Zfx genes for sex identification in humans, cattle, sheep and goats. *Nature Biotechnology* 8: 1279–1281.

Anderson, E.C. (2005). An efficient Monte Carlo method for estimating N_e from temporally spaced samples using a coalescent-based likelihood. *Genetics* 170: 955–967.

Anderson, E.C., Williamson, E.G., and Thompson, E.A. (2000). Monte Carlo evaluation of the likelihood for Ne from temporally spaced samples. *Genetics* 156: 2109–2118.

Austin, J.D., Lougheed, S.C., Neidrauer, L. et al. (2002). Cryptic lineages in a small frog: the post-glacial history of the spring peeper, *Pseudacris crucifer* (Anura: Hylidae). *Molecular Phylogenetics and Evolution* 25: 316–329.

Baillargeon, S. and Rivest, L. (2007). `Rcapture`: loglinear models for capture–recapture in R. *Journal of Statistical Software* 19: 1–31.

Balloux, F. (2004). Heterozygote excess in small populations and the heterozygote-excess effective population size. *Evolution* 58: 1891–1900.

Barrett, E.L.B. and Richardson, D.S. (2011). Sex differences in telomeres and lifespan. *Aging Cell* 10: 913–921.

Baums, I.B., Miller, M.W., and Hellberg, M.E. (2005). Regionally isolated populations of an imperiled Caribbean coral, *Acropora palmata*. *Molecular Ecology* 14: 1377–1390.

Beaumont, M.A. (1999). Detecting population expansion and decline using microsatellites. *Genetics* 153: 2013–2029.

Beaumont, M.A. (2003). Estimation of population growth or decline in genetically monitored populations. *Genetics* 164: 1139–1160.

Beaumont, M.A., Zhang, W., and Balding, D.J. (2002). Approximate Bayesian computation in population genetics. *Genetics* 162: 2025–2035.

Beerli, P. (2006). Comparison of Bayesian and maximum-likelihood inference of population genetic parameters. *Bioinformatics* 22: 341–345.

Beerli, P. and Felsenstein, J. (1999). Maximum-likelihood estimation of migration rates and effective population numbers in two populations using a coalescent approach. *Genetics* 152: 763–773.

Beerli, P. and Felsenstein, J. (2001). Maximum likelihood estimation of a migration matrix and effective population sizes in *n* subpopulations by using a coalescent approach. *Proceedings of the National Academy of Sciences USA* 98: 4563–4568.

Beja-Pereira, A., Oliveira, R., Alves, P.C. et al. (2009). Advancing ecological understandings through technological transformations in noninvasive genetics. *Molecular Ecology Resources* 9: 1279–1301.

Benson, D.A., Karsch-Mizrachi, I., Lipman, D.J. et al. (2005). GenBank. *Nucleic Acids Research* 33: D34–D38.

Bergl, R.A. and Vigilant, L. (2007). Genetic analysis reveals population structure and recent migration within the highly fragmented range of the Cross River gorilla (*Gorilla gorilla diehli*). *Molecular Ecology* 16: 501–516.

Berthier, P., Beaumont, M.A., Cornuet, J.-M., and Luikart, G. (2002). Likelihood-based estimation of the effective population size using temporal changes in allele frequencies: a genealogical approach. *Genetics* 160: 741–751.

Bertorelle, G., Benazzo, A., and Mona, S. (2010). ABC as a flexible framework to estimate demography over space and time: some cons, many pros. *Molecular Ecology* 19: 2609–2625.

Birky, C. (1995). Uniparental inheritance of mitochondrial and chloroplast genes: mechanisms and evolution. *Proceedings of the National Academy of Sciences USA* 92: 11331–11338.

Björklund, M., Bergek, S., Ranta, E., and Kaitala, V. (2010). The effect of local population dynamics on patterns of isolation by distance. *Ecological Informatics* 5: 167–172.

Bonte, D., Baert, L., Lens, L., and Maelfait, J. (2004). Effects of aerial dispersal, habitat specialisation, and landscape structure on spider distribution across fragmented grey dunes. *Ecography* 3: 343–349.

Bothwell, H., Bisbing, S., Therkildsen, N.O. et al. (2013). Identifying genetic signatures of selection in a non-model species, alpine gentian (*Gentiana nivalis* L.), using a landscape genetic approach. *Conservation Genetics* 14: 467–481.

Brooker, L.C. and Brooker, M.G. (2002). Dispersal and population dynamics of the Blue-breasted Fairy-Wren, *Malurus pulcherrimus*, in fragmented habitat in the Western Australian wheatbelt. *Wildlife Research* 29: 225–233.

Broquet, T. and Petit, E.J. (2009). Molecular estimation of dispersal for ecology and population genetics. *Annual Review of Ecology, Evolution, and Systematics* 40: 193–216.

Brown, A.H.D., Clegg, M.T., Kahler, A.L., and Weir, B.S. (1990). *Plant Population Genetics, Breeding, and Genetic Resources*. Sunderland, MA: Sinauer Associates Inc.

Burbrink, F.T. (2010). Historical versus contemporary migration in fragmented populations. *Molecular Ecology* 19: 5321–5323.

Carvajal-Rodríguez, A. (2008). Simulation of genomes: a review. *Current Genomics* 9: 155–159.

Cavalli-Sforza, L.L., Menozzi, P., and Piazza, A. (1994). *The History and Geography of Human Genes*. Princeton, New Jersey: Princeton University Press.

Cegelski, C.C., Waits, L.P., and Anderson, N.J. (2003). Assessing population structure and gene flow in Montana wolverines (*Gulo gulo*) using assignment-based approaches. *Molecular Ecology* 12: 2907–2918.

Charlesworth, B. (1998). Measures of divergence between populations and the effect of forces that reduce variability. *Molecular Biology and Evolution* 15: 538–543.

Charlesworth, B. (2009). Fundamental concepts in genetics: effective population size and patterns of molecular evolution and variation. *Nature Reviews Genetics* 10: 195–205.

Chen, C., Durand, E., Forbes, F., and François, O. (2007). Bayesian clustering algorithms ascertaining spatial population structure: a new computer program and a comparison study. *Molecular Ecology Notes* 7: 747–756.

Chesser, R.K., Rhodes, O.E., Sugg, D.W., and Schnabel, A. (1993). Effective sizes for subdivided populations. *Genetics* 135: 1221–1232.

Chiappero, M.B., Panzetta-Dutari, G.M., Gómez, D. et al. (2011). Contrasting genetic structure of urban and rural populations of the wild rodent *Calomys musculinus* (Cricetidae, Sigmodontinae). *Mammalian Biology* 76: 41–50.

Chiucchi, J.E. and Gibbs, H.L. (2010). Similarity of contemporary and historical gene flow among highly fragmented populations of an endangered rattlesnake. *Molecular Ecology* 19: 5345–5358.

Corander, J., Marttinen, P., Sirén, J., and Tang, J. (2008). Enhanced Bayesian modelling in BAPS software for learning genetic structures of populations. *BMC Bioinformatics* 9: 539.

Corander, J., Sirén, J., and Arjas, E. (2008). Bayesian spatial modeling of genetic population structure. *Computational Statistics* 23: 111–129.

Corander, J., Waldmann, P., Marttinen, P., and Sillanpää, M. J. (2004). BAPS 2: enhanced possibilities for the analysis of genetic population structure. *Bioinformatics* 20: 2363–2369.

Cormack, R. (2012). Log-linear models for capture-recapture. *Biometrics* 45: 395–413.

Cornuet, J.M. and Luikart, G. (1996). Description and power analysis of two tests for detecting recent population bottlenecks from allele frequency data. *Genetics* 144: 2001–2014.

Cornuet, J.M., Piry, S., Luikart, G. et al. (1999). New methods employing multilocus genotypes to select or exclude populations as origins of individuals. *Genetics* 153: 1989–2000.

Cornuet, J.-M., Ravigné, V., and Estoup, A. (2010). Inference on population history and model checking using DNA sequence and microsatellite data with the software DIYABC (v1.0). *BMC Bioinformatics* 11: 401.

Crandall, K., Posada, D., and Vasco, D. (2006). Effective population sizes: missing measures and missing concepts. *Animal Conservation* 3: 317–319.

Crother, B.I., White, M.E., Savage, J.M. et al. (2011). A reevaluation of the status of the foxsnakes *Pantherophis gloydi* Conant and *P. vulpinus* Baird and Girard (Lepidosauria). *ISRN Zoology* 2011: 1–15.

Crow, J.F. (1954). Breeding structure of populations. II. Effective population number. In: *Statistics and Mathematics in Biology* (eds. O. Kempthorne, T.A. Bancroft, J.W. Gowen and J.L. Lush), 543–560. Ames: Iowa State College Press.

Crow, J.F. and Denniston, C. (1988). Inbreeding and variance effective population numbers. *Evolution* 42: 482–495.

Csilléry, K., Blum, M.G.B., Gaggiotti, O.E., and François, O. (2010). Approximate Bayesian computation (ABC) in practice. *Trends in Ecology and Evolution* 25: 410–418.

Csilléry, K., François, O., and Blum, M.G.B. (2012). abc: an R package for approximate Bayesian computation (ABC). *Methods in Ecology and Evolution* 3: 475–479.

Dale, M.R.T. and Fortin, M.-J. (2010). From graphs to spatial graphs. *Annual Review of Ecology, Evolution, and Systematics* 41: 21–38.

De Barba, M., Waits, L.P., Garton, E.O. et al. (2010). The power of genetic monitoring for studying demography, ecology and genetics of a reintroduced brown bear population. *Molecular Ecology* 19: 3938–3951.

Dewey, S. and Heywood, J. (1988). Spatial genetic structure in a population of *Psychotria nervosa*. I. Distribution of genotypes. *Evolution* 42: 834–838.

Dias, P. (1996). Sources and sinks in population biology. *Trends in Ecology and Evolution* 11: 326–330.

van Dijk, E.L., Auger, H., Jaszczyszyn, Y., and Thermes, C. (2014). Ten years of next-generation sequencing technology. *Trends in Genetics* 30: 418–426.

DiLeo, M.F., Row, J.R., and Lougheed, S.C. (2010). Discordant patterns of population structure for two co-distributed snake species across a fragmented Ontario landscape. *Diversity and Distributions* 16: 571–581.

Do, C., Waples, R.S., Peel, D. et al. (2014). NeEstimator v2: re-implementation of software for the estimation of contemporary effective population size (N_e) from genetic data. *Molecular Ecology Resources* 14: 209–214.

Double, M.C., Peakall, R., Beck, N.R., and Cockburn, A. (2005). Dispersal, philopatry, and infidelity: dissecting local genetic structure in Superb Fairy-Wrens (*Malurus cyaneus*). *Evolution* 59: 625–635.

Dray, S., Legendre, P., and Peres-Neto, P.R. (2006). Spatial modelling: a comprehensive framework for principal coordinate analysis of neighbour matrices (PCNM). *Ecological Modelling* 196: 483–493.

Dudaniec, R.Y., Spear, S.F., Richardson, J.S., and Storfer, A. (2012). Current and historical drivers of landscape genetic structure differ in core and peripheral salamander populations. *PLoS One* 7: e36769.

Dyer, R.J. and Nason, J.D. (2004). Population graphs: the graph theoretic shape of genetic structure. *Molecular Ecology* 13: 1713–1727.

Elmer, K.R., Dávila, J., and Lougheed, S.C. (2007). Cryptic diversity and deep divergence in an upper Amazonian leaflitter frog, *Eleutherodactylus ockendeni*. *BMC Evolutionary Biology* 7: 247.

England, P.R., Cornuet, J.-M., Berthier, P. et al. (2006). Estimating effective population size from linkage disequilibrium: severe bias in small samples. *Conservation Genetics* 7: 303–308.

Epperson, B.K. (2003). *Geographical Genetics*. Princeton, New Jersey: Princeton University Press.

Epperson, B.K. (2005). Estimating dispersal from short distance spatial autocorrelation. *Heredity* 95: 7–15.

Estoup, A. and Angers, B. (1998). Microsatellites and minisatellites for molecular ecology: theoretical and empirical considerations. In: *Advances in Molecular Ecology* (ed. G. Carvalho), 55–86. Amsterdam: IOS Press.

Evanno, G., Regnaut, S., and Goudet, J. (2005). Detecting the number of clusters of individuals using the software STRUCTURE: a simulation study. *Molecular Ecology* 14: 2611–2620.

Ewens, W.J. (1982). On the concept of the effective population size. *Theoretical Population Biology* 21: 373–378.

Eytan, R.I. and Hellberg, M.E. (2010). Nuclear and mitochondrial sequence data reveal and conceal different demographic histories and population genetic processes in Caribbean reef fishes. *Evolution* 64: 3380–3397.

Falush, D., Stephens, M., and Pritchard, J.K. (2003). Inference of population structure using multilocus genotype data: linked loci and correlated allele frequencies. *Genetics* 164: 1567–1587.

Faubet, P. and Gaggiotti, O. (2008). A new Bayesian method to identify the environmental factors that influence recent migration. *Genetics* 178: 1491–1504.

Faubet, P., Waples, R., and Gaggiotti, O. (2007). Evaluating the performance of a multilocus Bayesian method for the estimation of migration rates. *Molecular Ecology* 16: 1149–1166.

Felsenstein, J. (2006). Accuracy of coalescent likelihood estimates: do we need more sites, more sequences, or more loci? *Molecular Biology and Evolution* 23: 691–700.

Figueira, W.F. (2009). Connectivity or demography: defining sources and sinks in coral reef fish metapopulations. *Ecological Modelling* 220: 1126–1137.

Fisher, R.A. (1930). *The Genetical Theory of Natural Selection*. Oxford, England: Clarendon Press.

Fortuna, M.A., Albaladejo, R.G., Fernández, L. et al. (2009). Networks of spatial genetic variation across species. *Proceedings of the National Academy of Sciences USA* 106: 19044–19049.

Fouquet, A., Ledoux, J., Dubut, V. et al. (2012). The interplay of dispersal limitation, rivers, and historical events shapes the genetic structure of an Amazonian frog. *Biological Journal of the Linnean Society* 106: 356–373.

François, O., Blum, M.G.B., Jakobsson, M., and Rosenberg, N.A. (2008). Demographic history of European populations of *Arabidopsis thaliana*. *PLoS Genetics* 4: e1000075.

Frankham, R. (1995). Effective population size/adult population size ratios in wildlife: a review. *Genetical Research* 66: 95–107.

Frankham, R. (2005). Genetics and extinction. *Biological Conservation* 126: 131–140.

Frantz, A.C., Cellina, S., Krier, A. et al. (2009). Using spatial Bayesian methods to determine the genetic structure of a continuously distributed population: clusters or isolation by distance? *Journal of Applied Ecology* 46: 493–505.

Frantz, A.C. and Roper, T.J. (2006). Simulations to assess the performance of different rarefaction methods in estimating population size using small datasets. *Conservation Genetics* 7: 315–318.

Freeland, J.R., Peterson, S.D., and Kirk, H. (2011). *Molecular Ecology*. Oxford, England: Wiley-Blackwell.

Galtier, N., Nabholz, B., Glémin, S., and Hurst, G.D.D. (2009). Mitochondrial DNA as a marker of molecular diversity: a reappraisal. *Molecular Ecology* 18: 4541–4550.

Garroway, C.J., Bowman, J., Carr, D., and Wilson, P.J. (2008). Applications of graph theory to landscape genetics. *Evolutionary Applications* 1: 620–630.

Garza, J.C. and Williamson, E.G. (2001). Detection of reduction in population size using data from microsatellite loci. *Molecular Ecology* 10: 305–318.

Gibbs, H.L., Corey, S.J., Blouin-Demers, G. et al. (2006). Hybridization between mtDNA-defined phylogeographic lineages of black ratsnakes (*Pantherophis* sp.). *Molecular Ecology* 15: 3755–3767.

Gilbert, K.J. and Whitlock, M.C. (2015). Evaluating methods for estimating local effective population size with and without migration. *Evolution* 69: 2154–2166.

Giordano, A.R., Ridenhour, B.J., and Storfer, A. (2007). The influence of altitude and topography on genetic structure in the long-toed salamander (*Ambystoma macrodactulym*). *Molecular Ecology* 16: 1625–1638.

Griffiths, R., Daan, S., and Dijkstra, C. (1996). Sex identification in birds using two CHD genes. *Proceedings of the Royal Society of London B* 263: 1251–1256.

Griffiths, R., Double, M.C., Orr, K., and Dawson, R.J.G. (1998). A DNA test to sex most birds. *Molecular Ecology* 7: 1071–1075.

Griffiths, R.C. and Tavaré, S. (1994). Sampling theory for neutral alleles in a varying environment. *Philosophical Transactions of the Royal Society of London B* 344: 403–410.

Guillot, G., Estoup, A., Mortier, F., and Cosson, J.F. (2005). A spatial statistical model for landscape genetics. *Genetics* 170: 1261–1280.

Guillot, G., Leblois, R., Coulon, A., and Frantz, A.C. (2009). Statistical methods in spatial genetics. *Molecular Ecology* 18: 4734–4756.

Guillot, G., Mortier, F., and Estoup, A. (2005). `Geneland`: a computer package for landscape genetics. *Molecular Ecology Notes* 5: 712–715.

Hájková, P., Zemanová, B., Roche, K., and Hájek, B. (2008). An evaluation of field and noninvasive genetic methods for estimating Eurasian otter population size. *Conservation Genetics* 10: 1667–1681.

Haldane, J.B.S. (1932). *The Causes of Evolution*. Princeton University Press.

Hardy, O.J. (2003). Estimation of pairwise relatedness between individuals and characterization of isolation-by-distance processes using dominant genetic markers. *Molecular Ecology* 12: 1577–1588.

Hardy, O.J. and Vekemans, X. (1999). Isolation by distance in a continuous population: reconciliation between spatial autocorrelation analysis and population genetics models. *Heredity* 83: 145–154.

Hardy, O. and Vekemans, X. (2002). `SPAGeDi`: a versatile computer program to analyse spatial genetic structure at the individual or population levels. *Molecular Ecology Notes* 2: 618–620.

Harris, R.B. (1989). Genetically effective population size of large mammals: an assessment of estimators. *Conservation Biology* 3: 181–191.

Hastings, A. (1993). Complex interactions between dispersal and dynamics: lessons from coupled logistic equations. *Ecology* 74: 1362–1372.

Hastings, W. (1970). Monte Carlo sampling methods using Markov chains and their applications. *Biometrika* 57: 97–109.

Hedgecock, D., Launey, S., Pudovkin, A.I. et al. (2007). Small effective number of parents (N_b) inferred for a naturally spawned cohort of juvenile European flat oysters *Ostrea edulis*. *Marine Biology* 150: 1173–1182.

Hedrick, P.W. (2005). A standardized genetic differentiation measure. *Evolution* 59: 1633–1638.

Hedrick, P.W. (2009). *Genetics of Populations*, 4e. Sudbury, MA: Jones and Bartlett Publishers.

Hedrick, P.W. (2011). *Genetics of Populations*. Burlington, Massachusetts: Jones and Bartlett Publishers.

van Heerwaarden, J., Doebley, J., Briggs, W.H. et al. (2011). Genetic signals of origin, spread, and introgression in a large sample of maize landraces. *Proceedings of the National Academy of Sciences USA* 108: 1088–1092.

van Heerwaarden, J., Ross-Ibarra, J., Doebley, J. et al. (2010). Fine scale genetic structure in the wild ancestor of maize (*Zea mays* ssp. *parviglumis*). *Molecular Ecology* 19: 1162–1173.

Hemmer-Hansen, J., Nielsen, E.E., Meldrup, D., and Mittelholzer, C. (2011). Identification of single nucleotide polymorphisms in candidate genes for growth and reproduction in a nonmodel organism; the Atlantic cod, *Gadus morhua*. *Molecular Ecology Resources* 11 (Suppl 1): 71–80.

Hey, J. (2010). Isolation with migration models for more than two populations. *Molecular Biology and Evolution* 27: 905–920.

Hill, W.G. (1981). Estimation of effective population size from data on linkage disequilibrium. *Genetical Research* 38: 209–216.

Hoehn, M., Gruber, B., Sarre, S.D. et al. (2012). Can genetic estimators provide robust estimates of the effective number of breeders in small populations? *PLoS One* 7: e48464.

Hoelzel, A., Hey, J., Dahlheim, M.E. et al. (2007). Evolution of population structure in a highly social top predator, the killer whale. *Molecular Biology and Evolution* 24: 1407–1415.

Holderegger, R. and Wagner, H.H. (2008). Landscape genetics. *BioScience* 58: 199.

Howes, B.J., Brown, J.W., Gibbs, H.L. et al. (2008). Directional gene flow patterns in disjunct populations of the black ratsnake (*Pantheropis obsoletus*) and the Blanding's turtle (*Emydoidea blandingii*). *Conservation Genetics* 10: 407–417.

Hudson, R. (2002). Generating samples under a Wright–Fisher neutral model of genetic variation. *Bioinformatics* 18: 337–338.

Huxley, J. (1942). *Evolution: The Modern Synthesis*. London, UK: George Alien and Unwin Ltd.

Jabot, F, Faure, T, Dumoulin, N, Carlo, A (2015) `EasyABC`: Efficient Approximate Bayesian Computation Sampling Schemes. CRAN.R-project.org/package=EasyABC.

Janssens, X., Fontaine, M.C., Michaux, J.R. et al. (2007). Genetic pattern of the recent recovery of European otters in southern France. *Ecography* 31: 176–186.

Johnson, J.A., Burnham, K.K., Burnham, W.A., and Mindell, D.P. (2007). Genetic structure among continental and island populations of Gyrfalcons. *Molecular Ecology* 16: 3145–3160.

Johnstone, I. (2001). On the distribution of the largest eigenvalue in principal components analysis. *The Annals of Statistics* 29: 295–327.

Jolivet, C., Höltken, A.M., Liesebach, H. et al. (2010). Spatial genetic structure in wild cherry (*Prunus avium* L.): I. Variation among natural populations of different density. *Tree Genetics and Genomes* 7: 271–283.

Jombart, T. (2008). `Adegenet`: a R package for the multivariate analysis of genetic markers. *Bioinformatics* 24: 1403–1405.

Jombart, T., Devillard, S., Dufour, A.-B., and Pontier, D. (2008). Revealing cryptic spatial patterns in genetic variability by a new multivariate method. *Heredity* 101: 92–103.

Jombart, T., Pontier, D., and Dufour, A.-B. (2009). Genetic markers in the playground of multivariate analysis. *Heredity* 102: 330–341.

Jorde, P.E. and Ryman, N. (2007). Unbiased estimator for genetic drift and effective population size. *Genetics* 177: 927–935.

Jost, L. (2008). GST and its relatives do not measure differentiation. *Molecular Ecology* 17: 4015–4026.

Kaitala, V., Ranta, E., and Stenseth, N.C. (2006). Genetic structuring in fluctuating populations. *Ecological Informatics* 1: 343–348.

Kalinowski, S.T. and Waples, R.S. (2002). Relationship of effective to census size in fluctuating populations. *Conservation Biology* 16: 129–136.

Kingman, J.F.C. (1982a). On the genealogy of large populations. *Journal of Applied Probability* 19A: 27–43.

Kingman, J.F.C. (1982b). The coalescent. *Stochastic Processes and their Applications* 13: 235–248.

Kingman, J.F.C. (2000). Origins of the coalescent: 1974–1982. *Genetics* 156: 1461–1463.

Kitano, J., Bolnick, D.I., Beauchamp, D.A. et al. (2008). Reverse evolution of armor plates in the threespine stickleback. *Current Biology* 18: 769–774.

Kobmoo, N., Hossaert-McKey, M., Rasplus, J.Y., and Kjellberg, F. (2010). *Ficus racemosa* is pollinated by a single population of a single agaonid wasp species in continental South-East Asia. *Molecular Ecology* 19: 2700–2712.

Koen, E.L., Bowman, J., Garroway, C.J. et al. (2011). Landscape resistance and American marten gene flow. *Landscape Ecology* 27: 29–43.

Kohn, M.H., York, E.C., Kamradt, D.A. et al. (1999). Estimating population size by genotyping faeces. *Proceedings of the Royal Society of London B* 266: 657–663.

Koumi, P., Green, H.E., Hartley, S. et al. (2004). Evaluation and validation of the ABI 3700, ABI 3100, and the MegaBACE 1000 capillary array electrophoresis instruments for use with short tandem repeat microsatellite typing in a forensic environment. *Electrophoresis* 25: 2227–2241.

Kuhner, M. (2006). LAMARC 2.0: maximum likelihood and Bayesian estimation of population parameters. *Bioinformatics* 22: 768–770.

Lambert, A. (2010). Population genetics, ecology and the size of populations. *Journal of Mathematical Biology* 60: 469–472.

Laval, G. and Excoffier, L. (2004). SIMCOAL 2.0: a program to simulate genomic diversity over large recombining regions in a subdivided population with a complex history. *Bioinformatics* 20: 2485–2487.

Lee, S., Wright, F.A., and Zou, F. (2011). Control of population stratification by correlation-selected principal components. *Biometrics* 67: 967–974.

Legendre, P. and Fortin, M.-J. (2010). Comparison of the Mantel test and alternative approaches for detecting complex multivariate relationships in the spatial analysis of genetic data. *Molecular Ecology Resources* 10: 831–844.

Liu, N., Chen, L., Wang, S. et al. (2005). Comparison of single-nucleotide polymorphisms and microsatellites in inference of population structure. *BMC Genetics* 6 (Suppl 1): S26.

Lougheed, S.C., Campagna, L., Dávila, J.A. et al. (2013). Continental phylogeography of an ecologically and morphologically diverse Neotropical songbird, *Zonotrichia capensis*. *BMC Evolutionary Biology* 13: 58.

Lowe, W.H. and Allendorf, F.W. (2010). What can genetics tell us about population connectivity? *Molecular Ecology* 19: 3038–3051.

Luikart, G., Allendorf, F.W., Cornuet, J.-M., and Sherwin, W. B. (1998). Distortion of allele frequency distributions

provides a test for recent population bottlenecks. *Journal of Heredity* 89: 238–247.
Luikart, G. and Cornuet, J.M. (1999). Estimating the effective number of breeders from heterozygote excess in progeny. *Genetics* 151: 1211–1216.
Luikart, G., Ryman, N., Tallmon, D.A. et al. (2010). Estimation of census and effective population sizes: the increasing usefulness of DNA-based approaches. *Conservation Genetics* 11: 355–373.
Lukacs, P.M. and Burnham, K.P. (2005). Review of capture-recapture methods applicable to noninvasive genetic sampling. *Molecular Ecology* 14: 3909–3919.
Lukacs, P.M., Eggert, L.S., and Burnham, K.P. (2007). Estimating population size from multiple detections with non-invasive genetic data. *Wildlife Biology in Practice* 3: 83–92.
Ma, J. and Amos, C.I. (2012). Principal components analysis of population admixture. *PLoS One* 7: e40115.
Madoui, M.-A., Engelen, S., Cruaud, C. et al. (2015). Genome assembly using Nanopore-guided long and error-free DNA reads. *BMC Genomics* 16: 327.
Madsen, T., Shine, R., Olsson, M., and Wittzell, H. (1999). Restoration of an inbred adder population. *Nature* 402: 34–35.
Magwene, P.M. (2001). New tools for studying integration and modularity. *Evolution* 55: 1734–1745.
Malécot (1948). *Les mathématiques de l'hérédité*, 63. Paris, France: Masson et Cie.
Malenfant, R.M., Coltman, D.W., and Davis, C.S. (2015). Design of a 9K illumina BeadChip for polar bears (*Ursus maritimus*) from RAD and transcriptome sequencing. *Molecular Ecology Resources* 15: 587–600.
Manel, S., Gaggiotti, O., and Waples, R. (2005). Assignment methods: matching biological questions with appropriate techniques. *Trends in Ecology and Evolution* 20: 136–142.
Manel, S., Poncet, B.N., Legendre, P. et al. (2010). Common factors drive adaptive genetic variation at different spatial scales in Arabis alpina. *Molecular Ecology* 19: 3824–3835.
Manel, S., Schwartz, M.K., Luikart, G., and Taberlet, P. (2003). Landscape genetics: combining landscape ecology and population genetics. *Trends in Ecology and Evolution* 18: 189–197.
Manier, M.K. and Arnold, S.J. (2006). Ecological correlates of population genetic structure: a comparative approach using a vertebrate metacommunity. *Proceedings of the Royal Society of London B* 273: 3001–3009.
Maruyama, T. and Fuerst, P.A. (1985). Population bottlenecks and nonequilibrium models in population genetics. II. Number of alleles in a small population that was formed by a recent bottleneck. *Genetics* 111: 675–689.
Mavárez, J., Salazar, C.A., Bermingham, E. et al. (2006). Speciation by hybridization in *Heliconius* butterflies. *Nature* 441: 868–871.
May, R.M. (1973). *Stability and Complexity in Model Ecosystems*. Princeton, New Jersey: Princeton University Press.
Meirmans, P.G. and Hedrick, P.W. (2011). Assessing population structure: FST and related measures. *Molecular Ecology Resources* 11: 5–18.
Metropolis, N., Rosenbluth, A., Rosenbluth, M., and Teller, A. (1953). Equation of state calculations by fast computing machines. *Journal of Chemical Physics* 21: 1087–1092.
Miller, C.R., Joyce, P., and Waits, L.P. (2005). A new method for estimating the size of small populations from genetic mark-recapture data. *Molecular Ecology* 14: 1991–2005.
Miller, C.R. and Waits, L.P. (2003). The history of effective population size and genetic diversity in the Yellowstone grizzly (*Ursus arctos*): implications for conservation. *Proceedings of the National Academy of Sciences USA* 100: 4334–4339.
Montarry, J., Andrivon, D., Glais, I. et al. (2010). Microsatellite markers reveal two admixed genetic groups and an ongoing displacement within the French population of the invasive plant pathogen *Phytophthora infestans*. *Molecular Ecology* 19: 1965–1977.
Murphy, M.A., Dezzani, R., Pilliod, D.S., and Storfer, A. (2010). Landscape genetics of high mountain frog metapopulations. *Molecular Ecology* 19: 3634–3649.
Nei, M. (1973). Analysis of gene diversity in subdivided populations. *Proceedings of the National Academy of Sciences USA* 70: 3321–3323.
Nei, M. (1975). *Molecular Population Genetics and Evolution*. North-Holland, Amsterdam: North-Holland Publishing Company.
Nei, M. and Tajima, F. (1981). Genetic drift and estimation of effective population size. *Genetics* 98: 625–640.
Neigel, J. (2002). Is F_{ST} obsolete? *Conservation Genetics* 3: 167–173.
NERC Centre for Population Biology, IC (2010) The Global Population Dynamics Database Version 2.
Neuhauser, C., Andow, D., and Heimpel, G. (2003). Community genetics: expanding the synthesis of ecology and genetics. *Ecology* 84: 545–558.
Nielsen, R. and Wakeley, J. (2001). Distinguishing migration from isolation – a Markov chain Monte Carlo approach. *Genetics* 158: 885–896.
Nikolic, N. and Chevalet, C. (2014). Detecting past changes of effective population size. *Evolutionary Applications* 7: 663–681.
Nosil, P., Egan, S.P., and Funk, D.J. (2008). Heterogeneous genomic differentiation between walking-stick ecotypes: "isolation by adaptation" and multiple roles for divergent selection. *Evolution* 62: 316–336.

Notohara, M. (1990). The coalescent and the genealogical process in geographically structured population. *Journal of Mathematical Biology* 29: 59–75.

Novembre, J. and Stephens, M. (2008). Interpreting principal component analyses of spatial population genetic variation. *Nature Genetics* 40: 646–649.

Nunney, L. and Elam, D. (1994). Estimating the effective population size of conserved populations. *Conservation Biology* 8: 175–184.

Ovenden, J.R., Peel, D., Street, R. et al. (2007). The genetic effective and adult census size of an Australian population of tiger prawns (*Penaeus esculentus*). *Molecular Ecology* 16: 127–318.

Paetkau, D., Calvert, W., Stirling, I., and Strobeck, C. (1995). Microsatellite analysis of population structure in Canadian polar bears. *Molecular Ecology* 4: 347–354.

Paetkau, D., Slade, R., Burden, M., and Estoup, A. (2004). Genetic assignment methods for the direct, real-time estimation of migration rate: a simulation-based exploration of accuracy and power. *Molecular Ecology* 13: 55–65.

Pakendorf, B. and Stoneking, M. (2005). Mitochondrial DNA and human evolution. *Annual Review of Genomics and Human Genetics* 6: 165–183.

Palsbøll, P.J., Bérubé, M., and Allendorf, F.W. (2007). Identification of management units using population genetic data. *Trends in Ecology and Evolution* 22: 11–16.

Palstra, F.P. and Ruzzante, D.E. (2008). Genetic estimates of contemporary effective population size: what can they tell us about the importance of genetic stochasticity for wild population persistence? *Molecular Ecology* 17: 3428–3447.

Paquette, S.R., Louis, E.E., and Lapointe, F.-J. (2010). Microsatellite analyses provide evidence of male-biased dispersal in the radiated tortoise *Astrochelys radiata* (Chelonia: Testudinidae). *Journal of Heredity* 101: 403–412.

Parisod, C. and Holderegger, R. (2012). Adaptive landscape genetics: pitfalls and benefits. *Molecular Ecology* 21: 3644–3646.

Paschou, P., Ziv, E., Burchard, E.G. et al. (2007). PCA-correlated SNPs for structure identification in worldwide human populations. *PLoS Genetics* 3: 1672–1686.

Pascual, M., Chapuis, M.P., Mestres, F. et al. (2007). Introduction history of *Drosophila subobscura* in the New World: a microsatellite-based survey using ABC methods. *Molecular Ecology* 16: 3069–3083.

Patterson, N., Price, A.L., and Reich, D. (2006). Population structure and eigenanalysis. *PLoS Genetics* 2: e190.

Peakall, R., Ruibal, M., and Lindenmayer, D. (2003). Spatial autocorrelation analysis offers new insights into gene flow in the Australian bush rat (*Rattus fuscipes*). *Evolution* 57: 1182–1195.

Peakall, R. and Smouse, P.E. (2006). `Genalex 6`: genetic analysis in Excel. Population genetic software for teaching and research. *Molecular Ecology Notes* 6: 288–295.

Peery, M.Z., Kirby, R., Reid, B.N. et al. (2012). Reliability of genetic bottleneck tests for detecting recent population declines. *Molecular Ecology* 21: 3403–3418.

Pennell, M.W., Stansbury, C.R., Waits, L.P., and Miller, C.R. (2013). `Capwire`: a R package for estimating population census size from non-invasive genetic sampling. *Molecular Ecology Resources* 13: 154–157.

Perry, J.C. and Rowe, L. (2011). Rapid microsatellite development for water striders by next-generation sequencing. *Journal of Heredity* 102: 125–129.

Peter, B.M., Wegmann, D., and Excoffier, L. (2010). Distinguishing between population bottleneck and population subdivision by a Bayesian model choice procedure. *Molecular Ecology* 19: 4648–4660.

Peters, J.L., Zhuravlev, Y.N., Fefelov, I. et al. (2008). Multilocus phylogeography of a holarctic duck: colonization of North America from Eurasia by Gadwall (*Anas strepera*). *Evolution* 62: 1469–1483.

Pierpaoli, M., Biro, Z.S., Herrmann, M. et al. (2003). Genetic distinction of wildcat (*Felis silvestris*) populations in Europe, and hybridization with domestic cats in Hungary. *Molecular Ecology* 12: 2585–2598.

Piry, S., Alapetite, A., Cornuet, J.-M. et al. (2004). `GENECLASS2`: a software for genetic assignment and first-generation migrant detection. *Journal of Heredity* 95: 536–539.

Poole, K. (1997). Dispersal patterns of lynx in the Northwest Territories. *Journal of Wildlife Management* 61: 497–505.

Pritchard, J., Stephens, M., and Donnelly, P. (2000). Inference of population structure using multilocus genotype data. *Genetics* 155: 945–959.

Provine, W.B. (1971). *The Origins of Theoretical Population Genetics*. Chicago: University of Chicago Press.

Pudovkin, A.I., Zaykin, D.V., and Hedgecock, D. (1996). On the potential for estimating the effective number of breeders from heterozygote-excess in progeny. *Genetics* 144: 383–387.

Pybus, O.G. and Rambaut, A. (2009). Evolutionary analysis of the dynamics of viral infectious disease. *Nature Reviews Genetics* 10: 540–550.

Rannala, B. and Mountain, J.L. (1997). Detecting immigration by using multilocus genotypes. *Proceedings of the National Academy of Sciences USA* 94: 9197–9201.

Reed, J., Mills, L.S., Dunning, J. et al. (2002). Emerging issues in population viability analysis. *Conservation Biology* 16: 7–19.

Reich, D., Price, A.L., and Patterson, N. (2008). Principal component analysis of genetic data. *Nature Genetics* 40: 491–492.

Riley, S.P.D., Pollinger, J.P., Sauvajot, R.M. et al. (2006). A southern California freeway is a physical and social barrier to gene flow in carnivores. *Molecular Ecology* 15: 1733–1741.

Robert, C.P., Cornuet, J.-M., Marin, J.-M., and Pillai, N.S. (2011). Lack of confidence in approximate Bayesian computation model choice. *Proceedings of the National Academy of Sciences USA* 108: 15112–15117.

Rockwood, L.L. (2006). *Introduction to Population Ecology*. Malden, MA: Wiley-Blackwell.

Rosenberg, N.A., Li, L.M., Ward, R., and Pritchard, J.K. (2003). Informativeness of genetic markers for inference of ancestry. *American Journal of Human Genetics* 73: 1402–1422.

Rosenberg, N.A. and Nordborg, M. (2002). Genealogical trees, coalescent theory and the analysis of genetic polymorphisms. *Nature Reviews Genetics* 3: 380–390.

Rousset, F. (2000). Genetic differentiation between individuals. *Journal of Evolutionary Biology* 13: 58–62.

Rousset, F. (2008). Dispersal estimation: demystifying Moran's I. *Heredity* 100: 231–232.

Row, J.R., Blouin-Demers, G., and Lougheed, S.C. (2010). Habitat distribution influences dispersal and fine-scale genetic population structure of eastern foxsnakes (*Mintonius gloydi*) across a fragmented landscape. *Molecular Ecology* 19: 5157–5171.

Row, J.R., Brooks, R.J., Mackinnon, C.A. et al. (2011). Approximate Bayesian computation reveals the factors that influence genetic diversity and population structure of foxsnakes. *Journal of Evolutionary Biology* 24: 2364–2377.

Row, J.R., Gomez, C., Koen, E.L. et al. (2012). Dispersal promotes high gene flow among *Canada lynx* populations across mainland North America. *Conservation Genetics* 13: 1259–1268.

Rozenfeld, A.F., Arnaud-Haond, S., Hernández-García, E. et al. (2008). Network analysis identifies weak and strong links in a metapopulation system. *Proceedings of the National Academy of Sciences USA* 105: 18824–18829.

Rudge, J.W., Lu, D.-B., Fang, G.-R. et al. (2009). Parasite genetic differentiation by habitat type and host species: molecular epidemiology of *Schistosoma japonicum* in hilly and marshland areas of Anhui Province, China. *Molecular Ecology* 18: 2134–2147.

Rueness, E.K., Stenseth, N.C., O'Donoghue, M. et al. (2003). Ecological and genetic spatial structuring in the Canadian lynx. *Nature* 425: 69–72.

Rutledge, L.Y., White, B.N., Row, J.R., and Patterson, B.R. (2012). Intense harvesting of eastern wolves facilitated hybridization with coyotes. *Ecology and Evolution* 2: 19–33.

Sacks, B.N., Brown, S.K., and Ernest, H.B. (2004). Population structure of California coyotes corresponds to habitat-specific breaks and illuminates species history. *Molecular Ecology* 13: 1265–1275.

Sanger, F., Nicklen, S., and Coulson, A.R. (2007). DNA sequencing with chain-terminating inhibitors. *Proceedings of the National Academy of Sciences USA* 74: 5463–5467.

Schmuki, C., Vorburger, C., Runciman, D. et al. (2006). When log-dwellers meet loggers: impacts of forest fragmentation on two endemic log-dwelling beetles in southeastern Australia. *Molecular Ecology* 15: 1481–1492.

Schwartz, M.K. and McKelvey, K.S. (2009). Why sampling scheme matters: the effect of sampling scheme on landscape genetic results. *Conservation Genetics* 10: 441–452.

Schwartz, M.K., Tallmon, D.A., and Luikart, G. (1998). Review of DNA-based census and effective population size estimators. *Animal Conservation* 1: 293–299.

Seber, G.A.F. (1973). *The Estimation of Animal Abundance and Related Parameters*. London: Edward Arnold.

Seeb, J.E., Carvalho, G., Hauser, L. et al. (2011). Single-nucleotide polymorphism (SNP) discovery and applications of SNP genotyping in nonmodel organisms. *Molecular Ecology Resources* 11 (Suppl): 1–8.

Selelstad, M. and Hebert, J. (1994). Construction of human Y-chromosomal haplotypes using a new polymorphic A to G transition. *Human Molecular Genetics* 3: 2159–2161.

Selkoe, K.A. and Toonen, R.J. (2006). Microsatellites for ecologists: a practical guide to using and evaluating microsatellite markers. *Ecology Letters* 9: 615–629.

Shaffer, M.L. (1981). Minimum population sizes for species conservation. *BioScience* 31: 131–134.

Shendure, J. and Ji, H. (2008). Next-generation DNA sequencing. *Nature Biotechnology* 26: 1135–1145.

Sjödin, P., Kaj, I., Krone, S. et al. (2005). On the meaning and existence of an effective population size. *Genetics* 169: 1061–1070.

Skrbinšek, T., Jelenčič, M., Waits, L. et al. (2012). Monitoring the effective population size of a brown bear (*Ursus arctos*) population using new single-sample approaches. *Molecular Ecology* 21: 862–875.

Slatkin, M. and Hudson, R. (1991). Pairwise comparisons of mitochondrial DNA sequences in stable and exponentially growing populations. *Genetics* 129: 555–562.

Smouse, P.E. and Peakall, R. (1999). Spatial autocorrelation analysis of individual multiallele and multilocus genetic structure. *Heredity* 82: 561–573.

Sokal, R.R. and Oden, N.L. (1978). Spatial autocorrelation in biology 1. Methodology. *Biological Journal of the Linnean Society* 10: 199–228.

Sousa, V.C., Fritz, M., Beaumont, M.A., and Chikhi, L. (2009). Approximate Bayesian computation without summary statistics: the case of admixture. *Genetics* 181: 1507–1519.

Stauffer, D. and Aharony, A. (1994). *Introduction to Percolation Theory*. London: Taylor and Francis.

Stephens, M., Smith, N.J., and Donnelly, P. (2001). A new statistical method for haplotype reconstruction from population data. *American Journal of Human Genetics* 68: 978–989.

Stevens, V.M. and Baguette, M. (2008). Importance of habitat quality and landscape connectivity for the persistence of endangered natterjack toads. *Conservation Biology* 22: 1194–1204.

Stewart, K.A., Austin, J.D., Zamudio, K.R., and Lougheed, S. C. (2016). Contact zone dynamics during early stages of speciation in a chorus frog (*Pseudacris crucifer*). *Heredity* 116: 239–247.

Storfer, A., Murphy, M.A., Evans, J.S. et al. (2007). Putting the “landscape” in landscape genetics. *Heredity* 98: 128–142.

Taberlet, P. and Luikart, G. (1999). Non-invasive genetic sampling and individual identification. *Biological Journal of the Linnean Society* 68: 41–55.

Takezaki, N. and Nei, M. (1996). Genetic distances and reconstruction of phylogenetic trees from microsatellite DNA. *Genetics* 144: 389–399.

Tallmon, D.A., Koyuk, A., Luikart, G., and Beaumont, M.A. (2008). ONeSAMP: a program to estimate effective population size using approximate Bayesian computation. *Molecular Ecology Resources* 8: 299–301.

Taylor, D.R. and Keller, S.R. (2007). Historical range expansion determines the phylogenetic diversity introduced during contemporary species invasion. *Evolution* 61: 334–345.

Thalmann, O., Fischer, A., Lankester, F. et al. (2007). The complex evolutionary history of gorillas: insights from genomic data. *Molecular Biology and Evolution* 24: 146–158.

Traill, L.W., Brook, B.W., Frankham, R.R., and Bradshaw, C. J.A. (2010). Pragmatic population viability targets in a rapidly changing world. *Biological Conservation* 143: 28–34.

Troupin, D., Nathan, R., and Vendramin, G.G. (2006). Analysis of spatial genetic structure in an expanding *Pinus halepensis* population reveals development of fine-scale genetic clustering over time. *Molecular Ecology* 15: 3617–3630.

Tung, J., Charpentier, M.J.E., Garfield, D.A. et al. (2008). Genetic evidence reveals temporal change in hybridization patterns in a wild baboon population. *Molecular Ecology* 17: 1998–2011.

Vandewoestijne, S. and Van Dyck, H. (2010). Population genetic differences along a latitudinal cline between original and recently colonized habitat in a butterfly. *PLoS One* 5: e13810.

Vekemans, X. and Hardy, O.J. (2004). New insights from fine-scale spatial genetic structure analyses in plant populations. *Molecular Ecology* 13: 921–935.

Veysey, J.S., Mattfeldt, S.D., and Babbitt, K.J. (2011). Comparative influence of isolation, landscape, and wetland characteristics on egg-mass abundance of two pool-breeding amphibian species. *Landscape Ecology* 26: 661–672.

Vignieri, S. (2005). Streams over mountains: influence of riparian connectivity on gene flow in the Pacific jumping mouse (*Zapus trinotatus*). *Molecular Ecology* 14: 1925–1937.

Wagner, H.H., Holderegger, R., Werth, S. et al. (2005). Variogram analysis of the spatial genetic structure of continuous populations using multilocus microsatellite data. *Genetics* 169: 1739–1752.

Wahlund, S. (1928). Zusammensetzung von Populationen und Korrelationserscheinungen vom Standpunkt der Vererbungslehre ausbetrachtet. *Hereditas* 11: 65–106.

Waits, L.P. (2004). Using noninvasive genetic sampling to detect and estimate abundance of rare wildlife species. In: *Sampling Rare or Elusive Species: Concepts, Designs, and Techniques for Estimating Population Parameters* (ed. W. Thompson), 211–228. Washington, D.C.: Island Press.

Waits, L. and Paetkau, D. (2005). Noninvasive genetic sampling tools for wildlife biologists: a review of applications and recommendations for accurate data collection. *Journal of Wildlife Management* 69: 1419–1433.

Wang, J. (1997a). Effective size and *F*-statistics of subdivided populations. I. Monoecious species with partial selfing. *Genetics* 146: 1465–1474.

Wang, J. (1997b). Effective size and *F*-statistics of subdivided populations. II. Dioecious species. *Genetics* 146: 1465–1474.

Wang, J. (2001). A pseudo-likelihood method for estimating effective population size from temporally spaced samples. *Genetical Research* 78: 243–257.

Wang, J. (2005). Estimation of effective population sizes from data on genetic markers. *Philosophical Transactions of the Royal Society of London B* 360: 1395–1409.

Wang, J. (2009). A new method for estimating effective population sizes from a single sample of multilocus genotypes. *Molecular Ecology* 18: 2148–2164.

Wang, J., Brekke, P., Huchard, E. et al. (2010). Estimation of parameters of inbreeding and genetic drift in populations with overlapping generations. *Evolution* 64: 1704–1718.

Wang, J. and Whitlock, M.C. (2003). Estimating effective population size and migration rates from genetic samples over space and time. *Genetics* 163: 429–446.

Wang, R., Compton, S.G., Shi, Y.-S., and Chen, X.-Y. (2012). Fragmentation reduces regional-scale spatial genetic structure in a wind-pollinated tree because genetic barriers are removed. *Ecology and Evolution* 2: 1–12.

Wang, T.P., Shrivastava, J., Johansen, M.V. et al. (2006). Does multiple hosts mean multiple parasites? Population genetic structure of *Schistosoma japonicum* between definitive host species. *International Journal for Parasitology* 36: 1317–1325.

Wang, Z., Gerstein, M., and Snyder, M. (2009). RNA-Seq: a revolutionary tool for transcriptomics. *Nature Reviews Genetics* 10: 57–63.

Waples, R.S. (1989a). Temporal variation in allele frequencies: testing the right hypothesis. *Evolution* 43: 1236–1251.

Waples, R.S. (1989b). A generalized approach for estimating effective population size from temporal changes in allele frequency. *Genetics* 121: 379–391.

Waples, R.S. (1991). Genetic methods for estimating the effective size of cetacean populations. *Report for the International Whaling Commission* (Special Issue 13): 279–300.

Waples, R.S. (2006). A bias correction for estimates of effective population size based on linkage disequilibrium at unlinked gene loci. *Conservation Genetics* 7: 167–184.

Waples, R.S. and Do, C. (2008). LDNE: a program for estimating effective population size from data on linkage disequilibrium. *Molecular Ecology Resources* 8: 753–756.

Waples, R.S. and Gaggiotti, O. (2006). What is a population? An empirical evaluation of some genetic methods for identifying the number of gene pools and their degree of connectivity. *Molecular Ecology* 15: 1419–1439.

Ward, J. (1963). Hierarchical grouping to optimize an objective function. *Journal of the American Statistical Association* 58: 236–244.

Wegmann, D. and Excoffier, L. (2010). Bayesian inference of the demographic history of chimpanzees. *Molecular Biology and Evolution* 27: 1425–1435.

Wegmann, D., Leuenberger, C., Neuenschwander, S., and Excoffier, L. (2010). ABCtoolbox: a versatile toolkit for approximate Bayesian computations. *BMC Bioinformatics* 11: 1–7.

Wells, J.V. and Richmond, M.E. (1995). Populations, metapopulations, and species populations: what are they and who should care. *Wildlife Society Bulletin* 23: 458–462.

Whitlock, M.C. and McCauley, D.E. (1999). Indirect measures of gene flow and migration: F_{ST} not equal to $1/(4N_m + 1)$. *Heredity* 82: 117–125.

Wilson, G. and Rannala, B. (2003). Bayesian inference of recent migration rates using multilocus genotypes. *Genetics* 163: 1177–1191.

Wright, S. (1931). Evolution in Mendelian populations. *Genetics* 16: 97–159.

Wright, S. (1943). Isolation by distance. *Genetics* 28: 114–138.

Wright, S. (1965). The interpretation of population structure by *F*-statistics with special regard to systems of mating. *Evolution* 19: 395–420.

Wright, S. (1978). *Evolution and the Genetics of Populations, Vol. 4: Variability within and among Natural Populations*. Chicago: University of Chicago Press.

Wright, S.I. and Gaut, B.S. (2005). Molecular population genetics and the search for adaptive evolution in plants. *Molecular Biology and Evolution* 22: 506–519.

Zhdanova, O.L. and Pudovkin, A.I. (2008). Nb_HetEx: a program to estimate the effective number of breeders. *Journal of Heredity* 99: 694–695.

Zink, R.M. and Barrowclough, G.F. (2008). Mitochondrial DNA under siege in avian phylogeography. *Molecular Ecology* 17: 2107–2121.

12

Spatial Structure in Population Data

Marie-Josée Fortin

Department of Ecology and Evolutionary Biology, University of Toronto, Toronto, Ontario, Canada

Summary

An understanding of population dynamics first requires the determination of a population's spatial structure. Knowledge about a population's spatial structure can be characterized and quantified using various spatial statistics. The selection of the appropriate spatial statistics will depend on the question investigated and the available data to address it. In this chapter, I first summarize the key spatial statistics that can be used as an exploratory spatial data analysis (ESDA) to quantify and test the significance of spatial patterns in population data based on: (i) point data using Ripley's *K*; or (ii) abundance data using spatial autocorrelation coefficients (Moran's *I*, Geary's *c*) or geostatistics (variograms). Then I present how density data can be estimated and mapped based on point data using kernel density functions. Similarly, spatially interpolated data can be estimated from quantitative data using spatial interpolation techniques such as Kriging. Next, spatial regressions can be used to relate spatially structured species abundance to environmental and ecological factors. Last, given that numerous processes may shape spatial distributions of populations, multiscale analysis (Moran's Eigenvector Maps [MEM]) should be performed to detect the key spatial scales of the data. I stress the advantages and limits of the presented spatial methods while highlighting how the methods can be used in tandem to gain knowledge about species ecology.

12.1 Introduction

Spatial distributions of species are the result of various biological, ecological, and environmental processes all occurring at different spatial and temporal scales (Levin 1992; Dungan et al. 2002; Fortin et al. 2012). Environmental and climatic conditions are exogenous processes that induce spatial structure in species distributions by affecting nutrient availability, habitat quality and quantity, and appropriate climatic physiological conditions that define the species' ecological niche. Exogenous conditions are expected to be spatially structured, and species distributions will mirror spatial patterns of resources at regional and landscape scales (Wagner and Fortin 2005; Fortin et al. 2012). At more local scales, biological endogenous (e.g. animal movement, species interactions, demographic processes, dispersal) and abiotic exogenous processes (e.g. environmental and climatic conditions, natural disturbances, human-made activities) within habitat patches also generate spatial patterns. All the processes and factors affect species spatial patterns at multiple spatial scales (Fortin et al. 2012). When spatial patterns are due to *endogenous processes* it is often referred to as **true spatial autocorrelation**, whereas when they are due to *exogenous processes* it is considered as **false spatial autocorrelation** or spatial dependency (Legendre 1993). The synergetic interactions between endogenous and exogenous processes result in embedded spatial patterns (e.g. trends and patchiness) at different spatial scales, all of which have their own magnitude, size, and shape. Species spatial distributions can therefore show different patterns depending on local, landscape, region, or continent scales (Fortin et al. 2012) as well as the level of organization studied being at the individual, population, metapopulation, or species level (Table 12.1).

Understanding species' responses to their environment requires quantitative characterization of spatial patterns of animal distribution and abundance. By determining spatial patterns, one can determine the key spatial scales and ecological processes that generate observed patterns (Table 12.1). There are many spatial statistics available to analyze spatial patterns in ecological data (Dale and Fortin 2014). In ecology, there is a long history of using and developing spatial statistics to relate complex spatial patterns to ecological processes and environmental factors (Watt 1947; Greig-Smith 1961; Lloyd 1967).

Population Ecology in Practice, First Edition. Edited by Dennis L. Murray and Brett K. Sandercock.

Companion website: www.wiley.com/go/MurrayPopulationEcology

Table 12.1 Spatial scales and spatial analyses.

Spatial scales	**Local/** ***Home range***	**Population/** ***Landscape***	**Metapopulation/** ***Region***	**Species range/** ***Continent***
Sampling unit	• Individuals (*x-y* coordinates) or features related to each individual (e.g., nest locations, home range centroids, etc.)	• Abundance data sampled at sampling units.	• Habitat patches (shape, size, and quality).	• Sampling units • Species range distribution
Processes	• Demography • Movement • Resource use	• Demography • Movement • Dispersal • Species interactions • Land use • Disturbance	• Demography • Dispersal • Species interactions • Land use • Climate change • Disturbance	• Speciation • Dispersal • Migration • Species interactions • Land use • Climate change • Disturbance
Spatial structure	• Patch size and shape • Spatial heterogeneity	• Patch configuration • Topography • Spatial heterogeneity	• Patch configuration • Patch isolation • Topography • Spatial heterogeneity	• Barriers • Topography • Spatial heterogeneity
Spatial analysis	• Home range delineation • Point pattern analysis	• Spatial autocorrelation • Spatial variance • Spatial interpolation • Multiscale analysis • Spatial regression	• Spatial autocorrelation • Spatial variance • Spatial interpolation • Multiscale analysis • Spatial regression • Gravity model	• Spatial autocorrelation • Spatial variance • Spatial interpolation • Multiscale analysis • Spatial regression • Species range delineation • Species distribution models

Furthermore, different spatial statistics have been developed in various disciplines ranging from geography, engineering, and epidemiology to economics. Yet, all spatial statistics are rooted in the statistical principles and assumptions of time series analysis, which assume that the processes that generated the pattern are stationary (i.e. same mean and variance through the study area).

The *spatial statistics* used by ecologists stem mostly from human geography (spatial statistics; Cliff and Ord 1981) and mining (geostatistics; Matheron 1970; Journel and Huijbregts 1978; Cressie 1993). The goal of the spatial statistics developed by geographers is to estimate the degree of **spatial autocorrelation** of quantitative data against various spatial lags. Such estimations can be assessed and their significance tested using Moran's *I* or Geary's *c* spatial autocorrelation coefficients (Cliff and Ord 1981). The original goal of the geostatistical methods (semivariance γ and spatial interpolation based on Kriging) was to infer the quantitative values of a variable at unsampled locations, incorporating knowledge of the spatial structure and autocorrelation present in the data with a spatial interpolator algorithm developed by the engineer Krige (Journel and Huijbregts 1978). Hence, the estimation of spatial autocorrelation, computed in term of spatial variance (see Section 12.4), was then not tested for its significance.

The spatial aggregation of point data was estimated using other methods based on *point pattern methods* such as Ripley's *K* statistics (Ripley 1981). With the availability of various types of sensors, data about individual's movement and species abundances are now possible to obtain over larger study areas, making the assumption of stationarity of the processes less likely. To deal with such regionally structured data, Local Indicator of Spatial Association methods (Local Moran's *I*, Local Geary's *c*; Local Getis' *G*) have been developed (Anselin 1995; Getis and Ord 1996).

Yet, the presence of spatial autocorrelated structure in ecological data invalidates the assumption of independence of parametric inferential tests such as correlation, regression, and analysis of variance (ANOVA, Cliff and Ord 1981). To address the statistical issue of dependence of the errors due to spatial autocorrelation, a series of corrections have been proposed for bivariate tests (Clifford et al. 1989; Dutilleul 1993). Recently, the acknowledgement of the importance of spatial scales in ecological studies has grown (Fortin et al. 2012). The determination of the key spatial scales at which processes shape the spatial patterns of ecological data can be done using spectral decomposition methods (wavelets; Keitt and Urban 2005) and eigenfunction methods (Moran Eigenvector Maps [MEM]; Dray 2011; Dray et al. 2012). Last, to relate how environmental factors and their spatial patterns influence species distributions, a wide range of spatial regressions have been developed by geographers (Haining 2003) and economists (Anselin 1988).

Before presenting these various spatial statistics in more detail, it is important to understand the various aspects of "space," how space affects the observed spatial patterns, and our ability to detect spatial relationships (Table 12.2). Four sets of issues are relevant.

First, one needs to decide how space will be conceptually measured. Indeed, space can be considered: (i) *implicitly*, where distance is measured in terms of relative topological nearest neighbors; (ii) *explicitly*, where distances are measured based on Euclidean distance; and (iii) *functionally*, which are weighted distances that account for species' behavioral dispersal responses to landscape heterogeneity.

Second, space is often used as a surrogate for unmeasured factors and processes. Indeed, in the presence of spatial patterns due to spatial autocorrelation, nearby values of a variable are most likely to be similar in terms of magnitude. Therefore, space recorded as x-y coordinates or measured as either Euclidean distance or functional connectivity can be used as a *spatial predictor* in a regression model.

Third, current processes and resulting spatial patterns can be influenced by historical events, habitat composition, or spatial configuration of the landscape. Such a phenomenon is referred to as a lag effect of spatial legacy or *ecological memory* (Peterson 2002; James et al. 2007).

Fourth, both spatial dependency and spatial autocorrelation act together so that the spatial pattern at a given location is affected by the values of the factors and processes from the immediate surroundings.

From these four issues, it is clear that space is important in structuring the spatial distribution of species (Table 12.2). Hence, one needs to be aware of how these aspects of space interplay in the spatial analysis of population data. In this chapter, I focus on the key methods used to quantify spatial patterns using either point data such as x-y coordinates of individuals or abundance data

Table 12.2 Multiple aspects of space: relative, absolute, and realistic.

	Implicit/Relative:	*Explicit/Absolute:*	*Functional/ Realistic:*
Conceptual representations of space	Relative position between sampling locations based on network topological neighborhoods.	Euclidean distances based on x-y coordinates of sampling locations.	Resistance (least-cost) distance proportional to landscape quality between the sampled sites.
Space as a surrogate for processes	Binary network topology is a surrogate for dispersal potential and any other ecological processes between spatial entities (individuals or sampling units).	Euclidean distance or x-y coordinates can be used in the context when the first law of geography prevails: nearby locations are more likely to have similar values. Hence "space" can be used as a surrogate for unsampled factors and processes that would generate spatial patterns in the observed data.	The actual position of the sampling locations is considered as well as the weights (landscape costs in terms of energetic costs, behavior limitation to dispersal and mortality risk, etc.) of area between the sampling locations. Hence, functional connectivity weights can be used as a surrogate for dispersal ability and movement impediment.
Spatial legacy	Influence of past spatial patterns within patches on current spatial patterns and processes.	Influence of past patch and matrix configuration on current spatial pattern and processes.	Influence of past spatial patterns of landscapes on current spatial pattern and processes.
Spatial contingency	Influence on targeted patch by its surrounding based on relative position of other landscape structures.	Influence on targeted patch by its surrounding based on Euclidean distance.	Influence on targeted patch by its surroundings based on both Euclidean distance and the resistance costs (based on telemetry data, genetic data, or expert knowledge) of the landscape structures.

such as count of individuals per quadrat, stressing the pros and cons of each method.

12.2 Data Acquisition and Spatial Scales

Species are constantly moving throughout their life cycle and, as a result, different life stages (e.g. juveniles vs. adults) contribute differentially to their spatial distribution. Detection of species spatial distributions is directly related to the spatial scales of three levels of investigation: (i) the spatial scales of the *processes* acting on a species; (ii) the spatial scale of the *sampling design*; and (iii) the spatial scale of the *statistical methods* used in an analysis (Dungan et al. 2002). The spatial scale of study area is referred to as the spatial extent and determines which of these spatial patterns can be detected. A good rule of thumb is to delimit the size of the study area so that it is three to five times greater than the spatial scale of the main process under study (O'Neill et al. 1996, 1999). The sampling design to be used within the study area should then be guided by prior knowledge about the species dispersal ability: the spacing among samples should be greater than the daily animal movement, such as greater than the size of a territory or home range, to ensure that the same individual is not recorded twice.

Then, the spatial analysis can be performed to estimate the spatial pattern: (i) for the entire extent of the study area using *global spatial analysis*; or (ii) for each sampling location using *local spatial analysis* (Anselin 1995). In the first case – global spatial analysis – the process that generates the spatial pattern is assumed to have the same intensity over the entire study area (i.e. mean and variance): this is the assumption of stationarity of the process that is required by most spatial statistics. Note that the assumption of stationarity is about the processes, not the data. Yet, such information is often unknown such that an *exploratory spatial data analysis* (ESDA) should be performed, computing the mean and variance of the data using a moving window over the entire extent: if the values are similar then stationarity can be assumed and hence the global spatial statistics can be used; if the values vary too much from one subregion to another then local spatial statistics should be used. In the second case – local spatial analysis – the extent may be large enough to include changes in the intensity of the process or a mixing of different processes. In such circumstances, the estimation of spatial pattern at each sampling location allows the detection of subareas that have comparable values.

12.3 Point Data Analysis

For sessile species (e.g. plants or sessile marine species), nests, dens, or territories, we can be interested in determining whether such point data are spatially *underdispersed* (aggregated) or *overdispersed* (segregated). Spatial patterns of point data can be tested using point data analysis methods such as Ripley's K statistic (Ripley 1981). The K statistic quantifies the degree of spatial aggregation of points by counting the number of points (e.g. individual positions) in increasing circular search areas, revealing either the presence of spatial clustering or overdispersion of points (Diggle 2003; Illian et al. 2008; Wiegand and Moloney 2014). Note that Ripley's K functions by counting the number of points within a radius and is not measuring the degree of spatial autocorrelation but rather the degree of spatial aggregation. Indeed, the degree of spatial autocorrelation is measured on quantitative values where local deviations in values of neighboring locations are more likely to be similar. Then, spatial aggregation is measured on point data while spatial autocorrelation can be estimated for quantitative data (see Section 12.4).

Ripley's K function is therefore the overall mean number of points present within a circular search window of radius r:

$$\hat{K}(r) = \lambda^{-1} \sum_{i=1}^{n} \sum_{j=1}^{n} I_r(p_i, p_j)/n, \; for\, i \neq j \text{ and } r > 0, \tag{12.1}$$

where the point intensity, λ, is estimated as the density n/A, where A is the area of the study area and n is the total number of points. I_r is an indicator function that takes value 1 when point p_j is within a radius distance r of point p_i, (and 0 otherwise). As the search window to compute the number of points is circular, the resulting Ripley's K value is an isotropic cumulative count of all points at distances from 0 to r. The expected number of events under a completely random spatial process (CSR) is the area of a circle, πr^2. To test whether the computed Ripley's K function values are significant, the null hypothesis uses the CSR position of the points in the entire study area, creating a confidence interval around the expected values according to the radius size. Based on a Poisson distribution of points, the expected number of points within a radius distance r of an arbitrary point is proportional to the area of a circle (πr^2) times the density of points, λ, over the study area. A plot of the K values against the radius values r provides insights on how the spatial aggregation of the points varies according to distance (Figure 12.1a and b). At short distances, it is common to have few points and therefore the spatial pattern cannot be distinguished from random dispersion

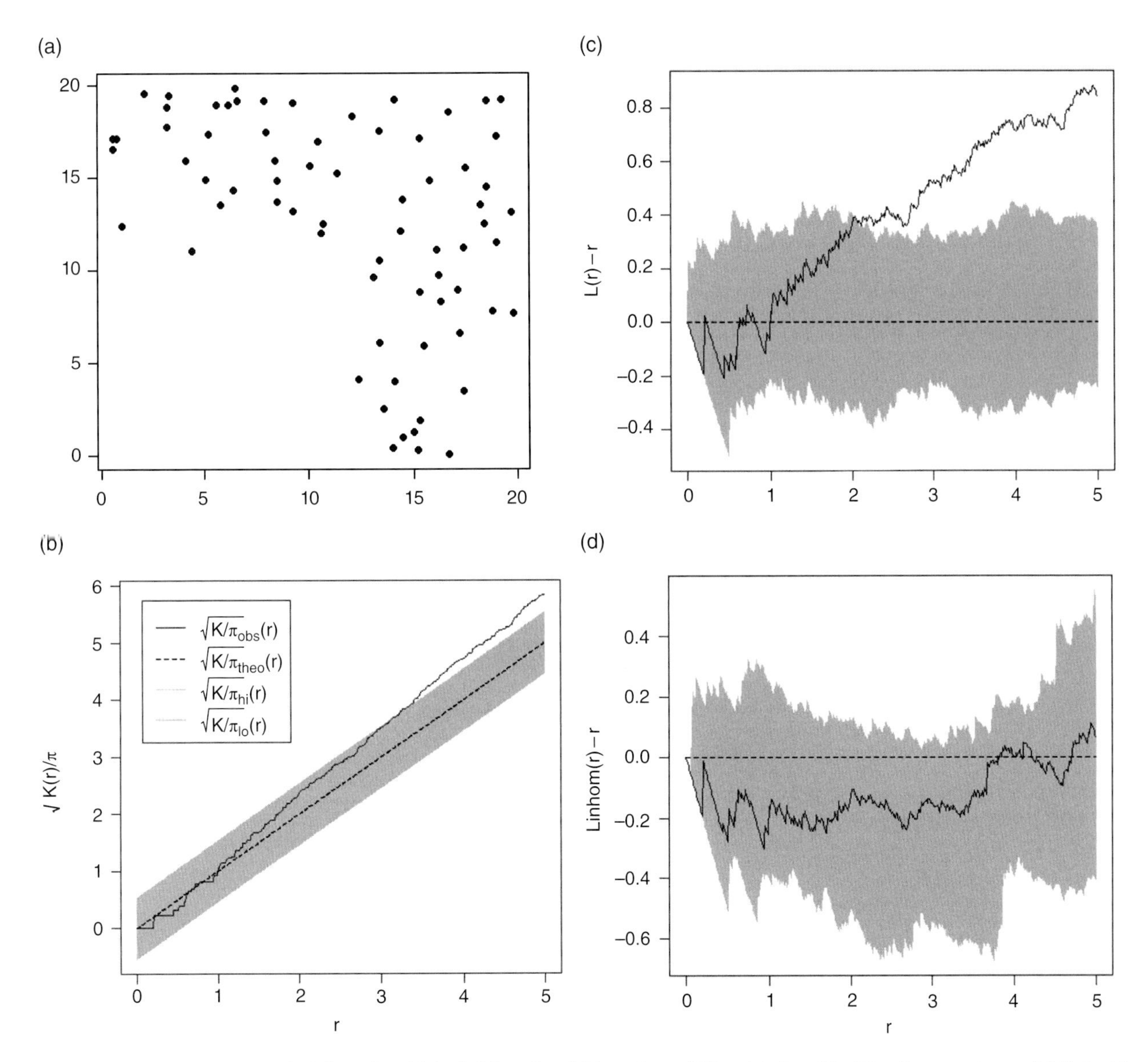

Figure 12.1 Point pattern analysis based on Ripley's K function. (a) Locations of 70 points in a 20×20 unit squared area. (b) Plot of the K values against the radius distance r where significant values are detected only at a large radius. (c) Plot of the K transformed into L values against the radius distance r where significant values are detected only at a large radius. (d) Plot of the L values against the radius distance r while correcting for the inhomogeneity of the study area as all the 70 points are in one sector. Based on the inhomogeneity test, the Ripley's values are not significant anymore for the large radius. The observed Ripley's values are indicated by the solid black lines, the expected values by the dashed lines, and the 95% confidence intervals by the gray region. To avoid edge effects, the maximum radius distance used is 5 units.

(CSR). As distance increases, either spatial aggregation (underdispersed) or segregation (overdispersed) can be significantly detected.

At large distances, the circular search window includes area outside the study area. To avoid such bias, it is recommended not to use a radius larger than half the smallest edge length of the study area. For example, if the study area is 10×20 m, then the maximum radius used to estimate Ripley's K should be at the most 5 m (10 m ÷ 2). In Figure 12.1, the study area is 20×20 units and to be conservative Ripley's K was only calculated up to a radius of 5 units.

Most software packages assume that the sampling frame for a study area is rectangular (or square). The `Programita` software accommodates irregularly shaped study areas using a mask to outline the study area (programita.org). Furthermore, as points next to the edge of the study area have fewer points around them than points in the middle of the study area, an edge effect that biases the measures of spatial aggregation may be present.

Various edge correction weights methods that account for the amount of the search area that is outside the study area are available (Haase 1995; Getzin et al. 2008; Schiffers et al. 2008).

To stabilize the variance of the K values and make the plot easier to interpret, the linear version of Ripley's K is commonly used: Ripley's L (Figure 12.1c). Under such a random spatial process, the expected value of $L(r)$ is zero (Ripley 1981; Diggle 2003). The CSR can be simulated using a Poisson distribution that assumes a complete census of all the points within the study area and a point process that is stationary and isotropic over the study area (Fortin and Dale 2005). Recording point data for sessile individuals is straightforward and feasible; it is less so for mobile individuals. For mobile individuals, surrogate information needs to be used to represent the location of an individual such as the centroid of an individual's territory (home range) or the location of a bird nest or den.

Often, Ripley's K function is used with sampled data that are a subset of all points instead of census data comprised of all points in a study area. Unfortunately, such use of sampled data will bias both the value of the Ripley's K statistic and its significance testing. While delineating the extent of the study area, one can then select an area that overlaps different types of soils or environmental conditions that affect the homogeneity of the spatial pattern. A lack of homogeneity of the study area will also result in biased Ripley's values (Wiegand and Moloney 2004; Schiffers et al. 2008). To correct the nonhomogeneity in the data, the modified inhomogeneous Ripley's function should be used (Getzin et al. 2008; Illian et al. 2008; Wiegand and Moloney 2014; Figure 12.1d).

Spatial pattern can be the result of several processes operating at different scales. Thus, a series of hierarchical null hypotheses, with their respective appropriate randomization procedures, allow for the testing of alternative hypotheses about the underlying processes that generate spatial patterns (Goovaerts and Jacquez 2004; Lancaster and Downes 2004). Such hypotheses can be hierarchically structured as illustrated by Melles et al. (2009) while studying nest patterns of Hooded Warblers (*Setophaga citrina*) in southern Ontario: (i) random within the study area such as complete spatial randomness based on Poisson distribution, the default in most spatial analysis; (ii) random within habitat with restricted randomization based on the spatial arrangement of land cover types; or (iii) random according to distance to mate with restricted randomization based on spatial arrangement of conspecifics.

Ripley's K is widely used even though its value is cumulative as the radius increases, so it can inhibit the scale-dependent patterns that one hopes to detect (Schurr et al. 2004). For example, spatial aggregation at short distances may generate above-average aggregation at larger distances. An alternative to Ripley's K is the pair-correlation function $g(r)$ (Stoyan and Stoyan 1994; Wiegand and Moloney 2004; Melles et al. 2009) which computes the mean number of points within a given ring radius rather than within a circle. Under complete spatial randomness, the $g(r)$ statistic is one; values above one indicate clumping or spatial aggregation, and values below one indicate segregation or uniformly spaced points.

12.4 Abundance Data Analysis

The presence of a spatial pattern implies that the values of the quantity measured are more similar at short distances than expected by chance, which is the definition of spatial autocorrelation. The processes that generate spatial autocorrelation are numerous and act on species at several spatial and temporal scales. While there are several processes influencing spatial patterns, the spatial statistics that estimate the degree of spatial autocorrelation cannot differentiate the source(s) of the spatial pattern. To determine the relative importance of various processes to the spatial pattern(s), one should either use an experimental design to separate the effects of some processes, or spatial regression (see Section 12.8). To detect significant spatial patterns from quantitative data (e.g. species abundance data, environmental data) it is recommended to have at least 30–50 samples and that the spacing between the samples be smaller than the habitat patch size or species' home range size (Dale and Fortin 2014).

The most commonly used measure of spatial autocorrelation is based on a Pearson correlation coefficient and has been proposed by Moran (1948): the Moran's I. The expected value of autocorrelation of absence of spatial autocorrelation is close to 0: $E(I) = -(n-1)^{-1}$ (Cliff and Ord 1981). The estimation of spatial autocorrelation computes deviation from the value at one location to the mean of the distribution of the variable using an increasing ring radius d (known as distance class) as follows:

$$I(d) = \left(\frac{n}{Wd}\right)\frac{\sum_{\substack{i=1\\i\neq j}}^{n}\sum_{\substack{j=1\\j\neq i}}^{n} w_{ij}(d)(x_i-\bar{x})}{\sum_{i=1}^{n}(x_i-\bar{x})^2}, \tag{12.2}$$

where x_i and x_j are the values of the variable x at location i and j. $w_{ij}(d)$ is a weight matrix that indicates if locations i and j are in the distance class d. $W(d)$ is the sum of $w_{ij}(d)$. Positive autocorrelation values imply that values of the variable at d distance are similar while negative autocorrelation values indicate that different values of the variable are at d distance. While the denominator should standardize and bound the range of the Moran's I values between −1 to +1, when there are too few pairs of locations in a distance class d the estimated value is unstable and can fall outside the range of −1 and +1. Such larger or smaller values than +1 or −1 occur mostly at the largest distances where there are the fewest pairs contributing to the statistic (Figure 12.2a).

Similarly to Ripley's K, it is recommended to not compute Moran's I for a larger d distance class than half the length of the smallest edge of the study area (Cliff and Ord 1981; Fortin and Dale 2005). Hence if the study area is 20 × 50 m, then the maximum radius used to estimate the degree of spatial autocorrelation should be 20 m divided by 2 (i.e. 10 m).

Moran's I is widely used for several reasons but the most important one is the ease of its interpretation as it is similar to the Pearson's correlation coefficient. Yet, Moran's I can produce biased estimates of spatial autocorrelation when the distribution of the values of the quantitative variable is not normally distributed due to skewed distribution. To avoid measuring the degree of spatial autocorrelation based on deviation to the mean of the variable, one can use Geary's c statistic, $c(d)$, where:

$$c(d) = \left(\frac{n-1}{2W(d)}\right) \frac{\sum_{\substack{i=1 \\ i \neq j}}^{n} \sum_{\substack{j=1 \\ i \neq j}}^{n} w_{ij}(d)\left(x_i - x_j\right)^2}{\sum_{i=1}^{n} \left(x_i - \bar{x}\right)^2}. \tag{12.3}$$

Geary's c is a distance measure based on the deviations between values at d distances. Geary's c varies from 0 which indicates positive autocorrelation up to the expected value of absence of spatial autocorrelation, $E(c) = 1$, while greater values than 1 up to 2 indicate negative autocorrelation (Figure 12.2c). With too few pairs of sampling locations or when outlier values due to skewed data are used to compute Geary's c, the deviations

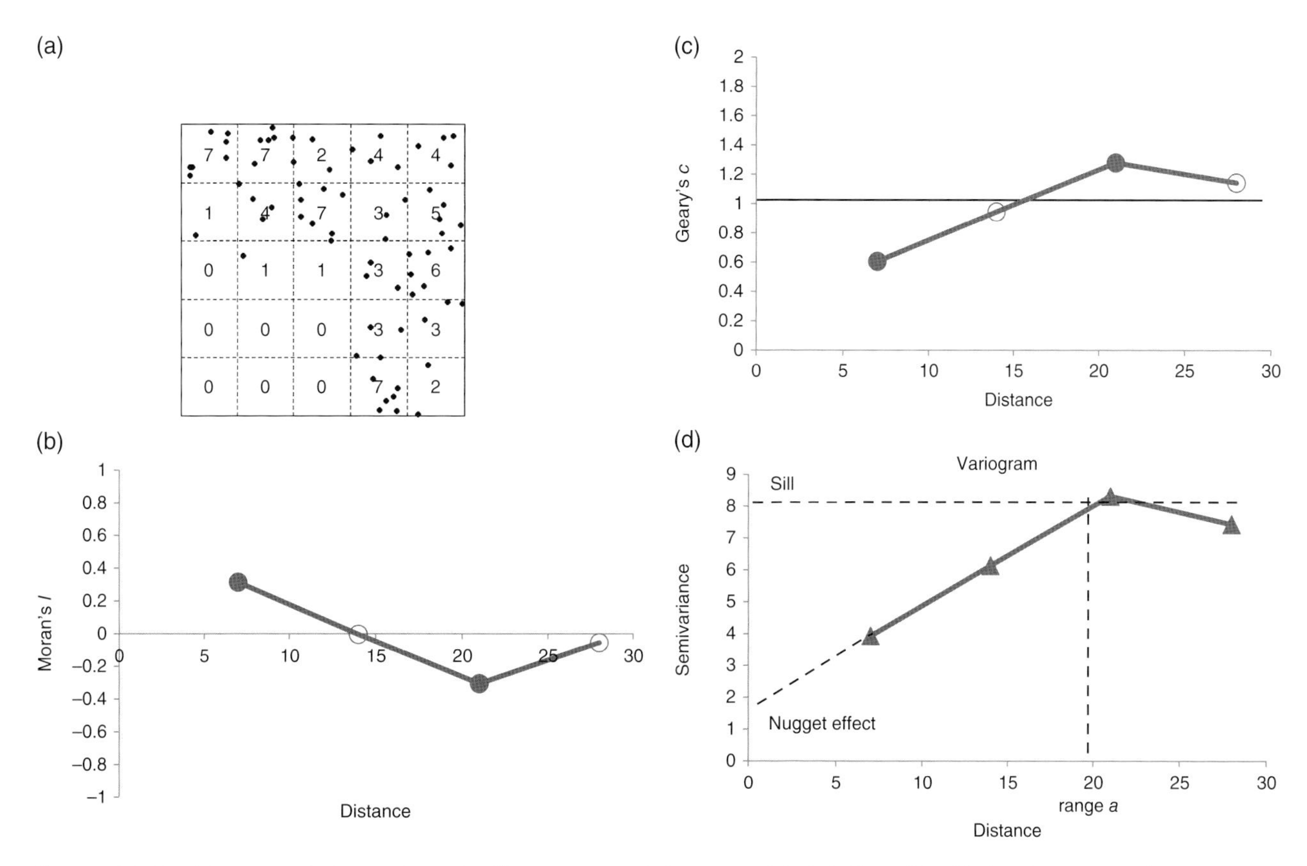

Figure 12.2 Spatial autocorrelation analyses: significant autocorrelation at $\alpha = 0.05$ is indicated by filled circles; nonsignificant autocorrelation is indicated by open circles. (a) Locations of $\lambda(s)$ resampled using 5 by 5 quadrats where numbers of points by quadrat are shown. (b) Moran's I correlogram where at the shortest distance the spatial autocorrelation is significantly positive moving to significantly negative at the third distance class. The two other values are not significant. (c) Geary's c correlogram is a mirror image of the Moran's I correlogram but is interpreted in term of distance so positive autocorrelation is close to 0 and negative autocorrelation is close to 2. (d) Variogram where the key parameters (spatial range a, sill, nugget effect) are shown.

between adjacent locations are squared, leading to extremely biased values of spatial autocorrelation.

Both Moran's I and Geary's c coefficient value of spatial autocorrelation against distance can be tested for their significance using either an asymptotic test or a randomization test (Cliff and Ord 1981; Legendre and Legendre 2012; Figure 12.2).

The degree of spatial autocorrelation can also be computed in terms of spatial variance using geostatistics (Matheron 1970; Journel and Huijbregts 1978; Cressie 1993). The estimation of spatial variance, $\hat{\gamma}(h)$ (also known as *semi-variance*), is:

$$\hat{\gamma}(h) = \frac{1}{2n(h)} \sum_{i=1}^{n(h)} (z(s_i) - z(s_i + h))^2, \qquad (12.4)$$

where z is the value of the variable at the ith sampling location s_i, and $n(h)$ is the number of pairs of sampling locations located at distance h apart. The equation of the semi-variance and Geary's c are similar except that Geary's c is standardized whereas the semi-variance is not. As such, the semi-variance values are in the same units as the data $z(s_i)$. The plot of the semi-variance values against the spatial lag h is a *variogram* (Figure 12.2d). At short distances the variance of a quantitative variable varies according to distance up to a given distance or "range" (a), where the variance plateau at the "sill" is no longer influenced by distance (Figure 12.2d). When the value of the semi-variance is interpolated to intercept the ordinate y-axis, the value at the intercept is the nugget effect (Figure 12.2d). The magnitude of the nugget effect is due to several issues related to the sampling design such as the size of the sampling units, the spatial distance among the sampling units, and the number of sampling units, as well as the degree of spatial autocorrelation of the variable (Fortin 1999; Fortin and Dale 2005). The estimated parameters from the variogram are subsequently used in a spatial interpolation technique known as Kriging (see Section 12.5).

The spatial autocorrelation and spatial variance measures as presented above are *isotropic* such that only distance classes affect the values of spatial variance and not directionality. Yet, all can be computed using both distance and angle (directionality) classes such that *anisotropic*, directional, and spatial patterns can be detected. When both distance and angle classes are used to compute anisotropic correlograms or variograms, fewer pairs of locations by distance classes are used as the pairs are divided into angle classes, potentially resulting in biased estimates of autocorrelation. To minimize bias, one should use larger spatial lags.

Last, depending on the spatial arrangement of the sampled locations and the species distribution, it is possible that the assumption of stationarity of the process does not prevail, such that mean and variance vary over the entire study area. In such circumstances, one should either subdivide the study area into homogeneous subareas using spatial clustering or boundary detection methods (Dale and Fortin 2014), or use local spatial statistics known as the *local indicator of spatial autocorrelation* (LISA, Anselin 1995; Sokal et al. 1998), which compute spatial autocorrelation for each sampling location. All of the above spatial autocorrelation measures can then be rewritten into local spatial statistics: local Moran's I_i and local Geary's c_i. One local spatial statistic that is often used when analyzing quantitative data is the local Getis' G_i statistic (Getis and Ord 1996; Dale and Fortin 2014). Unlike Moran's I and Geary's c, the local G_i statistic states the degree of *spatial clustering* of the quantitative variable at each sampling location i. In essence, local G_i is the ratio of the average value of a variable at the sampling location i given the values within a given spatial lag. If such ratio is greater than the global mean of the variable, it identifies clusters of high values known as *hot spots*. Alternatively, if the ratio is smaller than the mean, then clusters of low values or *cold spots* can also be identified. Local G_i can be used as an ESDA to determine if the assumption of stationarity about the mean of the processes is valid: if not significant, clusters (hot or cold spots) are not detected and the assumption of stationarity can be assumed; if significant hot and cold spots are detected and the assumption of stationarity should not be assumed.

12.5 Spatial Interpolation

Based on animal abundance having spatial pattern, it is possible to interpolate abundance at unsampled locations using techniques for *spatial interpolation*. The key underlying principle of spatial interpolation techniques is spatial autocorrelation among the values of a quantitative variable, such that nearby values are more similar in their magnitude than those further away. The task in spatial interpolation lies, therefore, in determining the spatial distance up to which values are spatially autocorrelated. Such a spatial distance is often referred to as the spatial range in geostatistics, or the zone of influence in spatial statistics literature (Cressie 1993; Legendre and Fortin 1989).

Each spatial interpolator has different properties ranging from *global* spatial interpolators such as polynomial regressions, *local* spatial interpolators such as the inverse distance weighting function, and geostatistical techniques such as Kriging (Cressie 1993; Haining 2003). Hence, the spatial interpolation methods to use depends on the goal of the analysis: when the goal is to produce a map for publication, one may use the inverse distance weighting function; when the goal is to predict values at unsampled locations, Kriging techniques should be used. Here, I review only the local spatial interpolator based on Kriging.

The first step of Kriging consists of fitting the best theoretical variogram model from a series of analytical models such as linear, spherical, exponential, and Gaussian to the observed variogram based on the sampled data. From a fitted theoretical variogram model, one can then estimate the three parameters: the spatial range a, the sill, and the nugget effect (Figure 12.2d). As Kriging aims to maximize the spatial interpolation based on the degree of spatial autocorrelation, the fitting of the parameters is often estimated for the shorter spatial lag distances h. Hence, variograms are usually computed up to two-thirds of the longest distance between the sampling locations, to avoid having too few pairs of locations to compute the semi-variance values as a function of distance. Based on the three estimated parameters from the fitted theoretical variogram, Kriging, which is a series of linear equations, optimizes the best combination of weights for each sampling location to interpolate values at unsampled locations. To solve this system of linear equations, the weights sum to 1 where there are more equations than unknown parameters to estimate.

There are three advantages to using Kriging to spatially interpolate data at unsampled locations: (i) it produces the *best linear unbiased predictions* (BLUP), implying that at the sampling locations the observed values are returned (Figure 12.3a); (ii) the predictions are based on local weights that maximize the degree of spatial autocorrelation with which the predictions are computed; and (iii) given that the interpolated data are based on a theoretical variogram model, each prediction has an associated standard error (Figure 12.3b). This last property of Kriging is to identify the areas of high standard error values. In a second phase of analysis, one can then use these maps to optimize the spatial sampling design where more sampling locations could be added to reduce the standard error. One needs to keep in mind that these standard errors are based on the theoretical model selected and the estimation of the parameters. If an incorrect theoretical variogram model is selected to fit the observed variogram or if parameter estimates are biased (e.g. strong value of sill, too-small or too-large spatial range a), then the interpolated values will be biased as well.

While Kriging was developed to model quantitative data, through the years the technique was expanded to include several variants, including: blocked Kriging, punctual Kriging, universal Kriging, multivariate Kriging, stratified Kriging, co-Kriging, and indicator Kriging (Cressie 1993). When Kriging is used to map species ranges in species atlases, stratified Kriging should be used as it masks the area where the probability of occurrence of the species is zero, such as open water or urban areas. Then to model qualitative data, an indicator Kriging algorithm can be used to obtain probability occurrence values. For example, to mitigate sampling bias of The Ontario Breeding Bird Atlas (OBBA), Cadman et al. (2007) based surveys on 10 × 10 km blocks and Polakowska et al. (2012) used indicator Kriging to model the probability of occurrence of birds in southern Ontario using only adjacent sampling units to account for location effects on species occurrence. Then the probability of occurrence of the birds was used to model the relationship between birds and landscape cover types.

(a)

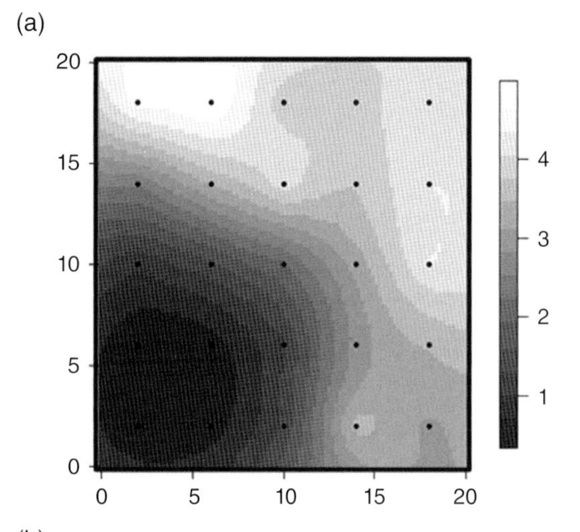

(b)

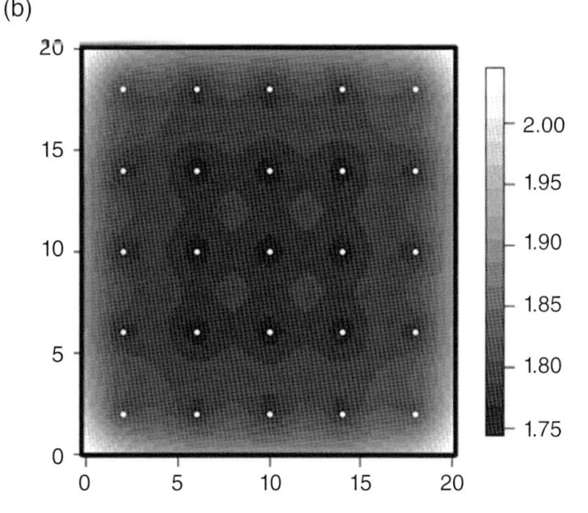

Figure 12.3 Spatially interpolated quantitative values developed by Kriging, using quantitative values based on the 25 quadrats as shown in Figure 12.2a. (a) Kriged (spatially interpolated) values based on a spherical model having a spatial range of 4.5 units, a nugget effect of 5.0, and a sill (including the nugget effect) of 10.5. (b) The standard errors associated with each predicted value where they are the highest in the region and where there are no points in the quadrats. The open circles indicate locations of the 70 points.

Last, when there are spatial patterns that are embedded in the data, it is recommended to first remove the large-scale trend before Kriging is performed, using a technique known as "universal Kriging."

12.6 Spatial Density Mapping

With the availability of novel sensors, it is possible to analyze individual movement using telemetry data. Several research questions can be addressed with such point data ranging from delineation of *home ranges* (Long and Nelson 2012) or up to *species ranges* (Fortin et al. 2005). Though, the golden rule of any spatial analysis is to plot the data first to have an understanding of the spatial distribution of individuals, populations, and species. Such visualization of the spatial aggregation of point data (as illustrated in Figure 12.1a) can be obtained using a *kernel density function* (Worton 1989). Kernel density functions estimate the density in term of intensity, $\lambda(s)$, for a given location s given the number of points within a kernel search area:

$$\hat{\lambda}_\tau(s) = \frac{1}{\delta_\tau(s)} \sum_{1=1}^{n} \frac{1}{\tau^2} k\left(\frac{s - s_i}{\tau}\right), \qquad (12.5)$$

where intensity $\lambda(s)$, at unsampled location s and given sampled locations s_i within the τ bandwidth of the k kernel function, and δ_τ is an edge correction factor. The shape of the kernel function can vary according to the weight one wants to apply to nearby points rather than those further away. The most commonly used kernel function, k, is a Gaussian function. Then, the size of the kernel search area is controlled by the value of τ bandwidth parameter: smaller bandwidth results in small concentrated intensity values around each observed point, and larger bandwidth results in a smooth gradient of intensity values (Figure 12.4).

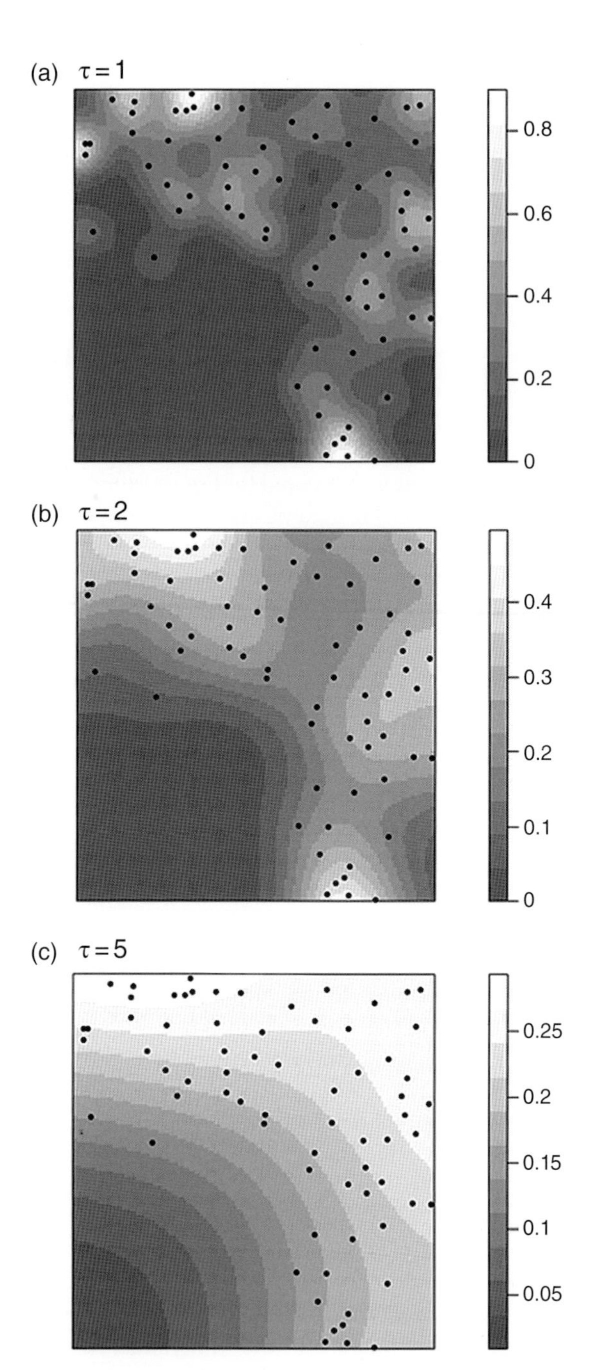

Figure 12.4 Kernel density maps based on the same 70 points as shown in Figure 12.1a. (a) Intensity map based on a τ bandwidth of 1. (b) Intensity map based on a τ bandwidth of 2. (c) Intensity map based on a τ bandwidth of 5.

12.7 Multiple Scale Analysis

Current ecological studies of populations can be carried out over larger regions due to the availability of novel data acquisition equipment and analytical tools. When populations are studied at large spatial scales, several processes may have shaped their spatial patterns (Table 12.1; Fortin et al. 2012). In such cases, one should carry out an ESDA aiming to identify whether there are spatial patterns at multiple spatial scales reflecting large-scale gradients and regional patchiness (see Chapter 2). *Multiscale analysis* can be performed using multiresolution techniques such as *wavelet decomposition* for lattice/grid data such as remotely sensed data (James and Fortin 2012), while MEM can be used with regularly or irregularly spaced spatial data (Dray 2011; Dray et al. 2012). The inherent principle of these decomposition methods is a spectral analysis at multiple frequencies and amplitudes (Griffith 1996; Borcard and Legendre 2002; Dray 2011). Here, I will summarize the key features of the MEM, which is an eigenfunction technique that can be used with sampled data.

The *spectral decomposition* was first proposed by Borcard and Legendre (2002) to be performed using a principal coordinates analysis (PCoA) on a Euclidean distance matrix based on the *x-y* coordinates of the sampling locations. The approach was first named the principal coordinates of neighbour matrices (PCNM) method, but nowadays it is also referred to as distance-based MEM (dbMEM). Another way to perform a spectral decomposition with the MEM approach is to use a $\mathbf{D_W}$ matrix that is the Hadamard product (i.e. element-to-element) of an undirected connectivity matrix, **C**, based on the relative position of the sampling location and a weight matrix, **W** (e.g. Euclidean distance). The $\mathbf{D_W}$ is a symmetric matrix that is centred. From the analysis of the $\mathbf{D_W}$, orthogonal eigenvectors can be obtained as well as eigenvalues that are related to Moran's *I* coefficient values. The largest eigenvalue corresponds to the largest value of Moran's *I*, while the last, smallest eigenvalue is the smallest value of Moran's *I*. The spatial eigenvectors are implicitly isotropic and orthogonal. Yet, there are $n - 1$ spatial eigenvectors so when the sample size n is high, there is little difference between subsequent spatial eigenvectors. The next step is therefore to identify the key spatial scales of the spatial pattern of the data using a forward selection procedure that relates species data and MEMs as spatial predictors (Legendre and Legendre 2012). Once key spatial scales are determined, they can provide insight on which factors and processes shape the spatial patterns. For example, Munoz (2009) used dbMEM to determine the key spatial scales in simulated metapopulation data where known habitat structures were combined.

For illustration purposes, I use estimates of abundance based on telemetry data from wolves (*Canis lupus*) in Algonquin Provincial Park in Ontario (Rutledge et al. 2009) to determine the key spatial scales using the dbMEM method (Figure 12.5a). Wolf abundance data were measured using a 5 × 5 km sampling unit and counting the number of distinct wolf individuals within those units. A total of 238 sampling units had non-zero values for a total 237 ($n - 1$) spatial eigenvectors. To reduce the number of spatial eigenvectors, a forward selection procedure relating the abundance data and 238 dbMEMs was computed. The forward selection procedure reduces the number of dbMEMs to 151. Figure 12.5b shows a few of the 151 dbMEMs to illustrate how they reflect large (e.g. dbMEM 1 is a large east–west trend; dbMEM 3 is a large north–south trend) and intermediate spatial scales (dbMEM 30 shows a series of intermingled small patches) of wolf abundance in Algonquin Provincial Park.

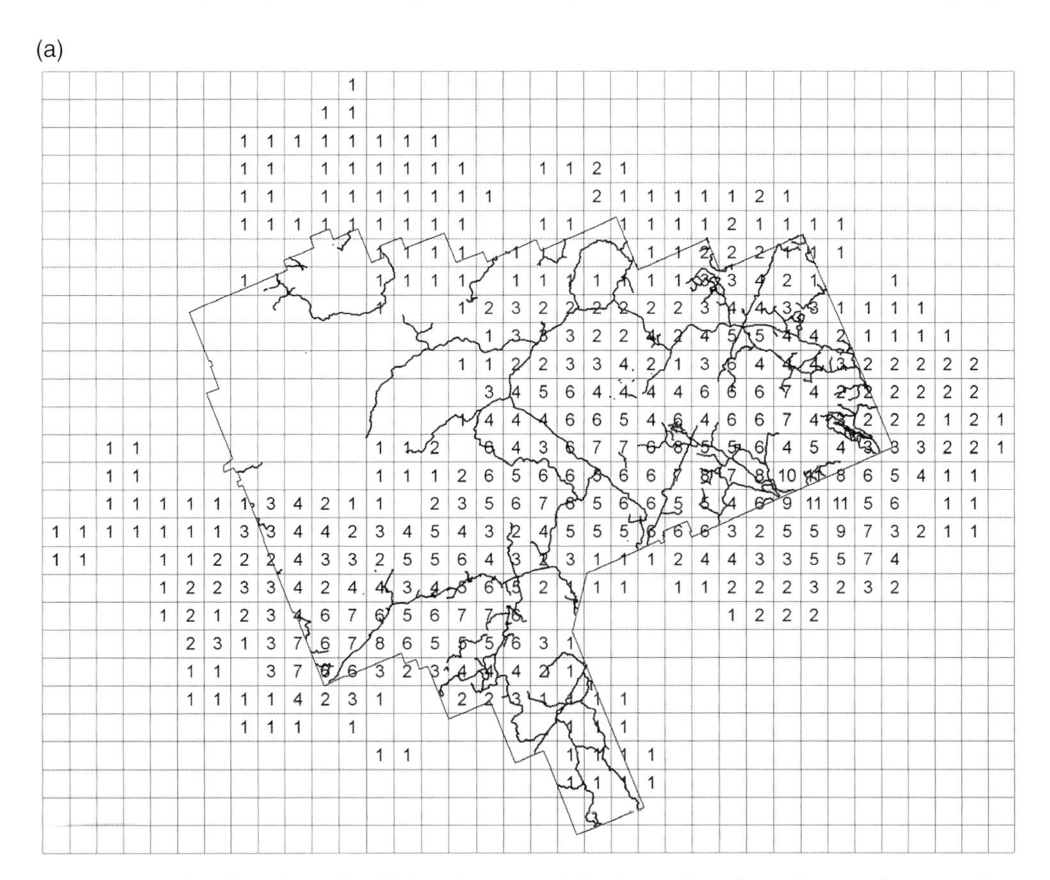

Figure 12.5 Multiscale analysis of wolf abundance spatial distribution based on telemetry data using distance-based MEM (dbMEM) (previously known as Principal Coordinates of Neighbour Matrices [PCNM]). (a) Study area in Algonquin Provincial Park in Ontario (as outlined) covering an area of 175 × 125 km where abundance is sampled in 5 × 5 km sampling units ($n = 440$ sampling units). (b) Example of dbMEM with large spatial scale patterns (two first rows: dbMEMs 1–10) and small spatial scale patterns (two lower rows: dbMEMs 21–118).

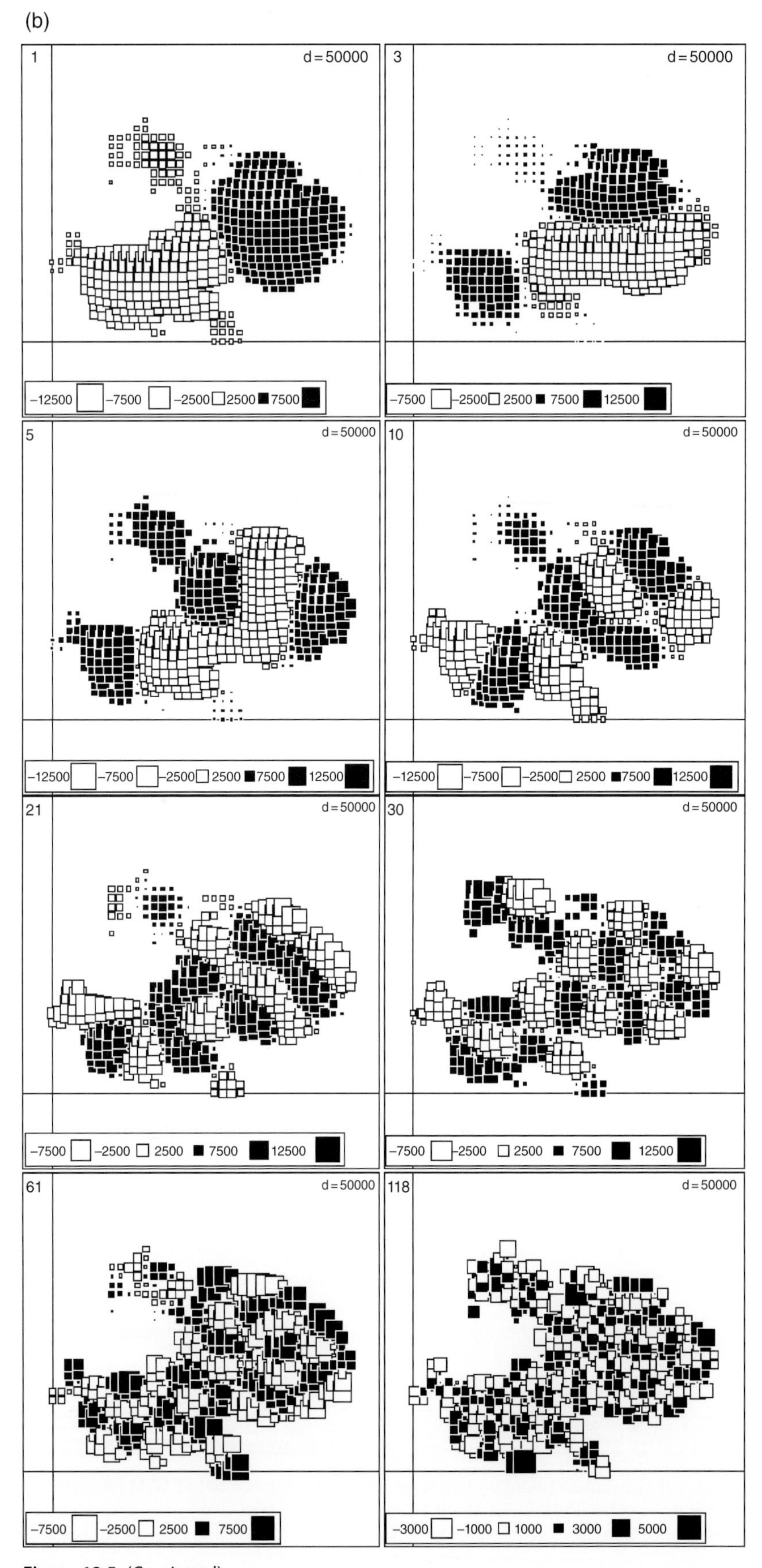

Figure 12.5 (Continued)

12.8 Spatial Regression

The detection of spatial patterns in species abundance is part of ESDA. Once the spatial pattern and scales of the data have been determined, the next step is to relate factors and processes that can explain such spatial patterns. When there is no spatial structure in the data, a linear regression framework is used:

$$y \sim \beta \mathbf{X} + \varepsilon, \tag{12.6}$$

where y is the response variable, β is a vector of regression coefficients of fixed predictors **X**, and ε are the random errors. In linear regression the errors are assumed to be independent such that $\text{var}(\varepsilon) = \sigma^2 \text{I}$, where I is the $n \times n$ identity matrix. However, the assumption of independence of the errors is often violated when analyzing ecological data. It is therefore important to perform a residual analysis to determine whether or not the residuals are significantly spatially autocorrelated. Such residual analysis can be performed by computing Moran's I correlogram with the residual values. Different reasons can explain spatially structured residuals: (i) the model is lacking some key covariates/predictors that could explain the spatial structure of the response (Melles et al. 2011); and (ii) there exist mismatches among the spatial patterns of the response and predictors.

If no new predictors can be added or if the scales match but the residuals are still spatially autocorrelated, one should use spatial regression models. There are three families of spatial regression methods: spatial filtering models, spatial error models (SEM), and local spatial regressions (Beale et al. 2010).

Spatial filtering models are also known as *spatial lag regressions*. In essence, spatial filtering regressions accounts for the effects of spatial structure in the predictor variables using adjacent neighboring samples based on either sample topology for irregularly spaced data or eight-neighbors for regularly spaced data. The most commonly used spatial filtering models are: *conditional autoregressive model* (CAR) and *simultaneous autoregressive model* (SAR). When the data show patterns at multiple spatial scales, however, one should use instead a spatial filtering model that includes spatial predictors based on MEM such that local, intermediate, and large spatial structures can be accounted in the regression (Bini et al. 2009; Beale et al. 2010; Manel et al. 2012). As there are as many MEM spatial predictors as there are sampling locations, one needs to perform a selection procedure (e.g. forward selection) to include in the regression only the key spatial scales that are having a significant influence of the values of the response variable. The selected MEM spatial predictors can be indicators of the spatial scales at which unsampled processes are shaping the values of the response variable.

For example, to relate adaptive genes to environmental conditions, Manel et al. (2012) used broad scaled MEMs to account for processes that affected the genetic structure of plants over the European Alps. We can also return to the earlier example of wolf abundance in Algonquin Provincial Park and the 151 dbMEMs obtained to compute a spatial filtering regression. When all 151 dbMEMs are used as spatial predictors, the adjusted R^2 of the linear regression is 0.87. Such high adjusted R^2 explaining most of the variability in wolf abundance solely based on spatial predictors is quite impressive. When only the 8 dbMEMs illustrated in Figure 12.5 are used, the adjusted R^2 drops to 0.27. This example illustrates the fact that spatial eigenvectors can improve the fit of the regression very well. Unfortunately, such spatial predictors do not provide any predictive power as they are specific to the study area analyzed. Hence, when possible, one should therefore include variables that correspond to factors and processes that can explain the response variables rather than using only the relative spacing between sampling locations in the regression.

The next type of spatial regression accounts for the spatial structure in the errors of the model directly and is called SEM. Generalized least-squares regression (GLS) is recommended as a good method to account for spatial errors (Beale et al. 2010). In GLS, the spatial structure is parameterized in the errors as a spatial covariance matrix using an inverse distance function, an autoregressive function, or a variogram model. The parameterization of the covariance matrix is achieved through an interactive process, between fitting the regression coefficients and then fitting the spatial structure of the residuals. Regression Kriging is also another spatial error model that is similar in essence to GLS (Hengl et al. 2004).

Last, local spatial regression models such *geographically weighted regressions* (GWR; Fotheringham et al. 2002) model as many regressions as there are sampled locations. GWR is the regression equivalent of local Moran's statistic. As GWR is always fitting the best model for each sampling location, the residuals are not spatially structured. The major drawback of GWR, however, is that they overfit the data as no single model exists for prediction purpose. Yet, GWR can be used as an ESDA tool to identify subregions where to relationship between the response and the predictors are comparable. For example, McNew et al. (2013) used GWR to model habitat selection of Greater Prairie-Chickens (*Tympanuchus cupido*) because the assumption of stationarity was not valid. The GWR analysis determined that nest site selection by female prairie-chickens occurred at multiple spatial scales. To determine how the distribution of Ovenbirds (*Seiurus aurocapilla*) in southern Ontario can be explained by forest cover, Fortin and Melles (2009) compared OLS, regression Kriging, and GWR. They found that although the R^2 of the OLS (ordinary least squares) was 0.43, the residuals of OLS showed significant spatial autocorrelation. Then with the GWR, the pseudo-R^2 ranged between 0.19 and 0.82 and there was no significant spatial autocorrelation in the residuals. Last, the R^2 of the regression Kriging was 0.68 and as for the GWR there was no significant spatial autocorrelation in the residuals. This example illustrates the importance of using spatial regression models to account for spatial structure in the data. Furthermore, the wide range of pseudo-R^2 values can be used as an indicator that some regionalization exists over southern Ontario and that the study area should be divided into sub-regions having comparable range of values. Hence in such a case, GWR can be used as ESDA tool.

12.9 Software Tools

Free software packages such as `PASSaGE` (Rosenberg and Anderson 2011) and `SAM` (Rangel et al. 2010) perform most of the spatial statistics and spatial regressions presented in this chapter. Spatial aggregation based on point data and kernel density functions can be computed using the `spatstat package` in `R`. The `Programita` software (http://programita.org) computed the univariate, bivariate and multivariate versions of point pattern analysis methods (e.g. Ripley's *K*, Ripley's *L*, pair correlation function). Furthermore, `Programita` accommodates irregularly-shaped study areas using a mask to outline the study area. Spatial autocorrelation (e.g. Moran's *I*) based on quantitative data can be computed through different `R` libraries: `ade4`, `ape`, `ncf`, `pgirmess`, `raster`, `spdep`, and `mpmcorrelogram`. Semivariance and Kriging can be estimated using `geoR` in `R` or `GS+` commercial software (https://www.gammadesign.com). Spatial regressions such as GLS can be computed using `gls` and `nlme` in `R` while GWR can be computed using `spdep` in `R` or with the `GWR4` software (https://geodacenter.asu.edu/software/downloads/gwr_downloads). Finally, dbMEM can be computed using `vegan` in `R`.

12.10 Online Exercises

The online exercises for this chapter illustrate the basics of point pattern analysis, area pattern analysis, and spatial interpolation. Exercise 1 provides an example of spatial data in a quadrat sampling scheme. The tools of the `spatstat package` in `R` are used to calculate different versions of Ripley's *K* and to calculate kernel density functions. The effect of outliers is illustrated by repeating the calculations on different subsets of the dataset. Exercise 2 illustrates spatial modeling where point data are transformed into abundance data. The `correlog` function in the `ncf` package is used to calculate Moran's *I* and to examine patterns of spatial autocorrelation. Last, functions in the `geoR`, `fields`, and `maps` packages are used to illustrate spatial interpolation based on kriging techniques. Potential useful R packages are also illustrated in Fletcher and Fortin (2019).

12.11 Future Directions

Understanding population spatial dynamics requires analyzing the spatial pattern of ecological data. I have presented an overview of key spatial statistics, providing their advantages and disadvantages. The most important take-home message from a statistical perspective is that all descriptive and inferential spatial statistics are sensitive to the assumption of stationarity, the sampling design used to collect the data, and the parameters selected to perform spatial statistics analyses. As larger regions are studied, the likelihood of violating the assumption of stationarity increases, requiring the use of a local version of the spatial statistics that will test for spatial structure at each sampling location. It is therefore important to perform an ESDA to first determine the degree of spatial autocorrelation in the data allowing one to establish which inferential model to use. Furthermore, the ESDA can reveal that either multiscale or a regionalized analysis is required before performing a spatial regression. Then with the availability of spatial and temporal data, more spatiotemporal analysis of ecological data will be become possible. Most spatial statistics have been modified to analyze bivariate data where the two variables can be measurements taken at different time periods. Cressie and Wikle (2011) presented various spatiotemporal statistics and how to use them. Yet, the power of detection of the signal will strongly depend on the ratio of data spacing in space and time, and the intensity of the structures analyzed.

Glossary

False spatial autocorrelation Spatial autocorrelation generated by exogenous processes such as disturbances and spatially structured environmental factors.

Spatial autocorrelation Spatial pattern where values of a variable at nearby locations are more similar (positive autocorrelation) or less similar (negative autocorrelation) than expected due to random chance.

True spatial autocorrelation Spatial autocorrelation generated by endogenous processes such as dispersal, species interaction, and species behavior.

References

Anselin, L. (1988). *Spatial Econometrics: Methods and Models*. New York: Kluwer Academic Publisher.

Anselin, L. (1995). Local indicators of spatial association: LISA. *Geographical Analysis* 27: 93–115.

Beale, C.M., Lennon, J.J., Yearsley, J.M. et al. (2010). Regression analysis of spatial data. *Ecology Letters* 13: 246–264.

Bini, L.M., Diniz, J.A.F., Rangel, T.F.L.V. et al. (2009). Coefficient shifts in geographical ecology: an empirical evaluation of spatial and non-spatial regression. *Ecography* 32: 193–204.

Borcard, D. and Legendre, P. (2002). All-scale spatial analysis of ecological data by means of principal coordinates of neighbour matrices. *Ecological Modelling* 153: 51–68.

Cadman MD, Sutherland DA, Beck GG, et al. (eds) (2007) *Atlas of the Breeding Birds of Ontario, 2001–2005*. Bird Studies Canada, Environment Canada, Ontario Field Ornithologists, Ontario Ministry of Natural Resources, Ontario Nature Toronto.

Cliff, A.D. and Ord, J.K. (1981). *Spatial Processes: Models and Applications*. London: Pion.

Clifford, P., Richardson, S., and Hémon, D. (1989). Assessing the significance of correlation between two spatial processes. *Biometrics* 45: 123–134.

Cressie, N.A.C. (1993). *Statistics for Spatial Data*, revised ed. New York: Wiley.

Cressie, N. and Wikle, C.K. (2011). *Statistics for Spatio-Temporal Data*. Hoboken NJ: Wiley.

Dale, M.R.T. and Fortin, M.-J. (2014). *Spatial Analysis: A Guide for Ecologists*. Cambridge: Cambridge University Press.

Diggle, P.J. (2003). *Statistical Analysis of Spatial Point Patterns*, 2e. London: Edward Arnold.

Dray, S. (2011). A new perspective about Moran's *I* coefficient: spatial autocorrelation as a linear regression problem. *Geographical Analysis* 43: 127–141.

Dray, S., Pélissier, R., Couteron, P. et al. (2012). Community ecology in the age of multivariate multiscale spatial analysis. *Ecological Monographs* 82: 257–275.

Dungan, J.L., Perry, J.N., Dale, M.R.T. et al. (2002). A balanced view of scale in spatial statistical analysis. *Ecography* 25: 626–640.

Dutilleul, P. (1993). Modifying the *t* test for assessing the correlation between two spatial processes. *Biometrics* 49: 305–314.

Fletcher, R. and Fortin, M.-J. (2019). *Spatial Ecology and Conservation Modeling: Applications with R*. Cham: Springer.

Fortin, M.-J. (1999). Effects of sampling unit resolution on the estimation of the spatial autocorrelation. *Écoscience* 6: 636–641.

Fortin, M.-J. and Dale, M.R.T. (2005). *Spatial Analysis: A Guide for Ecologists*. Cambridge: Cambridge University Press.

Fortin, M.-J., James, P.M.A., MacKenzie, A. et al. (2012). Spatial statistics, spatial regression, and graph theory in ecology. *Spatial Statistics* 1: 100–109.

Fortin, M.-J., Keitt, T., Maurer, B. et al. (2005). Species ranges and distributional limits: pattern analysis and statistical issues. *Oikos* 108: 7–17.

Fortin, M.-J. and Melles, S.J. (2009). Avian spatial responses to forest spatial heterogeneity at the landscape level: conceptual and statistical challenges. In: *Real World Ecology: Large-Scale and Long-Term Case Studies and Method* (eds. S. Miao, S. Carstenn and M. Nungesser), 137–160. New York: Springer.

Fotheringham, A.S., Brunsdon, C., and Charlton, M. (2002). *Geographically Weighted Regression: The Analysis of Spatially Varying Relationships*. Chichester: Wiley.

Getis, A. and Ord, J.K. (1996). Local spatial statistics: an overview. In: *Spatial Analysis: Modelling in a GIS Environment* (eds. P. Longley and M. Batty), 261–277. Cambridge: GeoInformation International.

Getzin, S., Wiegand, T., Wiegand, K., and He, F.L. (2008). Heterogeneity influences spatial patterns and demographics in forest stands. *Journal of Ecology* 96: 807–820.

Goovaerts, P. and Jacquez, G.M. (2004). Accounting for regional background and population size in the detection of spatial clusters and outliers using geostatistical filtering and spatial neutral models: the case of lung cancer in Long Island, New York. *International Journal of Health Geographics* 3: 1–14.

Greig-Smith, P. (1961). Data on pattern within plant communities. I. The analysis of pattern. *Journal of Ecology* 49: 695–702.

Griffith, D.A. (1996). Spatial autocorrelation and eigenfunctions of the geographic weights matrix accompanying geo-referenced data. *Canadian Geographer* 40: 351–367.

Haase, P. (1995). Spatial pattern-analysis in ecology based on Ripley *K*-function–introduction and methods of edge correction. *Journal of Vegetation Science* 6: 575–582.

Haining, R.P. (2003). *Spatial Data Analysis: Theory and Practice*. Cambridge: Cambridge University Press.

Hengl, T., Heuvelink, G.B.M., and Stein, A. (2004). A generic framework for spatial prediction of soil variables based on regression-Kriging. *Geoderma* 120: 75–93.

Illian, J., Penttinen, A., and Stoyan, D. (2008). *Statistical Analysis and Modeling of Spatial Point Patterns*. Chichester: Wiley.

James, P.M.A. and Fortin, M.-J. (2012). Ecosystems and spatial patterns. In: *Encyclopedia of Sustainability Science and Technology* (ed. R.A. Meyers), 3326–3342. Springer Science+Business Media.

James, P.M.A., Fortin, M.-J., Fall, A. et al. (2007). The effects of spatial legacies following shifting management practices and fire on boreal forest age structure. *Ecosystems* 10: 1261–1277.

Journel, A.G. and Huijbregts, C. (1978). *Mining Geostatistics*. London: Academia Press.

Keitt, T.H. and Urban, D.L. (2005). Scale-specific inference using wavelet. *Ecology* 86: 2497–2504.

Lancaster, J. and Downes, B.J. (2004). Spatial point pattern analysis of available and exploited resources. *Ecography* 27: 94–102.

Legendre, P. (1993). Spatial autocorrelation: trouble or new paradigm? *Ecology* 74: 1659–1673.

Legendre, P. and Fortin, M.-J. (1989). Spatial pattern and ecological analysis. *Vegetatio* 80: 107–138.

Legendre, P. and Legendre, L. (2012). *Numerical Ecology*, 3 English ed. Amsterdam: Elsevier.

Levin, S.A. (1992). The problem of pattern and scale in ecology. *Ecology* 73: 1943–1967.

Lloyd, M. (1967). Mean crowding. *Journal of Animal Ecology* 36: 1–30.

Long, J.A. and Nelson, T.A. (2012). Time geography and wildlife home range delineation. *Journal of Wildlife Management* 76: 407–413.

Manel, S., Gugerli, F., Thuiller, W. et al. (2012). Broad-scale adaptive genetic variation in alpine plants is driven by temperature and precipitation. *Molecular Ecology* 21: 3729–3738.

Matheron, G. (1970). La théorie des variables régionalisées, et ses applications, *Les Cahiers du Centre de Morphologie Mathématique de Fontainebleau*, 1–212. Fontainebleau: Fascicule 5.

McNew, L.B., Gregory, A.J., and Sandercock, B.K. (2013). Spatial heterogeneity in habitat selection: Nest site selection by Greater Prairie-Chickens. *Journal of Wildlife Management* 77: 791–801.

Melles, S.J., Badzinski, D., Fortin, M.-J. et al. (2009). Disentangling habitat and social drivers of nesting patterns in songbirds. *Landscape Ecology* 24: 519–531.

Melles, S.J., Fortin, M.-J., Lindsay, K., and Badzinski, D. (2011). Expanding northward: influence of climate change, forest connectivity, and population processes on a threatened species' range shift. *Global Change Biology* 17: 17–31.

Moran, P.A.P. (1948). The interpretation of statistical maps. *Journal of the Royal Statistical Society B* 10: 243–251.

Munoz, F. (2009). Distance-based eigenvector maps (DBEM) to analyse metapopulation structure with irregular sampling. *Ecological Modelling* 220: 2683–2689.

O'Neill, R.V., Hunsaker, C.T., Timmins, S.P. et al. (1996). Scale problems in reporting landscape pattern at the regional scale. *Landscape Ecology* 11: 169–180.

O'Neill, R.V., Riitters, K.H., Wickham, J.D., and Jones, K.B. (1999). Landscape pattern metrics and regional assessment. *Ecosystem Health* 5: 225–233.

Peterson, G.D. (2002). Contagious disturbance, ecological memory, and the emergence of landscape pattern. *Ecosystems* 5: 329–338.

Polakowska, A.E., Fortin, M.-J., and Couturier, A. (2012). Quantifying the spatial relationship between bird species' distributions and landscape feature boundaries in southern Ontario, Canada. *Landscape Ecology* 27: 1481–1493.

Rangel, T.F., Diniz, J.A.F., and Bini, L.M. (2010). SAM: a comprehensive application for Spatial Analysis in Macroecology. *Ecography* 33: 46–50.

Ripley, B.D. (1981). *Spatial Processes*. New York: Wiley.

Rosenberg, M.S. and Anderson, C.D. (2011). `PASSaGE`: pattern analysis, spatial statistics and geographic exegesis. Version 2. *Methods in Ecology and Evolution* 2: 229–232.

Rutledge, L.Y., Patterson, B.R., Mills, K.J. et al. (2009). Protection from harvesting restores the natural social structure of eastern wolf packs. *Biological Conservation* 143: 332–339.

Schiffers, K., Schurr, F.M., Tielborger, K. et al. (2008). Dealing with virtual aggregation—a new index for analysing heterogeneous point patterns. *Ecography* 31: 545–555.

Schurr, F.M., Bossdorf, O., Milton, S.J., and Schumacher, J. (2004). Spatial pattern formation in semi-arid shrubland: a priori predicted versus observed pattern characteristics. *Plant Ecology* 173: 271–282.

Sokal, R.R., Oden, N.L., and Thomson, B.A. (1998). Local spatial autocorrelation in biological variables. *Biological Journal of the Linnean Society* 65: 41–62.

Stoyan, D. and Stoyan, H. (1994). *Fractals, random shapes and point fields: Methods of geometrical statistics*. New York: Wiley.

Wagner, H.H. and Fortin, M.-J. (2005). Spatial analysis of landscapes: concepts and statistics. *Ecology* 86: 1975–1987.

Watt, A.S. (1947). Pattern and process in the plant community. *Journal of Ecology* 35: 1–22.

Wiegand, T. and Moloney, K.A. (2004). Rings, circles, and null models for point pattern analysis in ecology. *Oikos* 104: 209–229.

Wiegand, T. and Moloney, K.A. (2014). *Handbook of Spatial Point Pattern Analysis in Ecology*. Boca Raton: Chapman and Hall, CRC.

Worton, B.J. (1989). Optimal smoothing parameters for multivariate fixed and adaptive kernel methods. *Journal of Statistical Computation and Simulation* 32: 45–57.

13

Animal Home Ranges

Concepts, Uses, and Estimation

Jon S. Horne[1], *John Fieberg*[2], *Luca Börger*[3], *Janet L. Rachlow*[4], *Justin M. Calabrese*[5] *and Chris H. Fleming*[6]

[1] Idaho Department of Fish and Game, Boise, ID, USA
[2] Department of Fisheries, Wildlife and Conservation Biology, University of Minnesota, St. Paul, MN, USA
[3] Department of Biosciences, College of Science, Swansea University, Swansea, United Kingdom
[4] Department of Fish and Wildlife Sciences, University of Idaho, Moscow, ID, USA
[5] Department of Biology, University of Maryland College Park, College Park, MD, USA
[6] Smithsonian Conservation Biology Institute, Front Royal, VA, USA

Summary

Estimating animal home ranges and understanding the processes that influence home range behavior are important components of ecological investigations. While ecologists have been interested in the concept of home range for many decades, turning this concept into a usable statistical model or quantifiable metric for scientific purposes has been quite challenging. Keys to overcoming this challenge are (i) letting well-defined questions, and a solid understanding of the study system point to a specific home range metric to be quantified; (ii) choosing an appropriate estimator of this target metric; (iii) collecting location data during appropriate time periods and at the right sampling frequency; and (iv) understanding the trade-off between simplicity and complexity in modeling ecological data. If researchers follow these recommendations, many of the traditional challenges to studying animal home ranges will be alleviated due to technological advances in tracking systems and a solid foundation of models for understanding space use. However, we should continue to strive for more fundamental approaches to understanding animal movements. In particular, we see great opportunity for the continued development of agent-based models (ABM) of animal movements that allow for home range behavior to be an emergent property of the model instead of an a priori structure imposed on the data.

13.1 What Is a Home Range?

The important role of animal movements in natural selection and evolution was first discussed by Charles Darwin (1859) in *On the Origin of Species*. Individual movements are generally not random, but instead represent specific behavioral responses to certain needs or stimuli (Turchin 1998). Examples include movements to seek out resources or social partners, respond to environmental change, escape predators, or mitigate competitive pressures (Nathan et al. 2008). To accomplish these goals, many animals restrict their movements to certain confined areas that are smaller than the potential areas they could explore given their movement capacities – a behavior known as site fidelity, and the restricted areas used by animals known as the *home range* (Börger et al. 2008). Thus, home range behavior is fundamental to many ecological processes including resource selection (Moorcroft and Barnett 2008), social interactions (Emlen and Oring 1977; Wilson 1979), predator–prey interactions (Moorcroft and Lewis 2006), population (Gautestad and Mysterud 2005) and metapopulation dynamics (Matthiopoulos et al. 2005), and community structure (Holyoak et al. 2005). Consequently, investigating home ranges has been a dominant component of ecological research, allowing us to gain insight into important processes that shape the distribution and abundance of species.

Although not the first to use the term "home range," William Henry Burt has been credited with formally defining the concept in 1943. Prior to that time, the terms home range and territory were used loosely and often interchangeably. To set the record straight, Burt (1943) wrote in his seminal paper that "the home range concept is, in my opinion, entirely different from, although associated with, the territoriality concept" and suggested that the *home range* be defined as "the area, usually around a home site, over which the animal normally travels in search of food" (Burt 1943) whereas *territory* be restricted to "… the protected part of the home range." The key distinction is that a home range is the general area within which an animal lives, while a territory is that portion

Population Ecology in Practice, First Edition. Edited by Dennis L. Murray and Brett K. Sandercock.

Companion website: www.wiley.com/go/MurrayPopulationEcology

of the home range that is defended against conspecific or interspecific intrusion. Burt (1943) further narrowed the definition by suggesting that "...Occasional sallies outside the area, perhaps exploratory in nature, should not be considered part of the home range." Thus, home ranges need not incorporate the entire area an animal uses, but should be restricted only to those areas that we expect the animal to use repeatedly during its normal activities. Last, Burt commented on the concept of temporal stability in the area occupied, stating "it is only after they establish themselves, normally for the remainder of their lives, unless disturbed, that one can rightfully speak of the home range." Home ranges, therefore, are formed after animals have demonstrated *site fidelity*, defined as the tendency for an animal to remain in the same area for an extended period of time (White and Garrott 1990). Thus, three characteristics emerge from Burt's original definition and have provided the foundation for how ecologists view animal home ranges: (i) Home ranges are the consequence of behavioral and environmental processes such as forage acquisition, mating, predator avoidance, and conspecific attraction or avoidance that result in predictable patterns of restricted space use on the landscape; (ii) Animals might occasionally make exploratory forays, but home ranges should be those areas that are normally used; and (iii) Temporal stability in the area used is expected such that an animal will make repeated visits to places within the home range.

Burt's original definition of home range has been amazingly stable and resilient, but a great deal of work has been necessary to operationalize this definition as a usable statistical model or quantifiable metric for scientific purposes. For example, how do we conclude that an individual has displayed sufficient site fidelity to consider estimating a home range? Or, how should researchers separate exploratory movements from normal activities? Additionally, while it was clear that Burt thought animal home ranges should be described by the area in two-dimensional space, Aebischer et al. (1993) noted that an animal's movements actually define a line or "trajectory through space and time." Thus, the question arises, with complete knowledge of an animal's continuous path or trajectory over a period of time (i.e. a line), what should be the home range area? A major goal of this chapter is to explore these issues and suggest approaches for turning the behavioral concept of animal home range into a biologically informative model from which valuable metrics can be derived for use in scientific research. Keys to achieving this goal are (i) letting well-defined questions, and a solid biological understanding of the study system, point to a specific home range metric to be quantified; (ii) choosing an appropriate estimator of this target metric; (iii) collecting location data during appropriate time periods and at the right sampling frequency; and (iv) understanding the trade-offs between simplicity and complexity in modeling ecological data (Murtaugh 2007).

13.1.1 Quantifying Animal Home Ranges as a Probability Density Function

Not long after Burt promoted a formal definition of home range, others began working on ways to quantify it. Initially, home ranges were described by drawing polygons around observed locations (Figure 13.1a; Mohr 1947), but ecologists quickly realized that it would be better to view home ranges as a distribution describing the relative frequency of use across the landscape (Hayne 1949). Calhoun and Casby (1958) suggested defining home ranges as a *density function* describing the probability of an animal being present in some arbitrarily small area. This density function was subsequently deemed the *utilization distribution* by Jennrich and Turner (1969) and the term became synonymous with the mathematical expression of an animal's home range (Anderson 1982; Horne and Garton 2006a). Most contemporary definitions and descriptors of an animal's home range are formulated as a two-dimensional probability density function $UD(x, y)$, but can be extended to additional dimensions (Keating and Cherry 2009; Tracey et al. 2014). As with any probability density function, the probability associated with any point in continuous space is 0, but non-zero probabilities of occurrence in any spatial region (A) can be found by integrating $UD(x, y)$ over the region A:

$$P(A) = \int_A UD(x,y)dxdy \tag{13.1}$$

where $P(A)$ is the volume under $UD(x, y)$ and is defined as the probability that a randomly drawn "location" will occur in that area (Evans et al. 2000). The key concept, for any probability density function, including those describing animal home ranges, is the relative frequency with which a random event is expected to occur. Thus, the definition of home range is incomplete without explicit specification of what is to be considered the random event drawn from $UD(x, y)$. Below, we describe two specifications with each producing distinctly different estimates of an animal's home range (Fleming et al. 2015; Figures 13.1 and 13.2).

For our first specification, the *occurrence distribution*, the actual movement trajectory of the animal under investigation is the target of our model. In this case, we are interested in answering the question, "What proportion of time during the study was the animal in area A?" or, stated in terms of a probability function, "What is the probability that the animal was in A if we were to choose a time at random from the period of observation?". In the ideal case when the complete trajectory of an individual is

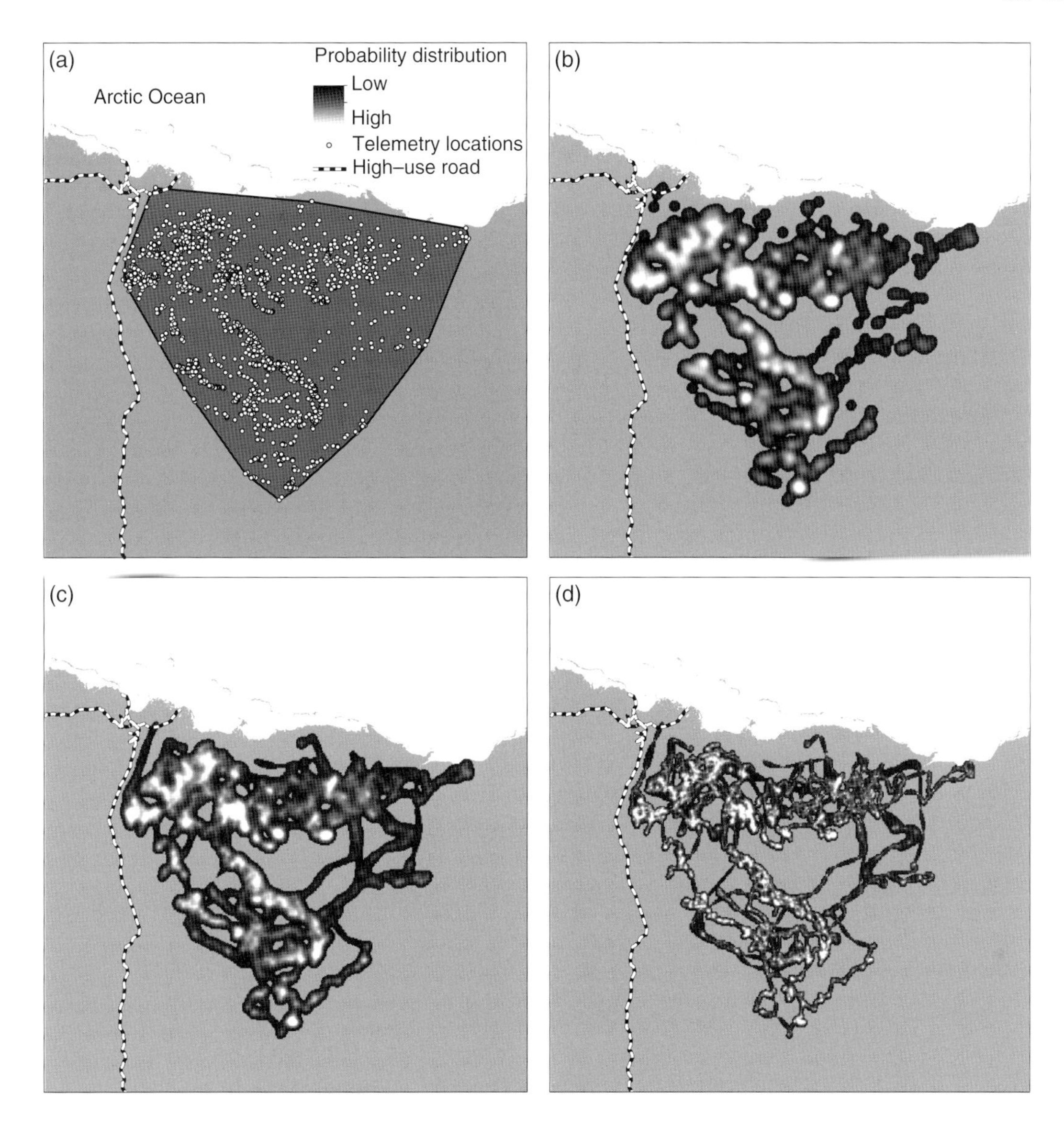

Figure 13.1 Home range estimates of an individual caribou in northern Alaska when the occurrence distribution is the target of estimation. Estimators include: (a) the minimum convex polygon (MCP) (open circles represent location data taken at approximately five-hour intervals), (b) kernel density estimate with smoothing parameter chosen using likelihood cross-validation (CVh), (c) the Brownian bridge movement model (BBMM), and (d) Kriging based on a Ornstein-Uhlenbeck with foraging movement model; Kriging based on the continuous-time correlated random walk model (CTCRW) was virtually indistinguishable from this estimate.

known, $P(A)$ equals the amount of time spent in A divided by the length of the observation period. Our second specification, the *range distribution*, considers animal movements more generally. For this specification, our goal is not to characterize where the animal went, but rather to describe the underlying stochastic process responsible for the animal's movement trajectory. By using a stochastic modeling framework, we allow for multiple possible trajectories and space-use patterns, all of which could have occurred during the observation period or perhaps some other time period of interest. For the range distribution, $P(A)$ maps the probability of finding the individual in A at a randomly chosen time, if we were to generate a novel movement path using the same rules that generated the original path. Hence, the range distribution seeks to answer the question, "Given a set of behavioral and environmental states, what proportion of time would we *expect* to find the animal in area A?". To reiterate, the occurrence distribution is focused on describing where the animal *went* (retrospective), while the range distribution describes where the animal *could have gone* or, in some instances, could go in the future (predictive). Although this might appear to be a subtle difference, the appropriate choice of a home range

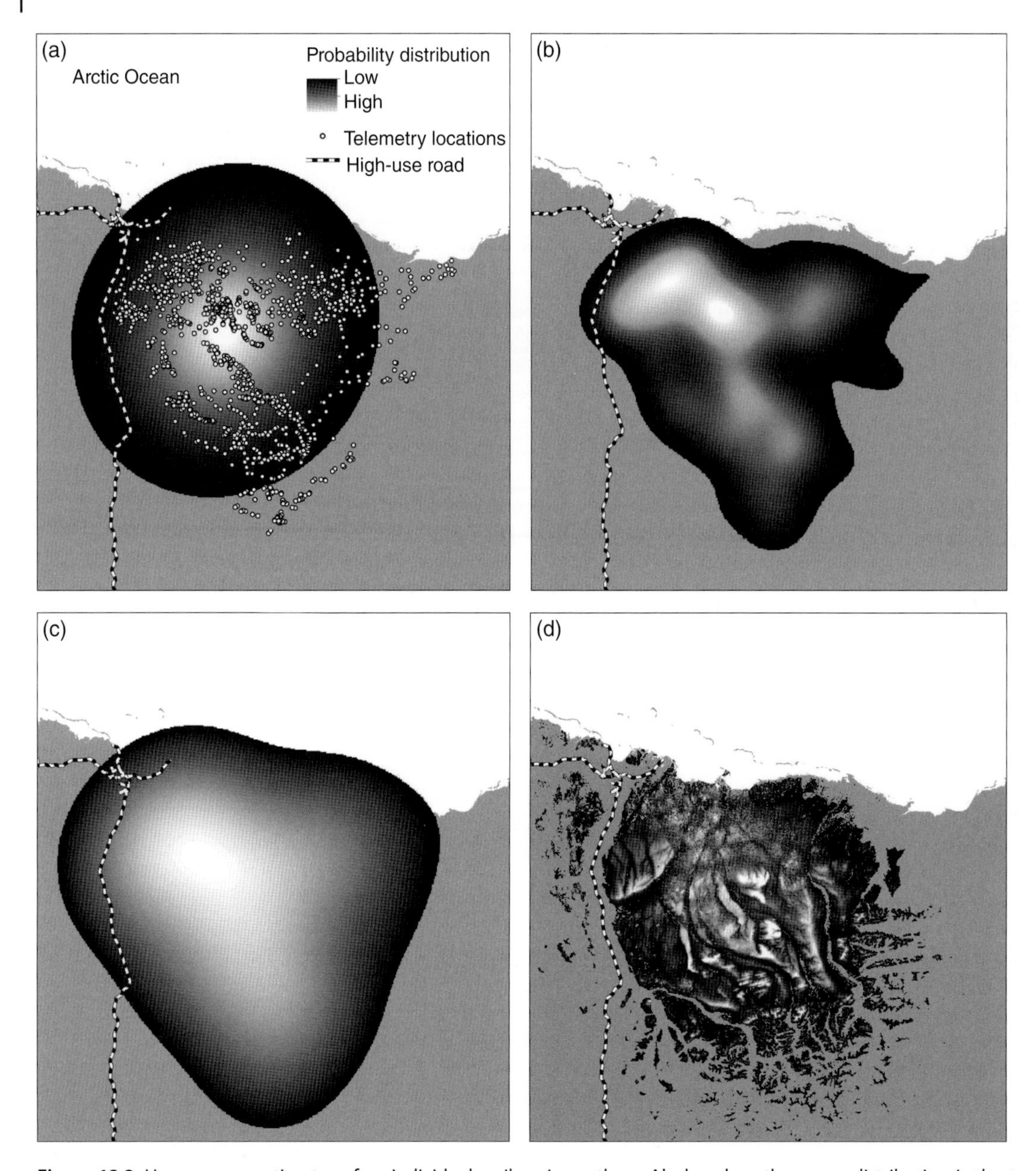

Figure 13.2 Home range estimates of an individual caribou in northern Alaska when the range distribution is the target of estimation. Estimators include: (a) the bivariate normal distribution based on the Ornstein–Uhlenbeck (OU) diffusion process (open circles represent location data taken at approximately five-hour intervals), (b) kernel density estimate with h_{opt} (Eq. 13.3) for the smoothing parameter, (c) autocorrelated kernel density estimate, and (d) the synoptic model. The synoptic model included covariates related to landscape cover, roads, and topographic position index (TPI). Selection for TPI was found to vary with the amount of mosquito activity and the distribution depicted here represents expected space use when mosquito activity is high.

estimator or model will depend on the specification that most appropriately answers the research question of interest.

13.1.2 Why Estimate Animal Home Ranges?

Estimates of home range have been used to address a wide variety of ecological questions. Many early natural history studies described patterns of space use by animals seeking to simply establish the existence of home ranges. Seton (1920) was instrumental in elevating the "home range of the individual" to a life history parameter of biological interest, and studies published during subsequent decades provided basic descriptors of home ranges focused primarily on size (Hall and Linsdale 1929; Hamilton 1937; Lay and Baker 1938). Following these studies, others began to look for relationships between fundamental ecological or biological processes and home range characteristics. For example, McNab (1963) demonstrated that the area used by mammals increased with body mass at a rate consistent with metabolism, suggesting that energetic requirements determine the size of areas used by individuals.

Questions related to the structuring of animal populations and their management have also relied on analyses of home ranges. For example, estimates of the size of either individual or group home ranges, together with information about overlap among adjacent ranges, commonly have been used to estimate population densities (Fuller et al. 1992). In addition, the location and size of home ranges are sometimes estimated to guide species control or removal programs (Sparklin et al. 2009; Bengsen et al. 2012) and to delineate areas for habitat conservation or restoration (Johnson et al. 2010).

One of the most prominent uses of home range estimates in the ecological literature is due to the hierarchical framework for habitat selection outlined by Johnson (1980). Following this publication, decades of research have delineated the home range as an independent step in the process of quantifying resource selection: choice of the geographic range (first order), selection of a home range from a larger surrounding area (second order), and then selection of general areas or habitats within the home range (third order). A more recent line of investigation has sought to examine the effects of landscape patterns and habitat features on metrics of animal space use, including home range size (Walter et al. 2009; Massé and Côté 2012; Naidoo et al. 2012) and movement patterns (Mueller et al. 2011).

Last, the recognition that intrinsic as well as extrinsic factors also affect decisions about space use has motivated studies investigating the influence of sex, age, reproductive state, and dominance status on movement behaviors (Börger et al. 2006a; Murray et al. 2007; Anadón et al. 2012; van Beest et al. 2011). A variety of methods have been proposed for quantifying overlap in use of space among individuals of the same species (Fieberg and Kochanny 2005), and home range overlap indices have been used to evaluate the potential for disease transmission (Marsh et al. 2011), habitat quality (McLoughlin et al. 2000), and social associations relative to kinship (Frère et al. 2010). Similarly, estimates of spatial interactions between species have been employed to evaluate mechanisms of coexistence (Edelman 2012), predation (Whittington et al. 2011), and competition (Grassel et al. 2015).

13.2 Estimating Home Ranges: Preliminary Considerations

As with any scientific study, it is important to have clearly defined research questions and objectives (see Chapter 1 and 2). In Section 13.1.2, we provided a small sample of studies that have estimated some aspect of animal home ranges to answer questions about the behavior, ecology, and management of species. However, even within this small sample, it is clear that there is a great breadth of questions we might ask related to home range, and these questions should influence what we choose to estimate, how precise the estimate needs to be, and the most cost-effective methods of data collection. Thus, we strongly urge researchers to guard against a "one size fits all" approach to studying home ranges and instead take an approach that allows research questions to drive decisions related to various design and analysis strategies (Fieberg and Börger 2012).

To begin with, well-defined research objectives should identify an appropriate target population, defined by its geographic range and possibly limited to one or more unique population segments such as a particular age or sex class. Substantial thought and effort should be expended to ensure a representative sample of individuals is obtained from this target population (Morrison et al. 2001; Millspaugh et al. 2012). Researchers should also be cognizant that monitoring techniques have the potential to interfere with the natural behavior of individuals in the study (Dennis and Shah 2012; Rachlow et al. 2014; Coughlin and van Heezik 2015) and plan to account for or mitigate these effects. Finally, research objectives should identify whether the range distribution or occurrence distribution best answers the research question(s). This estimation target will play an important role in determining the frequency and duration for obtaining location data (Börger et al. 2006a, 2006b; Fieberg and Börger 2012) as well as guide the choice of an appropriate home range estimator.

With regard to how often location data should be collected on individuals, observations taken close in time will generally be close in space, thus they will be *autocorrelated*. The shorter the time interval separating two observed locations, the more strongly autocorrelated they will be. While autocorrelation was traditionally viewed as a significant obstacle to home range estimation (Harris et al. 1990), modern analyses recognize that autocorrelation is actually beneficial for estimating the occurrence distribution and can be either accommodated with an appropriate sampling schedule (Fieberg 2007a), or directly modeled (Fleming et al. 2015) when estimating the range distribution. For practical purposes, it will often be beneficial to sample frequently when estimating the occurrence distribution, whereas infrequent locations collected over an extended period of time might be sufficient for estimating the range distribution. Once a schedule has been determined, researchers should be aware that some methods of obtaining telemetry data have the potential to miss locations due to topography, vegetation cover, or individual behaviors (Frair et al. 2010). Similar to bias in selecting individuals, researchers should plan to mitigate observation bias or correct for it when estimating the home range (Horne et al. 2007b).

With regard to how long individuals should be monitored, it is important to remember that one of the defining tenets of animal home ranges is site fidelity. With an understanding of the behavior of the target species, researchers should plan a study that will be long enough such that they can expect temporal stability in the estimate of space use (Fieberg and Börger 2012). Following data collection, this concept should be revisited by plotting the area occupied by an individual versus time, which should asymptote as the area occupied becomes temporally stable (Laver and Kelly 2008).

To determine time periods during which the individual is demonstrating home range behavior versus other movement modes such as dispersal or migration, researchers will need to do some preliminary analyses of their location data. We strongly recommend visualizing each individual's location data within a GIS (geographic information system) to get familiar with their data and better understand the species' behavior and ecology. Visualizing location data is also helpful for finding "outliers" in space-use patterns, such as when an individual might have taken an occasional sally (Figure 13.3); how one treats these outliers will depend on one's study objectives as well as the reason for that particular behavior. Like statistical outliers more generally, we suggest researchers critically evaluate these locations as they may be some of the most interesting points offering insight into rare, but potentially critical, behaviors.

In addition to visualizing location data, a convenient metric for understanding movement modes is an animal's squared displacement over time (Börger and Fryxell 2012). Specifically, for any organism or object moving for a certain interval, the straight-line distance from the start to the end point is called the net displacement, and the square of this value is the *net squared displacement* (NSD; Turchin 1998). NSD is a fundamental quantity for studying movements of organisms or particles (Skellam 1951; Turchin 1998; Nouvellet et al. 2009) and is defined as:

$$NSD_{(t)} = (x_t - x_0)^2 + (y_t - y_0)^2, \tag{13.2}$$

where (x_0, y_0) is the location of the animal's first observation, and (x_t, y_t) is the location of the animal at time t. Different patterns of NSD can be predicted a priori for individuals exhibiting different movement behaviors (Börger and Fryxell 2012). By plotting NSD over time, researchers can gain insights into periods when animals are wandering nomadically, dispersing, displaying site fidelity within a home range, or migrating between seasonal ranges (Figure 13.4). For a *nomadic* or dispersing animal, it can be demonstrated that the value of the NSD will continuously increase over time. For a *resident* animal that moves within a home range, the rate of increase of NSD over time will decrease and approach an asymptote (Turchin 1998). We also note that a related approach to identify stationary movement periods that makes more complete use of the location data was recently developed based on the *empirical variogram* of the data (Fleming et al. 2014a; implemented in the `ctmm` package of `R`). Similar to the NSD, if the empirical variogram does not show a clear asymptote, then home range analysis of any kind is tenuous because

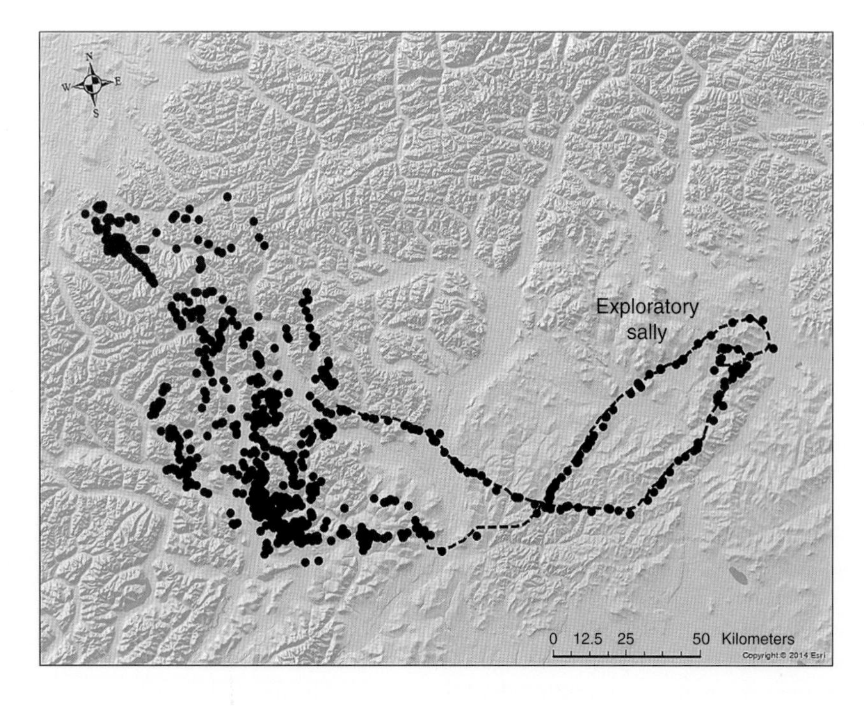

Figure 13.3 Telemetry locations for an individual caribou during four winters illustrating an "exploratory sally" (dashed line), as defined in early descriptions of home range behavior (Burt 1943).

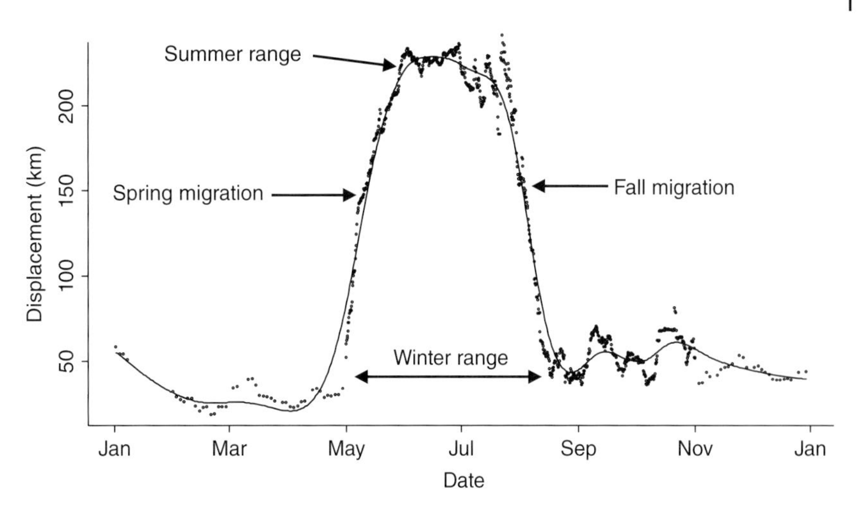

Figure 13.4 Analysis of net squared displacement (NSD) used to identify movement behaviors of an individual caribou on the north slope of Alaska. For easier interpretation, we report the square root of NSD. Small open circles are the actual values calculated for each observation and the solid line are these values smoothed with a spline function in R.

the individual's movement did not show evidence of site fidelity.

Last, with regard to choosing an estimator, contemporary ecologists are faced with a daunting array of possible home range models, but many of the models are better suited for characterizing the occurrence distribution versus the range distribution, and vice versa. To assist with this dichotomy, we describe several home range estimators by placing each under the estimation target to which we think they are best suited, while fully recognizing that the distinction is not always clear. Thus, our goal is not to necessarily force a particular model into a particular estimation target but to encourage researchers to critically think about the connections among research questions, estimation targets, and home range estimators.

13.3 Estimating Home Ranges: The Occurrence Distribution

The *occurrence distribution* provides estimates of where an animal went based on a set of known locations (Figure 13.1). It is the most common conceptualization of home range and has the longest history of use by ecologists. Occurrence models are well-suited to addressing questions that require fine-scale information about where the animals went (Figures 13.1c and d). For example, they have been used to quantify the area used by individuals for evaluating resource selection (Kittle et al. 2015), delineate movement corridors (Sawyer et al. 2009), and define space use for animals that wander over great extents such as pelagic mammals (Lowther et al. 2013) and birds (Adams et al. 2012).

While estimating the occurrence distribution from limited data can be quite challenging, modern telemetry devices often enable researchers to collect location data over larger spatial extents, with greater frequency, and with much better accuracy than traditional methods. We now have the ability to estimate the occurrence distribution with greater accuracy, and the development of models for estimating an animal's actual movement path from location data is an active area of ecological research. Thus, we begin by describing models that simply define an outer boundary of the occurrence distribution; then, we describe popular methods based on kernel smoothing; and end this section with more contemporary models that recognize the time series nature of telemetry data and model the occurrence distribution based on animal movement processes.

13.3.1 Minimum Convex Polygon

The *minimum convex polygon* (MCP) utilized by Mohr (1947) is the oldest and most commonly used method for estimating animal home ranges (Laver and Kelly 2008). It is simply the smallest convex polygon that contains all locations (Figure 13.1a). MCP is a crude estimate of the occurrence distribution that is sensitive to extreme data points and sample size, ignores the interior data points, and implies uniform use within the boundary (White and Garrott 1990; Powell 2000). Nevertheless, MCPs may provide an adequate estimate of home range in some cases such as when (i) comparing sizes across taxa where large differences in range size mask smaller differences due to the choice of home range estimator (Nilsen et al. 2008); (ii) generating abundance estimates based on territory occupation and size (Adams et al. 2008; Suwanrat et al. 2015); or (iii) defining used habitats for second-order resource selection (Horne et al. 2009) or an outer boundary of availability for third-order resource selection (Zeale et al. 2012).

13.3.2 Kernel Smoothing

Kernel density estimation (KDE) has been used across scientific disciplines as a nonparametric approach to estimate an unknown probability distribution (Silverman 1986). KDE works by placing small kernels or bumps over each data point and then averaging the contribution from all kernels to obtain an estimate of the probability distribution at any point in space (Figure 13.5). A key component of KDE is adjusting the width of the kernels, often called the *smoothing parameter* or bandwidth, to best represent the target probability distribution. The bandwidth can have a substantial impact on the resulting estimate and there are several methods available for choosing its value based on statistical properties of the data (Silverman 1986 Jones et al. 1996). If the true underlying distribution is assumed to be a bivariate normal, the optimal level of smoothing is calculated using

$$h_{opt} = \sqrt{\sigma^2} \times n^{-1/6}, \tag{13.3}$$

where σ^2 is the average marginal covariance estimated from the x and y coordinates of the locations and n is the sample size (Silverman 1986). If the true underlying distribution is multimodal, other data-based methods can be used such as *least-squares cross-validation* (LSCVh) or *likelihood cross-validation* (CVh) which seek to minimize the "distance" between the KDE estimate and the true distribution that generated the data. For LSCVh, distance is measured based on the integrated squared error, and for CVh, it is measured based on the Kullback–Leibler discrepancy (Silverman 1986; Horne and Garton 2006b).

As for home range estimation, KDE was first suggested by the seminal work of Bruce Worton (1989). It is important to understand that when KDE was introduced for home range estimation, VHF telemetry dominated ecological studies, so most location data were spaced relatively far apart in time and could generally be considered independent observations from an animal's range distribution. Indeed, when there is little autocorrelation or a relatively large smoothing parameter is used, traditional KDE can be a convenient approach for estimating the range distribution of animals as we discuss in Section 13.4.4; but what about using KDE to estimate the occurrence distribution with autocorrelated data? In fact, as the time between locations goes to zero, many of the data-based choices of bandwidth selection, for example LSCVh or CVh, will also approach zero. This means that as location data are acquired at smaller and

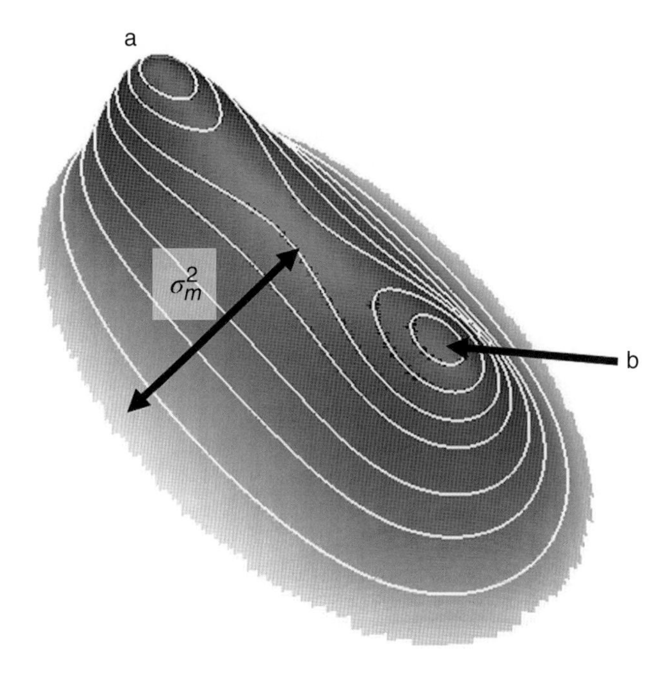

Figure 13.6 Probability density based on the Brownian bridge movement model (BBMM) describing where the animal could have been knowing that it had to get from location (a) to location (b) within the observed time interval and that its possible movement path, represented by the spread of the distribution, is constrained by its mobility (governed by parameter σ_m^2).

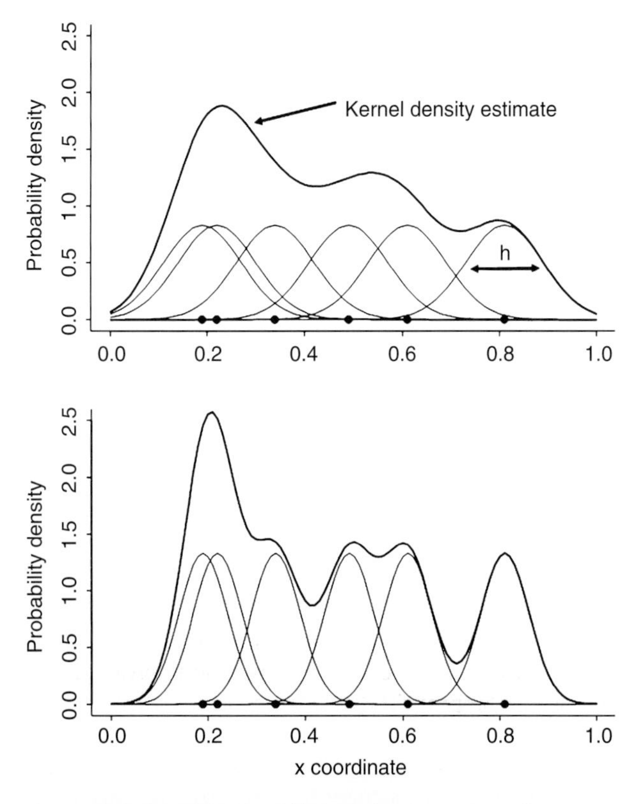

Figure 13.5 One-dimensional kernel density estimate from six observations (closed circles) and two smoothing parameters (top: $h = 0.8$; bottom: $h = 0.5$). Estimated density at a position x is obtained by placing a kernel over each observation and then summing the contributions of all kernel.

smaller intervals, KDE will become closer and closer to the animal's movement path. Therefore, with frequently sampled location data and a small enough smoothing parameter, KDE can be an adequate estimator of the occurrence distribution (Figure 13.1c; also see Figure 3 of Horne et al. 2007a). However, if traditional KDE is to be used for estimating the occurrence distribution, researchers must give careful thought to the sampling schedule as well as how to choose the smoothing parameter, because as time intervals between locations become larger or bandwidths increase, KDE will tend to produce estimates more characteristic of the range distribution versus the occurrence distribution (compare Figure 13.1b versus Figure 13.2b).

13.3.3 Models Based on Animal Movements

The increasing availability of high-frequency and accurate location data has spurred several advances that allow researchers to model the occurrence distribution of animals based on actual movement processes. As part of his Master's thesis, Floyd Bullard (1999) recognized that previous approaches to home range estimation ignored the time component of location data despite the fact that locations were being taken with increasing frequency. Bullard thought that it would be far better to base home range estimation on a movement process that considers the temporal sequence of locations, and suggested estimating the unknown movement path between observed locations by assuming animals moved according to *Brownian motion*, characterized by unpredictable or random movement, between any pair of consecutive locations. Additionally, the potential movement path would be constrained by the fact that the animal had to get from one location to the next within the observed time interval and the animal's mobility, represented by the parameter σ_m^2, would restrict the potential paths it could have taken (Figure 13.6). Subsequent to Bullard's work, Horne et al. (2007a) further developed the *Brownian bridge movement model* (BBMM) by describing an approach for estimating σ_m^2. Kranstauber et al. (2012) then generalized the BBMM by allowing the mobility parameter to vary over the movement path. More recently, Pozdnyakov et al. (2014) derived the exact full-dataset likelihood function for the BBMM which provided a method for unbiased and more efficient estimation of σ_m^2 and allowed the observation error to be simultaneously estimated from the data when it is unknown.

A key characteristic of the BBMM is that Brownian motion has no directional persistence, while real animals may exhibit a tendency to keep moving in a particular direction and speed. Thus, an alternative to the BBMM is the *continuous-time correlated random walk model* (CTCRW; Johnson et al. 2008a), more generally known as the Integrated Ornstein–Uhlenbeck (IOU) process. The CTCRW adds a layer of biological realism by incorporating directional persistence in movement via autocorrelated velocities, where velocity of movement includes both direction of travel *and* movement speed. Thus, autocorrelation in velocity means that an individual's direction and speed at one point in time will be similar to those same quantities at other neighboring times.

The BBMM and the CTCRW models will often be adequate for estimating the movement path (Figure 13.1c and d). However, depending on the frequency of the location data and the behavior of the animals under investigation, alternative movement models might need to be considered. For example, if location data are taken at long enough time intervals such that home range behavior or the tendency to remain in a restricted area, needs to be incorporated, the Ornstein–Uhlenbeck (OU) process (Uhlenbeck and Ornstein 1930; Dunn and Gipson 1977) or the recently introduced Ornstein–Uhlenbeck with foraging (OUF) process (Figure 13.1d; Fleming et al. 2014a, 2014b) could be used.

Any of the movement models described in this section could be used to provide a probabilistic estimate of an animal's movement path. However, a priori choice of a particular movement model requires an understanding of how autocorrelation in velocities is affected by the interaction between the time interval between location data and the characteristics of real animals moving through their environment. In recent work, Fleming et al. (2016) described a more general approach to probabilistic path reconstruction based on the time series analog of *Kriging* from geostatistics (Chapter 12; Cressie 1993; Diggle and Ribeiro 2007). Under their approach, researchers are free to hypothesize and fit a variety of movement models to the location data, including all of the models described in this section; then use model selection criteria to help decide which movement model is most appropriate (Fleming et al. 2014a, 2014b); and finally, estimate the occurrence distribution using the movement model that best fits the observed location data. Thus, the general Kriging approach alleviates the need to choose a movement model a priori, a significant benefit to those researchers who do not have the knowledge and expertise to appropriately match a particular movement model to the behavior of the focal individuals and their location data. The downside to this approach is that there is added time and effort required for implementation and depending on the objectives of the study and the particular dataset, the practical difference between alternative movement models might be insignificant (for example, see Figures 13.1c and d).

13.3.4 Estimation from a One-dimensional Path

The occurrence distribution quantifies our uncertainty about where the animal was located during the observation period, including times when no locations were observed. In the limit, when the time between location observations and the magnitude of location errors both go to zero, the occurrence distribution will collapse to the individual's movement path. Ongoing advances in tracking technology are beginning to make this mathematical limit a reality (Li et al. 2015), and even if we never get there, it is still an interesting methodological exercise to contemplate this situation. Thus, to conclude this section, we return to a question posed at the beginning of our chapter: "With complete knowledge of an animal's continuous path or trajectory over a period of time, what should be the home range area?" Indeed, this question should give the reader pause as the answer is not obvious. However, we can look to a field study by Ostro et al. (1999a) to help guide our thinking. While studying the spatial ecology of a translocated population of black howler monkeys (*Alouatta pigra*) in Belize, Ostro et al. (1999a) confronted this question after following groups of monkeys continuously during their daily movements and recording their position without error. Thus, they had observed a *one-dimensional path* for animal movements. To estimate a home range, they developed a technique that assumes animals actually utilize an area around the traveled path (Ostro et al. 1999b). Defining the occurrence distribution was a simple process involving three steps: (i) map the observed trajectory in a GIS; (ii) buffer the observed trajectory by a distance appropriate for the individual(s) under investigation and superimpose a MCP on the resulting map; and (iii) incorporate lacunae, the gaps or spaces within the outer boundary not filled by the buffer, smaller than a designated minimum size into the digitized polygon home range. While straightforward, the approach requires two somewhat subjective decisions. First is the choice of a buffer width for the observed trajectory. For Ostro et al.'s (1999a) study of howler monkeys, mean group spread was used. For other situations, Ostro et al. (1999b) suggested it might be more appropriate to use the area visually surveyed, or otherwise "perceived," by the study animals. Second, researchers must choose the size of lacunae to be incorporated into the home range. For this, Ostro et al. (1999b) incorporated lacunae for which the area was <1% of the MCP.

13.4 Estimating Home Ranges: The Range Distribution

Recall that the range distribution aims to predict where an animal might go rather than define where it went (Figure 13.2). Historically, one of the most common uses of the range distribution was to define availability for third-order resource selection (Shirk et al. 2014). But because these models describe the process resulting in a particular home range, they allow researchers to ask questions about why animals move the way they do. For example, study questions might include: Which behaviors lead to the emergence of a home range?; How will the structure of an individual's home range vary under changing environmental conditions or in response to variation in the density of neighboring individuals (e.g. competitors, mates, predators, prey)?; or How do resource-use strategies structure animal home ranges? Using a process-based approach, hypotheses for these questions can be formulated as mathematical models and evaluated for their predictive ability. Models that provide good predictions can then be used not only to describe the home range, but also to provide a greater understanding of the ecological processes driving home range behavior.

13.4.1 Bivariate Normal Models

Some of the early models of animal home ranges were *focal point attraction models* that depicted space use of animals as having a centralized area where activity was focused. The first of these, introduced by Calhoun and Casby (1958), was the circular *bivariate normal model*. The circular distribution was later generalized by Jennrich and Turner (1969) to allow elongate or elliptical home ranges (Figure 13.2a). Recognizing the need to accommodate the fact that telemetry data are autocorrelated, Dunn and Gipson (1977) developed a bivariate normal home range model based on the OU diffusion process, which allowed for bivariate normal parameters to be estimated from autocorrelated location data typical of telemetry studies. More importantly, the OU model was the first that linked individual behavior with a statistical model of home range. For example, the focal attraction point may be due to a den or nest site from which movements radiate out, or it could be the result of behavior designed to retain familiarity of resources within the home range. Subsequent to the work of Dunn and Gipson (1977), Blackwell (1997, 2003) introduced a broad generalization of the OU model that addresses many of the criticisms leveled against earlier models. In particular, it allows an individual animal to switch between different movement processes or behavioral states within the home range.

13.4.2 The Synoptic Model

One of the main criticisms of the focal point attraction models described in Section 13.4.1 is that they are too unrealistic for describing the complex space use of animals in real landscapes. However, these models can provide the basis for more realistic descriptions of the home range if additional information, in the form of spatiotemporal covariates such as habitat or locations of other organisms, is included in the model. This is the approach taken by Horne et al. (2008) who developed the *synoptic model* that estimates the range distribution as an explicit function of environmental and behavioral variables in a model comparison framework (see also Horne and Garton 2006a). Different sets of covariates represent different hypotheses about the ecological processes determining animal space use allowing one to simultaneously estimate the home range while evaluating the set of covariates that most likely determine that form. Interestingly, an analogous model of space use was simultaneously proposed by Johnson et al. (2008b) from the perspective of estimating resource selection. The two avenues of inquiry converged because home ranges emerge, in part, as a consequence of resource selection or avoidance (Horne et al. 2008; Powell and Mitchell 2012), and resource selection cannot be evaluated independently of home range behavior (Rhodes et al. 2005). The synoptic model is defined as

$$g_u(s,t) = \frac{g_a(s) \times w_t(s)}{\int g_a(z) \times w_t(z)dz}, \quad (13.4)$$

where

$g_u(s,t)$ the range distribution defined as the probability density of use at spatial location s at time t

$w_t(s)$ a weighting function describing habitat preferences at time t,

$g_a(s)$ the probability of use in the absence of habitat selection, also viewed as null model of the home range in a homogeneous environment.

The denominator of Eq. (13.4) integrates over space and is simply the normalizing constant for a weighted distribution that ensures $g_u(s,t)$ integrates to 1 and thus, is a proper probability density function. Under a resource selection paradigm, the form of g_a is an explicit definition of what is considered "available" for subsequent resource selection. In the context of home range estimation, g_a is viewed as a null model of the home range in the absence of any effect from habitat selection. Thus far, g_a has taken the form of a bivariate normal distribution, which is characteristic of animals as having a centralized area of activity, or a bivariate exponential power distribution allowing for a more uniform distribution of use characteristic of territorial animals (Horne et al. 2008). The weighting function for habitat selection is based on a proportional change in g_a,

$$w_t(s) = Exp[\boldsymbol{\beta}_t{}'\mathbf{X}_t(s)], \quad (13.5)$$

where $\boldsymbol{\beta}$ is a vector of selection coefficients to be estimated and $\mathbf{X}$ is a vector of habitat covariates. The subscript t on $\mathbf{X}$ and $\boldsymbol{\beta}$ allows for temporal changes in habitat as well as changes in selection over time.

The synoptic model is particularly well suited when the study has dual goals of both providing a description of areas that could be utilized within a home range as well as investigating the environmental characteristics that influence space use (Horne et al. 2014; Wilson et al. 2014). Furthermore, the ability of this model to incorporate temporally dynamic selection or habitat conditions makes it particularly useful for describing space use under variable environmental conditions. For example, as part of a study to investigating space use of caribou (*Rangifer tarandus*) in northern Alaska, synoptic models were developed to estimate individual home ranges as a function of habitat type and amount of human infrastructure. Combined with selection for other landscape covariates, the best synoptic model suggested that selection for ridge tops versus valley bottoms depended on the level of mosquito activity, a temporally dynamic process with high activity driven by a combination of warm temperature and low-velocity wind speed. By allowing selection for topographic position to vary temporally as a function of temperature and wind speed, researchers were able to derive a general estimate of home range that could be used to predict where an individual would be expected to occur within its home range on days of high versus low mosquito activity (Figure 13.2d).

13.4.3 Mechanistic Models

A powerful approach to answer questions about the ecological processes that determine and structure animal home ranges is to use what are called *mechanistic modeling approaches* (reviewed in Börger et al. 2008), that is, analytical or individual-based modeling approaches that aim to develop spatially explicit predictions of animal space use by modeling specific individual movement processes (Mitchell and Powell 2012; Moorcroft 2012). The resulting probability density functions from contrasting models are then compared to the observed distribution of location data to evaluate alternative hypotheses of space use. Mechanistic models are directly based on the processes generating patterns of space use. Thus, they are especially useful for predicting space-use responses to environmental change, as well as for understanding the evolutionary significance of movement behaviors (Moorcroft and Lewis 2006). In general, two approaches have been used for mechanistic modeling of home ranges: an *analytical approach* stemming from the field of statistical physics and random walks (Turchin 1998; Okubo and Levin 2001; Moorcroft and Lewis 2006), and an *individual-based approach* developed from optimal foraging theory (Mitchell and Powell 2004). Interestingly, a convergence is developing between these two approaches as both are increasingly focusing on including resource selection processes (Mitchell and Powell 2012; Moorcroft 2012).

The analytic approach uses computer-based simulations of a random walk model to translate movement behavior to space-use predictions. Specifically, the trajectory of an individual animal is modeled through a sequence of moves with a certain length, duration, and orientation, all drawn from a redistribution kernel that describes the probability of moving from one location to the next (Moorcroft and Lewis 2006). Movement can then be directly simulated on a computer, and the outcome of multiple simulations compared to observed space-use patterns. For example, Van Moorter et al. (2009) used simulations to show that a complex combination of short-term and long-term memory processes might be necessary to lead to the emergence of home range patterns for noncentral place foraging animals.

An alternative to using simulated random walks is to formulate partial differential equations to model the spread of individual locations within a home range or territory (Lewis and Murray 1993; Blackwell 2003; Moorcroft and Lewis 2006). Here, movement processes are translated into a set of partial differential equations composed of two parts, one modeling diffusive movement component(s) and one modeling directed movement component(s) – a simple example would be a focal point of activity to which an animal returns repeatedly (Moorcroft and Lewis 2006). Other behaviors can be incorporated into the diffusive or directed movement part, such as scent marking and olfactory orientation (Benhamou 1989), area-restricted foraging (Benhamou 1994), conspecific avoidance and prey availability (Lewis and Murray 1993; Moorcroft and Lewis 2006), and memory processes (Moorcroft 2012).

A different set of mechanistic models has been developed by Mitchell and Powell (2004, 2007, 2012). These behavior-based models of animal movement attempt to develop an understanding of the processes that determine the structure and size of home ranges and which resource-use strategies are most profitable to an individual. The models are derived from optimal foraging theory which is based on the assumption that animals make space-use decisions by evaluating trade-offs between patch profitability versus travel costs. Each model represents contrasting strategies that animals might use to decide which resource patches to include within the home range under different spatiotemporal resource distribution scenarios.

13.4.4 Kernel Smoothing

We previously described how KDE can be used to estimate the occurrence distribution if location data are highly autocorrelated (Section 13.3.2). KDE can also be used to estimate the range distribution provided that the location data can be considered independent observations from the range distribution. In this case, one only needs to choose an appropriate smoothing parameter: h_{ref} if the range distribution is assumed to be a bivariate normal or one of the data-based methods, such as LSCV or CV, that seek to minimize the difference between KDE and the true range distribution. Of course, the trend in location data is toward increasingly finely sampled observations through time (Li et al. 2015). For these types of locations, Fleming et al. (2015) rederived the KDE explicitly assuming that the data are autocorrelated. The resulting *autocorrelated kernel density estimate* (AKDE) is based on finding the optimal smoothing bandwidth for estimating the range distribution when data are autocorrelated.

An AKDE analysis proceeds in similar fashion to the Kriging approach (Section 13.3.3). Like Kriging, the AKDE conditions upon both the data and a previously selected movement model that accurately represents the autocorrelation structure of the data. This means that first, one must select an appropriate movement model for the data using the model identification, fitting, and selection tools described in Fleming et al. (2014a, 2014b). Once an appropriate movement model has been fitted and selected, it can be used within the AKDE to yield an estimate of the range distribution (Figure 13.2d). Importantly, only models that feature restricted space use, such as the OU and OUF models, should be used in AKDE.

13.5 Software Tools

Contemporary home range analysis requires significant effort devoted to data management, visualization, and manipulation. Software such as ESRI's `ArcGIS` software, or open-source alternatives such as `GRASS` or `QGIS`, can assist with these tasks. Additionally, there are at least two `ArcGIS` extensions that implement several home range analyses including `HRT` (Home Range Tools for ArcGIS) and `ArcMET` (Movement Ecology Tools for ArcGIS). A powerful set of tools for conducting a wide variety of movement and home range analyses is the Geospatial Modeling Environment (`GME`) which is stand-alone software that utilizes both `ArcGIS` and `Program R` for functionality. Last, there is a relatively new software program called `Agent Analyst for ArcGIS` that can be used to facilitate agent-based modeling of animal movements.

Specific packages for Program R related to analysis of animal space use include the ctmm package for calculating movement variograms, for fitting several movement models, and for implementing the Kriging and AKDE approaches. The BBMM can be estimated with the BBMM, move or smam packages, as well as within the Kriging framework by using a Brownian motion movement model. The CTCRW model can be implemented within the package crawl or within the Kriging framework by using a movement model based on the IOU process. Last, the adehabitat and rhr packages implement many of the recent and traditional home range models, including KDE and MCP.

13.6 Online Exercises

The online exercises for this chapter illustrate representative analyses of animal movement data based on functions in the R packages adehabitHR and ctmm. Exercise 1 presents movement data from two caribou and shows how movement modes can be calculated from location data with estimates of NSD and variograms. Exercise 2 presents movement data from three caribou and illustrates alternative methods for estimating occurrence distributions, including MCPs, kernel density estimators, BBMMs, and Kriging as a generalization to non-Brownian movement models. Exercise 3 uses the same dataset on summer locations of caribou to estimate predicted range distributions based on a synoptic model of home range, or an ADKE.

13.7 Future Directions

13.7.1 Choosing a Home Range Model

The study of animal home ranges has been, and will continue to be, integral to advancing our ecological understanding. Indeed, the complexity and generality of the home range concept has allowed for fruitful studies of a myriad of ecological processes including habitat selection, sociality, predation, competition, and population dynamics, among others. But it is this same breadth and generality of the home range concept that has led to a variety of estimation targets and models to estimate these targets, presenting a persistent challenge for practicing ecologists needing to choose among home range estimators.

Often, researchers will promote a particular approach or method as "one size fits all," as if there is a single best model for answering all questions involving animal home ranges. We think this viewpoint is counterproductive and emphasize that the first and most important step in choosing a home range model is to have well-defined research questions that point to a particular estimation target, for example the occurrence or range distribution. This initial step is critical because it will help direct researchers toward a particular class of home range estimators, and will assist with development of the sampling design for location data. After ensuring that a group of home range estimators will provide the appropriate metric to answer your research question, the next step is to evaluate available choices, paying particular attention to sampling considerations, robustness to misspecification of model parameters, and the availability of software tools or computer code to implement the method.

In general, models that allow for more accurate estimation of the home range also require better location data both in terms of quantity and quality, greater understanding of complex modeling approaches, and more time to compile and analyze the data. Thus, as is true for ecological modeling in general, there are trade-offs between simple models that are easy to implement but are crude approximations of the real process versus complex models that have greater potential to accurately depict reality but are more difficult to implement. Simulation studies can be a powerful approach for vetting different estimators but to be useful, they should attempt to mimic the true underlying processes that generated the location data. These goals can be met by simulating animals moving on realistic landscapes, with locations collected according to a prespecified sampling schedule (Siniff and Jessen 1969; Fieberg 2007a, 2007b; Signer et al. 2015). Additionally, simulation studies should evaluate the implications of using alternative home range estimators on actual research questions. In other words, estimators should be compared in terms of their ability to capture meaningful biological signals, such as trends in space use, rather than simply address whether estimates include or exclude particular areas that are used or unused (Fieberg and Börger 2012; Signer et al. 2015). Recent studies by Börger et al. (2006b), Nilsen et al. (2008), and Signer et al. (2015) are examples that we hope will become more commonplace in the future, because their work demonstrates that oftentimes conclusions will be robust to the choice of analytical method even when absolute estimates differ considerably (Signer et al. 2015). As such, it may be reasonable to pick a simple method that is easy to understand and implement.

13.7.2 The Future of Home Range Modeling

Historically, studies of animal home ranges involved efforts to describe patterns of space use and then use subsequent analyses to try to explain these observed patterns. In the future, productive analysis of home range behavior will be facilitated by research that seeks a more direct understanding of the physiological, behavioral, and ecological processes that influence animal movements. By modeling these fundamental processes, we should be able to estimate home ranges as an *emergent property* instead of simply a description of observed location data. Over five decades ago, Sanderson (1966) encouraged researchers to focus less on calculating home range metrics and more on asking questions about why animals move the way they do, a call that has been frequently repeated (Fieberg and Börger 2012; Powell 2012). We believe there is still ample room for progress toward this goal. The recent surge of process-based models coupled with finer resolution and accuracy of location and environmental data has opened the door for ecologists to link ecological processes with statistical models more directly (Börger et al. 2008; See also Chapter 2). We view these process-based models as important advancements in the study of animal home ranges and believe they will lead to a greater understanding of the ecology of animal space use.

A relatively new and promising approach to gain a deeper understanding of ecological processes, including animal movements and space use, is the use of *agent-based models* (ABM; McLane et al. 2011). In fact, the analytical approach to mechanistic home range modeling we described earlier is an example of a relatively simple ABM. The goal of an ABM is to explicitly consider the "agents" of a system, in our case the individuals moving about a landscape, and use the fundamental processes influencing these agents to understand and predict the emergent properties of the system. For example, an ABM could be used to simulate an individual's movements by incorporating processes such as memory, learning, energy gain versus expenditure, predator avoidance, interactions with other individuals, and reproductive strategies. The emergent properties might include predictions related to site fidelity, habitat selection, and social interactions, among others; all of which are important components and processes related to home range behavior. Thus, instead of using location data to describe home range behavior, an ABM would predict home-range behavior based on an understanding of the interactions between individuals and their environment. These predictions could then be compared to observed location data to evaluate alternative research hypotheses. While this approach is still in its infancy, we strongly believe that it will constitute the next frontier in the study of animal home ranges.

Last, it is important to recognize that home range behavior is not isolated from other processes affecting animal movements. For example, in the past, research on animal movements often split into separate interest groups, including home range, dispersal, or migration. Currently, most researchers agree that animal movements encompass a continuum of different movement modes, from sedentarism (home range) to migration, dispersal, and nomadism (Börger et al. 2011; Mueller et al. 2011). Thus, a realistic representation of the population effects of animal movements might require explicit consideration of the dynamic connections among different movement modes (Wang and Grimm 2007).

References

Adams, J., MacLeod, C., Suryan, R.M. et al. (2012). Summertime use of west coast U.S. National Marine Sanctuaries by migrating Sooty Shearwaters (*Puffinus griseus*). *Biological Conservation* 156: 105–116.

Adams, L.G., Stephenson, R.O., Dale, B.R. et al. (2008). Population dynamics and harvest characteristics of wolves in the Central Brooks Range, Alaska. *Wildlife Monographs* 170, 25 pp.

Aebischer, N.J., Robertson, P.A., and Kenward, R.E. (1993). Compositional analysis of habitat use from animal radio-tracking data. *Ecology* 74: 1313–1325.

Anadón, J.D., Wiegland, T., and Giménez, A. (2012). Individual-based movement models reveals sex-biased effect of landscape fragmentation on animal movement. *Ecosphere* 3: art64.

Anderson, D.J. (1982). The home range: a new nonparametric estimation technique. *Ecology* 63: 103–112.

van Beest, F.M., Rivrud, I.M., Loe, L.E. et al. (2011). What determines variation in home range size across spatiotemporal scales in a large browsing herbivore? *Journal of Animal Ecology* 80: 771–785.

Bengsen, A., Butler, J., and Masters, P. (2012). Estimating and indexing feral cat population abundances using camera traps. *Wildlife Research* 38: 732–739.

Benhamou, S. (1989). An olfactory orientation model for mammal movements in their home ranges. *Journal of Theoretical Biology* 139: 379–388.

Benhamou, S. (1994). Spatial memory and searching efficiency. *Animal Behaviour* 47: 1423–1433.

Blackwell, P.G. (1997). Random diffusion models for animal movement. *Ecological Modelling* 100: 87–102.

Blackwell, P.G. (2003). Bayesian inference for Markov processes with diffusion and discrete components. *Biometrika* 90: 613–627.

Börger, L., Dalziel, B.D., and Fryxell, J.M. (2008). Are there general mechanisms of animal home range behaviour? A review and prospects for future research. *Ecology Letters* 11: 637–650.

Börger, L., Franconi, N., De Michele, G. et al. (2006a). Effects of sampling regime on the mean and variance of home range estimates. *Journal of Animal Ecology* 75: 1393–1405.

Börger, L., Franconi, N., Ferretti, F. et al. (2006b). An integrated approach to identify spatiotemporal and individual-level determinants of animal home range size. *American Naturalist* 168: 471–485.

Börger, L. and Fryxell, J. (2012). Quantifying individual differences in dispersal using net squared displacement. In: *Dispersal Ecology and Evolution* (eds. J. Clobert, M. Baguette, T.G. Benton and J.M. Bullock), 222–230. Oxford, UK: Oxford University Press.

Börger, L., Matthiopoulos, J., Holdo, R.M. et al. (2011). Migration quantified: constructing models and linking them with data. In: *Animal Migration—A Synthesis* (eds. E.J. Milner-Gulland, J.M. Fryxell and A.R.E. Sinclair), 111–128. Oxford, UK: Oxford University Press.

Bullard, F. 1999. Estimating the home range of an animal: A Brownian bridge approach. M.S. thesis. University of North Carolina, Chapel Hill, NC.

Burt, W.H. (1943). Territoriality and home range concepts as applied to mammals. *Journal of Mammalogy* 24: 346–352.

Calhoun, J. B., and Casby, J. U. 1958. *Calculation of home range and density of small mammals*. US Dept. of Health, Education, and Welfare, Public Health Service Publication No. 592. 24 pp.

Coughlin, C.E. and van Heezik, Y. (2015). Weighed down by science: do collar-mounted devices affect domestic cat behaviour and movement? *Wildlife Research* 41: 606–614.

Cressie, N. (1993). *Statistics for Spatial Data*, Revised ed. New York: Wiley.

Darwin, C. (1859). *On the Origin of Species by Means of Natural Selection, or Preservation of Favoured Races in the Struggle for Life*. London, UK: John Murray.

Dennis, T.E. and Shah, S.F. (2012). Acute effects of trapping, handling, and tagging on the behavior of wildlife using GPS telemetry: a case study of the common brushtail possum. *Journal of Applied Animal Welfare Science* 15: 189–207.

Diggle, P.J. and Ribeiro, P.J. (2007). *Model-based Geostatistics*. New York: Springer.

Dunn, J.E. and Gipson, P.S. (1977). Analysis of radiotelemetry data in studies of home range. *Biometrics* 33: 85–101.

Edelman, A. (2012). Positive interactions between desert granivores: localized facilitation of harvester ants by kangaroo rats. *PLoS One* 7: e30914.

Emlen, S.T. and Oring, L.W. (1977). Ecology, sexual selection, and the evolution of mating systems. *Science* 197: 215–223.

Evans, M., Hastings, N., and Peacock, B. (2000). *Statistical Distributions*, 3e. New York, NY: Wiley.

Fieberg, J. (2007a). Kernel density estimators of home range: smoothing and the autocorrelation red herring. *Ecology* 88: 1059–1066.

Fieberg, J. (2007b). Utilization distribution estimation using weighted kernel density estimators. *Journal of Wildlife Management* 71: 1669–1675.

Fieberg, J. and Börger, L. (2012). Could you please phrase "home range" as a question ? *Journal of Mammalogy* 93: 890–902.

Fieberg, J. and Kochanny, C.O. (2005). Quantifying home-range overlap: the importance of the utilization distribution. *Journal of Wildlife Management* 69: 1346–1359.

Fleming, C.H., Calabrese, J.M., Mueller, T. et al. (2014a). From fine-scale foraging to home ranges: a semi-variance approach to identifying movement modes across spatiotemporal scales. *American Naturalist* 183: E154–E167.

Fleming, C.H., Calabrese, J.M., Mueller, T. et al. (2014b). Non-Markovian maximum likelihood estimation of autocorrelated movement processes. *Methods in Ecology and Evolution* 5: 462–472.

Fleming, C.H., Fagan, W.F., Mueller, T. et al. (2015). Rigorous home range estimation with movement data: a new autocorrelated kernel-density estimator. *Ecology* 96: 1182–1188.

Fleming, C.H., Fagan, W.F., Mueller, T. et al. (2016). Estimating where and how animals travel: an optimal framework for path reconstruction from autocorrelated tracking data. *Ecology* 97: 576–582.

Frair, J.L., Fieberg, J., Hebblewhite, M. et al. (2010). Resolving issues of imprecise and habitat-biased locations in ecological analyses using GPS telemetry data. *Philosophical Transactions of the Royal Society of London B* 365: 2187–2200.

Frère, C.H., Krützen, M., Mann, J. et al. (2010). Home range overlap, matrilineal and biparental kinship drive female associations in bottlenose dolphins. *Animal Behaviour* 80: 481–486.

Fuller, T.K., Berg, W.E., Radde, G.L. et al. (1992). A history and current estimate of wolf distribution and numbers in Minnesota. *Wildlife Society Bulletin* 20: 42–55.

Gautestad, A.O. and Mysterud, I. (2005). Intrinsic scaling complexity in animal dispersion and abundance. *American Naturalist* 165: 44–55.

Grassel, S.M., Rachlow, J.L., and Williams, C.J. (2015). Spatial interactions between sympatric carnivores: asymmetric avoidance of an intraguild predator. *Ecology and Evolution* 5: 2762–2773.

Hall, E.R. and Linsdale, J.M. (1929). Notes on the life history of the kangaroo mouse (*Microdipodops*). *Journal of Mammalogy* 10: 298–305.

Hamilton, W.J. Jr. (1937). Activity and home range of the field mouse, *Microtus pennsylvanicus pennsylvanicus*. *Ecology* 18: 255–263.

Harris, S., Cresswell, W.J., Forde, P.G. et al. (1990). Home range analysis using radio-tracking data – a review of problems and techniques particularly as applied to the study of mammals. *Mammal Review* 20: 97–123.

Hayne, D.W. (1949). Calculation of size of home range. *Journal of Mammalogy* 30: 18.

Holyoak, M., Leibold, M.A., and Holt, R.D. (eds.) (2005). *Metacommunities: Spatial Dynamics and Ecological Communities*. Chicago, IL: University of Chicago Press.

Horne, J.S., Craig, T., Joly, K. et al. (2014). Population characteristics, space use and habitat selection of two non-migratory caribou herds in Central Alaska, 1994-2009. *Rangifer* 34: 1–20.

Horne, J.S. and Garton, E.O. (2006a). Selecting the best home range model: an information theoretic approach. *Ecology* 87: 1146–1152.

Horne, J.S. and Garton, E.O. (2006b). Likelihood cross-validation versus least-squares cross-validation for choosing the smoothing parameter in kernel home-range analysis. *Journal of Wildlife Management* 70: 641–648.

Horne, J.S., Garton, E.O., Krone, S.M., and Lewis, J.S. (2007a). Analyzing animal movements using Brownian bridges. *Ecology* 88: 2354–2363.

Horne, J.S., Garton, E.O., and Rachlow, J.L. (2008). A synoptic model of animal space use: simultaneous estimation of home range, habitat selection, and inter/intra-specific relationships. *Ecological Modelling* 214: 338–348.

Horne, J.S., Garton, E.O., and Sager-Fradkin, K.A. (2007b). Correcting home-range models for observation bias. *Journal of Wildlife Management* 71: 996–1001.

Horne, J.S., Tewes, M.E., Haines, A.M., and Laack, L.L. (2009). Habitat partitioning of sympatric ocelots and bobcats: implications for ocelot recovery in southern Texas. *Southwestern Naturalist* 54: 119–126.

Jennrich, R.I. and Turner, F.B. (1969). Measurement of non-circular home range. *Journal of Theoretical Biology* 22: 227–237.

Johnson, D.H. (1980). The comparison of usage and availability measurements for evaluating resource preference. *Ecology* 61: 65–71.

Johnson, R.R., Granfors, D.A., Niemuth, N.D. et al. (2010). Delineating grassland bird conservation areas in the U.S. Prairie Pothole region. *Journal of Fish and Wildlife Management* 1: 38–42.

Johnson, D.S., London, J.M., Lea, M.A., and Durban, J.W. (2008a). Continuous-time correlated random walk model for animal telemetry data. *Ecology* 89: 1208–1215.

Johnson, D.S., Thomas, D.L., Ver Hoef, J.M., and Christ, A. (2008b). A general framework for the analysis of animal resource selection from telemetry data. *Biometrics* 64: 968–976.

Jones, M.C., Marron, J.S., and Sheather, S.J. (1996). A brief survey of bandwidth selection for density estimation. *Journal of the American Statistical Association* 91: 401–407.

Keating, K.A. and Cherry, S. (2009). Modeling utilization distributions in space and time. *Ecology* 90: 1971–1980.

Kittle, A.M., Anderson, M., Avgar, T. et al. (2015). Wolves adapt territory size, not pack size to local habitat quality. *Journal of Animal Ecology* 84: 1177–1186.

Kranstauber, B., Kays, R., LaPoint, S.D. et al. (2012). A dynamic Brownian bridge movement model to estimate utilization distributions for heterogeneous animal movement. *Journal of Animal Ecology* 81: 738–746.

Laver, P.N. and Kelly, M.J. (2008). A critical review of home range studies. *Journal of Wildlife Management* 72: 290–298.

Lay, D.W. and Baker, R.H. (1938). Notes on the home range and ecology of the Attwater wood rat. *Journal of Mammalogy* 19: 418–423.

Lewis, M.A. and Murray, J.D. (1993). Modeling territoriality and wolf-deer interactions. *Nature* 366: 738–740.

Li, J., Brugere, I., Ziebart, B., et al. (2015). Social information improves location prediction in the wild. *AAAI Workshop*.

Lowther, A.D., Harcourt, R.G., Page, B., and Goldsworthy, S. D. (2013). Steady as he goes: At-sea movement of adult male Australian sea lions in a dynamic marine environment. *PLoS One* 8: e74348.

Marsh, M.K., Hutchings, M.R., McLeod, S.R., and White, P. C.L. (2011). Spatial and temporal heterogeneities in the contact behaviour of rabbits. *Behavioural Ecology and Sociobiology* 65: 183–195.

Massé, A. and Côté, S.D. (2012). Linking habitat heterogeneity to space use by large herbivores at multiple scales: from habitat mosaics to forest canopy openings. *Forest Ecology and Management* 285: 67–76.

Matthiopoulos, J., Harwood, J., and Thomas, L. (2005). Metapopulation consequences of site fidelity for colonially breeding mammals and birds. *Journal of Animal Ecology* 74: 716–727.

McLane, A.J., Semeniuk, C., McDermid, G.J., and Marceau, D.J. (2011). The role of agent-based models in wildlife ecology and management. *Ecological Modelling* 222: 1544–1556.

McLoughlin, P.D., Ferguson, S.H., and Messier, F. (2000). Intraspecific variation in home range overlap with habitat quality: a comparison among brown bear populations. *Evolutionary Ecology* 14: 39–60.

McNab, B.K. (1963). Bioenergetics and the determination of home range size. *American Naturalist* (894): 133–140.

Millspaugh, J.J., Kesler, D.C., Kays, R.W. et al. (2012). Wildlife radiotelemetry and remote monitoring. In: *Wildlife Techniques Manual* (ed. N.J. Silvy), 258–283. Johns Hopkins Press.

Mitchell, M.S. and Powell, R.A. (2004). A mechanistic home range model for optimal use of spatially distributed resources. *Ecological Modelling* 177: 209–232.

Mitchell, M.S. and Powell, R.A. (2007). Optimal use of resources structures home ranges and spatial distribution of black bears. *Animal Behaviour* 74: 219–230.

Mitchell, M.S. and Powell, R.A. (2012). Foraging optimally for home ranges. *Journal of Mammalogy* 93: 917–928.

Mohr, C.O. (1947). Table of equivalent populations of North American small mammals. *American Midland Naturalist* 37: 223–249.

Moorcroft, P.R. (2012). Mechanistic approaches to understanding and predicting mammalian space use: recent advances, future directions. *Journal of Mammalogy* 93: 903–916.

Moorcroft, P.R. and Barnett, A. (2008). Mechanistic home range models and resource selection analysis: a reconciliation and unification. *Ecology* 89: 1112–1119.

Moorcroft, P.R. and Lewis, M.A. (2006). *Mechanistic Home Range Analysis*. Princeton, NJ: Princeton University Press.

Morrison, M.L., Block, W.M., Strickland, M.D., and Kendall, W.L. (2001). *Wildlife Study Design*. New York, New York, USA: Springer-Verlag New York, Inc.. 210 pp.

Mueller, T., Olson, K.A., Dressler, G. et al. (2011). How landscape dynamics link individual- to population-level movement patterns: a multispecies comparison of ungulate relocation data. *Global Ecology and Biogeography* 20: 683–694.

Murray, C.M., Mane, S.V., and Pusey, A.E. (2007). Dominance rank influences female space use in wild chimpanzees, *Pan troglodytes*: towards an ideal despotic distribution. *Animal Behaviour* 74: 1795–1804.

Murtaugh, P.A. (2007). Simplicity and complexity in ecological data analysis. *Ecology* 88: 56–62.

Naidoo, R., Du Preez, P., Stuart-Hill, G. et al. (2012). Factors affecting intraspecific variation in home range size of a large African herbivore. *Landscape Ecology* 27: 1523–1534.

Nathan, R., Getz, W.M., Revilla, E. et al. (2008). A movement ecology paradigm for unifying organismal movement research. *Proceedings of the National Academy of Science USA* 105: 19052–19059.

Nilsen, E.B., Pedersen, S., and Linnell, J.D.C. (2008). Can minimum convex polygon home ranges be used to draw biologically meaningful conclusions? *Ecological Research* 23: 635–639.

Nouvellet, P., Bacon, J.P., and Waxman, D. (2009). Fundamental insights into the random movement of animals from a single distance-related statistic. *American Naturalist* 174: 506–514.

Okubo, A. and Levin, S. (2001). *Diffusion and Ecological Problems: Modern Perspectives*, 2e. Berlin, Germany: Springer-Verlag.

Ostro, L.E.T., Silver, S.C., Koontz, F.W. et al. (1999a). Ranging behavior of translocated and established groups of black howler monkeys *Alouatta pigra* in Belize, Central America. *Biological Conservation* 87: 181–190.

Ostro, L.E.T., Young, T.P., Silver, S.C., and Koontz, F.W. (1999b). A geographic information system method for estimating home range size. *Journal of Wildlife Management* 63: 748–755.

Powell, R.A. (2000). Animal home ranges and territories and home range estimators. In: *Research Techniques in Animal Ecology: Controversies and Consequences* (eds. L. Boitani and T.K. Fuller), 65–100. New York, NY: Columbia University Press.

Powell, R.A. (2012). Diverse perspectives on mammal home ranges or a home range is more than location densities. *Journal of Mammalogy* 93: 887–889.

Powell, R.A. and Mitchell, M.S. (2012). What is a home range? *Journal of Mammalogy* 93: 948–958.

Pozdnyakov, V., Meyer, T., Wang, Y.B., and Yan, J. (2014). On modeling animal movements using Brownian motion with measurement error. *Ecology* 95: 247–253.

Rachlow, J.L., Peter, R.M., Shipley, L.A., and Johnson, T.R. (2014). Sub-lethal effects of capture and collaring on wildlife: experimental and field evidence. *Wildlife Society Bulletin* 38: 458–465.

Rhodes, J.R., McAlpine, C.A., Lunney, D., and Possingham, H.P. (2005). A spatially explicit habitat selection model incorporating home range behavior. *Ecology* 86: 1199–1205.

Sanderson, G.C. (1966). The study of mammal movements: a review. *Journal of Wildlife Management* 30: 215–235.

Sawyer, H., Kauffman, M.J., Nielson, R.M., and Horne, J.S. (2009). Identifying and prioritizing ungulate migration routes for landscape-level conservation. *Ecological Applications* 19: 2016–2025.

Seton, E.T. (1920). For a methodic study of life-history of mammals. *Journal of Mammalogy* 1: 67–69.

Shirk, A.J., Raphael, M.G., and Cushman, S.A. (2014). Spatiotemporal variation in resource selection: insights from the American marten (*Martes americana*). *Ecological Applications* 24: 1434–1444.

Signer, J., Balkenhol, N., Ditmer, M., and Fieberg, J. (2015). Does estimator choice influence our ability to detect changes in home range size? *Animal Biotelemetry* 3: 1–9.

Silverman, B.W. (1986). *Density Estimation for Statistics and Data Analysis*. London, United Kingdom: Chapman and Hall.

Siniff, D.B. and Jessen, C.R. (1969). A simulation model of animal movement patterns. In: *Advances in Ecological Research*, vol. 6 (ed. J.B. Cragg), 185–219. London, UK: Academic Press.

Skellam, J.G. (1951). Random dispersal in theoretical populations. *Biometrika* 38: 196–218.

Sparklin, B.D., Mitchell, M.S., Hanson, L.B. et al. (2009). Territoriality of feral pigs in a highly persecuted population on fort Benning, Georgia. *Journal of Wildlife Management* 73: 497–502.

Suwanrat, S., Ngoprasert, D., Sutherland, C. et al. (2015). Estimating density of secretive terrestrial birds (Siamese Fireback) in pristine and degraded forest using camera traps and distance sampling. *Global Ecology and Conservation* 3: 596–606.

Tracey, J.A., Sheppard, J., Zhu, J. et al. (2014). Movement-based estimation and visualization of space use in 3D for wildlife ecology and conservation. *PLoS One* 9: e109065.

Turchin, P. (1998). *Quantitative Analysis of Movement.* Sunderland, MA: Sinauer Associates, Inc.

Uhlenbeck, G.E. and Ornstein, L.S. (1930). On the theory of the Brownian motion. *Physical Review* 36: 823.

Van Moorter, B., Visscher, D., Benhamou, S. et al. (2009). Memory keeps you at home: a mechanistic model for home range emergence. *Oikos* 118: 641–652.

Walter, W.D., VerCauteren, K.C., Campa, H. III et al. (2009). Regional assessment on influence of landscape configuration and connectivity on range size of white-tailed deer. *Landscape Ecology* 24: 1405–1420.

Wang, M. and Grimm, V. (2007). Home range dynamics and population regulation: an individual-based model of the common shrew *Sorex araneus*. *Ecological Modelling* 205: 397–409.

White, G.C. and Garrott, R.A. (1990). *Analysis of Wildlife Radio-Tracking Data*. San Diego, CA: Academic Press.

Whittington, J., Hebblewhite, M., DeCesare, N.J. et al. (2011). Caribou encounters with wolves increase near roads and trails: a time-to-event approach. *Journal of Applied Ecology* 48: 1535–1542.

Wilson, E.O. (1979). *Sociobiology: The New Synthesis*. New York, NY: Wiley.

Wilson, R.R., Horne, J.S., Rode, K.D. et al. (2014). Identifying polar bear resource selection patterns to inform offshore development in a dynamic and changing Arctic. *Ecosphere* 5: art136.

Worton, B.J. (1989). Kernel methods for estimating the utilization distribution in home-range studies. *Ecology* 70: 164–168.

Zeale, M.R., Davidson-Watts, I., and Jones, G. (2012). Home range use and habitat selection by barbastelle bats (*Barbastella barbastellus*): implications for conservation. *Journal of Mammalogy* 93: 1110–1118.

14

Analysis of Resource Selection by Animals

Joshua J. Millspaugh[1], Christopher T. Rota[2], Robert A. Gitzen[3], Robert A. Montgomery[4], Thomas W. Bonnot[5], Jerrold L. Belant[6], Christopher R. Ayers[7], Dylan C. Kesler[8], David A. Eads[9] and Catherine M. Bodinof Jachowski[10]

[1] Wildlife Biology Program, Department of Ecosystem and Conservation Sciences, University of Montana, Missoula, MT, USA
[2] Wildlife and Fisheries Resources Program, School of Natural Resources, West Virginia University, Morgantown, WV, USA
[3] Forestry and Wildlife Sciences, Auburn University, Auburn, AL, USA
[4] Department of Fisheries and Wildlife, Michigan State University, East Lansing, MI, USA
[5] Department of Fisheries and Wildlife Sciences, University of Missouri, Columbia, MO, USA
[6] Camp Fire Program in Wildlife Conservation, College of Environmental Science and Forestry, State University of New York, Syracuse, NY, USA
[7] Wildlife, Fisheries, and Aquaculture, Mississippi State University, Starkville, MS, USA
[8] The Institute for Bird Populations, Point Reyes Station, CA, USA
[9] U.S. Geological Survey, Fort Collins Science Center, Fort Collins, CO, USA
[10] Department of Forestry and Environmental Conservation, Clemson University, Clemson, SC, USA

Summary

The study of resource selection by animals has been of great interest to ecologists for many decades. A renewed sense of urgency has arisen in studies of resource selection given new environmental challenges such as climate change, conservation planning, and habitat loss. Technological and analytical advancements have furthered our ability to collect data and model resource selection over large areas. Here, we discuss several important considerations in resource selection studies including sampling strategies, study design, and potential sources of bias. We also provide advice on selecting analytical methods, linking resource selection to fitness, and on incorporating animal behavior and biological seasons in studies of resource selection. All inference begins and ends with sampling so we stress the importance of applying rigorous study designs and sampling strategies while avoiding the temptations of convenience sampling and the idea that some information is better than none. We emphasize the use of radio-tracking technologies because such tools have greatly enhanced studies of resource selection, but also include other methods of data collection. A theme linking analytical methods is the development of resource selection functions (RSFs) and resource selection probability functions (RSPFs), which are the primary means to assess selection of resources by animals. We describe many of the most common procedures to analyze resource selection data, including compositional analysis, logistic regression, discrete choice models, Poisson regression, ecological niche factor analysis, and mixed-effect models. We describe some of the key assumptions and highlight the necessary data. We generally recommend process-based resource selection models because the statistical properties of these models are better understood, they have a long history of application and development in the statistical literature, and because they make good biological sense. We caution that global recommendations on analytical methods and study designs are not possible given such choices should be aligned with study objectives, data types, and assumptions inherent in different sampling protocols. Looking ahead, we anticipate several new developments in the future of resource selection studies, including continued development of individual-animal-based models which are scaled up to the population level, applications of new technology, and increased reliance on mixed models and Bayesian methods which handle highly complex models and appropriately propagate uncertainty.

14.1 Introduction

Resource selection is the study of resource "choices" made by animals (Manly et al. 2002; Lele et al. 2013). Resources can include environmental features such as vegetation, food, and water; anthropogenic disturbance; and landscape attributes. Resource selection has often been viewed in a cost–benefit framework, where costs and benefits derived from different resources are weighed by an animal, consciously or instinctively, to determine which resources should be used (Hildén 1965; Jones 2001). Factors such as competition (MacArthur and Levins 1964; Martin 1996), predation (Creel et al. 2005; White et al. 2008), parasitism (Stamp et al. 2002), temporal changes in the availability of resources (Arthur et al. 1996), and climate (Mysterud 1996; Martin 2001) can all affect patterns of resource selection.

Population Ecology in Practice, First Edition. Edited by Dennis L. Murray and Brett K. Sandercock.

Companion website: www.wiley.com/go/MurrayPopulationEcology

There are many practical and theoretical motivations to study resource selection. From a practical perspective, natural resource managers seek to provide resources such as food and cover that promote survival of species (Manly et al. 2002), map the occurrence of species for planning purposes (Dzialak et al. 2013; Sawyer and Brashares 2013), conduct quantitative risk assessments for wildlife populations (McDonald and McDonald 2002), delineate conservation corridors (Chetkiewicz and Boyce 2009; Squires et al. 2013), prioritize animal translocation sites (Bodinof et al. 2012), and understand the effects of human activities and development on wildlife (Jobes et al. 2004; Sawyer et al. 2006; McNew et al. 2014). At other times, such as in the case of wildlife damage management, it is useful to know which resources might dissuade animals from using certain areas (McArthur et al. 2000). Theoretical interest in resource selection addresses fundamental questions pertaining to species distributions (Fretwell and Lucas 1970; Fretwell 1972), the evolutionary costs and benefits of differential habitat use (Martin 1996), the role of ontogeny in resource selection (Wiens 1972), interspecific competition and niche overlap (MacArthur and Levins 1964; Rosenzweig 1991), and the interplay between use of habitats and genetic flow (Scribner et al. 2005). Regardless of the motivations for a resource selection study, it should be carried out with knowledge of the state of the art of design elements and considerations, data types, and methods of analysis.

The study of resource selection has strong historical roots, but has developed tremendously in recent years. Resource selection by birds has been studied for many decades (Kendeigh 1945; Svärdson 1949) and key concepts identified 50 years ago remain widely applicable today (Hildén 1965). Issues of spatial scale (Wiens 1973; Orians and Wittenberger 1991; DeCesare et al. 2012), the relative role of proximate and ultimate factors (Hutto 1985), and prior experience (Wiens 1972) have been discussed for many decades and remain critically important to contemporary conservation issues involving resource selection. Modern studies, involving the development of resource selection functions (RSFs), can be traced to applications in the early 1990s (McDonald et al. 1990). However, renewed interest in the study of resource selection has grown (Jones 2001), due to interest in conservation planning initiatives (Johnson et al. 2004; Millspaugh and Thompson III 2009) and the need to understand how species may respond to climate change (Martin 2001; Durner et al. 2009; Wiens et al. 2009) and ongoing habitat loss (Oppel et al. 2004; Sawyer et al. 2006; Dzialak et al. 2011). The past 20 years have seen a surge in the number of resource selection studies and the development and application of sophisticated technological tools and analytical techniques (Aarts et al. 2012; McDonald et al. 2013; Renner et al. 2015).

New technologies have facilitated the recent growth in resource selection research. For instance, subminiaturization of radio tags, the development of new global tracking technologies, accessibility of camera traps, and the availability of new sensors that allow for more detailed studies of wild animals have greatly expanded the breadth and scope of resource selection research (Moll et al. 2007; Cagnacci et al. 2010; Rota et al. 2016). Researchers have the ability to collect more detailed information from a greater diversity of species for longer periods of time than ever before (Bridge et al. 2011; Kranstauber et al. 2011). Technological advances are important to studies of resource selection, but such developments carry new assumptions with analysis and interpretation and potential issues such as spatial and temporal autocorrelation, which must be addressed using modern techniques (Johnson et al. 2008; Brost et al. 2015). Technological advancements also involve trade-offs regarding financial costs and reduced field effort, which can be critically important in providing context for data.

Advances in statistical methodology have closely tracked technological advancements (Gillies et al. 2006). As with all statistical analyses, one needs to carefully consider methodological assumptions when modeling resource selection. In studies involving wild animals in field environments, some otherwise reasonable assumptions may be difficult to meet, or the effort of meeting these assumptions may not sufficiently minimize biases. Further, careful interpretation of model output requires researchers to consider whether one can interpret the result as a *relative* probability of resource use, or as an *absolute* probability of resource use – a fundamental distinction in resource selection studies.

Our chapter focuses on the design, analysis, and interpretation of resource selection studies. Radio-tracking technology is a primary method to study resource selection, so much of this chapter discusses telemetry-based designs and analysis. However, many of these designs and analyses are well-suited to other data collection efforts. We focus on practical motivations for studies of resource selection, such as mapping the occurrence of species or understanding which resources are associated with the probability an animal will use a particular area. As a theme for this chapter, we make a distinction between RSFs (McDonald et al. 1990; Manly et al. 2002) and *resource selection probability functions* (RSPFs; Sólymos and Lele 2016). We have attempted to highlight the most commonly used procedures with the greatest value for studying resource selection by animals. We first introduce readers to basic terminology, which is followed by important considerations in resources selection studies, before ending with descriptions of analytical options.

Throughout this chapter, we emphasize that choice of an appropriate analytical method and the quality of a resource selection study should be driven by study objectives and fundamental principles of experimental design.

14.2 Definitions

14.2.1 Terminology and Currencies of Use and Availability

The study of resource selection comes with its own formalized terminology. As you will note later, this terminology describes concepts that should drive study design, choice of analytical methods, and interpretation of results. We focus on seven primary terms: *resource, use, availability, non-use, selection, preference*, and *avoided.*

A *resource* is a discrete unit an animal may encounter (Lele et al. 2013). The exact definition of a resource depends on the context of a study, but could include discrete food items, a vegetation cover type, or other features expressed in a pixel in a geographic information system (GIS) map. Resource units typically have a set of one or more attributes that animals may use to discriminate among resource units, such as the type of food item, the specific cover type, or the elevation a particular pixel represents. These attributes are commonly referred to as *covariates* or *environmental variables.* An *available* resource unit is one that can potentially be encountered by an animal, whereas a *used* resource unit is one that has received some investment by an animal. The exact nature of this investment depends on the context of the study, but can include the quantity of resource units obtained in a fixed period (Johnson 1980); or the time spent, distance traveled, energy expended, or presence within a resource unit (Buskirk and Millspaugh 2006). In contrast, *non-use* considers resources at sample units where no use occurred. It is necessary to confirm the lack of use to declare that a site or resource was not used, since failure to detect an animal does not mean use did not occur and lack of use may be confounded with imperfect detection (MacKenzie et al. 2002; Chapter 3).

Because non-use of a resource is often impossible to determine, many studies compare used resources to available resources (Figure 14.1). Note that available resources are called *background* or *pseudo-absences* in some contexts (Phillips and Dudik 2008; Renner et al. 2015; Chapter 15). In some studies of resource selection, such as food choice studies where animals are offered different food types, availability of resources may be known precisely (Carter et al. 1999). However, in field studies, resource availability is often difficult to determine

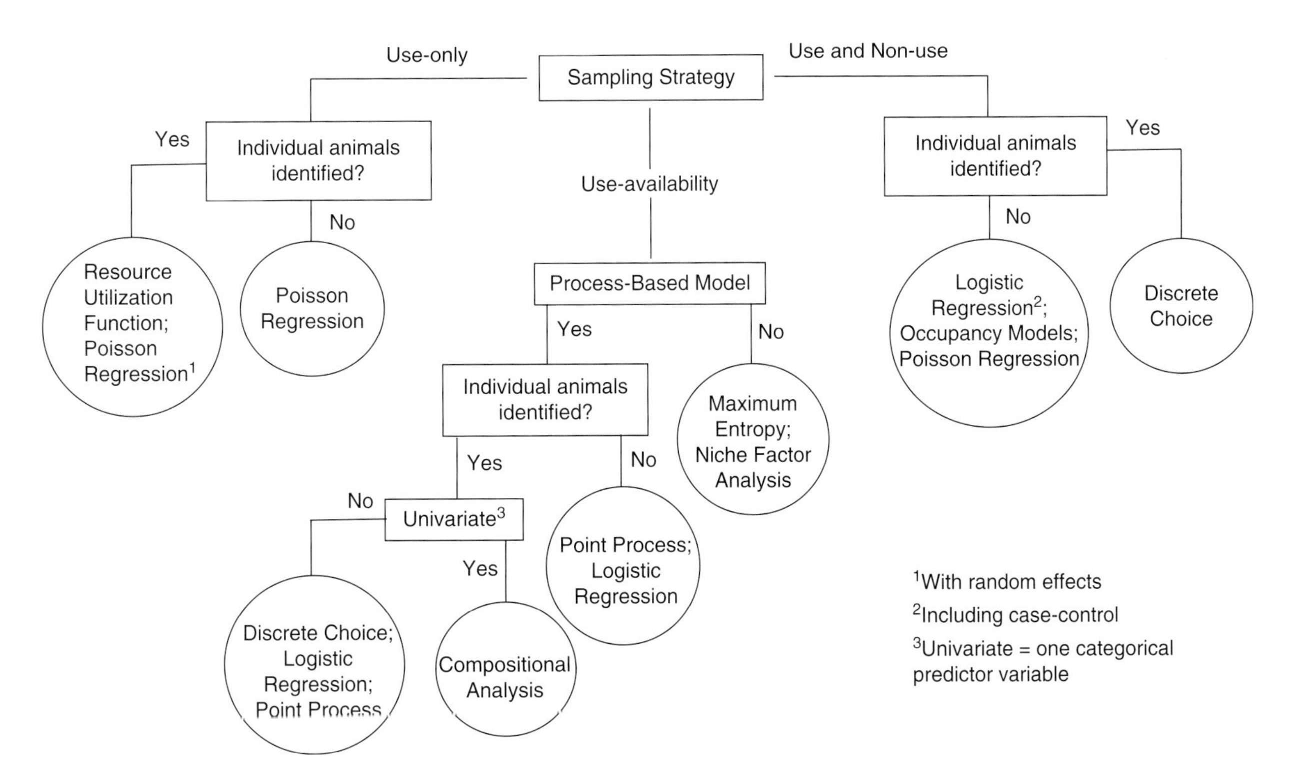

Figure 14.1 Conceptual diagram of several commonly used methods to fit resource selection models. Many of these categories are not mutually exclusive, but rather represent commonly used methods given sample strategy, whether individuals are identified, and whether they use process-based models.

because it can be affected by social hierarchies, risk of predation, and other uncontrolled factors (Otis 1997, 1998; Mysterud and Ims 1998). Thus, available resources are often defined within specific spatial bounds determined by the researcher. The definition of availability is critically important in resource selection studies because it determines the distribution of resources against which use is compared (Buskirk and Millspaugh 2006; Benson 2012; Northrup et al. 2013).

Comparing used resource units to unused or available resource units allows inference regarding preference and selection. *Preference* for a resource reflects the relative likelihood that a resource is chosen if offered on an equal basis with others (Johnson 1980). Inferring preference from field studies is difficult because resources are not equally and randomly available (Marzluff et al. 2001). For this reason, the term preference is not appropriate in most studies of resource selection and use of the term should be generally restricted to controlled feeding trials. The term "selection" is used when resource use is disproportionately greater than resource availability (Johnson 1980). Conversely, resources are "avoided" when resource use is disproportionately less than would be expected based on the availability of that resource (Johnson 1980). Avoidance can be a misleading term because even avoided resources may still be used by animals.

14.2.2 Use-availability, Paired Use-availability, Use and Non-use (Case-control), and Use-only Designs

A common goal of many resource selection studies is to predict either the relative or absolute probability a resource unit is used, depending on how sample units were selected and categorized (Keating and Cherry 2004). There are many different methods for sampling used, unused, and available resources, each of which implies its own statistical model. The first online exercise accompanying this chapter demonstrates how modeling and interpretation of RSFs and RSPF scan depend on sampling methodologies.

Perhaps the simplest conceptual design in a resource selection study is to collect a random sample of available resource units from the population of interest and determine if each unit was used or unused. While useful as a starting point in considering different sampling designs, this deceptively simple sampling protocol is usually difficult to implement. In most practical settings, *non-use* cannot be determined with certainty, and replicate surveys to determine detection and nondetection are often necessary to account for imperfect detection. In many other practical settings, a random sample of available resource units will result in few or no observations of use. This situation may occur, for example, with rare or difficult-to-detect species. In these settings, a random sample of used resources units may be obtained such as with radio telemetry and a separate sample of unused but available resources may be obtained, which is a so-called *case-control design*. Alternatively, a randomly sampled used resource unit could be paired with one or more resource units that were unused at that time such as sampling a choice set in *discrete choice analysis* (Cooper and Millspaugh 1999). Such an approach avoids ambiguity in establishing non-use, since an animal cannot use two resource units simultaneously, but also limits the types of predictions that can be made. The difficulty of establishing non-use leads many to collect a random sample of used resource units, and a separate random sample of available resource units, acknowledging that available units may or may not be used themselves. Last, the difficulty of establishing non-use leads to some sampling schemes that only collect random samples of used resources.

14.2.3 Differences Between RSF, RSPF, and RUF

When describing resource selection by animals, it is important to distinguish among three distinct categories of analytical methods: RSF, RSPF, and *resource utilization functions* (RUF). RSFs are proportional to RSPFs up to an arbitrary constant and can be used to estimate the relative probability of use. In contrast, the estimated parameter coefficients in a RSPF allow prediction of the absolute probability that a particular sampling unit will be used by an animal (Manly et al. 2002; Lele and Keim 2006; Sólymos and Lele 2016). RSFs may not be insightful if the absolute probabilities of use are close to 0 or 1. For example, consider a case where use of one resource is 10× more likely than another. However, if the absolute probability of using one resource is 0.01 and the other resource is 0.10, neither resource is likely to be used. For this reason, RSPFs offer a more intuitive and useful interpretation. Furthermore, we caution users of RSFs against a probabilistic interpretation, even when predictions are scaled between 0 and 1, because the relative probability might not be proportional to the absolute probability of use. The underlying data, choice of analytical methods, and study objectives ultimately determine whether a RSF or RSPF is used, but we recommend using RSPFs whenever possible, given their ability to estimate absolute probability of use and the resulting more-direct interpretation of outputs.

RSFs and RSPFs can both be used when the response variable is discrete: used versus unused, or use versus availability. In contrast, RUFs can be used to model resource selection when a continuous density estimate derived from a *utilization distribution* (UD) is treated as the response variable (Marzluff et al. 2004; Millspaugh et al. 2006; Figure 14.2).

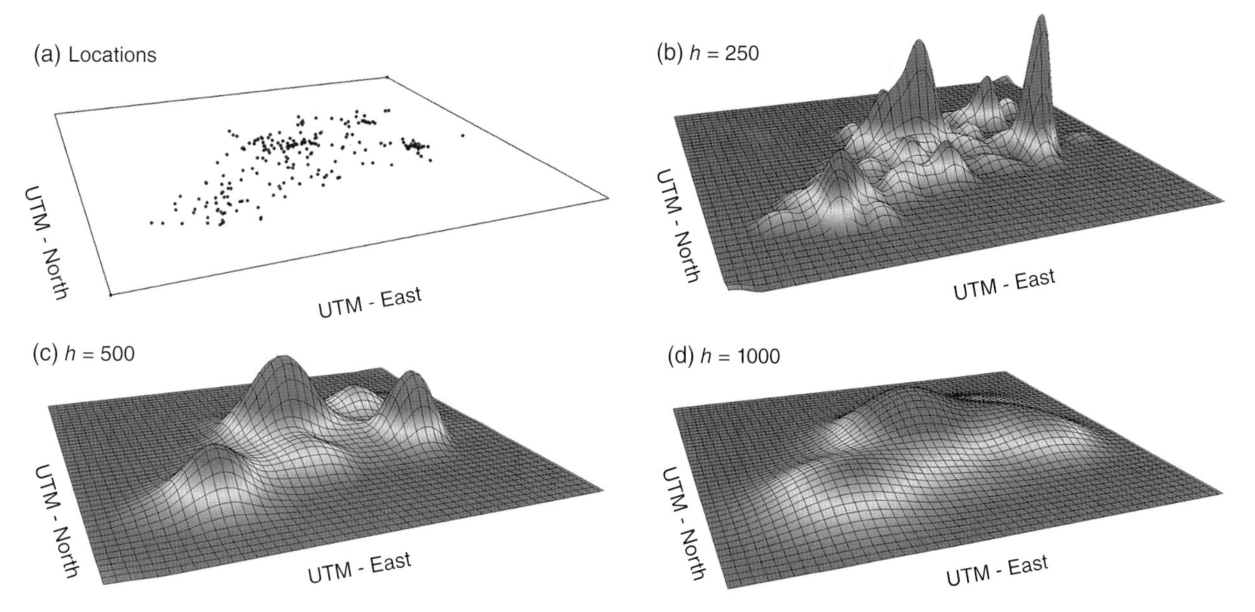

Figure 14.2 Kernel-based utilization distributions (UD) quantify the probability of use computed from locations obtained from radio tracking of white-tailed deer (*Odocoileus virginianus*) in Missouri. Locations of the animal (a) and choice of bandwidth (*h*) aka smoothing factor has a substantial impact on the shape of the UD surface. (b) $h = 250$, (c) $h = 500$, (d) $h = 1000$.

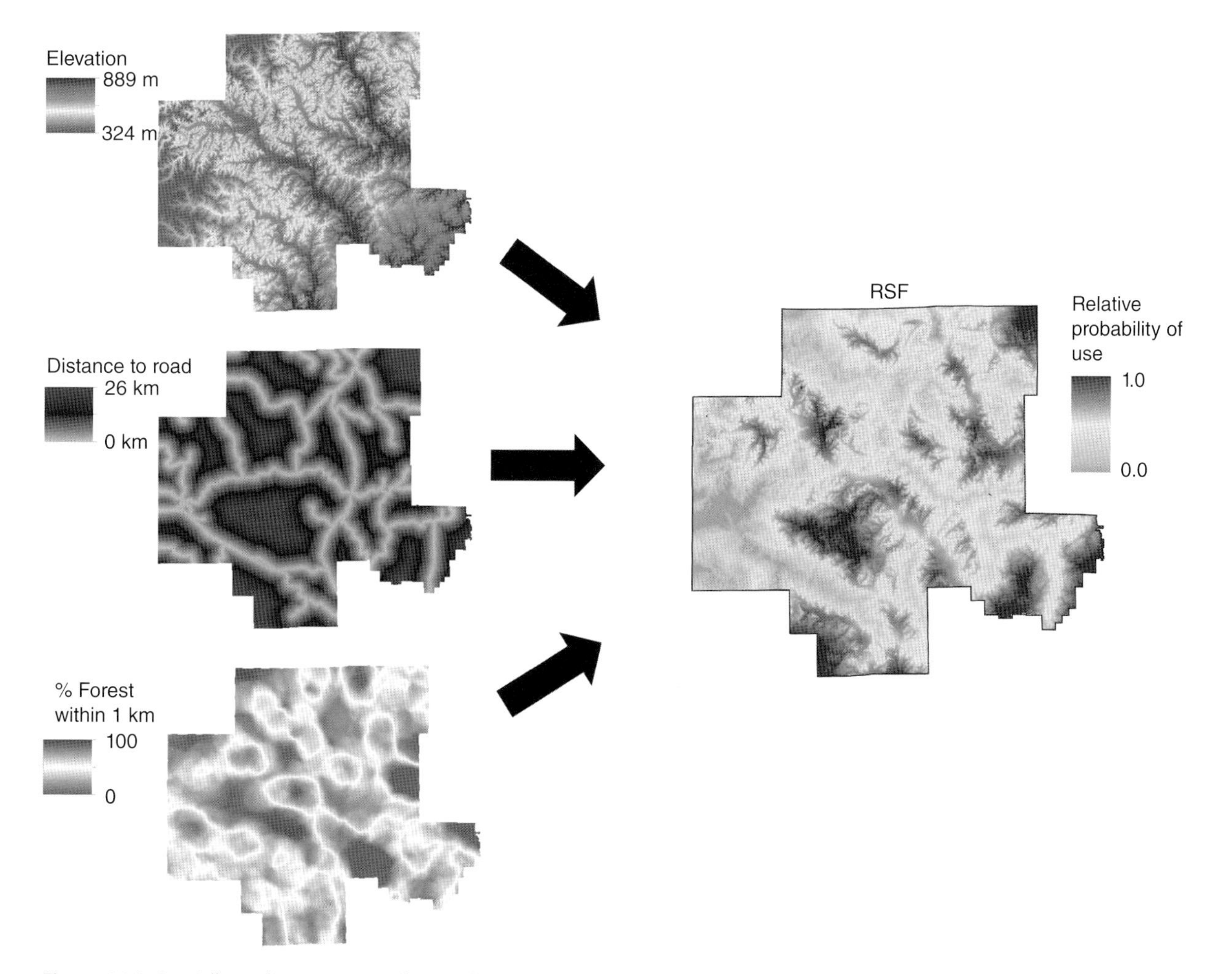

Figure 14.3 Spatially explicit resource selection function (RSF). If a RSF is developed based on various resource characteristics that can be mapped across the study area, then the relative probability of use can be estimated for the study area as shown here. In this example, animals selected for higher elevations, increased distances from roads, and increased surrounding forest cover.

14.3 Considerations in Studies of Resource Selection

In studies of resource selection, there is no "one size fits all" strategy to design and analysis, in part because each study has unique objectives, sampling requirements, and factors that might affect the accuracy of the data including telemetry location error and fix rate bias. The choice of design and analysis options should be driven by careful consideration of fundamental issues such as the desired temporal and spatial scale of inference, the value of linking resource use to fitness effects, the benefits of considering behavior associated with resource use, and a clear question needed to inform management decision-making. Here, we discuss some of the fundamental considerations in studies of resource selection and how to address them.

Before any decision is made about design or analysis, it is important to clarify the scope of inference. In studies of resource selection, there are several critical questions about the animals under study and data to be collected. For example, is the intent to describe the ecology of males or females or both? Is inference desired for all time periods? What is the temporal extent: a year, a season, a month? What is the biological period of interest: breeding season, nesting period, winter? What is the spatial extent that the study intends to characterize? These are all important considerations that ultimately affect how, when, and where data are collected.

14.3.1 Two Important Sampling Considerations: Selecting Sample Units and Time of Day

A critical step when considering sampling design is to clearly identify the level at which replicate observations are made. Throughout, we use the term *sample unit* to identify the unit from which a response and predictor variables are measured. Sample units can be defined in several ways in resource selection studies. Studies that monitor animals using radio tracking, where the animal is repeatedly observed through time, can define either the individual animal or the observed use sites as the sample unit, depending on how they are applied in the analysis (Aebischer et al. 1993; Marzluff et al. 2004). For example, if individual-based models are developed, the animal is the sample unit, but if all observations across multiple individuals are pooled, then the individual animal locations are the sample unit. Alternatively, events such as nest visits may be considered sample units. In this example, nest survival or failure, which is the response variable, and associated vegetation characteristics, which are the predictor variables, are recorded during each nest visit. Last, spatial locations based on coordinates or GIS pixels may be considered sample units, where use and non-use is recorded as the response variable, and the associated resource attributes are the predictor variables. In our discussion below we attempt to set context for these different approaches to selecting sample units. Because the goal of field studies is accurate and defensible inference about a target population, it is important that sampling be conducted appropriately.

Selection of sample units should reflect random, probabilistic sampling to the fullest extent possible. However, logistical and biological realities sometimes prohibit completely probabilistic sampling practices, such as behavioral differences among animals that influence susceptibility to capture or observability (Lundy et al. 2012). Logistical and biological constraints are not excuses to apply convenience sampling, and researchers should make every attempt to minimize sources of bias that are at least partially controllable.

Another important consideration is the timing of data collection. Biologists need to clearly consider the desired temporal scope of inference and should consider the biology of the animal and activity patterns when designing a resource selection study (Beyer and Haufler 1994). In telemetry-based studies, it is imperative that the underlying sampling strategy fully covers the temporal window of interest because resource selection could differ by time of day or season. Either all portions of this window need to be represented equally, or unequal allocation of effort should reflect a predetermined scope of inference (Fieberg 2007). Our recommendation can be met most easily with a systematic schedule for data collection across both a 24-hour cycle, or the period of activity if diurnal, and the longer-term period of interest (Otis and White 1999), though more complex sampling strategies can also be considered if appropriate and efficient. There are other advantages to applying standard sampling designs such as the ability to assess economic costs of alternative sampling strategies.

14.3.2 Estimating the Number of Animals and Locations Needed

It is always advisable to consider the sample sizes necessary to achieve study objectives. Calculations of sample sizes necessary to achieve a desired level of precision of parameter estimates, statistical power, or ability to discriminate among competing models requires initial information about expected variability within and among animals, effect sizes, and the desired significance level. Pilot data can often explore potential variability in parameter estimates, which can be used to evaluate sample size trade-offs. For any individual study, the most useful sample size guidance comes from simulations or analyses tailored to the specific designs being considered and the focal study area (Baasch et al. 2010). We encourage others to use a study-specific approach when considering sample size requirements for particular methods of analysis.

Resource selection studies based on radio tracking contain two levels of sampling. One level is the sampling of individual animals from the population of interest. The other level is the sampling of sites from all possible sites at which each study animal could have been recorded while radio-tagged. The appropriate allocation of survey effort between these two levels must be considered when evaluating sample sizes. For example, if *population-level inference* on resource selection coefficients is desired, then many individuals should be included with relatively few observations per individual. Conversely, if *individual-level inference* is desired, more observations should be made from fewer individuals. In general, population-level inference is more useful for answering management-oriented and theoretical resource selection questions whereas individual-level inference is more useful for questions in behavioral ecology.

There is some general guidance available to estimate the number of animals and observations needed per animal when using observational or radio-tracking data to estimate the parameters of an RSF or RSPF. Most of the simulation work involving statistical models that consider only one categorical resource feature, such as vegetation cover type, suggests that 20–25 animals with 30–50 observations per animal each is adequate (Alldredge and Ratti 1986, 1992; Leban et al. 2001). Other methods, such as *species distribution models*, have their own guidelines (Wisz et al. 2008; Chapter 15). However, we caution against broad application of these recommendations to studies involving multiple resource features or studies with great individual variation in selection patterns. Baasch et al. (2010) provide a useful framework for addressing site-specific sample size requirements in studies of resource selection because they are tailored to the study area and animals under study.

14.3.3 Location Error and Fix Rate Bias Resource Selection Studies

An important source of bias in resource selection studies arises if location error varies systematically with terrain or habitat. For example, bias will be introduced if GPS locations cannot be obtained from particular areas within animal's home range such as low elevation sites with poor signal transmission. The issue becomes particularly important with interactions among habitat use (Frair et al. 2004, 2010; Hebblewhite et al. 2007), time of day, or the ability of a biologist to collect observations during those periods. The two main sources of bias that influence the reliability of telemetry technology are fix rate and locational error. *Fix rate* is the probability of obtaining a location (Frair et al. 2004, 2010; Lewis et al. 2007) whereas *locational error* is the distance between the estimated and actual location of an animal. GPS locational errors can range into the 100s of meters, though errors averaging less than 12 m are common (Rempel et al. 1995; D'Eon et al. 2002; Cargnelutti et al. 2007). In contrast, very high frequency (VHF) locational error from triangulation can range into the thousands of meters (White 1985; Nams and Boutin 1991) and geolocators can be ±30 km which can preclude an assessment of resource selection. In light of these potential errors, we encourage researchers to evaluate the accuracy of location data in each study situation. Fix rate and locational error are problematic for resource selection models because they can lead to misidentification of resource selection (Saltz 1994; Johnson and Gillingham 2008). The potential for misidentification of resource selection may be particularly acute for species that select edge habitats, small habitat patches, or rare habitats (McKenzie et al. 2009; Nielson et al. 2009).

Once locational accuracy is assessed, analytical approaches can be applied to minimize the effects of locational error (Frair et al. 2004, 2010). Methods include data screening for fixes with poor precision (Lewis et al. 2007), modeling resource selection in relation to continuous covariates rather than categorical covariates (Montgomery et al. 2010), altering the scale of inference associated with the model results according to the measurement error inherent to the data (Findholdt et al. 2002), explicitly incorporating the error into the modeling design (Montgomery et al. 2010, 2011; Brost et al. 2015), or using techniques such a kernel density estimates which are more robust to lack of precision with locations (Moser and Garton 2007).

14.3.4 Consideration of Animal Behavior in Resource Selection Studies

With the exception of event site studies of resource selection, which consider specific behavioral events such as nesting, den sites, and roosting, we may not know why an animal uses a resource. Visual confirmation of activity (Eads et al. 2011, 2012), remote monitoring of behavior via sensors (Moll et al. 2009), or still images can be collected (O'Connell et al. 2010), but most often we only know the timing of use within a sample unit. Patterns of resource selection are likely to vary depending on the particular behavior an animal is engaged in. For example, Cooper and Millspaugh (2001) found that parameter coefficients for RSFs varied depending on the behavior of elk (*Cervus canadensis*). In particular, they found canopy cover was important only for particular behaviors, such as bedding, which did not adequately capture resource needs for foraging. Indeed, Marzluff et al. (2004) cautioned that pooling animal use data in studies of resource selection might mask the importance of some resources should resource use differ by behavior. Additionally, knowledge of animal behavior when using specific resource units improves our understanding of why the animal uses certain units and the potential importance of those resources to animals (Marzluff et al. 2004; Forester et al. 2007; Lundy et al. 2012).

14.3.5 Biological Seasons in Resource Selection Studies

Researchers should consider the phenology and seasonality of environmental conditions when designing resource selection studies. For many species and in many regions, organisms respond to seasonal variation in temperature and precipitation with large-scale and long-term phenomena like migration, breeding, postbreeding, or dispersal. Phenological timing can be driven by endogenous and exogenous factors as diverse as hormone cycles, resources, climate, and mate availability. Biological seasons often alter the relationship between animals and resource use because there are seasonal changes in the resource needs of study animals and in the availability of resources in the environment.

Biological seasons can be defined with a number of criteria, but most approaches use the ecology and climatic patterns of the study system in light of the natural history of focal species. For example, the energy needed by a 21 g White-breasted Nuthatch (*Sitta carolinensis*) to survive at winter temperatures of 0 °C is approximately twice what is required during summer conditions at 30 °C (Kendeigh 1970; Grubb and Pravosudov 2008). Changes in energy needs translate to substantially altered food resource requirements during each season. An investigator wishing to evaluate resource selection in nuthatches would thus be wise to carefully define the biological seasons of interest before embarking on an investigation.

Alternatively, investigators could consider potential biological influences by aligning data to account for differences in timing among individuals. For example, an investigator might have resource-use data for 20 pairs of birds, collected across a 60-day period encompassing prebreeding, incubation, nestling-rearing, and fledging. If nest initiation dates for those birds were spread across a 15-day period, ordinal dates would serve as poor bounds for resource selection analyses. Rather, the investigator might classify resource-use data for each study animal in the context of its reproductive timing. Data collected before the first egg was laid on each territory would be considered prebreeding and analyzed separately from resource-use data collected during incubation and brood-rearing phases. Resource selection analyses could then be aimed at natural history phases, despite slightly shifted seasonal phases among individual birds. The interpretation of results from such carefully designed studies may be much more straightforward when placed in the context of biological seasons, and inference will ultimately tie to the biology of the study animal much better.

14.3.6 Scaling in Resource Selection Studies

Scale is a ubiquitous issue in ecology, and the heterogeneity and patterns of resources across landscapes is inherently *scale dependent* (Turner 1990; Levin 1992; Boyce 2006). Within the context of resource selection, scale refers to the spatial and temporal extent at which organisms make decisions and researchers evaluate resource selection (Boyce 2006). Thomas and Taylor (2006) described four general designs for resource selection studies based on whether use and availability data are collected at the individual or population scale (Figure 14.4).

Johnson's (1980) classification scheme for orders of selection illustrates that resource selection, whether at the individual, population, or species level is inherently hierarchical (Johnson 1980; Manly et al. 2002; Figure 14.5). Thus, organisms are adapted to niches of varying breadth at the species range, at the home range

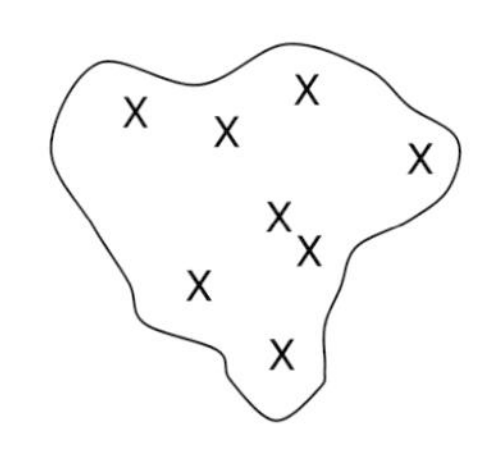

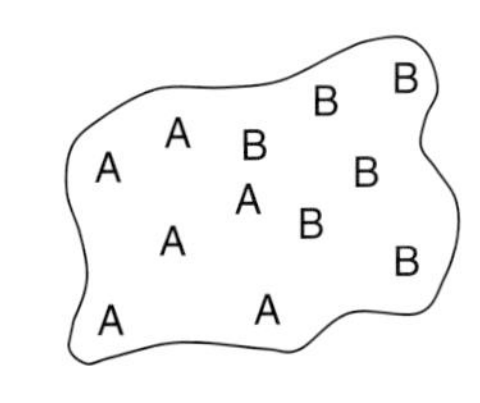

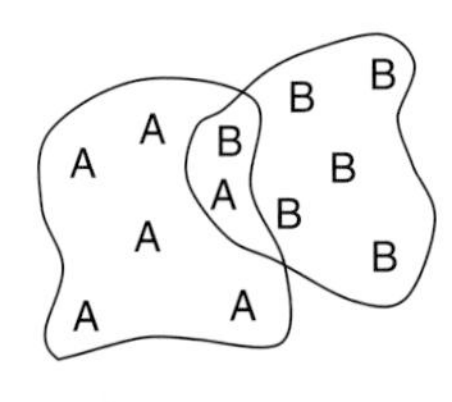

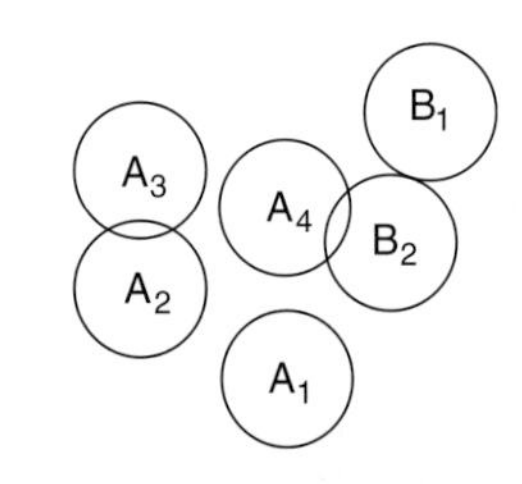

Figure 14.4 Four study designs for resource selection studies. The first design (I) collects use and defines availability at the population level. Under this design, observations of individual animals (the Xs) are not distinguished. Under study design (II), use is recorded for individual animals (denoted by "A" and "B"), but availability is still defined at the population level. Study design (III) estimates use and availability separately for each animal. Design (IV) summarizes use of individual animals, but availability is defined per individual use location or event (designated by subscripts).

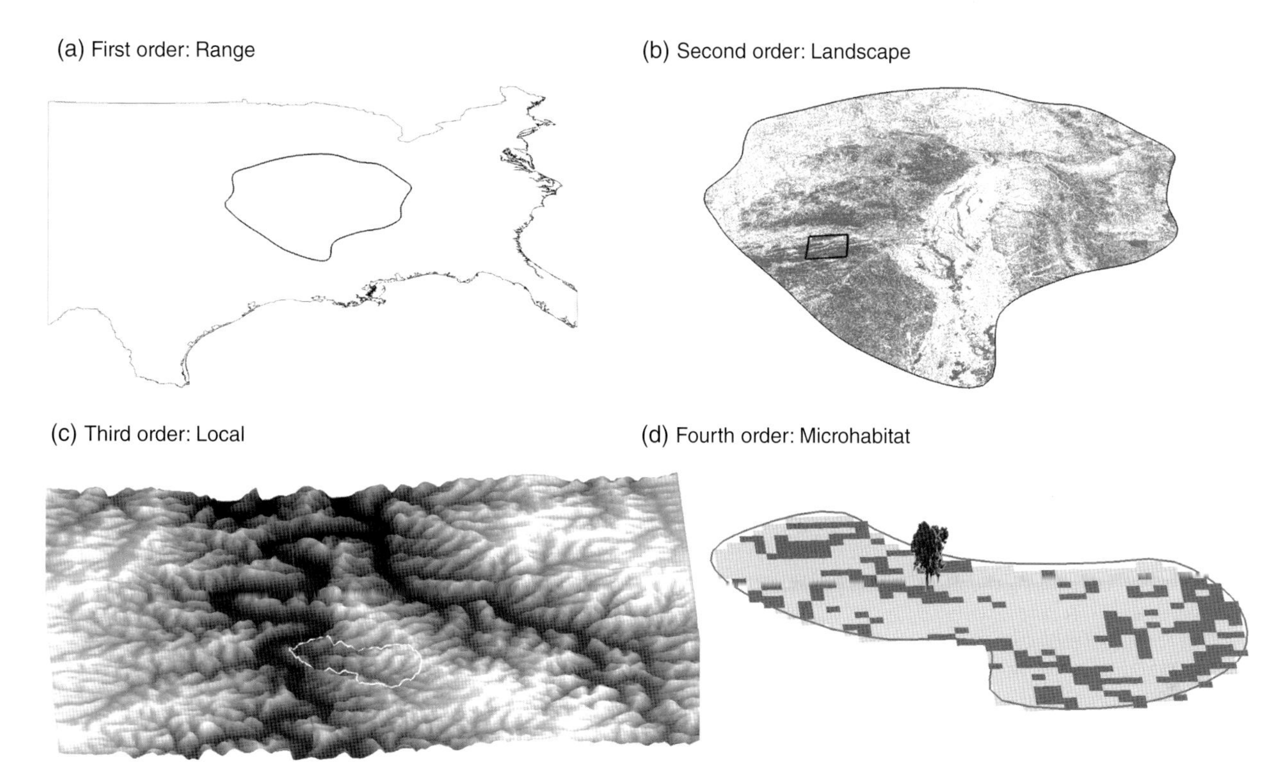

Figure 14.5 An example of the orders of resource selection for a representative songbird in the Midwestern U.S.A. (Johnson 1980). The first order of selection is the geographical range of a species (a). Within that range the bird selects a home range with heavily forested landscapes (second-order selection; b). Within the home range, the bird selects heavily dissected southwest facing slopes (third-order selection; c). For the fourth order of selection, the bird chooses to nest in a tree (d) having greater than 50% canopy closure (green cells).

scale, and among habitat patches within the home range (Kie et al. 2002).

Assessing an ecological question at an inappropriate scale increases the probability of misidentifying resource selection (Boyce 2006). At the core of this issue is the recognition that scale directly influences the relationship between response variables such as abundance, occupancy, and the predictor variables, such as biotic and abiotic environmental characteristics of the landscape. For example, Orians and Wittenberger (1991) described how Yellow-headed Blackbirds (*Xanthocephalus xanthocephalus*) used cues about insect availability to determine where they would locate a territory (second-order selection). However, nest sites were selected based on vegetation density (third-order selection), not on insect availability, so if the scale of analysis had only considered nest site selection, they might have underestimated the importance of insect availability in the overall pattern of resource selection. When considering spatial and temporal scale it is important to consider both extent due to the range of data and an organism's distribution, as well as resolution due to the finest grain of the analysis (Turner et al. 2001).

Identifying an appropriate spatiotemporal scale is based on several factors: including the biology of the species, population, and/or individual organism, the resolution and extent of the covariate data, and the temporal resolution and extent required to evaluate the question of interest. For example, if a species is particularly adapted to and dependent upon a rare habitat type, then the analysis must be designed at a spatial scale that will capture that landscape heterogeneity. A critical issue is that it may be necessary to evaluate ecological relationships at finer resolutions to comprehend ecological processes occurring at broader spatial scales (Turner et al. 1989). It is always possible to scale up, but not to scale down: meaning that inferences cannot be made at resolutions finer than the grain of the analysis.

14.3.7 Linking Resource Selection to Fitness

An implied assumption is that resource selection is correlated with fitness of an animal. However, such an association is not guaranteed (Van Horne 1983; Buskirk and Millspaugh 2006). While it is often possible to relate resources to the distribution of a species in time and space, the ultimate consequences of resource selection to the animal often remain unknown. Few studies have tested the assumption that resource selection is associated with fitness (Both and Visser 2000; Franklin et al.

2000; McLoughlin et al. 2007). Linking resource selection to biological outcomes such as survival and reproduction will improve the value of resource selection studies tremendously (Garshelis 2000; Garton et al. 2001). Recent developments begin to explicitly link resource selection and demography (Matthiopoulos et al. 2015), and *spatial capture-recapture models* are an area of active research that offer exciting opportunities to link demography and resource selection (Royle et al. 2013).

14.4 Methods of Analysis and Examples

Choice of methods to analyze resource selection data is diverse and dependent on many factors such as research objectives, study design, sampling strategies, and available data (Figure 14.1). It is important to note that several analytical techniques *could* be used to analyze the same data in a resource selection study and Figure 14.1 is intended to serve as only a general guide to common methods used for criteria shown.

14.4.1 Compositional Analysis

Aebischer et al. (1993) introduced the use of *compositional analysis* (Aitchison 1986) for studies of resource selection to overcome issues with earlier methods (Neu et al. 1974; Quade 1979; Johnson 1980) such as the unit-sum constraint, nonindependence of sample units, and consideration of differential resource selection by individuals. Compositional analysis was recommended by Aebischer et al. (1993) who suggested a two-stage analysis, first examining home range selection within the study area and then examining resource selection within the home range (Figure 14.5). The response variable is the proportion of each discrete level of a resource variable within the home range of an individual animal, and must sum to one over all discrete vegetation types, defined as the "unit-sum constraint." Thus, the individual animal is the sample unit, which also helps overcome problems of nonequal weighting among animals when the number of locations across animals is not equal (Pendleton et al. 1998). Availability is similarly defined by the proportional occurrence of discrete levels of a resource variable, but at a scale that is appropriate to the question being posed (Figure 14.6).

Compositional analysis uses *multivariate analysis of variance* (MANOVA) to compare proportional use of discrete vegetation types to proportional availability (Aebischer et al. 1993). Once the proportional use and availability data are obtained, these data are transformed to ln-ratios $\boldsymbol{y}$ (used) and $\boldsymbol{y_o}$ (available), and the difference in the ln transformed data for each pairwise comparison (i.e., animal) is calculated ($\mathbf{d} = \boldsymbol{y} - \boldsymbol{y_o}$) to test the hypothesis that $\mathbf{d} = \mathbf{0}$ (use = availability). The overall test for selection is derived from the MANOVA statistics. If $\mathbf{d} \neq \mathbf{0}$,

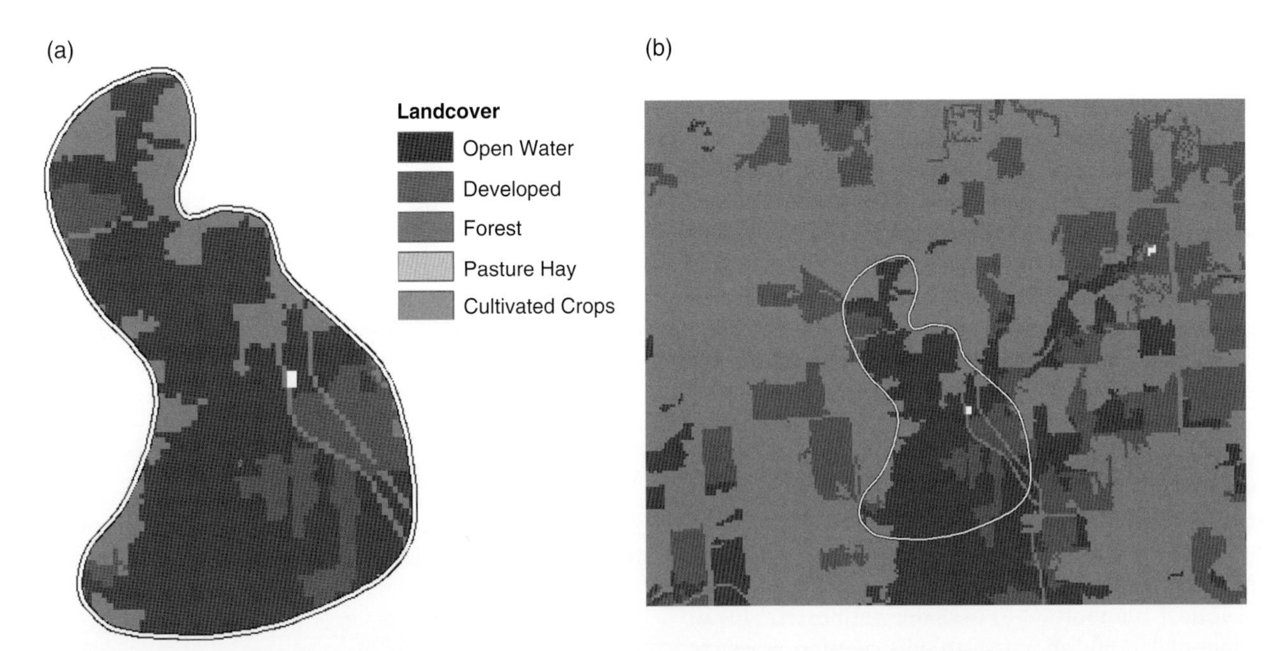

Figure 14.6 An example of compositional analysis. Compositional analysis uses a paired used-availability design to rank categories of a resource variable with discrete levels (e.g. different cover types of vegetation such as forest, pasture hay, cultivated crops). In this example, the proportion of land cover used by a white-tailed deer (*Odocoileus virginianus*) in Missouri is quantified using the animal's home range (a). The used proportions are compared to those found in the study area (b) to determine that forested areas are used in greater proportion than their availability in the larger area.

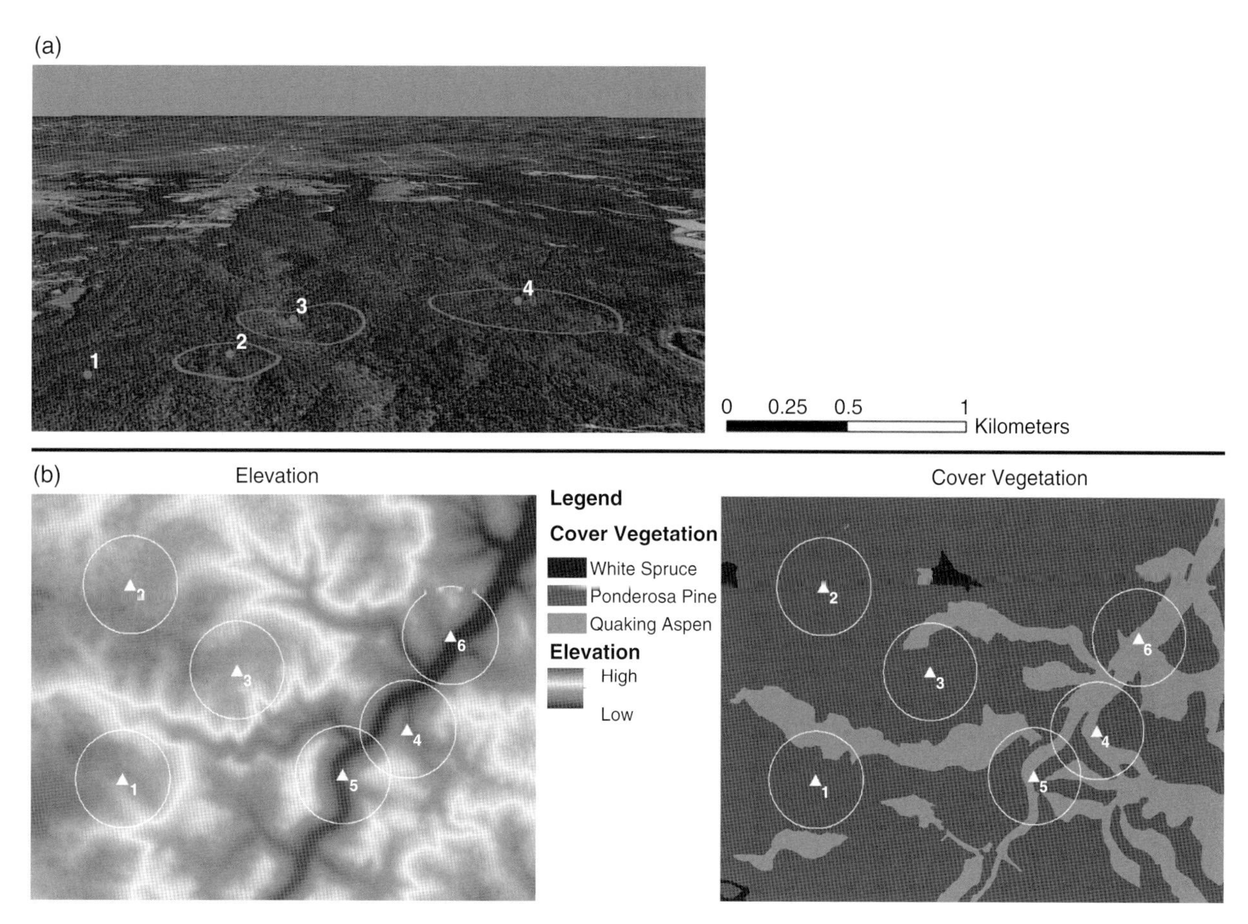

Figure 14.7 Defining choice sets for discrete choice models. Using the paired use-availability design within discrete choice allows for the use of choice sets to characterize availability. Variation in resource availability can be incorporated on a per use location basis and in a biologically informed way. For example, the extent of availability for successive elk (*Cervus elaphus*) radio-tracking locations in a forested landscape (a) could be dependent on its movement rates between successive use locations. As distances between locations 2, 3, and 4 increase, a larger spatial area might be considered available. For an elk whose successive locations demonstrate a seasonal movement to lower elevations (b), individual choice sets capture the increased availability of quaking aspen surrounding locations 4, 5, and 6 after the animal has moved to lower elevations.

separate *t*-tests (Aebischer et al. 1993) or randomization tests (Manly 1991) are used to determine how selection differs by resource pairs (Erickson et al. 2001). Using this information, Aebischer et al. (1993) describe ranking each discrete level based on the number of times it was selected over another level.

Despite these advantages, a few challenges complicate use of compositional analysis. First, ln-ratios are undefined for zeros, which are observed when a discrete vegetation type remained unused. Aebischer et al. (1993) suggested that zeros be replaced by a small constant (e.g. 0.01) but this solution is problematic because the model assumes this value is a real number, so misclassification is highly probable (Bingham and Brennan 2004; Bingham et al. 2007). To address this issue, researchers can drop vegetation categories that are not used by an animal, reclassify discrete vegetation categories by grouping types to eliminate unused types, or omit individual animals that have not used all available vegetation types. Although we agree that reclassifying habitat types can be sensible, we do not recommend the omission of habitat types or individual animals because it results in loss of observation data. Last, when using the animal's home range boundary to quantify proportional use, there is no consideration of the amount of time spent in each habitat type. Millspaugh et al. (2006) address this problem by incorporating use intensity into estimates of proportional use of habitats through incorporation of the UD. Using this approach, the study showed selection results that were biased when random use was assumed through the range.

14.4.2 Logistic Regression

In studies of resource selection, we are often interested in estimating the probability that a particular sample unit is used by an animal as a function of the resources present in that sample unit. Perhaps the most straightforward approach is to divide an area of interest into a population of N discrete sample units such as GIS pixels, and assess use and non-use at a random collection of sample units. Importantly, sample units are selected without regard to use. We can then correlate the use or non-use of each sample unit with associated covariates to infer the strength and direction of selection of particular resources. *Logistic regression* is a commonly used analytical method for such data.

In the simplest applications of logistic regression, we are interested in estimating the probability that a species uses a discrete sampling unit, and in determining the effects that a suite of m resources, written as $x_1, \ldots, x_m$, may have on the probability of using each sample unit. The probability that a species uses sample unit i can be written as:

$$\psi_i = \frac{e^{\beta_0 + \beta_1 x_{i1} + \ldots + \beta_m x_{im}}}{1 + e^{\beta_0 + \beta_1 x_{i1} + \ldots + \beta_m x_{im}}}, \tag{14.1}$$

where ψ_i is the probability that a species occurs in sampling unit i, β_0 is the intercept parameter, $\beta_1, \ldots, \beta_m$ are parameter coefficients associated with the M resources, and x_{im} is the value of resource unit m at sample unit i.

If we survey n sample units, recording use or non-use of the species of interest and measuring m resources at each of these sample units, we can write the log-likelihood as:

$$\ln(L(\beta_0,\beta_1,\ldots,\beta_m \mid \mathbf{y},\mathbf{x})) = \sum_{i=1}^{n} y_i \ln(\psi_i) + (1-y_i)\ln(1-\psi_i). \tag{14.2}$$

Here $y_i = 1$ if a species used sample unit i, and $y_i = 0$ if a species did not use sample unit I. *Maximum* likelihood techniques can then be used to estimate parameters β_0, $\beta_1, \ldots, \beta_m$. With a basic understanding of logistic regression, we now consider a series of sampling protocols that represent variations on the general model described above. The appropriateness and interpretation of logistic regression models, and the choice between logistic regression and other similar models, depends on the sampling design.

14.4.3 Sampling Designs for Logistic Regression Modeling

14.4.3.1 Random Sampling of Units within the Study Area

A common sampling design in resource selection studies involves sampling a collection of n sample units, usually discrete sites, within a study area and measuring use versus non-use at each sample unit. The researcher can use many methods to select the n sample units, such as *simple random sampling*, *systematic sampling*, or *stratified random sampling*, though methods other than simple random sampling may add complexity to analytical methods. Most resource selection studies use model-based inference, but with probability sampling of discrete sites from a finite population of sites, inference can be either design- or model-based.

An important distinction between a sampling design which involves random sampling of units within the study area and those discussed later, is that each sample unit is selected without regard to whether the unit has been used. Once the n sample units are selected, survey methods such as point-counts or camera traps are used to detect use and non-use. Additionally, a suite of m covariates, the resources, are measured at each sample unit.

An important assumption of logistic regression is that use and non-use are detected without error. However, many animal species are elusive and difficult to detect, leading to imperfect detection probabilities. This limitation applies to virtually all forms of use versus non-use analysis, whether it be via radio telemetry, visual observation, or passive monitoring. A variation on the basic logistic regression model allows researchers to account for imperfect detection when estimating the probability a sample unit is used, which can be accounted for through multiple surveys at each sample unit and careful recording of detection and nondetection events (MacKenzie et al. 2002). Sampling units must be closed to changes in occurrence between surveys so that a species that is present at a sample unit during one survey must be present during all surveys and multiple repeated surveys at each sample unit can be used to estimate detection probability. Under this sampling protocol, we can write the log-likelihood as:

$$\ln(L(\beta_0,\beta_1,\ldots,\beta_m,p)) = \sum_{i=1}^{n} y_i \ln\left(\psi_i p^{j_i}(1-p)^{J-j_i}\right) + (1-y_i)\ln\left(\psi_i(1-p)^{J} + (1-\psi_i)\right), \tag{14.3}$$

where p is the probability of detecting a species during a single survey, J is the number of replicate surveys conducted at each sample unit, j_i is the number of sampling occasions at site i when a species was detected, and $y_i = 1$ if $j_i > 0$, or $y_i = 0$ if $j_i = 0$. In this model, detection probability, p, can be modeled as a function of covariates in the manner described for ψ_i in Eq. (14.1) above (MacKenzie 2006).

14.4.3.2 Random Sampling of Used and Unused Units

Instead of drawing a single sample from the overall population in the study area and then determining use for each selected survey unit, it may be desirable to sample directly from the two populations of used and unused sample units within a study area. The case-control design is commonly considered when use is rare across the study area and a random sample from the overall population may lead to few observations of use (Hosmer and

Lemeshow 2000; Keating and Cherry 2004). With the case-control design, a random sample of used sites, denoted n_1, is collected from the population of all used sites throughout the study area, denoted as N_1. Next, a random sample of unused sites, denoted as n_0, is collected from the population of all unused sites throughout the study area (N_0). In practice, a random sample of used sites is generally obtained through use of radio telemetry or a similar method to determine which sites are used, and a random sample of unused sites is then selected from the population of sites within the study area where the animal was never observed. The case-control design described below assumes that use and non-use can be determined with certainty, an assumption we will later relax.

Under a case-control sampling design, the probability that site i is occupied is:

$$\psi_i^{cc} = \frac{e^{\beta_0 + \ln\left(\frac{S_1}{S_0}\right) + \beta_1 x_{i1} + \ldots + \beta_m x_{im}}}{1 + e^{\beta_0 + \ln\left(\frac{S_1}{S_0}\right) + \beta_1 x_{i1} + \ldots + \beta_m x_{im}}}, \qquad (14.4)$$

where $S_1 = n_1/N_1$ and $S_0 = n_0/N_0$ (Keating and Cherry 2004). Note that this equation is similar to Eq. 14.1, except that now the intercept parameter β_0 is adjusted by the natural log of the ratio S_1/S_0, which accounts for the disproportionate occurrence of used and unused samples in the dataset. The log-likelihood can then be written as:

$$\ln\left(L(\beta_0, \beta_1, \ldots, \beta_m)\right) = \sum_{i=1}^{n} y_i \ln\left(\psi_i^{cc}\right) + (1 - y_i)\ln\left(1 - \psi_i^{cc}\right). \qquad (14.5)$$

Note that this equation is identical to the log-likelihood for the basic logistic regression model described in equation 14.2. If the proportion S_1/S_0 is known, then the case-control logistic regression model can be used to estimate a RSPF. In this situation, the case-control model can be fit with standard logistic regression software routines and the estimated value of the intercept parameter, β_0, can be obtained by subtracting $\ln(S_1/S_0)$ from the estimate of β_0 or by including an offset term.

In most field studies, the proportion S_1/S_0 is rarely known. In this situation, case-control logistic regression still provides unbiased estimates of parameter coefficients $\beta_1, \ldots, \beta_m$, but estimates of the intercept parameter β_0 are confounded with $\ln(S_1/S_0)$. In this situation, logistic regression techniques can only be used to estimate a RSF. However, unbiased estimates of linear model slope parameters can still be obtained if none of the unused sample units are contaminated because imperfect knowledge of the number of used and unused samples only introduces bias into estimates of the linear model intercept parameter (Keating and Cherry 2004). In such settings, estimates of odds ratios can be obtained to rank the relative probability a sample unit is used relative to a user-defined baseline, but inference cannot be obtained on the absolute probability a sample unit is used. To demonstrate more precisely, let:

$$\pi_x = \frac{e^{(\beta_0 + x\beta_1)}}{1 + e^{(\beta_0 + x\beta_1)}} \qquad (14.6)$$

denote the absolute probability a sample unit is occupied when a specific covariate takes the value x, with β_0 and β_1 denoting intercept and slope parameters of linear predictors, respectively. The odds of use when the covariate takes value x is defined as $\text{odds}_x = \pi_x/(1 - \pi_x) = \exp(\beta_0 + x\beta_1)$. Finally, the odds ratio when the covariate takes the value $x + z$, relative to baseline value x, is defined as:

$$\text{OR}_{x+z} = \frac{e^{(\beta_0 + (x+z)\beta_1)}}{e^{(\beta_0 + x\beta_1)}} = e(z\beta_1). \qquad (14.7)$$

There are several important items to note here. First, interpretation of model predictions as odds ratios does not depend on the intercept parameter, which makes case-control logistic regression useful even if the number of used and unused sample units is unknown. Second, interpretation of odds ratios is relative to a baseline set of covariates – it is important to explicitly acknowledge such baselines when reporting odds ratios. Third, the resulting odds ratios are not directly proportional to the absolute probability of use. In other words, the absolute probability of use is not simply a constant multiplied by the odds ratio. Practically, this issue means predictions from case-control models when the number of used and unused samples in the population is unknown can only be used to rank sample units from low to high probability of use relative to an explicit baseline. Last, the expected value for an odds ratio under no resource selection is one.

For case-control sampling to be appropriately used, the researcher must know with certainty which sample units are used versus unused. While use can usually be determined with certainty in many animal studies, non-use can rarely be determined with certainty. This situation may lead to cases where sample units classified as unused may actually have been used at some point, but this use is not observed by the researcher. This situation is commonly referred to as "contamination" (Lancaster and Imbens 1996). Case-control logistic regression should not be used when contamination is possible, as is most commonly the case with animal field research. Instead, researchers should use a use-availability sampling scheme, which is described in the next section.

14.4.3.3 Random Sample of Used and Available Sampling Units

Like the case-control sampling scheme above, a use and availability sampling scheme draws samples from two populations: all sample units that were used and all available sample units. Under this sampling design, a random sample of used sample units, denoted n_1, is collected from the population of all used sample units throughout the study area, denoted as N_1. A second random sample of available sample units, denoted n_a, is then drawn from the entire population of all sample units, denoted as N. Note that the random sample of available sample units is made without regard to whether a site is used or unused, since non-use is not known with certainty. Random sampling of used and available sampling units is a common design when animals are fitted with radio-transmitters or GPS units. In these situations, only use is observed with certainty, and units where use was not observed may or may not have remained unused; therefore these units are classified simply as available.

Logistic regression is commonly used when data are collected with this random use-availability sampling scheme (Manly et al. 2002). Once a model is fit to use-availability data, estimated slope parameters are used to construct a RSF, typically with an exponential form:

$$RSF = \exp\left(\mathbf{x}'\hat{\boldsymbol{\beta}}\right), \tag{14.8}$$

where $\mathbf{x}$ is a vector of covariates and $\hat{\boldsymbol{\beta}}$ is as associated vector of estimated slope coefficients, excluding an intercept term. The slope coefficients estimated under such a model are not guaranteed to be unbiased, and the resulting RSF is neither guaranteed to be proportional to the absolute probability of use, nor a reasonable ranking of relative probabilities of use (Keating and Cherry 2004). Despite these shortcomings, logistic regression is still a commonly used and useful tool for modeling resource selection with use-availability data (Johnson et al. 2006; see also its use as an approximation to point process models below).

When estimating an RSPF with random use-availability data, we are interested in estimating the probability that a sample unit is observed to be used, conditional on that unit being included in the sample (ϕ_i, Keating and Cherry 2004). This probability can be written as:

$$\phi_i = \frac{\frac{h}{q(1-h)}\psi_i}{1+\frac{h}{q(1-h)}\psi_i}, \tag{14.9}$$

where $h = n_1/n$ (with $n = n_a + n_1$) is the ratio of used sample units to the total number of sample units, ψ_i is the probability of use defined in Eq. (14.1), and q is the unconditional probability of use, or prevalence (see online exercises).

Manly et al. (2002) suggested, for convenience of fitting a use and availability model, that the probability of use, ψ_i, take an exponential form. Assuming the exponential form for ψ_i, one could rewrite the equation above as:

$$\phi_i^{\mathrm{RSF}} = e^{\beta_0+\ln\left(\frac{h}{q(1-h)}\right)+\beta_1 x_{i1}+\ldots+\beta_m x_{im}} \Big/ \left(1+e^{\beta_0+\ln\left(\frac{h}{q(1-h)}\right)+\beta_1 x_{i1}+\ldots+\beta_m x_{im}}\right). \tag{14.10}$$

Assuming this form for ϕ_i, one can use readily available logistic regression software packages to estimate the regression coefficients $\beta_1, \ldots, \beta_m$. However, β_0 is confounded with the ratio $\ln\left(\frac{h}{q(1-h)}\right)$, and assuming the exponential form for ψ_i will only yield an RSF, where estimated probability of use is proportional, but not equal, to the "true" probability of use.

Assuming the exponential form for ψ_i, the log-likelihood of the use-availability model can be written as:

$$\ln(L(\beta_0,\beta_1,\ldots,\beta_m)) = \sum_{i=1}^{n} z_i \ln(\phi_i) + (1-z_i)\ln(1-\phi_i), \tag{14.11}$$

where $z_i = 1$ if sampling unit i was observed to be used, 0 otherwise. While this equation looks similar to the logistic regression log-likelihood above, use of logistic regression to estimate parameter coefficients in this context has been heavily criticized (Keating and Cherry 2004). Several authors have proposed techniques closely related to case-control logistic regression that enable users to estimate absolute probabilities of use from use-availability data (Lele and Keim 2006; Royle et al. 2012; Rota et al. 2013). These techniques, however, require strong assumptions regarding the functional form of the absolute probability of use (Fithian and Hastie 2013), precluding their use in many practical settings. Indeed, estimating absolute probabilities of use from use-availability data remains an important area of methodological research.

Last, the use of point process models represents an exciting area of development in the analysis of use-availability data (Warton and Shepherd 2010; Renner et al. 2015). *Point-process models* assume the location and number of used points is random and can be used to estimate the intensity of presences as a function of covariates. In fact, in many cases, point-process models have been found to be equivalent to many of the logistic regression and maximum entropy techniques ecologists have been using for years (Warton and Shepherd 2010; Renner and Warton 2013).

14.4.4 Discrete Choice Models

Case-control logistic regression requires one to assume there is little or no contamination in unused sample units, which is not met in many practical field settings. An alternative approach to identifying unused sample units is to pair used sample units with one or more sample units that were not used at the same time. For example, Rota et al. (2014) paired trees used for foraging by Black-backed Woodpeckers (*Picoides arcticus*) with trees that woodpeckers could have used, but did not use at the time of sampling. Unused sample units may have been used at a previous or future time, but they were not used at the same time as the used sample unit. Additionally, there may be other advantages to pairing used and unused sample units, such as when available resources are not uniformly distributed in space, perhaps because of changes in resource characteristics through time, differences in spatial availability, or restrictions in access to resources (see online exercises). In these settings, the used and associated unused sample units are commonly referred to as a *choice set*, and comprise a single replicate.

The approach offers great flexibility in defining availability, which can change spatially and temporally (Figure 14.6). For example, Cooper and Millspaugh (1999) defined availability as a 1 km circle centered on a point between a telemetry location and a used bed site. Durner et al. (2009) defined the center of a choice set on an animal's location at the previous time step with the radius of that circle, dependent on the time between locations and the distance an animal could move in that interval. They computed a radius of available habitat as $[a + (b \times 2)] \times c$ where a is the mean hourly movement rate for all animals being sampled, b is the standard deviation of that movement rate, and c is the number of hours between locations. We advocate the use of these biologically appropriate measures of defining availability to guide realistic definitions of resource availability provided they meet objectives of the study.

Discrete choice models are commonly applied when used sample units are paired with one or more unused sample units (Ben-Akiva and Lerman 1985; McCracken et al. 1998; Cooper and Millspaugh 1999; Figure 14.1). Predictions from discrete choice models can be made in terms of odds ratios. Additionally, predictions can be made in terms of the probability a sample unit is used, relative to the other sample units within the choice set. To demonstrate, consider the probability of selecting a sample unit with covariate value $x + z$, when an animal has a choice between sample units with covariates x and $x + z$. We can write this probability as:

$$\pi_{x+z} = \frac{e^{((x+z)\beta_1)}}{e^{((x+z)\beta_1)+e(x\beta_1)}} = \frac{e(z\beta_1)}{1+e(z\beta_1)}. \tag{14.12}$$

Notice that this probability depends strongly on the denominator and will change depending on the number of choices available. To convince yourself, consider a choice set with a third alternative that has covariate value $x + y$ and calculate π_{x+z}. Notice also that discrete choice models do not require an intercept parameter (you could place an intercept parameter in the linear predictors above, but it will cancel from all terms), again indicating that predictions of odds ratios and probabilities of use are made relative to a set of defined conditions.

14.4.5 Poisson Regression

In some resource selection studies, investigators are interested in understanding relative intensity of use across sample units, with use measured as discrete counts, typically from animal encounters or sign counts, in each cell. For example, one could grid a study area into sample units, and count the number of times use occurred in each sample unit based on a camera trap array, counts of animal tracks, scat, burrows, or other sign or a tally of direct visual observations (Eads et al. 2011; Kays et al. 2016). In these situations, *Poisson regression models* and other extensions to this analysis use counts as the response variable for modeling resource selection. How sample units are delineated is important because the size of the units should not be arbitrary, but might consider the scale of resource choices made by an animal and whether variation in the intensity of animal use can be expected. The size of sample units is analogous to bandwidth smoothing in the RUF analysis described in Section 14.4.6 (Millspaugh et al. 2006). Poisson regression could be considered a use-only method if only sample units where the animal is observed are included in the analysis or it could be a use-availability method if areas where animals are not observed are included in the analysis (see online exercises).

Poisson regression with an offset term can estimate an RSF, the relative frequency of use within sample units. Count c_i, collected at sample unit i can be modeled as a Poisson random variable:

$$c_i \sim Poisson(\lambda_i), \tag{14.13}$$

where $\ln(\lambda_i) = \beta_0 + \beta_1 x_{1,\,i} + \ldots + \beta_p x_{p,i} + \ln(total)$. Here, ln(total) is an offset term equal to the sum of all counts across the study area. In this manner, $\beta_0 + \beta_1 x_{1,i} + \ldots + \beta_p x_{p,i}$, represents the relative intensity of use at sample unit i. Thus, Poisson regression is a logical extension from the logistic regression approach described above with increased complexity in the response variable reflecting intensity rather than simply presence or absence. Poisson regression is a generalized linear model with the distributional assumption that c_i comes from a Poisson distribution with parameter $\lambda = E(c_i)$. The offset term $\ln(total)$ is a quantitative variable whose regression coefficient is set to 1.0. Because the offset in this model is the same for all sampled habitat units within the study area or home range for a particular animal, it is absorbed into the estimate of β_0 and ensures that we are modeling relative frequency of use, instead of counts of use, although the response variable specified is the counts of use (Millspaugh et al. 2006). That is, subtracting the offset term from both sides of the equation above, the left hand side becomes $\ln[E(r_i)]$-$[\ln(total)]$ = $\ln[E(r_i/total)]$ = $\ln[E$ (Relative Frequency$_i$)].

Advantages of Poisson regression include flexibility in study design for use-only, use-availability, and use-non-use (Figure 14.1) and a response variable based on intensity of use. Additionally, extensions to the basic Poisson model allow modeling of overdispersed and zero-inflated data (Zuur et al. 2009). However, additional evaluations of such applications of Poisson models, particularly accuracy and precision of estimates of probability of resource use with varying densities of animal locations, are warranted.

14.4.6 Resource Utilization Functions

RUF are an alternative analytical technique that represents a departure from the binary use and non-use and use-availability techniques described in previous sections (Marzluff et al. 2004; Millspaugh et al. 2006). The RUF method is a logical extension of techniques that use individual animal observations as the sample unit by summarizing those individual observations as a probabilistic and continuous measure of activity in an animal's home range. The RUF method first requires a collection of spatially referenced used sample units. Once obtained, coordinates of the used sample units are used to obtain a two-dimension UD, which approximates the probability density function giving rise to the observed use locations. Last, the UD is treated as the response variable in a regression model with spatial covariates as explanatory variables, often with spatially correlated errors. Many decisions must be made when fitting RUF models, including choice of UD estimation method (Kernohan et al. 2001; Getz and Wilmers 2004), the bandwidth to use when estimating UDs (Seaman et al. 1999; Kie et al. 2010; Lichti and Swihart 2011), and what covariance function, if any, to assume when fitting the regression model. Furthermore, because the method is a two-stage analysis, additional effort is required if one wishes to propagate uncertainty in the estimated UD. The method is attractive because it does not require collection of unused or available sample units, and Hooten et al. (2013) suggest this model is most useful when telemetry points are subject to location error.

The process of relating the UD density estimate to underlying resources is typically done at the individual level, using multiple linear regression. It is possible to account for spatial autocorrelation induced by smoothing the UD (Marzluff et al. 2004), but recent work indicates bias in parameter estimates is most substantially reduced by applying the same level of smoothing to the predictor variables (Hooten et al. 2013). More work is needed to address the spatial autocorrelation function in RUF models and to determine whether other functions are more appropriate than the spatial covariance model used by Marzluff et al. (2004).

Once individual models are fit, it is possible to scale up to a population-level model (see online exercises). First, fit the same model to each of the n individuals. Then, estimate population-level model coefficients as the empirical mean and variance of individual-level regression coefficients (Marzluff et al. 2004; Sawyer et al. 2006). In principle, this approach could be extended to a hierarchical modeling framework where individual-level slope coefficients are themselves assumed to be random variables drawn from a higher-level distribution, but we are unaware of any such developments. Standard or bootstrap test statistics or confidence intervals could be used to assess the statistical significance of the population-level model coefficients. The equations give equal weight to each study animal, but can be adjusted to incorporate sampling weights.

There are several advantages to using a RUF. The method is intuitive in that animal space use can be generalized as a continuous variable. By using the UD as the response variable, problems with relating estimated locations with specific resource covariates are avoided. This issue can make analyses based on a model-estimated UD more appropriate than analysis of individual used locations when telemetry error is an issue (Moser and Garton 2007; Hooten et al. 2013). Further, because the animal is the sample unit and the "height" of the UD at each cell is used as the response variable, issues such as independence of observations become less of a concern. Instead, additional location data, which might be correlated, offer a better UD estimate, provided data were collected in a systematic manner.

14.4.7 Ecological Niche Factor Analysis

An important sampling design issue facing investigators conducting resource selection studies is the difficulty in determining absence of animals from areas or habitats of interest. Although presence can be readily quantified, absence is challenging to confirm given the difficulty in knowing whether the habitat is unsuitable, has not yet been occupied (e.g. by colonizing species), or use was not detected (Basille et al. 2008). Multiple analytical approaches are available to evaluate resource selection using presence and "background" data on resource availability (Elith et al. 2006). Here, we note the functional equivalence among terms like "background," "pseudo-absence," and "available." One of the most common methods is *ecological-niche factor analysis* (ENFA; Hirzel et al. 2002), which is based on the ecological niche concept (Hutchinson 1957), and is meant to assess habitat suitability for the individual, population, or species, rather than the probability of respective occurrences (Hirzel et al. 2001). Therefore, an *ecological niche* is quantified using suitability functions that compare distributions of environmental characteristics between sites occupied by the species and the available area of interest.

ENFA models species presence data and ecogeographical variables, which are the *resources*. Ecogeographical variables are environmental characteristics presumed to describe the ecological niche for a species. The ENFA uses a geometric mean distance algorithm to model suitability based on the ecogeographical variables (Cianfrani et al. 2010). Thus, the niche is defined by uncorrelated measures of marginality and specialization factors that are calculated for each environmental characteristic. Marginality defines the deviation between values of a given environmental characteristic using sites with known animal use or presence and values of available sites without known use. Marginality is measured as the squared distance between the mean available sampling unit to the mean used sampling unit. Thus, with characteristics of each sampling unit, marginality calculates the magnitude of deviation of the niche relative to available sampling units. Higher marginality scores represent greater deviations of the occupied niche relative to available resources (Basille et al. 2008).

Once marginality is calculated, ENFA calculates a *specialization factor*. Specialization measures dispersion of the ecological niche, or niche width, and is the ratio of the variation of the distribution of an environmental variable at all used sample units to the variation of the distribution of that same variable at all available sample units (Hirzel et al. 2001; Unger et al. 2008; Cianfrani et al. 2010). As specialization values increase, the niche becomes more restricted along the axis relative to the reference. Applications of ENFA have included development of predictive habitat suitability maps (Dettki et al. 2003; Sattler et al. 2007), and recent improvements have been made to identify mechanistic explanations of observed space use. For example, Basille et al. (2008) described the use of biplots with marginality and specialization axes to characterize relative contributions of ecogeographical variables in defining habitat selection. For reviews of ENFA and similar modeling approaches, see Hirzel et al. (2001), Pearce and Boyce (2006), and Elith et al. (2011).

14.4.8 Mixed Models

Mixed models, consisting of both fixed and random effects, are often appropriate for RSF or RSPF analyses. For example, consider a Poisson regression analysis in which the same set of sample units were observed in each of two years, with n unique sites visit once per year. Assuming the $2*n$ counts were independent would ignore potential correlation between the repeated counts at each site. Instead, a *random site effect* could be added, which would account for dependence between multiple observations at the same site. Most commonly, the random site effect would be assumed to come from a normal distribution with mean 0 and variance parameter σ^2_{site}, although random slope models can be specified differently, with the mean of the higher-level distribution equal to the population-level mean of the slope parameter.

The use of random effects also automatically accounts for differences in the number of observations per site, individual, or other factor modeled with a random effect. The power of incorporating random effects into resource selection models should not be seen simply as their ability to account for dependence among observations. Most importantly, in the resource selection context, random effects can be used to model variation in parameter coefficients among individuals or subpopulations (Gillies et al. 2006; Thomas et al. 2006). In turn, higher-order covariates can be incorporated to examine hypotheses about factors associated with variation among individuals in selection (Wagner et al. 2011), such as functional responses. Further, such procedures are useful for examining both population-level questions and consequences of individual-level resource selection on fitness (Gillies et al. 2006), while effectively treating individual animals as the sample units. Random effects are tied to variance parameters, which may be of high interest in themselves for assessing the magnitude of different sources of variation.

The past decade has seen increased use of mixed-effects models in the study of resource selection (Gillies et al. 2006; Thomas et al. 2006; Hebblewhite and Merrill 2008). Hebblewhite and Merrill (2008) found that mixed-effects models in studies of resource selection received much stronger support than models with fixed effects only. Gillies et al. (2006) provided similar evidence of the utility of random effects and noted improvements to model fit and the ability to discern patterns of resource selection. Given such benefits, we have observed increase application of mixed-effects models in the study of resource selection (Rota et al. 2014; Jachowski et al. 2016). Random effects can be directly integrated in any of the procedures described above for studies of resource selection in wildlife species (Thomas et al. 2006).

14.5 Software Tools

Many statistical techniques for the analysis of resource selection data are available within Program R (R Core Team 2016). All of the common off-the-shelf methods, such as linear models, generalized linear models, and generalized linear mixed-effects models, are available within the basic installation of Program `R`. Many specialized packages are also available for analysis of resource selection data. Useful packages include `ResourceSelection` (Lele et al. 2016), `adehabitat` (including `adehabitatMA`, `adehabitatHS`, `adehabitatLT`, and `adehabitatLT`; Calenge 2006), `mlogit` (Croissant 2013), and `unmarked` (Fiske and Chandler 2011), with custom functions to fit a variety of specialized models described in this chapter. Program `R` also interfaces with several stand-alone programs useful for conducting resource selection analyses, including `MaxEnt` (Elith et al. 2011, with the `dismo` package [Hijmans et al. 2016]) and Program `MARK` (with the "`RMark`" package [Laake 2013]). Program `R` also interfaces with software designed to fit Bayesian models, including `Stan` (Carpenter et al. 2017, with `rstan` [Stan Development Team 2016]) and `OpenBUGS` (Lunn et al. 2000, with `R2OpenBUGS` [Sturtz et al. 2005]). Additionally, there are many stand-alone software packages that are completely independent of `R`, such as Program `PRESENCE` for fitting occupancy models (www.mbr-pwrc.usgs.gov/software/presence.html), `ENFA` in `Biomapper` for Ecological Niche Factor Analysis (http://www2.unil.ch/biomapper/enfa.html), and programs such as `FRAGSTATS` which compute habitat metrics within a GIS framework.

14.6 Online Exercises

We include five online exercises intended to provide users with practical applications of the analysis of resource selection data. Throughout this chapter we have highlighted the importance of sampling design and how different strategies to collect data affect the appropriateness of analysis choices. Exercise 1 uses a simulation approach reflecting different common sampling methods. This exercise highlights the importance of simulation work for evaluating bias and precision of parameter estimates and different common designs used in studies of resource selection. Exercise 2 demonstrates the use of discrete choice models to identify important resource features for reintroduced hellbenders (*Cryptobranchus alleganiensis bishopi*) in the Ozark Mountains of Missouri. The example highlights the use of discrete choice models which can be used when there are explicitly paired used and available resources, which was important for this species given that resource availability changed over short temporal scales. Exercise 3 demonstrates the development of a RUF which demonstrates the application of a continuous response variable as estimated by a UD. It also demonstrates how to fit individual animal models and scale up for population-level inference. Exercise 4 provides an application of Poisson Regression to investigate resource selection by black-footed ferrets (*Mustela nigripes*). This method applies count data of individually identifiable ferrets and relates those observations to resource features associated with their prey, prairie dogs (*Cynomys* spp.). Exercise 5 assesses the effects of resource use on percentage body fat of black bears (*Ursus americanus*). This exercise highlights how resource use can be tied with a species condition.

14.7 Future Directions

Resource selection studies are a cornerstone of ecological research as fundamental models for understanding species distributions and animal–habitat relationships. Indeed, many new applications of RSFs and RSPFs demonstrate their utility in addressing urgent conservation issues (Sawyer et al. 2006; Bodinof et al. 2012; Squires et al. 2013). For most species, we have moved beyond the need for simple, observational studies that describe basic resource relationships. Placing radio tags on a small sample of animals is not likely to yield robust data (Lindberg and Walker 2007). The wealth of available tools and guidance means that there is no excuse for not developing robust and defensible projects. These expectations, coupled with increasing availability of statistical tools and advanced technology, have increased the quantitative value of resource selection studies.

A theme for our chapter has been RSFs and RSPFs and we see their continued application and development. These types of models are intuitively appealing, flexible, and robust for addressing resource-animal questions. These techniques are also increasingly available to ecologists through specialized software packages (Elith et al. 2011; Fiske and Chandler 2011). While we promote the availability of software through open sharing of analysis code, we caution choices of analytical options based on ease of analysis. Instead, it is of paramount importance to line up objectives, hypotheses, and data types with the choice of analysis. Nearly all of the methods we have discussed can produce a colorful predictive map, but the utility of that map should be robust relative to the objectives and data in hand. Further, such maps should explicitly report accuracy and not obscure the uncertainty in projections on the landscape.

We see continued development and application of models that take full advantage of GPS equipment and other technological advancements in field techniques (Benson 2010). Even relatively small animals can carry radio-tracking devices and GPS units now contain proximity sensors and remote cameras (Moll et al. 2007; Rutz et al. 2007; Kays et al. 2016). Such advancements continue to expand the range of species, questions, and available analytical options. Further, the availability of sensors such as accelerometers will provide greater insights into animal behavior which will likely be more cleverly integrated into resource selection analyses. Challenges of integrating emerging data streams in existing data analyses has been an ongoing issue in resource selection studies for decades. Despite advancements in available data, we cannot rely on technology to make up for pitfalls in study design and the need to continue collecting in-field data and observations on animals.

Great attention has been paid to new analytical techniques that, at their core, are RSFs or RSPFs because they seek to relate occurrence or abundance of animals to resources. We see species distribution models collectively falling under the RSF and RSPF umbrella. The models have some differences in assumptions, data requirements, and interpretation, but ultimately the general objectives of both approaches are to understand how resources affect the occurrence of wildlife. Terminology has been an impediment when trying to interpret model results because each approach has its own jargon. It is important we avoid getting caught up in semantics, and instead take care to delineate assumptions, appropriate application, and interpretation of alternative models.

We see great utility in sharing data to promote increased comparisons of animal-resource relationships across time and space. For example, the Movebank data model for radio tracking and similar approaches is an important new resource (Kranstauber et al. 2011). Movebank is an online database which stores, processes, analyzes, and allows sharing of animal radio-tracking data. Contributors have control over sharing options, which helps ensure proper use of data. The availability of shared data facilitates comparisons across a species range and improves opportunities for meta-analyses. In a resource selection context, rather than individual studies of species x at location y, Movebank provides an opportunity for more comprehensive assessments of species x or location y. It also facilitates reanalysis if new models become available in the future. While site- and species-specific studies will always be needed to address many of the pressing resource selection issues such as climate change and habitat loss, additional spatial and temporal replication will be advantageous.

We expect continued development of individual-animal-based models which are scaled up to the population level and an increased reliance on mixed models and Bayesian methods. Bayesian statistical techniques are becoming increasingly popular in resource selection studies because of their ability to fit complex models that are often unavailable within a frequentist paradigm (Hobbs and Hooten 2015). In many cases, the field and analytical technology is available to estimate individual-animal-based models within a hierarchical modeling framework. Novel analytical techniques will prove helpful in addressing issues such as spatial and temporal autocorrelation (Gillies et al. 2006; McLoughlin et al. 2010), dependence between two or more species (Rota et al. 2016), the link between resource selection and demography (Matthiopoulos et al. 2015), and numerous other ecological processes that influence patterns of resource selection.

References

Aarts, G., Fieberg, J., and Matthiopoulos, J. (2012). Comparative interpretation of count, presence-absence and point methods for species distribution models. *Methods in Ecology and Evolution* 3: 177–187.

Aebischer, N.J., Robertson, P.A., and Kenward, R.E. (1993). Compositional analysis of habitat use from animal radio-tracking data. *Ecology* 74: 1313–1325.

Aitchison, J. (1986). *The Statistical Analysis of Compositional Data*. New York, NY: Chapman and Hall.

Alldredge, J.R. and Ratti, J.T. (1986). Comparison of some statistical techniques for analysis of resource selection. *Journal of Wildlife Management* 50: 157–165.

Alldredge, J.R. and Ratti, J.T. (1992). Further comparison of some statistical techniques for analysis of resource selection. *Journal of Wildlife Management* 56: 1–9.

Arthur, S.M., Manly, B.F.J., McDonald, L.L., and Garner, G. W. (1996). Assessing habitat selection when availability changes. *Ecology* 77: 215–227.

Baasch, D.M., Tyre, A.J., Millspaugh, J.J. et al. (2010). An evaluation of three statistical methods used to model resource selection. *Ecological Modelling* 221: 565–574.

Basille, M., Calenge, C., Marboutin, E. et al. (2008). Assessing habitat selection using multivariate statistics: some refinements of the ecological-niche factor analysis. *Ecological Modeling* 211: 233–240.

Ben-Akiva, M. and Lerman, S.R. (1985). *Discrete Choice Analysis: Theory and Application to Travel Demand*. Cambridge, MA: The MIT Press.

Benson, E. (2010). *Wired Wilderness: Technologies of Tracking and the Making of Modern Wildlife*. Baltimore, MD: Johns Hopkins University Press.

Benson, J.F. (2012). Improving rigour and efficiency of use-availability habitat selection analyses with systematic estimation of availability. *Methods in Ecology and Evolution* 4: 244–251.

Beyer, D.E. Jr. and Haufler, J.B. (1994). Diurnal versus 24-hour sampling of habitat use. *Journal of Wildlife Management* 58: 178–180.

Bingham, R.L. and Brennan, L.A. (2004). Comparison of type I error rates for statistical analyses of resource selection. *Journal of Wildlife Management* 68: 206–212.

Bingham, R.L., Brennan, L.A., and Ballard, B.M. (2007). Misclassified resource selection: compositional analysis and unused habitat. *Journal of Wildlife Management* 71: 1369–1374.

Bodinof, C.M., Briggler, J.T., Junge, R.E. et al. (2012). Habitat attributes associated with short-term settlement of Ozark hellbender (*Cryptobranchus alleganiensis bishopi*). *Freshwater Biology* 57: 178–192.

Both, C. and Visser, M.E. (2000). Breeding territory size affects fitness: an experimental study on competition at the individual level. *Journal of Animal Ecology* 69: 1021–1030.

Boyce, M.S. (2006). Scale for resource selection functions. *Diversity and Distributions* 12: 269–276.

Bridge, E.S., Thorup, K., Bowlin, M.S. et al. (2011). Technology on the move: recent and forthcoming innovations for tracking migratory birds. *BioScience* 61: 689–698.

Brost, B.M., Hooten, M.B., Ephriam, M.H., and Small, R.J. (2015). Animal movement constraints improve resource selection inference in the presence of telemetry error. *Ecology* 96: 2590–2597.

Buskirk, S.W. and Millspaugh, J.J. (2006). Metrics for studies of resource selection. *Journal of Wildlife Management* 30: 358–366.

Cagnacci, F., Boitani, L., Powell, R.A., and Boyce, M.S. (2010). Animal ecology meets GPS-based radiotelemetry: a perfect storm of opportunities and challenges. *Philosophical Transactions of the Royal Society of London B* 365: 2157–2162.

Calenge, C. (2006). The package `adehabitat` for the R software: a tool for the analysis of space and habitat use by animals. *Ecological Modeling* 197: 516–519.

Cargnelutti, B., Coulon, A., Hewison, A.J.M. et al. (2007). Testing global positioning system performance for wildlife monitoring using mobile collars and known reference points. *Journal of Wildlife Management* 71: 1380–1387.

Carpenter, B., Gelman, A., Hoffman, M. et al. (2017). `Stan`: a probabilistic programming language. *Journal of Statistical Software* 76: https://doi.org/10.18637/jss.v076.i01.

Carter, S.K., Rosas, F.C.W., Cooper, A.B., and Corderio-Duarte, A.C. (1999). Consumption rate, food preferences and transit time of captive giant otters *Pteronura brasiliensis*: implications for the study of wild populations. *Aquatic Mammals* 25: 79–90.

Chetkiewicz, C.L.-B. and Boyce, M.S. (2009). Use of resource selection functions to identify conservation corridors. *Journal of Applied Ecology* 46: 1036–1047.

Cianfrani, C., Le Lay, G., Hirzel, A., and Loy, A. (2010). Do habitat suitability models reliably predict the recovery areas of threatened species. *Journal of Applied Ecology* 47: 421–430.

Cooper, A.B. and Millspaugh, J.J. (1999). The application of discrete choice models to wildlife resource selection studies. *Ecology* 80: 566–575.

Cooper, A.B. and Millspaugh, J.J. (2001). Accounting for variation in resource availability and animal behavior in resource selection studies. In: *Radio Tracking and Animal Populations* (eds. J.J. Millspaugh and J.M. Marzluff), 243–274. San Diego, CA: Academic Press.

Creel, S., Winnie, J. Jr., Maxwell, B. et al. (2005). Elk alter habitat selection as an antipredator response to wolves. *Ecology* 86: 3387–3397.

Croissant, Y (2013) `mlogit`: multinomial logit model. R package version 0.2–4. https://CRAN.R-project.org/package=mlogit

DeCesare, N.J., Hebblewhite, M., Schmiegelow, F. et al. (2012). Transcending scale dependence in identifying habitat with resource selection functions. *Ecological Applications* 22: 1068–1083.

D'Eon, R.G., Serrouya, R., Smith, G., and Kochanny, C.O. (2002). GPS radiotelemetry error and bias in mountainous terrain. *Wildlife Society Bulletin* 30: 430–439.

Dettki, H., Lofstrand, R., and Edenius, L. (2003). Modeling habitat suitability for moose in coastal northern Sweden: empirical vs. process-oriented approaches. *Ambio* 32: 549–556.

Durner, G.M., Douglas, D.C., Nielson, R.M. et al. (2009). Predicting 21st-century polar bear habitat distribution from global climate models. *Ecological Monographs* 79: 25–58.

Dzialak, M.R., Olson, C.V., Harju, S.M. et al. (2011). Identifying and prioritizing Greater Sage-Grouse nesting and brood rearing habitat for conservation in human-modified landscapes. *PLoS One* 6: e26273.

Dzialak, M.R., Webb, S.L., Harju, S.M. et al. (2013). Greater Sage-Grouse and severe winter conditions: identifying habitat for conservation. *Rangeland Ecology and Management* 66: 10–18.

Eads, D.A., Jachowski, D.S., Biggins, D.E. et al. (2012). Resource selection models are useful in predicting fine-scale distributions of black-footed ferrets in prairie dog colonies. *Western North American Naturalist* 72: 206–215.

Eads, D.A., Millspaugh, J.J., Biggins, D.E. et al. (2011). Post-breeding resource selection by adult black-footed ferrets in the Conata Basin, South Dakota. *Journal of Mammalogy* 92: 760–770.

Elith, J., Graham, C.H., Anderson, R.P. et al. (2006). Novel methods improve prediction of species' distributions from occurrence data. *Ecography* 29: 129–151.

Elith, J., Phillips, S.J., Hastie, T. et al. (2011). A statistical explanation of `MaxEnt` for ecologists. *Diversity and Distributions* 17: 43–57.

Erickson, W.P., McDonald, T.L., Gerow, K.G. et al. (2001). Statistical issues in resource selection studies with radio-marked animals. In: *Radio Tracking and Animal Populations* (eds. J.J. Millspaugh and J.M. Marzluff), 211–242. San Diego, CA: Academic Press.

Fieberg, J. (2007). Kernel density estimators of home range: smoothing and the autocorrelation red herring. *Ecology* 88: 1059–1066.

Findholdt, SL, Johnson, BK, McDonald LL, et al. (2002) Adjusting for radiotelemetry error to improve estimates of habitat use. USDA Forest Service Pacific Northwest Research Station General Technical Report PNW-GTR-555.

Fiske, I. and Chandler, R. (2011). `unmarked`: an R package for fitting hierarchical models of wildlife occurrence and abundance. *Journal of Statistical Software* 43: 1–23.

Fithian, T. and Hastie, W. (2013). Inference from presence-only data; the ongoing controversy. *Ecography* 36: 864–867.

Forester, J.D., Ives, A.R., Turner, M.G. et al. (2007). State-space models link elk movement patterns to landscape characteristics in Yellowstone National Park. *Ecological Monographs* 77: 285–299.

Frair, J.L., Fieberg, J., Hebblewhite, M. et al. (2010). Resolving issues of imprecise and habitat-biased locations in ecological analyses using GPS telemetry data. *Philosophical Transactions of the Royal Society of London B* 365: 2187–2200.

Frair, J.L., Nielsen, S.E., Merrill, E.H. et al. (2004). Removing GPS collar bias in habitat selection studies. *Journal of Applied Ecology* 41: 201–212.

Franklin, A.B., Anderson, D.R., Gutiérrez, R.J., and Burnham, K.P. (2000). Climate, habitat quality, and fitness in Northern Spotted Owl populations in Northwestern California. *Ecological Monographs* 70: 539–590.

Fretwell, S.D. (1972). Populations in a seasonal environment. *Monographs in Population Biology* 5: 1–219.

Fretwell, S.D. and Lucas, H.L. (1970). On territorial behavior and other factors influencing habitat distribution in birds. *Acta Biotheoretica* 19: 16–36.

Garshelis, D.L. (2000). Delusions in habitat evaluation: measuring use, selection, and importance. In: *Research Techniques in Animal Ecology: Controversies and Consequences* (eds. L. Boitani and T.K. Fuller), 111–164. New York, NY: Columbia University Press.

Garton, E.O., Wisdom, M.J., Leban, F.A., and Johnson, B.K. (2001). Experimental design for radiotelemetry studies. In: *Radio Tracking and Animal Populations* (eds. J.J. Millspaugh and J.M. Marzluff), 16–42. New York, NY: Academic Press.

Getz, W.M. and Wilmers, C.C. (2004). A local nearest-neighbor convex-hull construction of home ranges and utilization distributions. *Ecography* 27: 489–505.

Gillies, C.S., Hebblewhite, M., Nielsen, S.E. et al. (2006). Application of random effects to the study of resource selection by animals. *Journal of Animal Ecology* 75: 887–898.

Hebblewhite, M. and Merrill, E. (2008). Modelling wildlife-human relationships for social species with mixed-effects resource selection models. *Journal of Applied Ecology* 45: 834–844.

Hebblewhite, M., Percy, M., and Merrill, E.H. (2007). Are all global positioning system collars created equal? Correcting habitat-induced bias using three brands in the Central Canadian Rockies. *Journal of Wildlife Management* 71: 2026–2033.

Hijmans, RJ, Phillips, S, Leathwick, J, Elith, J (2016) `dismo`: Species Distribution Modeling. R package version 1.1–1. https://CRAN.R-project.org/package=dismo

Hildén, O. (1965). Habitat selection in birds. *Annales Zoologici Fennici* 2: 53–75.

Hirzel, A., Hausser, J., Chessel, D., and Perrin, N. (2002). Ecological-niche factor analysis: how to compute habitat-suitability maps without absence data? *Ecology* 83: 2027–2036.

Hirzel, A.H., Helfer, V., and Metral, F. (2001). Assessing habitat-suitability model with a virtual species. *Ecological Modelling* 145: 111–121.

Hobbs, N.T. and Hooten, M.B. (2015). *Bayesian Models: A Statistical Primer for Ecologists*. Princeton, NJ: Princeton University Press.

Hooten, M.B., Hanks, E.M., Johnson, D.S., and Alldredge, M.W. (2013). Reconciling resource utilization and resource selection functions. *Journal of Animal Ecology* 82: 1146–1154.

Hosmer, D.W. and Lemeshow, S. (2000). *Applied Logistic Regression*. New York, NY: Wiley.

Hutchinson, G.E. (1957). Concluding remarks. *Cold Springs Harbor Symposium on Quantitative Biology* 22: 277–297.

Hutto, R.L. (1985). Habitat selection by nonbreeding, migratory land birds. In: *Habitat Selection in Birds* (ed. M. L. Cody), 455–476. New York, NY: Academic Press.

Jachowski, D.S., Rota, C.T., Dobony, C.A. et al. (2016). Seeing the forest through the trees: considering roost site selection at multiple spatial scales. *PLoS One* 11: e0150011.

Jobes, A.P., Nol, E., and Voigt, D.R. (2004). Effects of selection cutting on bird communities in contiguous eastern hardwood forests. *Journal of Wildlife Management* 68: 51–60.

Johnson, C.J. and Gillingham, M.P. (2008). Sensitivity of species-distribution models to error, bias, and model design: an application to resource selection functions for woodland caribou. *Ecological Modelling* 213: 143–155.

Johnson, C.J., Nielsen, S.E., Merrill, E.H. et al. (2006). Resource selection functions based on use-availability data: theoretical motivation and evaluation methods. *Journal of Wildlife Management* 70: 347–357.

Johnson, C.J., Seip, D.R., and Boyce, M.S. (2004). A quantitative approach to conservation planning: using resource selection functions to map the distribution of mountain caribou at multiple spatial scales. *Journal of Applied Ecology* 41: 238–251.

Johnson, D.H. (1980). The comparison of usage and availability measures for evaluating resource preference. *Ecology* 61: 65–71.

Johnson, D.S., Thomas, D.L., Ver Hoef, J., and Christ, A. (2008). A general framework for the analysis of animal resource selection from telemetry data. *Biometrics* 64: 968–976.

Jones, J. (2001). Habitat selection studies in avian ecology: a critical review. *Auk* 118: 557–562.

Kays, R., Parsons, A.W., Baker, M.C. et al. (2016). Does hunting or hiking affect wildlife communities in protected areas? *Journal of Applied Ecology* 54: 242–252.

Keating, K.A. and Cherry, S. (2004). Use and interpretation of logistic regression in habitat-selection studies. *Journal of Wildlife Management* 68: 774–789.

Kendeigh, S.C. (1945). Community selection by birds on the Helderberg Plateau of New York. *Auk* 62: 418–436.

Kendeigh, S.C. (1970). Energy requirements for existence in relation to size of bird. *Condor* 72: 60–65.

Kernohan, B.J., Gitzen, R.A., and Millspaugh, J.J. (2001). Analysis of animal space use and movements. In: *Radio Tracking and Animal Populations* (eds. J.J. Millspaugh and J.M. Marzluff), 125–166. San Diego, CA: Academic Press.

Kie, J.G., Bowyer, R.T., Nicholson, M.C. et al. (2002). Landscape heterogeneity at differing scales: effects on spatial distribution of mule deer. *Ecology* 83: 530–544.

Kie, J.G., Matthiopoulos, J., Fieberg, J. et al. (2010). The home-range concept: are traditional estimators still relevant with modern telemetry technology. *Philosophical Transactions of the Royal Society of London B* 365: 2221–2231.

Kranstauber, B., Cameron, A., Weinzerl, R. et al. (2011). The Movebank data model for radio tracking. *Environmental Modelling and Software* 26: 834–835.

Laake, JL (2013). `RMark`: An R Interface for Analysis of Capture-Recapture Data with MARK. AFSC Processed Rep 2013-01, 25p. Alaska Fish. Sci. Cent., NOAA, Natl. Mar. Fish. Serv., 7600 Sand Point Way NE, Seattle, WA.

Lancaster, T. and Imbens, G. (1996). Case-control studies with contaminated controls. *Journal of Econometrics* 71: 145–160.

Leban, F.A., Wisdom, M.J., Garton, E.O. et al. (2001). Effect of sample size on the performance of resource selection analyses. In: *Radio Tracking and Wildlife Populations* (eds. J.J. Millspaugh and J.M. Marzluff), 291–307. San Diego, CA: Academic Press.

Lele, S.R. and Keim, J.L. (2006). Weighted distributions and estimation of resource selection probability functions. *Ecology* 87: 3021–3028.

Lele, S.R., Keim, J.L., Solymos, P. (2016). `ResourceSelection`: Resource Selection (Probability) Functions for Use-Availability Data. R package version 0.2-6. https://CRAN.R-project.org/package=ResourceSelection

Lele, S.R., Merrill, E.H., Keim, J., and Boyce, M.S. (2013). Selection, use, choice and occupancy: clarifying concepts in resource selection studies. *Journal of Animal Ecology* 82: 1183–1191.

Levin, S.A. (1992). The problem of pattern and scale in ecology: the Robert H. MacArthur award lecture. *Ecology* 73: 1943–1967.

Lewis, J.S., Rachlow, J.L., Garton, E.O., and Vierling, L.A. (2007). Effects of habitat on GPS collar performance: using data screening to reduce location error. *Journal of Applied Ecology* 44: 663–671.

Lichti, N.I. and Swihart, R.K. (2011). Estimating utilization distributions with kernel versus local convex hull methods. *Journal of Wildlife Management* 75: 413–422.

Lindberg, M.S. and Walker, J. (2007). Satellite telemetry in avian research and management: sample size considerations. *Journal of Wildlife Management* 71: 1002–1009.

Lundy, M.G., Buckley, D.J., Boston, E.S.M. et al. (2012). Behavioural context of multi-scale species distribution models assessed by radio-tracking. *Basic and Applied Ecology* 13: 188–195.

Lunn, D.J., Thomas, A., Best, N., and Spiegelhalter, D. (2000). `WinBUGS` – a Bayesian modelling framework: concepts, structure, and extensibility. *Statistics and Computing* 10: 325–337.

MacArthur, R. and Levins, R. (1964). Competition, habitat selection, and character displacement in a patchy environment. *Proceedings of the National Academy of Sciences USA* 51: 1207–1210.

MacKenzie, D.I. (2006). Modeling the probability of resource use: the effect of, and dealing with, detecting a species imperfectly. *Journal of Wildlife Management* 70: 367–374.

MacKenzie, D.I., Nichols, J.D., Lachman, G.B. et al. (2002). Estimating site occupancy rates when detection probabilities are less than one. *Ecology* 83: 2248–2255.

Manly, B.F.J. (1991). *Randomization and Monte Carlo Methods in Biology*. London, UK: Chapman and Hall.

Manly, B.F.J., McDonald, L.L., Thomas, D.L. et al. (2002). *Resource Selection by Animals: Statistical Design and Analysis for Field Studies*. Dordrecht, The Netherlands: Kluwer Academic Publishers.

Martin, T.E. (1996). Fitness costs of resource overlap among coexisting bird species. *Nature* 380: 338–340.

Martin, T.E. (2001). Abiotic vs. biotic influences on habitat selection of coexisting species: climate change impacts? *Ecology* 82: 175–188.

Marzluff, J.M., Knick, S., and Millspaugh, J.J. (2001). High-tech behavioral ecology: modeling the distribution of animal activities to better understand animal space use. In: *Radio Tracking and Animal Populations* (eds. J.J. Millspaugh and J.M. Marzluff), 309–328. San Diego, CA: Academic Press.

Marzluff, J.M., Millspaugh, J.J., Hurvitz, P., and Handcock, M.S. (2004). Relating resources to a probabilistic measure of space use: forest fragments and Steller's Jays. *Ecology* 85: 1411–1427.

Matthiopoulos, J., Fieberg, J., Aarts, G. et al. (2015). Establishing the link between habitat selection and animal population dynamics. *Ecological Monographs* 85: 413–436.

McArthur, C., Goodwin, A., and Turner, S. (2000). Preferences, selection and damage to seedlings under changing availability by two marsupial herbivores. *Forest Ecology and Management* 139: 157–173.

McCracken, M.L., Manly, B.F.J., and Vander Heyden, M. (1998). The use of discrete-choice models for evaluating resource selection. *Journal of Agricultural, Biological, and Environmental Statistics* 3: 268–279.

McDonald, L., Manly, B., Huettmann, F., and Thogmartin, W. (2013). Location-only and use-availability data: analysis methods converge. *Journal of Applied Ecology* 82: 1120–1124.

McDonald, L.L., Manly, B.F., and Raley, C.M. (1990). Analyzing foraging and habitat use through selection functions. *Studies in Avian Biology* 13: 325–331.

McDonald, T.L. and McDonald, L.L. (2002). A new ecological risk assessment procedure using resource selection models and geographic information systems. *Wildlife Society Bulletin* 30: 1015–1021.

McKenzie, H., Jerde, C., Visscher, D. et al. (2009). Inferring linear feature use in the presence of GPS measurement error. *Environmental and Ecological Statistics* 16: 531–546.

McLoughlin, P.D., Gaillard, J.M., Boyce, M.S. et al. (2007). Lifetime reproductive success and composition of the home range in a large herbivore. *Ecology* 88: 3192–3201.

McLoughlin, P.D., Morris, D.W., Fortin, D. et al. (2010). Considering ecological dynamics in resource selection functions. *Journal of Animal Ecology* 79: 4–12.

McNew, L.B., Hunt, L.M., Gregory, A.J. et al. (2014). Effects of wind energy development on nesting ecology of Greater Prairie-Chickens in fragmented landscapes. *Conservation Biology* 28: 1089–1099.

Millspaugh, J.J., Nielson, R.M., McDonald, L. et al. (2006). Analysis of resource selection using utilization distributions. *Journal of Wildlife Management* 70: 385–395.

Millspaugh, J.J. and Thompson, F.R. III (eds.) (2009). *Models for Planning Wildlife Conservation in Large Landscapes*. San Diego, CA: Elsevier Science.

Moll, R.J., Millspaugh, J.J., Beringer, J. et al. (2007). A new "view" of ecology and conservation through animal-borne video systems. *Trends in Ecology and Evolution* 22: 660–668.

Moll, R.J., Millspaugh, J.J., Beringer, J. et al. (2009). A terrestrial animal-borne video system for large mammals. *Computers and Electronics in Agriculture* 66: 133–139.

Montgomery, R.A., Roloff, G.J., and Ver Hoef, J.M. (2011). Implications of ignoring telemetry error on inference in wildlife resource use models. *Journal of Wildlife Management* 75: 702–708.

Montgomery, R.A., Roloff, G.J., Ver Hoef, J., and Millspaugh, J.J. (2010). Can we accurately characterize wildlife resource use when telemetry data are imprecise? *Journal of Wildlife Management* 74: 1917–1925.

Moser, B.W. and Garton, E.O. (2007). Effects of telemetry location error on space-use estimates using a fixed-kernel density estimator. *Journal of Wildlife Management* 71: 2421–2426.

Mysterud, A. (1996). Bed-site selection by adult roe deer *Capreolus capreolus* in southern Norway during summer. *Wildlife Biology* 2: 101–106.

Mysterud, A. and Ims, R.A. (1998). Functional responses in habitat use: availability inferences relative use in trade-off situations. *Ecology* 79: 1435–1441.

Nams, V.O. and Boutin, S. (1991). What is wrong with error polygons? *Journal of Wildlife Management* 55: 172–176.

Neu, C.W., Byers, C.R., and Peek, J.M. (1974). A technique for analysis of utilization-availability data. *Journal of Wildlife Management* 38: 541–545.

Nielson, R.M., Manly, B.F.J., McDonald, L. et al. (2009). Estimating habitat selection when GPS fix success is less than 100%. *Ecology* 90: 2956–2962.

Northrup, J.M., Hooten, M.B., Anderson, C.R. Jr., and Wittemyer, G. (2013). Practical guidance on characterizing availability in resource selection functions under a use-availability design. *Ecology* 94: 1456–1463.

O'Connell, A.F., Nichols, J.D., and Karanth, K.U. (2010). *Camera Traps in Animal Ecology: Methods and Analyses*. New York, NY: Springer.

Oppel, S., Schaefer, H.M., Schmidt, V., and Schröder, B. (2004). Habitat selection by the Pale-headed Brush-Finch (*Atlapetes pallidiceps*) in southern Ecuador: implications for conservation. *Biological Conservation* 118: 33–40.

Orians, G.H. and Wittenberger, J.F. (1991). Spatial and temporal scales in habitat selection. *American Naturalist* 137: S29–S49.

Otis, D.L. (1997). Analysis of habitat selection studies with multiple patches within cover types. *Journal of Wildlife Management* 61: 1016–1022.

Otis, D.L. (1998). Analysis of the influence of spatial pattern in habitat selection studies. *Journal of Agricultural, Biological, and Environmental Statistics* 3: 254–267.

Otis, D.L. and White, G.C. (1999). Autocorrelation of location estimates and the analysis of radiotracking data. *Journal of Wildlife Management* 63: 1039–1044.

Pearce, J.L. and Boyce, M.S. (2006). Modelling distribution and abundance with presence-only data. *Journal of Applied Ecology* 43: 405–412.

Pendleton, G.W., Titus, K., DeGayner, E. et al. (1998). Compositional analysis and GIS for study of habitat selection by Goshawks in Southeast Alaska. *Journal of Agricultural, Biological, and Environmental Statistics* 3: 280–295.

Phillips, S.J. and Dudik, M. (2008). Modeling of species distributions with Maxent: new extensions and a comprehensive evaluation. *Ecography* 31: 161–175.

Quade, D. (1979). Using weighted rankings in the analysis of complete blocks with additive block effects. *Journal of the American Statistical Association* 74: 680–683.

R Core Team. 2016. R: A language and environment for statistical computing. R Foundation for Statistical Computing, Vienna, Austria. URL https://www.R-project.org.

Rempel, R.S., Rodgers, A.R., and Abraham, K.F. (1995). Performance of a GPS animal location system under boreal forest canopy. *Journal of Wildlife Management* 59: 543–551.

Renner, I.W., Elith, J., Baddeley, A. et al. (2015). Point process models for presence-only analysis. *Methods in Ecology and Evolution* 6: 366–379.

Renner, I.W. and Warton, D.I. (2013). Equivalence of MAXENT and Poisson point process models for species distribution modeling in ecology. *Biometrics* 69: 274–281.

Rosenzweig, M.L. (1991). Habitat selection and population interactions: the search for mechanism. *American Naturalist* 137: S5–S28.

Rota, C.T., Millspaugh, J.J., Kesler, D.C. et al. (2013). A re-evaluation of a case-control model with contaminated controls for resource selection studies. *Journal of Animal Ecology* 82: 1165–1173.

Rota, C.T., Rumble, M.A., Millspaugh, J.J. et al. (2014). Space-use and habitat associations of Black-backed Woodpeckers occupying recently disturbed forests in the Black Hills, South Dakota. *Forest Ecology and Management* 313: 161–168.

Rota, C.T., Wikle, C.K., Kays, R.W. et al. (2016). A two-species occupancy model accommodating simultaneous spatial and interspecific dependence. *Ecology* 97: 48–53.

Royle, J.A., Chandler, R.B., Sollmann, R., and Gardner, B. (2013). *Spatial Capture–Recapture*. Waltham, Massachusetts, USA: Academic Press.

Royle, J.A., Chandler, R.B., Yackulic, C., and Nichols, J.D. (2012). Likelihood analysis of species occurrence probability from presence-only data for modelling species distributions. *Methods in Ecology and Evolution* 3: 545–554.

Rutz, C., Bluff, L.A., Weir, A.A.S., and Kacelnik, A. (2007). Video cameras on wild birds. *Science* 318: 765.

Saltz, D. (1994). Reporting error measures in radio location by triangulation: a review. *Journal of Wildlife Management* 58: 181–184.

Sattler, T., Bontadina, F., Hirzel, A.H., and Arlettaz, R. (2007). Ecological niche modeling of two cryptic bat species calls for a reassessment of their conservation status. *Journal of Applied Ecology* 44: 1188–1199.

Sawyer, H., Nielson, R.M., Lindzey, F., and McDonald, L.L. (2006). Winter habitat selection of mule deer before and during development of a natural gas field. *Journal of Wildlife Management* 70: 396–403.

Sawyer, S.C. and Brashares, J.S. (2013). Applying resource selection functions at multiple scales to prioritize habitat use by the endangered Cross River gorilla. *Diversity and Distributions* 19: 943–954.

Scribner, K.T., Blanchong, J.A., Bruggeman, D.J. et al. (2005). Geographical genetics: conceptual foundations and empirical applications of spatial genetic data in wildlife management. *Journal of Wildlife Management* 69: 1434–1453.

Seaman, D.E., Millspaugh, J.J., Kernohan, B.J. et al. (1999). Effects of sample size on kernel home range estimates. *Journal of Wildlife Management* 63: 739–747.

Sólymos, P. and Lele, S.R. (2016). Revisiting resource selection probability functions and single-visit methods: clarification and extensions. *Methods in Ecology and Evolution* 7: 196–205.

Squires, J.R., DeCasare, N.J., Olson, L.E. et al. (2013). Combining resource selection and movement behavior to predict corridors of Canada lynx at their southern range periphery. *Biological Conservation* 157: 187–195.

Stamp, R.K., Brunton, D.H., and Walter, B. (2002). Artificial nest box use by the North Island Saddleback: effects of nest box design and mite infestations on nest site selection and reproductive success. *New Zealand Journal of Zoology* 29: 285–292.

Stan Development Team (2016) `RStan`: the `R` interface to `Stan`, Version 2.10.1. http://mc-stan.org

Sturtz, S., Ligges, U., and Gelman, A. (2005). `R2WinBUGS`: a package for running `WinBUGS` from `R`. *Journal of Statistical Software* 12: 1–16.

Svärdson, G. (1949). Competition and habitat selection in birds. *Oikos* 1: 157–174.

Thomas, D.L., Johnson, D., and Griffith, D.B. (2006). A Bayesian random effects discrete-choice model for resource selection: population-level selection inference. *Journal of Wildlife Management* 70: 404–412.

Thomas, D.L. and Taylor, E.J. (2006). Study designs and tests for comparing resource use and availability II. *Journal of Wildlife Management* 70: 324–336.

Turner, M.G. (1990). Spatial and temporal analysis of landscape patterns. *Landscape Ecology* 4: 21–30.

Turner, M.G., Gardner, R.H., and O'Neill, R.V. (2001). *Landscape Ecology in Theory and Practice: Pattern and Process*. New York, NY: Springer Verlag.

Turner, M.G., O'Neill, R.V., Gardner, R.H., and Milne, B. T. (1989). Effects of changing spatial scale on the analysis of landscape pattern. *Landscape Ecology* 3: 153–162.

Unger, D.E., Fei, S., and Maehr, D. (2008). Ecological niche factor analysis to determine habitat suitability of a recolonizing carnivore. In: *Proceedings of the 6th Southern Forestry and Natural Resources GIS Conference* (eds. P. Bettinger, K. Merry, S. Fei, et al.), 237–249. Athens. GA: Warnell School of Forestry and Natural Resources, University of Georgia.

Van Horne, B. (1983). Density as a misleading indicator of habitat quality. *Journal of Wildlife Management* 47: 893–901.

Wagner, T., Diefenbach, D.R., Christensen, S.A., and Norton, A.S. (2011). Using multilevel models to quantify heterogeneity in resource selection. *Journal of Wildlife Management* 75: 1788–1796.

Warton, D.I. and Shepherd, L.C. (2010). Poisson point process models solve the "pseudo-absence problem" for presence-only data in ecology. *Annals of Applied Statistics* 4: 1383–1402.

White, G.C. (1985). Optimal locations of towers for triangulation studies using biotelemetry. *Journal of Wildlife Management* 49: 190–196.

White, P.J., Garrott, R.A., Cherry, S. et al. (2008). Changes in elk resource selection and distribution with the reestablishment of wolf predation risk. *Terrestrial Ecology* 3: 451–476.

Wiens, J.A. (1972). Anuran habitat selection: early experience and substrate selection in *Rana cascadae* tadpoles. *Animal Behaviour* 20: 218–220.

Wiens, J.A. (1973). Pattern and process in grassland bird communities. *Ecological Monographs* 43: 237–270.

Wiens, J.A., Stralberg, D., Jongsomjit, D. et al. (2009). Niches, models, and climate change: assessing the assumptions and uncertainties. *Proceedings of the National Academy of Sciences USA* 106: 19729–19736.

Wisz, M.S., Higmans, R.J., Li, J. et al. (2008). Effects of sample size on the performance of species distributions models. *Diversity and Distributions* 14: 763–773.

Zuur, A.F., Ieno, E.N., Walker, N.J. et al. (2009). *Mixed Effects Models and Extensions in Ecology with R*. New York, NY: Springer Verlag.

15

Species Distribution Modeling

Daniel H. Thornton[1] *and Michael J.L. Peers*[2]

[1] *School of Environment, Washington State University, Pullman, WA, USA*
[2] *Department of Biological Sciences, University of Alberta, Edmonton, Alberta, Canada*

Summary

Species distribution modeling is one of the fastest growing areas of ecology. Although most research in population ecology occurs at relatively small spatial scales, species distribution models (SDMs) are often applied to understand large-scale patterns in the distribution and abundance of species. SDMs may be used to address a myriad of different issues, ranging from the predicted response of species to environmental change to the discovery of new biodiversity. The goal of this chapter is to provide a brief overview of SDMs and their implementation, as well as detail current developments and advances in the field. The literature on species distribution modeling is vast and growing rapidly, but our chapter provides a starting point to better understand these methods. At its core, species distribution modeling entails using a statistical model to relate locations of species occurrence (and perhaps locations of confirmed absence) with environmental variables. While a simple concept, SDMs require complex theoretical and methodological considerations. Development of an SDM starts with the nontrivial steps of collecting and processing species and environmental data, then choosing an appropriate statistical model while making important decisions related to the application of the model (such as how to deal with sampling bias), and then fitting and potentially evaluating the model. Interpretation of SDMs is complex and requires a careful consideration of niche ecology, the nature of the analysis employed, and the meaning of the derived output. As the field of distribution modeling progresses, new techniques are being utilized to address key limitations of existing methods. Innovations include incorporation of dispersal constraints and biotic interactions into SDMs, and coupling of SDMs with more complex population models. New methods hold great promise for developing a new generation of SDMs that will further advance our understanding of species–environment relationships at large spatial scales.

15.1 Introduction

Species distribution models (SDMs) are being applied to address some of the most important issues in ecology and conservation, including niche characterization and conservatism, predictions of species response to climate and land use change, and characterization of species invasive potential, disease dynamics, and conservation planning (Box 15.1; Guisan and Thuiller 2005; Elith and Leathwick 2009). The application of species distribution modeling has increased markedly over the past 20 years, and is a current "hot topic" in ecological and conservation research (Lobo et al. 2010). Although this rapid growth has led to a degree of methodological and conceptual uncertainty in the application and interpretation of SDMs (Peterson et al. 2011; Warren 2012), the use of these techniques shows no signs of abating, and they remain a powerful, and sometimes the only available tool, to address a variety of theoretical and applied questions in ecology. For the purposes of this chapter, we define SDMs as large-scale correlative models of the environmental tolerances of species based on presence-only, presence-background, or presence-absence data (Franklin 2009; Peterson et al. 2011). By large-scale, we are referring to analyses conducted at large spatial extents that span a good portion of, or the entirety of, a species' range, and that typically consist of species or environmental data collected at a relatively coarse grain or resolution. As such, SDMs ideally represent some of the limiting environmental factors that are important in determining range-wide distribution patterns. SDMs provide parameter estimations used typically to map probability of occurrence,

Population Ecology in Practice, First Edition. Edited by Dennis L. Murray and Brett K. Sandercock.

Companion website: www.wiley.com/go/MurrayPopulationEcology

Box 15.1 Applications of SDMs to Ecology and Conservation Biology

Species distribution models can be applied to provide insight into some of the most important questions in ecology and conservation. These include, among others, predicting species invasions, assessing impact of climate change on species, discovering new populations or new species, supporting conservation planning efforts, disease mapping, and testing biogeographical and evolutionary hypotheses. We highlight a few of these here.

1) *Climate change*: One of the most common applications of species distribution modeling is the prediction of the impacts of climate change on species. For example, SDMs can indicate future range shifts of species as the climate warms, identifying areas of contraction and expansion (Figure B15.1.1). Although such extrapolation may be a tenuous application of species distribution modeling, forward prediction of SDMs is rapidly becoming a key tool in management and conservation of species. A recent application of this kind of analysis comes from Benito et al. (2014), who used SDMs to predict future range shifts in 176 plant species in the Mediterranean Basin. They found that although extinction and quasi-extinction would occur for a variety of species in the twenty-first century, accounting for dispersal in SDMs decreased the expected number of species experiencing substantial range contractions. Working at an even larger scale, Lawler et al. (2013) applied SDMs to 2903 vertebrates to model shifts in suitable climates and identify likely routes of movement for these species across the Americas. Their work revealed regions of high densities of climate-driven movements of species that also were heavily impacted by people, highlighting the need for protection efforts in those areas.
2) *Discovering new biodiversity*: Because SDMs often predict habitat beyond the known distribution limits of species, they can be used to indicate suitable areas for future surveys of poorly studied species. For example, Guisan et al. (2006) used SDMs to indicate potentially suitable areas for rare and endangered plants in Switzerland, improving their sampling efficiency when searching for new populations. Raxworthy et al. (2003) developed SDMs for a variety of chameleon species in Madagascar, overlapped the models, and then used areas that were highly suitable for multiple species as targets of likely occurrence of undiscovered species (this approach resulted in the discovery of 7 new chameleon species).
3) *Conservation planning*: SDMs have a large potential to inform conservation planning and decision-making, but to date, examples of the use of SDMs for addressing such issues remain scarce in the peer-review literature (Guisan et al. 2013). However, some examples

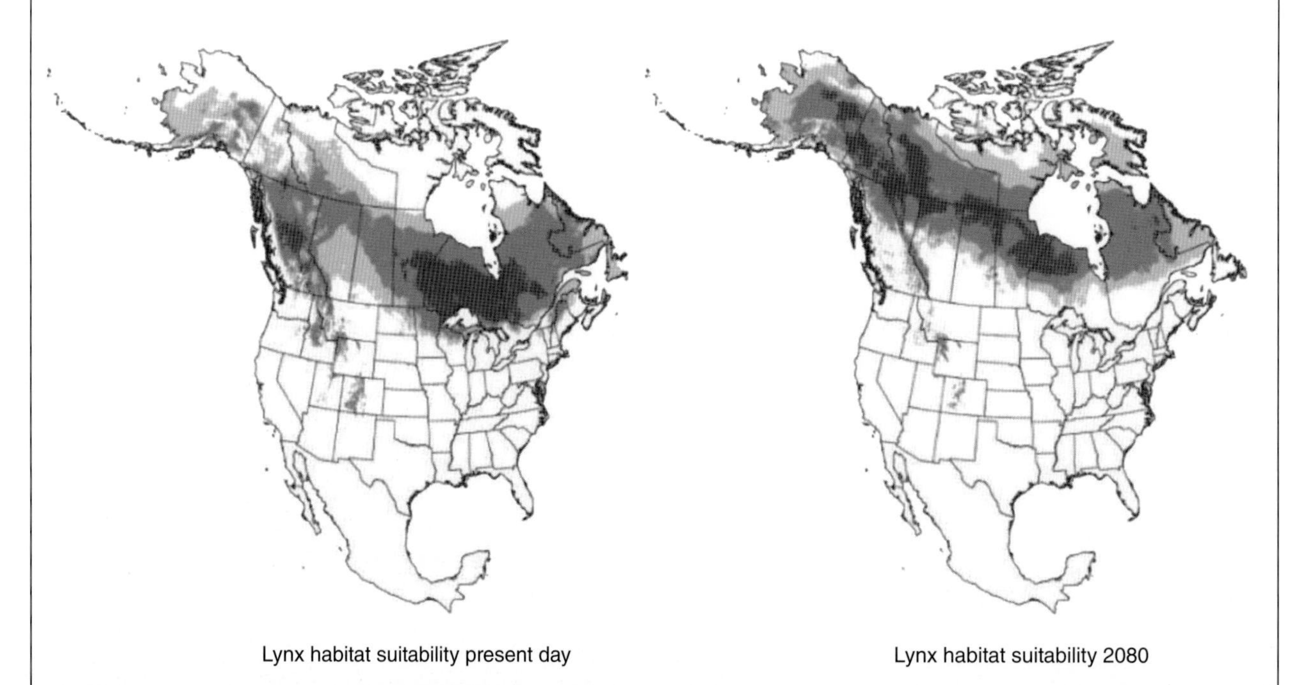

Figure B15.1.1 Predicted species distribution of Canada lynx based on habitat suitability in present day versus possible changes in 2080.

indicate the high potential of SDMs in this area. Araújo et al. (2004) developed SDMs for 1200 plant species across Europe in current and future time steps, and used those data to compare the performance of different reserve selection algorithms in conserving biodiversity. They conclude that new approaches are needed for reserve selection in the face of climate change. SDMs are also being applied to aid in planning reintroduction sites for species. Given that one of the most important criteria for reintroduction success is suitable habitat, SDMs can be used to indicate areas that most likely meet this criteria. Kuemmerle et al. (2010) used SDMs to predict areas of suitable habitat for European bison (*Bison bonasus*) and were able to confirm the suitability of a proposed Romanian reintroduction site. In an innovative application, we have used SDMs to predict the past, current, and future suitability of protected areas for Passenger Pigeons (*E. migratorius*) as part of assessment of the feasibility of this species as a de-extinction candidate (Peers et al. 2016). Our results reveals that many of the areas that were historically suitable for pigeons are no longer suitable, but areas outside the current range may become suitable in the future (Figure B15.1.2).

4) *Evolutionary history*: SDMs are increasingly being combined with genetic and phylogeographic data to provide insight into past evolutionary history of species. For example, Hugall et al. (2002), integrated molecular phylogeography with predictions from SDMs that were projected backward in time using paleoclimatic data to examine the evolutionary history of an endemic snail. The authors conclude that combining these two approaches provides a powerful means to assess location of past refugia and how species responded to them. Schorr et al. (2013) combined SDMs with phylogeographic approaches to study past refugia of alpine *Primula* species during the last glacial maximum. They conclude that the two sources of data were complementary in providing a fuller picture of past ranges: the phylogeographic data permitted the estimation of likely source areas for recolonization after range contraction, and the SDMs predicted larger and more divergent refugia than could have been accomplished with phylogeographic data alone.
5) *Niche conservatism*: SDMs are also being applied to address questions related to niche conservatism, or the degree to which niches are conserved in space and time. This question has been addressed largely in literature on species invasions, where niche conservatisms of an invader between its native and non-native range is key to understanding their invasion potential and the successful application of SDMs to predict invasion spread. Although niche shifts have been documented in some cases, debate remains regarding the ubiquity of niche shift vs. conservatism. For example, Strubbe et al. (2015) document conservatism between native and invasive range niches for 29 vertebrate species in Europe and North America, whereas Early and Sax (2014) document a large number of niche shifts in 51 plant species between their native (European) and introduced ranges (USA).

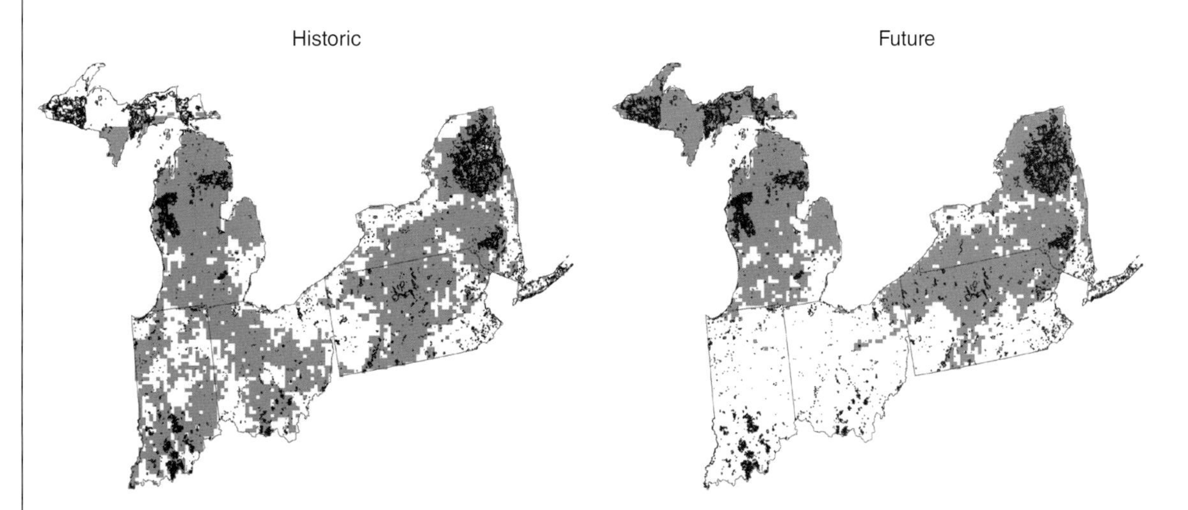

Figure B15.1.2 Species distribution models (SDMs) for the extinct Passenger Pigeon. We examined how a changing climate would change suitable habitats (gray) to unsuitable habitats (white) between historic versus future periods, and in relation to distributions of protected areas in the northeast (black).

or habitat suitability, across geographic space. These models may be more appropriately called *ecological niche models* (ENM), because they implicitly or explicitly estimate some components of a species' niche (Warren 2012). We refer to them as SDMs given the common usage of this term in the literature, and because both terms are used interchangeably to address similar or identical questions (Peterson and Soberon 2012). Previous chapters have considered occupancy models (Chapter 3) and resource selection functions (Chapter 14). These two techniques sometimes fall under the general rubric of species distribution modeling, but we did not include them here because they are usually conducted at relatively smaller spatial scales, in both extent and resolution.

15.1.1 Relationship of Distribution to Other Population Parameters

Given the topic of this book, a useful starting point is to examine the relationship between the distribution of a species and other population parameters such as abundance, density, and population variability. A positive relationship between distribution and abundance of species is well documented as one of the most common patterns in macroecology (Brown 1984; Gaston and Blackburn 2000), and is usually referred to as an *occupancy-abundance* or *distribution-abundance* relationship. The relationship refers to the fact that widely distributed species also tend to be locally abundant, and has been documented in a variety of forms for a diversity of taxa (Borregaard and Rahbek 2010). Occupancy-abundance relationships are often nonlinear (Gaston et al. 2000), and depend critically on a number of factors, including how distribution and abundance are measured (Blackburn et al. 2006). For example, if distribution is measured as the extent of occurrence or range size, the positive distribution-abundance relationship is much less strong than when distribution is measured as the proportion of sites occupied (Blackburn et al. 2006). Interspecific patterns in the distribution-abundance relationships are well supported: species that occupy a high proportion of sites across their range tend to be more abundant than those that have low rates of occupancy. However, intraspecific patterns also exist, where areas of high levels of occupancy within a species' range tend to be the same areas with high mean abundance (Borregaard and Rahbek 2010). Positive intraspecific distribution-abundance patterns have important conservation implications: as a species range contracts due to human pressures, we expect the average density within sites to also decline, and so the proportional loss of individuals is greater than expected based on range contraction alone (Gaston et al. 2000). The exact mechanisms behind the distribution-abundance relationship remain poorly understood, and likely involve a variety of processes acting at different scales. Processes may include niche-related mechanisms, such that species with greater niche breadth can occupy more sites at higher average abundance than those with lesser niche breadth, or demographic mechanisms, such that colonization and persistence within regions is positively linked to average abundance due to metapopulation processes, among others (Holt et al. 2002; Borregaard and Rahbek 2010). Regardless of the mechanism, the link between distribution and abundance suggests that patterns in distribution among or within a species can reveal important information about other components of its population ecology.

In a related question, intraspecific links between distribution and abundance have also been studied to determine if there are consistent patterns in how abundance varies across the geographic distributional range of a species. The *abundant-center hypothesis* states that abundance declines as one moves away from the center of a species' range and approaches the edge (Hengeveld and Haeck 1982). The hypothesis is a widely cited concept in ecological literature, but empirical support remains equivocal (Sagarin and Gaines 2002). A recent analysis by Martínez-Meyer et al. (2013) suggests that abundance declines, not with distance from the center of the distributional range, but rather with distance from the centroid of the species' ecological niche, which has been termed the *distance to the niche centroid* hypothesis (DNC). Under this formulation, distance in ecological space is the most important driver of abundance patterns across a species range, not distance in geographical space. Given that distance from the geographic center of a species' range will sometimes, but not always, correspond to distance from the centroid of the niche, the DNC hypothesis provides an explanation for the disparate results of empirical tests of the abundant-center hypothesis to date. The idea of abundance declining away from the niche centroid has received some empirical support (Yañez-Arenas et al. 2012; Lira-Noriega and Manthey 2013; Martínez-Meyer et al. 2013), but remains little explored. Recent simulation work suggests that DNC estimates account for 52–99% of geographic variability in density, with the utility of DNC being heavily influenced by the bias and sample size of presence data used to estimate niche parameters (Yañez-Arenas et al. 2014). The DNC hypothesis provides a rationale for linking SDMs and population processes, given that SDMs can potentially be used to characterize some components of the niche of a species (Section 15.1.2). Along those lines, the relationship between SDM model output – probability of occupancy or habitat suitability maps – and population density or abundance, has also been explored recently. To date, strong, weak, and no relationships have been observed between SDM model output and other population metrics (Thuiller et al. 2010; Oliver et al. 2012; Yañez-Arenas et al. 2012; Gutiérrez et al. 2013; Van Couwenberghe et al. 2013). The relationship between SDM output and population processes may be complex, and can vary with the type of population metric (such as stability vs. density), species identity, and adequacy of the SDM. For example, Törres et al. (2012) developed SDMs for jaguars and correlated SDM output (habitat suitability) with independent density estimates for the species, and found that SDMs were better at predicting locations of high density vs. low density. Taken together, the above evidence is suggestive that range-wide estimates of occupancy, suitability, or niche characteristics from SDMs may go beyond merely providing information on distribution, but rather inform about other population processes. The importance of the possible relationship between distribution and other population processes cannot be overstated. If distribution or habitat suitability relates strongly to other population metrics, our ability to investigate macroecological questions and design conservation strategies will be greatly enhanced, given the relative ease with which distribution data can be collected compared to population parameters such as abundance, density, or demography. Thus, the relationship between distribution and population processes remains an important area of future research in population ecology.

15.1.2 Species Distribution Models and the Niche Concept

The basic process for developing most SDMs involves collecting data on species presence (or presence-absence) across a geographic area, developing a statistical model that relates those presences to a suite of environmental variables by comparing environmental characteristics at presence sites with those from the available background (or absence) sites, and projecting the resultant model across the study area to develop maps of habitat suitability or probability of occurrence (Figure 15.1). SDMs thus integrate across both *geographic space* (*G*-space) and *environmental space* (*E*-space). The initial data processing is carried out in *G*-space, model fitting occurs in *E*-space, and model outputs are projected and visualized in *G*-space (Peterson and Soberon 2012). However, note that SDMs can also be developed based solely on spatial relationships between points of species occurrence, without reference to environmental variables. Although these spatial interpolation methods can produce accurate models of occupied distributional area (Bahn and McGil 2007), they are not widely implemented and suffer some restrictions due to their ignorance of environmental constraints on distribution patterns, such as an inability to project the models to other time periods or geographic areas (Warren 2012). Most types of SDMs relate species occurrence to environmental variables in the model fitting process, and identify the environmental domain permitting existence of the species (Figure 15.2), leading to a natural connection between SDMs and niche concepts (Peterson et al. 2011; Higgins et al. 2012). Indeed, the development and wide implementation of SDMs has led to a resurgence of "niche ecology."

However, the idea that SDMs can be used to estimate the characteristics of a species' niche is fraught with difficulties. To begin with, niches may be defined in various ways that have stronger and weaker connections with SDMs. One of the most obvious distinctions is between Grinnellian and Eltonian niche concepts, where a *Grinnellian niche* refers to a coarse-scale definition of a niche that typically considers only *abiotic* limiting factors, whereas an *Eltonian niche* refer to local-scale definitions of niches that can include *biotic* interactions between species (Peterson et al. 2011). Because Grinnellian niches operate at larger scales, they are thought to provide a more natural connection with SDMs (Peterson et al. 2011), and most distribution modeling exercises rely on abiotic factors such as climate to explain occurrence patterns. However, whether or not SDMs should incorporate more Eltonian concepts, such as biotic interactions, as part of the modeling process, remains heavily debated (Araújo and Rozenfeld 2014; de Araújo et al., 2014). Regardless, further complexities confront the application of SDMs to niche ecology. Species may not occupy all suitable habitats, defined as areas within their environmental tolerance, due to restrictions on dispersal or movement. Therefore, presence records of the species will not occur in those areas, and SDMs developed from the full set of records may underestimate environmental tolerance. The same issue can arise due to biotic interactions: a species may be absent from environmentally suitable areas due to lack of a key resource or presence of a competitor. For example, Canada lynx (*Lynx canadensis*) may not occur in areas without their key prey species, snowshoe hares (*Lepus americanus*), even if the unoccupied areas are climatically suitable. On the other hand, some presence records from species may occur in sink habitats, which are not in environmentally suitable habitats, overestimating environmental tolerances. Complications with presence records, and how they relate to the output of SDMs, have been conceptualized through a simple heuristic called a *Biotic Abiotic Movement diagram* (BAM; Box 15.2), depicting the interactions among abiotic, biotic, and movement factors (Peterson et al. 2011). Various distributional areas and their corresponding environmental niches can be defined based on this diagram. The bottom line of this heuristic is that several different components of a niche can be estimated from

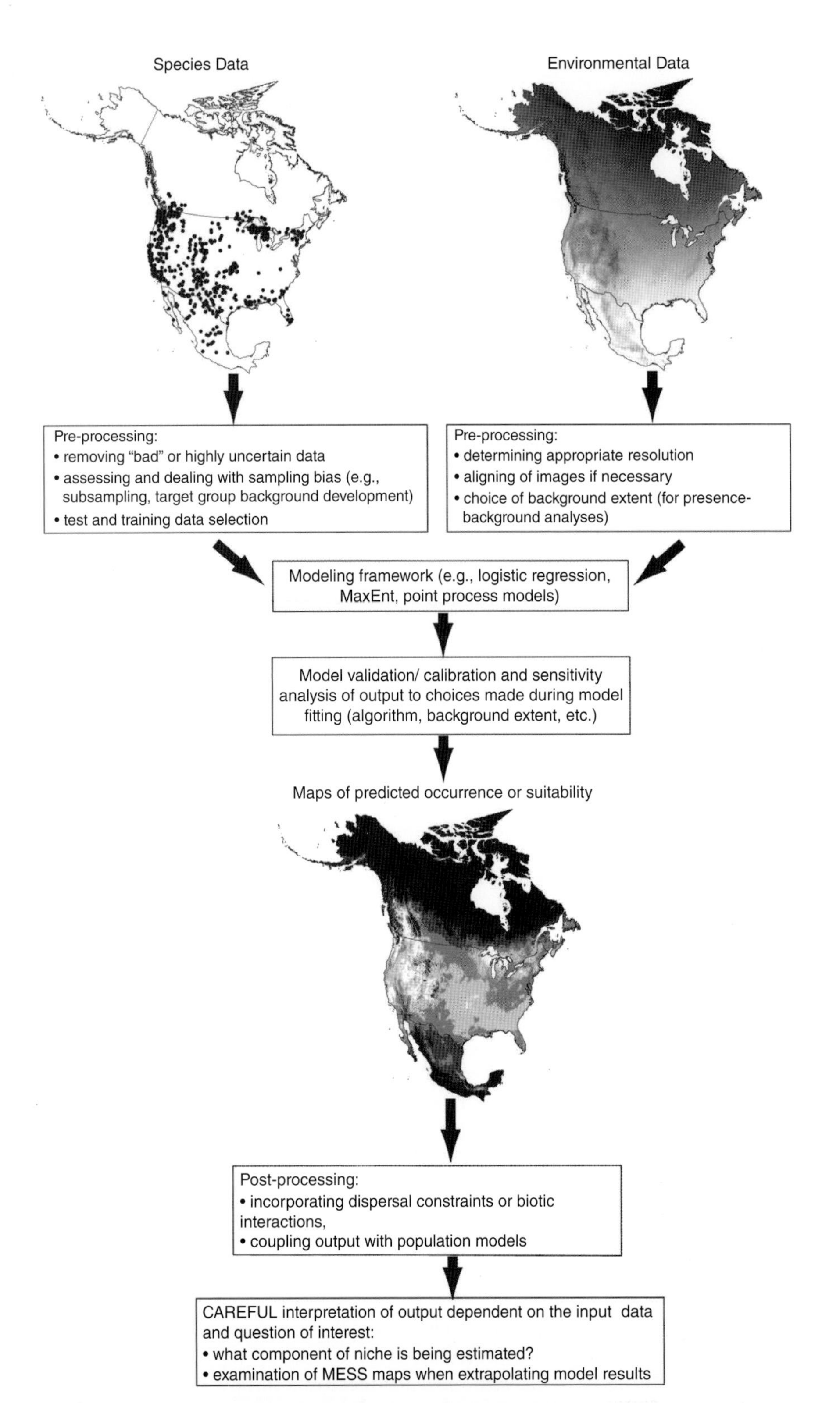

Figure 15.1 Diagram of the essential steps in species distribution modeling. Species and environmental data may need to be pre-processed in various ways prior to analysis. Once the input data are ready, model fitting can proceed, with the algorithm chosen dependent on the data types (e.g. presence-only vs. presence-absence), as well as the nature of the study question. After model fitting, models should be evaluated for discrimination and calibration, and sensitivity to modeling choices should also be assessed as a means to explore the impact of methodology on model results (e.g. choice of background size). If satisfied with modeling outcomes, results can be used to create maps predicting occurrence/suitability, or address questions of interest. Some post-processing may be necessary, particularly if trying to incorporate added complexity such as dispersal constraints. Note that adding complexity can also occur during the model-fitting stage (for example, including biotic interactions as covariates in the models). Post-processing could include a number of different tasks, including but not limited to: constraining distributions to biotically suitable or accessible areas, and ensemble forecasting of species distribution (i.e. combining output from multiple species distribution models (SDMs); Araújo and New 2007). Finally, model results must be interpreted cautiously, taking into account the assumptions and limitations of SDMs, particularly as it relates to interpretation of model output and extrapolation to new regions/time periods.

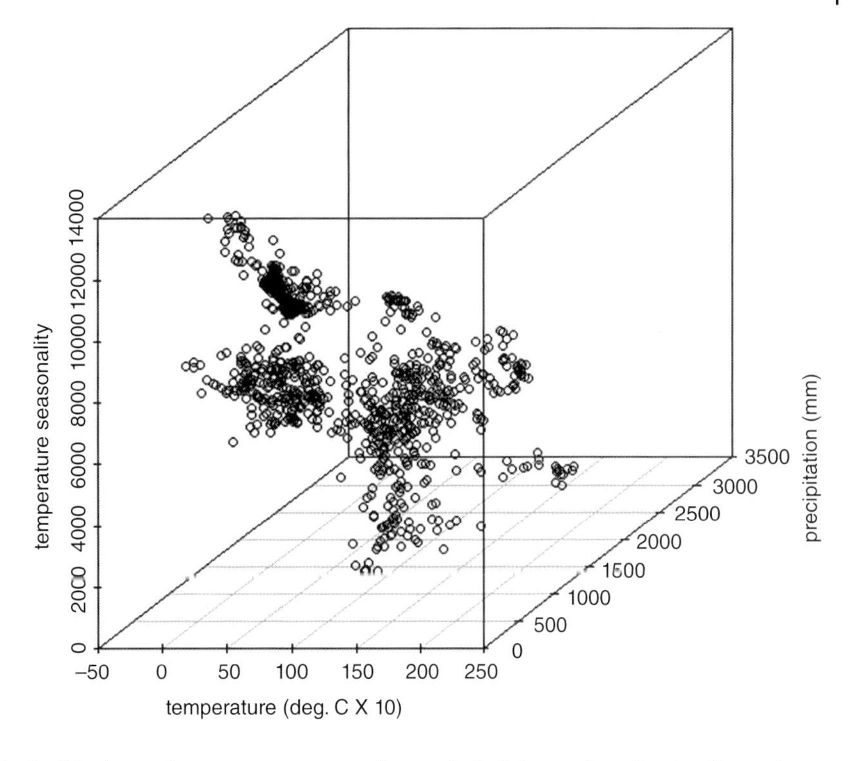

Figure 15.2 Scatterplot of environmental characteristics of bobcat presence records. Points indicate location of presence records along three environmental axes. This multidimensional volume indicates, to some degree, the environmental tolerance of the species. For example, bobcats (*Lynx rufus*) do not appear to occupy areas where the annual temperature is below 0 °C.

distribution data, dependent on how abiotic, biotic, and movement factors have shaped the distribution of a particular species (see Box 15.2 for further explanation). Recognizing that not all SDMs will estimate the same components of a species' niche, and understanding where limitations in our interpretation of SDM model output may occur, is important to keep in mind when using SDMs to characterize species niche space (Warren 2012).

Box 15.2 Species Distribution Models Versus the Niche – The BAM Diagram

The BAM diagram is a useful heuristic for thinking about how SDMs relate to the niche of a species (Figure B15.2.1). The three domains of the Venn diagram include: the biotically suitable areas for the species (*B*), the abiotically suitable areas for the species (*A*), and the areas accessible to the species by movements (*M*). Given these three intersecting domains, several distinct distributional areas can be defined, including the occupied distributional area (G_0, those areas that are simultaneously abiotically and biotically suitable, and accessible to the species), the invadable distributional area (G_I, those areas that are abiotically and biotically suitable but are not reachable by the species), and the abiotically suitable area (G_A, those areas within the geographic region that are abiotically suitable). The geographic spaces (*G*) can be translated to environmental space (*E*) with the corresponding niches: E_0 = occupied niche (the set of environmental conditions currently occupied by the species), E_I = invadable niche (regions in environmental space that the species can tolerate, but which are not represented within the set of conditions that are reachable by the species), and E_A = existing fundamental niche (the existing subset of E that is within the tolerance limits of the species).

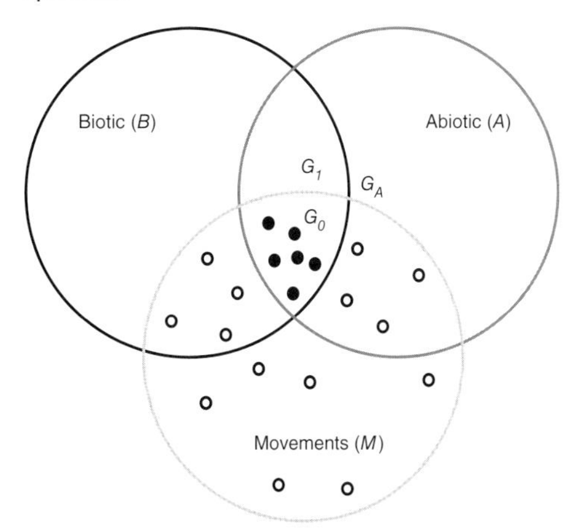

Figure B15.2.1 The *Biotic Abiotic Movement* (BAM) diagram for visualizing the geographic and environmental space occupied by species with domains for biotic (*B*), abiotic (*A*), and movements (*M*).

Which components of *G*- and *E*-space do SDMs actually estimate? The answer depends in part on the form of the BAM diagram for a given species and study area, the methodology used, and the nature of the presence records collected (Peterson et al. 2011). For instance, it is often assumed that biotic interactions influence species distribution and abundance at finer scales than abiotic characteristics such as climate. If this is true, then large-scale distribution patterns which are the focus of many SDMs will not be strongly influenced by biotic interactions. This would correspond to a BAM diagram where the '*A*' circle was contained completely within the '*B*' circle, and hence distribution models would likely estimate a niche that was closer to E_A than one where biotic interactions restricted distribution patterns (reviewed by Peterson et al. 2011). Even when studies are conducted across the range of a species, and using an adequate set of presence records, limited dispersal or biotic interactions usually prevent a species from occupying certain areas, thus restricting the estimate of the species niche. As a general rule, SDMs probably estimate some component of the niche of a species that falls along a continuum from E_O to E_A, and hence predict suitable habitat outside of the currently occupied area and somewhat into G_I (Peterson et al. 2011). However, the choice of method and type of data will determine the usefulness of the SDM output to model distinct components of geographic distribution and niches (Jiménez-Valverde et al. 2008), and thus care needs to be taken in interpretation if SDMs are to be used as a basis for conservation or ecological research (Hortal et al. 2012; Kamino et al. 2012).

15.2 Building a Species Distribution Model

15.2.1 Species Data

SDMs require information on species distribution patterns at large spatial extents, represented as grid cells or as georeferenced point locations on a map. Species distribution data are usually *presence-only* where data exist for the presence of the species, but not its absence, or alternatively, *presence-absence* data where data include location of both species presence and absence. Distinctions between the two categories have implications for the fitting and interpretation of SDMs, which will be discussed later in the chapter. Various sources of pre-existing information on species distribution patterns exist (Table 15.1), and prior data often serve as the basis for the development of SDMs. Data sources can include museums or herbaria, biodiversity inventories, atlases, large-scale field surveys, or citizen-science initiatives. The vast majority of pre-existing data on species distributions at large scales only provides information on species presence but not species absence. However, rigorously designed surveys do exist that provide presence-absence data at large spatial extents, such as the Breeding Bird Survey in North America.

Museum and herbaria records have become increasingly accessible as collection information is made available through data portals on the internet (Graham et al. 2004). However, many museum collections remain to be digitized, and many existing records of species presence are still widely scattered and difficult to access (Newbold 2010; Beck et al. 2013). Furthermore, locations with large positional uncertainties are fairly common with biodiversity inventory data from online repositories, as are gross errors in location. For example, a location for a specimen record may be recorded as the centroid of a state or province, and the actual location may not be known more precisely. *Location errors* can create a substantial amount of work that must be done to "clean" a dataset prior to use in distribution modeling. Positional uncertainty is an aspect of species presence data that is commonly ignored or unreported in SDM literature. However, positional uncertainty of location records has the possibility of substantially reducing the accuracy of SDMs, particularly in areas where environments are spatially heterogeneous (Naimi et al. 2014). Removing highly uncertain locations within heterogeneous environments is one strategy to reduce problems of positional uncertainty (Naimi et al. 2014). Another approach is to ensure that positional accuracy is not greater than the resolution of the environmental layers used in modeling.

Ideally, information from these pre-existing data sources should provide a fairly complete picture of the distribution patterns of a variety of taxa around the globe. Unfortunately, many of the databases of species distribution contain information that is incomplete or spatially biased. Comparably few records exist for invertebrates or from less-developed countries (Kamino et al. 2012; Beck et al. 2014), whereas at-risk species or developed regions may be disproportionately represented in inventory data (Schmidt-Lebuhn et al. 2013). Taxonomic biases in coverage may be particularly vexing when attempting to model entire guilds or communities of species, as data variability will constrain the number of species that can be modeled. Spatial bias is also problematic

Table 15.1 Examples of common sources of species distributional data (note that this is not an exhaustive list, and most of these sources do not have data on species absence).

Full Name	Taxonomic Group	Geographic Coverage	Web Address
Global Biodiversity Information Facility	Broad	Global	www.data.gbif.org/welcome.htm
The World Information Network on Biodiversity	Broad	146 Countries	www.conabio.gob.mx/remib/cgi-bin/clave_remib.cgi?lengua=EN
European Natural History Specimen Information Network	Broad	Europe	cordis.europa.eu/projects/rcn/52229_en.html
Canadian Biodiversity Information Facility	Broad	Canada	www.cbif.gc.ca/eng/home/?id=1370403266262
Distributed Information Network For Biological Collections	Broad	Brazil	www.splink.org.br/index?lang=en
Instituto Nacional de Biodiversidad	Broad	Costa Rica	atta.inbio.ac.cr
FishNet	Fish	North America	fishnet2.net
HerpNET	Reptiles/ Amphibians	Global	herpnet2.org
Missouri Botanical Garden (Tropicos)	Plants	Global	www.tropicos.org
Ornithology Information System	Birds	Global	ornis2.ornisnet.org
Biodiversity Information Serving Our Nation	Broad	United States	bison.usgs.gov/#home
INaturalist	Broad	Mainly United States	www.inaturalist.org/observations
North American Breeding Bird Survey	Birds	North America	www.pwrc.usgs.gov/bbs/RawData/Choose-Method.cfm
Christmas Bird Count	Birds	North America	birds.audubon.org/christmas-bird-count
eBird	Birds	Global	ebird.org/ebird/map
Biological Records Center/ National Biodiversity Network	Broad	United Kingdom	nbn.org.uk/

when considering any given species. *Spatially biased data* can arise from a number of sources: certain areas may be more likely to be sampled due to accessibility, certain areas may be targets for extensive surveys, or certain areas may be more heavily sampled if data sources are mixed (e.g. mixing museum records that have few locations in any given region with radio telemetry data that have many locations in a few regions (Kramer-Schadt et al. 2013). Problems with spatial bias in distribution modeling, and potential corrections, are discussed in Section 15.3.2.

Although biodiversity inventories contain a wealth of information that is easily accessible, it is prudent to scrutinize these records carefully, and understand potential biases in the data (Beck et al. 2014). As part of our own research utilizing museum records for SDM development for well-known mammal species, we have found it necessary to individually contact museums or utilize other sources of information to increase the sample size and geographic spread of locations for a variety of species. For example, to obtain sufficiently large datasets of presence records for our work on distribution modeling of North American mesocarnivores (Peers et al. 2013; Thornton and Murray 2014), we combined information from freely accessible biodiversity databases (e.g. Global Biodiversity Information Facility), data from numerous museums that we contacted individually, and data on harvest records obtained from government agencies in over 20 states and provinces. Increasing recognition of problems with biased or incomplete data, and other limitations of biodiversity inventories, will hopefully lead to various methods to reduce these problems, either through increased digitization and error checking of inventory records, use of additional sources (such as private collections), development of more systematic methods such as targeted surveys, or increased use of large-scale citizen science efforts (Cases-Marce et al. 2012; Schmidt-Lebuhn et al. 2013; Duputié et al. 2014).

15.2.2 Environmental Data

Selecting which *environmental variables* to include in SDMs depends in large part on the availability of data sources and scale of the analysis (Peterson et al. 2011). For most large-scale SDMs, researchers choose a variety of abiotic variables such as temperature or precipitation, and therefore work within the framework of estimating Grinnellian niches. However, even within this framework, choices must be made about the types of variables to include. Austin (2007) provides a breakdown of environmental variables into proximal versus distal variables based on their degree of causality. Species respond directly to *proximal variables* whereas they respond in an indirect manner to *distal variables* via a correlation with other proximate variables. For example, a drought-intolerant species may respond directly to an environmental variable representing the number of days without rain as a proximate variable, whereas it would respond indirectly to mean annual precipitation as a distal variable, via the negative relationship between annual precipitation and days without rain. Most readily available environmental layers at large spatial extents are likely distal in nature, which requires care in the selection of environmental variables and in the interpretation of their influence. For researchers working at smaller spatial scales, other types of environmental variables representing biotic interactions with predators or competitors, resources, and microclimate can also be included. Disagreement persists regarding the importance of these variables, and how best to incorporate their influences, at the larger extents and resolutions that characterize the correlative SDMs discussed in this chapter.

Various sources exist for environmental data commonly employed in large-scale distribution modeling, and more sources are regularly appearing as remote-sensing databases become more accessible and additional sensors are employed (Franklin 2009). Commonly used sources include climatic, topographical, and land cover layers (Table 15.2). Franklin (2009) gives an excellent overview of the types of environmental data used in SDMs. As with any analysis utilizing large-scale environmental coverages, care must be taken that information from spatial layers is accurate and of good quality. Ensuring quality may not be difficult when dealing with "preprocessed" environmental coverages such as WorldClim data (Hijmans et al. 2005), but may involve substantial effort if more specialized coverages must be obtained and analyzed.

Often one of the most difficult choices in developing an SDM is deciding which environmental layers to use as the predictor variables. Variable selection should ideally be based on a priori knowledge of species ecology (Chapter 2), and include variables that most closely reflect environmental gradients important to the species as either proximate or distal factors. In practice, key factors are usually not known, and a wide selection of variables is included. By including a large number of variables in an SDM, another problem is induced: *multicollinearity* among predictors. The problem of correlated predictor variables in a regression setting is discussed elsewhere (Chapter 2). Here, we emphasize that correlated predictors make interpretations regarding the most important environmental drivers of species distribution patterns problematic, regardless of the procedure used for model fitting. Correlations also create problems when extrapolating SDMs to different time periods or locations, because correlations among predictors may change in scale and direction. Therefore, a judicious selection of predictor variables, with some pre-processing to remove highly correlated environmental variables, is usually good practice (Elith et al. 2011).

One final consideration is the *temporal matching* of environmental and species data. The timing of the collection of the species data should reflect the timing of the collection of the environmental data, to a reasonable degree (Peterson et al. 2011). For example, if one wanted to develop distribution models for extinct Passenger Pigeons (*Ectopistes migratorius*), using presence records collected from 1850 through 1910, it would be prudent to use historic climate averages from a similar time range in an attempt to match presence records with appropriate environmental data. There are no hard and fast rules for what corresponds to an adequate degree of match between presence records and environmental data, and consideration of how fast environmental layers are likely to change over time is warranted. Land cover layers may change quite dramatically over short time periods compared to climatic layers, and therefore require more precise matching to species data for unbiased inference.

15.2.3 Model Fitting

Once the species data and environmental data have been obtained and preprocessed, model fitting can proceed (Figure 15.1). There are three main types of SDMs, distinguished by how they deal with locality data for species: *presence-only*, *presence-background*, and *presence-absence* (Figure 15.3). A variety of modeling algorithms can be used to fit these data, but an overview of all these different analytical methods is beyond the scope of this

Table 15.2 Sample of various sources of large-scale environmental data commonly used in distribution modeling.

Full Name	Geographic Coverage	Time Frame /Resolution	Web Address
Climate			
WorldClim	Global	1950–2000/ 30 arc-seconds	www.worldclim.org
CliMond	Global	1950–2000/ 10 arc-minutes	www.climond.org
PRISM	United States	1971–2010/ 30 arc-seconds	www.prism.oregonstate.edu/
North American Regional Reanalysis	North America	1979–Present/ 32 km	nomads.ncdc.noaa.gov/data.php?name=access#narr_datasets
Canadian Center for Climate Modeling and Analysis	Global	2020–2080/ 30 arc-seconds	www.ccafs-climate.org/data
NASA Earth Exchange Downscaled Project	United States	1950–2100/ 800 m	theclimatedatafactory.com/?gclid=EAIaIQobChMI1bGSjuHp4AIVgR-tBh3q4gIQEAAYASAAEgIskfD_BwE
Land Cover			
GlobCover	Global	2009/ 300 m	due.esrin.esa.int/page_globcover.php
National Gap Analysis Program (GAP)	United States	1999–2001/ 30 m	gapanalysis.usgs.gov/gaplandcover/data/download
U. S. Environmental Protection Agency	North America	1987 (continuously updated)/ shapefile	www.epa.gov/wed/pages/ecoregions/na_eco.htm
Global Land Cover Facility	Global	Broad	www.landcover.org/data
GeoBase	Canada	1996–2005/ 30 m	www.geobase.ca/geobase/en/find.do?produit=csc2000v&language=en
Soils/Topography			
Harmonized World Soil Database (HWSD)	Global	2008/ 30 arc-seconds	webarchive.iiasa.ac.at/Research/LUC/External-World-soil-database/HTML
Shuttle Radar Topography Mission	Global	2000/ 30 arc-seconds	www.worldclim.org/current

chapter. We will introduce data requirements, analytical processes, and pros and cons of a subset of these methods. Elith et al. (2006), Elith and Leathwick (2009), and Franklin (2009) provide a more comprehensive overview of many different algorithms for fitting species distribution data.

1) *Presence-only models* make use of data on species presence alone. Therefore, these SDMs characterize the locations of known presence, without reference to locations of absence or what locations might be available to the species in the study area. Environmental envelope and environmental distance analyses are commonly used with presence-only data. For example, envelope methods such as BIOCLIM typically calculate the range (or 95% CI) of the values of environmental variables at presence locations of a species, and then compare all points on the landscape to see if those points fall within this multidimensional envelope to produce a binary map of suitable/nonsuitable habitat (Beaumont et al. 2005). In contrast, environmental distance analyses (e.g. DOMAIN, Mahalanobis distance) classify locations on a landscape in terms of their suitability based on their distance in environmental space to known presence records. Output from these models shows how similar each area is to environments of known presence and is used to determine how favorable the habitat is for the species. Generally, presence-only SDMs have less discriminative capacity, and lower performance, than presence-background or presence-absence models (Elith et al. 2006), and have been

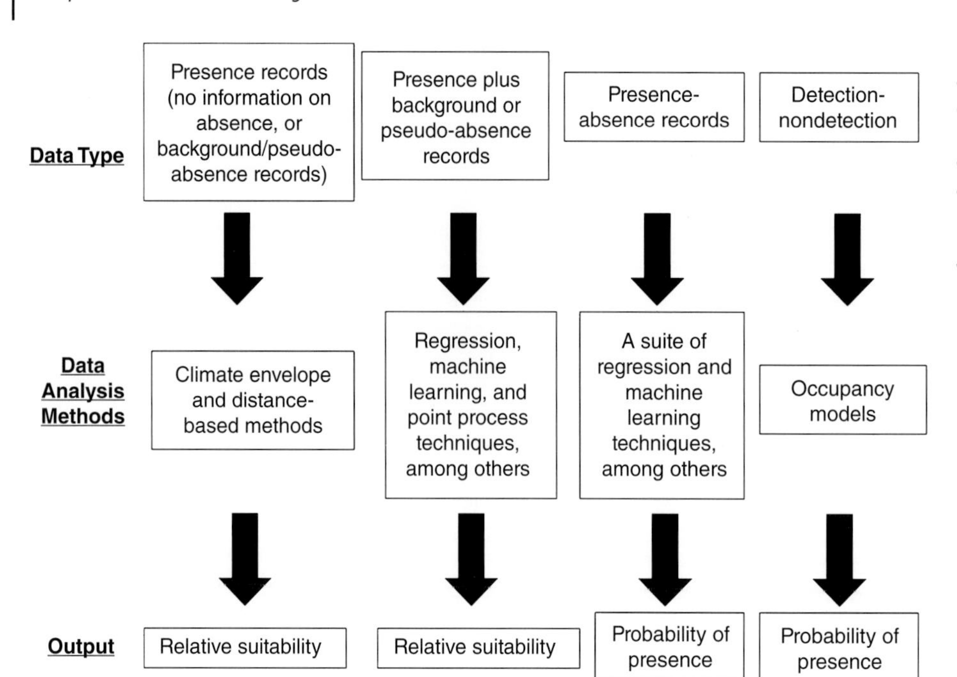

Figure 15.3 Flowchart for determining the methods and output of species distribution model (SDMs) that result from different data types. For presence-absence models, the model output may be probability of presence or relative suitability depending on whether or not detectability was near 100%.

supplanted by these improved methods. However, presence-only methods can be valuable in situations where available data are limited, or when the underlying assumptions of more complex methods cannot be met. For example, Falcucci et al. (2013) used presence-only data and Mahalanobis distance analyses to model the potential distribution of wolves (*Canis lupus*) recolonizing the European Alps. The authors argued that more complex presence-absence or presence-background methods were not advisable for their study due to a lack of absence data and the inability to define a background area in a rapidly expanding population of wolves.

2) *Presence-background models* make use of data on species presence, and data related to the background environment in which the species is located. These SDMs compare the environments of locations of species presence versus the "background" of the available environment in the study area, to try and determine which environments are favored by the species. Presence-background techniques are equivalent to use-availability designs in resource selection studies, but the two lines of investigation have been developed independently (McDonald 2013; Warton and Aarts 2013). Here, we discuss techniques commonly applied in the SDM literature, and we refer the reader to Chapter 14 for a discussion of use-availability methods.

Generally in presence-background SDMs, a large number of randomly selected background points are selected for comparison with the presence records. The number of background points selected, and whether or not to select background points in a biased manner, depends on the type of analysis employed (Barbet-Massin et al. 2012). Note that the background points that characterize the *available environment* may include known presence records, or could fall in areas that may or may not hold populations of the species of interest. We combine presence-background SDMs with presence-pseudo-absence SDMs. In the latter case, the background locations are selected to explicitly exclude locations of known presence, but otherwise, the techniques are similar and can be fit using the same statistical models. The types of algorithms that can be used to fit presence-background datasets are various and include: (i) common regression methods (e.g. generalized linear models [GLMs]; generalized additive models [GAMs]; and multivariate adaptive regression splines [MARS]); (ii) machine learning techniques (e.g. boosted regression trees [BRT]; random forests [RF]; artificial neural networks [ANNs]; and maximum entropy models [MaxEnt]); (iii) ecological niche-factor analysis (ENFA); (iv) genetic algorithms for rule prediction (GARP); and (v) spatial point process models.

Regression techniques are well established in the literature and relate the response variables (presence and background or pseudo-absence points) to single or multiple environmental predictors. GLMs are able to model relationships between response and predictor that generally include linear, cubic, or quadratic terms. GAMs and MARS are able to fit more complex relationships between the response variable and the environmental predictors (e.g. complex curves), via smoothing functions (GAMs) or piecewise linear fits (MARS). MaxEnt models are one of the more popular machine learning methods for fitting SDMs. These models essentially minimize the entropy between two probability densities, one for the occurrence records and one for the background records, defined in environmental space (Elith et al. 2011). Maximum entropy models can also fit complex response curves. Spatial point process models have recently been utilized to fit SDMs. These algorithms model the spatial location and number of presence records jointly, as an intensity function (or expected number of presence records per unit area). Covariates can influence this intensity function much like they can in regression models (Renner and Warton 2013).

The output of presence-background models is the relative suitability of a site for occupancy – in other words, a ranking of the sites in terms of their favorability to the species. This fact has been underappreciated in the literature, where the output of presence-background SDMs is often referred to incorrectly as probability of occurrence (Yackulic et al. 2013), and can have major implications for the utility of SDMs to answer a variety of questions (Section 15.2.4).

3) *Presence-absence models* make use of data on species presence and species absence (areas that have been searched for the species, but where the species was not detected). Note that these SDMs are appropriate when absence locations are not "false absences" that result from a failure to detect a species at a site. If imperfect detection is a problem, then occupancy modeling study design and statistical analysis should be performed (Chapter 3). By comparing environmental conditions at species presence records versus environmental conditions of confirmed absence records, these models seek to identify how habitats that are occupied differ from those that are not occupied. Presence-absence models can be fit using many of the same techniques listed above for presence-background techniques (e.g. GLMs, GAMS, MARS, regression trees). Output from presence-absence methods can be interpreted as probability of occurrence for a given location on the landscape – in other words, the proportional likelihood that a site is occupied, given its location and the suite of environmental predictors. For example, a typical model for a GLM with presence-absence data employs the logistic link to model probability of presence as a function of environmental predictors:

$$\ln\left(\frac{\pi(x)}{1-\pi(x)}\right) = \alpha + \beta x, \quad (15.1)$$

where π is probability of presence, x is one or more covariates, α is an intercept and β is the coefficient for the effect of the covariate on the log odds of presence. Rearranging this equation gives the probability of presence as a function of the intercept and covariates:

$$\pi(x) = \frac{\exp(\alpha + \beta x)}{1 + \exp(\alpha + \beta x)}. \quad (15.2)$$

Given that a statistical algorithm has been selected to model presence-only, presence-background, or presence-absence data, methods must be used to compare multiple models and determine the influence of predictor variables on the response. We will not engage in a detailed discussion of these issues, but refer the reader to Chapter 2, which deals with the topic of model selection and predictor influence in detail.

15.2.4 Interpretation of Model Output

The output of statistical models based on presence-absence data can typically be used to predict occurrence probability across geographic space. Thus, output from these models predicts the absolute probability that a species occurs at a particular site or location, dependent on the suite of environmental covariates at that location. In contrast, output of statistical models based on presence-background data typically can only be used to predict relative occurrence probability across geographic space. Output from these models provides information on the suitability of sites for the species that is proportional, but not equivalent to, the probability of occupancy (Hastie and Fithian 2013). To convert output of these models to absolute probability of occupancy, information on the prevalence of the species (total no. of sites occupied in the study area) is required, and that information is not retrievable from presence-only or presence-background data (Phillips and Elith 2013). For example, if we have presence records of a species from 20 of a possible 200 sites in a study area but no data on absence, we do not know if this species is rare and well surveyed (with low prevalence), or common but poorly surveyed (with high prevalence).

Some statistical methods can make use of supplementary information on prevalence, such as counts or presence-absence surveys in other locations, or expert opinion, to estimate occupancy probability from presence-background data (Phillips and Elith 2013; Dorazio 2014). Similarly, output from presence-background algorithms can be post-processed to provide estimates of absolute probability of occupancy, but information must be available to inform a parameter in the model that gives the probability of presence at sites with "typical" conditions for the species. However, this information is lacking for most species (Elith et al. 2011). Several authors have suggested that presence-only data can indeed be used to estimate the *absolute* probability of presence (Lele and Keim 2006; Royle et al. 2012; Rota et al. 2013). However, debate remains regarding the usefulness and applicability of these methods, due to their strict assumptions (Hastie and Fithian 2013; Phillips and Elith 2013; Guillera-Arroita et al. 2015). Therefore, unless supplementary information is available, the modeler must assume that output from presence-background analysis gives a *relative* probability of occupancy, and not an absolute estimate. Similarly, large-scale presence-absence data, when detectability is less than one, also may only be able to estimate relative probability of presence (Guillera-Arroita et al. 2015, Figure 15.3).

Significantly, the inability to estimate absolute probability of presence from presence-only or presence-background data greatly reduces the ability to answer many key conservation and management questions, such as estimates of area of occupancy or changes in occupied area over time (Guillera-Arroita et al. 2015). These limitations demonstrate the value of well-designed surveys that record presence-absence data (or detection-nondetection) for many applications of SDMs. However, relative probability of presence is still informative for a number of important ecological and conservation questions such as examining relative changes in suitability over time, or spatial prioritization exercises.

Indeed, the subject of data types and how it relates to SDM output and usefulness is receiving increasing attention (Dorazio 2014; Guillera-Arroita et al. 2015). Rules of thumb emerging from research on distribution models generally use presence-absence data when available for development of SDMs, as these are more likely to result in data that provide absolute probability of presence, and are more robust to sampling bias and reduce problems associated with selecting a background extent (Section 15.3.3). Unfortunately, the vast majority of large-scale data that are used for SDMs only have information on species presence, greatly limiting the options of the modeler. Recent critiques of presence-only or presence-background models (Yackulic et al. 2013), and our own experience with numerous reviewers, highlight the skepticism with which many researchers approach the results of such analyses. From a practical standpoint, the availability of data on species presence and the urgent need for large-scale modeling of distribution necessitate presence-background models for the near future. Moreover, output from these models has been useful for informing about distribution patterns and range shifts under a variety of circumstances (Peterson 2003; Peterson et al. 2011), and thus has considerable merit. Considering all these issues together, we recommend a careful reading of recent literature on interpretation of different kinds of SDMs, using the most informative data possible for each species, and consideration of the question to be addressed, prior to engaging in modeling.

15.2.5 Model Accuracy

Given that an SDM has been developed, one of the most important procedures is to evaluate *model accuracy* (Figure 15.1). Best practices for species distribution modeling require fitting the model using a "training" set of presence or presence-absence records, and then using a "testing" set of records to validate model accuracy. In any exercise of model evaluation with SDMs, decisions regarding how to select the test data with which to evaluate the model must be considered carefully. The alternatives include *model interpolation* where the same data is used to train and test the mode,l versus *model extrapolation* where separate datasets are used to train and test the model. For example, the data can be split into single training and testing sets, split into *k*-fold partitions of testing and training data, or the test data can be a completely independent dataset in space or time. The latter approach may be best for obtaining realistic understanding of model accuracy, however, ensuring similar time spans for training and testing data is crucial in case suitability has changed through time (Chapter 2).

We focus here on two facets of model accuracy for SDMs: *discrimination capacity* and *calibration* (or reliability). Discrimination refers to the ability of models to correctly predict the status of known presence and absence locations – in other words, the agreement between known observations and model predictions for occurrence. Discrimination is the most commonly used approach to evaluate SDM accuracy. Discrimination is usually assessed via a *confusion matrix* that cross-tabulates the predicted and observed presence-absence values (Table 15.3). Two types of errors are possible: false positive errors of *commission*, where the model predicts presence when in fact the species was absent, and false negative errors of *omission*, where the model predicts absence when in fact the species was present. Using the confusion matrix, a variety of accuracy statistics can be calculated, including sensitivity and specificity, overall accuracy, the Kappa statistic, the true skill statistic, and

Table 15.3 Confusion matrix for presence-absence data.

Prediction from model	Observed response	
	Presence	**Absence**
Presence	TP (true positive)	FP (false positive)
Absence	FN (false negative)	TN (true negative)

the F-measure, among others (Fielding and Bell 1997; Liu et al. 2011). *Sensitivity* is the probability that a particular known presence location is correctly classified, and is given by the formula:

$$TP/(TP + FN), \quad (15.3)$$

where TP is proportion of true positives and FN the proportion of false negatives (Table 15.3). *Specificity* is the probability that an absence location is correctly classified, and is given by the formula:

$$TN/(FP + TN), \quad (15.4)$$

where TN is the proportion of true negatives and FP the proportion of false positives. The true skill statistic (TSS) provides another measure of accuracy and is calculated as sensitivity + specificity − 1. Generally, several different measures of accuracy should be computed to test model performance, as each measure quantifies a different aspect of predictive power (Elith and Graham 2009). However, the choice of statistic employed should also depend on the goal of the analysis: to emphasize reduction of commission errors, omission errors, or both. For example, in using SDMs to guide reserve design selection, it may be prudent to weight misclassifications of absences – commission errors – as more detrimental than misclassifications of presences (Lobo et al. 2008). A weighting procedure would help guard against unrealistically optimistic scenarios, where models overpredict presence of species in reserves, and would be especially recommended in situations where the cost of protecting an unsuitable reserve or area is quite substantial.

All of the above metrics of model discrimination are considered *threshold dependent*. The modeling results, which are usually continuous measures of suitability or probability of occurrence, must be converted to binary measures of predicted presence-absence before the metrics can be calculated. The researcher must determine a cut-off point above which continuous output from an SDM is converted to "presence," and below which the output is converted to "absence." The threshold used to determine predicted presence locations is therefore an important choice for the modeler to make. One objective approach to threshold selection is to choose the threshold value as the value that can maximize predictive accuracy of an independent evaluation dataset (Liu et al. 2005). Other ad hoc approaches are also possible, such as the minimum predicted probability value of a training or test presence location.

Threshold-independent measures of model accuracy also exist (Liu et al. 2011), and one of the most widely used measures is the *Area Under the Receiver Operating Characteristic curve* (AUC). The AUC value of an SDM is interpreted as the probability that the model will rank a randomly chosen presence site higher that a randomly chosen absence site. AUC values range from 0.5 to 1, with values of 0.5 indicating discrimination no better than random, values of greater than 0.7 indicating reasonable discriminative capacity, and values of 1.0 indicating perfect discrimination.

Measures of discriminative capacity work well and are easy to interpret when SDMs have been developed from presence-absence data. When SDMs are developed from presence-background data, discriminative capacity is harder to measure. The problem arises because true absence is not known, and background locations may contain presence or absence of the species. For threshold-dependent measures of discrimination, the omission error can still be calculated from the observed presence data and hence measures such as model sensitivity can be estimated, but the commission error cannot be calculated (Li and Guo 2013). Lack of commission errors limits the types of accuracy assessments that can be employed. Li and Guo (2013) have recently recommended the use of a modified F-measure for use with accuracy assessment of presence-background data that performs similarly to the traditional F-measure used with presence-absence data. More developments in this area are needed as presence-background modeling continues to grow in application.

Threshold-independent measures are also compromised by the use of presence-background data for modeling. Although AUC is commonly employed as an accuracy measure for presence-background data and is integrated into common software packages such as `MaxEnt`, it suffers from several drawbacks that limit its application (Lobo et al. 2008; Peterson et al. 2008; Jiménez-Valverde et al. 2013). AUC values will be highly sensitive to the extent of the background used to create the models (Peterson et al. 2011). For example, larger background areas used to compare with presence locations will often artificially inflate AUC values due to the fact that background locations will be drawn from areas that may not have been accessible, or from which the species is excluded by other factors. The problem is easy to envision: the environment of nearby sites is expected to be more similar than the environment of distant sites, and selecting background locations from areas that are farther away from the presence locations will inflate AUC values (Hijmans 2012). For example, in developing distribution models for Canada lynx, selecting a background size equivalent to all of North America results in AUC values 0.10 higher than selecting a background size just a little larger than the current range of the species. However, the exact same presence dataset is used in both exercises. Similar reasoning holds when looking at AUC values for species with different range sizes: AUC values from SDMs will be artificially higher for species with small range, because test and training presence data will tend to be closer together (Hijmans 2012). Sampling bias in presence data can also inflate AUC values. In this case, if test data for model evaluation are drawn from a dataset that suffers from sampling bias, they will suffer from the same bias and tend to reflect modeling outputs, inflating AUC values. Hijmans (2012) has recently proposed a procedure to account for some of these problems through adjustment of AUC values by use of pairwise distance sampling of presence and background locations, and use of null models.

Calibration (reliability) is another means of examining model accuracy, and refers to the agreement between predicted probabilities of occurrence and the observed proportion of sites occupied (Li and Guo 2013). For example, if a species occupies 25 of 50 sites in a certain type of environment, a well-calibrated model would predict close to 50% probability of presence in that environment. Calibration is much more rarely employed as a means to evaluate SDMs, but is independent of discrimination as a measure of model quality, in that a poorly calibrated model can have excellent discrimination, and vice versa (Phillips and Elith 2010). The discrepancies arise because discrimination measures the ability to separate presence and absence based on predictions of the model, but the absolute values of those predictions are unimportant (Lawson et al. 2014). Calibration is sometimes assessed with binned calibration plots, where model output is partitioned into a number of bins, and then the fraction of true presences is plotted against the average model output value in each bin (Pearce and Ferrier 2000). Once again, presence-background data present added difficulties in estimating calibration of a model, but recently developed presence-only calibration plots (POC plots) can be used (Phillips and Elith 2010).

Evaluation of SDMs remains an active area of research. However, a cautious approach is to examine several different metrics for evaluating a model's accuracy, and note potential drawbacks to any one approach. For presence-absence SDMs, combining calibration and discrimination metrics is recommended for gaining a broader understanding of model performance. For presence-background models, employing some combination of recently developed techniques such as the adjusted F-measure, adjusted AUC, or POC plots, may be recommended as a method to reduce bias and provide better estimates of model accuracy. Sensitivity is also a reliable metric from presence-background methods because it does not rely on pseudo-absence in the calculation. Use of a spatially independent dataset to test predictions cannot be overestimated as an important step to take with presence-absence or presence-background SDMs if the model will be used to extrapolate results (Muscarella et al. 2014; Radosavljevic and Anderson 2014).

15.3 Common Problems when Fitting Species Distribution Models

15.3.1 Overfitting

As with many regression techniques, *overfitting* is a concern with species distribution modeling. Overfitting occurs when model complexity is too high and a large number of predictor variables are used in the same model, or the response curves are too complex, so that the model becomes too tightly constrained to the input data used to train or fit the model (Chapter 2). Models that are overfit will perform poorly when extrapolated to new environments or locations. The increasing availability of software tools to fit SDMs using a large number of predictors has led to an explosion of studies that are likely overfit to the training data (Peterson et al. 2011). Overfitting can best be detected by testing models on completely independent datasets (in space or time) of presence, or presence-absence, records. Careful a priori selection of predictor variables to avoid collinearity before modeling can help alleviate this problem (Chapter 2). The simple step of examining pairwise correlation between predictor variables and eliminating one of the variables with high correlation ($r > 0.70$) may decrease the possibility of overfitting, and aid interpretation of model output. Usually, the variable that has the most proximate role in determining distribution patterns, based on a priori knowledge, is retained.

Various means of model selection can be used to compare performance of models of differing complexity, thus reducing the possibility of producing overly complex models. Akaike's Information Criterion (AIC) is commonly used with regression methods fit to presence-absence data to help in producing parsimonious models (Chapter 2), but cannot be used with all SDM modeling algorithms. AIC can also be estimated from some presence-background models (Warren and Siefert 2011), and is increasing in use as a method to select among models (Radosavljevic and Anderson 2014; Warren et al. 2014). Many of the metrics mentioned previously for assessing model accuracy (such as AUC) have also been used to select models of differing complexity in species distribution modeling. For example, overfitting can be identified by models that perform much better in terms of AUC values on training data than test data (Warren and Siefert 2011), particularly if the held out data are spatially independent (Merow et al. 2014). Merow et al. (2014) provide an excellent overview of the trade-off between simple vs. complex models, and how these trade-offs relate to a variety of different study objectives for SDMs. We recommend judicious a priori selection of a limited number of predictor variables, followed by a performance comparison among simple and complex models based on AIC, AUC, or other performance metrics, to select a parsimonious but informative model.

15.3.2 Sample Selection Bias

Spatial sampling bias is one of the more difficult issues to deal with in species distribution modeling, and can severely affect model quality (Phillips et al. 2009). Reductions in model quality due to sampling bias occur because geographically biased data are likely to also exhibit environmental bias, and thus, environments that are surveyed heavily will appear to be more suitable to a species. A fitted SDM therefore may provide a model of sampling effort, rather than species distribution (Figure 15.4). Note that sampling bias is more of a problem with presence-only or presence-background techniques; presence-absence SDMs are generally more robust to sampling bias (Phillips et al. 2009) because the absence and presence data will show the same pattern of spatial bias (Elith et al. 2011). Spatial sampling bias is the rule rather than the exception when dealing with datasets of species presence records (Yackulic et al. 2013), and thus demands careful consideration in the modeling process.

Despite its prevalence and importance, adjusting for sampling bias remains a poorly studied area of species distribution modeling and is often overlooked when fitting and interpreting models. Correcting sampling bias when only data on species presence are available can occur in several ways, none of which are completely satisfactory. Subsampling or spatially filtering records by reducing the number of occurrence records in over-sampled regions is a means to reduce clumping of the presence records and thus reduce the effect of spatial bias in the data (Veloz 2009). Subsampling can improve predictive abilities of SDMs (Kramer-Schadt et al. 2013; Boria et al. 2014; Fourcade et al. 2014), and our own work has shown that subsampling produces much more realistic models than raw, uncorrected data. Indeed, for models of coyote distribution, we found that subsampling the presence records provided a much more realistic fit to the known distribution of the species than the full, unfiltered dataset (Thornton and Murray 2014). In a comparison of multiple bias correction techniques, Fourcade et al. (2014) found *subsampling* to be one of the most consistently effective techniques to adjust SDMs for sampling bias. Filtering presence records in environmental, as opposed to geographic, space is also possible and has been found effective (Varela et al. 2014). However, subsampling may not be practical when dealing with small datasets, may not entirely eliminate the clumping in the presence records, and may degrade model performance if the

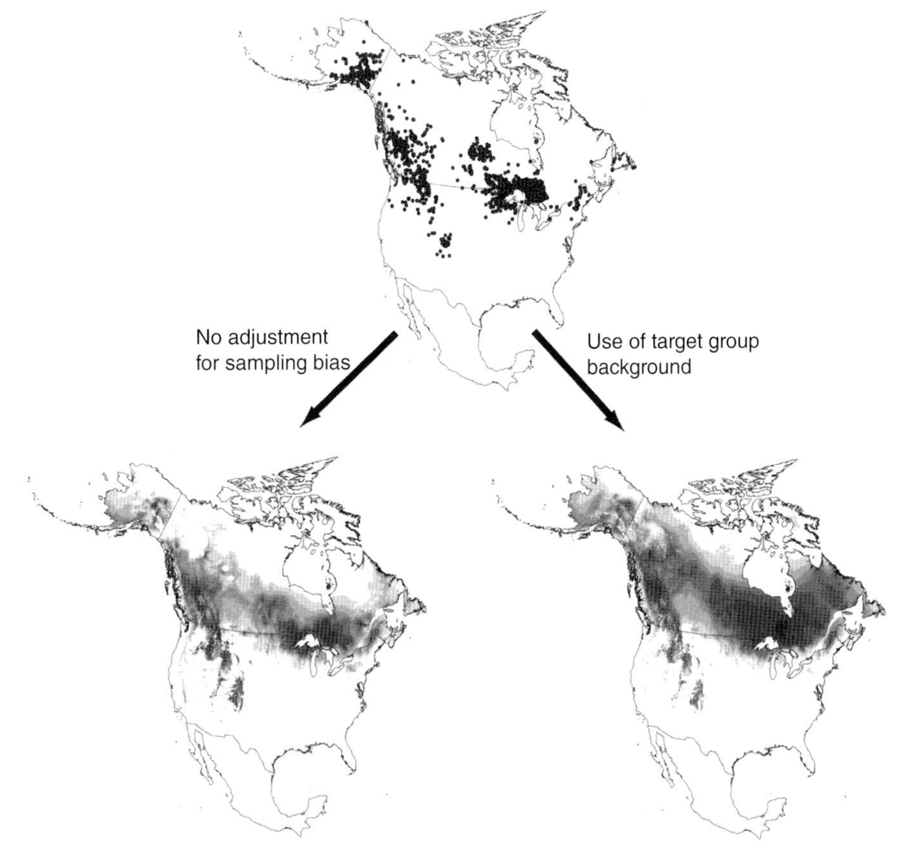

Figure 15.4 Example of the problem of sampling bias with presence-only data. Top image shows the location of presence records for Canada lynx (*Lynx canadensis*) in North America. A clumping of location records can be seen in several regions (e.g. Ontario, British Columbia) due to the uneven availability of high-resolution harvest records and collection bias from museums. A habitat suitability model from `MaxEnt` with no adjustment for the spatial bias in occurrence records is shown at the bottom left, and a habitat suitability model using a target group background is shown on the bottom right (increasingly dark colors indicate higher predicted suitability). Note that when no adjustment is made, the predicted areas of high habitat suitability are tightly constrained to the areas of high clumping in the presence records, indicating that the model reflects sampling bias, not real patterns in habitat suitability. The target group background approach helps alleviate this problem, providing a more realistic suitability model.

spatial clumping of presence records is due to the actual distribution patterns of the species (e.g. a species prefers a particular soil type that is clumped in space), as opposed to being driven by sampling bias. Another approach is to include covariates in the model that may reflect the sampling bias, such as distance to roads. However, this may only be useful when the bias is known a priori and can be reasonably modeled with additional covariates. Spatial point process models may lend themselves particularly well to this type of correction. In a recent paper by Renner et al. (2015) that sought to model incidental sightings of an endemic tree (*Eucalyptus sparsifolia*) using point process models, spatial bias was modeled by including distance from roads and distance from urban areas, operating under the assumption that these areas would be more likely to be traveled, and thus, more likely to result in opportunistic sightings of the species.

A third approach for presence-background techniques is to bias the selection of background or pseudo-absence data in the same manner that the presence data are biased. *Data selection* can occur in one of two ways: (i) developing a bias grid that reflects either a priori knowledge of sampling bias, or is based on densities of presence records, where background records are sampled preferentially from those areas that have high densities of presence records (Elith et al. 2010); or (ii) use of a target group background (Phillips et al. 2009). We have used the former approach to model distribution patterns of Canada lynx and bobcats (*Lynx rufus*, Peers et al. 2013, 2014), where subsampling produced unrealistic models based on a priori knowledge of distribution patterns, and the use of a bias file produced much more realistic models. However, the bias file approach, particularly when based on density of presence records, is an ad hoc approach, and some comparisons suggest it does not perform as well as basic spatial filtering (Fourcade et al. 2014). The use of target group background involves selecting background locations only from areas that have been surveyed for other taxonomically similar species to the focal species, and using a similar sampling strategy as the protocols to generate presence records for the species of interest (Figure 15.4). For example, if the goal was to develop a model of the distribution of a mesocarnivore based on occurrence locations obtained from harvest records, background locations would be sampled only from those areas in the study region that contained at least one occurrence location from harvest records of a similar mesocarnivore species. The approach assumes that the "target group" appropriately represents the spatial bias in the presence records for the species of interest, so care must be taken in selecting this group. Target group background approaches have been found to improve model quality in multiple studies but may not eliminate bias entirely (Phillips et al. 2009; Syfert et al. 2013). Phillips et al. (2009) developed presence-background SDMs using a variety of modeling algorithms and datasets, and found that target group approaches improved AUC values from independent presence-absence test data for most groups and models. Last, spatial point process models may advance our understanding in this area, as these models offer diagnostic tools for checking for spatial bias/nonindependence of presence-only records, which is currently lacking from analyses such as MaxEnt. If nonindependence is detected, area-interaction models can be fit to adjust for this bias (Renner and Warton 2013).

Good practice suggests that presence records should be examined visually for spatial bias at the outset of any analysis, and the likelihood that the bias is driven by sampling artifacts or real patterns in distribution should be assessed. If the records are likely impacted by sampling bias, methods should be used to correct for spatial bias in the occurrence records during the modeling process. Given recent research, and our own experience, as a general rule of thumb we suggest first considering subsampling or filtering of the presence records, or explicitly modeling the bias with additional covariates. If those methods are not reasonable to employ for a particular dataset, target group approaches may be most appropriate. Sensitivity analyses are also recommended, to determine whether results change drastically depending on the type of bias correction used.

15.3.3 Background Selection

When engaging in presence-background modeling, one of the key decisions that modelers must make is choice of the background extent. The *background extent* defines the available environment: the environmental attributes that will be used to compare against species' presence records to develop a model of the species distributional limits. Decisions on an appropriate background extent and the number of random samples from that background to select for comparison with presence records may be a function of the software or statistical procedure used (Giovanelli et al. 2010), and can lead to both over- and underestimating predictions regarding model performance (Lobo and Tognelli 2011). Furthermore, changing the background used in the model may alter the ecological question that is being addressed. For example, using a background of all Australia for a species endemic to the southwest poses the question: why is the species only in southwestern environments (Elith et al. 2011)?

In general, the background area should include the full environmental range of the species, excluding areas that have not been surveyed, or where the species does not occur due to barriers in dispersal, such as elevation. If selecting background points from environments well beyond species' dispersal potential, the modeler is operating under the assumption that the area is unsuitable without the species having sampled this region. Therefore, large background extents tend to produce apparently better models in terms of discrimination, but that are largely uninformative (Lobo et al. 2010; Acevedo et al. 2012).

Decisions on background extents may need to be species-specific, taking into account dispersal attributes and geographical characteristics as well as the time span of the species' presence on the landscape, and environmental changes that have occurred during that period (Barve et al. 2011). For example, invasive species that recently arrived could have a background extent consisting of the present distribution buffered by an estimate of the maximum dispersal distance achieved since their time of arrival, if the rate of spread is predictable (Barve et al. 2011). A framework can also be established to define the area that has the highest probability of being accessible to a species while simultaneously avoiding regions that are uninformative for an ecological model (Acevedo et al. 2012). General rules of thumb would suggest selecting a background that is not too far removed from the range limits of the species, and perhaps doing a sensitivity analysis of different background extents when developing models.

15.3.4 Extrapolation

Many common applications of SDMs require extrapolation of the model predictions to new regions or time periods, for example, using SDMs to predict how the range of a species will shift due to climate change, or predicting suitable habitat in a region that has never been surveyed. Transfer of model predictions can be problematic if extrapolating outside of the training data, where the model is used to predict responses or distribution under novel environmental conditions. As a simple example of this, imagine that an SDM is developed for a given place and time that indicates how suitability of a site will change as the temperature varies within the range of 0–20 °C. What happens when that model is used to predict to a location/time where the temperatures increase to 25 °C? What will be the form of the response curve? Not only is extrapolation to values outside of that used to calibrate the initial model problematic, but problems also arise with respect to extrapolation to new combinations of multiple predictors (Owens et al. 2013). Regardless of the nature of the novel conditions, assumptions must be made about the shape of the response curves of probability of presence or habitat suitability as it relates to environmental variables that extend beyond the training data, which adds considerable uncertainty to the predictions. There are options for how to extrapolate responses to novel environments, including truncation

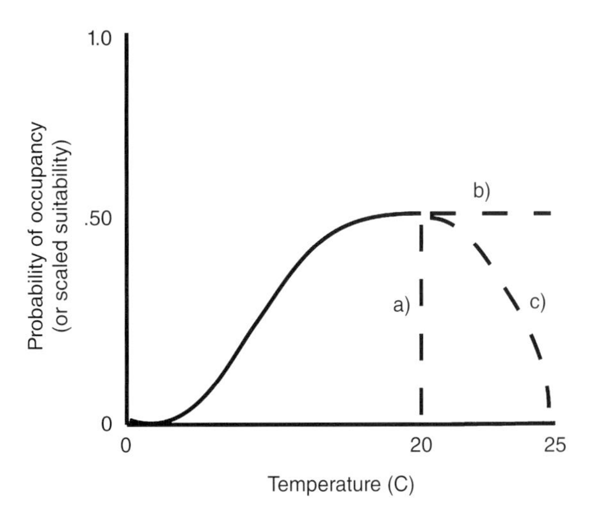

Figure 15.5 Hypothetical response curve obtained from a species distribution model. The solid curve shows the model fit for how probability of occupancy (or environmental suitability) changes as a function of temperature in the study area. Note that the response curve only extends to the limits of the environmental data in the study area (which in this case, reached 20 °C). If this model is used to extrapolate to time periods or places where temperature reaches 25 °C, the response curve must be extended in some way: a) represents *truncation*, where all temperatures beyond 20 °C are considered unsuitable; b) represents *clamping*, where all temperatures beyond 20 °C are considered to stay at the same level of suitability as the highest temperature in the training data; and c) represents *extrapolation*, where response to temperatures beyond 20 °C is assumed to fit some sort of more complex model (e.g. a curvilinear or linear rise or decline).

and clamping (Figure 15.5). When engaging in extrapolation, it is highly recommended to employ *multivariate environmental similarity surfaces* (MESS) or *mobility-oriented parity* (MOP). These two tools compare the environmental space of the data used to develop models versus the environmental space into which the model is extrapolated, to develop maps that indicate where the most novel environmental conditions exist in the extrapolation space (Elith et al. 2010; Owens et al. 2013). Use of these tools can be a way to visualize areas where model results likely have high uncertainty because they are extrapolating well outside the training range.

For presence-background methods, projecting models into future climates or new geographic space can add uncertainty to decisions regarding an appropriate size for the background extent. Owens et al. (2013) note that for species that have presence records that fall toward the edge of the available environment space, extrapolation is likely to be more uncertain than for species where presence records are centrally located within the available environmental space. The discrepancy arises because algorithms for the central locations will be better able to model necessarily unsuitable environments. Even if models are developed using the full range of the species, including areas within their dispersal limits, projections well outside this area will still involve considerable extrapolation into environments beyond the training data (Elith et al. 2011). Thus, the degree of uncertainty when extrapolating SDMs may vary according to the intricacies of sampling and environmental space for a particular species.

15.3.5 Violation of Assumptions

Correlative species distribution modeling has at least three implicit assumptions that are necessary to consider in the modeling process: (i) *Equilibrium*: When developing SDMs, the species of interest is assumed to be in equilibrium with its environment. The assumption may not be met for invasive species that are expanding their range, or for species that are experiencing range contraction due to land use or climate change. Again, this is primarily a problem with presence-background methods, and not presence-absence approaches. Elith et al. (2010) detail some possible techniques to deal with violations of this assumption, for example, by restricting the background extent to reflect the fact that a species is currently expanding its range and has not yet sampled the entire study region. (ii) *Niche conservatism*: SDMs also assume niche conservatism, in that the relationship between environmental predictors and species occurrence is not variable in space or time. The prevalence of niche conservatism is an ongoing debate, and one that has been generated in large part over concerns about how well SDMs can be translated to new time periods or spatial locations (Pearman et al. 2008; Peterson 2011). Modeling subspecies independently may be one way to account for spatial variability in niche characteristics (Pearman et al. 2010). (iii) *Microrefugia*: Microenvironments are often ignored, due to the relatively large spatial scales at which SDMs are often applied. However, such environments may be critical to the persistence of certain species, but this information will not be captured in most models based on coarse-resolution environmental predictors. The issue may be addressed in the near future, as the resolution of environmental data obtainable with remote sensing increases and there is a push to collect datasets of presence (or presence-absence) records at a fine resolution over large spatial extents (Hannah et al. 2014).

15.4 Recent Advances

SDMs are continuing to increase in sophistication and the pace by which SDMs are becoming increasingly complex is among the most rapidly evolving areas of population ecology today. Outside of continuing developments in statistical methodology and niche ecology, SDMs may advance in three other emerging areas with new models that include dispersal, population dynamics, or biotic interactions.

15.4.1 Incorporating Dispersal

Correlative SDMs are commonly used for extrapolation of distribution patterns in space or time. Predicting a shift in the occupied range of a species, however, requires that the dispersal potential of the species be considered. *Dispersal potential* has been suggested as the most important determinant for the expansion rate of invasive species and demands full consideration when understanding the ability of species to keep pace with moving suitable environments (Schloss et al. 2012; Travis et al. 2013). It is imperative that SDMs extrapolating distribution patterns also incorporate measures of a species' dispersal potential.

Initially, SDMs used either unlimited dispersal or no dispersal as possible scenarios in their projections (Thuiller et al. 2004). In this case, species were assumed to be able to occupy all available suitable habitat no matter how far outside the current range, or a species could not occupy any suitable habitat outside the current range. Recent studies have incorporated some aspect of dispersal into their predictions to generate *partial dispersal models* (Bateman et al. 2013). The simplest methods for incorporating dispersal constraints into SDMs have used standard dispersal distances for a group of species multiplied over the time of projection to restrict future habitat suitability (Williams et al. 2005; Midgley et al. 2006; Reside et al. 2012). In other words, suitable habitat beyond the dispersal distance of the species of interest is considered unsuitable, and is then removed from the map of suitable habitat during post-processing. For example, Reside et al. (2012) developed future SDMs for over 200 species of birds inhabiting tropical savannas of Australia. To incorporate dispersal limitations, the authors limited suitable habitat to within 30 km of current range limits for every decade of change, based on average range shifting rates in birds. Such approaches may be more accurate than unlimited or no dispersal models, but still ignore important factors influencing a species' dispersal potential (Bateman et al. 2013). A slightly more sophisticated approach is to use dispersal kernels, where the probability of dispersing and occupying new habitat is a function of distance to currently occupied habitat. For example, Summers et al. (2012) modeled response of plant species to climate change by using SDMs to indicate areas of potential future suitability. Dispersal into future habitat was modeled based on a negative exponential function, whereby cells close to currently occupied habitat were much more likely to be colonized than those farther away. Recently, models have been developed that allow the incorporation of multiple factors that impact a species' ability to track suitable habitat through dispersal (Iverson et al. 2004; Midgley et al. 2010; Engler et al. 2009, 2012). Models with movement can all be termed *hybrid models*, as they combine SDM output with simple or complex models of dispersal and demography. By integrating population dynamics and dispersal into the predictions of future ranges, these models may improve predictions of future distribution patterns. However, they do require a greater amount of information, and thus may only be applicable to a small number of species.

The above approaches have combined dispersal with SDMs to produce spatially explicit predictions of extrapolated distribution patterns in space and time, but other approaches address the question of whether or not a species can keep up with its shifting suitable climate (Brotons et al. 2012). For example, Leroux et al. (2013) used reaction–diffusion equations and information on the intrinsic rate of increase and rate of shift of climatic conditions to estimate the dispersal distance needed to keep pace with shifting habitat suitability. Other studies have combined SDMs with metapopulation models to project species distributions through time (Fordham et al. 2012; Naujokaitis-Lewis et al. 2013). The latter models represent the most inclusive modeling attempts to date, but key assumptions regarding the link between SDM predicted probability of occurrence with population dynamics used in metapopulation models remain largely untested (Bateman et al. 2013).

The more complex partial dispersal models are an improvement from the more simplistic projections that ignore dispersal limitations, however, caution is warranted. Accurate information for dispersal parameters is not available for most species. Minor alterations in values used for various dispersal parameters, which represent conservative uncertainties in species dispersal potential, can lead to drastically different projections of range shifts (Peers et al., unpublished data). The role of rare, long-distance dispersal events in tracking suitable habitat through time may be important, but remains unexplored. Furthermore, species dispersal rates may not be fixed through time but subject to evolution (Travis et al. 2013), which has been shown at the range-edge of invasive species (Lindström et al. 2013). Regardless of the method used and potential lack of information, incorporating dispersal constraints into SDMs is becoming a common practice of the "next-generation" approaches for understanding species' responses to environmental change.

15.4.2 Incorporating Population Dynamics

Hybrid SDMs have also started to be used in combination with sophisticated, spatially explicit population modeling to incorporate additional demographic complexity beyond just movement. Here, SDMs are used to initially define habitat patches and perhaps also initial patch-carrying capacity, and then stage- or age-structured matrix population models are applied to simulate change in distribution or abundance over time or across space (Keith et al. 2008; Anderson et al. 2009; Fordham et al. 2013). A *metapopulation approach* allows incorporation of local population dynamics such as density dependence and demographic and environmental stochasticity, as well as spatial population dynamics such as dispersal between patches, to provide a more realistic understanding of how distributions may shift over time. These *coupled niche-population models* are better able to capture details such as time lags and thresholds in response of species to climate change that are not included in simple correlative SDMs, and can incorporate added complexities such as land-use changes (Anderson et al. 2009; Franklin 2010). The SDM alone, and more complex coupled-niche population models, have not been extensively compared. Limited data suggests similar performance in predicting current distributions, but divergent future predictions (Fordham et al. 2013). Other types of hybrid models include the coupling of SDMs with spatially explicit individual-based population models, and with occupancy dynamics (Franklin 2010; De Caceres and Brotons 2012). Working with 12 species of open-habitat bird species in Catalonia, De Caceres and Brotons (2012) used correlative SDMs to develop an initial distribution map and future habitat suitability maps, and then occupancy models to examine colonization processes in grid cells.

Although hybrid models incorporating dispersal or local population dynamics may address some of the most critical limitations of correlative SDMs, particularly when extrapolating distribution patterns across space or time, there is an obvious trade-off between complex, more mechanistic models and simple, correlative models. Franklin (2010) provides a rule of thumb in this regard, suggesting that at a minimum, dispersal constraints should be accounted for in some way in SDM predictions, either through simply post-processing models to eliminate habitat beyond the migration rate of the species, or applying more complex grid-based simulations of dispersal. If knowledge of a species' life history and habitat requirement is available, more complex coupled-niche population models may be employed to improve predictions.

15.4.3 Incorporating Biotic Interactions

One implicit assumption of many SDMs is that biogeographic patterns at a range-wide scale are responsive to abiotic factors, with biotic interactions only influential at smaller spatial scales or through effects on local patterns of abundance. Debate remains regarding the importance of predator–prey dynamics, facilitation, competition, and other *biotic interactions* for large-scale SDMs (Kissling et al. 2012; de Araújo et al. 2014; Araújo and Rozenfeld 2014). Although it could be argued that biotic interactions are implicitly incorporated in the modeling process because species presence or presence-absence records will hold a signal of this interaction, such an argument does not consider extrapolations in space or time which require information on all important predictors – thus, it remains important to parse the abiotic environmental signal from the biotic one, if the biotic signal is indeed influential. The weight of recent work appears to support the idea that large-scale species distributions can also be altered by biotic interactions (Wisz et al. 2013; Araújo and Rozenfeld 2014; de Araújo et al. 2014). For example, distribution and abundance of several taxa have been strongly linked to the presence of plant species (Novotny et al. 2006; Kissling et al. 2007). Predators can control the range limits of prey species in numerous ecosystems (Estes et al. 2011) and predator and prey richness are strongly linked at macroscales (Sandom et al. 2013). Displacement from competitors can also alter broad-scale relationships between species and their environment (Peers et al. 2013). Given this growing body of evidence, inclusion of biotic interactions in species distribution modeling is potentially desirable, especially when estimating range shifts resulting from invasions or anthropogenic change.

The proper way of incorporating biotic interactions in SDMs remains an active area of study. One simple method is to apply post-processing to the output of an SDM, whereby the distribution of one species is restricted in some way based on knowledge of the distribution of another species. For example, a butterfly distribution might be restricted to areas that contain suitable host plants, no matter how suitable the environment may be in other locations. Such an approach is only feasible when a priori information exists regarding a strong interaction between two or more species. Another possible method for incorporating biotic information is to include interacting species or dietary resources as predictor variables in model development (Heikkinen et al. 2007; Pellissier et al. 2010; Araújo et al. 2014). Information on the distribution, abundance, or habitat suitability of a potentially interacting species is included in the model-fitting process in the same manner as other environmental predictor variables. Studies employing this technique have demonstrated more realistic or better-fitting models of distribution patterns when incorporating biotic information compared to models based solely on environmental variables (Heikkinen et al. 2007; Preston et al. 2008; Aragón and Sánchez-Fernández 2013). However, the approach has been criticized because biotic interactions in the model may be correlated with abiotic factors, and the causal links among biotic interactions are uncertain. Moreover, SDMs developed with biotic predictor variables may fit better than models from abiotic variables alone, but improvement in SDM fit can also be achieved by including noninteracting species in SDMs (Giannini et al. 2013), suggesting that improvement in fit with biotic variables could be a statistical artifact. It may also be problematic to include biotic interactions in SDMs, if the biotic variable is affected by its interactions with the species of interest, such as resources that are affected by consumptive activities (Anderson 2013).

More complex methods for modeling biotic interactions have recently been suggested. Pollock et al. (2014) recommend the application of *joint SDM*, whereby distributions of multiple species are modeled simultaneously based on environmental variables, and correlated residuals between species can reveal information on positive or negative interactions. Trainor et al. (2014) presented another approach, where data on the spatial location of trophic interactions between species is modeled explicitly for the example of predation by lynx on snowshoe hares. They found that a normal SDM for lynx, where locations of lynx presence from telemetry data were related to environmental variables, differed from a *trophic interaction distribution model* (TIDM) where the spatial location of predation events was related to environmental variables. The results suggest that SDMs may under- or over-represent conditions needed to maintain key interactions. However, both Pollock et al. (2014) and Trainor et al. (2014) focused their studies at relatively small scales – their application to large-scale models, or models that rely on presence-only or presence-background data, may be limited. Studies have also begun using surrogates for biotic interactions, where variables indirectly reflecting gradients in biotic interactions across a geographic extent are incorporated into models. For example, variables reflecting vegetation height or biomass could be applied as surrogates of competitive intensity in plants (Midgley et al. 2010).

The importance of considering biotic interactions in SDMs is perhaps most pressing with models of species range shifts in response to climate change. In fact, biotic interactions may be more important than physiological tolerance of temperature and precipitation, in predicting response to climate change (Urban et al. 2013). Limited work incorporating biotic interactions into future predictions has been done to date, and has generally revolved around use of biotic predictor variables or post-processing of distribution patterns (Hof et al. 2012; Peers et al. 2014). However, considerable uncertainty remains regarding our ability to model interactions that are not well understood using SDMs, and in predicting how these interactions will behave in novel abiotic/biotic environments that are driven by climate change (Anderson 2013). Some of the more complex interaction models may be an improvement in modeling distributions, but uncertainty within each parameter on model performance will be an important insight moving forward (Naujokaitis-Lewis et al. 2013). Given the weight of evidence showing the importance of biotic interactions in mediating macroscale distribution patterns, methods for incorporating this information in SDMs will remain a key goal of future research.

15.5 Software Tools

15.5.1 Fitting and Evaluation of Models

The two most commonly used software tools for fitting distribution models are packages in R (hwww.r-project.org) and MAXENT (Ahmed et al. 2015). For presence-only models, the dismo package in R provides functions for fitting climate envelope and distance-based models (Hijmans et al. 2015). For presence-background models, the stand-alone software MAXENT will fit models of maximum entropy (www.cs.princeton.edu/~schapire/maxent), and these models can also be fit within the dismo package. MAXENT is user-friendly, provides numerous options, evaluation methods, and mapping tools (e.g. MESS maps), and is used for two of the online exercises in this chapter. Software tools for implementing many other methods for modeling presence-background, or presence-absence data, such as GLM, GAM, and BRT, can be found in a variety of packages within R. In particular, the biomod2 package allows for the fitting of ten different algorithms (Thuiller et al. 2012), including regression and machine learning techniques, as well as ensemble modeling abilities. Many of the R packages have functions for pre-processing species and environmental data. In addition, there are some useful stand-alone packages in this regard, such as spThin, which spatially subsets species occurrence data to reduce clumping. SDMtoolbox in ArcMap allows for several different functions for pre-processing species and environmental data for analysis in SDMs (Brown 2014). GARP models of distribution can be fit using DesktopGarp, and an ENFA can be conducted within the Biomapper software. Many of the software packages for fitting distribution models also offer a variety of options for model evaluation, including dismo, biomod2, and MAXENT. Another R package, ENMEval, provides additional tools for model evaluation of MAXENT models (Muscarella et al. 2014), including AIC statistics that are absent from the stand-alone software. In addition, ENMEval provides automated splitting of testing and training data in different ways, such as in spatial blocks by latitude and longitude, that may be useful for evaluating overfitting and model extrapolation performance of MAXENT models. More recently, two online sources allow for the development of SDMs using several different modeling techniques. These tools include SPACES: Spatial Portal for Analysis of Climatic Effects on Species, and ModEco: Integrated Software for Species Distribution Analysis and Modeling.

15.5.2 Incorporating Dispersal or Population Dynamics

The Program SHIFT was developed for tree species and incorporates dispersal probabilities into a model (Iverson et al. 2004), taking into account the influence of abundance of the species near the range limit, the forest density within and beyond this limit, and the distance between cells. The Program BioMove was developed for plant species, and integrates SDMs with demographic rates, dispersal parameters, and landscape-level processes including species' responses to vegetation structure (Midgley et al. 2010). The Program MigClim uses output from current and projected SDMs, along with multiple demographic- and dispersal-related parameters that can be easily manipulated to generate predictions of future distribution patterns (Engler et al. 2009, 2012; see the online exercises for a detailed description). Although designed with plants in mind, MigClim is flexible enough to be used to model dispersal dynamics for a variety of species.

15.6 Online Exercises

The online exercises for our chapter illustrate species distribution modeling based on locality records for the striped skunk (*Mephitis mephitis*). Exercise 1 uses the tools of MaxEnt to develop a range-wide model of the probability of occurrence. MaxEnt assumes that sampling of presence records is unbiased and we illustrate how spatial rarification of occurrence records can be used to account for spatial bias. Exercise 2 extends the model to predict the impacts of climate change on the species distribution. We explore changes in species distributions under different scenarios of climate change. Species distributional responses could be affected by other factors, including dispersal capability. Exercise 3 illustrates use of the Program MigClim to combine scenarios of climate change with information on dispersal distances to predict future species distributions.

15.7 Future Directions

The rapid growth in species distribution modeling is likely to continue over the coming years, as new methods and techniques begin to address concerns that have arisen from the initial work in this field. Indeed, a new generation of models will likely be needed to deal with the stickiest issues that plague many distribution models, such as incorporation of biotic interactions. However, progress in the field will be hampered by continued lack of comprehensive large-scale datasets of species distribution (Duputié et al. 2014). Continued digitization of museum records and other sources of occurrence data will aid in the development of more comprehensive datasets of species occurrence, and represents an important step forward, but there is a pressing need to generate more presence-absence (or detection-nondetection) data for species at large scales. The issue is particularly pertinent given the advantages of presence-absence distribution models over presence-only models that have been discussed in this chapter. Citizen science efforts may offer a way forward and are gaining increasing attention as an approach for generating better ecological data on species distributions and abundance (Dickinson et al. 2012).

Outside of generating better distribution data, methods for combining different data sources, in an effort to correct for the deficiencies of presence-only data, represent another promising approach to improving data quality and interpretation from SDMs. Fithian et al. (2015) provides a method to account for sampling bias in presence-only datasets by modeling multiple species that suffer from similar bias, and where a subset of the species have presence-absence data available. Dorazio (2014) details an approach that combines large-scale presence-only data with smaller-scale count or occupancy to correct for bias and imperfect detection in presence-only models. These kinds of innovative models may allow for more flexible application of presence-background modeling efforts.

On the other side of the modeling equation, increased use of remote sensing to generate environmental predictors for use in SDMs is another area that we expect will contribute to enhancing the utility of these models over the coming decades. The variety of environmental predictors that can be obtained from remote sensing data is a vast but still relatively underutilized resource in distribution modeling. In particular, remote sensing platforms can be used to generate large-scale proximal predictors for use in modeling, such as the Normalized Difference Vegetation Index (NDVI) or vegetation structure and seasonality, with great potential to improve SDMs given their likely strength of influence on distribution patterns (le Roux et al. 2013; Jarnevich et al. 2014). Indeed, addition of these variables to topographic or climatic predictors have been found to improve predictions from SDMs (Cord and Rödder 2011; Parvianinen et al. 2013; Wilson et al. 2013). Moreover, remote sensing may contribute to the development of future predictions of land-use change to match with future climate predictions (Halmy et al. 2015) – the inability to model land use in the future is a current deficiency in our ability to predict future range limits of species with SDMs.

We also expect increasing integration of SDMs with population demographic approaches (Fordman et al. 2014). Although these are only applicable to a subset of species which have considerable demographic data, the advantages in obtaining a more realistic understanding of mechanisms that are driving distributional changes is considerable. Indeed, the union of population ecology and demography with large-scale niche models may be one way to address disadvantages with either approach – the lack of realism in large-scale distribution models, and inability to scale up for demographic models.

Improvements in methodology and data are certainly important in pushing the field of distribution modeling forward, but an increasing emphasis on using SDMs to address key outstanding ecological and evolutionary questions is also an emerging area. Already, an increasing number of studies are using SDMs to test ecological questions such as niche conservatism and species' evolutionary history. Many of the ecological questions that SDMs may illuminate are of keen interest to the field of population ecology, such as biogeographical patterns in distribution and abundance, controls on range limits, and competitive interactions. Similarly, the promise of using SDMs to address key conservation problems is only in its infancy, and as these models become more rigorous, and our interpretations become more nuanced, SDMs will become a key part of the tool box for managers and conservation biologists. SDMs are already used extensively to examine impacts of climate change on species and for priority-setting exercises for protected areas, but we envision a larger role. For example, we recently used SDMs to forecast suitable present-day habitat for de-extinction candidates such as the Passenger Pigeon, in an effort to assess the feasibility of this potentially transformative conservation strategy. In another fascinating conservation application, SDMs are being used to model high-use areas for poachers of tigers (*Panthera tigris*) to aid law enforcement efforts in deterrence efforts.

One avenue of development for SDMs that we expect will contribute to their ability to address key questions in ecology and evolution is the increased integration of distribution models with genetic, phylogeographic, isotopic, or other forms of data. For example, the combination of SDMs with stable isotope data has revealed novel insights into the distribution and origin of migratory species (Flockhart et al. 2013; Cardador et al. 2015). SDMs are also starting to be used to address questions outside the ecological disciplines. SDMs are being combined with human demography and paleogeography data to examine hypotheses regarding past human distribution, impacts on the environment, and resource use (Franklin et al. 2015; Tallavaara et al. 2015). Therefore, as SDMs become more sophisticated and increasingly integrated with other forms of data, our ability to tackle a wide variety of questions in a diverse number of fields and subfields will be greatly enhanced, highlighting the tremendous potential and flexibility of these techniques for research in the twenty-first century.

References

Acevedo, P., Jiménez-Valverde, A., Lobo, J.M. et al. (2012). Delimiting the geographical background in species distribution modelling. *Journal of Biogeography* 39: 1383–1390.

Ahmed, S.E., McInerny, G., O'Hara, K. et al. (2015). Scientists and software - surveying the species distribution modelling community. *Diversity and Distributions* 21: 258–267.

Anderson, B.J., Akçakaya, H.R., Araújo, M.B. et al. (2009). Dynamics of range margins for metapopulations under climate change. *Proceedings of the Royal Society of London B* 276: 1415–1420.

Anderson, R.P. (2013). A framework for using niche models to estimate impacts of climate change on species distributions. *Annals of the New York Academy of Sciences* 1297: 8–28.

Aragón, P. and Sánchez-Fernández, D. (2013). Can we disentangle predator-prey interactions from species distributions at a macro-scale? A case study with a raptor species. *Oikos* 122: 64–72.

Araújo, M.B., Cabeza, M., Thuiller, W. et al. (2004). Would climate change drive species out of reserves? An assessment of existing reserve-selection methods. *Global Change Biology* 10: 1618–1626.

Araújo, M.B. and New, M. (2014). Ensemble forecasting of species distributions. *Trends in Ecology and Evolution* 22: 42–47.

Araújo, M.B. and Rozenfeld, A. (2014). The geographic scaling of biotic interactions. *Ecography* 37: 406–415.

Austin, M. (2007). Species distribution models and ecological theory: a critical assessment and some possible new approaches. *Ecological Modelling* 200: 1–19.

Bahn, V. and McGil, B.J. (2007). Can niche-based distribution models outperform spatial interpolation? *Global Ecology and Biogeography* 16: 733–742.

Barbet-Massin, M., Jiguet, F., Albert, C.H. et al. (2012). Selecting pseudo-absences for species distribution models: how, where and how many? *Methods in Ecology and Evolution* 3: 327–338.

Barve, N., Barve, V., Jiménez-Valverde, A. et al. (2011). The crucial role of the accessible area in ecological niche modeling and species distribution modeling. *Ecological Modelling* 222: 1810–1819.

Bateman, B.L., Murphy, H.T., Reside, A.E. et al. (2013). Appropriateness of full-, partial- and no-dispersal scenarios in climate change impact modelling. *Diversity and Distributions* 19: 1224–1234.

Beaumont, L.J., Hughes, L., and Poulsen, M. (2005). Predicting species distributions: use of climatic parameters in BIOCLIM and its impact on predictions of species' current and future distributions. *Ecological Modelling* 186: 250–269.

Beck, J., Ballesteros-Mejia, L., Nagel, P. et al. (2013). Online solutions and the 'Wallacean shortfall': what does GBIF contribute to our knowledge of species' ranges? *Diversity and Distributions* 19: 1043–1050.

Beck, J., Böller, M., Erhardt, A. et al. (2014). Spatial bias in the GBIF database its effect on modeling species geographic distributions. *Ecological Informatics* 19: 10–15.

Benito, B.M., Lorite, L., Perez-Perez, R. et al. (2014). Forecasting plant range collapse in a Mediterranean hotspot: when dispersal uncertainties matter. *Diversity and Distributions* 20: 72–83.

Blackburn, T.M., Cassey, P., and Gaston, K.J. (2006). Variations on a theme: sources of heterogeneity in the form of the interspecific relationship between abundance and distribution. *Journal of Animal Ecology* 75: 1426–1439.

Boria, R.A., Olson, L.E., Goodman, S.M. et al. (2014). Spatial filtering to reduce sampling bias can improve the performance of ecological niche models. *Ecological Modelling* 275: 73–77.

Borregaard, M.K. and Rahbek, C. (2010). Causality of the relationship between geographic distribution and species abundance. *Quarterly Review of Biology* 85: 3–25.

Brotons, L., De Cáceres, M., Fall, A. et al. (2012). Modeling bird species distribution change in fire prone Mediterranean landscapes: incorporating species dispersal and landscape dynamics. *Ecography* 35: 458–467.

Brown, J.H. (1984). On the relationship between abundance and distribution of species. *American Naturalist* 124: 255–279.

Brown, J.L. (2014). `SDMtoolbox`: a Python-based GIS toolkit for landscape genetic, biogeographic and species distribution model analyses. *Methods in Ecology and Evolution* 5: 694–700.

Cardador, L., Navarro, J., Forero, M.G. et al. (2015). Breeding origin and spatial distribution of migrant and resident harriers in a Mediterranean wintering area: insights from isotopic analyses, ring recoveries and species distribution modelling. *Journal of Ornithology* 156: 247–256.

Cases-Marce, M., Revilla, E., Fernandes, M. et al. (2012). The value of hidden scientific resources: preserved animal specimens from private collections and small museums. *BioScience* 62: 1077–1082.

Cord, A. and Rödder, D. (2011). Inclusion of habitat availability in species distribution models through multi-temporal remote-sensing data? *Ecological Applications* 21: 3285–3298.

de Araújo, C.B., Marcondes-Machado, L.O., and Costa, G. C. (2014). The importance of biotic interactions in species distribution models: a test of the Eltonian noise hypothesis using parrots. *Journal of Biogeography* 41: 513–523.

De Caceres, M. and Brotons, L. (2012). Calibration of hybrid species distribution models: the value of general-purpose vs. targeted monitoring data. *Diversity and Distributions* 18: 977–989.

Dickinson, J.L., Shirk, J., Bonter, D. et al. (2012). The current state of citizen science as a tool for ecological research and public engagement. *Frontiers in Ecology and the Environment* 10: 291–297.

Dorazio, R.M. (2014). Accounting for imperfect detection and survey bias in statistical analysis of presence-only data. *Global Ecology and Biogeography* 23: 1472–1484.

Duputié, A., Zimmermann, N.E., and Chuine, I. (2014). Where are the wild things? Why we need better data on species distribution. *Global Ecology and Biogeography* 23: 457–467.

Early, R. and Sax, D.F. (2014). Climatic niche shifts between species' native and naturalized ranges raise concern for ecological forecasts during invasions and climate change. *Global Ecology and Biogeography* 23: 1356–1365.

Elith, J. and Graham, C.H. (2009). Do they? How do they? WHY do they differ? On finding reasons for differing performances of species distribution models. *Ecography* 32: 66–77.

Elith, J., Graham, C.H., Anderson, R.P. et al. (2006). Novel methods improve prediction of species' distributions from occurrence data. *Ecography* 2: 129–151.

Elith, J., Kearney, M., and Phillips, S. (2010). The art of modelling range-shifting species. *Methods in Ecology and Evolution* 1: 330–342.

Elith, J. and Leathwick, J.R. (2009). Species distribution models: ecological explanation and prediction across space and time. *Annual Review of Ecology, Evolution, and Systematics* 40: 677–697.

Elith, J., Phillips, S.J., Hastie, T. et al. (2011). A statistical explanation of MaxEnt for ecologists. *Diversity and Distributions* 17: 43–57.

Engler, R., Hordijk, W., and Guisan, A. (2012). The MIGCLIM R package - seamless integration of dispersal constraints into projections of species distribution models. *Ecography* 35: 872–878.

Engler, R., Randin, C.F., Vittoz, P. et al. (2009). Predicting future distributions of mountain plants under climate change: does dispersal capacity matter? *Ecography* 32: 34–45.

Estes, J.A., Terborgh, J., Brashares, J.S. et al. (2011). Trophic downgrading of planet Earth. *Science* 333: 301–306.

Falcucci, A., Maiorano, L., Tempio, G. et al. (2013). Modeling the potential distribution for a range-expanding species: wolf recolonization of the alpine range. *Biological Conservation* 158: 63–72.

Fielding, A.H. and Bell, J.F. (1997). A review of methods for the assessment of prediction errors in conservation presence/absence models. *Environmental Conservation* 24: 38–49.

Fithian, W., Elith, J., Hastie, T. et al. (2015). Bias correction in species distribution models: pooling survey and collection data for multiple species. *Methods in Ecology and Evolution* 6: 424–438.

Flockhart, D.T.T., Wassenaar, L.I., Martin, T.G. et al. (2013). Tracking multi-generational colonization of the breeding grounds by monarch butterflies in eastern North America. *Proceedings of the Royal Society of London B* 280: art20131087.

Fordham, D.A., Akçakaya, H.R., Araújo, M.B. et al. (2012). Plant extinction risk under climate change: are forecast range shifts alone a good indicator of species vulnerability to global warming? *Global Change Biology* 18: 1357–1371.

Fordham, D.A., Mellin, C., Russell, B.D. et al. (2013). Population dynamics can be more important than physiological limits for determining range shifts under climate change. *Global Change Biology* 19: 3224–3237.

Fordman, D.A., Shoemaker, K.T., Schumaker, N.H. et al. (2014). How interactions between animal movement and landscape processes modify local range dynamics and extinction risk. *Biology Letters* 10: art20140198.

Fourcade, Y., Engler, J.O., Rodder, D. et al. (2014). Mapping species distributions with MAXENT using a geographically based sample of presence data: a performance assessment of methods for correcting sampling bias. *PLoS One* 9: e97122.

Franklin, J. (2009). *Mapping Species Distributions: Spatial Inference and Prediction*. Cambridge: Cambridge University Press.

Franklin, J. (2010). Moving beyond static species distribution models in support of conservation biogeography. *Diversity and Distributions* 16: 321–330.

Franklin, J., Potts, A.J., Fisher, E.C. et al. (2015). Paleodistribution modeling in archaeology and paloeanthropology. *Quaternary Science Reviews* 110: 1–14.

Gaston, K.J. and Blackburn, T.M. (2000). *Pattern and Process in Macroecology*. Oxford: Blackwell Science.

Gaston, K.J., Blackburn, T.M., Greenwood, J.J.D. et al. (2000). Abundance-occupancy relationships. *Journal of Applied Ecology* 37: 39–59.

Giannini, T.C., Chapman, D.S., Saraiva, A.M. et al. (2013). Improving species distribution models using biotic interactions: a case study of parasites, pollinators and plants. *Ecography* 36: 649–656.

Giovanelli, J.G.R., de Siqueira, M.F., Haddad, C.F.B. et al. (2010). Modeling a spatially restricted distribution in the Neotropics: how the size of calibration area affects the performance of five presence-only methods. *Ecological Modelling* 221: 215–224.

Graham, C.H., Ferrier, S., Huettman, F. et al. (2004). New developments in museum-based informatics and applications in biodiversity analysis. *Trends in Ecology and Evolution* 19: 497–503.

Guillera-Arroita, G., Lahoz-Monfort, J.J., Elith, J. et al. (2015). Is my species distribution model fit for purpose? Matching data and models to applications. *Global Ecology and Biogeography* 24: 276–292.

Guisan, A., Broennimann, O., and Engler, R. (2006). Using niche-based models to improve sampling of rare species. *Conservation Biology* 20: 501–511.

Guisan, A. and Thuiller, W. (2005). Predicting species distribution: offering more than simple habitat models. *Ecology Letters* 8: 993–1009.

Guisan, A., Tingley, R., Baumgartner, J.B. et al. (2013). Predicting species distributions for conservation decisions. *Ecology Letters* 16: 1424–1435.

Gutiérrez, D., Harcourt, J., Díez, S.B. et al. (2013). Models of presence–absence estimate abundance as well as (or even better than) models of abundance: the case of the butterfly *Parnassius apollo*. *Landscape Ecology* 28: 401–413.

Halmy, M.W.A., Gessler, P.E., Hicke, J.A. et al. (2015). Land use/land cover change detection and prediction in the north-western coastal desert of Egypt using Markov-CA. *Applied Geography* 63: 101–112.

Hannah, L., Flint, L., Syphard, A.D. et al. (2014). Fine-grain modeling of species' response to climate change: holdouts, stepping-stones, and microrefugia. *Trends in Ecology and Evolution* 29: 390–397.

Hastie, T. and Fithian, W. (2013). Inference from presence-only data; the ongoing controversy. *Ecography* 36: 864–867.

Heikkinen, R.K., Luoto, M., Virkkala, R. et al. (2007). Biotic interactions improve prediction of boreal bird distributions at macro-scales. *Global Ecology and Biogeography* 16: 754–763.

Hengeveld, R. and Haeck, J. (1982). The distribution of abundance. I. Measurements. *Journal of Biogeography* 9: 303–316.

Higgins, S.I., O'Hara, R.B., and Römermann, C. (2012). A niche for biology in species distribution models. *Journal of Biogeography* 39: 2091–2095.

Hijmans, R.J. (2012). Cross-validation of species distribution models: removing spatial sorting bias and calibration with a null model. *Ecology* 93: 679–688.

Hijmans, R.J., Cameron, S.E., Parra, J.L. et al. (2005). Very high resolution interpolated climate surfaces for global land areas. *International Journal of Climatology* 25: 1965–1978.

Hijmans RJ, Phillips S, Leathwick J et al. (2015) Package 'dismo'. R package version 1.0–12. Available at http://CRAN.R-project.org/package= dismo

Hof, A.R., Jansson, R., and Nilsson, C. (2012). How biotic interactions may alter future predictions of species distributions: future threats to the persistence of the arctic fox in Fennoscandia. *Diversity and Distributions* 18: 554–562.

Holt, A.R., Gaston, K.J., and He, F.L. (2002). Occupancy-abundance relationships and spatial distribution: a review *Basic and Applied Ecology* 3: 1–13.

Hortal, J., Lobo, J.M., and Jimenez-Valverde, A. (2012). Basic questions in biogeography and the (lack of) simplicity of species distributions: putting species distribution models in the right place. *Natureza and Conservacao* 10: 108–118.

Hugall, A., Moritz, C., and Moussalli, A. (2002). Reconciling paleodistribution models and comparative phylogeography in the Wet Tropics rainforest land snail *Gnarosophia bellendenkerensis* (Brazier1875). *Proceedings of the National Academy of Sciences USA* 99: 6112–6117.

Iverson, L.R., Schwartz, M.W., and Prasad, A.M. (2004). How fast and far might tree species migrate in the eastern United States due to climate change? *Global Ecology and Biogeography* 13: 209–219.

Jarnevich, C.S., Esaias, W.E., Ma, P.L.A. et al. (2014). Regional distribution models with lack of proximate predictors: Africanized honeybees expanding north. *Diversity and Distributions* 20: 193–201.

Jiménez-Valverde, A., Acevedo, P., Barbosa, A.M. et al. (2013). Discrimination capacity in species distribution models depends on the representativeness of the environmental domain. *Global Ecology and Biogeography* 22: 508–516.

Jiménez-Valverde, A., Lobo, J.M., and Hortal, J. (2008). Not as good as they seem: the importance of concepts in species distribution modelling. *Diversity and Distributions* 14: 885–890.

Kamino, L.H., Stehmann, Y.J.R., Amaral, S. et al. (2012). Challenges and perspectives for species distribution modelling in the neotropics. *Biology Letters* 8: 324–326.

Keith, D.A., Akçakaya, H.R., Thuiller, W. et al. (2008). Predicting extinction risks under climate change: coupling stochastic population models with dynamic bioclimatic habitat models. *Biology Letters* 4: 560–563.

Kissling, W.D., Dormann, C.F., Groeneveld, J. et al. (2012). Towards novel approaches to modelling biotic interactions in multispecies assemblages at large spatial extents. *Journal of Biogeography* 39: 2163–2178.

Kissling, W.D., Rahbek, C., and Böhning-Gaese, K. (2007). Food plant diversity as broad-scale determinant of avian frugivore richness. *Proceedings of the Royal Society of London B* 274: 799–808.

Kramer-Schadt, S., Niedballa, J., and Pilgrim, J.D. (2013). The importance of correcting for sampling bias in MaxEnt species distribution models. *Diversity and Distributions* 19: 1366–1379.

Kuemmerle, T., Perzanowski, K., Chaskovskyy, O. et al. (2010). European bison habitat in the Carpathian Mountains. *Biological Conservation* 143: 908–916.

Lawler, J.J., Ruesch, A.S., Olden, J.D. et al. (2013). Projected climate-driven faunal movement routes. *Ecology Letters* 16: 1014–1022.

Lawson, C.R., Hodgson, J.A., and Wilson, R.J. (2014). Prevalence, thresholds and the performance of presence-absence models. *Methods in Ecology and Evolution* 5: 54–64.

le Roux, P.C., Aalto, J., and Luoto, M. (2013). Soil moisture's underestimated role in climate change impact modelling in low-energy systems. *Global Change Biology* 19: 2965–2975.

Lele, S.R. and Keim, J.L. (2006). Weighted distributions and estimation of resource selection probability functions. *Ecology* 87: 3021–3028.

Leroux, S.J., Larrivée, M., Boucher-Lalonde, V. et al. (2013). Mechanistic models for the spatial spread of species under climate change. *Ecological Applications* 23: 815–828.

Li, W. and Guo, Q. (2013). How to assess the prediction accuracy of species presence-absence models without absence data? *Ecography* 36: 788–799.

Lindström, T., Brown, G.P., Sisson, S.A. et al. (2013). Rapid shifts in dispersal behavior on an expanding range edge. *Proceedings of the National Academy of Sciences USA* 110: 13452–13456.

Lira-Noriega, A. and Manthey, J.D. (2013). Relationship of genetic diversity and niche centrality: a survey and analysis. *Evolution* 68: 1082–1093.

Liu, C., White, M., and Newell, G. (2011). Measuring and comparing the accuracy of species distribution models with presence-absence data. *Ecography* 34: 232–243.

Liu, C.R., Berry, P.M., Dawson, T.P. et al. (2005). Selecting thresholds of occurrence in the prediction of species distributions. *Ecography* 28: 385–393.

Lobo, J.M., Jiménez-Valverde, A., and Hortal, J. (2010). The uncertain nature of absences and their importance in species distribution modelling. *Ecography* 33: 103–114.

Lobo, J.M., Jimenez-Valverde, A., and Real, R. (2008). AUC: a misleading measure of the performance of predictive distribution models. *Global Ecology and Biogeography* 17: 145–151.

Lobo, J.M. and Tognelli, M.F. (2011). Exploring the effects of quantity and location of pseudo-absences and sampling biases on the performance of distribution models with limited point occurrence data. *Journal for Nature Conservation* 19: 1–7.

Martínez-Meyer, E., Díaz-Porras, D., Peterson, A.T. et al. (2013). Ecological niche structure and rangewide abundance patterns of species ecological niche structure and rangewide abundance patterns of species. *Biological Letters* 9: art20120637.

McDonald, T.L. (2013). The point process use-availability or presence-only likelihood and comments on analysis. *Journal of Animal Ecology* 82: 1174–1182.

Merow, C., Smith, M.J., Edwards, T.C. et al. (2014). What do we gain from simplicity versus complexity in species distribution models? *Ecography* 37: 1267–1281.

Midgley, G.F., Davies, I.D., Albert, C.H. et al. (2010). BioMove - an integrated platform simulating the dynamic response of species to environmental change. *Ecography* 36: 1058–1069.

Midgley, G.F., Hughes, G.O., Thuiller, W. et al. (2006). Migration rate limitations on climate change-induced range shifts in cape Proteaceae. *Diversity and Distributions* 12: 555–562.

Muscarella, R., Galante, P.J., Soley-Guardia, M. et al. (2014). ENMeval: an R package for conducting spatially independent evaluations and estimating optimal model complexity for MAXENT ecological niche models. *Methods in Ecology and Evolution* 5: 1198–1205.

Naimi, B., Hamm, N.A.S., Groen, T.A. et al. (2014). Where is positional uncertainty a problem for species distribution modelling? *Ecography* 37: 191–203.

Naujokaitis-Lewis, I.R., Curtis, J.M.R., Tischendorf, L. et al. (2013). Uncertainties in coupled species distribution – metapopulation dynamics models for risk assessments under climate change. *Diversity and Distributions* 19: 541–554.

Newbold, T. (2010). Applications and limitations of museum data for conservation and ecology, with particular attention to species distribution models. *Progress in Physical Geography* 34: 3–22.

Novotny, V., Drozd, P., Miller, S.E. et al. (2006). Why are there so many species of herbivorous insects in tropical rainforests? *Science* 313: 1115–1118.

Oliver, T.H., Gillings, S., Girardello, M. et al. (2012). Population density but not stability can be predicted from species distribution models. *Journal of Applied Ecology* 49: 581–590.

Owens, H.L., Campbell, L.P., Dornak, L.L. et al. (2013). Constraints on interpretation of ecological niche models by limited environmental ranges on calibration areas. *Ecological Modelling* 263: 10–18.

Parvianinen, M., Zimmermann, N.E., Heikkinen, R.K. et al. (2013). Using unclassified continuous remote sensing data to improve distribution models of red-listed plant species. *Biodiversity and Conservation* 22: 1731–1754.

Pearce, J. and Ferrier, S. (2000). Evaluating the predictive performance of habitat models developed using logistic regression. *Ecological Modelling* 133: 225–245.

Pearman, P.B., D'Amen, M., Graham, C.H. et al. (2010). Within-taxon niche structure: niche conservatism, divergence and predicted effects of climate change. *Ecography* 33: 990–1003.

Pearman, P.B., Guisan, A., Broennimann, O. et al. (2008). Niche dynamics in space and time. *Trends in Ecology and Evolution* 23: 149–158.

Peers, M.L.J., Thornton, D.H., and Murray, D.L. (2013). Evidence for large-scale effects of competition: niche displacement in Canada lynx and bobcat. *Proceedings of the Royal Society of London B* 280: art20132495.

Peers, M.L.J., Thornton, D., Majchrzak, Y.N., Bastille-Rousseau, G., and Murray, D.L. (2010). De-extinction potential under climate change: Extensive mismatch of historic and future habitat suitability for three candidate bird species. *Biological Conservation* 197: 164–170.

Peers, M.L.J., Wehtje, M., Thornton, D. et al. (2014). Prey switching as a means of enhancing persistence in predators at the trailing southern edge. *Global Change Biology* 20: 1126–1135.

Pellissier, L., Bråthen, K.A., Pottier, J. et al. (2010). Species distribution models reveal apparent competitive and facilitative effects of a dominant species on the distribution of tundra plants. *Ecography* 33: 1004–1014.

Peterson, A.T. (2003). Predicting the geography of species' invasions via ecological niche modeling. *Quartely Review of Biology* 78: 419–433.

Peterson, A.T. (2011). Ecological niche conservatism: a time-structured review of evidence. *Journal of Biogeography* 38: 817–827.

Peterson, A.T., Papes, M., and Soberon, J. (2008). Rethinking receiver operating characteristic analysis applications in ecological niche modeling. *Ecological Modelling* 213: 63–72.

Peterson, A.T. and Soberon, J. (2012). Species distribution modeling and ecological niche modeling: getting the concepts right. *Natureza and Conservacao* 10: 102–107.

Peterson, A.T., Soberon, J., Pearson, R.G. et al. (2011). *Ecological Niches and Geographic Distributions.* Princeton: Princeton University Press.

Phillips, E. and Elith, J. (2010). POC plots: calibrating species distribution models with presence-only data. *Ecology* 91: 2476–2484.

Phillips, S.J., Dudík, M., Elith, J. et al. (2009). Sample selection bias and presence-only distribution models: implications for background and pseudo-absence data. *Ecological Applications* 19: 181–197.

Phillips, S.J. and Elith, J. (2013). On estimating probability of presence from use-availability or presence-background data. *Ecology* 94: 1409–1419.

Pollock, L.J., Tingley, R., Morris, W.K. et al. (2014). Understanding co-occurrence by modeling species simultaneously with a joint species distribution model (JSDM). *Methods in Ecology and Evolution* 5: 397–406.

Preston, K.L., Rotenberry, J.T., Redak, R.A. et al. (2008). Habitat shifts of endangered species under altered climate conditions: importance of biotic interactions. *Global Change Biology* 14: 2501–2515.

Radosavljevic, A. and Anderson, R.P. (2014). Making better MAXENT models of species distributions: complexity, overfitting and evaluation. *Journal of Biogeography* 41: 629–643.

Raxworthy, C.J., Martinez-Meyer, E., Horning, N. et al. (2003). Predicting distributions of known and unknown reptile species in Madagascar. *Nature* 426: 837–841.

Renner, I.W., Elith, J., Baddeley, A. et al. (2015). Point process models for presence-only analysis. *Methods in Ecology and Evolution* 6: 366–379.

Renner, I.W. and Warton, D.I. (2013). Equivalence of MAXENT and Poisson point process models for species distribution modeling in ecology. *Biometrics* 69: 274–281.

Reside, A.E., Vanderwal, J., and Kutt, A.S. (2012). Projected changes in distributions of Australian tropical savanna birds under climate change using three dispersal scenarios. *Ecology and Evolution* 2: 705–718.

Rota, C., Millspaugh, J.J., Kesler, D.C. et al. (2013). A re-evaluation of a case-control model with contaminated controls for resource selection studies. *Journal of Animal Ecology* 82: 1165–1173.

Royle, J.A., Chandler, R.B., Yackulic, C. et al. (2012). Likelihood analysis of species occurrence probability from presence-only data for modelling species distributions. *Methods in Ecology and Evolution* 3: 545–554.

Sagarin, R.D. and Gaines, S.D. (2002). The 'abundant centre' distribution: to what extent is it a biogeographical rule? *Ecology Letters* 5: 137–147.

Sandom, C., Dalby, L., Fløjgaard, C. et al. (2013). Mammal predator and prey species richness are strongly linked at macroscales. *Ecology* 94: 1112–1122.

Schloss, C.A., Nuñez, T.A., and Lawler, J.J. (2012). Dispersal will limit ability of mammals to track climate change in the Western hemisphere. *Proceedings of the National Academy of Sciences USA* 109: 8606–8611.

Schmidt-Lebuhn, A.N., Knerr, N.J., and Kessler, M. (2013). Non-geographic collecting biases in herbarium specimens of Australian daisies (Asteraceae). *Biodiversity and Conservation* 22: 905–919.

Schorr, G., Pearman, P.B., Guisan, A. et al. (2013). Combining paleodistribution modelling and phylogeographical approaches for identifying glacial refugia in Alpine Primula. *Journal of Biogeography* 40: 1947–1960.

Strubbe, D., Beauchard, O., and Matthysen, E. (2015). Niche conservatism among non-native vertebrates in Europe and North America. *Ecography* 38: 321–329.

Summers, D.M., Bryan, B.A., Crossman, N.D. et al. (2012). Species vulnerability to climate change: impacts on spatial conservation priorities and species representation. *Global Change Biology* 18: 2335–2348.

Syfert, M.M., Smith, M.J., and Coomes, D.A. (2013). The effects of sampling bias and model complexity on the predictive performance of MaxEnt species distribution models. *PLoS One* 8: e55158.

Tallavaara, M., Luoto, M., Korhonen, N. et al. (2015). Human population dynamics in Europe over the last glacial maximum. *Proceedings of the National Academy of Sciences USA* 112: art1503784112.

Thornton, D.H. and Murray, D.L. (2014). Influence of hybridization on niche shifts in expanding coyote populations. *Diversity and Distributions* 20: 1355–1364.

Thuiller, W., Albert, C.H., Dubuis, A. et al. (2010). Variation in habitat suitability does not always relate to variation in species' plant functional traits. *Biology Letters* 6: 120–123.

Thuiller, W., Brotons, L., Araujo, M.B. et al. (2004). Effects of restricting environmental range of data to project current and future species distributions. *Ecography* 27: 165–172.

Thuiller W, Georges D, Engler R (2012) Package 'biomod2'. R package version 3.1–64. Available at http://CRAN.R-project.org/package= biomod2

Törres, N.M., Júnior, P.D.M., Santos, T. et al. (2012). Can species distribution modelling provide estimates of population densities? A case study with jaguars in the Neotropics. *Diversity and Distributions* 18: 615–627.

Trainor, A.M., Schmitz, O.J., Ivan, J.S. et al. (2014). Enhancing species distribution modeling by characterizing predator-prey interactions. *Ecological Applications* 24: 204–216.

Travis, J.M.J., Delgado, M., Bocedi, G. et al. (2013). Dispersal and species' responses to climate change. *Oikos* 122: 1532–1540.

Urban, M.C., Zarnetske, P.L., and Skelly, D.K. (2013). Moving forward: dispersal and species interactions determine biotic responses to climate change. *Annals of the New York Academy of Sciences* 1297: 44–60.

Van Couwenberghe, R., Collet, C., Pierrat, J.-C. et al. (2013). Can species distribution models be used to describe plant abundance patterns? *Ecography* 36: 665–674.

Varela, S., Anderson, R.P., Garcia-Valdes, R. et al. (2014). Environmental filters reduce the effects of sampling bias and improve predictions of ecological niche models. *Ecography* 37: 1084–1091.

Veloz, S.D. (2009). Spatially autocorrelated sampling falsely inflates measures of accuracy for presence-only niche models. *Journal of Biogeography* 36: 2290–2299.

Warren, D.L. (2012). In defense of 'niche modeling'. *Trends in Ecology and Evolution* 27: 497–500.

Warren, D.L. and Siefert, S.N. (2011). Ecological niche modeling in MaxEnt: the importance of model complexity and the performance of model selection criteria. *Ecological Applications* 21: 335–342.

Warren, D.L., Wright, A.N., Siefert, S.N. et al. (2014). Incorporating model complexity and spatial sampling bias into ecological niche models of climate change risks faced by 90 California vertebrate species of concern. *Diversity and Distributions* 20: 334–343.

Warton, D. and Aarts, G. (2013). Advancing our thinking in presence-only and used-available analysis. *Journal of Animal Ecology* 82: 1125–1134.

Williams, P., Hannah, L., Andelman, S. et al. (2005). Planning for climate change: identifying minimum-dispersal corridors for the Cape Proteaceae. *Conservation Biology* 19: 1063–1074.

Wilson, J.W., Sexton, J.O., Jobe, R.T. et al. (2013). The relative contribution of terrain, land cover, and vegetation structure indices to species distribution models. *Biological Conservation* 164: 170–176.

Wisz, M.S., Pottier, J., Kissling, W.D. et al. (2013). The role of biotic interactions in shaping distributions and realised assemblages of species: implications for species distribution modelling. *Biological Reviews* 88: 15–30.

Yackulic, C.B., Chandler, R., Zipkin, E.F. et al. (2013). Presence-only modelling using MAXENT: when can we trust the inferences? *Methods in Ecology and Evolution* 4: 236–243.

Yañez-Arenas, C., Guevara, R., Martínez-Meyer, E. et al. (2014). Predicting species' abundances from occurrence data: effects of sample size and bias. *Ecological Modelling* 294: 36–41.

Yañez-Arenas, C., Martínez-Meyer, E., and Mandujano, S. (2012). Modelling geographic patterns of population density of the white-tailed deer in Central Mexico by implementing ecological niche theory. *Oikos* 121: 2081–2089.

Part V

Software Tools

16

The R Software for Data Analysis and Modeling

Clément Calenge

Direction de la Recherche et de l'Expertise, Office National de la Chasse et de la Faune Sauvage, Saint Benoist, Auffargis, France

Summary

Due to the large number of available statistical methods, the R software has become a central tool for ecologists. This software was originally designed to "turn ideas into software," according to John Chambers. Users can easily program their own methods for the R environment and bundle them into a package; the "Comprehensive R Archive Network" (CRAN) hosts more than 7500 packages. Thus, R has become the lingua franca of statistics. In this chapter, I describe the basic concepts that will allow readers to learn R. I first present how to work with vectors, matrices/arrays, and lists in R. I also describe how to write a function, as well as the basic control-flow constructs that facilitate programming. I finally present how to define the class of an object. The primary message of this chapter is that the main skill required to use R is to know how to find, read, and understand help in R. Many resources are given in this chapter to develop this skill. A web exercise, available at the URL of the book, allows the reader to practice these concepts.

16.1 An Introduction to R

16.1.1 The Nature of the R Language

During the S-plus user conference held in New Orleans in 1999, the biostatistician Frank Harrell posed the question, "Can one be a good data analyst without being a half-good programmer? The short answer to that is, 'No.' The long answer to that is, 'No.'" Indeed each case has its specifics, and therefore each dataset requires its own data analysis strategy, as the chapters of this book clearly illustrate. The analyst must be able to implement techniques on a computer, which requires basic programming skills.

Common low-level programming languages, such as C or Fortran, are not convenient for data analysis. Such compiled languages are efficient for intensive calculation, but require a considerable knowledge in computer science: one has to know how to manage the computer memory, and how the different types of data (e.g. double, integer, long integer) are handled by the computer. The learning curve of these languages is steep, and data analysts often do not have enough time to invest in it. In addition, even when the analyst has strong programming skills, writing a program using such languages is usually cumbersome, with considerable time spent in debugging, and does not fit well with the needs of the analyst to quickly implement a statistical technique during the step of data exploration. Exploratory analysis is where the R environment is especially useful.

Ross Ihaka and Robert Gentleman, from Auckland University (New Zealand), developed the R environment to provide a statistical environment to their laboratory in 1992, and based it on the S language (Ihaka and Gentleman 1996). S was invented at the AT&T Bell laboratories by John Chambers and his colleagues during the mid-1970s, and gained success in the community of applied statisticians, due to its simplicity and its flexibility. The S language has been especially designed to allow the "half-good programmer" to easily implement statistical solutions to particular problems, without requiring a deep understanding of computer programming.

With the help of Martin Mächler (ETH Zürich, Switzerland), Ihaka and Gentleman released the R environment as open-source software in 1995. Presently, the Comprehensive R Archive Network (CRAN) is the core of an increasingly growing R community, contributing to the development of the functionalities of this language.

The R environment is extendable: the S language has been designed to encourage the user to "slide into

Population Ecology in Practice, First Edition. Edited by Dennis L. Murray and Brett K. Sandercock.

Companion website: www.wiley.com/go/MurrayPopulationEcology

programming, perhaps without noticing," and the R language, which is a dialect of S, shares this characteristic. Therefore, it is easy for users to program their own functions, and to bundle them into a package. Packages built by the users can then be uploaded on CRAN (cran.r-project.org), and thereby be made available for all R users. At the time of writing, there are more than 7500 available packages, and this number is continually growing. R users can therefore have access to most statistical methods, whatever the scientific field in which they have been developed including statistics and biological science, but also geography, geology, forensic science, political science, and economy. The R environment also provides numerous graphic functions and is suitable for data exploration, as I demonstrate later in this chapter.

R is a scripting language: users can program their data analysis in a text file and source it in the R environment to allow its execution. However, the R environment in itself does not allow the writing of scripts. It is possible, although inconvenient, to write such scripts in any text editor available on the computer (e.g. on Microsoft Windows, Notepad, Wordpad, or even Microsoft Word). However, it is preferable to write such scripts in an Integrated Development Environment (IDE), a software application designed for the writing of R scripts. Such IDEs provides functionalities that facilitate the development of R programs such as parentheses matching, indentation of the code, or syntax coloring, and interact deeply with the R environment (typically, the R code can be sourced to R directly from the IDE, allowing a quick debugging of the code). Many IDEs are available for the R environment. Thus, the software GNU Emacs (www.gnu.org/software/emacs) is highly customizable, and thanks to the ESS extension (ess.r-project.org), it can be used to write R scripts and source. However, despite its high flexibility, new users prefer more user-friendly software to write their R programs, such as the commonly used RStudio (www.rstudio.com).

16.1.2 Qualities and Limits

The R language has many qualities. It is a simple, intuitive, and elegant language. The R functions usually have a name that can be found easily (`t.test()` for a *t*-test, `anova()` for an Analysis of Variance, `lm()` for a linear model, and `glm()` for a general linear model). As already noted, no deep knowledge in computer programming is required to use R. In addition, numerous graphical functions are available, allowing the user to easily design publication-quality figures. Moreover, since it is used by data analysts in many fields, it is one of the software where the number of available statistical methods is the largest.

However, R also has characteristics that can be considered as drawbacks. Statistical analysts may be used to different types of statistical software, where statistical methods are accessed through mouse-clickable menus and carried out with the help of dialog boxes. New users are often dubious when they face the R prompt for the first time. Whereas "traditional software" does not require any learning phase and can be used almost immediately, the learning curve of the language R is steeper. However, as noted in Frank Harrell's comments highlighted earlier, a sensible data analysis usually cannot be conducted by someone who is not at least a half-good programmer.

Another criticism of R stems from a fundamental property of this language: R is an interpreted language. In other words, every command typed at the R prompt has to be interpreted by the R software prior to its execution. On the other hand, programs written using compiled languages, such as C or Fortran, can be executed directly. When a statistical method requires a large number of calculation steps such as Monte Carlo Markov chains, the corresponding R program can be slow. Several R functions can be used to circumvent this drawback to some extent (e.g. the function `lapply()`), but such functions cannot be used in all contexts. For this reason, when an intensive calculation is implemented in R by a programmer, the core of the approach is usually programmed in a low-level language (usually C or Fortran), and interfaced in R. However, when R users want to implement an intensive calculation in pure R, they should be prepared to wait a long time before the calculation ends.

16.1.3 R for Ecologists

Because of the central place of statistics in ecological studies, the availability of all common statistical methods makes R an ideal tool for ecologists. Thus, many seminal books describing the use of common statistical methods in ecological studies explain how to implement them in R, for example Bolker (2007) on statistical modeling in ecology, Zuur et al. (2009) on mixed models, and Borcard et al. (2011) on how to use common factor analyses and spatial statistics in ecological studies.

But R also provides tools that are used only in ecology (see for example the primer to ecology by Stevens 2009). Thus, ecological population models can be developed in R with the package popbio (Stubben and Milligan 2007); the packages adehabitatLT, adehabitatHR, and adehabitatHS provide methods to analyze animal home ranges, animal movement data, and habitat selection by the wildlife (Calenge 2006); and the package ResourceSelection can be used to fit resource selection functions (Lele et al. 2015). The CRAN website contains a web page that describes many packages of interest to ecologists (at the URL: cran.r-project.org/web/views/Environmetrics.html).

As noted by many authors (e.g. De Leeuw et al. 2007), R has become the lingua franca of statistics; any new statistical method proposed in the literature is now available in R, no matter the particular scientific field where it has been originally developed. Thus, although the previous chapters in the present book are not specific to R, most of the methods can be performed in R, and R serves as the basis for the online exercises that accompany each chapter.

16.1.4 R is an Environment

It should now be clear that R is not simply a software package. It is also an environment designed to enhance the communication between users and developers across the disciplines of statistics, ecology, and computer science. John Chambers and his colleagues designed the S language "to turn ideas into software" (Chambers 1998), and the R environment also shares this property.

Thus, some functions are implemented only to illustrate an idea. For example, the help page of the function `pie()`, which draws pie charts, clearly advises the user against its use because of the low readability of such displays (alternative methods are proposed on the help page); the help page of the function `SnowsPenultimateNormalityTest()` in the package TeachingDemos contains interesting information regarding the uselessness of normality tests. Therefore, there is much more to be gained from the R environment than simply the execution of statistical tests.

Furthermore, R packages always propose a point of view on the methods that they implement. Therefore, the same method can be implemented in different packages, but with different points of view. For example, the bivariate kernel method is a general method that smooths a point pattern in a two-dimensional space (Silverman 1986). This approach is implemented in several R packages, including:

1) the function `kde2d()` (package MASS): can be used to smooth a point distribution on a scatterplot relating two random variables;
2) the function `kernel2d()` (package splancs): can be used to smooth a point pattern in geographical space;
3) the function `kernelUD()` (package adehabitatHR): can be used to estimate the utilization distribution from animal radio-tracking data.

The implemented method is the same for all these functions. But these functions are different because they are designed to work on different types of data, in different contexts. Of course, it is possible to use, say, the function `kde2d()` to estimate the utilization distribution of an animal, but the use of `kernelUD()` is more convenient here: the R object returned by the function `kernelUD()` can be used by other functions of the package adehabitatHR for further analysis, such as the estimation of the home range or the study of the interaction between animals. Other R packages do not provide functions for the use of the kernel method in this context, because their author(s) did not design the function for this kind of use. The philosophy underlying a package is usually described in the documentation of the package (help pages, tutorials, articles in scientific journals, mailing lists, etc.), and the package user should be aware of the purpose and goals.

Any user can write a package for the R software and submit it to CRAN. The package should respect several guidelines: the functions available in a package should be thoroughly documented, no compiled code should be included in the sources, and the licensing information should be clearly indicated. When a package is submitted, the CRAN system carries out a large number of automatic tests to check that these guidelines are respected. However, CRAN does not test the soundness of the approaches implemented in a package. It is not required to beta-test a package prior to its submission to CRAN. It is therefore up to the users to decide whether they should use a given function implemented in a given package; they can decide to use it because they know and trust the programmer, or because the package has been thoroughly tested in the scientific literature, or because they have checked the code themselves. It is, however, risky to blindly use any function available in any random package, without knowing the context in which it has been proposed.

The main message of this chapter is that R should never be used uncritically. There is no excuse for failing to read thoroughly the documentation prior to the use of a function.

16.2 Basics of R

In this section, I describe the basic commands that the new user needs to know in order to use R efficiently. The R software can be downloaded from CRAN at the following URL: cran.at.r-project.org. The reader should also install an IDE (see section 16.1.1 for a list of programs) to facilitate code writing. Usually, organizing the code in an IDE prior to its execution in R is a means to keep track of the analyses performed on a dataset, which should be the norm to allow reproducible research. The code presented in this chapter is also available as a text file on the website of the book, and can be opened in an IDE prior to its execution.

When we launch R for the first time, this software shows the symbol ">." This symbol is named the *prompt*; it indicates that R is ready to take a command. R can be used to perform simple calculations. For example, if you type 3*6 at the prompt and then type enter:

```
> 3*6
[1] 18
```

R calculates the result of this product, and displays the result. R detects when a command is not complete. For example, if you do not close a parenthesis, R will assume that the command that you have typed is not complete. R will then ask the user to complete the command by replacing the prompt by the symbol "+." For example:

```
> (3*6
+ )
[1] 18
```

In the rest of this chapter, when I show some R code, I also include the prompt, so that the user can see exactly what R returns after the execution of the code.

16.2.1 Several Basic Modes of Data

Everything in R is an object. Each object has a name that can be called by the user, and contains data. Open R, and type the following commands at the command prompt:

```
> foo <- 3
> foo
[1] 3
> # R is case sensitive: the following
> # returns an error
> Foo
Error: object 'Foo' not found
> foo + 2
[1] 5
> bar <- foo + 2
> bar
[1] 5
```

Note that any text following a dash # is considered as a comment and is not interpreted by R. The symbol <– is used in R to assign a value to an object. Thus, the first command assigns the value 3 to the object named foo. Calling foo at the command line prints its value. In the fourth command, we add the value 2 to the object foo. As foo contains the number 3, R prints the object 5. The last command illustrate how the results of this operation can be stored in a new object bar. And again, the object bar contains the value 5. This example also illustrates how to use the common basic operations (+, −, *, /) in R.

The objects foo and bar are now stored in the R environment. The function ls() can be used to list the objects in the environment, and the function rm() can be used to remove some of them:

```
> ls()
[1] "bar" "foo"
> rm(bar)
> ls()
[1] "foo"
```

The object foo stores a numeric value. However, we can store values of other modes in an object. There are many storage modes that can be handled by R (integer, raw, etc.), but basically, the new R user needs to know only three storage modes: (i) the mode *numeric* is used to store real numbers; (ii) the mode *character* is used to store character strings; and (iii) the mode *logical* is used to store logical values (TRUE or FALSE).

The mode character is illustrated in the following self-explanatory example:

```
> myString <- "A character string must be
enclosed in single or double quotes"
> myString
[1] "A character string must be enclosed in
simple or double quotes"
```

Common operations with the logical mode are illustrated in the following example:

```
> ## Is 3 greater than 5 ? The following
> ## operation returns a logical value
> Is3GreaterThan5 <- (3 > 5)
> Is3GreaterThan5
[1] FALSE
> ## Then 3 is lower than 5 ?
> 3 < 5
[1] TRUE
> ## Is 3 < 5 and 4 < 3 ?
> (3 < 5)&(4 < 3)
[1] FALSE
> ## Is 3 < 5 or 4 < 3 ?
> (3 < 5)|(4 < 3)
[1] TRUE
> ## Equality can be tested with ==, and
difference with !=
> 3 == 5
[1] FALSE
> 3 != 5
[1] TRUE
> ## Last, the sign ! is used to denote the
> ## logical complement
> ## (i.e., if A is TRUE, then !A, 'not A',
> ## is FALSE)
> !Is3GreaterThan5
[1] TRUE
```

Logical values can be coerced to numerical values, and vice versa:

```
> as.numeric(TRUE)
[1] 1
> as.numeric(FALSE)
[1] 0
> as.logical(1)
[1] TRUE
> as.logical(0)
[1] FALSE
```

Note the value NA is used by R to store missing values for all modes (NA stands for "Not Available"). For the numeric mode, when a missing value is used in a calculation, the result is also a missing value (e.g. 3 + NA returns a missing value). For the logical mode, R respects the rules of logical calculus, as illustrated below:

```
> ## Will it rain tomorrow? We do not know, so
> ## we use NA
> RainTomorrow <- NA
> ## Sunshine at the time of writing? We
> ## know it:
> SunshineNow <- TRUE
> ## Snow at the time of writing? We also
> ## know it:
> SnowNow <- FALSE
>
> ## Now consider the two conditions:
> ## RainTomorrow and SunshineNow
> ## Since one of these conditions is NA, we
> ## do not know whether both are TRUE
> RainTomorrow&SunshineNow
[1] NA
> ## But we know that at least one of these two
> ## conditions is TRUE
> RainTomorrow|SunshineNow
[1] TRUE
> ## We know that it is not snowing now, but we
> ## do not know if it will rain
> ## tomorrow, so that we cannot be sure that
> ## at least one of these conditions
> ## is TRUE. Therefore, the result is NA
> RainTomorrow|SnowNow
[1] NA
```

16.2.2 Several Basic Types of Objects

We now describe more precisely the R objects themselves. The new R user really needs to know only four basic types of objects: (i) vectors; (ii) matrices and arrays; (ii) lists; and (iv) functions.

Vectors: vectors are one-dimensional objects used in R to store several values *of the same mode* in one object. Thus, it is not possible to store both character and numeric values in the same vector. Vectors are most commonly created using the function c():

```
> # Create a vector containing 6 numeric
> # values
> myVector <- c(2, 1, 3, 8, 9, 2)
> myVector
[1] 2 1 3 8 9 2
> # The length of this vector is:
> length(myVector)
[1] 6
```

Note that in the previous section, the objects foo, bar, myString, and myLogical were vectors of length one. The elements of a vector can be recovered with the bracket notation:

```
> # Prints the second element of the vector
> myVector[2]
[1] 1
> # Creates another numeric vector
> myIndex <- c(2,4)
> # Prints the second and fourth elements
> # from myVector
> myVector[myIndex]
[1] 1 8
> # Or, more directly:
> myVector[c(2,4)]
[1] 1 8
> # stores the result in another vector
> myOtherVector <- myVector[c(2,4)]
> myOtherVector
[1] 1 8
```

It is sometimes useful to generate vectors of consecutive integers. This is possible with the help of the colon ":". For example:

```
> # generates a vector of consecutive
integers from 2 to 5
> 2:5
[1] 2 3 4 5
> # Prints the elements 2 to 5 from myVector
> myVector[2:5]
[1] 1 3 8 9
```

Negative indices can be used to remove elements from a vector:

```
> ## prints myVector without the first
element
> myVector[-1]
[1] 1 3 8 9 2
> ## prints myVector without the first and
> ## fifth elements
> myVector[c(-1, -5)]
[1] 1 3 8 2
```

In this example, specific elements are extracted from the vector by passing another vector of integers inside the brackets. However, R also allows us to extract elements with the help of a vector, as illustrated in the following example:

```
> GreaterThan4 <- (myVector > 4)
> GreaterThan4
[1] FALSE FALSE FALSE  TRUE  TRUE FALSE
>
> # The logical vector GreaterThan4 has the
> # same length as myVector
> # It contains TRUE if the corresponding
> # element in myVector is > 4
> # and FALSE otherwise
>
> # This vector can be used to print the
> # elements of myVector that are
> # greater than 4:
> myVector[GreaterThan4]
[1] 8 9
> # Or, more simply:
> myVector[myVector > 4]
[1] 8 9
```

Last, it is also possible to redefine the value of one or several vector elements using the bracket notation, and even to extend the length of a vector:

```
> ## define a vector of length two
> myvec <- c(10, 20)
> ## add a third element to this vector
> myvec[3] <- 40
> myvec
[1] 10 20 40
> ## Substitute new values into the first
> ## two elements:
> myvec[1:2] <- c(80, 90)
> myvec
[1] 80 90 40
```

Matrices and arrays: a *matrix* is a collection of data elements, all of the same mode, arranged in a two-dimensional rectangular layout. A matrix is typically created from a vector with the help of the function matrix, as in the example below:

```
> myMatrix <- matrix(myVector, nrow = 3)
> myMatrix
     [,1] [,2]
[1,]    2    8
[2,]    1    9
[3,]    3    2
> # number of columns
> ncol(myMatrix)
[1] 2
> # number of rows
> nrow(myMatrix)
[1] 3
```

The vector myVector created previously is of length six. When we create the matrix, we indicate here how many rows the matrix should contain, in this case, three. Because the vector is of length six, the resulting matrix has two columns. The matrix is then filled with the elements of the vector (filling the rows first and then the columns). Elements of the matrix can be recovered or assigned with the help of the bracket notation [row_number, column_number]. If the element row_number (or column_number) is missing, then the result is a vector containing all the rows (or columns) for the given column (resp. row). This operation is illustrated in the example below:

```
> ## Recover the element located in the
> ## third row and the second column
> myMatrix[3,2]
[1] 2
> ## Recover the second and third rows in the
> ## first column
> myMatrix[2:3, 1]
[1] 1 3
> ## Recover the second row
> myMatrix[2,]
[1] 1 9
> ## Recover the second column
> myMatrix[,2]
[1] 8 9 2
> ## Example of assignation. We change the
> ## first row of the matrix
> myMatrix[1,] <- c(9, 5)
> myMatrix
     [,1] [,2]
[1,]    9    5
[2,]    1    9
[3,]    3    2
```

Again, negative indices and logical vectors can be used for assignation and subsetting:

```
> myMatrix[-1,]
     [,1] [,2]
[1,]    1    9
[2,]    3    2
>
> myMatrix[c(FALSE, TRUE, TRUE),]
     [,1] [,2]
[1,]    1    9
[2,]    3    2
```

A matrix organizes data in a two-dimensional structure. R also provides a broader type of multidimensional structure, named “array.” A matrix is a special type of array, with two dimensions. It is possible (though not common) to organize data in arrays with three or more dimensions. Arrays are created with the help of the function `array()`:

```
> ar <- array(1:24, dim=c(4,3,2))
> ar
, , 1

     [,1] [,2] [,3]
[1,]    1    5    9
[2,]    2    6   10
[3,]    3    7   11
[4,]    4    8   12

, , 2

     [,1] [,2] [,3]
[1,]   13   17   21
[2,]   14   18   22
[3,]   15   19   23
[4,]   16   20   24
```

In this example, we create a three-dimensional array containing the integers 1 to 24. The first dimension contains four elements, the second one contains three elements, and the last one contains two elements. Similarly, extraction can be done with the help of the bracket notation:

```
> ar[4,2,1]
[1] 8
```

And similarly, subsetting and assignation can be done with the help of vectors of positive integers, negative integers, or logical values.

Lists: a list is a collection of R objects. A list can store any type of R object (vectors, matrices or arrays, functions, and even other lists). One given list can store elements of different modes. It is typically created with the function `list()`. For example here is a list with a vector, a matrix, a string, and a logical vector:

```
> myList <- list(myVector, myMatrix,
+  "a character string", myVector > 4)
> myList
[[1]]
[1] 2 1 3 8 9 2

[[2]]
     [,1] [,2]
[1,]    2    8
[2,]    1    9
[3,]    3    2

[[3]]
[1] "a character string"

[[4]]
[1] FALSE FALSE FALSE  TRUE  TRUE FALSE
```

Lists are considered internally as generic vectors. Therefore, it is possible to recover or modify subparts of the list using the bracket notation:

```
> myList[2:3]
[[1]]
     [,1] [,2]
[1,]    2    8
[2,]    1    9
[3,]    3    2

[[2]]
[1] "a character string"
> myList[1]
[[1]]
[1] 2 1 3 8 9 2
> ## Example of assignation
> myList[3] <- list("Another string")
```

Similarly to vectors and matrices, it is possible to subset or assign elements in a list using a vector of positive integers, negative integers, or logical values. Note that the use of the single bracket notation on a list returns a list. Similarly, only another list can be assigned as a given element of a list using this notation. If we want to extract or change the object stored in a particular element of the list, we have to use the double bracket notation. For example, to recover or change the object stored in the first element of the list:

```
> myList[[1]]
[1] 2 1 3 8 9 2
> myList[[1]] <- "yet another character
string"
```

Since the double bracket notation does not necessarily return a list, it is not possible to use it to extract several elements of the list. Therefore, the following command is invalid:

```
> myList[[1:3]]
Error in myList[[1:3]] : recursive
indexing failed at level 2
```

Note that the elements of a list can be named. This is also true for vectors and matrices, but it is more important for lists, as the elements can be recovered with their name, using the $ notation. For example:

```
> ## This list has two named elements
> myList2 <- list(theVector = myVector,
+                 TheMatrix = myMatrix)
> ## Call the element theVector
> myList2$theVector
```

```
[1] 2 1 3 8 9 2
> ## Same result as
> myList2[[1]]
[1] 2 1 3 8 9 2
```

Functions: functions are R structures designed to perform operations on other R objects. Each time that the user writes a command at the R prompt and hits the "Enter" key, one or several functions are executed. An R function is typically called by its name followed by parentheses. Inside the parentheses, the user must pass the arguments required by the function. We have already described several functions previously, such as c(), matrix(), and list(). Let us consider the function round(), which rounds the values in a numeric vector. The list of arguments of this function can be printed with the function args():

```
> args(round)
function (x, digits = 0)
NULL
```

The function round() takes two arguments: x contain the vector of value(s) to be rounded, and digits is the number of decimal places to be used (an integer is expected). The function defines a default value digits = 0, which means that by default, this function rounds x to the nearest integer. The following commented example illustrates the use of this function:

```
> ## Rounding using the default value for
> ## digits, i.e. digits = 0
> round(x = 3.78)
[1] 4
> ## Note that it is not necessary to specify
> ## x explicitly (see below)
> round(3.78)
[1] 4
> ## Now, round to the first decimal place
> round(x = 3.78, digits = 1)
[1] 3.8
> ## When the arguments are named, the order
> ## of the arguments is not important
> round(digits = 1, x = 3.78)
[1] 3.8
> ## However, when the arguments are not
> ## named, R interprets the arguments
> ## in the same order as the one indicated by
> ## args()
> ## In other words, The following command
> ## is identical to the previous one:
> round(3.78, 1)
[1] 3.8
> ## But this one is not:
> round(1, 3.78)
[1] 1
> ## Here, R understands x = 1 and digits
> ## = 3.78
> ## (as digits should be an integer,
> ## R interprets it as 4)
```

To find out which type of objects is expected by a function, we can use the important function help():

```
> help(round)
```

The function help() is so intensively used that a shortcut is available in R: typing the name of the function after a question mark is an alias to call the help page:

```
> ?round
```

The help page describes the arguments of the function and gives details on how the function should be used, as well as implementation details and examples of use. The next section describes more precisely how to find help in R.

16.2.3 Finding Help and Installing New Packages

The CRAN provides more than 7500 R packages, with each package generally providing numerous functions. As the reader can imagine, even the most experienced R users do not know all the functions available in all packages. The main skills required to use R are: (i) the ability to find the function suitable to perform a particular operation; and (ii) the ability to find help on how to use this function. In this section, I focus on these skills.

Numerous functions are intended to help the user to find help within the R software. The user already knows the function help(). However, the use of help() supposes that the user already knows the name of the function that he wants to use. In most cases, the new R user will know the name of the statistical method that he wants to use, but not the name of the function(s) that implement it. This is where the function help.search() is the most useful (see the help page of the function help.search() for a detailed description of this function, by typing ?help.search at the command prompt).

For example, imagine that we have two vectors a and b, representing the value of two variables measured on seven animals:

```
> a <- c(3, 2, 5, 7, 1.5, 12, 4.1)
> b <- c(17, 18, 12, 20, 3, 8, 40)
```

We would like to calculate a Spearman correlation coefficient between these two variables, and test its significance. However, we do not know the name of the function that we need to call to perform these operations. We use the function help.search() to search a character string in the help pages of the functions available on your system. For example:

```
> help.search("correlation")
```

Note that the quotes are important (as indicated on the help page, the main argument of the function `help.search()` should be a character string, therefore enclosed in quotes). On my system, this command displays a list of 115 functions with a help page where the word "correlation" has been found. This result will likely be different on another system: because this function searches patterns on the R help pages available on a computer, the number of matches will depend on the set of packages installed on the system. In my case, it will be too long to study the help pages of 115 functions to identify the function that I need. We may therefore refine the search:

```
> help.search("Spearman correlation")

Help files with alias or concept or title
matching 'Spearman correlation' using
fuzzy matching:

stats::cor.test        Test for Association/
                       Correlation Between
                       Paired Samples

Type '?PKG::FOO' to inspect entries 'PKG::
FOO', or 'TYPE?PKG::FOO' for
entries like 'PKG::FOO-TYPE'.
```

There is only one function that matches the pattern, the function `cor.test()`, available in the package `stats`, which is loaded by default. We have more information about this function by typing `?cor.test` at the command prompt. Take the time to read the help page of a function that you do not know! The author of the function generally presents important information on the help page of a function, and it is a common error among new R users to overlook this step. Failing to read the documentation may lead to a misuse of the function, and therefore, to incorrect results.

On this help page, we see that the function `cor.test()` tests "for association between paired samples, using one of Pearson's product moment correlation coefficient, Kendall's tau, or Spearman's rho". This function takes an argument named `method`:

```
method: a character string indicating
        which correlation coefficient is
        to be used for the test. One of
        '"pearson"', '"kendall"', or
        '"spearman"', can be abbreviated.
```

This help page provides all the required information to run this function. We are now able to calculate the Spearman correlation coefficient between a and b, and to test whether it is significantly different from 0 (two-sided test, see the help page):

```
> cor.test(a, b, method="spearman")

      Spearman's rank correlation rho

data: a and b
S = 48, p-value = 0.7825
alternative hypothesis: true rho is not
equal to 0
sample estimates:
   rho
0.1428571
```

It turns out that the coefficient is rather low at rho = 0.14, and not significantly different from 0, with a *P*-value of *P* = 0.78. This example is a simple one, because the function that we were looking for was already present on our system. However, in many cases, we will need functions implemented in packages available on CRAN, but not on our system. For example, imagine that we want to fit a random forest to our data for predictive purposes (Breiman 2001). Using the approach described above:

```
> help.search("random forests")
No vignettes or demos or help files found
with alias or concept or
title matching 'random forests' using
fuzzy matching.
```

However, this approach is common in statistics, and it would be surprising if it was not available in R. The function `RSiteSearch()` can be used in the same way as `help.search()` to find the function implementing it on CRAN:

```
> RSiteSearch("random forests")
```

This opens a browser window listing all the web pages on CRAN where the search pattern is present. Looking at the results, we can see that there are many packages that propose an implementation of methods related to random forests (nonexhaustive list):

1) The package `bigrf` implements the fit of random forests on large datasets.
2) The package `randomForestSRC` implements the fit of random forests for survival, regression, and classification.
3) The package `randomSurvivalForest` implements the fit of random survival forests.
4) The package `varSelRF` implements the variable selection in a random forest fit.
5) The package `randomForest` implements "Breiman and Cutler's random forests for classification and regression."

The function `RSiteSearch()` returned 537 results at the time of writing! The list of results illustrates an important characteristic of the R software that we have already discussed. Each function implements a particular point of view on the approach, and the user should be aware of the differences between the points of view. As for most topics in science, we can see from these results that there are many points of views on random forests: on the algorithms that can be used to implement a given approach, on the kind of data required as input, or on the kind of results expected. It is dangerous to randomly pick one of these functions, hoping that we make the correct choice. Users are expected to know exactly what they want to do.

For example, imagine that you want to fit a random forest as described by Breiman (2001). Browsing the results of `RSiteSearch()`, we can see that the package `randomForest` seems to implement this approach, as and cites explicitly the Breiman and Cutler's approach. Browsing the web (Google "`randomForest R`"), we learn that this package includes the original C code of the Program "`Random Forests`" written by Breiman (2002), and provides an R interface to this reference program (see for example finzi.psych.upenn.edu/R/library/randomForest/DESCRIPTION). Browsing the web further, we discover a scientific paper describing the package, and how it should be used (Liaw and Wiener 2002). This function seems to be the best choice, given our objectives. Browsing the mailing lists hosted on the R website (especially the R-help mailing list), we also obtain additional information on the use of this approach (required sample sizes, common errors in using this function). Background research is part of the scientific work expected before the analysis, and cannot be overlooked.

Now, imagine that we are satisfied with the package `randomForest`. We can download and install it in R, using the function `install.packages()`:

```
> install.packages("randomForest")
```

The package is then loaded into R with the function `library()`:

```
> library(randomForest)
```

Then, we can see the list of functions available in that package:

```
> help(package = "randomForest")
```

The function `randomForest()` of this package seems to perform the desired operation. As previously, you can obtain additional information on this function by typing `?randomForest` at the command prompt.

I stress that it is important to read thoroughly all the material related to the programs that you want to use, and the help page generally provides useful references related to the method. The reader should be aware that there is absolutely no warranty that the methods available in a given R package are sound for the issue at hand.

16.2.4 How to Write a Function

Writing a custom R function is easy. The definition of a function has the following structure:

```
functionName <- function(arguments)
{
    body of the function
    ...
    return(result)
}
```

The first line of the code defines the function, and curly brackets are used to group together the command lines that build the body of the function. We consider the classical example of a function that converts a number C of Celsius degrees into a number F of Fahrenheit degrees. The conversion formula is the following:

$$F = C \times 9/5 + 32$$

We would like to create a function that takes a number of Celsius degrees as an argument and returns a number of Fahrenheit as output. You can write the following function in an IDE (Section 16.1.1):

```
Celsius2Fahrenheit <- function(Celsius)
{
    Fahrenheit <- Celsius * 9/5 + 32
    return(Fahrenheit)
}
```

Just copy and paste this function into R (or source the file in R with the function `source()`). You can now use it for conversion. For example:

```
> Celsius2Fahrenheit(40)
[1] 104
```

It is important to understand the scope of variables in R. In the above example, the object Celsius was defined in the arguments of the function. Now, what would happen if we had an R object named Celsius in the R global environment? Would there be confusion between the object Celsius in the function and the object Celsius in the global environment? This can be easily checked:

```
> Celsius <- 200
> Celsius2Fahrenheit(40)
[1] 104
```

This does not change the result. Actually, R defines a temporary local environment for the function to store the value of the variables passed as arguments (here `Celsius = 40`), as well as the value of the variables created in the body of the function (here Fahrenheit). This environment is deleted when the function returns its value. When an object name is called within a function, R first searches this object in this temporary local environment, and, if it does not find it, R searches for it in the global environment. For example, consider the following example:

```
> foo <- function(x)
+ {
+   return(x + y)
+ }
> foo(3)
Error in foo(3) : object 'y' not found
```

Here, we define a function taking only one argument x. The body of the function contains only one command: the sum x + y. However, the object y has not been defined in the arguments or in the body of the function. In addition, there is no object y in the global environment. Therefore, calling foo(3) assigns three to x, but y has no value. The function therefore returns an error. Now consider this example:

```
> y <- 2
> foo(3)
[1] 5
```

We define an object y in the global environment. The function foo() finds the object x in the temporary environment of the function, and the object y in the global environment.

Basic control-flow constructs are available to facilitate programming. We present the most common construct, but there is a help page that lists them all (type ?Control at the command prompt). The if... else... construct can be used to test a condition. For example, consider the function Celsius2Fahrenheit() programmed previously. We could program another function to convert Fahrenheit degrees to Celsius degrees. Or, we could program a more general function that converts between Celsius and Fahrenheit or conversely, depending on the value of an argument. The following function illustrates the use of the if... else... construct for this function. You can write the following function in an IDE:

```
ConvertDegrees <- function(Degrees,
conversion="CelsiusToFahrenheit")
{
  if (conversion == "CelsiusToFahrenheit") {
        DegreesOut <- Degrees * 9/5 + 32
    } else {
        DegreesOut <- (Degrees - 32) * 5/9
    }
    return(DegreesOut)
}
```

This function takes two arguments: the first one is a number of degrees, and the second one indicates the type of desired conversion. By default, the argument conversion is set to "CelsiusToFahrenheit." In the body of the function, we first test the type of expected conversion. If the user asks for "CelsiusToFahrenheit," then we use the formula to convert Celsius degrees to Fahrenheit degrees. If conversion takes any other value, we use the formula to convert Fahrenheit to Celsius. Write this function in an IDE, then copy it and paste it in R (or source it in R with the function source()). You can then use it for conversion:

```
> ConvertDegrees(40)
[1] 104
> ConvertDegrees(104,
+       conversion="Fahrenheit2Celsius")
[1] 40
> ConvertDegrees(ConvertDegrees(40),
+       conversion="Fahrenheit2Celsius")
[1] 40
```

16.2.5 The `for` loop

Another useful construct is the for loop. It can be used to repeat an operation numerous times: The example below illustrates the use of the for loop:

```
> vec <- 1:5
> for (i in vec) {
+ cat("The variable i now takes the value",
+       i, "\n")
+ }
The variable i now takes the value 1
The variable i now takes the value 2
The variable i now takes the value 3
The variable i now takes the value 4
The variable i now takes the value 5
```

The function cat() collapses its arguments into a character string and prints it in the console (note that \n is the character used for the newline). The for loop is here repeated on a vector of length 5, containing the integers from 1 to 5. At first, the variable i takes the first value of this vector, i.e. the value 1. The body of the loop is executed and all the commands in this loop are executed with the value i = 1; the function cat() therefore prints the value of the variable i = 1. Then, i is incremented: it takes the second value in this vector, i.e. the value 2. This operation is repeated for all the values of the vector.

The for loop is extremely useful when an operation is to be repeated many times. For example, imagine that you have a folder with many files that you wish to import in R. Rather than importing them manually one after the other, you can define a vector containing the names of the files in that folder (there is a function to build such a vector:

list.files()) and build a for loop to import them in a list, i.e. something like:

```
lifi <- list.files()
results <- list()
for (i in 1:length(lifi)) {
   results[[i]] <- read.table(lifi[[i]])
}
```

However, the for loop has a drawback: relying too heavily on for loops can make the script run slow: at each iteration, the code inside the for loop must be interpreted and then executed. The interpretation step can be slow when the number of commands or the number of iterations in the for loop is large. In such situations, it is often preferable to rely on vectorized functions such as lapply(), which perform the interpretation step only once.

16.2.6 The Concept of Attributes and S3 Data Classes

Any R object has an attribute list that can optionally be used to store additional information about the object. The attributes list can be accessed with the function attributes(), and any particular attribute can be defined and set with the function attr(). The following example illustrates this concept:

```
> age <- c(3,2,5,6)
> attr(age, "units") <- "Months"
> age
[1] 3 2 5 6
attr(,"units")
[1] "Months"
> attributes(age)
$units
[1] "Months"
```

In this example, the vector age contains the age of a sample of four animals. We have defined an attribute named "units" to store the measurement units for the age (we could of course have used a completely different name). This attribute is considered as meta-information on the object.

Why should we use attributes? The average R user will rarely define the attributes of the objects. However, it is important to understand them, as they are intensively used by R programmers to store information in the output of a function. For example, consider the function na.omit(), that removes missing values from a vector:

```
> vec <- c(3, NA, NA, 5, 6)
> vecWithoutNA <- na.omit(vec)
> vecWithoutNA
[1] 3 5 6
attr(,"na.action")
[1] 2 3
attr(,"class")
[1] "omit"
```

The function na.omit() returns the desired vector, without missing values, but also defines two attributes. We will describe later the special attribute "class". The attribute "na.action" stores the position in the original vector of the removed missing values, so that it is still possible to rebuild the original vector from vecWithoutNA, if desired.

There is an attribute that has a special meaning in R: the attribute "class". In the previous sections, we have described the basic object types available in R (vectors, matrices and arrays, lists, and functions). Actually, it is possible to extend this list by defining your own object classes based on these basic types. R provides a simple class definition mechanism. The classes created using this approach are named S3 classes (standing for language S version 3, where this mechanism has been developed).

For example, imagine a study of the relationship between the age of an animal (in months) and its weight. We have a small sample of animals, each animal being captured several times. For each animal, we have the following information:

1) Whether the animal is a female (TRUE) or a male (FALSE)
2) A vector of the animal age at each capture (of variable length depending on the number of capture)
3) A vector of weights at each capture (same length as the vector of ages)

The elements are of different modes (logical, numeric), so that we need to use the only object type allowing to store different modes, i.e. a list. We could define a list to store the information for a given animal. For example, for the first animal:

```
> # The vector of ages for animal 1
> vecage <- c(2, 4, 7, 12, 32)
> # The vector of weights for animal 1
> vecweight <- c(5, 25, 32, 50, 120)
> Animal1 <- list(Female = TRUE,
+       age = vecage, weight = vecweight)
```

The object Animal1 stores all the information related to animal 1. Most R users will be satisfied with this approach, and will not go further. However, R programmers think differently. If this type of study is common, then for every animal in this type of study, we will have this type of information. Therefore, we could define a special type of list designed to store this information. More formally, we define a *class* of objects. We could name this class "Animal", for example. The class "Animal" that we define is a list with three elements: (i) a logical value named Female; (ii) a numeric vector named age; and (iii) a numeric vector named weight. To define the class of an object in R, we use the function class():

```
> class(Animal1) <- "Animal"
> Animal1
$Female
[1] TRUE

$age
[1]  2  4  7 12 32

$weight
[1]  5 25 32 50 120

attr(,"class")
[1] "Animal"
```

The function `class()` just defines an additional attribute to the object `Animal1`, the attribute "`class`". This function does nothing more than:

```
> attr(Animal1 , "class") <- "Animal"
```

Why should we bother with this particular attribute? And if we really want to define a new attribute to define the type of an object, why should we use the attribute "`class`" rather than any user-defined attribute, like "`type`" or "`AnyNameIWishToRepresentACategory`"?

The point is that this particular attribute "`class`" is important because it can be used to define the behavior of some special functions, named *generic functions*. Generic functions have a behavior that depends on the class of the object passed as argument. There are many generic functions available in R, especially graphic functions (e.g. `plot()`, `hist()`). We will focus of the function `summary()`, which provides a summary of an object. If you look at the help page of this function, it is clearly indicated:

```
'summary' is a generic function used to
  produce result summaries of the results
  of various model fitting functions
```

Actually, when the function `summary()` is called on an object, this function first checks the class attribute of the object. In our case, the class of the object `Animal1` is "`Animal`". Then, this function searches in the global environment a function named `summary.Animal()`. If there is such a function in the environment, the generic function `summary()` will call this other function `summary.Animal()`, and pass the object as argument. If there is no function `summary.Animal()`, the generic function `summary()` will use the function `summary.default()` (this default method is generally described on the help page of the generic function, in this case `summary()`).

In our case, we have defined a class named "`Animal`", but we have not defined any function named `summary.Animal()`. Therefore, when we call the function summary of this object, it calls the function `summary.default()`:

```
> summary(Animal1)
       Length Class  Mode
Female 1      -none- logical
age    5      -none- numeric
weight 5      -none- numeric
>
> ## This is the same as:
> summary.default(Animal1)
       Length Class  Mode
Female 1      -none- logical
age    5      -none- numeric
weight 5      -none- numeric
```

However, it is possible to define a method to the generic function `summary()` for the class "`Animal`." For example, you can write the following function in an IDE:

```
summary.Animal <- function(object, ...)
{
cat("Object of Class 'Animal'\nThere are",
    length(object$age),
    "captures of this animal\n")
}
```

The function `summary.Animal()` just prints some information on the object. Copy and paste this function to the command prompt. Now, when you will call the function `summary()`, it will call the method `summary.Animal()`:

```
> summary(Animal1)
Object of Class 'Animal'
There are 5 captures of this animal
```

An important concept when working with object classes is the inheritance. In our example, we may wish to define a finer class "`FrenchAnimal`" that contains all the elements of the parent class "`Animal`" (a logical element named `Female`, and two numeric vectors named `age` and `weight`), but which also contains an additional element "`Location`" giving the official code from the French institute for Statistics, Institut national de la statistique et des études économiques (INSEE), of the municipality where it has been captured. Because this code is specific to France, it would be pointless for animals captured in other countries. However, it could possibly be useful for mapping purposes when we have this information. Because this new class has the same elements as the class "`Animal`", all the functions designed to work with the class "`Animal`" would also work with the subclass "`FrenchAnimal`". We say that the class "`FrenchAnimal`" *extends* the class "`Animal`". And we say that *the object inherits* both classes. This inheritance can be stated in the class attribute, by providing a character vector containing all the classes that a given object inherits. So, for example:

```
> # The vector of ages for animal 2
> vecage <- c(2, 4, 7, 12, 32)
> # The vector of weights for animal 2
> vecweight <- c(5, 25, 32, 50, 120)
> Animal2 <- list(Female = TRUE,
+    age = vecage, weight =
+    vecweight, location=71270)
> class(Animal2) <- c("FrenchAnimal",
+                         "Animal")
```

Then, generic functions having methods for these two classes can be used. Note that any generic function `fun()` will first search a function named `fun.FrenchAnimal()` and pass the object to it. If there is no such function, it will search a function named `fun.Animal()` and pass the object to it. And if there is no such function, it will pass the object to the function `fun.default()`.

Note there is a special generic function that the user should know: the function `print()`. This function is called invisibly each time that the name of the object is typed at the command prompt. That is:

```
> 3
[1] 3
> # is the same as
> print(3)
[1] 3
```

Therefore, it is possible to define a `print()` method for any object class. For example, we could define a `print()` method for the class "`Animal`". For the sake of simplicity, in our example it is identical to the function `summary.Animal()`:

```
print.Animal <- function(x, ...)
{
    cat("Object of Class 'Animal'\n",
        length(x$age)
        "captures of this",
        "animal\n")
}
```

Write this function in an IDE and copy-paste it to the command prompt. Then, each time you will simply call `Animal1`, it will call the function `print.Animal()`:

```
> Animal1
Object of Class 'Animal'
5 captures of this animal
```

Note that the availability of the function `print.Animal()` does not change the content of the object `Animal1`; it changes only the way it is printed. Defining print methods is common in the R environment. For many functions, the object returned contains numerous large elements (e.g. a list of vectors containing thousands of values), and it is not convenient to print the whole content of an object each time the object is called. The availability of print methods allows us to print only a short description of the object.

Thus, the content of the object may differ from what is printed. When a function returns an object, and when the user needs to use components of this object, it is always useful to know the structure of the class to which this object belongs. The structure is usually described on the help page of the function (section "value" of the help pages). The user may also look at the structure of the object, thanks to the function `str()`:

```
> str(Animal1)
List of 3
 $ Female: logi TRUE
 $ age   : num [1:5] 2 4 7 12 32
 $ weight: num [1:5] 5 25 32 50 120
 - attr(*, "class")= chr "Animal"
```

Another way is to use the function `unclass()`, that deletes the attribute "`class`" of an object:

```
> unclass(Animal1)
$Female
[1] TRUE

$age
[1]  2  4  7 12 32

$weight
[1]   5  25  32  50 120
```

16.2.7 Two Important Classes: The Class `factor` and the Class `data.frame`

Section 16.2.6 described the S3 class-definition mechanism available in R. Most packages in R define their own classes and methods for these classes. It is therefore virtually impossible to describe all of them. However, there are two classes that are so commonly used in R, that deserve their own section: the class "`factor`" and the class "`data.frame`".

The class "`factor`" is used to handle qualitative variables in R. Objects of class "`factor`" are defined with the function `factor()` (see the help page of this function). The following example shows an example of use of this function:

```
> fac <- c("Adult", "Young", "Young",
+         "Adult", "Young")
> age <- factor(fac)
> age
[1] Adult Young Young Adult Young
Levels: Adult Young
> ## The function attributes()
> ## lists all attributes of object
> attributes(age)
$levels
[1] "Adult" "Young"

$class
[1] "factor"
>
> unclass(age)
[1] 1 2 2 1 2
attr(,"levels")
[1] "Adult" "Young"
```

In this example, the vector `fac` is a character vector containing the age class of several animals. The animal may be either young or adult. The function `factor()` returns an object of class `"factor"`, with an attribute named `"levels"`. As the name indicates, this attribute defines the two levels of the factor. The returned vector is itself a vector of integers, each integer corresponding to one particular level. Thus, the value 1 is associated with the first level, `"Adult"`, and the value 2 is associated to the second level, `"Young."` The use of the class `"factor"` is preferred when dealing with qualitative variables in the `R` functions, for several reasons. First, there are many functions designed to work on this class. For example, you can get or redefine the list of levels of the factor with the function `levels()`; there is a method for the function `summary()`; etc.

```
> levels(age)
[1] "Adult" "Young"
> levels(age) <- c("Old", "Baby")
> age
[1] Old  Baby Baby Old  Baby
Levels: Old Baby
> summary(age)
 Old Baby
   2    3
```

In addition, factors need less computer memory than character strings. Memory management is often a problem in `R` when working with large datasets. Last, the use of this function allows us to make sure that the levels of the factor are those that we expect. Consider the following example:

```
> fac <- c("Adult", "Young", "young",
+         "Adult", "Young")
> age <- factor(fac)
> age
[1] Adult Young young Adult Young
Levels: Adult young Young
```

It is clear that there is a typo here, as three levels are defined instead of two. In this case, it is easy to spot the difference between `"Young"` and `"young"` because there are only five elements in the vector, but error checking is harder when the dataset is large.

There is another important class that the reader should know: the class `"data.frame"`. This class is designed to store data tables. Basically, the class `"data.frame"` is a list of vectors, all the vectors having the same length. Objects of this class are created with the help of the function `data.frame()`. Note that functions for data import into R, such as `read.table()` or `read.csv()`, also return data frames. We illustrate this class in the example below:

```
> # The vector of ages for animal 1
> vecage <- c(2, 4, 7, 12, 32)
> # The vector of weights for animal 1
> vecweight <- c(5, 25, 32, 50, 120)
> df <- data.frame(age=vecage,
+                  weight=vecweight)
> df
  age  weight
1   2       5
2   4      25
3   7      32
4  12      50
5  32     120
>
> unclass(df )
$age
[1]  2  4  7 12 32

$weight
[1]   5  25  32  50 120

attr(,"row.names")
[1] 1 2 3 4 5
```

Objects of the class `"data.frame"` can be handled in the same way as list. Therefore, to get the second value from the variable `age`, just type:

```
> df$age[2]
[1] 4
> # Or
> df[[1]][2]
[1] 4
```

The class `"data.frame"` can also be considered as a two-dimensional object, in the same way as a matrix.

Therefore, it is possible to extract values from the object using the matrix bracket notation:

```
> df[1,2]
[1] 4
```

Note the difference between the matrix and the `data.frame`. The matrix is a two-dimensional structure, but the different columns of the object should be of the same mode (all numeric, all logical, all character). Because a data frame is basically a list, the different columns may belong to different modes.

16.2.8 Drawing Graphics

One of the great strengths of the R environment is the ability to easily draw insightful graphics. R developers have provided many ways to build graphics. The interested reader may have a look at the packages `ggplot2` and `lattice`, which contain numerous interesting functions for plotting data. In this section we will present only the base graphics, the most common approach to plot data.

The base R environment provides numerous functions to plot data. For example, the function `plot()` can be used to display scatterplots, the function `hist()` can be used to draw histograms, the function `boxplot()` can be used to draw boxplots, the function `barplot()` can be used to display bar charts, and the function `dotchart()` can be used to display dot charts.

We will not illustrate all these functions here (we recommend reading the help page of these functions; see also the excellent book by Beckerman and Petchey 2012). Rather, we will focus on one of these functions, the function `plot()`, and illustrate how customized graphics can be done in R. To illustrate these functions, we will work on an example dataset already available by default in the R environment, the dataset `trees` (see the help page of this dataset by typing `?trees` at the command prompt). This dataset is a data frame with three variables, giving the girth, the height, and the volume of 31 black cherry trees. The function `head()` displays the first six rows of this data frame:

```
> head(trees)
  Girth Height Volume
1   8.3     70   10.3
2   8.6     65   10.3
3   8.8     63   10.2
4  10.5     72   16.4
5  10.7     81   18.8
6  10.8     83   19.7
```

First, we could have a look at the relationship between the height and the volume, without taking the girth into account. As noted previously, this is achieved with the help of the function `plot()`. Note that `plot()` is a generic function; therefore the use of plot in this case is identical to the use of the function `plot.default()`, as the objects given as arguments do not have any class:

```
plot(trees$Height, trees$Volume, xlab =
"Height of the tree",
    ylab = "Volume of the tree",
    main = "Relationship between height and
volume",
    pch = 3)
```

The arguments of the function `plot()` allow control on the various elements of the plot (see the help page of the function `plot.default()`, which gives many details on how to use this function). The arguments `xlab` and `ylab` set axis labels, `main` sets a main title, and `pch = 3` set crosses for points. The results are displayed in Figure 16.1.

We first use the function `cut()` to divide the range of the variable `Girth` into three classes:

```
> girthcla <- cut(trees$Girth, 3)
> girthcla
 [1] (8.29,12.4] (8.29,12.4] (8.29,12.4]
     (8.29,12.4] (8.29,12.4] (8.29,12.4]
 [7] (8.29,12.4] (8.29,12.4] (8.29,12.4]
     (8.29,12.4] (8.29,12.4] (8.29,12.4]
[13] (8.29,12.4] (8.29,12.4] (8.29,12.4]
     (12.4,16.5] (12.4,16.5] (12.4,16.5]
[19] (12.4,16.5] (12.4,16.5] (12.4,16.5]
     (12.4,16.5] (12.4,16.5] (12.4,16.5]
[25] (12.4,16.5] (16.5,20.6] (16.5,20.6]
     (16.5,20.6] (16.5,20.6] (16.5,20.6]
[31] (16.5,20.6]
Levels: (8.29,12.4] (12.4,16.5] (16.5,20.6]
```

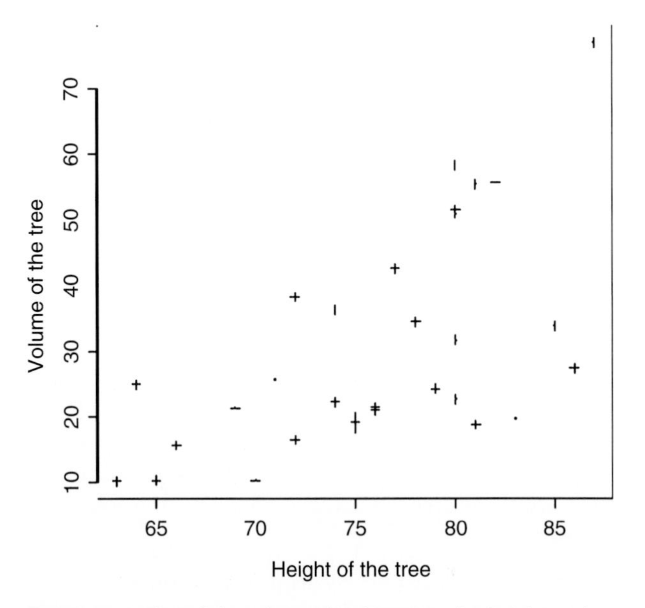

Figure 16.1 Plot of the relationship between the height and volume of 31 black cherry trees.

The resulting vector is a factor with three levels. Now, we will display the relationship between the height and the volume for the different girth classes. We have many ways to do so, but we will try two different approaches.

We first draw a blank graphics with the function plot() and then add the points with the function points(). We will use different points symbols for the three classes:

```
> ## Note the use of type = "n", which does not
> ## draw the points
> plot(trees$Height, trees$Volume, xlab =
"Height of the tree",
+    ylab = "Volume of the tree",
+    main = "Relationship between height
and volume",
+    type = "n")
>
> ## and different point shapes
> shap <- c(8, 17, 1)
>
> ## Then use a for loop to draw the points on
> ## this device
> for (i in 1:3) {
+ points(trees$Height[girthcla ==
+                      levels(girthcla)[i]],
+        trees$Volume[girthcla ==
+                      levels(girthcla)[i]],
+        pch=shap[i], cex = 1.5)
+}
```

Here, we used the function points() to add points to a pre-existing graphical window. We have increased the size of the points by defining cex = 1.5 (the default value for cex is 1). There are many functions that can be used to add graphical elements to a graphical window. Thus, the function lines() can be used to add lines; the function polygon() can be used to add a polygon to a plot; the function symbols() can be used to draw various symbols on a plot (circles, square, etc.). Last, we can add a legend to the plot with the function legend():

```
> legend(65,70, levels(girthcla),
+         pch=shap, cex = 1.5)
```

The resulting plot is displayed on Figure 16.2.

Now, we discuss the important function par(). The reader is strongly encouraged to spend some time in reading the help page of this function, as its use is required to produce publication-quality figures. This function can be used to control graphical parameters, including the size of margins in a plot, the font of axes labels and of the title, the length of the ticks on the axes, etc. It can also be used to draw multi-panel figures, thanks to the argument mfrow. For example, we could plot the relationship between the height and the volume for the three girth classes separately (i.e. one panel for each girth class):

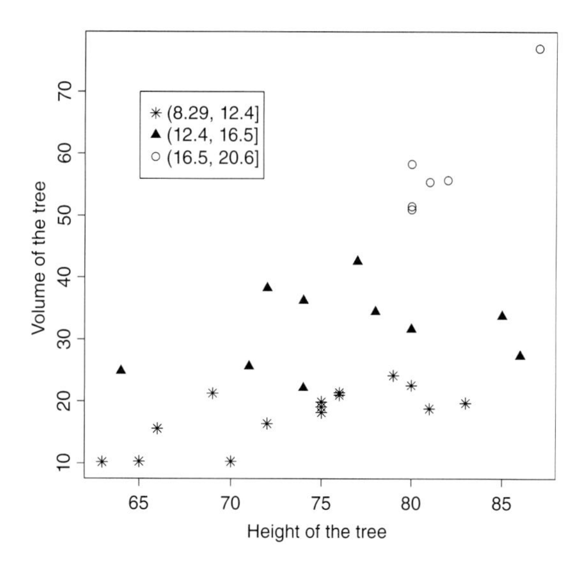

Figure 16.2 Plot of the relationship between the height and the volume of 31 black cherry trees. The different symbols correspond to different classes of girths.

```
> # Define a multi-panel figure made of 2
> # columns and 2 rows:
> par(mfrow=c(2,2))
> # Then, a for loop to draw the three plots
> for (i in 1:3) {
+  plot(trees$Height[girthcla ==
+                      levels(girthcla)[i]],
+       trees$Volume[girthcla ==
+                      levels(girthcla)[i]],
+       pch=3, xlab="Height",
  ylab="Volume",
+       main = levels(girthcla)[i])
+}
```

The resulting plot is displayed on Figure 16.3.

16.2.9 S4 Classes: Why It Is Useful to Understand Them

In Section 16.2.5, I have described the basic S3 class-definition mechanism, relying on the setting of an attribute named "class". This mechanism is simple, but it has drawbacks: the only thing that defines a class is the attribute "class" of the object. Therefore, nothing prevents the user to set a given class attribute to the wrong type of object, leading to further errors, sometimes unnoticed. For example, try to create an object of the class "data.frame" without the function data.frame():

```
> JustATry <- list(var1 = c(1,3,2),
+                  var2 = c(3,2,1))
> class(JustATry) <- "data.frame"
> JustATry[1,2]
[1] 3
> nrow(JustATry)
[1] 0
```

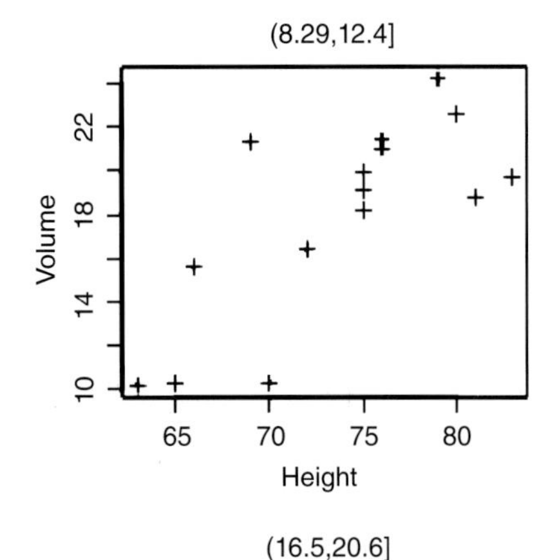

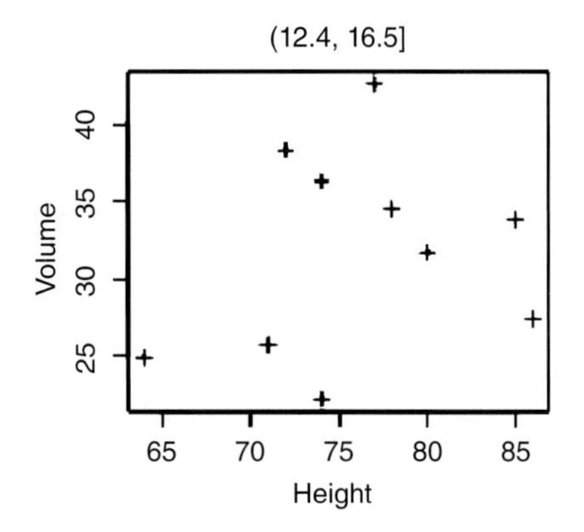

Figure 16.3 Plot of the relationship between the height and the volume of 31 black cherry trees. The different panels present this relationship for the different classes of girths.

Remember that a data frame is a list of vectors, all of them having the same length. The object `JustATry` that we have created satisfies these criteria. However, there are also other criteria that define a data frame. Thus, a data frame should have an attribute named `"row.names"`, corresponding to a character vector of the same length as the elements of the data frame, and containing the names of the rows. We forgot to define this attribute, and the function `nrow()`, that relies on it, indicates erroneously that the data frame is empty. For this reason, it is always recommended to rely on the functions designed to create the class rather than trying to create it ourselves. We should have used the function `data.frame()`, that creates all the required elements:

```
> JustATry2 <- data.frame(var1=c(1,3,2),
+                         var2 = c(3,2,1))
> nrow(JustATry2)
[1] 3
```

Moreover, generic functions work in a simplistic way with `S3` classes: they check the attribute `"class"` of the first argument to select the suitable method. However, it would be interesting to define methods that depend on the class of all the objects passed as arguments. That is, if we have a function `MyFunction()` taking two arguments `x` and `y`, we may wish to define four different methods for the four *signatures*:

```
myFunction(x = Object_of_class_A, y =
Object_of_class_B)
myFunction(x = Object_of_class_B, y =
Object_of_class_A)
myFunction(x = Object_of_class_B, y =
Object_of_class_B)
myFunction(x = Object_of_class_A, y =
Object_of_class_A)
```

Of course, it is possible to create two methods `myFunction.A()` and `myFunction.B()` and to use a `if... else...` condition in these functions to check the class of `y` and use appropriate commands. However, programming may quickly become cumbersome in such situations.

This is where the `S4` classes are the most useful. The `S4` class-definition mechanism is intended to provide *formal classes* to the `R` environment. More precisely, when a `S4` class is defined by `R` programmers, they create a *prototype* of the object that stores all the features that characterize it. This prototype is located somewhere in the global environment. The prototype also contains a formal inheritance hierarchy stored in it. The user does not have to know where this prototype is, but only has to know that when an object of class `S4` is created, the function that creates it checks that all the requirements of the class are satisfied. Also, generic functions can be defined and the method is selected *depending on the signature of the function*, and not only on the class of the first argument.

For example, consider the useful package sp, which provides S4 classes for spatial data. First load the package, and load the example dataset meuse, which is a data frame containing the value of several variables mapped over a grid:

```
> ## First install the package sp if it is not
> ## available
> install.packages("sp")
> ## loads the package sp
> library(sp)
> data(meuse)
> head(meuse)
```

```
       x      y cadmium copper lead zinc  elev       dist   om ffreq soil lime
1 181072 333611    11.7     85  299 1022 7.909 0.00135803 13.6     1    1    1
2 181025 333558     8.6     81  277 1141 6.983 0.01222430 14.0     1    1    1
3 181165 333537     6.5     68  199  640 7.800 0.10302900 13.0     1    1    1
4 181298 333484     2.6     81  116  257 7.655 0.19009400  8.0     1    2    0
5 181307 333330     2.8     48  117  269 7.480 0.27709000  8.7     1    2    0
6 181390 333260     3.0     61  137  281 7.791 0.36406700  7.8     1    2    0
  landuse dist.m
1     Ah     50
2     Ah     30
3     Ah    150
4     Ga    270
5     Ah    380
6     Ga    470
```

We create an object of the class "SpatialPoints" to store the coordinates of the points:

```
> sppo <- SpatialPoints(meuse[,1:2])
> class(sppo)
[1] "SpatialPoints"
attr(,"package")
[1] "sp"
```

We can look at the structure of this object:

```
> str(sppo)
Formal class 'SpatialPoints' [package
"sp"] with 3 slots
  ..@ coords      : num [1:155, 1:2] 181072
181025 181165 181298 181307 ...
  .. ..- attr(*, "dimnames")=List of 2
  .. .. ..$ : NULL
  .. .. ..$ : chr [1:2] "x" "y"
  ..@ bbox        : num [1:2, 1:2] 178605 329714
181390 333611
  .. ..- attr(*, "dimnames")=List of 2
  .. .. ..$ : chr [1:2] "x" "y"
  .. .. ..$ : chr [1:2] "min" "max"
  ..@ proj4string:Formal class 'CRS'
[package "sp"] with 1 slots
  .. .. ..@ projargs: chr NA
```

The function str() indicates that the object belongs to a formal class. An S4 object is an empty list, and all the data are stored in the attribute list. For S4 objects, these attributes are named "slots," and are accessed with the at sign, @. For example, to get the bounding box in this object, you can type:

```
> sppo@bbox
     min    max
x 178605 181390
y 329714 333611
```

However, it is usually not recommended to handle these slots directly. Programmers usually provide functions to work with these slots. Thus, to get the bounding box of an object of class "SpatialPoints" it is preferable to use the function bbox(). The structure of the class is documented on the help page of the class, available by typing:

```
> help("SpatialPoints-class")
```

More generally, for a class "X", information on the class can be found by typing help("X-class"). This help page describes the structure of the class, its position in the inheritance hierarchy, and the methods associated to the different signatures implying an object of the class "SpatialPoints". For example, the help page indicates a method for the function plot:

```
 plot 'signature(x = "SpatialPoints",
y = "missing")': plot points
```

Therefore, to plot the points, we can type:

```
> plot(sppo)
```

We do not show the resulting plot, to be concise.

We have noted previously that for each S4 class, there is a prototype somewhere in the environment. Although this is not of prime interest for the R user, who will never have to use it, we show how this prototype is stored in the environment:

```
> ## The function apropos() finds all the
> ## elements in the global
> ## environment with a name matching a
> ## given pattern
```

```
> apropos("SpatialPoints")
[1] ".__C__SpatialPoints"
          ".__C__SpatialPointsDataFrame"
[3] "rbind.SpatialPoints"
          "rbind.SpatialPointsDataFrame"
[5] "SpatialPoints"
                 "SpatialPointsDataFrame"
>
> ## The object .__C__SpatialPoints
> ## contains the prototype for the class:
> .__C__SpatialPoints
Class "SpatialPoints" [package "sp"]

Slots:

Name:    coords      bbox  proj4string
Class:   matrix    matrix          CRS

Extends: "Spatial"

Known Subclasses:
Class "SpatialPointsDataFrame", directly
Class "SpatialPixels", directly
Class "SpatialPixelsDataFrame", by class
"SpatialPixels", distance 2
```

Thus, the prototype is an object of class "classRepresentation" and contains all the relevant information for the class: (i) what are the required slots; (ii) what class of objects these slots must be; (iii) which class it extends; and (iv) which classes extend it (subclasses). The interested user can look at the structure of this object:

```
> str(.__C__SpatialPoints)
```

We omit the output of this function and do not describe the S4 classes further. The interested reader will find all relevant information in Bates (2003).

16.3 Online Exercises

Three online exercises for this chapter will provide an introduction to Program R. Exercise 1 is an example of a graphical exploration of data on the ecological niches of two species of grouse in France: Rock Ptarmigan (*Lagopus muta*) and Black Grouse (*Tetrao tetrix*). The example illustrates base functions for importing and summary of data, creating graphs, and programming loops. Exercise 2 illustrates steps for using a generalized linear model (GLM) to model species occurrence as a function of habitat variables. Exercise 3 shows how predicted values from the GLM can be plotted to visualize the model results.

16.4 Final Directions

In this chapter, we have described the basics of R. As we have indicated many times, the main skill required to use R is to know how to find help (Section 16.2.3). The reader should also know that there are numerous R resources in all possible formats available on the internet:

1) The website of the R software (www.r-project.org) contains numerous manuals describing particular approaches related with R (how to make a package, internal structures in R, etc.). Many tutorials and wikis in various languages are also available on this site.
2) The R Journal (journal.r-project.org/current.html) publishes papers on new approaches implemented in the software and changes in the software. It supersedes the former journal R News (www.r-project.org/doc/Rnews), which also provided interesting information on the R software.
3) A central hub of content collected from bloggers who write in English about R is available at the URL: www.r-bloggers.com (573 bloggers contribute to this hub at the time of writing).
4) The R website hosts numerous mailing lists related to this software. The reader will be interested to know that this sites also hosts thematic lists, such as the list r-sig-ecology related to the use of R in ecology, and r-sig-Geo related to the use of R with geographical data and mapping, etc. (see a complete list at the following URL: www.r-project.org/mail.html).
5) There are also R user groups connecting local people interested with R (see a list at rwiki.sciviews.org/doku.php?id=rugs:r_user_groups).

We finish the chapter with a final humorous note:

```
> ## Install the package fortunes
> install.packages("fortunes")
> ## loads the package
> library(fortunes)
> ## Never forget: ?fortune to find help on
> ## the function
> fortune("Yoda")


Evelyn Hall: I would like to know how (if)
I can extract some of the information from
the summary of my nlme.
Simon Blomberg: This is R. There is no if.
Only how.
  - Evelyn Hall and Simon 'Yoda' Blomberg
    R-help (April 2005)
```

References

Bates, D. (2003). Converting packages to S4. *R News* 3: 6–8.

Beckerman, A.P. and Petchey, O.L. (2012). *Getting started with R: an introduction for biologists.* Oxford, UK: Oxford University Press.

Bolker, B. (2007). *Ecological Models and Data in R.* Princeton: Princeton University Press.

Borcard, D., Gillet, F., and Legendre, P. (2011). *Numerical Ecology with R.* New York: Springer Science and Business Media.

Breiman, L. (2001). Random forests. *Machine Learning* 45: 5–32.

Breiman, L. (2002) *Manual on Setting Up, Using, and Understanding Random Forests V3.1.* Statistics Department University of California Berkeley, California, USA.

Calenge, C. (2006). The package adehabitat for the R software: a tool for the analysis of space and habitat use by animals. *Ecological Modelling* 197: 516–519.

Chambers, J.M. (1998). *Programming with data: A guide to the S language.* New York: Springer.

De Leeuw, J., Mair, P. et al. (2007). An introduction to the special volume on "Psychometrics in R.". *Journal of Statistical Software* 20: 1–5.

Ihaka, R. and Gentleman, R. (1996). R: A Language for Data Analysis and Graphics. *Journal of Computational and Graphical Statistics* 5: 299–314.

Lele, S. R., Keim, J. L. and Solymos, P. (2015). ResourceSelection: Resource Selection (Probability) Functions for Use-Availability Data. R package version 0.2–5.

Liaw, A. and Wiener, M. (2002). Classification and regression using randomForest. *R News* 2: 18–22.

Silverman, B. (1986). *Density Estimation for Statistics and Data Analysis.* London: Chapman and Hall.

Stevens, H. (2009). *A Primer of Ecology with R.* Berlin: Springer.

Stubben, C.J. and Milligan, B.G. (2007). Estimating and analyzing demographic models using the popbio package in R. *Journal of Statistical Software* 22: 1–23.

Zuur, A., Ieno, E., Walker, N. et al. (2009). *Mixed Effects Models and Extensions in Ecology with R.* New York: Springer.

Index

Population Ecology in Practice, First Edition. Edited by Dennis L. Murray and Brett K. Sandercock.

Companion website: www.wiley.com/go/MurrayPopulationEcology

q

r

t